DEVELOPMENTS IN SEDIMENTOLOGY 47

Diagenesis, III

FURTHER TITLES IN THIS SERIES

VOLUMES 1-11, 13-15, 17, 21-25A, 27, 28, 31, 32 and 39 are out of print

12 *R.G.C. BATHURST*
CARBONATE SEDIMENTS AND THEIR DIAGENESIS
16 *H.H. RIEKE III and G.V. CHILINGARIAN*
COMPACTION OF ARGILLACEOUS SEDIMENTS
18A *G.V. CHILINGARIAN and K.H. WOLF, Editors*
COMPACTION OF COARSE-GRAINED SEDIMENTS, I
18B *G.V. CHILINGARIAN and K.H. WOLF, Editors*
COMPACTION OF COARSE-GRAINED SEDIMENTS, II
19 *W. SCHWARZACHER*
SEDIMENTATION MODELS AND QUANTITATIVE STRATIGRAPHY
20 *M.R. WALTER, Editor*
STROMATOLITES
25B *G. LARSEN and G.V. CHILINGAR, Editors*
DIAGENESIS IN SEDIMENTS AND SEDIMENTARY ROCKS
26 *T. SUDO and S. SHIMODA, Editors*
CLAYS AND CLAY MINERALS OF JAPAN
29 *P. TURNER*
CONTINENTAL RED BEDS
30 *J.R.L. ALLEN*
SEDIMENTARY STRUCTURES
33 *G.N. BATURIN*
PHOSPHORITES ON THE SEA FLOOR
34 *J.J. FRIPIAT, Editor*
ADVANCED TECHNIQUES FOR CLAY MINERAL ANALYSIS
35 *H. VAN OLPHEN and F. VENIALE, Editors*
INTERNATIONAL CLAY CONFERENCE 1981
36 *A. IIJIMA, J.R. HEIN and R. SIEVER, Editors*
SILICEOUS DEPOSITS IN THE PACIFIC REGION
37 *A. SINGER and E. GALAN, Editors*
PALYGORSKITE-SEPIOLITE: OCCURRENCES, GENESIS AND USES
38 *M.E. BROOKFIELD and T.S. AHLBRANDT, Editors*
EIOLIAN SEDIMENTS AND PROCESSES
40 *B. VELDE*
CLAY MINERALS—A PHYSICO-CHEMICAL EXPLANATION OF THEIR OCCURRENCE
41 *G.V. CHILINGARIAN and K.H. WOLF, Editors*
DIAGENESIS, I
42 *L.J. DOYLE and H.H. ROBERTS, Editors*
CARBONATE-CLASTIC TRANSITIONS
43 *G.V. CHILINGARIAN and K.H. WOLF, Editors*
DIAGENESIS, II
44 *C.E. WEAVER*
CLAYS, MUDS, AND SHALES
45 *G.S. ODIN, Editor*
GREEN MARINE CLAYS
46 *C.H. MOORE*
CARBONATE DIAGENESIS AND POROSITY
48 *J.W. MORSE and F.F. MACKENZIE*
GEOCHEMISTRY OF SEDIMENTARY CARBONATES
49 *K. BRODZIKOWSKI and A.J. VAN LOON*
GLACIGENIC SEDIMENTS
50 *J.L. MELVIN*
EVAPORITES

DEVELOPMENTS IN SEDIMENTOLOGY 47

Diagenesis, III

Edited by

K.H. Wolf
18, Acacia Street, Eastwood, Sydney, N.S.W. 2122 (Australia)

and

G.V. Chilingarian
Civil Engineering Department, University of Southern California, Los Angeles, CA 90089 – 1211 (USA)

ELSEVIER
Amsterdam—London—New York—Tokyo 1992

ELSEVIER SCIENCE PUBLISHERS B.V.
Sara Burgerhartstraat 25
P.O. Box 211, 1000 AE Amsterdam, The Netherlands

ISBN 0-444-88516-1

This book has been printed on acid-free paper.

Printed in The Netherlands

DEDICATION

This book is dedicated to the following geologists who have made fundamental contributions to geology:

Yakov Eventov, for his important contributions to basin analysis, exploration, geochemistry, origin of oil, plate tectonics, and salt basins and tectonics;
Albert V. Carozzi, who contributed to carbonate petrology;
E.C. Dapples, his stratigraphic-sedimentary-tectonic frameworks put diagenesis into a regional context;
H.P. Eugster, who contributed to saline lake diagenesis;
Konrad B. Krauskopf and **Robert M. Garrels,** who have contributed to low-T/low-P geochemical (diagenetic, etc.) information;
J.B. Maynard, for his work on low-T/low-P sedimentary rock-hosted ore genesis;
S.J. Mazzullo, for his contributions to carbonate sedimentology, diagenesis, source rocks and reservoir rocks;
Arthur A. Meyerhoff, for his ideas on plate tectonics, basin analysis, oil genesis — where diagenesis finds an unequivocal context;
E. Roedder, who has urged the application of fluid-inclusion studies to diagenesis, among others;
Richard C. Selly, for his useful overviews of sedimentology, stratigraphy, basin analysis, among others;
Bern P. Tissot and **Dietrich H. Welte,** who have made many invaluable contributions to maturation, burial diagenesis, etc., related to oil genesis and exploration; and
I. Valeton, for her work in pedogenesis-related secondary changes (e.g., laterite- and bauxite-related diagenesis).

We also dedicate this book to **H. (HERBIE) S. ARMSTRONG** (Emeritus Dean, University of Guelph, Ontario), whose superb first-year geology lectures "compelled" the senior editor (K.H. Wolf) to transfer to geology; and to **MIHRAN AGHBABIAN,** President of American University in the Republic of Armenia, who inspired the second editor (George V. Chilingarian) to write many books.

LIST OF CONTRIBUTORS

G.V. CHILINGARIAN, Civil Engineering Dep., University of Southern California, Los Angeles, CA 90089-1211, USA

P.K. DUTTA, Dep. of Geography and Geology, Indiana State University, Terre Haute, IN 47809, USA

M.R. GIBLING, Dep. of Geology, Dalhousie University, Halifax, N.S. B3H 3J5, Canada

R.V. HURST, Chempet Research Corporation, 330 N Zachary Avenue, Suite 107, Moorpark, CA 93021, USA

H.L. JAMES, 1320 Lakeway Drive, Foothills 121, Bellingham, WA 98228, USA

M.P.R. LIGHT, ECL Petroleum Technologies, Henley-on-Thames, Oxon RG9 4PS, Great Britain

A.H. MOHAMAD, NAM (Shell/Exxon), Schepersmaal 1, 9405 TA Assen, The Netherlands

P.K. MUKHOPADHYAY, Global Geoenergy Research Ltd., P.O. Box 23070, Dartmouth, S.C., Nova Scotia, B3A 4S9, Canada

H.H. POSEY, Consulting Geologist, 2020 Routt Street, Lakewood, CO 80215, USA

W. RICKEN, Geologisches Institut, Universität Tübingen, Sigwartstrasse 10, 74 Tübingen, Germany

B.R. RUST† (Dep. of Geology, University of Ottawa, and Ottawa – Carleton Geoscience Centre, Ottawa, Ont. K1N 6N5, Canada)

E.V. TUCKER, School of Engineering, Geomaterials, Queen Mary College, Mile End Road, London E1 4NS, Great Britain

C.H. VAN DER WEIJDEN, Dep. of Geochemistry, Inst. of Earth Sciences, Rijksuniversiteit Utrecht, P.O. Box 80021, 3508 TA Utrecht, The Netherlands

A.J. VAN LOON, Julianaweg 5, 6862 ZN Oosterbeek, The Netherlands

K.H. WOLF, P.O. Box 909, Woden, Canberra, A.C.T. 2606, Australia, and 18 Acacia Street, Eastwood, Sydney, N.S.W. 2122, Australia

V.P. WRIGHT, Postgraduate Research Institute for Sedimentology, The University, P.O. Box 227, Whiteknights, Reading RG6 2AB, Great Britain

YOUNG IL LEE† (Dep. of Geological Sciences, College of Natural Sciences, Seoul National University, Seoul 151, Korea)

CONTENTS

Chapter 1

INTRODUCTION: FROM MARINE INTERSTITIAL FLUIDS TO PALEOSOLS – A REVIEW

GEORGE V. CHILINGARIAN AND KARL H. WOLF

The present volume, Diagenesis III, is an intellectual agglomeration covering a variety of topics to please everyone. It starts with the diagenesis of marine pore waters and soft-sediment deformations, followed by two chapters on sandstones – one on climatic influence in terrestrial sandstone diagenesis and the other on the deep-sea volcaniclastic sandstones. Diagenesis of carbonates follows next – one chapter is on compactional diagenesis, whereas the other is devoted to a case study (Aymestry Limestone Beds in the United Kingdom). Next, there are two chapters on the origin and migration of oil: (a) maturation of organic matter, and (b) relation of diagenesis to mineralization and hydrocarbon reservoir development. Then follows a chapter on sedimentary ore genesis – banded iron-formation. Finally, the book concludes with two chapters on paleosols. These twelve chapters are summarized here.

CHAPTER 2 – EARLY DIAGENESIS AND MARINE PORE WATER, BY C. VAN DER WEIJDEN

This chapter offers an extensive overview of the developments during the last decade in the understanding and modeling of pore-water chemistry as related to early diagenesis. The decomposition of organic matter is the major driving force of early diagenesis in marine sediments. Organic matter reaching the sea floor consists of a mixture of organic compounds having different decomposition reactivities. Very labile organic compounds are rapidly mineralized* in the water column and at the sediment – water interface. Labile organic compounds are decomposed at more moderate rates and become buried in the sediment. The accumulation of organic matter in the sediments is a function of: (1) primary production in the photic zone above; (2) the total sedimentation rate; and (3) the porosity of the sediment.

Decomposition occurs mainly by micro-biota under oxic, suboxic, and anoxic conditions. Oxygen is used by aerobic bacteria first under oxic conditions. Consumption of oxygen within the top of the sediment can be measured in-situ by oxygen microprobes, even at great water depth. Suboxic processes take over at greater depths in the sediment where oxygen is depleted. Nitrate reduction (denitrification) is the second process involved in the mineralization of organic matter. On a global scale, oxygen consumption by breakdown of organic matter is about 5 – 10 times

* Mineralization of organic matter can be defined as the microbially-mediated oxidative breakdown of organic matter into the inorganic compounds – carbon dioxide, ammonium/nitrate, phoshates, plus trace constituents.

higher than nitrate consumption. The other processes involved in the decomposition of organic matter are reduction of solid manganese and iron oxides. Especially in the latter case, the extent to which this occurs depends on the content of easily-reducible ferric oxides, mostly present as coatings on sediment particles. Finally, under anoxic conditions, sulphate takes over the role of electron acceptor. In shallow-marine sediments, the contribution of sulphate reduction to the mineralization of organic matter is roughly an order of magnitude greater than oxygen consumption, whereas in deep-sea sediments the reverse is true. Methanogenesis starts when the sulphate pool is practically exhausted. During the upward diffusion, the produced methane acts as a carbon source for sulphate reduction at higher levels. Sediments with appreciable methane production usually are: (1) shallow-marine sediments with high organic matter content; and (2) sediments deposited under anoxic conditions.

As a result of decomposition of organic matter, carbonic acid may be either produced or consumed and bicarbonate is formed. This affects the pH of the interstitial water. The calculation of pH is not simple, because it depends on the buffer capacity of the bulk sediment. Nitrate that is produced by oxic mineralization is partly removed by denitrification as dinitrogen. Ammonium formed by ammonification can be partly adsorbed by clay minerals, or, after upward diffusion, is oxidized by nitrifying bacteria. Phosphate may be partly precipitated in the form of apatite, adsorbed onto carbonates or ferric oxides, and partly removed by diffusion in anoxic sediments. Depending on the content of easily-reducible ferric oxide in the sediment, the produced hydrogen sulphide and ferrous ions are removed as ferrous sulphides or pyrite. The relation between organic carbon and sulphidic S (C/S ratio) can be used as a diagnostic tool in the reconstruction of sedimentary environments. Any hydrogen sulphide that diffuses into suboxic/oxic sediment zones is readily oxidized into sulphate. Manganous ions may be either precipitated as carbonate (pure or mixed), or, during upward diffusion, become oxidized again. This produces a small zone of manganese oxide enrichment in the sediment.

Studies of pore-water chemistry are used to estimate the consumption of oxygen, nitrate, and sulphate, as well as to estimate the methane production involved in the breakdown of organic matter. Pore-water profiles for various constituents are the base for the estimation of fluxes across the sediment – water interface. Such fluxes play a role in the (bio)geochemical cycles of nutrients.

In the second part of Chapter 2, Van der Weijden presents some basic concepts and equations relevant to the modeling of profiles of pore-water constituents. Diffusion, advection, and production/consumption of constituents can be formulated by the general diagenetic equation of Berner (1980). The concept of steady-state is usually applied as a first, and mostly last, approximation in modeling profiles of dissolved constituents in pore water. The role of coupling of ion fluxes, which has been analyzed by Van der Weijden, plays a role for large concentration gradients. Diffusion coefficients of ion-pairs are of the same order of magnitude as those of the constituting ions. Although sophisticated models of enhanced dispersion in the benthic layer by bioturbation have been successfully applied, it has been shown that they can be replaced by more simple models. The lability of organic matter or compounds, which has been taken into account, generally decreases with depth in the sediment. A continuous rather than a stepwise reduction in lability of organic matter with depth is the most realistic assumption.

In the last part of Chapter 2, some models are presented that have been applied to measured profiles for oxygen, nitrate, manganese, sulphate, sulphide, methane, bicarbonate, phosphate, and silica in pore water. The selection was made in such a manner that it shows a broad spectrum of parameters and boundary conditions. Model-fits were performed by analytical as well as numerical solutions of the appropriate diagenetic equations, parameters, and pertinent boundary conditions. These examples demonstrated the power of realistic models in the understanding of early diagenetic processes and element fluxes across the sediment – water interface.

In the opinion of the editors, one of the main unresolved problems is obtaining representative pore-water samples, because the chemistry of solutions squeezed from marine muds changes with the magnitude of applied pressure (see Rieke and Chilingarian, 1974), and with time of storage prior to analysis.

CHAPTER 3 – THE RECOGNITION OF SOFT-SEDIMENT DEFORMATIONS AS EARLY-DIAGENETIC FEATURES – A LITERATURE REVIEW, BY A.J. VAN LOON

The deformations, which occur during early diagenesis, have attracted the attention of earth scientists for over a hundred years. The periods from 1950 to 1960 and from 1970 to 1980 have been particularly fruitful with regard to the understanding of these early-diagenetic phenomena. During these periods, new insights into their genesis were gained and new approaches in their analysis were started. The study of these early-diagenetic features is now considered to be almost as important as that of primary sedimentary structures in the reconstruction of the geological history.

Some types of soft-sediment deformations aroused great interest. This is particularly true for the wide variety of deformations formed under glacigenic conditions and for the group of deformations related to mass transport. Some specific structures have been (and still are) studied intensively, e.g., clastic dikes, load casts and convolutions. Other structures, such as rain-drop imprints, desiccation cracks, and sole markings received somewhat less attention – although the descriptive literature of these structures is vast. A third group of soft-sediment deformations was described and analyzed only occasionally, in spite of their most interesting genesis; examples of this category are gravifossum and kink structures.

This chapter by Van Loon shows how soft-sediment deformations were gradually recognized as early-diagenetic features, which cannot be neglected if the geological history of a sediment is to be accurately reconstructed.

In this context, the following ought to be pointed out. The authors have frequently directed the attention to one (at least) important enigma that needs to be further investigated: How to distinguish between "soft-sediment deformation" structures and "solid-rock-stage, tectonically-deformed" structures that are often very similar! The similarity between these two types of genetically-different structures can be so great that up to this day no method has been found that would permit a distinction. Many of the "soft-sediment" versus "hard-rock tectonic" deformation problems have been encountered in sedimentary-rock-hosted laminated and bedded ore deposits, such as at the world-renown Mount Isa mine, Australia (among many others on various continents).

CHAPTER 4 – CLIMATIC INFLUENCE ON DIAGENESIS OF FLUVIAL SANDSTONES, BY P.K. DUTTA

As pointed out by Dutta, significant progress has been made in many aspects of sandstone diagenesis. This has helped to explore and exploit sandstone reservoirs which produce nearly half of the world's petroleum. In spite of such progress, there is no understanding of the source of cement and mass-transfer mechanisms in the diagenesis of siliciclastic sediments. Whereas the earlier works on diagenesis have focussed on postdepositional factors/processes that turn "loose sand" into "indurated sandstone", the processes/factors controlling the precursor materials have received only little attention. Climate in the source area is one factor that has influenced the precursor components which, in turn, determined the type of sandstone diagenesis.

In continents, except along the high-relief mountainous regions, climate controls the intensity of chemical weathering of the lithosphere and this, in turn, controls the nature and abundance of the precursor materials. Chemical weathering is most intense in a warm humid climate where unstable minerals are chemically decayed, leaving behind relatively stable minerals in the zone of weathering. As a result, sediments derived from such a source are mineralogically mature. As a consequence of high atmospheric precipitation, groundwater within shallow depths is also low in dissolved mineral content. In a cold climate, the rates of chemical weathering are extremely slow, whereas in arid regions chemical weathering is insignificant owing to the lack of moisture. Under both cold and arid conditions, therefore, the products in the weathering profiles are characterized by the abundance of unstable minerals; i.e., "unstable" when moved to other depositional environments. Detritus derived from cold and arid terrains yields immature sediments in these new regimes.

During shallow burial of sediments with high initial porosity and permeability, groundwater moves fast and dissolution of detrital minerals is minimal. Groundwater at this stage is mainly controlled by climate. In both warm and temperate, but humid, climates, the early authigenic minerals are cation-poor silicates. Under extremely warm and humid conditions, gibbsite may also form as an early cement. In arid and cold climates, cation-rich silicates and carbonates constitute the early cements.

During deep burial, the volume of groundwater is restricted and moves very slowly. Because of high burial temperature and sluggish movements of groundwater flow, the detrital minerals tend to react through dissolution and alteration mechanisms. This results in making the pore water highly concentrated with respect to total dissolved material. The sediments derived from intensely weathered humid regions are rich in silica, and dissolution of these materials gives rise to a pore solution deficient in most mobile metallic cations. Even during deep burial, therefore, such sediments will have cation-poor cements. On the other hand, the sediments derived from regions with an arid and cold climate will be rich in immature detrital minerals. Dissolution of such minerals, which are enriched in most common cations, will give rise to cation-rich interstitial solutions and, consequently, cements.

Relating the nature and abundance of authigenic minerals to climate is of great importance to the understanding and reconstruction of any diagenetic history in both space and time: it will assist in characterizing and assessing the potential of petroleum and groundwater reservoirs, for example.

CHAPTER 5 – DIAGENESIS OF DEEP-SEA VOLCANICLASTIC SANDSTONES, BY YONG IL LEE

In deep-sea environments of the active plate margins, volcaniclastic sandstones occur abundantly in both back-arc and fore-arc basins. The sandstones in both of these basins have a large amount of basaltic and andesitic rock (= lithic) fragments and glass matrix, which are subject to intensive diagenetic alteration during burial.

Sandstones in the back-arc basins show a range of differences in the degree of diagenesis depending on the times of deposition with respect to rifting in the basin. Major diagenetic changes are associated with sandstones accumulated at a slow burial rate during intensive heat-flow events associated with early basinal rifting, whereas sandstones deposited at a more rapid burial rate when rifting ceased and heat flow from the basement was normal show least-intensive diagenetic alteration. Similar diagenetic alterations are expected in the fore-arc basin sandstones, with potential minor differences.

As pointed out by Lee, such factors as sandstone age, sandstone composition, and burial rate, as well as heat flow from basement should be considered in the diagenesis of back-arc and fore-arc basin sandstones. According to Lee, other factors being equal, lesser thermal influence from the basement is expected in the fore-arc basins.

CHAPTER 6 – A VOLUME AND MASS APPROACH TO CARBONATE DIAGENESIS: THE ROLE OF COMPACTION AND CEMENTATION, BY W. RICKEN

In this chapter, Ricken introduces a new concept for the quantification of carbonate diagenesis which describes diagenetic processes in terms of changing sediment and rock volumes. A basic expression of this concept is a numerical relationship between compaction, porosity, and the carbonate content, i.e., Ricken's carbonate compaction equation. Compaction measurements prove that this relationship is well-documented in the rock record – thus, a new approach to carbonate diagenesis is possible: the various types of diagenetic sediment-to-sedimentary rock transformations can be distinguished, and the diagenetic histories of given calcareous rock samples can be simulated. Using Ricken's compaction equation, the amount of compaction, the cement content, and the quantity of carbonate dissolution can be predicted. Carbonate mass balances and related numerical decompaction show the primary composition, porosity, and degree of closure of the diagenetic system. The compactional enrichment of minor constituents, such as organic carbon and trace elements, is also demonstrated in this chapter.

Application of the volume approach to diagenetic processes is demonstrated by Ricken for interbedded marl – limestone sequences. He also quantified the diagenetic influence on bedding rhythms, i.e., the enhancement in difference in carbonate content.

CHAPTER 7 – DIAGENETIC HISTORY OF THE AYMESTRY LIMESTONE BEDS (HIGH GORSTIAN STAGE), LUDLOW SERIES, WELSH BORDERLAND, U.K., BY A.H. MOHAMAD AND E.V. TUCKER

Drs. Mohamad and Tucker provide a diagenetic history of the Aymestry Limestone, which represents the only significant carbonate facies in the Ludlow Series of the British Isles. Shallowing and shoaling sequences on a narrow shelf sea in the Welsh Borderland preserved two distinctive megafacies: (1) nodular limestones (wackestones/packstones), which are genetically related to burrow-fills, cut-and-fill sedimentary structures and post-diagenetic pressure-solution phenomena; and (2) calcareous mudstones and siltstones. Storm-generated shell bank deposits succeed the mudstones and siltstones near the shelf edge. The temporal and spatial distribution of these facies reflect the progressive infill of a semi-protected embayment, proximal to the shelf edge. Within the carbonate-rich sequence, sediment lithification originated from early submarine cementation, modified progressively as the salinity of the pore fluid changed in response to comingling of marine and freshwater phreatic phases associated with shallowing. Subsequent diagenetic changes are associated with phreatic to vadose phases and burial diagenesis.

Early-formed marine cements are associated with micrite envelopes, which are ascribed to repeated boring – infilling by endolithic algae (*Girvanella* sp.) at or near the water – sediment interface. In addition, syntaxial micrite cement (high-Mg precursor) also occurs as pore-fill within the stomapores of crinoid ossicles. Second-generation cements probably originated within the mixing zone environment and are preserved as syntaxial fibrous and/or botryoidal calcite, and isopachous rims comprising high-Mg calcite precursors. Other characteristics of these early-formed cements are multiple domal growth fronts associated with competitive growths on free surfaces, dusty inclusions as a result of rapid crystallization, and microboring traces on fibrous crusts. The occurrence of micro-omission surfaces, truncating the syntaxial calcite crusts related to chondritic burrowing, provide circumstantial evidence of early cementation. Inclusion-rich, neomorphosed granular cements, which postdate the syntaxial fibrous calcite, probably crystallized in the freshwater end of the mixing zone diagenetic spectrum. Incipient dolomitization is present, primarily preserved within crinoid stomapores as microdolomite rhombs.

Inclusion-free drusy calcite, which forms the subsequent void-filling, grew and enlarged centripetally. These paraxial blocky calcites are coarsely crystalline with rhombic to scalenohedral shapes and planar crystalline boundaries with a characteristic enfacial junction. Staining by alizarin red – potassium ferricyanide revealed multiphase zonation (rhombocentric) of ferroan and non-ferroan calcite compositions. These calcites are associated with slow rate of crystallization in a freshwater phreatic environment saturated with $CaCO_3$.

According to Mohamad and Tucker, vadose diagenetic fabrics are mainly confined to incipient hardground per se, characterized by poikilotopic calcite engulfing pre-existing cement fabrics. Typically, the poikilotopic cements are very coarsely crystalline (1000 μm in size) having planar boundaries with enfacial junctions. The inclusions or poikilotopes comprise recrystallized syntaxial overgrowths and pseudomorphs of former dolomites. Both recrystallization and dedolomitization of the poikilotopes (which predate the genesis of poikilotopic cement) are due to flushing of magnesium by undersaturated meteoric waters. Another notable feature associated with hardground is the character of the wall lining of *Trypanite* borings: the sparry calcites that line the wall have been degraded to spotty micrite. The genesis of such secondary micrite is ascribed to dissolution – reprecipitation processes. Other subordinate fabrics, such as fracture healing, pressure solution, and silicification, are generally associated with deep-burial diagenesis.

The types and fabric relationship of the various cements and their chemical/mineralogical characteristics indicate a progressive change from marine through mixing zone to phreatic and vadose diagenetic environments. The dominant diagenetic setting of the Aymestry Limestone, however, was the mixing zone environment.

CHAPTER 8 – GEOCHEMICAL AND ISOTOPIC CONSTRAINTS ON SILICA AND CARBONATE DIAGENESIS IN THE MIOCENE MONTEREY FORMATION, SANTA MARIA AND VENTURA BASINS, CALIFORNIA, BY R.W. HURST

As pointed out by Hurst, silica and carbonate diagenesis provide important controls on petroleum production and migration in the Miocene Monterey Formation in California. Before introducing new trace element and isotopic (Sr, O) data, which bear on silica and carbonate diagenesis, he reviewed numerous contributions by his predecessors covering various aspects of Monterey Formation geology and geochemistry. These areas include: paleoceanography; the oxygen minimum layer, its relation to laminated diatomites, and resulting source-bed potential; depositional history; climate and tectonic events and their influence on Monterey Formation diagenesis; Monterey Formation lithofacies; diagenesis and physical properties of siliceous sediments; secondary carbonates and their geologic – geochemical characteristics; and more recent experimental and oceanographic observations.

The Sr isotopic evolution of seawater and its utility as a chronostratigraphic tool are discussed in this chapter. Results of Sr isotopic analyses of dolomites from the Ventura Basin (opal-A diagenetic grade) and Santa Maria Basin (opal-CT to quartz chert diagenetic grades) are also presented. The data indicate that dolomites subjected to longer diagenetic histories and higher silica grade record the geochemical evolution of the enclosing sediment, including events such as the silica polymorphic transformations (opal-A to opal-CT to quartz).

The Rb/Sr systematics of silica polymorphic transformations indicate that well-mixed diatomaceous sediments expel fluid during the transition to opal-CT. The

$^{87}Sr/^{86}Sr$-ratio of the expelled fluid is identical to that of the opal-CT which precipitates from this fluid. This dehydration reaction also accompanies the opal-CT to quartz transition. Hence, the higher-grade silica polymorph, formed during solution – precipitation reactions, is in Sr isotopic equilibrium with the fluid expelled. This equilibrium is recorded as the Sr initial ratio of linear arrays defined by the opal-CT and quartz cherts and marks the Sr isotopic evolution of interstitial waters in the Monterey Formation.

The evolution of the Rb/Sr system in Monterey siliceous sediments can be used to determine the timing of the silica polymorphic transformations and to correlate episodes of dolomitization with the transformations. Based upon the Sr isotopic data from dolomites in the Santa Maria Basin, Hurst concluded that Monterey Formation sediments approached mineralogic and chemical homogeneity during early diagenesis. This controlled the Sr isotopic evolution of interstitial waters on a basin-wide scale. As a result, dolomites occurring within fractures in dilation breccias have Sr isotopic compositions which are identical to the Sr initial ratios of the silica polymorphic linear arrays. According to Hurst, the association of hydrocarbons with the dolomite in the dilation breccias suggests that fractured dolomites, produced during silica polymorphic dehydration reactions, may be important reservoir rocks in the Monterey Formation. In this connection, the editors would like to mention that if oil is present only in the fractures, then the total quantity of oil is indeed very small, because porosity (ϕ) due to the fractures alone is less than 1% (see Chilingarian and Yen, 1986).

CHAPTER 9 – MATURATION OF ORGANIC MATTER AS REVEALED BY MICROSCOPIC METHODS: APPLICATIONS AND LIMITATIONS OF VITRINITE REFLECTANCE, AND CONTINUOUS SPECTRAL AND PULSED LASER FLUORESCENCE SPECTROSCOPY, BY P.K. MUKHOPADHYAY

Incident-light microscopic measurements of vitrinite reflectance, continuous spectral fluorescence, and pulsed laser fluorescence reveal various aspects of organic matter evolution.

As pointed out by Mukhopadhyay, vitrinite reflectance measurement is often oversimplified resulting in major confusion in solving geological problems. His chapter deals with the concepts, principles, and fundamental problems regarding vitrinite reflectance measurements. The potential of vitrinite reflectance is fully revealed when: (a) standardization methods are properly followed; (b) choice of vitrinite macerals or submacerals are fully understood; (c) the effects of heat flux, organic facies, and kerogen type on vitrinite reflectance are known; (d) the relation between the increase in vitrinite reflectance and hydrocarbon generation from liptinites is fully established; and (e) the kinetics of vitrinite reflectance is known to some extent.

Spectral fluorescence parameters (lambdamax, *Q*-value, and alteration) are found to be more useful in determining maturity and characterize genetic types of kerogen (especially Types I and IIA in source rocks), bitumen, crude oil, and condensate in a single microscopic system, provided chemical kinetics of fluorescence is fully understood. Fluorescence parameters are more reliable in documenting "oil window" in a sedimentary sequence than vitrinite reflectance in some cases, because hydrocarbon generation can be correlated with the fluorescence red shift of some specific liptinite macerals.

Pulsed laser fluorescence, which utilizes fluorescence decay time and component spectra of at least three fluorophores, was documented for the first time as a useful technique in delineating maturation of crude oil (especially biodegraded) and condensates as well as for oil – oil correlation by microscopic means.

CHAPTER 10 – DIAGENESIS AND ITS RELATION TO MINERALIZATION AND HYDROCARBON RESERVOIR DEVELOPMENT: GULF COAST AND NORTH SEA BASINS, BY M.P.R. LIGHT AND H.H. POSEY

Mineral and fluid diagenesis and low-rank metamorphic reactions in sedimentary basins are controlled, in the simplest sense, by little more than variations in burial temperatures and original mineral and fluid composition. However, in basins that undergo fluid expulsion as a result of overpressuring, and particularly in basins where oil and gas are generated and moved by overpressuring, these diagenetic and metamorphic reactions respond to more complex controls. Multiple generations of mineral deposition and destruction develop because fluids and hydrocarbon-associated gases are driven from high-temperature environments, where they form, to shallower, lower-temperature regimes where they are out of equilibrium and, thus, cause minerals to dissolve or form. In basins where salinities vary widely during burial, these reactions are even more complex.

For the U.S. Gulf Coast Basin, which is an evaporite – sediment – hydrocarbon-generating basin, Light and Posey developed an integrated hydrothermal model to explain relationships between basin structures, authigenic mineral composition, salinity, hydrocarbon maturation, salt dome caprock formation, and salt dome mineralization. Their integrated hydrothermal model applies to the North Sea Basin. Throughout the chapter, the possible, though unknown, roles that metamorphic reactions, particularly the important reactions that release water and carbon dioxide, play on diagenesis in the shallower sedimentary column are discussed.

Salt domes, their associated caprocks, minerals, and hydrocarbons are a normal aspect of evaporite basin evolution that involves either the destruction and transformation of minerals by temperature and chemical changes associated with burial diagenesis, material flow within and around the diapir, oil and gas maturation and migration, and fresh-water intrusion in the shallow environment. Inasmuch as diapirs intrude over relatively long periods, the amount of piercement during any single stage is small compared with the total amount of apparent upward migration of the salt mass. Halite diapirs "intrude" sedimentary cover probably with the aid of warm fluid that invades the evaporite body, probably from beneath the evaporite body.

Halite diapirism and anhydrite caprock formation begin soon after deposition of the mother salt and cease either when the supply of halite from mother salt is exhausted, the evaporite plug is cut off from the mother salt, or when rapid burial forms too thick a cover for the diapir to penetrate. Calcite caprocks may form any time after anhydrite caprock, but require mature hydrocarbons for their formation. Base metal sulfides and barite form during the entire caprock forming event and must involve fluids both from below and above the mother salt. Fluids involved in anhydrite caprock formation are of unknown salinity, slightly undersaturated with respect to anhydrite, and are slightly reducing. Calcite caprock-forming fluids are mixtures of formation and meteoric waters, the mixture of which is probably relatively warm. Gypsum and sulfur, which form late in the caprock sequence, form in the presence of substantially lower temperature, lower salinity, and, probably, meteoric fluids.

Caprock metals are derived during burial diagenesis principally from the breakdown of plagioclase and clay, albitization of feldspar, transformation of smectite to illite, and, possibly, the destruction of anhydrite and calcite. Reduced sulfur for metal sulfides is derived from two sources: thermochemically-reduced and biochemically-reduced end-members or deep and shallow sources, respectively. Sulfur for native sulfur deposits is probably derived through biogenic reduction of anhydrite sulfate, the unreduced portions of which may form barite.

CHAPTER 11 – PRECAMBRIAN IRON-FORMATIONS: NATURE, ORIGIN, AND MINERALOGIC EVOLUTION FROM SEDIMENTATION TO METAMORPHISM, BY H.L. JAMES

The Precambrian iron-formations of the world have been the focus of thousands of individual studies and surveys over the past hundred years, and they continue to be a vital target for further detailed examinations. They are the source of the bulk of iron ore mined, and the reserves, even in the face of an extraction rate approaching a billion tons per year, remain enormous. Scientifically, the distinctive chemical compositions, the virtual limitation to the Precambrian, and the contained evidence of biological activity in some provide fertile grounds for speculation on the evolution of the earth's hydrosphere, atmosphere, and biosphere, as well as the lithosphere.

Iron-formation is a general term that covers a variety of iron-rich, thinly-bedded or laminated rocks of chemico-sedimentary origin and varying metamorphic grade. Typical compositions are 25 – 35% Fe and 35 – 50% SiO_2. The latter occur most commonly as interlayered chert or its metamorphic equivalent.

In a quantitative sense (i.e., based on total tonnage), three-fourth (or more) of the known iron-formations is found in eight major districts, distributed over five continents, within shelf-type sequences of early Proterozoic age (2500 – 1900 m.y.). These particular deposits of iron-formation, as much as 1000 m thick, are considered to be products of a reaction between upwelling anaerobic deep ocean waters, in which iron and silica had been accumulating throughout much of the Archean time, and oxidic surface waters of continental shelves and marginal basins that developed in early Proterozoic time, following world-wide Archean cratonization. Thousands of other deposits, most of lesser dimension, are spatially associated with, and genetically related to, volcanic rocks of contemporary age. A large percentage of these occur in greenstone belts of late Archean age, but some are as young as early Paleozoic. A third group, relatively small in number, but including some of substantial size, are of late Proterozoic age, deposited along rifted continental margins.

The mineralogical composition of existing iron-formations, all of which are metamorphosed to some degree, was set initially by the physicochemical properties of the depositional environment, which had a possible range from strongly oxidizing to strongly reducing. Initial deposits, depending upon local conditions, consisted variously of iron oxide hydrates, iron-rich carbonate, and iron-rich silicate mud, all interbedded with silica gel. These materials were converted by sea-bottom reactions and ensuing diagenesis to stable assemblages of iron oxides, siderite, iron silicates, and chert. Under extreme reducing conditions, reactions between organic material, entrapped sulfate-bearing seawater, and initial iron precipitates yielded pyritic iron-formation. The metamorphic imprints on these deposits are many and varied. Of particular significance is the persistence of stable assemblages consisting of quartz (chert, initially) and iron oxides (magnetite and/or hematite) even under extreme conditions of temperature and pressure. Oxidic iron-formation remains a recognizable constituent in Precambrian metasedimentary sequences, even in those of great age and high metamorphic grade.

In this context, James emphasized that many genetic interpretations of the iron-formations have been offered by numerous researchers – and no full agreement has as yet been achieved.

CHAPTER 12 – PALEOSOL RECOGNITION: A GUIDE TO EARLY DIAGENESIS IN TERRESTRIAL SETTING, BY V.P. WRIGHT

Most terrestrial sediments are likely to have undergone some pedogenic alteration. It is essential to recognize such changes, not only to differentiate them from later shallow or deep-burial alteration, but also because paleosols can provide a wealth of fine detail about ancient environments.

This chapter, by Wright, aims to provide an introduction to the recognition of paleosols especially in outcrop. Criteria such as horizonation, color, and the nature of boundaries are considered especially useful, as also are distinctive pedogenic structures. Confirmation of extent and type of the pedogenic processes usually requires granulometric, chemical, mineralogical, and micromorphological studies.

Having identified pedogenically-altered zones in the rock record, there are several complicating factors to be aware of before reliable interpretations can be made. Soils form slowly – whereas environmental changes, as a consequence of climatic, geomorphic and vegetation changes, can be rapid and frequent.

As pointed out by Wright, not only will these changes impart complexity to any soil, but the profile itself will naturally change as the soil evolves from an immature to mature form. In aggrading situations, any point on the soil will pass through various soil levels as the profile aggrades. Erosion of the profiles will result in horizons being overprinted by higher ones. Many paleosols reported from the geological record so far are simple, single-phase profiles, which is clear evidence of the episodic nature of sedimentation, that occurs in discrete, short-lived phases separated by long periods of no net sedimentation.

CHAPTER 13 – SILICA-CEMENTED PALEOSOLS (GANISTERS) IN THE PENNSYLVANIAN WADDENS COVE FORMATION, NOVA SCOTIA, CANADA, BY M.R. GIBLING AND B.R. RUST

Ganister-bearing paleosols in the Waddens Cove Formation of Nova Scotia, Canada, formed within crevasse-splay, levee, and channel deposits following landform abandonment under a seasonal, tropical climate. According to Gibling and Rust, the ganisters were substantially lithified at or just below the floodplain surface, as shown by channel margins stepped over ganisters and ganister slump-blocks and fragments in channel deposits.

The ganisters contain 81 – 86% silica, with aluminous and ferruginous material, and up to 1% titanium oxides. Microquartz replaced clays interstitial to framework quartz grains and filled vugs, but overgrowths on quartz grains are rare. Illuviation and embedded-grain cutans (argillans and sesquans) are common, along with sideritic rhizoliths and hematitic glaebules. Authigenic titanium oxides are disseminated within clay and microquartz patches. Silica was derived from dissolution of feldspar and embayed quartz grains, in addition to clays.

Weathering and cementation of sandy parent material took place under low and, probably, fluctuating pH conditions associated with a seasonally variable groundwater level and abundant vegetation. Poor development of paleosol profiles probably reflects the aggradational, proximal floodplain setting and cumulative profile formation. Gibling and Rust pointed out that unlike ganisters elsewhere, these ganister-bearing paleosols are not overlain by coal, and a subsequent rise in relative base-level probably would have been required for peat accumulation.

The ganisters show petrographic features, including microquartz cement and evidence for titanium oxide authigenesis, which are analogous to those of some silcretes.

REFERENCES

Berner, R.A., 1980. *Early Diagenesis.* Princeton Univ. Press, Princeton, N.Y., 241 pp.

Chilingarian, G.V. and Yen, T.F., 1986. Notes on carbonate reservoir rocks, No. 3: Fractures. *Energy Sources,* 8 (2/3): 261 – 275.

Rieke, H.H., III and Chilingarian, G.V., 1974. *Compaction of Argillaceous Sediments.* Developments in Sedimentology, 16. Elsevier, Amsterdam, 424 pp.

Chapter 2

EARLY DIAGENESIS AND MARINE PORE WATER

CORNELIS H. VAN DER WEIJDEN

INTRODUCTION

This chapter is restricted to the study of the degradation of organic matter by a suite of redox processes in marine sediments, with an emphasis on the effect on pore water chemistry, the modeling of concentration profiles, and the calculation of the concomitant fluxes across the sediment – water interface. The accumulation of organic matter in marine sediments is highly relevant to actuo- and paleo-oceanography. In the last decade important progress has been made toward the qualitative and quantitative understanding of the processes and conditions that determine the preservation of organic matter in marine sediments. Geochemists, sedimentologists, chemical oceanographers, and geomicrobiologists have made important contributions to this progress of the present-day understanding of biogeochemical cycles in the oceans. The number of papers and reviews on marine pore-water chemistry, and changes therein in relation to diagenetic reactions in the sediments, has increased dramatically during the last decade. It is not possible to refer to all of them in the context of this chapter. The major cause of changes in the pore-water chemistry is no doubt the breakdown of organic matter during early diagenesis, with its most pronounced activity in the top layers of the sediments. Although such activities can be studied in a black box approach, in which the marine geochemist or sedimentologist only studies the net effects of this breakdown, it certainly adds much to our understanding of the apparent processes when attention is paid to the role of macro-, meio- and micro-organisms in this breakdown. Excellent reviews and books are dedicated to this subject (e.g., Nedwell, 1984).

Accumulation of organic matter in marine sediments, i.e., the organic matter that escapes (micro)biological breakdown and becomes permanently buried in the sediment column, has always been the subject of intensive study from the point of view of fossil fuel genesis. But the role that this accumulation plays in the geochemical cycles of carbon, nitrogen, oxygen and phosphorus, and their role in global change is by now the most important motivation for detailed studies of the processes that govern the breakdown of organic matter and recycling of the mineralized products. International cooperation in the next decade in global ocean flux studies will certainly bring about a still better understanding of these cycles and of the role of early diagenesis in marine sediments and their pore waters in these cycles.

Apart from the geomicrobial point of view, the geochemical studies on the breakdown of organic matter and its products has also contributed much to a better understanding of the processes (Degens and Mopper, 1975, 1976). The introduction and increasing use of sediment traps has provided a method to relate the fluxes of organic matter toward the sediment – water interface (sedimentation), into the top sediment layers with high microbiological activity, and, finally, into the deep sediment layers with low microbiological activity (accumulation). Pore-water studies, with emphasis on the consumption of oxygen and other electron acceptors and production of conjugate electron donors accompanied by the production of carbon dioxide and nutrient species, have contributed much to the understanding of the kinds and rates of degradation processes affecting organic matter within the sediment.

In addition to the role played by the degradation of organic matter, other reactions may also have an impact on changes in pore-water chemistry. Sediments consist of a mixture of solid particles immersed in pore waters. Secondary reactions will occur because particles of one kind or in combination with other different types, can be unstable in their sedimentary habitat, giving rise to dissolution, recrystallization, or neoformation of mineral phases. Because most reactions have an impact on the pore-water chemistry, the studies of changes therein often provide evidence of the occurrence of such diagenetic reactions. Diagnosis of pore-water chemistry profiles can also reveal the presence of sources (e.g., deep-seated evaporites) or sinks (ocean-floor basalts) for pore-water constituents and record sedimentological events (e.g., turbidites). Reviews on pore-water chemistry and their use in the interpretation of the sedimentological and paleo-oceanographic record and processes can be found in Rieke and Chilingarian (1974), Gieskes (1975, 1983), Manheim (1976), Price (1976), and Hesse (1986).

The aims of studies on interstitial water (as mentioned by Gieskes, 1975) have broadened since then and now may be restated as follows:

(1) To study early diagenesis of organic matter, its concomitant chemical reactions and reaction rates that are involved, and its role in the burial or recycling of solid and dissolved constituents in the sediments.

(2) To give insight into secondary diagenetic reactions and their character (dissolution, recrystallization, and neoformation of mineral phases), occurring within the sedimentary column in a variety of depositional regimes.

(3) To provide qualitative and quantitative information on element fluxes into or out of the sediment column at its outside boundaries (sediment – water or sediment – ocean-floor interfaces).

(4) To reveal events that changed the boundaries or boundary conditions (e.g., salinity, turbidites, intrusion of sills), the presence of deep-seated sources (e.g., evaporites), or externally-induced advection processes (e.g., hydrothermal activity).

The separation of pore waters from the bulk sediment is usually done in one of the following ways (in order of the most frequent use). (1) The squeezing of an aliquot of the sediment under pressure in a cylinder equipped with a membrane filter on top of a supporting screen located above the outlet through which the squeezed-out solution can flow into sampling bottles. Because it is possible that the magnitude of compaction pressure used changes the chemistry of the squeezed-out solutions, the accuracy of this method has been questioned by Rieke and Chilingarian (1974). (2) Whole-sediment squeezing: a recently developed method (Bender et al., 1987). (3) High-speed centrifugation of an aliquot of the sediment, and siphoning-off of the supernatant expelled solution into a sampling bottle. (4) In-situ dialysis (or filtration) of pore water through membranes placed in fixed positions in a tube that encloses the sediment core after penetration of the corer into the sediment. The latter method is in principle the most reliable for getting samples of pore water with a minimum of deviations from their actual chemistry. For the same reason, some concentrations or parameters can be measured by in-situ probes (e.g., oxygen, pH, Eh, and formation factor) with a minimum of disturbance. For short cores (box cores) this method can be done in a continuous profile starting at the sediment – water interface up to a depth of some tens of centimeters. For longer cores (piston cores), the probes have to be installed at fixed depths. For better interpretation of these in-situ measurements, it is generally necessary to retrieve the sediment core after the measurements and/or after the completion of the pore water collection, to further examine the recovered sediment column, shipboard or in the home-based laboratory. Prior to the shipboard sqeezing or centrifuging to obtain samples of pore water, the sediment cores have to be transferred from the sea floor to sea level. This always means, to a minor or major extent, depressurization of the sediment and, in many cases, also a change in temperature, before the sediment can be further processed. A change in pressure brings about a shift in chemical equilibria governing the pore-water chemistry. The example, most often given, of this change is the occurrence of precipitation of calcium carbonate, which changes the calcium concentration and carbonate alkalinity of the pore water to an extent that is dictated by the change in solubility. Relative to a change in pressure, a change in temperature has an even greater effect on chemical equilibria in the bulk sediment. This is the reason why the sediments are squeezed or centrifuged in an environment that is held at its original, in-situ temperature.

A problem one has to be aware of, is that the sediments are usually depleted in oxygen, consumed in oxidative diagenetic reactions. Sampling and squeezing/centrifuging has in these cases to be carried out under an artificial atmosphere (in glove bags, boxes or centrifuge bodies) with negligible oxygen content. If temperature and atmosphere, under which the expression of pore water is carried out, are not kept under proper control, the resulting pore-water chemistry will, for a number of elements, differ from its true (in-situ) composition. For some chemical constituents the results will be (and have been, unfortunately, in the past) so erroneous that they should be discarded.

Other problems that one has to be aware of in the interpretation of pore-water data are: (1) loss of the top of the sediment in the cores, (2) distortion of sediment layers, even with cores that are filled only with material from one horizon, and (3) shortening of the recovered sediment column in comparison to its thickness (e.g., Lebel et al., 1982). Coring techniques with a minimum of distortion, e.g., those carried out by divers and from submersibles, are generally not realistic. The improvement of corers launched from research vessels is still in progress. Thus, for the meaningful study of pore-water chemistry, an experienced team of geochemists or geomicrobiologists, sedimentologists/paleontologists, and marine technicians is required. In order to be able to evaluate the results of pore-water studies, a full description of the applied techniques and methods should be included in the published papers.

The major framework of early diagenetic processes is discussed in this chapter at two levels. Part I presents a largely qualitative description of the processes. In Parts

II and III, a more quantitative approach of the processes is presented in physico-chemical models. The choice of the examples in Part III is made in such a manner that different and typical applications of modeling on the basis of diagenetic equations are presented.

PART I: EARLY DIAGENETIC PROCESSES

Organic carbon, diagenesis and related impact on pore-water chemistry

Most estimates of the burial rate of organic carbon (C-org) in marine sediments range between 40 and 180 Tg a^{-1} (Williams, 1975; Holland, 1978; Berner, 1982; Romankevich, 1984; Lasaga et al., 1985). Typical C-org contents of marine sediments are 0.3% for deep-sea sediments, 2% for geosynclinal sediments, and 1 – 5% for shelf sediments (Degens and Mopper, 1976). The fraction of C-org that becomes permanently buried in the sediments is about 0.4 – 1% of the mean annual primary production in the oceans. Most of the primary production is consumed and recycled in the food chain in the euphotic zone. According to De Vooys (1979), the range of estimates of the relative amounts of C-org raining out from the euphotic zone into deep waters is 1 – 15%. Pelet (1981) estimated that 1 – 2% of the formerly-living organic matter escapes biological recycling and he stated that the accumulation of organic carbon correlates with the oxygen content of the water column.

Degens and Mopper (1976) estimate that 3 – 7% of the annual global p.p. (primary production) leaves the euphotic zone, 1.3 – 2.4% (relative to p.p.) of which is solubilized/oxidized in the water column and 1 – 4% (relative to p.p.) in the sediments, leaving 0.2 – 0.6% (relative to p.p.) to become trapped within the sediments. Such figures illustrate the importance of the degradation of organic matter within the sediments. This degradation is a biological process in which macro-, meio- and microfauna are involved. The complexity of these systems is discussed by Fenchel and Jørgensen (1977), who focus on the role of bacteria in the metabolism of organic detritus. Degens and Mopper (1975, 1976) noted that metazoan activity, and not bacterial activity, seems to be responsible for the breakdown of organic matter in oxic surface sediments. This may still be in accord with Fenchel and Jørgensen's (1977) conclusion that grazing (the constant removal of individuals) of the bacterial population increases its productivity. The role of bacteria was also discussed by Rowe and Denning (1985). They stated that, although the effectiveness of shallow-water bacteria decreases drastically when brought under hydrostatic pressure, microorganisms typical for the deep-sea environment have been found, that show the reverse behavior (barophiles) (Yayanos et al., 1979, 1981). This illustrates the ability of living organisms to adjust to changes in habitat. According to Rowe and Denning (1985), bacterial counts in Atlantic abyssal plains are about $5\ (\pm\ 3) \times 10^8$ per gram of dry sediment in the top 5 cm, with a rapid decrease by one order of magnitude over the next 10 cm. For the bacterial biomass in these A.P.'s, Rowe and Denning give an average of 142 mg C m^{-2}. The calculated C-org utilization was 0.3 mg C m^{-2} day^{-1} (Biscay abyssal plain) and 1.25 mg C m^{-2} day^{-1} (Demarara abyssal plain), which, on average, amounted to about 60% of the organic matter that accumulated at the top of the sediments. This, in turn, amounted to only 16% of the flux of organic matter to the top of the sediments as measured by sediment traps. In other words, 6.5% of the total flux of C-org becomes permanently buried and 93.5% of this flux is utilized, 84% of which is consumed by organisms on or above the sediment – water interface. Jahnke et al. (1982a) found for MANOP sites C and S in the Central Pacific that 98.5 – 100% of the flux of C-org, as measured by sediment traps, was decomposed on top of or within the sediments, leaving only 1.5 – 0% to be buried. This high degradation is confirmed by Emerson et al. (1985), who found that this occurs mainly in the uppermost few centimeters of the sediments. Berelson et al. (1987) concluded on the basis of chemical balances and total carbon dioxide fluxes, that roughly 1/3 of the C-org arriving at the floor of San Pedro (about 850 m depth) and San Nicolas Basin (about 1550 m depth) off the Californian coast at Los Angeles is recycled. Henrichs and Farrington (1984) estimated that in the highly productive region near 15°S off the Peru coast, 20 – 30% of C-org flux leaving the euphotic zone is oxidized at the sediment – water interface and in the surface sediments, and that 50 – 80% of this flux is accumulated in the sediments. The mineralization in the water column is thought to be relatively small in this region because most of the flux is composed of large pellets. Reimers and Suess (1983) calculated that between 37 and 83% of the sediment flux (as measured by sediment traps) was mineralized on top of the sediments, with further 47 – 11% diagenetically mineralized within the sediments on the Pacific – Antarctic Ridge, leaving 6 – 16% organic matter (relative to the flux) to become permanently buried. They also found, in agreement with Müller and Mangini (1980), that the rate constants for degradation of organic matter are several orders of magnitude higher for oxic than for anoxic sediments with comparable sedimentation rates. They attribute this to the macrobenthic community, which exists only under oxic conditions. Henrichs and Reeburgh (1987), however, concluded on the basis of microcosm and laboratory studies, that anaerobic degradation rates are not intrinsically lower than aerobic rates; fresh organic matter degrades at similar rates under oxic and anoxic conditions. Middelburg (1989) found that the rate constants for degradation of organic matter gradually decrease with time described by a simple power function. Pedersen and Calvert (1990) also concluded that the degradation rates of organic matter are equal for oxic and anoxic conditions. In sediments with high sedimentation rates labile organic matter is rapidly buried before it is degraded, whereas in sediments with low sedimentation rates appreciable amounts of labile organic matter can be destroyed at or above the sediment – water interface. In the deep ocean water column, particulate organic matter is continuously ingested, digested, and mineralized, depleting the organic matter in labile (i.e., most readily metabolizable) constituents. Organic matter entering the sediment in shallow water will tend to contain a greater portion of labile components (Nedwell, 1984). The shallow-water areas are also often the environments where the sedimentation rates are high. Labile organic compounds are consumed by reactions based on the elimination of functional groups by deamination, decarboxylation and condensation, depolymerization, isomerization, and certain intermolecular redox reactions (Price, 1976). It is commonly found that the C/N ratio in the remaining organic matter increases upon diagenesis, which indicates that proteins and the constituing aminoacids are preferentially decomposed in the (micro)biological processes, albeit not all to the same extent. Bacterial breakdown of aminoacids may be prevented by adsorption of these acids on clay minerals. Likewise, phosphorus-containing functional groups are more easily decomposed during diagenesis, so that the C/P ratio in the remaining organic matter will increase.

Heath et al. (1976) calculated a relation between the accumulation of C-org and sedimentation rate for deep ocean sediments:

$$\text{accumulation rate} = 0.01 \times (\text{sedimentation rate})^{1.4} \tag{2-1a}$$

Müller and Suess (1979) proposed the following expression for the accumulation (= permanent burial) of C-org in marine sediments:

$$C\text{-org}\ (\%) = 3 \times 10^{-3}\ (R_c \times s^{0.3})/[d_s\ (1 - \phi)] \tag{2-1b}$$

where:

C-org (%) = concentration of organic carbon in dry sediment;
R_c = primary production in surface water above (g m^{-2} a^{-1});
s = sedimentation rate (cm ka^{-1});
d_s = bulk sediment density (mass per unit of bulk sediment volume); and
ϕ = porosity (volume of pore water per unit of bulk volume of sediment).

Henrichs and Reeburgh (1987) found a similar relationship and could fit their data by using the following simplified equation:

$$C\text{-org}\ (\%) = s^{0.4}/2.1 \tag{2-1c}$$

where s is sedimentation rate (cm a^{-1}).

Such simple correlations between preservation of organic matter and primary productivity have been questioned by Emerson (1985). He stressed that the availability of O_2 in bottom waters is at least one important factor in the degradation of organic matter. The diagenetic chemical reactions that are involved are mostly written in a manner which takes into account the C:N:P ratios in organic matter. The average atomic ratios for pelagic phytoplankton are 106:16:1 (Redfield et al., 1963). But that ratio changes for bulk organic matter during sedimentation toward the sediment – water interface and further degradation at the top of the sediments. In coastal seas, the atomic ratio in plankton may be different from the Redfield ratio, for instance, C/N is about 5 (Walsh et al., 1985). The degradation of organic matter is mostly an oxidative process, for which electron acceptors must be present. The sequence in which the couples with the highest potentials relative to the reductants (organic groups/substances) act before the couple with the next lower potential becomes active, etc., are all relative to the in-situ pH. This sequence in marine sediments is, usually:

$$O_2,\ MnO_2 \approx NO_3^-,\ Fe_2O_3/FeOOH,\ SO_4^{2-},\ HCO_3^-$$

Bacteria, specialized in using the energy released in each of these oxidation – reduction (redox) reactions, obey this sequence. So, upon burial of organic matter, firstly oxygen will become depleted, then nitrate and manganese (III, IV)-oxide take over as oxidants, followed by reactive iron (III)-oxide. Then, the sulphate reservoir will become used and, finally, when at greater depths sulphate becomes eventually depleted before metabolizable C-org, bicarbonate may take over the role of oxidant. In general, not counting sediments with anoxic bottom waters, O_2 plays by far the most dominant role in the degradation of organic matter. According to Aller et al. (1983), the redox reactions can be written in the following manner:

$$(CH_2O)_x\,(NH_3)_y\,(H_3PO_4)_z + x\,H_2O \rightarrow$$
$$x\,CO_2 + y\,NH_4^+ + z\,HPO_4^{2-} + (4x - y + 2z)\,H^+ + 4x\,e^- \quad \text{(2-2a)}$$

where:

$(CH_2O)_x\,(NH_3)_y\,(H_3PO_4)_z$ = organic matter

e^- = electron (charge minus one)

For each redox level during diagenesis, this reaction can be combined with one of the pertinent following reactions:

$$x\,O_2 + 4x\,H^+ + 4x\,e^- \rightarrow 2x\,H_2O \quad \text{(aerobic respiration)} \quad \text{(2-2b)}$$

$$0.8x\,NO_3^- + 4.8x\,H^+ + 4x\,e^- \rightarrow 0.4x\,N_2 + 2.4x\,H_2O \quad \text{(denitrification)} \quad \text{(2-2c)}$$

$$2x\,MnO_2 + 8x\,H^+ + 4x\,e^- \rightarrow 2x\,Mn^{2+} + 4x\,H_2O \quad \text{(2-2d)}$$

$$4x\,FeOOH + 1.2x\,H^+ + 4x\,e^- \rightarrow 4x\,Fe^{2+} + 8x\,H_2O \quad \text{(2-2e)}$$

$$0.5x\,SO_4^{2-} + 4.5x\,H^+ + 4x\,e^- \rightarrow 0.5x\,HS^- + 2x\,H_2O \quad \text{(sulphate reduction)} \quad \text{(2-2f)}$$

$$0.5x\,CO_2 + 4x\,H^+ + 4x\,e^- \rightarrow 0.5x\,CH_4 + x\,H_2O \quad \text{(2-2g)}$$

Some overlap in consecutive reactions is possible in this sequence. The combination of reaction 2-2a with reactions 2-2b – 2-2f, substituting the Redfield ratio $x{:}y{:}z$ = 106:16:1, is discussed in great detail by Froelich et al. (1979). They visualized the sequence of redox reactions as shown in Fig. 2-1. The term "suboxic" is used for processes that take place when all O_2 has been depleted and reactions 2-2c – 2-2e are involved.

Because the Gibbs free energy yields for Mn-oxide reduction and denitrification do not differ much, it is not surprising that spatial overlap of these reactions occurs. Accepting fixed and known carbon over nutrient element ratios in the degradable organic matter, the stoichiometry of these reactions can be used advantageously to describe and explain diagenetic reactions in the sediments by careful analysis of changes in the pore-water chemistry (e.g., Emerson et al., 1980; Elderfield et al., 1981a,b; Anderson et al., 1986; De Lange, 1986).

The pore-water chemistry follows quite rapidly changes in sedimentary conditions. Anderson et al. (1986) give a characteristic time of one year for a pore-water profile to adjust to a 10-cm shift in conditions. This still allows for seasonal changes to show up in pore-water profiles, which will affect mostly the profiles in the top of the sediment with high organic sedimentation rates (e.g., Rutgers van der Loeff, 1980; Elderfield et al., 1981a,b; Jørgensen and Sørensen, 1985; Martin and Bender, 1988).

Oxygen consumption

Ammonium as a degradation product from organic matter as formulated in Eq. 2-2a will, under aerobic conditions, be oxidized by nitrifying bacteria (nitrification), according to the reaction:

$$NH_4^+ + 2\,O_2 \rightarrow NO_3^- + 2\,H^+ + H_2O \qquad (2\text{-}3)$$

Combination of Eqs. 2-2a, 2-2b and 2-3 then leads to the reaction describing aerobic degradation of organic matter:

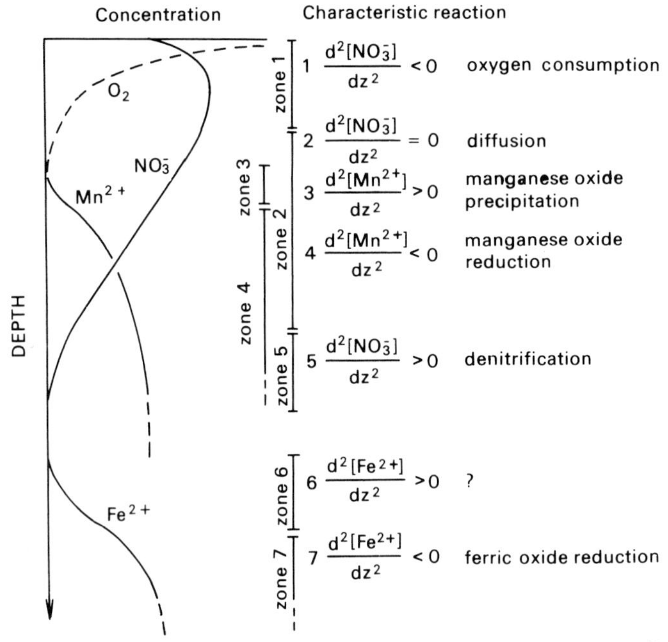

Fig. 2-1. Schematic representation of trends in pore-water profiles, showing the sequence of redox reactions involved in oxidative degradation of organic matter. Axes are in arbitrary units. The zones, characteristic curvature of the gradients, and reactions are discussed in the following sections. (Redrawn and slightly modified from Froelich et al., 1979.)

$$(CH_2O)_x\,(NH_3)_y\,(H_3PO_4) + (x + 2y)\,O_2 \rightarrow$$
$$x\,CO_2 + y\,NO_3^- + z\,HPO_4^{2-} + (y + 2z)\,H^+ + (x + y)\,H_2O \qquad (2\text{-}4a)$$

Or, in the Redfield stoichiometry:

$$(CH_2O)_{106}\,(NH_3)_{16}\,(H_3PO_4) + 138\,O_2 \rightarrow$$
$$106\,CO_2 + 16\,NO_3^- + HPO_4^{2-} + 18\,H^+ + 122\,H_2O \qquad (2\text{-}4b)$$

This reaction assumes a concerted breakdown of organic molecules containing amino- and phosphogroups, in which O_2 is consumed and acidity is produced. Part of these reactions occur at the sediment – water interface, and will scarcely affect pore-water chemistry. The O_2 respiration of the sediment community, which constitutes the sedimentary part of the benthic boundary layer, was formulated by Smith and Hinga (1983) for the Pacific Ocean (Eq. 2-5a) and Atlantic Ocean (Eq.2-5b), respectively, as follows:

$$OC = 0.3508 - 1.142 \times 10^{-4}\,D + 7.68 \times 10^{-3}\,R_c \qquad (2\text{-}5a)$$

$$OC = 0.9421 - 1.621 \times 10^{-4}\,D - 1.25 \times 10^{-3}\,R_c \qquad (2\text{-}5b)$$

and for the whole ocean (Eq. 2-5c) as:

$$OC = 0.3789 + 7.577 \times 10^{-3}\,R_c - 0.14692\,(O_2) \qquad (2\text{-}5c)$$

where:

OC	respiration rate (ml O_2 m^{-2} h^{-1})
R_c	primary production in overlying surface waters (g C m^{-2} a^{-1})
D	depth (m)
(O_2)	oxygen concentration of bottom water (ml l^{-1})

These authors concluded that Eqs. 2-5a,b give better estimates than Eq. 2-5c. The number of observations that can be used is still too small to allow for general predictive equations. This explains the discrepancies between the data for O_2 respiration given by Hinga et al. (1979), Smith and Hinga (1983), and Jørgensen (1983), as compiled in Table 2-1.

Agreement exists that the deep-sea sediments play only a minor role in the global uptake of O_2. Organic detritus raining down from the euphotic zone into pelagic regions is already largely oxidized in the water column and, therefore, only a small part of it will reach the sediment – water interface. The shelf regions (less than 200 m water depth) are responsible for some 60 – 85% of the total O_2 consumption. These areas will also have the largest benthic activity, because of the relatively large food supply, with bioturbation enhancing the contact between organic matter and oxygenated bottom water. The effect of aerobic degradation of organic matter within the sediment will be a decrease of pH of the pore water, to an extent deter-

TABLE 2-1

Oxygen consumption as estimated by (1) Jørgensen (1983); (2) Hinga et al. (1979), and (3) Smith and Hinga (1983)

Location	Depth interval (m)	Area (%) of whole ocean			O_2 uptake rate ($mM\ cm^{-2}\ a^{-1}$)			O_2 uptake (%) total benthic	
		1	*2*	*3*	*1*	*2*	*3*	*1*	*2*
Shelf	0 – 200	8.6	7.0		0.52	0.18		83	61
Upper slope	200 – 1000	4.2	5.2		0.11	0.09		9	22
Lower slope	1000 – 4000	29	32.2		0.01	0.01		6	14.5
Deep-sea	>4000	58	55.6		0.002	0.001		2	2.5
Global average	> 200			93			0.7		

mined by the availability of labile organic matter, its rate of degradation, the presence of buffering solid and dissolved constituents, and the rate of diffusion of the various species involved in the control of pH. For instance, calcium carbonate, if present, will partially dissolve until a new chemical equilibrium is established in the system $CO_2 - H_2O - Ca^{2+}$ (Emerson et al., 1980, 1982a). Drops in pH below the sediment – water interface are a common feature. Measurements of pH can be made in pore waters separated from the bulk sediment by dialysis, squeezing, or centrifugation. The last two methods bear the problem of possible escape of CO_2. The pH can also be measured by punching directly a proper set of electrodes into the sediment. This may cause a problem because of the so-called "suspension effect", that in principle can cause difference between the measured pH and the real pH of 0.1 to 0.2 of a pH unit.

Because of the secondary reactions and the concomitant complex response of pH making it difficult, if not impossible, to use pH changes as a means to calculate the extent of early diagenetic reactions, an easier way is to measure the O_2 profile within the sediment. Before discussing this, however, it must be emphasized that O_2 consumption is not only due to the reaction 2-4a, but also to the oxidation of reduced constituents in the right-hand side (rhs) of Eqs. 2-2c to 2-2f that move upward by diffusion. In shallow waters, photosynthesis may produce O_2 in the top of the sediment during daytime. Also seasonal changes in the temperature regime will affect the consumption rates and profiles of O_2 in pore water. The O_2 profile is, therefore, generated in principle by a number of processes taking place within the top of the sediment.

The introduction of microelectrodes for measuring O_2 concentrations has made it possible to measure the O_2 profiles in marine sediments very precisely (submillimeter scale) (Revsbech et al., 1980a,b, 1981, 1986; Reimers et al., 1984; Jørgensen and Revsbech, 1985; Emerson et al., 1985; Reimers and Smith, 1986; Silverberg et al., 1987). The in-situ O_2 measurements in the deep sea with the use of microelectrodes, as carried out by Reimers et al. (1986), are an important step forward in receiving the most reliable and precise measurements of undisturbed sediments at great depths. Revsbech et al. (1980a) compared millimeter-scale measurements of dissolved O_2 and of redox potential (Fig. 2-2) in sediments at water depths of 4–44 m. They demonstrated that estimates of the depth of penetration of O_2 into sediments, based on the thickness of the brown, oxidized surface layer, are too high. The anoxic part within a brown layer is often much thicker than the oxic part. Redox potential readings can still be positive in layers below the horizon where the oxygen concentration is zero. This is not surprising when one is familiar with the stability fields of dissolved and solid constituents in sediments, but it is a warning that the terms "oxic" and "anoxic" can be misleading. Another, still more erroneous concept often used, is that positive redox potentials indicate oxidizing, and negative potentials show reducing conditions.

Some typical O_2 profiles will be shown and used in model calculations in the last part of this chapter. The stoichiometry concept can also be used to assess consumption of O_2 (e.g., Emerson et al., 1980).

Nitrate consumption (denitrification)

Knowles (1982) defined denitrification as the dissimilatory reduction by essentially aerobic bacteria, of one or both ionic nitrogen oxides (nitrate or nitrite) to the gaseous oxides [nitric oxide (NO) and nitrous oxide (N_2O)], which may themselves be further reduced to dinitrogen (N_2). Denitrification rates are positively related to pH with an optimum rate at pH 7–8, i.e., exactly within the typical range of marine sediments. Denitrification starts to become dominant at low O_2 concentrations ($< 6\ \mu mol\ O_2\ l^{-1}$). Measured denitrification rates range from 3–10 $\mu mol\ N\ m^{-2}\ day^{-1}$ for aerobic deep-sea sediments of the Eastern Atlantic Ocean to values of about 7 mmol N m^{-2} day^{-1} for sediments of relatively rich eutrophic coastal systems (e.g., Jenkins and Kemp, 1984; Horrigan and Capone, 1985). Jahnke et al. (1982b) predicted, based on a model, that the maximum amount of organic matter that can be oxidized by denitrification is only 30% of that oxidized by O_2 respiration in the Pacific Ocean and only 13% in the Atlantic Ocean. Bender and Heggie (1984) revised these values at 9% for the Pacific Ocean and a lower value for the

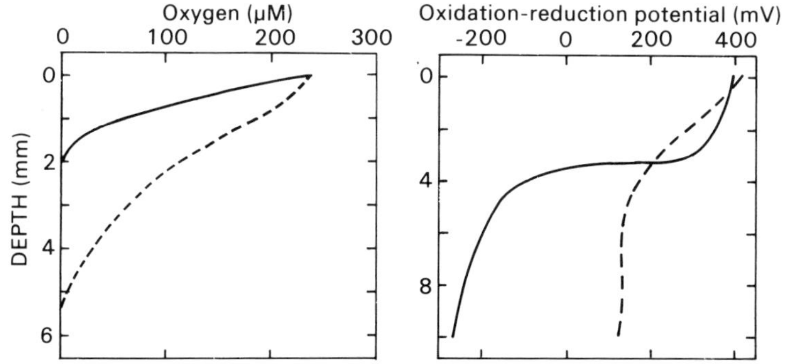

Fig. 2-2. O_2 and Eh profiles from two different sediments. The full O_2 curve displays the normal, nearly parabolic shape; the sigmoidal dashed curve is caused by turbulent O_2 transport in the upper 1 mm of sediment. The concomitant Eh curves show that the "oxidized" sediment layer having a positive Eh was much thicker than the oxic layer. (Redrawn and modified after Revsbech et al., 1980.)

Atlantic Ocean. Seitzinger et al. (1984) measured the denitrification in Narragansett Bay as the flux of N_2 and N_2O from the sediments into bottom water. They reported that about 50% of the inorganic combined N, entering the bay by rivers and sewers and firm land, is removed by denitrification, and that about 35% of organic N that is mineralized in the sediments is removed by denitrification. The role of N_2O in the removal is only of minor importance (< 10%). Jørgensen and Sørensen (1985) found for the marine part of a Danish estuary that O_2 uptake and denitrification are 65% and 3%, respectively, of the total electron flux, the remainder being due to nitrate reduction to ammonium ("nitrate ammonification"; Sørensen, 1987) plus sulphate reduction. The annual emission of N_2O amounted to only 1 – 5% of the measured denitrification. The annual loss of combined N by denitrification in the estuary as a whole corresponds to 5% of the nitrate from the river. Inorganic N in pore waters is partly supported by the mineralization of amino acids from organic matter. Degens and Mopper (1975) mentioned that the composition of amino acids in average sediments is very similar to that of marine plankton, with some enrichment in serine, glycine, hexosamine, and some depletion in glutamic acid, methionine, and arginine. This, apparently, is due to selective breakdown in the food chain before burial. Dominant species are: glycine, alanine, aspartic and glutamic acids. Müller (1975) studied the relative diagenetic mineralization rates of amino acids in two sediments in the Eastern Atlantic Ocean and found that arginine and lysine are least readily mineralized, whereas cysteine and methionine are most readily mineralized. Combination of Eqs. 2-2a and 2-2c gives the denitrification reaction:

$$(CH_2O)_x\ (NH_3)_y\ (H_3PO_4)_z\ +\ 0.8x\ NO_3^-\ \rightarrow\ x\ CO_2\ +\ y\ NH_4^+\ +\ 0.4x\ N_2\ +\ z\ HPO_4^{2-}\ +\ (-0.8x\ -\ y\ +\ 2z)\ H^+\ +\ 1.4x\ H_2O \qquad (2\text{-}6a)$$

or, in Redfield stoichiometry, and combining it with the familiar equations for dissociation reactions for water and carbonic acid:

$$(CH_2O)_{106}\ (NH_3)_{16}\ (H_3PO_4)\ +\ 84.8\ NO_3^-\ \rightarrow\ 7.2\ CO_2\ +\ 98.8\ HCO_3^-\ +\ 16\ NH_4^+\ +\ 42.4\ N_2\ +\ HPO_4^{2-}\ +\ 49.6\ H_2O \qquad (2\text{-}6b)$$

Both equations assume that ammonium, as a product of degradation of N-org, is not oxidized to N_2 in this bacterial process. Emerson et al. (1980) found no trace of NH_4^+ in the zone of denitrification and they, therefore, assumed that it is oxidized either to N_2 or N_2O, or is quantitatively adsorbed. In the case that all NH_3 is completely transformed into N_2 in the process, the equation is:

$$(CH_2O)_{106}\ (NH_3)_{16}\ (H_3PO_4)\ +\ 94.4\ NO_3^-\ \rightarrow\ 13.6\ CO_2\ +\ 92.4\ HCO_3^-\ +\ 55.2\ N_2\ +\ 84.8\ H_2O \qquad (2\text{-}7)$$

Fenchel and Jørgensen (1977) noted that denitrification and nitrification (= bacterial oxidation of ammonium to nitrate via nitrite ions by nitrifying bacteria) occur closely together in surface sediments and that this, in combination, may lead to the conversion of NH_4^+ via NO_3^- to N_2 (cf., Berelson et al., 1987).

Suess et al. (1980) state that nitrification occurs in the biologically-active oxic surface layer in ocean sediments. This means that ammonification (= conversion of N-org into NH_4^+) produces NH_4^+ that is added to the nitrate pool by nitrification and subsequently reduced to dinitrogen by denitrification. This means that the overall stoichiometry would be as given in Eq. 2-7. This reaction produces alkalinity. Again, the pH change depends on the bulk sediment chemistry with its buffering properties (Emerson et al., 1980, 1982a,b).

Goloway and Bender (1982) discussed three basic models for nitrate profiles. These profiles are shown in Fig. 2-3. The O_2 consumption model assumes conditions with a sequence of oxidative deamination/ammonification, oxidation of ammonium, and oxidation of nitrite (Suess et al., 1980), all occurring in the same layer with a rate high enough not to cause separation of these processes with depth. The two-layer model (Fig. 2-3) assumes a combination of O_2 consumption in the top of the sediment and denitrification below the horizon where O_2 is practically all consumed. The boundary between the zones of nitrification and denitrification is indicated by the stippled horizontal line where the second derivative of the NO_3^- concentration with depth is zero. The three-layer model (Fig. 2-3) shows a linear decrease in NO_3^- concentration between the zones of O_2 and NO_3^- reduction; it is assumed that the only process occurring here is a downward diffusion of NO_3^-. Such profiles are seen in sediments where a lithological change from calcareous to

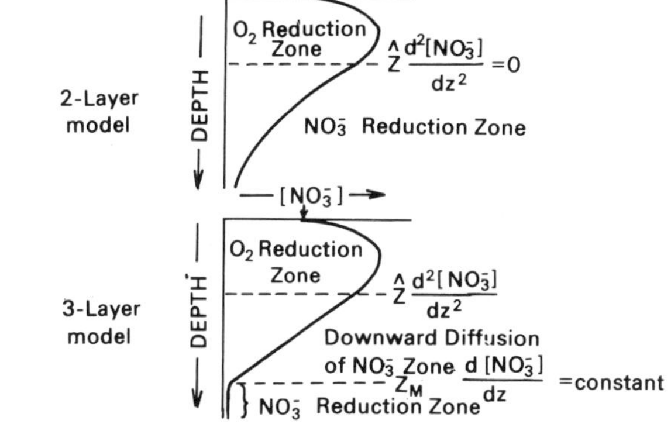

Fig. 2-3. Hypothetical pore water NO_3^- profiles. In the top figure, asymptotic NO_3^- content is defined as $[NO_3^-]^{\infty}$. (Modified after Goloway and Bender, 1982.)

terrigenous sediments is accompanied by a drastic decrease in the nitrate reduction rate. Jahnke et al. (1982b) showed the effect of denitrification rates on the profiles for a particular set of parameter values in a two-layer model (Fig. 2-4).

Jahnke (1985) discussed the effects of suboxic microenvironments, i.e., suboxic zones within the particles that are surrounded by oxygenated pore water. This has the effect of decreasing the magnitude of the NO_3^- maximum, shifting the depth of the maximum to a shallower horizon, and decreasing the buildup of N_2 in pore waters. The existence of such microenvironments is favored by organic detritus of large particle sizes. Wilson (1978) presented evidence that denitrification can take place in the oxic zone of pelagic sediments and explained this by assuming that zooplankton faecal pellets constitute a temporarily-isolated microenvironment. Christensen and Rowe (1984) estimated that about two-fifth of the NH_4^+ produced in the oxygenated layer is derived from microniches with anaerobic respiration. Rutgers van der Loeff et al. (1981) discussed nitrification and denitrification that occur very closely together and maybe even simultaneously in sediments with a high content of organic matter. Where both processes act together, nitrogen may be lost as N_2 or N_2O without any apparent consumption of NO_2^- or NO_3^-. In-faunal activity may also lead to a direct input of NH_4^+ into bottom water, thus causing a loss in the N budget that cannot be accounted for (Henriksen et al., 1983). Not all NH_4^+ produced in the anoxic zone will diffuse toward the sediment – water interface and become oxidized by nitrifying bacteria. The NH_4^+ has a high affinity for reactive exchange sites in clay minerals and, therefore, will become adsorbed readily. This sink has to be taken into account in the N budgets and N models for marine sediments. Nitrification will produce a downward positive gradient in pore-water profiles; therefore, in sediments with an oxygenated top, a flux of NO_3^- (and some

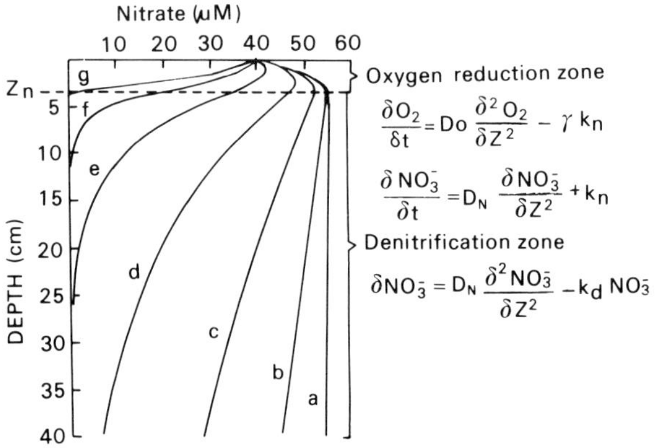

Fig. 2-4. Example model profiles calculated for bottom water concentrations of $O_2 = 110\ \mu M\ kg^{-1}$ and $NO_3^- = 40\ \mu M\ kg^{-1}$. Whole sediment diffusion coefficients of NO_3^-: $D_N = 3.5 \times 10^{-6}\ cm^2\ s^{-1}$ and of O_2: $D_O = 4.2 \times 10^{-6}\ cm^2\ s^{-1}$; k_N is the zero-order production function for NO_3^- production during O_2 reduction and k_d is the first-order denitrification rate constant; γ is the O_2/N ratio in Eq. 2-4b; and Z_n is the depth at which O_2 is depleted. M = moles. The family of curves marked by *a* – *g* represents denitrification rates (*a*) $10^{-12}\ s^{-1}$, (*b*) $10^{-10}\ s^{-1}$, (*c*) $10^{-9}\ s^{-1}$, (*d*) $10^{-8}\ s^{-1}$, (*e*) $10^{-7}\ s^{-1}$, (*f*) $10^{-6}\ s^{-1}$, (*g*) $10^{-4}\ s^{-1}$. (Modified after Jahnke et al., 1982b.)

NO_2^-) into the bottom water will occur. In anoxic sediments this will be the case for NH_4^+. Bender et al. (1977) studied cores in the Eastern Equatorial Atlantic and calculated that of the NO_3^- produced in nitrification about 96% returns to the bottom water and 4% is consumed by denitrification. Still, their conclusion is that the flux of nitrate from the sediments into bottom waters is of minor importance in the overall budget of deep ocean-water nitrate. Christensen et al. (1987a,b) compiled measurements/calculations of the nitrate consumption rate in a variety of sedimentary environments (Table 2-2). As compared to the O_2 consumption rates (Table 2-1), these data are much lower for deep-sea sediments, but higher for sediments that are very rich in organic matter (fjords; coastal and continental shelf areas). On a global scale, the annual denitrification rates are estimated to be equal to 4 – 7 Tg N for deep-sea sediments (Liu and Kaplan, 1984) and 75 Tg N in continental shelf areas including the Baltic Sea (Christensen et al., 1987b).

Manganese and iron oxide reduction

In contrast to other oxidants, Mn and Fe oxides are present in solid phases. Reduction of these phases renders the elements soluble as Mn^{2+} and Fe^{2+}, respectively. Consequently, no supply from bottom waters occurs by downward diffusion in solution. The oxidizing capacity of these oxides is, therefore, primarily determined by their quantity as incorporated in the sediment during its desorption. Under anoxic conditions no such accumulation is possible, because the oxides will dissolve in the water column or at the sediment – water interface; the oxides will be deposited in oxygenated waters and the elements will be preserved in sediment columns with oxic top layers. This can have a bearing on the growth of ferromanganese encrustrations and nodules at the ocean floor, where lateral supply is a partial or even the only source of metals (Calvert and Piper, 1984). Also, periodical changes from an

TABLE 2-2

Sedimentary nitrate consumption (after compilations by Christensen et al., 1987a,b)

Locations	Rate (μM NO_3 cm^{-2} a^{-1})
Eastern Atlantic	0.026
Northwest Atlantic	0.05 – 1.5
Equatorial Atlantic	0.3 – 0.9
Equatorial Pacific	0.2 – 3
Santa Barbara Basin	0.2 – 0.5
Washington Continental Slope	6
Coastal Equatorial Africa	9
San Clemente Basin	9.8
North Sea	25
Bering Sea	19 – 60
Coastal North Sea	7 – 110
Narragansett Bay	150
Washington Shelf	180

oxic to an anoxic state of the top of the sediment, as can be expected in shallow-water sediments with a high influx of organic matter, may cause a flux of dissolved Mn and Fe from the sediment into bottom waters (Eaton, 1979; Aller, 1980c; Elderfield et al., 1981a; Graybeal and Heath, 1984; Sundby et al., 1986).

Manganese oxide has seldom the stoichiometry of pure MnO_2. A mixture of divalent, perhaps trivalent, and tetravalent Mn is more common. This is often indicated by an average stoichiometric formula MnO_x (with $1.1 \leq x \leq 2$). Klinkhammer and Bender (1980) assumed a stoichiometry of $MnO_{1.33}$ for solid Mn oxide in the water column of the Pacific Ocean, whereas Murray et al. (1984) reported $MnO_{1.90-2.00}$ for the Eastern Tropical Pacific and Equatorial Pacific. The latter authors, noted, however, that for increased Mn^{2+} concentrations in the pore waters, the oxidation state decreases to as low as $MnO_{1.40}$. Kalhorn and Emerson (1984) reported $MnO_{1.65-1.95}$ for MANOP sites M and H in the Eastern Pacific Ocean. Emerson et al. (1982b) reported an average oxidation state of Mn for Saanich Inlet, a partially anoxic fjord, as low as $MnO_{1.16-1.36}$. Murray et al. (1984) discussed the transient mineral phases observed in the oxidation of Mn^{2+}, with a final phase to be manganite (γ-MnOOH). All MnO_x is a highly reactive oxidant. The reduction of these phases is usually mediated by bacteria (Ehrlich, 1981; Nealson, 1983). Only a small fraction of Fe(III)-oxide present in the solid phase of sediments is usually available for bacterial reduction. The most reactive phases are those with the highest solubility, i.e., amorphous and poorly crystallized Fe-oxyhydroxides (cf., Lovley and Phillips, 1986). These phases are usually present as coatings on sediment particles. For the crystallized phases, the order is: lepidocrocite (γ-FeOOH), goethite (α-FeOOH), hematite (α-Fe_2O_3) (Ehrlich, 1981). As is the case for MnO_x, reduction of Fe(III)-oxide can occur via enzymatic or nonenzymatic processes. As a prominent example of the latter, the reduction by hydrogen sulphide produced by sulphate-reducing bacteria can be mentioned.

The overall reactions for oxidation of organic matter by Mn oxide can be obtained by combination of Eqs. 2-2a and 2-2d, but allowing for the partial oxidation of NH_4^+ by Mn oxide according to the reaction:

$$2\,NH_4^+ \rightarrow N_2 + 8\,H^+ + 6\,e^- \tag{2-8a}$$

which gives:

$$(CH_2O)_x\,(NH_3)_y\,(H_3PO_4)_z + (2x + 3/2\,y)\,MnO_2 \rightarrow x\,CO_2 + y/2\,N_2 + z\,HPO_4^{2-} + (2x + 3/2\,y)\,Mn^{2+} + (-4x - 3y + 2z)\,H^+ + (3x + 3y)\,H_2O \tag{2-8b}$$

or, in Redfield stoichiometry, in combination with the dissociation reactions of water and dissolved carbon dioxide:

$$(CH_2O)_{106}\,(NH_3)_{16}\,(H_3PO_4) + 236\,MnO_2 + 364\,CO_2 + 104\,H_2O \rightarrow 470\,HCO_3^- + 8\,N_2 + 236\,Mn^{2+} + HPO_4^{2-} \tag{2-8c}$$

In these reactions, it is taken into account that at the Eh – pH boundary of Mn oxide, N_2 is the stable N species. This would also apply for the redox boundary of Fe(III) oxide reduction (cf., Breck, 1974). Usually, however, it is assumed, in keeping with the observed NH_4^+ profiles, that NH_4^+, released from organic matter breakdown, is not oxidized, directly or indirectly, by Fe(III)-oxide (Froelich et al., 1979; Emerson et al., 1980). The reaction for oxidation of organic matter by Fe-oxide can be obtained by combination of Eqs. 2-2a and 2-2e:

$$(CH_2O)_x\,(NH_3)_y\,(H_3PO_4)_z + 4x\,FeOOH \rightarrow x\,CO_2 + y\,NH_4^+ + z\,HPO_4^{2-} + 4x\,Fe^{2+} + (-8x - y + 2z)\,H^+ + 7x\,H_2O \qquad (2\text{-}9a)$$

or, in Redfield stoichiometry, in combination with the dissociation reactions of water and carbonic acid:

$$(CH_2O)_{106}\,(NH_3)_{16}\,(H_3PO_4) + 424\,FeOOH + 756\,CO_2 \rightarrow 862\,HCO_3^- + 16\,NH_4^+ + HPO_4^{2-} + 424\,Fe^{2+} + 120\,H_2O \qquad (2\text{-}9b)$$

Reactions 2-8c and 2-9b produce alkalinity and consume carbon dioxide. This would result in an increase of pH to an extent controlled by the buffering of the total sediment. Emerson et al. (1982a) show that, for a closed system, the alkalinity increase only becomes apparent when solid calcium carbonate is absent. When pore water is already (super)saturated with respect to $CaCO_3$, however, the alkalinity actually decreases because of precipitation according to:

$$Ca^{2+} + 2\,HCO_3^- \rightarrow CO_2 + H_2O + CaCO_{3(s)} \qquad (2\text{-}10)$$

which provides the CO_2 and takes away the produced alkalinity in the foregoing reactions as well.

The sequence in which oxidative organic matter mineralization takes place by denitrification or by Mn oxide reduction depends on the standard free energy of the latter (Froelich et al., 1979). In sediments with high input of organic matter the depth boundaries between these two processes will be hard to determine and will practically have an overlap. De Lange (1986) showed, for a situation of slow oxidative breakdown of organic matter, that denitrification preceded Mn oxide reduction. But, as demonstrated in Fig. 2-5, the observed redox boundaries indeed practically coincide, whereas the following step, i.e., reduction of Fe-oxide, lies, as expected, at a significantly lower redox boundary (Fig. 2-5c).

The reduction of MnO_x and concomitant production of Mn^{2+} below the zone of O_2 consumption, brings about a concave-down profile of increasing dissolved Mn with depth. The reduction depletes the solid sediment in reactive solid Mn-oxide and the manganous species will diffuse according to the actual concentration gradients. The upward diffusion is toward the horizon where the downward diffusive flux of CO_2 is just enough to re-oxidize Mn^{2+} into MnO_x. When this is a continuous process at steady-state, the sediment column will exhibit an horizon that is enriched in solid Mn-oxide. This is an often-described feature in marine sediments. Sometimes more than one of such horizons are found in sediments. In combination with the profile of dissolved Mn in pore waters, it will be apparent which one of the horizons is active and which one is fossil. Ideally, an active horizon would look like the one shown in Fig. 2-6A, but in reality the locus of complete Mn^{2+} oxidation may be situated above the oxide peak and this peak may be smeared out in an upward direction (Froelich et al., 1979; Burdige and Gieskes, 1983). The position of a fossil horizon would not be related to the apex of the dissolved Mn profile. The occur-

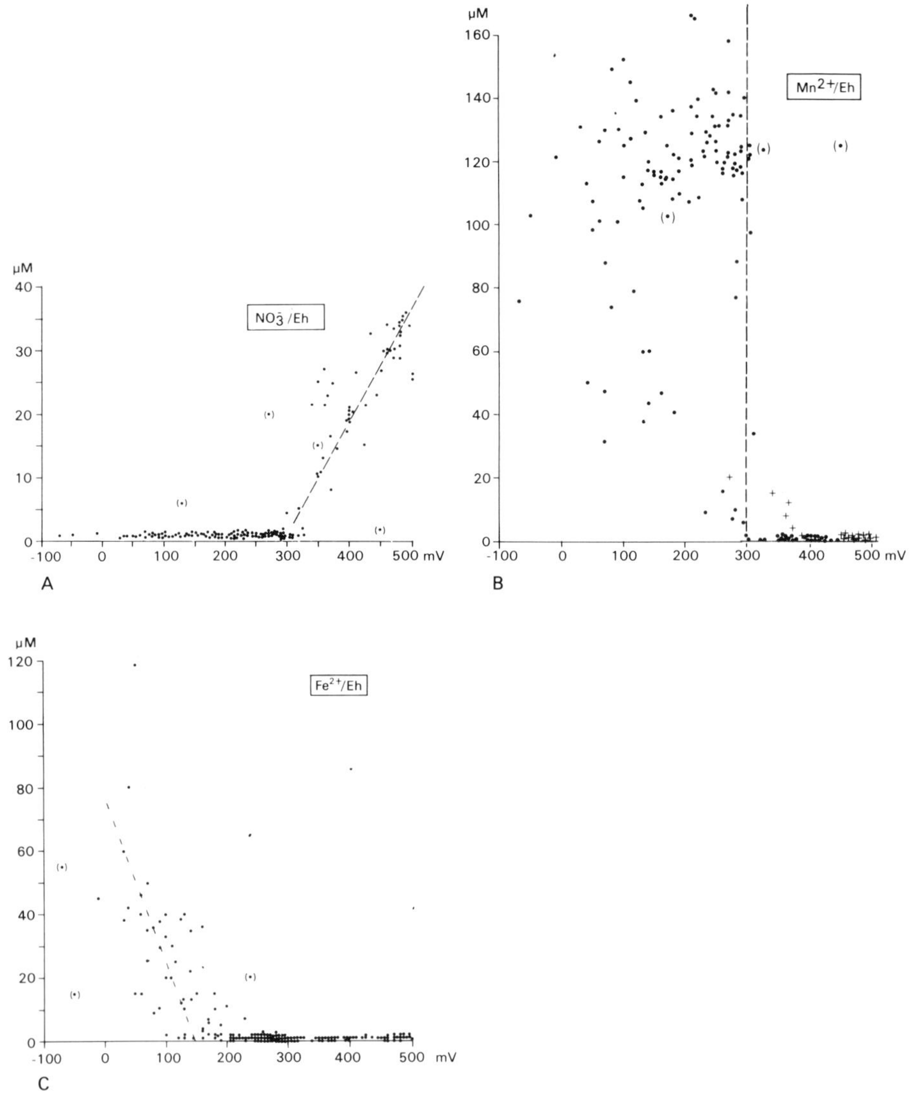

Fig. 2-5. Concentrations of NO_3^-, Mn^{2+} and Fe^{2+} in sediment pore water from the Nares Abyssal Plain versus Eh. (Modified after De Lange, 1986.)

rence of more than one horizon enriched in MnO_x has been related to changes in the regime of sedimentation, e.g., from glacial to interglacial (Berger et al., 1983; Thomson et al., 1984). Such changes disrupt steady-state and consequently it takes time to reach a new steady-state. For the dissolved Mn profile the time for attainment of a new pseudo steady-state is on the order of 10^0 years, but for the solid Mn oxide profile it is on the order of 10^2 to 10^3 years, depending on the actual degree of enrichment and its thickness (Froelich et al., 1979). When the sudden change in sedimentary condition involves an increased rain of organic matter toward the sediment, O_2 will become depleted at a shallower depth in the sediment and Mn oxide reduction will become active in a zone above the earlier steady-state horizon of Mn enrichment. The whole coupled system of oxidation and reduction will have to shift in an upward direction, but the actual fluxes due to concentration gradients in pore waters are low and, therefore, the upward transport of solid Mn-oxide will be slow. The sequence of events can be depicted as shown in Fig. 2-6B, C. A steady-state system at high input of C-org will have a shallow Mn oxide peak. A change toward a regime with low C-org rain will initially result in O_2 depletion below the Mn oxide peak and will leave this peak in place. Upon growth of the sediment column, the horizon of O_2 depletion will pass upward through the Mn-oxide peak; Mn-oxide then becomes metastable and will start to dissolve. Because of the already lower O_2

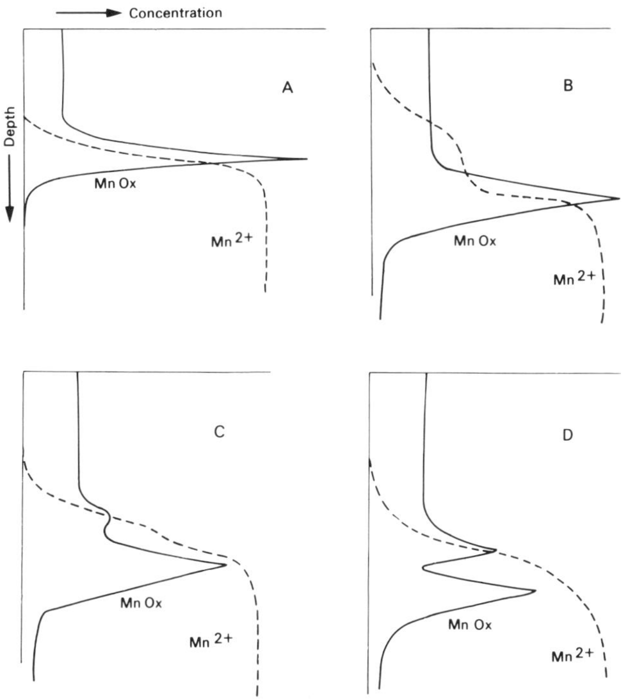

Fig. 2-6. (A) Steady state and (B, C, D) transient profiles of dissolved and solid Mn showing progressive stages after a change in the sedimentary regime with higher C-org content in the top sediment.

demand and the possible excess of Mn-oxide over labile C-org, this dissolution may only be partial and double peaks can develop (Thomson et al., 1984).

Hartmann (1979) and Kalhorn and Emerson (1984) reported that Mn oxide reduction takes place within the oxic zone in microniches rich in C-org, where O_2 is rapidly exhausted. This can create small yellowish discolorations in the brown oxic zone. The production of Mn^{2+} in the suboxic zone will not lead to unlimited concentrations, because of secondary reactions serving as sinks for Mn^{2+}. Holdren et al. (1975) mentioned that rhodochrosite ($MnCO_3$) and reddingite [$Mn_3(PO_4)_2 \cdot 3H_2O$] can be the controlling solid phases, but calculated that pore waters in Chesapeake Bay were supersaturated only with respect to rhodochrosite. In fact, this is the most frequently reported situation (Middelburg et al., 1987, and references therein). A calculated (super)saturation does not imply that a mineral is actually precipitating, and only mineralogical evidence of its existence can prove beyond doubt that this occurs. It is almost impossible to detect rhodochrosite because of the small amounts that eventually form. Suess (1979) used a leaching scheme applying increasingly higher acidity on sediments from the Baltic Sea, in combination with analysis of distinct mineral phases. The presence of a mixed Mn carbonate ($Mn_{0.85}Ca_{0.10}Mg_{0.05}CO_3$), a Mn-sulphide ($\gamma$-MnS), and a Mn phosphate [$Mn_3(PO_4)_2$] was found in this way. Mn-carbonates containing also Ca, Mg, and Fe in varying proportions have been identified more often (Pedersen and Price, 1982). These authors reported the occurrence of a mixed carbonate phase having composition $(Mn_{0.48}Ca_{0.47}Mg_{0.05})CO_3$ in Panama Basin sediments. Elderfield et al. (1981a) inferred from their pore-water studies in Narragansett Bay the presence of a mixed carbonate phase of composition $(Mn_{0.49}Ca_{0.51})CO_3$. Middelburg et al. (1987) calculated that for a general range of pore water compositions with respect to dissolved Mn^{2+} and Ca^{2+}, kutnahorite $(Mn_{0.5}Ca_{0.5})CO_3$ will be the stable phase. Comans and Middelburg (1987) showed that solubility of Mn^{2+} in the presence of $CaCO_3$ is controlled by a continuous sequence of adsorption, solid solution, and surface precipitation. In the oxic top of the sediments, the solubility of Mn(III,IV) oxide is so low that this phase will control the low Mn^{2+} concentrations. In the suboxic and anoxic part of the sediments solid carbonate phases as discussed will control these concentrations.

The abiological oxidation of Mn^{2+} is an autocatalytic process, not only depending on the concentrations of dissolved O_2 and Mn^{2+}, but also on the "concentration" of the solid phase [MnO_x]:

$$d(Mn^{2+})/dt = k\,(Mn^{2+})\,[MnO_{x(s)}]\,(OH^-)^2\,(O_2) \tag{2-11}$$

When pH as well as the concentrations of MnO_x and O_2 remain constant because of constant aeration in a steady-state profile, the rates will become pseudo first-order (Elderfield et al., 1981a). Burdidge and Kepkay (1983) found also that for concentrations of $Mn^{2+} < 15\ \mu M$, microbial Mn-binding and oxidation will be first-order in the Mn^{2+} concentration in most sedimentary environments. The oxidation of Mn^{2+} is microbially catalyzed (Ehrlich, 1981; Nealson, 1983; Tebo, 1983; Tebo and Emerson, 1985). The oxidation rate does depend on the O_2 concentration and on the concentration of microbial binding sites. This would mean that the rate of oxidation increases from the O_2 minimum in the sediment column upward due to the increasing O_2 concentration, but the population of Mn-oxidizing bacteria may well be favored in the niche at the boundary between oxic and suboxic conditions, so as to offset this.

The reduction of Fe(III)-oxide is apparent in the increase of dissolved Fe^{2+} species at a certain depth (cf., Fig. 2-5C). Fe^{2+} species will diffuse according to the concentration gradient in pore water and be reprecipitated oxidatively at a higher level. But a peak similar to that of MnO_x is generally not observed. This fact may be explained by the high noise in the Fe(III)-oxide profile and a signal-to-noise ratio that is unfavorable for detection of such patterns. Another explanation is that secondary reactions within or below the Fe-oxide zone that acts as sinks for the produced reduced Fe species, can effectively decrease the buildup of high upward concentration gradients. The precipitation may consume any O_2 if left over from Mn^{2+} oxidation, or NO_3^- (Klinkhammer, 1980). The redox boundary between Fe(III) and Fe(II) is usually visible in the sediment where the color changes from brown to green; this transition marks the reversible reduction of Fe(III) to Fe(II) in smectites (Lyle, 1983). This boundary also marks the precise depth of disappearance of NO_3^- from pore water. The concentration of ferrous ions in pore waters is controlled by the solubility of Fe(II)-minerals. Among these the hydrogenous ferrous monosulphides and pyrite (FeS_2) are the most important ones, but vivianite [$Fe_3(PO_4)_2 \cdot 8\ H_2O$; Elderfield et al., 1981b] or with a mixed composition [e.g., $(Fe_{0.86}Ca_{0.14})_3(PO_4)_2$ as reported by Suess, 1979] may also play a role. The latter phase can also be important as a control of phosphate in the pore waters.

From the modeling point of view, much work has been done on the Mn profiles in pore waters, scarcely on the Fe profiles. Therefore, only the modeling of Mn oxide reduction and the resulting Mn profiles in marine pore waters are discussed here.

Sulphate reduction

After O_2 respiration, SO_4^{2-} reduction is, on a global scale, the most important process in the diagenesis of organic matter. The sulphidic primary and secondary reaction products are very conspicuous in modern and ancient sediments and sedimentary rocks. Sulphide is not just produced by respiratory reduction of SO_4^{2-}, but also to some extent by the hydrolysis of proteins in organic detritus. Fresh plankton has an organic sulphur content of 1 – 2% or 0.3 – 0.6 mM S-org per gram on a dry weight basis (Jørgensen, 1977). Sulphate-reducing bacteria of the genus *Desulfovibrio* only can utilize low-molecular-weight organic molecules, like lactate, pyruvate, and malate, that are oxidized into acetate and subsequently excreted. But newly isolated strains of sulphate-reducing bacteria can oxidize all the major fermentation products to carbon dioxide and water (Jørgensen, 1982).

The major effect of the foregoing processes of O_2 and NO_3^- respiration is the combustion of C-org before it can become buried in the deeper anoxic sediment layers. Sulphate respiration supplies the energy to metabolize labile organic matter that is left over. This means that SO_4^{2-} reduction is prominent in environments of high sedimentation rates of both organic and inorganic matter, mostly in shallow coastal water with a high primary productivity and muddy sedimentation. Sørensen and Jørgensen (1987) found that sulphate reduction can take place within the microenvironment of organic aggregates in the oxic zone. Reduction of SO_4^{2-} can be formulated by combination of Eqs. 2-2a and 2-2f:

$$(CH_2O)_x\,(NH_3)_y\,(H_3PO_4)_z + 0.5x\,SO_4^{2-} \rightarrow x\,CO_2 + y\,NH_4^+ + z\,HPO_4^{2-} + (-0.5x - y - 2z)\,H^+ + 0.5x\,HS^- + x\,H_2O \qquad (2\text{-}12a)$$

or, in Redfield stoichiometry, and in combination with the dissociation reactions for water and carbonic acid:

$$(CH_2O)_{106}\,(NH_3)_{16}\,(H_3PO_4) + 53\,SO_4^{2-} \rightarrow 39\,CO_2 + 67\,HCO_3^- + 16\,NH_4^+ + HPO_4^{2-} + 53\,HS^- + 39\,H_2O \qquad (2\text{-}12b)$$

The reaction produces alkalinity (bicarbonate, phosphate, sulphide), causing an increase in pH that is partly offset by the simultaneous increase of the CO_2 concentration. The actual increase of pH depends on the buffering action of the other bulk sediment constituents (Emerson et al., 1980) and on the extent of the formation of sulphide minerals. Berner (1984, 1985) schematized the steps in the formation of sulphide minerals ending with the most stable pyrite, as shown in Fig. 2-7. Precipitation of mackinawite (FeS) would reduce the alkalinity increase drastically, as shown by the reaction:

$$Fe^{2+} + HS^- \rightarrow FeS + H^+ \qquad (2\text{-}13a)$$

Formation of pyrite, involving partial oxidation of sulphide, enhances the increase of alkalinity:

$$2\,FeOOH + 4\,H^+ + 2\,HS^- \rightarrow FeS_2 + Fe^{2+} + 4\,H_2O \qquad (2\text{-}13b)$$

In the latter overall reaction, $S^{(-2)}$ is oxidized to $S^{(-1)}$. Instead of FeOOH, also Fe^{3+} adsorbed on clays or MnO_x can serve as an oxidant in the suboxic zone. Although not visible in Eq. 2-13b the precipitation of pyrite generally seems to occur via FeS as a precursor phase. But direct precipitation has been reported fot Gotland Deep (Boesen and Postma, 1988). Pyrite in contact with O_2 is unstable, which would imply that pyrite formation actually occurs at the boundary layer between suboxic and anoxic conditions (Giblin and Howarth, 1984; Sørensen and Jørgensen, 1987; Feijtel et al., 1988; Oenema, 1988). Depending on the actual bacterial oxidation process, other oxidants may be involved, but all of them would produce an equivalent amount of alkalinity. The stoichiometry described by reaction 2-12 predicts an inter- and intrarelationship of the amounts of reduced sulphate and of produced total carbon dioxide, carbonate alkalinity, ammonium, phosphate, and sulphide. The ratios of the concentrations of the dissolved products can vary for different sedimentary regimes, because of (1) differences in the original C:N:P ratios, (2) differential diffusion, and (3) adsorption (NH_4^+, HPO_4^{2-}) or precipitation (HCO_3^-, HS^-, HPO_4^{2-}). Plots of the concentrations of SO_4^{2-} versus produced

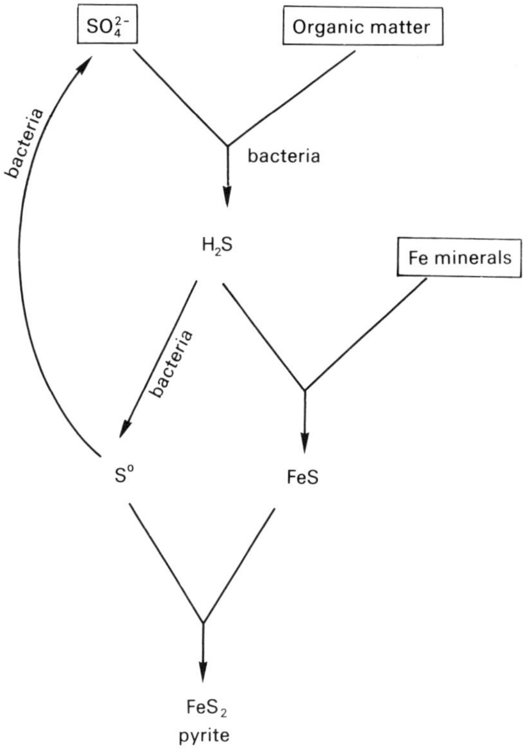

Fig. 2-7. Schematic diagram summarizing the major steps in sedimentary pyrite formation. (Modified after Berner, 1985.)

species is often used to decipher the reactions and processes (e.g., Hartmann et al., 1973, 1976). Apart from its role in the carbon cycle, sulphate reduction plays an important role in the sulphur cycle.

The presence of pyrite in recent and old sediments can be used in the diagnosis of sedimentary environments. Berner (1985) stated that in normal marine sediments, deposited in oxygenated water, the formation of pyrite is limited by the concentration and reactivity of organic matter, whereas in euxinic basins the formation of pyrite is limited by the abundance and reactivity of detrital phases of Fe (e.g., Boesen and Postma, 1988). The amount of labile C-org is, in a roundabout manner, related to the amount of the total C-org. Also, the amount of pyrite-S, formed in a more or less homogeneous sediment, is related to the HS^- produced in SO_4^{2-} reduction used to metabolize the labile organic compounds. When these processes begin within the sediment column, the relation between pyrite and C-org contents can be expected to be represented by a straight line through the origin. When, however, pyrite is already formed in the water column and on top of the sediments, as will be the case in euxinic sediments, this straight line will have an intercept with the FeS_2 axis. Such plots can thus be useful for distinguishing between different sedimentary environments, including sedimentation in fresh water (Leventhal, 1983;

Berner and Raiswell, 1983, 1984; Raiswell and Berner, 1985; Sheu, 1987; Boesen and Postma, 1988). In present-day marine sediments, deposited in an oxygenated water column, the rates of burial of C-org and pyrite-S correlate positively and have a constant ratio (C/S $\approx$ 3 on a weight basis). Deviations toward higher C/S ratios can be explained by burial in fresh water, and toward lower C/S ratios by burial in a euxinic environment (Berner and Raiswell, 1983). An idealized plot of the relation is given in Raiswell and Berner (1985), as shown in Fig. 2-8. Berner (1984) warned that this relation can be far from ideal, because pyrite can form in the water column and be laterally transported to other localities, thus upsetting the reliability of determining the slopes and the intercepts of the curves. Berner and Raiswell (1986) showed that the C/S ratio in marine sediments has changed in the geological history of the last 600 Ma. They attributed this to changes in the euxinic environments and to the rise of the land plants. Not all Fe present in the solids that rain out of the water column is available for the formation of Fe-sulphides. The most reactive phases are the Fe(III)-oxyhydroxide coatings on mineral grains. In most terrigenous sediments, enough reactive Fe(III)-phases are available to sustain the formation of pyrite. But even though the pyrite formation is not limited by the amount of reactive Fe(III)-mineral phases, their reactivity itself may well be limiting. This means that only part of the produced HS^- will be fixed in the neoformation of sulphidic minerals. In predominantly biogenous sediments (calcareous or siliceous), a shortage of reactive Fe(III)-phases is likely and HS^- can build up without being removed by precipitation (Berner, 1984). In general, these observations mean that the amount of C-org that is metabolized by sulphate-reducing bacteria is considerably greater than the pyrite that is formed, because a high percentage of HS^- diffuses out of the anoxic zone toward the oxic zone where it is oxidized to SO_4^{2-} again.

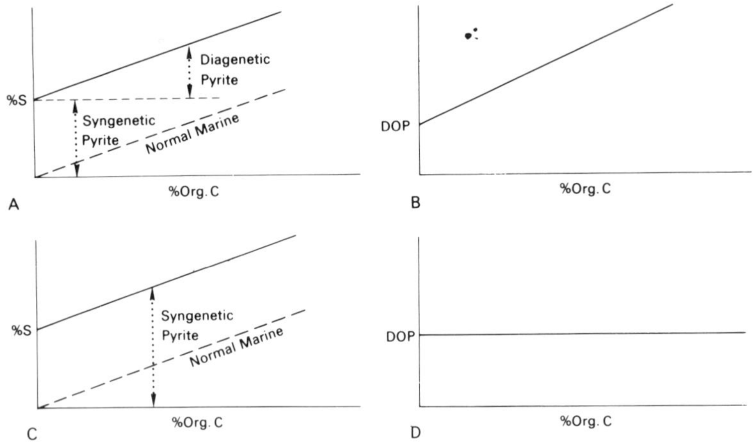

Fig. 2-8. Idealized plots of pyrite-S and degree of pyritization of iron (DOP) versus C-org for a hypothetical euxinic sediment. (A) The results for the formation of extra C-org-limited diagenetic pyrite, indicated by the increase in DOP with C-org content as shown in (B). The dashed curve is that expected for a normal marine sediment.
(C) The results for the formation of Fe-limited syngenetic pyrite alone, as indicated by the uniform DOP with increasing C-org content. (Modified after Raiswell and Berner, 1985.)

Jørgensen (1982) estimated that for sediments deposited in water depths in the range of 0 – 20 m, 90 – 95% of the sulphide produced is re-oxidized by O_2, equivalent to about 50% of the total O_2 uptake. For water depths between 20 and 200 m, his estimate is that 80% of the sulphide is re-oxidized, amounting to about 25% of the total O_2 consumption. Because sulphate reduction plays a much smaller role in deep-sea sediments and the slow rates of sedimentation, the reaction is such that a buildup of HS^- within the anoxic part of the sediment (if existing) is unlikely because of conditions favorable for pyrite formation. Sulphate reduction depends on the availability of labile organic matter as long as the SO_4^{2-} concentration < 5 mM. At lower SO_4^{2-} concentrations, the rates will become dependent on the SO_4^{2-} concentration (Berner, 1984). In the upper parts of the sedimentary column, more labile organic matter is present than in the lower parts, and a layer-by-layer measurement of the reduction rate will reflect this, as shown in Fig. 2-9.

In rapidly deposited sediments, more labile organic matter will be available for sulphate reduction than in slowly-deposited sediments (Middelburg, 1989). This leads to a positive relation between the average rate constant and the rate of sedimentation. According to Berner (1980), this relation is:

$$k_c \approx A\, w^2 \tag{2-14}$$

where:

k_c = rate constant for degradation of labile organic matter (Ma^{-1});
A ≈ 0.04 (a cm^{-2}); and
w = sedimentation rate (cm ka^{-1}).

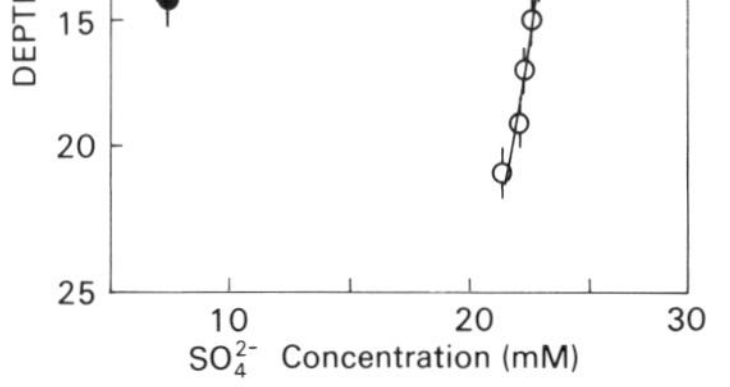

Fig. 2-9. Plots of SO_4^{2-} reduction rate by using a ^{35}S tracer (dots) and concentration of dissolved SO_4^{2-} (open circles) versus depth in a sediment at FOAM site from Long Island Sound. The low value for the reduction rate in the top 2 cm is due to the constant reoxygenation of large proportions of sediment within this depth range by burrowing infauna. (Modified after Berner, 1985.)

The values of k_c show a worldwide range of more than six orders of magnitude, reflecting the large range in sedimentation rates. Another consequence of the same features is that the depth, at which sulphate reduction starts in the sediment, varies from the uppermost layers for rapid sedimentation and high C-org content to below one meter in the continental slope (c.q. slope and rise), and is absent for many deep-sea sediments. Westrich and Berner (1984) compared the first-order rates for decomposition by O_2 and SO_4^{2-} respiration. According to them, the rate constant k_1 for labile, readily metabolizable organic matter compounds, and the rate constant k_2 for the less reactive compounds are as follows:

	O_2	SO_4^{2-}
k_1 (y^{-1})	24	8.8 − 7.2
k_2 (y^{-1})	1.4	0.84 − 1.02

They reported similar results obtained in other studies, indicating roughly a tenfold decrease in the ratio from the highly labile to the less reactive fraction of C-org. Because the bacterial activity is influenced by the ambient temperature, it is not surprising that seasonal changes in rates of metabolism occur in sediments at shallower water depths. Examples of this phenomenon are presented by Elderfield et al. (1981b), Klump and Martens (1981), Jørgensen and Sørensen (1985), and Oenema (1988). Some published depth-integrated rates of sulphate reduction are given in Table 2-3. An extensive overview of published reduction rates in coastal areas is given by Skyring (1987).

Consumption of SO_4^{2-} sets up a concentration gradient, which means that diffu-

TABLE 2-3

Some published depth-integrated rates of sulphate reduction

Location	Rate (mM cm^{-2} y^{-1})	References
MANOP site M	0.07×10^{-3}	Bender and Heggie (1984)
East Eq. Atlantic	0.11×10^{-3}	
Kattegat/Skagerak	0.32 − 0.45	Iversen and Jørgensen (1985)
Limfjorden	0.43	Howarth and Jørgensen (1984)
Coast of Peru (high productivity, 245 m depth)	0.18	Rowe and Howarth (1985)
Saanich Inlet (anoxic)	0.48	Devol and Ahmed (1981)
Skan Bay	0.43 (1980)	Reeburgh (1983)
	1.37 (1979)	
Eastern Scheldt:		
Creeks and channels	0.7 ± 0.5	Oenema (1988)
Mussel banks	1.0 ± 0.5	
Cape Lookout Bight	1.82 ± 0.16	Chanton et al. (1987a)
Chesapeake Bay	2.26	Reeburgh (1983)
Salt Marsh (New England)	3.28	Howes et al. (1984)

sion of SO_4^{2-} from bottom water into the sediment must occur as well as diffusion of HS^- (if not trapped in sulphidic minerals) toward to oxic zone. The re-oxidation of HS^- to SO_4^{2-} after diffusion into the oxic zone is not given the same amount of attention as paid to the oxidation of NH_4^+ in the oxic zone. Jørgensen (1977) quantified the fluxes of HS^- in Limfjorden (Denmark) and mentioned that 10% of all the HS^- produced, was precipitated as Fe(II)-sulphides and 90% was re-oxidized at the oxic surface. Berner and Westrich (1985) calculated losses of HS^- produced by SO_4^{2-} reduction for four stations in Long Island Sound, ranging from 25 to 94%. Bioturbation plays an important role in this process by the introduction of O_2-rich water into anoxic layers, and the enhancement of HS^- transport via benthic irrigation. Also, H_2S (in equilibrium with HS^-) may be stripped by upward-moving methane bubbles, a process also described by Klump and Martens (1981) and Chanton et al. (1987a). Christensen et al. (1984) argued that the downward transport of SO_4^{2-} from bottom water into the zone of sulphate reduction, was in large part due to benthic irrigation. Because this also brings O_2 directly to deeper horizons, this transport can be partially explained by re-oxidation of HS^- at these depths. Sweeney and Kaplan (1980) and Goldhaber and Kaplan (1980) addressed the evidence at hand for diffusion of SO_4^{2-} into sediments; their conclusions are not in line with each other. One of the reasons is that, in rapidly depositing sediments, the incorporation of syngenetic pyrite obfuscates the contribution of enrichment in total sulphur by diffusion. Study of the isotopic composition of the S pool can be helpful to elucidate the processes that brought about this enrichment. Bacterial sulphate reduction favors the reduction of $^{32}SO_4$ over $^{34}SO_4$; therefore, the sulphide pool will be lighter and the remaining sulphate pool heavier with respect to the S isotopes. Description of the diffusion of the dissolved S species has, therefore, to be broken down in two coupled processes, one for ^{32}S and one for ^{34}S isotopes. Taking this into account, Goldhaber and Kaplan (1980) quantified the amount of S added by diffusion for sediments in Pescadora Basin, as compared to the contribution of other processes (bioturbation and burial). Chanton et al. (1987b) did a similar study in Cape Lookout Bight and presented a good isotopic mass balance between input and output. A general treatment of the modeling of such coupled processes was also offered by Jørgensen (1979) and is discussed later in this chapter.

Methane production

A general consensus exists that the production of methane (CH_4) in marine sediments becomes prominent when practically all sulphate is exhausted. The bacterial conversion of suitable substrates in marine sediments by so-called methanogens seems to be predominantly based on the following reactions:

$$\text{fermentation: } CH_3COOH \rightarrow CH_4 + CO_2 \quad (2\text{-}15a)$$

$$\text{reduction: } CO_2 + 4\,H_2 \rightarrow CH_4 + 2\,H_2O \quad (2\text{-}15b)$$

In Fig. 2-10, the processes for marine environments are schematized. Ehrlich (1981)

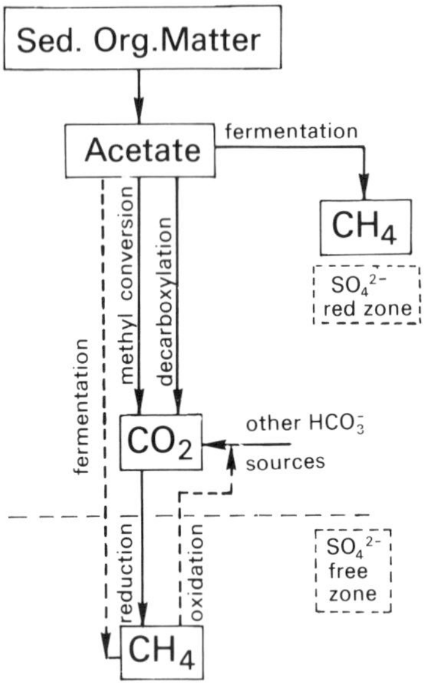

Fig. 2-10. Flow diagram comparing the methanogenesis pathway by acetate fermentation and CO_2 reduction in marine sediments. (Modified after Whiticar et al., 1986.)

mentioned that specialized strains of bacteria use the fermentation reaction, whereas others use the reduction reaction (Eq. 2-15b). Sansone and Martens (1982) concluded from their literature survey that in marine sediments the fermentation and reduction reactions describe the actual microbially-mediated processes. Crill and Martens (1986) reported for Cape Lookout Bight sediments that CH_4 production by reduction occurred in the whole sediment column with the exception of the top 2 cm, whereas fermentation only occurred below 10 cm depth where all SO_4^{2-} is exhausted. Fermentation provided about 1/3 of the total produced CH_4, whereas reduction provided the other 2/3.

Whiticar et al. (1986) reviewed existing data on the CH_4 production and found that reduction is the dominant process in the SO_4^{2-}-free zone of marine sediments. All methanogens use molecular hydrogen as an energy source and, therefore, this H_2 is an important agent in this conversion. Sansone and Martens (1982) discussed the role of H_2 produced by anaerobic heterotrophs by fermentation of carbohydrates. Acid-forming fermentative bacteria degrade large molecules to smaller molecules, e.g., to volatile fatty acids (aliphatic carboxylic acids with less than six carbon atoms per molecule) and can transfer H_2 derived from the oxidation of C-org to other bacteria capable of anaerobic respiration, such as sulphate or carbon dioxide reducers. The H_2 can also be transferred from sulphate-reducing bacteria to methanogens when SO_4^{2-} concentrations are low. These processes, called interspecies hydrogen transfer, lead to a rapid consumption of H_2 in organic-rich anaerobic environments ($< 10^2$ Pa) (Rudd and Taylor, 1980). Apart from hydrogen and acetate only formate, methanol and methylamines are suitable substrates for use by methanogens (Rudd and Taylor, 1980; Sansone and Martens,

1982; Crill and Martens, 1986). The reason why the presence of SO_4^{2-} tends to suppress methanogenesis is thought to be that sulphate-reducing bacteria are much more effective in the competition for H_2 and acetate produced by acid-forming bacteria than are methanogens. It must be mentioned, however, that the effectiveness of this competition can be the result of the typical bacterial communities and their populations in marine sediments. Oremland and Taylor (1978) showed that sulphate reduction and methanogenesis are not mutually exclusive under favorable laboratory conditions.

Sansone and Martens (1982) summarized the steps in the anaerobic decomposition of organic matter in the model presented in Fig. 2-11. These authors pointed out that the complex-organic substrate must at least be partially respired, before it can be completely oxidized to CO_2. In an anaerobic environment, SO_4^{2-} or CO_2 must be used as electron acceptors. The bacteria responsible for these respirations must rely on fermentative microbes to provide them with the utilizable simple substrates. The combination of Eqs. 2-2a and 2-2g gives the following overall process:

$$(CH_2O)_x\ (NH_3)_y\ (H_3PO_4)_z \rightarrow$$
$$0.5x\ CO_2 + 0.5x\ CH_4 + y\ NH_4^+ + z\ HPO_4^{2-} + (-y + 2z)\ H^+ \qquad (2\text{-}16a)$$

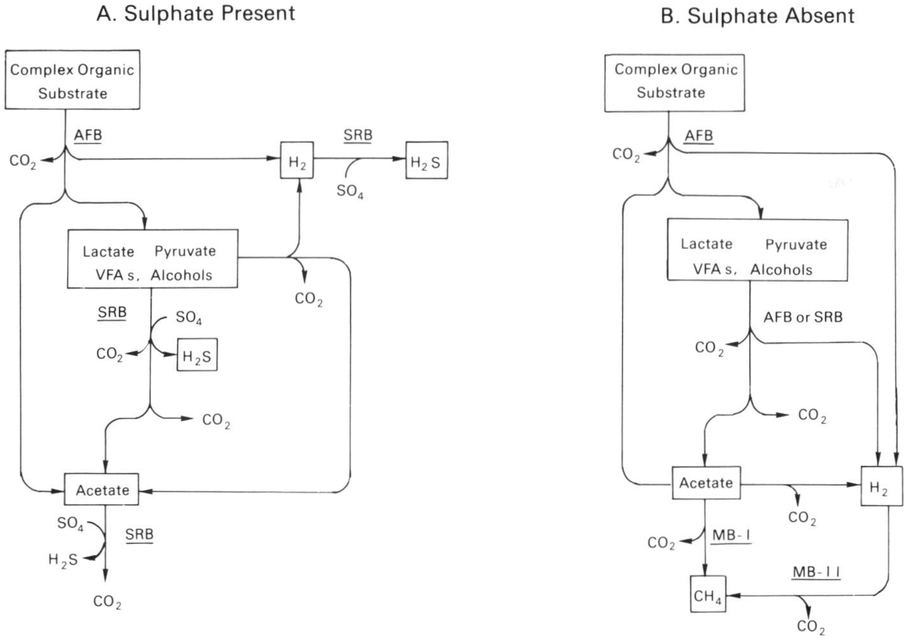

Fig. 2-11. Model of the terminal steps of anaerobic decomposition in: (A) sulphate-containing, and (B) sulphate-depleted environments. *AFB* = acid-forming bacteria, *SRB* = sulphate-reducing bacteria, and *MB* = methanogenic bacteria. (Modified after Sansone and Martens, 1982.)

or, in Redfield stoichiometry, in combination with the dissociation reactions for water and dissolved carbon dioxide:

$$(CH_2O)_{106}\ (NH_3)_{16}\ (H_3PO_4) + 14\ H_2O \rightarrow$$
$$39\ CO_2 + 14\ HCO_3^- + 53\ CH_4 + 16\ NH_4^+ + HPO_4^{2-} \qquad (2\text{-}16b)$$

Methane production increases alkalinity and total inorganic dissolved CO_2, the latter in equal amounts as the produced CH_4. Because of the increase of carbonate alkalinity, precipitation of $CaCO_3$ may occur (Claypool and Kaplan, 1974). The solubility of this solid phase, however, will be higher at higher CO_2 concentration that is produced simultaneously. For the calculation of the actual effect, the total sediment chemistry has to be taken into account.

Shaw et al. (1984) and Sansone and Martens (1982) reported that the 1:1 stoichiometry of produced total CO_2 and CH_4, suggested by reactions 2-16a,b, seems not to be observed when a labeled substrate ($^{14}CH_3COOH$) is used to study methanogenesis. They explain this by the interspecies hydrogen transfer mechanism, which converts acetate into $CO_2 + H_2$ by one organism. The products are excreted and the H_2 scavenged by another bacterial species for recombination with CO_2 to form CH_4. The CO_2 used in this second step, however, contains stable carbonate species from the much larger pore water carbonate pool, that dilutes the labeled CO_2. Thus methanogenesis could escape detection by relying on ^{14}C. These, as well as other complications, lead Iversen and Jørgensen (1985) to advocate the use of CH_4 oxidation rates as a combined measure of CH_4 production. The produced CH_4 can stay in the dissolved phase until a certain concentration is reached depending on temperature and pressure. Under conditions of high pressure and low temperature, the formation of solid CH_4 hydrate is possible, but this can eventually be expected in deep-sea sediments (Claypool and Kaplan, 1974), that are usually not the logical environment for methanogenesis during early diagenesis.

According to Kvenfolden (1988), methane hydrates are, however, likely to occur in outer continental margins and the amount of methane at subsurface depths of < 2 km is potentially very large. More likely, especially in shallower water environments, is the formation of gaseous CH_4. Methane as a hydrate and as a gas can be observed in seismic profiles. Methane can diffuse upward by molecular diffusion in the dissolved state or quasi-advectively by ebullition. The rates of these processes have been studied by Sansone and Martens (1981) and Kipphut and Martens (1982) in Cape Lookout Bight (North Carolina) sediments. Seasonal variations are observed, due to the strong temperature dependence of acetate utilization by methanogens, leading to much greater fluxes in the summer season. The measured quasi-advective flux arising from the release of gas bubbles was approximately six times larger than the diffusive flux. Ebullition may cause a direct injection of CH_4 into the overlying bottom water, even if the top of the sediment is oxygenated. But a large part of CH_4 that diffuses upward is oxidized. Most workers are convinced that this oxidation takes place even anaerobically, by or in connection with sulphate reduction, although the microbial processes that are involved are not yet known (Alperin and Reeburgh, 1985; Iversen and Jørgensen, 1985, Whiticar and Faber,

1985). In Table 2-4 some integrated rates of sediment methane oxidation are given.

Martens and Berner (1977) modeled the CH_4 profile in interstitial waters in a core retrieved from Long Island Sound, the result of which is shown in Fig. 2-12. Their conclusion is that the observed data points can be fitted satisfactorily by assuming a first-order consumption rate for anaerobic methane oxidation in the transition zone where SO_4^{2-} is still present. As a reference, the theoretical profile is shown for the case that no CH_4 is oxidized within the sediment column. This profile shows the typical features as reported by a number of workers (e.g., Warford et al., 1979; Sansone and Martens, 1981; Reeburgh, 1982; Devol et al., 1984; Iversen

TABLE 2-4

Some integrated oxidation rates of methane in a variety of sedimentary environments

Locality	Integrated oxidation rates (μM cm^{-2} a^{-1})	References
Santa Barbara Basin	0.011	Barnes and Goldberg (1976)
Kysing Fjord	0.12 – 0.63	Iversen and Blackburn (1981)
Kattegat (43 m)	0.4	Iversen and Jørgensen (1985)
Eckernfördner Bay	1.16 – 4.76	Whiticar (1978, 1982)
Cape Lookout Bight	5.6	Martens (1984)
Guinea Basin	8.8	Miller (1980)
Saanich Inlet	25 – 71	Reeburgh (1983)
Skan Bay	23.6	Reeburgh (1983)
Kattegat (65 m)	30	Iversen and Jørgensen (1985)
Skagerak (200 m)	42	Iversen and Jørgensen (1985)
Chesapeake Bay	220 – 360	Reeburgh (1983)

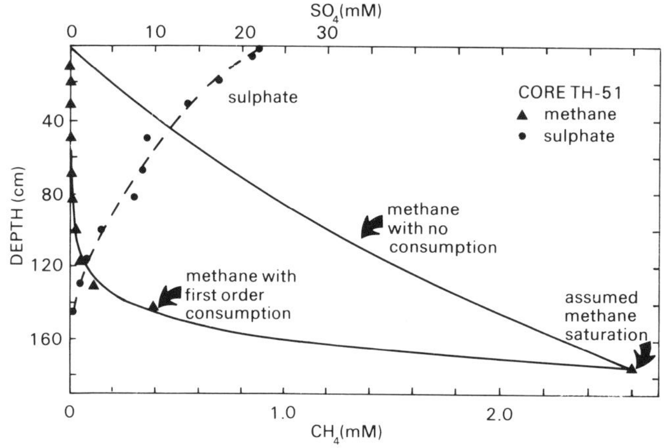

Fig. 2-12. CH_4 and SO_4^{2-} concentrations versus depth for a core from Long Island Sound. Full lines represent plots of theoretical curves for methane with no consumption and for methane with consumption by SO_4^{2-} via first-order kinetics of oxidation ($k = 8 \times 10^9$ s^{-1}); the dashed line is an exponential fit to SO_4^{2-} data. (Modified after Martens and Berner, 1977.)

and Jørgensen, 1985). The concave-upward profiles of CH_4 are indicative of consumption within the sediment column above the zone of methanogenesis. Both Alperin and Reeburgh (1985) and Iversen and Jørgensen (1985) have shown that the CH_4 oxidation rate is greatest just in the zone where SO_4^{2-} is still present. This means also that a second maximum in the sulphate reduction rate is present just above the boundary where SO_4^{2-} becomes exhausted. Alperin and Reeburgh (1984) used three approaches (diagenetic models, quasi in-situ rate measurements, and stable isotopes) to demonstrate that anaerobic CH_4 oxidation occurs in anoxic sediments. Because this overview stresses the power of modeling in deciphering diagenetic processes, the findings, of the last-mentioned authors, using the model approach, are shown in Fig. 2-13. This figure shows that CH_4 is being oxidized in the upper part of the sediment, with the highest rates occurring in the transition zone between sulphate reduction and methanogenesis. The upward migration of CH_4 and its subsequent oxidation at higher horizons has an effect on the isotopic distribution of carbon.

Reeburgh (1982) and Alperin and Reeburgh (1984) discussed the effect of methanogenesis and CO_2 oxidation on the carbon isotopes of total dissolved inorganic CO_2. The $^{13}CO_2$ profile typically shows a minimum approximately at the depth of the onset of CH_4 oxidation. Metabolism of C-org, with a typical marine $\delta^{13}C$ value of $-(5-17)$‰, causes the dissolved inorganic carbon pool to become lighter (more negative with depth). Methanogenesis in marine sediments produces very light CH_4 with $\delta^{13}C$ of $-(60-110)$‰ (Whiticar et al., 1986), making the remaining dissolved inorganic carbon pool heavier. Diffusion of the light CH_4 and its subsequent oxidation, makes the dissolved inorganic carbon pool lighter in this oxidation zone, giving rise to a minimum in the $\delta^{13}C$ profile. This is shown for Skan Bay sediments in Fig. 2-14 (cf. also Fig. 2-13).

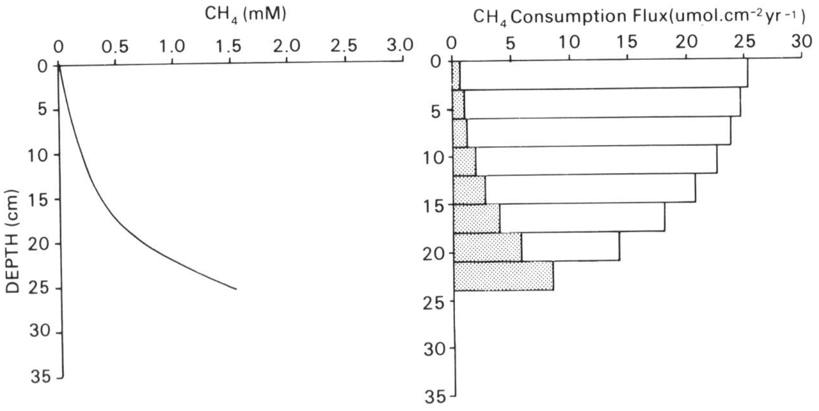

Fig. 2-13. Distribution of CH_4 with depth in Skan Bay sediments. Left: data for the upper 25 cm were fit to an exponential curve. Right: absolute (hatched) and cumulative (unhatched) CH_4 consumption fluxes are obtained by fitting the CH_4 concentration data. These fluxes show that CH_4 consumption is highest at depth and that consumption of the upward CH_4 flux is complete. (After Alperin and Reeburgh, 1984.)

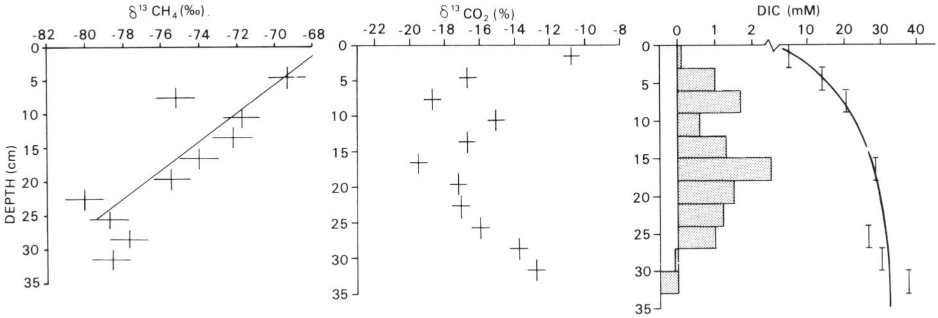

Fig. 2-14. (Left) Distribution of $\delta^{13}CH_4$ with depth in Skan Bay sediments, showing sample depth intervals and error bars; the curve shows results of a Rayleigh distillation model applied to the data. (Middle) Distribution of $\delta^{13}CO_2$ with depth. (Right) Results of a mass and stable carbon isotope balance model, showing variation with depth of dissolved inorganic carbon (DIC) and CH_4-derived CO_2 (hatched). (After Alperin and Reeburgh, 1984.)

Production of carbon dioxide and alkalinity

All reactions involving oxic, suboxic, and anoxic metabolism of organic matter produce CO_2 and/or (bi)carbonate (cf., Eqs. 2-6, 8, 9, 12, 13, 16). In the foregoing sections it was assumed that the pore water pH did not change and that the dominant ionic species of the oxidation products are in the typical pH-range between 7 and 8: HCO_3^-, HS^-, HPO_4^{2-}, NH_4^+. The pH can be buffered by $CaCO_3$ and by ion-exchange reactions on active surface sites of, for instance, clay minerals.

Dickson (1981) proposed the following definition for alkalinity: "The total or titration alkalinity is the number of moles of hydrogen ion equivalent to the excess of proton acceptors (bases formed from weak acids with a dissociation constant $K = 10^{-4.5}$ at 25°C and ionic strength = 0) over proton donors (acids with $K > 10^{-4.5}$) in one kilogram of sample". Production of weak acids, therefore, does not produce alkalinity, because upon their dissociation, equal amounts of protons and conjugate titratable bases are produced ($CO_2 + H_2O \rightleftarrows H^+ + HCO_3^-$; $H_2S \rightleftarrows H^+ + HS^-$; $H_3PO_4 \rightleftarrows 2\,H^+ + HPO_4^{2-}$). But when protons are subsequently consumed by other reactions (redox reactions or dissolution reactions), the conjugate bases become part of the total alkalinity. On the other hand, when protons are consumed, the alkalinity decreases. An example of a reaction consuming protons is denitrification (Eq. 2-6), whereas an example of production of protons is nitrification (Eq. 2-3).

Determination of the alkalinity by titration not only measures contributions of dissolved anionic carbonate species (carbonate alkalinity), but also of ammonia (NH_3), phosphate, (bi)sulphide, and borate. In order to calculate the carbonate alkalinity from the titration alkalinity, separate determinations are necessary to calculate total dissolved inorganic combined nitrogen, inorganic phosphate, hydrogen sulphide, and boric acid. In combination with the pH of the solution and with the values of the relevant apparent acid – base dissociation constants, the carbonate contribution to the titration alkalinity can be obtained by subtraction of the contributions of all other solute species (Dickson, 1981). The carbonate alkalinity in marine pore waters is usually much greater than those of the other species. At the pH of marine pore waters, carbonate alkalinity is mainly in the form of the HCO_3^- ion plus its complexes, c.q. ion pairs, because the carbonate species in that pH-range are relatively negligible. The equations for the total dissolved carbonate species (ΣCO_2) and carbonate alkalinity, are therefore:

$$\Sigma CO_2 \approx (H_2CO_3^*) + (HCO_3) + M_i(HCO_3)_j \leftrightarrows (H_2CO_3^*) + (HCO_3)_T \quad (2\text{-}17a)$$

$$CA \approx (HCO_3)_T \quad (2\text{-}17b)$$

or

$$\Sigma CO_2 \approx (H_2CO_3) + CA \quad (2\text{-}17c)$$

where:

$H_2CO_3^*$ = $CO_{2(l)} + H_2CO_3$ = total dissolved free carbon dioxide
M = cations in solution
i, j = $0, \ldots, n$
T = total analytical concentration

CO_2 can be determined by infrared spectrometry (or other methods), after acidification of the sample and driving out the CO_2 in a gas flow. Carbonate alkalinity can only be determined by titration, after proper correction for the contributions of other species to the total alkalinity. The relation between CO_2 and CA is via pH. When the dissociation reaction is written as:

$$H_2CO_3^* \rightleftarrows H^+ + HCO_3^- \quad (2\text{-}18)$$

then the apparent dissociation constant for seawater, including all HCO_3 species, is:

$$K_1' = (HCO_3^-)_T \, a_{H^+} \, (H_2CO_3^*)^{-1} \quad (2\text{-}19)$$

This gives the following relation:

$$CO_2 \approx CA \, (1 + a_{H^+}/K_1') \quad (2\text{-}20)$$

where:

a_{H^+} = hydrogen ion activity in appropriate scale = 10^{-pH};
K_1' = apparent first dissociation constant of $H_2CO_3^*$ in seawater.

The problem with the relation 2-20 is that K_1' depends on the bulk composition of the pore waters and is in fact not a constant when deviations from average seawater composition occur. This means that the conversion of CO_2 into CA, or vice versa, by the use of pH is not as straightforward as one would desire. These complications have to be realized when, as is often the case, the degree of saturation of pore waters with respect to solid carbonate phases is calculated.

The diagenesis of C-org can be followed by the concomitant change in dissolved CO_2. Anderson et al. (1986) used this approach and concluded that this is suitable in sediments where methane production can be neglected after proper correction for dissolution or precipitation of $CaCO_3$. It also offers a possibility to measure directly the rates of oxic and anoxic mineralization. A quantitative treatment of the relation between the production of CO_2 and metabolism of C-org is also given by Henrichs and Farrington (1984). The measured increase of ΣCO_2 is often used as a parameter to relate to the increase or decrease of oxidants (O_2, NO_3^-, SO_4^{2-}) or mineralization products (NH_4^+, HS^-, HPO_4^{2-}). When plots of these constituents in pore waters versus ΣCO_2 exhibit linear relationships, stoichiometric modeling is feasible. The ratios of ΣCO_2 and titration alkalinity (TA) that are produced in relation to the oxidants that are consumed, can be calculated from the pertinent mineralization reactions for sediments with or without $CaCO_3$ on the basis of an initial Redfield ratio, as given in the first four columns of Table 2-5. As pointed out by Emerson et al. (1982a), however, these stoichiometric ratios are based on the assumption that Eq. 2-10 does not represent an equilibrium reaction. They use a model to calculate the net effect of the mineralization on the carbonate chemistry of the pore waters:

$$d\Sigma CO_2/dOx = [\delta\Sigma CO_2/\delta Ox]_{Ca} + [\delta\Sigma CO_2/\delta Ox]_{OM} \qquad (2\text{-}21a)$$

$$dTA/dOx = [\delta TA/\delta Ox]_{Ca} + [\delta TA/\delta Ox]_{OM} \qquad (2\text{-}21b)$$

TABLE 2-5

Ratios of total CO_2 and alkalinity produced by various oxidants (Ox) in the presence or absence of $CaCO_3$

Reactions and equations	Absence of $CaCO_3$		Presence of $CaCO_3$			
	(1)*	(2)**	(1)	(2)	(1)	(2)
O_2 respiration (Eqs. 2-4b, 2-10)	−0.77	+0.12	−1.67	−1.67	−1.58	−1.50
Denitrification (Eqs. 2-6c, 2-10)	−1.12	−0.98	−1.28	−1.28	−1.21	−1.15
Mn oxide reduction (Eqs. 2-8c, 2-10)	−0.45	−2.00	+1.09	+1.09	+1.05	+0.96
Fe oxide reduction (Eqs. 2-9b, 2-10)	−0.25	−2.04	+1.53	+1.53		
Sulphate reduction (Eqs. 2-12b, 2-10)	−2.00	−2.28	−2.55	−3.57		

* (1) = $\partial\Sigma CO_2/\partial Ox$.
** (2) = $\partial TA/\partial Ox$.

where the first partial derivatives on the right-hand side indicate the change during organic matter degradation with the Ca concentration held constant (no $CaCO_3$ dissolution), whereas the second partial derivatives indicate the change as a result of $CaCO_3$ reaction in the absence of organic matter degradation. The latter derivatives follow from the equilibrium considerations in the system $CO_2 - Ca^{2+} - H_2O$, in combination with the definitions of CO_2 and *TA*. Emerson et al. (1982a) presented the results (Fig. 2-15) for their model calculations for oxic and suboxic diagenesis together with the stoichiometric predictions given in the first four columns of Table 2-5. Their conclusion is that the stoichiometric ratios in the

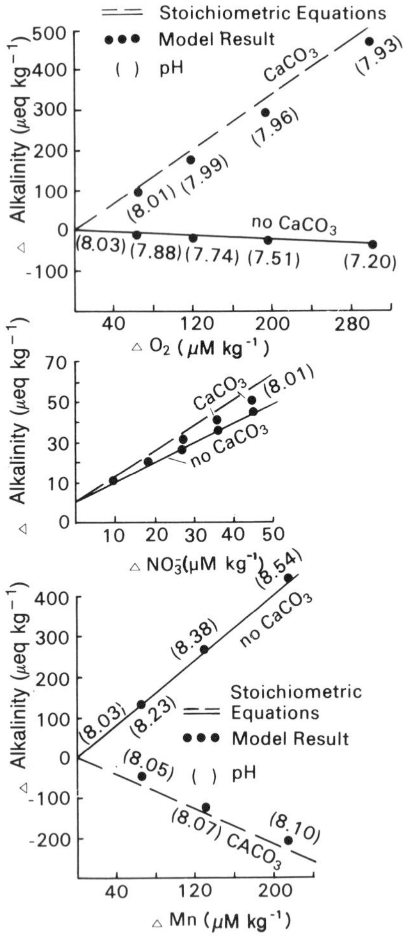

Fig. 2-15. The changes in alkalinity during labile organic matter degradation in a closed system in the presence and absence of $CaCO_3$. Lines represent the values predicted by the stoichiometry of the oxidative degradation reactions for C-org with O_2 (top), NO_3^- (middle) and MnO_2 (bottom) as electron acceptors, and dissolution of $CaCO_3$. The symbols are model-derived results with pH values in parentheses. The initial conditions are: total alkalinity = 2.446×10^{-3}, total dissolved CO_2 = 2.278×10^{-3}, pH = 8.03, T = 1.5°C, P = 437 bars. (After Emerson et al., 1982a.)

third and fourth column, are about 10% in error for oxic and suboxic oxidation. The last two columns give their model results. These data can be used for closed systems, but for open systems one has to take into account the fluxes of the various constituents due to concentration gradients. Boudreau (1987) used a model in which the production of CO_2 by decay of organic matter, the production of carbonate alkalinity by suboxic and anoxic processes, and acid – base equilibria and diffusion of all carbonate species are considered. He concluded that suboxic decay can lead to supersaturation of pore waters with respect to $CaCO_3$. Emerson et al. (1982a) also addressed the problem of the loss of alkalinity caused by precipitation of $CaCO_3$ due to depressurization while bringing the sediment core from in-situ pressure to 1 bar. Their conclusion is that this precipitation appears to be dependent on the abundance of $CaCO_3$ in the sediments. They believe that calculation of this effect of pressure on the alkalinity of pore waters, proposed as a method of correction by Murray et al. (1980), on the basis of assumption of ideal chemical equilibrium in the carbonate system, is not warranted and cannot replace the values that are obtained on pore water samples that are collected in-situ.

The main solid phase to control the carbonate concentration in pore waters is $CaCO_3$. But, as already mentioned, precipitation of pure or mixed Mn carbonates can also occur. Another complication is that aragonite and high-Mg calcites, when present, have a higher solubility than do low-Mg calcites. Finally, upward diffusion of CH_4 can lead to the production of CO_2 in the sulphate reduction zone and so enhance this reduction as compared to normal mineralization of solid labile C-org.

The isotopic composition of dissolved carbonate in marine pore waters changes, because plankton has low $\delta^{13}C$ values (typically in the range of $-(18-23)$‰ for the northern hemisphere, and $-(18-29)$‰ for the southern hemisphere) with values becoming lower with increasing latitudes (Rau et al., 1982). These values may become still a few per mille lower in the pycnocline, because of biological reworking (Jeffrey et al., 1983). Biogenic $CaCO_3$ has $\delta^{13}C$ values generally in the range of $0-2$‰ (all values relative to the standard PDB). Upon diagenesis, relatively lighter carbon is added to the inorganic carbon pool. As already discussed, the anoxic oxidation of CH_4 with its isotopically even lighter carbon, will have an even greater effect on the isotopic composition of dissolved carbonate at that depth. The effect of mineralization of C-org on the isotopic carbonate composition of pore waters can already be significant in the top few centimeters of the sediments, with $\delta^{13}C$ gradients as steep as -1.0‰ cm^{-1}. McCorkle et al. (1985), Corliss (1985), and McCorkle and Emerson (1988) pointed out that this will have an effect on the isotopic composition of benthic infauna. This effect has to be considered when the stable isotope record of benthic foraminifera tests is studied and interpreted.

Production of dissolved phosphate and silica

Phosphorus and silicon are both essential nutrients for aquatic life. Phosphorus is incorporated in cells and plays a key role in the energy transfer necessary for growth and functioning of living species. Phosphate may also be incorporated in protective hard tissue of aquatic flora and fauna (tests, shells). Silicon is an essential nutrient for the growth of a number of conspicuous aquatic organisms: diatoms, radiolaria, silicoflagellates, and sponges. The standing crop of marine organisms can, by far, not be maintained by the input of phosphate and silica from the continents. This means that considerable internal cycling of these nutrients takes place. Regeneration of the nutrients occurs in the food web within the euphotic zone with short turnover times, as well as beneath this zone by (micro)biological mineralization of soft tissue and chemical dissolution of hard tissue. The oceanic circulation brings deep water, with high dissolved nutrient levels, back to the surface, with circulation times in the order of 500 – 1500 years. Assuming a steady-state of the oceans, the permanent burial of the nutrients in marine sediments has to be equal to their input from the continents. A requirement of this input is that it partially takes place in a chemical form that can be utilized by marine organisms. Dissolved inorganic phosphate and silicium, and organic phosphorus in rivers, are readily available biologically. Reversibly adsorbed phosphate on solid particles may be utilized directly by algae or becomes desorbed in suboxic diagenesis and subsequently transferred to the water column. Amorphous silica can be solubilized in the water column or in the sediments and thus become, directly or after diffusion, available. But the input of practically insoluble forms of phosphate and silica, e.g., apatites and many aluminosilicate minerals, has no bearing on the maintenance of marine aquatic life.

Phosphate, for certain areas, and silica, for certain species, are growth-limiting nutrients. This means that these nutrients are among the first to become exhausted in the euphotic zone, so that further production of the species that depend on these nutrients comes to a halt. This leads sometimes to seasonal explosive blooms in a time of optimal conditions for growth, followed by a long time of zero growth during which the conditions favorable for production are restored. The main concern of this chapter is the role played by sediments to partially recycle or bury these nutrients. Froelich et al. (1982) identified the major fluxes of phosphorus to the sediments. About 90% is in the form of biogenic debris or its regeneration products, mainly in C-org, in $CaCO_3$ (mostly coccoliths), and in authigenic phosphorites. Studies by Palmer (1985) and Sherwood et al. (1987) have shown that the P content of calcareous sediments and sediment particles is in fact largely located in Fe and Mn oxide coatings of the particles, with only a small P content in the $CaCO_3$ phase proper. The extraction methods used by these authors show that the P content in the Mn and Fe oxide phases is not in a readily available form, but may be liberated under reducing, suboxic conditions. Production of phosphorites is restricted to areas of high productivity in some upwelling areas, e.g., off the coast of Peru. The burial of organic phosphorus (P-org) is related to the burial of C-org. The data for the $(P/C)_{org}$ ratio in marine sediments vary widely, depending on the sedimentary regime and geographical area. Mach et al. (1986) analyzed data that were published and found a tendency of higher $(P/C)_{org}$ ratios in sediments with low sedimentation rates. This is in contrast to what is usually assumed, namely, a preferential release of P-org from organic matter during early stages of diagenesis. Their findings may be warped by the techniques used by various workers in determining the P-org content in sediments. But if these findings are approximately correct, it has to be assumed that after the initial stages of diagenesis, C-org is more labile than P-org, which means that the $(P/C)_{org}$ ratio has a minimum at a specific depth and then increases again at greater depths. The Redfield P/C ratio for marine plankton is 9.4×10^{-3}. Peng and Broecker (1987) stated that this ratio in marine detritus differs from the commonly accepted Redfield ratio and proposed a value of $7.9\ (\pm 0.4) \times 10^{-3}$. Mach et al. (1986) proposed a global $(P/C)_{org}$ ratio for buried organic matter of 5.3×10^{-3} or roughly 40% preferential mineralization of P-org over C-org in comparison to phytoplankton composition. The rain of organic matter toward the sediment, as collected in sediment traps, has a $(P/C)_{org}$ ratio of $(1-5) \times 10^{-3}$ (Suess and Müller, 1980), with the higher values in rapidly accumulating sediments. This means that, in general, the preferential mineralization of P-org over C-org at the top or within the sediment is in the order of 50%.

The stoichiometric reactions for mineralization of P-org are given in Eqs. 2-4a, b; 2-6a, b; 2-7a, b; and 2-11a, b. But the amounts of phosphate that are liberated by these reactions are often not equal to those predicted by the equations. The reason can be that the $(P/C)_{org}$ ratio differs from the ideal Redfield ratio. But there are more explanations for observed profiles. Van Cappellen and Berner (1988) mentioned the possibility of dissolution of inorganic fish hard parts and the precipitation of carbonate fluorapatite. Furthermore, adsorption of phosphate plays a major role in the control of its concentration in natural waters. Ferric oxihydroxides are recognized as the most important mineral phases for such adsorption (e.g., Billen, 1982a). This means that, as long as sediments are oxic, such coatings are a preferred substrate for adsorption of phosphate. A flux of phosphate from the anoxic zone upward cannot reach the bottom waters because of this sorption taking place in the oxic top of the sediments. If, however, the oxic top is seasonally or more permanently situated at or very close to the sediment – water interface, phosphate will not be trapped and can diffuse out of the sediment into the overlying water (e.g., Klump and Martens, 1981; Fisher et al., 1982; Sundby et al., 1986). Such areas are, therefore, important for the re-injection of this nutrient into the water column. In general, these areas will have a high input of organic matter, bringing the horizon of O_2 depletion close to the sediment – water interface. Some published exchange rates are given in Table 2-6. On the other hand, suboxic oxidation of organic matter will set free the gradually accumulated adsorbed phosphate in this zone. This will give rise to a maximum in the dissolved phosphate concentrations at this depth, that is not due to mineralization of P-org (Froelich et al., 1979; Krom and Berner, 1981;

TABLE 2-6

Fluxes of phosphate from the sediment into the bottom water (s = summer; f = fall; w = winter)

Locality	Flux of phosphate (mM m^{-2} day^{-1})	References
La Jolla Bight	0.08	Hartwig (1974)
Narragansett Bay	0.23	Hale (1975)
Long Island Sound (FOAM)	0.10(s), 0.06(f), 0(w)	Aller (1980b)
(NWC)	0.32(s), 0.40(f), 0.02(w)	
(DEEP)	0.20(s), 0.02(f), 0.01(w)	
Cape Lookout Bight	2.4(s), ≈0(w)	Klump and Martens (1981)
Gulf of Mexico (outer shelf)	0.005 – 0.05	Filipek and Owen (1981)

Filipek and Owen, 1981; Anderson et al., 1986).

Sinks for phosphate in the anoxic zone are adsorption and precipitation of mineral phases. The linear adsorption coefficients for adsorption onto mineral phases in the anoxic zone are much lower than they are for Fe(III)-oxyhydroxides (Billen, 1982a). Among distinct mineral phases that are often proposed as neoformations in the anoxic zone are: vivianite $[Fe_3(PO_4)_2 \cdot 8H_2O]$, struvite $(MgNH_4PO_4 \cdot 6H_2O)$, and marine apatite $[Ca_5(PO_4)_{3-x}(CO_3)_xF_{1+x}]$. Even if the pore water turns out to be supersaturated with respect to one of these phases, that does not imply that these phases are actually forming and controlling the pore-water phosphate concentration (cf., Murray et al., 1978). Often, the dissolved phosphate profiles are quite irregular, especially in comparison to other products of microbial metabolism, e.g., ammonium. The PO_4/NH_4 ratio is not as constant as one would expect. Berner (1977) pointed out that nonlinear plots of PO_4 versus SO_4 for anoxic pore waters indicate that stoichiometric modeling is impossible, presumably because of preferential breakdown of different types of organic matter or precipitation of phosphate minerals. One should also be aware of possible sampling and analytical errors: phosphate is rapidly adsorbed on filters and walls of tubes and bottles when the anoxic pore water comes in contact with air, causing rapid oxidation of ferrous ions into amorphous Fe(III)-oxyhydroxide, which coats these surfaces and is a highly reactive sorbent for phosphate. Jahnke et al. (1982a) reported significant losses of phosphate using squeezing and centrifuging techniques in comparison to in-situ sampling of pore water. Other causes of irregular profiles of dissolved P can be changes in the sedimentary regime. For sediments from the Madeira and Nares A.P.'s with turbidites (having a higher C-org content) intercalated in normal pelagic sediments (with a low C-org content), De Lange (1984, 1986) found a good correlation between dissolved phosphate and C-org contents. But he showed that the profiles could not be explained by this correlation alone, because of the excessively high degradation rates that would be necessary to maintain the profiles as measured. To reconcile his findings and their explanations, he had to assume active precipitation of a phosphate phase, but without identifying which one. In both cases, the pore water profiles have not reached a steady-state.

Dissolved Si is predominantly produced by dissolution of biogenic siliceous tests and frustrules. Production of these materials does not occur evenly in the oceans. Apparently, they are favored in areas of potentially high productivity; in areas of average to low productivity, calcareous organisms can compete much better for nutrients (Bogdanov et al., 1980a,b). In the Antarctic regions, the production of diatoms dominates, whereas in tropical zones, radiolaria are the important species. Schink et al. (1975) estimated that, for the marine plankton, the Si/C molar ratio of 1/4.6 can be used as a global average in the euphotic zone, with a broad range of ratios for the various regions. The ocean waters and pore water are undersaturated with respect to solid amorphous silica. This means that huge quantities of biogenic silica dissolve in the water column and within the sediment. Broecker and Peng (1982) estimated that only about 5% of the biogenic silica produced in the euphotic zone becomes permanently buried in the sediments. The burial of this silica is not controlled by chemical equilibria, but probably by dissolution kinetics in comparison to rates of sedimentation. Accumulation of siliceous sediments occurs where the production and, consequently, the rate of sedimentation of biogenic silica are high. The solubility of amorphous silica decreases with temperature, about 50% from 25° to 0°C (Wollast, 1974). The solubility increases from sea-level pressure to a depth of 5 km by some 15% (Willey, 1974). The overall effect is that the solubility of siliceous debris will generally decrease from surface-water conditions to deep-water conditions, except at high latitudes. The rate of dissolution will also decrease, because this rate depends linearly on the degree of undersaturation of the solution (Vanderborght et al., 1977a). This dissolution rate also depends on the protection of the silica surface by adsorbed inhibitors, that drastically reduce this rate (Berner, 1980). In the water column this difference between equilibrium and actual concentrations is depth-dependent (i.e., increase of Si concentration with depth), as well as ocean-dependent (i.e., the lowest Si concentrations in the North Atlantic Ocean and the highest in the Pacific Ocean). The concentration in marine pore waters becomes higher with depth in the sediment to an extent that depends on (1) the amorphous silica content rate of bioturbation (Schinck et al., 1975), (2) the mechanical disturbation (Vanderborght et al., 1977a), and (3) the molecular diffusion back into the bottom water. The dissolved Si profiles mostly have a concave-down shape, indicating production at all depths, gradually arriving at a certain depth at a constant value, still below the chemical equilibrium but determined by dynamic equilibrium (Schinck et al., 1975). But advection (Lerman, 1975) or secondary reactions can change the profile. So-called "reverse weathering", that is the reconstitution of aluminosilicate minerals, was thought to be one of the possible sinks, and neoformation (e.g., of sepiolite) another sink. Sayles (1981), in his study of Atlantic sediments, argued that sepiolite probably does not control interstitial Si concentrations at the low temperature of deep-sea sediments, but may eventually become a controlling factor at temperatures higher than 20°C that can be expected at a depth of several hundreds of meters in the sediments. He also could not find evidence for control by neoformation of simple cation silicate phases ("reverse weathering") in keeping with his observations.

Mackin and Aller (1984a,b) presented arguments for the occurrence of authigenesis of clay minerals in marine sediments in the very early stages of sedimentation. Mackin (1987) presented a model in which the dissolved silica concentration is conditioned by equality of the Si/Al ratio of authigenic clays on the one hand and dissolving minerals on the other hand. This would explain why dissolved silica concentrations never quite reach the saturation value for amorphous silica. Williams and Crerar (1985) discussed the transformation of biogenic opal-A, via opal-CT, into authigenic quartz. The concomitant dissolved silica concentrations in "equilibrium" with these phases decrease in the same order. This may also explain why concentration profiles of dissolved silica in pore waters do not attain the concentrations expected for equilibrium with metastable opal-A. Borel-Curial and Rio (1988) found that the dissolved silica concentrations of pore water in radiolarian sediments are generally lower than those in the diatomaceous sediments; adsorption of dissolved silica onto clays and authigenesis of clays on the surface of existing (aluminium-)silicate minerals will decrease the Si concentrations in the pore waters. Higher Si concentrations can build up in almost pure carbonate sediments and this can give rise to incipient opal-CT nucleation (probably heterogeneously, on organic matter) and precipitation. Because the newly-formed minerals are, quantitatively, orders of magnitude smaller than the detrital ones, their existence and formation are almost impossible to detect without the help of electron microscopy. De Lange and Rispens (1986) found evidence for the neoformation of Fe(II)-silicate in cores from the Nares A.P.

The fluxes of Si out of the sediments into bottom waters can vary seasonally because of changes in bioturbation or physical perturbation rates (e.g., Elderfield et al., 1981b). Aller et al. (1985) also discussed the role played by a benthic community in generating the effective flux from sediments. Rutgers van der Loeff et al. (1984) demonstrated a drastic decrease in this flux upon the onset of anoxic conditions in the overlying bottom water, ending all benthic activity.

PART II: DIAGENETIC EQUATIONS

Introductory remarks

For an excellent treatment of the mathematical equations describing diagenesis and their use in chemical sedimentology, reference can be made to Berner's (1980) book. Extensive use of his key equations, their restrictions, and their applications, is no doubt the best way to present in this chapter the most important mathematical models used in the study of diagenesis. When not stated, this classical book will be the reference to most equations. Reference should also be made to Lerman (1975, 1979) and Lerman and Lietzke (1977).

The problem in mathematical modeling of diagenesis is that almost all parameters change in space and time. Sedimentation usually is a continuous process and, therefore, the boundary formed by the interface of the sediment and overlying water is moving upward in space with time, changes taking place at a certain horizon of a given age in the sediment column are not the same as the changes taking place at a certain depth beneath the sediment – water interface. For practical purposes, the most obvious choice of the origin in a one-dimensional description of the sediment column is at this sediment – water interface, at least for processes occurring near the top of the sediment column as is the case in early diagenesis. Mathematically, this means that a transformation is necessary to relate changes in a fixed layer to changes at a fixed depth. For any property of function f this relation is:

$$(\partial f/\partial t)_z = (\partial f/\partial t)_H - w\,(\partial f/\partial z)_t \tag{2-22}$$

where:

z = depth below sediment – water interface, positive downward;
H = considered sediment layer or horizon;
w = burial rate of layer below the sediment – water interface; and
t = absolute time.

The subscripts indicate the variables that are considered to be constant for the pertinent partial derivative. The properties or functions that are most important in describing diagenesis are porosity and concentration.

Porosity

Porosity shows up in many diagenetic equations to account for the differences in reference volumes. In terms of the sediment, the appropriate reference is a unit of bulk sediment. Concentrations are, however, usually given in mass per unit volume of solution or per unit mass of solid material. Recalculations on the base of a unit volume of sediment can be done as follows:

$$C_{bd} = \phi \, C_d \tag{2-23a}$$

$$C_{bs} = (1 - \phi) \, C_s \, d_s \tag{2-23b}$$

where:

C_{bd} = concentration of dissolved constituent per unit volume of sediment;
C_d = concentration of dissolved constituent per unit volume of pore water;
C_{bs} = concentration of solid component per unit volume of sediment;
d_s = mean density of total solids; and
ϕ = porosity in volume of interconnected pore water per unit of bulk volume (total sediment = pore water + solids).

Equation 2-22 can be used to explore the changes of porosity under different sedimentary regimes. In the absence of compaction within the sediment column, the porosity is constant in a given horizon:

$$(\partial\phi / \partial t)_H = 0 \tag{2-24a}$$

and, consequently:

$$(\partial\phi / \partial t)_z = -w \, (\partial\phi / \partial z)_t \tag{2-24b}$$

This means that, in the absence of compaction, the porosity at a given depth is, through the burial rate, related to the changes in the initial porosities of the sediment layers. Such changes can occur when the sedimentation rates and/or the type of sediments change. Another possibility is that the porosity profile, measured from the water – sediment interface, does not change in time, which can represent steady-state compaction. In that case the equations are:

$$(\partial\phi / \partial t)_z = 0 \tag{2-24c}$$

and, consequently:

$$(\partial\phi / \partial t)_H = w \, (\partial\phi / \partial z)_t \tag{2-24d}$$

This means that, in the case of steady-state compaction, the porosity of a given horizon changes with the rate of burial and with the porosity gradient. When both Eqs. 2-24a and c are true, then:

$$(\partial\phi/\partial z)_t = 0 \tag{2-24e}$$

which means that porosity is constant throughout the sediment column.

Changes in porosity upon compaction cause an advective flux of water. Considering a unit volume of sediment, mass balance requires that the change in this flux of water is due to the change in porosity (continuity of fluid), with the implicit assumption that the densities of solids and solutions do not really increase with depth. Also, the compaction of total solids within this unit volume of sediment is related to the burial rate:

$$\text{for solution:} \quad (\partial\phi/\partial t)_z = -(\partial\phi v/\partial z)_t \tag{2-25a}$$

$$\text{for solids:} \quad [\partial(1-\phi)/\partial t]_z = -[\partial(1-\phi)\, w/\partial z]_t \tag{2-25b}$$

where v = velocity of flow relative to the sediment – water interface.
Combination of Eqs. 2-25a and b gives:

$$(\partial\phi v/\partial z)_t = -[\partial(1-\phi)\, w/\partial z]_t \tag{2-25c}$$

Transformation of the left-hand side of Eq. 2-25a into layer-based coordinates via Eq. 2-22 and for the right-hand side substituting the right-hand side of Eq. 2-25c results in:

$$(\partial\phi/\partial t)_H = -(\partial\phi v/\partial z)_t + w\,(\partial\phi/\partial z)_t = (1-\phi)\,(\partial w/\partial z)_t \tag{2-25d}$$

The actual velocity of flow of pore water relative to a fixed horizon, v_H, equals the difference in velocities, relative to the sediment – water interface, of pore-water flow (v) and burial rate (w) at this horizon. Equation 2-25c can then be written in terms of velocity of pore-water flow relative to that horizon, v_H, as:

$$(\partial\phi/\partial t)_H = -(\partial\phi v_H/\partial z)_t - \phi\,(\partial w/\partial z)_t \tag{2-25e}$$

which relates the change of porosity of a horizon with the effective pore-water flow and burial rate relative to that horizon. In the absence of compaction, the left-hand side of Eqs. 2-25c and d are equal to zero, in which case the burial rate $w(z,t)$ equals the rate at which new sediment is deposited on top of the existing sediment column at time t.

Steady-state compaction means that the left-hand side of Eqs. 2-25a, b are equal to zero, or:

$$(\partial w/\partial z)_t = [w/(1-\phi)]\,(\partial\phi/\partial z)_t \tag{2-25f}$$

$$(\partial v/\partial z)_t = -(v/\phi)\,(\partial\phi/\partial z)_t \tag{2-25g}$$

This shows that the usual negative porosity gradient is related to a negative gradient of the burial rate and a positive gradient of the pore-water velocity, both relative to the sediment – water interface.

In the case of changes of the mineral composition of layers buried in the sediment column, the initial porosities differ as well, and so does the porosity profile. Imboden (1975) analyzed this situation mathematically. He assumed that the porosity in each layer, H, is a unique function of pressure, P, alone, pressure being due to the weight of the overlying sediment mass, M. He presented an empirical relation for the porosity as a function of the pressure due to the overlying solids alone:

$$E = E^0 - b \log (P/P^0) = E^0 - b \log[P^0 + gM(z)]/P^0 \tag{2-26a}$$

where:

E = $\phi/(1 - \phi)$ = relative pore volume;
E^0 = relative pore volume for $P^0 = 1$ bar (E^0 and b are characteristic parameters for a given sediment composition);
g = gravity acceleration; and
$M(z)$ = mass of overlying sediment (function of depth z).

This then leads to the relation:

$$\phi = \frac{\phi^0 - (1 - \phi^0)\, b \log [1 + gM(z)/P^0]}{1 - (1 - \phi^0)\, b \log [1 + gM(z)/P^0]} \tag{2-26b}$$

where: ϕ^0 = initial porosity.

M can be calculated from:

$$M = \int_0^H (m/s)\mathrm{d}z \tag{2-26c}$$

where:

m = mass sedimentation rate (solid mass per unit time); and
s = sedimentation rate in length per unit time.

Another useful expression is derived by Imboden (1975) for non-constant sedimentation with steady-state porosity and compaction. He assumed that the sediment column consists of only one mineral and has constant initial density, which means that porosity becomes a function of z alone $[(\partial\phi/\partial t)_z = 0]$:

$$\phi = 1 - [m^0(t)/wd_s] \tag{2-27}$$

where:

$m^0(t)$ = mass sedimentation rate or accumulation of solid sediment mass per unit time, assumed to be variable in time; and

w = velocity of burial of sediment particles below the sediment – water interface.

Concentrations

There are three potential processes that affect the concentrations of dissolved components in sediment pore waters:

(1) **Diffusion,** which arises from random motions of individual components and, because of these motions, acts to erase concentration differences in physically-connected compartments.

(2) **Advection,** which is a unidirectional flow as a result of an impressed internal force (compaction) or external force.

(3) **Reaction,** which can occur as consumption or production (e.g., adsorption/desorption, precipitation/dissolution, consumption of oxidants and production of reduced components during mineralization of organic matter, and decay/generation of radionuclides).

The mathematical formulation is:

$$(\partial C_{\mathrm{b}i} / \partial t)_z = - (\partial F_i / \partial z) + \Sigma R_i = \partial[D_i\, (\partial C_{\mathrm{b}i} / \partial z)_t] / \partial z - (\partial v c_{\mathrm{b}i} / \partial z)_t + \Sigma R_i \qquad \text{(2-28a)}$$

or, in layer-based coordinates:

$$(\partial C_{\mathrm{b}i} / \partial t)_H = + \partial[D_i\, (\partial C_{\mathrm{b}i} / \partial z)_t] / \partial z - (\partial v c_{\mathrm{b}i} / \partial z)_t + w\, (\partial C_{\mathrm{b}i} / \partial z)_t + \Sigma R_i \qquad \text{(2-28b)}$$

where:

F_i = flux of component i in mass per unit area of total sediment per unit of time (positive upward);

$C_{\mathrm{b}i}$ = concentration of component i in mass per unit volume of total (bulk) sediment;

D_i = diffusion coefficient of component i in area of total sediment per unit of time; and

ΣR_i = combined production and consumption reactions within the sediment column affecting the concentration of component i, in mass per unit volume of total sediment per unit of time.

Berner (1980) called Eqs. 2-28a and b the "general diagenetic equations". Basically, they can be applied to the solid as well as to the liquid phases, but the main interest in this chapter is on pore waters, i.e., the liquid phase. Because practically all parameters in these equations are variable in time and space, the differential equations are nonlinear and their solutions are quite complicated. Reference works presenting analytical solutions of these equations under various initial and boundary conditions are, among others: Bouldin (1968), Crank (1975) and Van Genuchten and Alves (1982). Berner (1980) restricted himself to steady-state diagenesis, which means that the concentration profiles relative to the sediment – water interface remain unaltered with time. One then may ask under what circumstances the assumption of steady state is justified.

Steady state

When can steady state be assumed? Lerman (1975) answered this question as follows: A steady-state model is an acceptable approximation, if the ages of the sediment column are comparable to the time scales imposed by rates of sedimentation, of diffusion, and of reactions taking place in the sediment column. Lerman and Lietzke (1977) phrased this alternatively: If diffusional fluxes and chemical reaction rates are fast in comparison with the rate of growth of the sediment – pore water column, the concentration profiles may be expected to be near steady state at all times during the continuous growth at the sediment column with constant boundary concentrations. They showed that at least no gross errors are introduced by a steady-state model under these conditions. For typical deep-sea conditions of low sedimentation rates the response time of the sedimentary system is roughly proportional to the square of the length of the sediment column and inversely proportional to the sediment diffusion coefficient (McDuff, 1978). The length of the sediment column under consideration can vary, depending on the length of the retrieved sediment core, for instance. Requirements are: (1) that concentration levels at the bottom of the column have stabilized, (2) that changes in porosity are negligible at that depth, and (3) that sedimentation rate, type of sediment, and boundary concentrations have been maintained for lengths of time comparable to the calculated time scale. For deep-sea sediments with a typical thickness of 400 m and a diffusion coefficient of 2×10^{-6} cm^2 s^{-1}, McDuff (1978) calculated a response time of 12 Ma, which is of the same order of magnitude as calculated by Lerman (1975, 1979) for similar cases.

Lasaga and Holland (1976), using a different approach, analyzed under which conditions variations in the sedimentary regime, both in the rate and composition, would be preserved in a concentration profile different from the steady-state profile. Again, the ratio between the rates of sedimentation and diffusion play a role in the preservation of the signal of initial variations with time of the input into the sediment. Slow sedimentation rates relative to the diffusion rate tend to dampen the positive or negative signals due to changes in the sedimentation regime. Working with a time scale of 1 ka and with a frequency of oscillations in the input of 5 per ka, the values of the ratio between burial rate squared and the diffusion coefficient in bulk sediment (w^2/D_s) has to be > 5, in order to be observed in the pore water chemistry. In general, the maximum frequency observable for given sedimentation and diffusion rates is *equal* to w^2/D_s. More frequent oscillations in the input will be averaged out completely in the pore-water profiles, and the profiles become indistinguishable from steady-state ones. Because D_s varies only slightly between different ionic species, the most important parameter is w. High sedimentation rates (in x m per 10^3y) are favorable for preservation of input periodicity in the pore water profile, whereas low sedimentation rates, typical for most marine sediments, are not. The concept of steady-state profiles, therefore, will be applicable in many studies of marine pore waters.

Advection

In Berner's (1980) general diagenetic equations (Eqs. 2-28a and b) the flux of component i is composed of a Fickian diffusion term and an advection term. Advection is brought about by internal compaction (i.e., the loss of water from a sediment layer due to compression by the overburden) or from externally impressed hydrostatic gradients. The latter situation is not common in the domain of marine sediments, but has to be kept in mind when dealing with sediments bordering the continents (for instance in areas of subduction), or in active spreading zones. Most commonly, however, advection is due to compaction only.

The concept of continuity of fluid was used in the derivation of Eqs. 2-25a and b. For steady-state compaction, i.e., $(\partial\phi/\partial t)_z = 0$, Eqs. 25f and g were derived. At a certain depth, z_d, below the sediment – water interface, the porosity gradient, $(\partial\phi/\partial z)_t$, will become negligibly small and the porosity will become constant ($= \phi_d$). Then the left-hand side of Eqs. 2-15f and g approach zero. This means that $w(z,t) \rightarrow w(z_d,t)$ and $v(z,t) \rightarrow v(z_d,t)$. When $w_d = w(z_d,t)$ and $v_d = v(z_d,t)$, the equations for advective velocity become:

$$v = \phi_d \, w_d / \phi \tag{2-29a}$$

$$w = (1 - \phi_d) \, w_d / (1 - \phi) \tag{2-29b}$$

The pore-water flux relative to a fixed layer of solid particles can be calculated by subtraction of Eq. 2-29b from Eq. 2-29a to give:

$$v_H = (\phi_d - \phi_H) \, w_d / \phi_H \, (1 - \phi_H) \tag{2-29c}$$

Because w_d is considered positive and because $1 > \phi_H > \phi_d > 0$, v_H is negative, which means that the advective flux relative to that horizon is upward, as expected. Equation 2-29c can be rewritten in terms of the upward flux of water through that horizon, $\phi_H v_H$, or:

$$\phi_H v_H = \phi_H \, (v - w) = (\phi_d - \phi_H) \, w_d / (1 - \phi_H) = \phi_d w_d - \phi_H w_H \tag{2-29d}$$

In general, the advective flux through a certain horizon relative to that horizon, corresponds to the change of pore-water volume below that horizon. In the case of non steady-state compaction, $(\partial\phi/\partial t)_z \neq 0$, this flux can be formulated as follows (Imboden, 1975):

$$\phi_H v_H = \int_H^{z_d} (\partial\phi/\partial t)_z \, dz' + \phi_d w_d - \phi_H w_H \tag{2-29e}$$

where dz' is the depth derivative in the interval from H down to z_d.

One is easily tempted to believe that, inasmuch as compactive flow is upward, pore water is expelled from the sedimentary column by compaction. But under conditions of ongoing sedimentation and normal porosity profiles, the rate of the advective pore-water flux cannot keep pace with the rate at which pore water is buried. Einsele (1977), preferring a graphical over an analytical approach in order to describe advection of pore water in sediments, showed that under normal continuing sedimentation, vertically-ascending pore waters do not reach the sediment – water interface. This was also stressed by Imboden (1975), Schink and Guinasso (1978), and by Berner (1980) based on analytical evidence. The actual vertical distance of advective pore-water flow depends on the difference between initial porosity before burial and the porosity at the base of the sequence or at a depth where porosity does not change further. The distance of movement and the velocity of upward-moving pore water upon continuous sedimentation decreases from top to bottom in a sequence. Pore water may be lost to the overlying water in a special situation where the top of a sediment column is eroded and when consolidation of the remaining sediment is still proceeding towards an equilibrium state. Negative (= downward) advection of overlying water into the remaining sediment column may occur when consolidation in the remaining sediment column had surpassed the equilibrium state relative to new static conditions. Other situations in which pore water can be expelled from the sediment is when, due to external conditions, upwelling occurs in the sediment column (e.g., along the Oregon/Washington Margin; Suess et al., 1985; Ritger et al., 1987).

How important is advective flow relative to diffusive flow in determining profiles of dissolved components? Lerman (1975) used the criterion that diffusion dominates the redistribution of components if $D_s >> hv$, and advection is more important if $hv >> D_s$ (where h = length of sediment column within which a concentration profile is observed, and v = rate of advection). Berner (1980) calculated that for $D_s = 100\ cm^2\ a^{-1}$ and for realistic advection rates, the considered length has to exceed 1 m in order for the advective flow to be more effective in the redistribution than the diffusive flow. For most deep-sea sediments ($v < 0.01\ cm\ a^{-1}$) this length would be at least several hundreds of meters, which means that diffusion is usually the important process.

Diffusion

The diffusive fluxes of solutes in pore waters depend on the concentration gradients and on the mobilities (diffusion coefficient) in the sedimentary column. Contrary to the advective flux in which both solvent and solutes move relative to a given boundary or layer, the diffusive flux of solutes assumes a stationary solvent. Concentration gradients are brought about by production or consumption of solutes in the sediments. Interesting boundaries are the base of the sedimentary column and, usually even more so, the sediment – water interface. The concentration gradient at the sediment – water interface determines the flux of the dissolved components to the overlying bottom water, whereas the concentration gradient at the base of the sedimentary column determines the rate of input or output due to weathering conditions on the ocean floor. The steepest gradients are usually present at this very interface. It is, therefore, important and difficult to determine precisely the concentration profile close to these interfaces. The diffusion that one is concerned with mostly, is that of ionic species. The mobility of an ion (u_i), defined as the velocity of an ion under a unit driving force (f_i), is called the absolute mobility (u_i^0) for very dilute solutions. The driving force acting on an ion (charge z_i) is the sum of the gradient of the chemical (μ_i) and electrical (Φ) potential, or, in one dimension (z):

$$f_i = -(\partial\mu_i/\partial z) + z_i(\partial\Phi/\partial z) \quad (2\text{-}30)$$

The Nernst – Einstein relation between the mobility and the diffusion coefficient of an ion is:

$$D_i^0 = RT\, u_i^0 \quad (2\text{-}31a)$$

The limiting diffusion coefficient (D_i^0) depends on the temperature and viscosity of the solution, which can be written as the Stokes – Einstein equation:

$$D_i^0 = -\, kT/6\pi\eta r \quad (2\text{-}31b)$$

where:

T = temperature in K;
η = viscosity;
k = Boltzman's constant; and
r = radius of the ionic sphere consisting of the ion plus hydrate layer.

Based on this relation, temperature corrections can be calculated using:

$$[D_i^0(T_1)\,\eta^0(T_1)]/T_1 = [D_i(T_2)\,\eta^0(T_2)]/T_2 \quad (2\text{-}31c)$$

Li and Gregory (1974) mentioned that this equation can be used to correct diffusion coefficients for temperature differences for ions with coefficients lower than that of the fluoride ion. For ions with coefficients higher than that of F^- (e.g., Cl^-, Br^-, I^-, HS^-, K^+, NH_4^+) the following equation fits the data better:

$$D_i^0(T_1)\,\eta^0(T_1) = D_i^0(T_2)\,\eta^0(T_2) \quad (2\text{-}31d)$$

The ratios they reported are:

$$D^0(25°C)/D^0(0°C) \simeq 2.2\ (< F^-) \text{ and } \simeq 2.0\ (> F^-)$$

The viscosity corrections for seawater in relation to pore water are:

$$\eta_0/\eta_{sw} = 0.95\ (0°C) \text{ and } = 0.92\ (25°C)$$

The pressure effect on η and D_i is very small, resulting in an increase of D_i at a depth of 6 km of no more than 8%.

The diffusion coefficients of dissolved species in sediments (D_s) differ from those in free solution (D) of identical composition as the pore water, in that the random movement of the species is restricted by the geometry, more specifically the tortuosity of the pore space. Tortuosity (τ) is defined as:

$$\tau = d_{\text{pore water}} / d_{\text{sediment}}$$

where d = mean free path of dissolved species, and relates D_s with D through the equation:

$$D_s = D / \tau^2 \tag{2-32}$$

The diffusive flux in Fick's first law, which is part of Eqs. 2-22a, b is valid in this form only for concentration gradients in solution. In order to enter D_s into these equations, one, therefore, has to multiply D_s by the porosity (ϕ):

$$F_i = - \phi D_s (\partial c_i / \partial z)_t \tag{2-33}$$

where:

F_i = diffusive flux of component i in terms of mass per area of total sediment per unit time; and

c_i = concentration of dissolved component in terms of mass per unit volume of pore solution.

Because tortuosity cannot be measured or calculated directly, this parameter is indirectly determined by measurement of the so-called formation factor, F_0. The latter is determined by the measurement of the resistivities (McDuff and Ellis, 1979) or of porosity (Manheim, 1970) as follows:

$$\tau^2 = \phi F = \phi \Omega / \Omega_0 \approx \phi^{1-n} \tag{2-34}$$

where:

Ω, Ω_0 = electrical resistivity of sediment and of pore water alone, respectively; and

n = exponent, depending on the type of sediment.

The latter equation is based on the experimental relationship formulated by Archie (1942). Ullman and Aller (1982) evaluated this relationship for different types of sediments. The published ranges of n are as follows: for sands and sandstones, n = 1.3 – 2, for clays n = 2.5 – 5.4, and for compacted sediments with $0.2 < \phi < 0.7$, $n \approx 2$. They suggested that for near-shore muddy sediments with high porosities ($\phi > 0.7$), the best estimate for n is 2.5 – 3.

Coupled fluxes and ion-pairs

A charged specie, moving in a vertical direction according to the prevailing concentration gradient, cannot move without upsetting the requirement of electroneutrality throughout the system. This means that fluxes of ions have to be coupled in order to maintain electroneutrality. For a simple 1:1 electrolyte, this means that the actual fluxes of cation and anion are equal in direction and in magnitude. For a multicomponent electrolyte, the relations become more complicated, because the diffusion of a cation is not necessarily matched by diffusion of an anion, but also can be compensated for by back-diffusion of other cations, including protons. Thorough analyses of the theoretical background and mathematical treatment are given by Lasaga (1979) and by McDuff and Ellis (1979). During early diagenesis, the absolute concentration gradients are usually so low that cross-coupling can be ignored. But, in cases where large salinity gradients exist within the sediment column, for example, cross-coupling has to be considered.

Another problem that has to be considered is that of ion-pairing. Several reasons exist to expect that ion-pairs behave differently in diffusion than the constituting free ions. One reason is that the absolute formal charge of the ion-pair is lower. The ion-pair is less hydrated and, thus, actually has a smaller size than the two single constituting ions including their water mantles (Lasaga, 1979; Katz and Ben-Yaakov, 1980). This idea was criticized by Johnson (1981), who argued that the concept of hydration as water molecules physically bound to an ion, is probably not correct. Instead, he supported the view of a difference in the residence time of solvent molecules adjacent to the ions compared to the solvent molecules in the bulk. In any case, he does not believe in diffusion of ions having to carry along a water mantle. Another reason to believe different diffusional behavior between ion pairs and their constituting free ions is that the hindrance by the electrical field becomes smaller or becomes even zero when the charge (z_i) and the ion potential (z_i/r_i) (r_i being the ionic radius) decrease or become zero. The effect of ion-pairing has also been discussed by Applin and Lasaga (1984). These authors refer to the theoretical values for diffusion coefficients of ion-pairs as proposed by Pikal (1971), but prefer to use a different model for calculating the association constants than the one used by Pikal. They proposed a slightly different equation to calculate the diffusion coefficients of ion-pairs (ions *m* and *n*):

$$D^0_{mn} = 1/2\ \{(D^0_m\ D^0_n)^{1/2} + (kT/2\eta a_0)\ [(b + 3)/b^2]\} \qquad (2\text{-}35)$$

where:

$D^0_{mn,m,n}$ = tracer diffusion coefficients of ion-pairs (*mn*), cation (*m*) and anion (*n*);
a_0 = size parameter of ion-pair;
b = $|z_m z_n| e^2/\epsilon k T a_0$;
e = electron charge;
ϵ = dielectric constant of water; and
z_m, z_n = charges of cation and anion.

The calculated values agree very closely with the experimental values referred to in their paper. The important conclusion drawn by Johnson (1981) and confirmed by Applin and Lasaga (1984) is that the diffusion coefficients of ion-pairs are of the same order of magnitude as those of the constituting ions and not greater than usually assumed. The latter authors presented a summary of the tracer diffusion coefficients for free ions and ion pairs in seawater, including the choice of the stoichiometric (in $1\ mole^{-1}$) association constants for seawater. This is shown in Table 2-7. The ion-pairing of alkali and earth alkali ions with chloride is taken into account, as strongly suggested by Johnson (1981; and his preceding papers), but which is ignored in most other models of ion association. The calculated diffusion coefficients can be used in the mathematical model of Applin and Lasaga (1984) to calculate the fluxes of major and minor components.

Enhanced mass transport across the sediment – water interface

The top of the sediment column is generally (when at least the overlying bottom water is oxygenated) subject to bioturbation by deposit-feeders and/or incidentally (mostly in shallow waters) subject to physical perturbation due to wave and bottom current action. Much attention has been paid to the physical – mathematical modeling of the profiles of solid substances and of pore-water components that are altered by such processes. The nature of the processes is, however, so complex that gross simplifications are necessary. Two approaches are discussed here.

The first approach assumes that the apparent diffusion coefficients in the sediment from the very top down to the horizon ($z = L$) where the disturbances are effective (Zone I), are higher than those below (Zone II). This model is still one-dimensional (depth) and requires formulation of continuity of properties and mass transport across Zones I and II. Inasmuch as deposit-feeders mix the bulk sediment in a variety of ways and some pump overlying seawater into the burrows they inhabit, Berner (1980) distinguished between a biodiffusion coefficient (D_B) for the

TABLE 2-7

Tracer diffusion coefficients D (in $10^{-5}\ cm^2\ s^{-1}$) for free ions and ion pairs of seawater (after Applin and Lasaga, 1984)

Ion	D	Ion-pair	D	Ion-pair	D
Mg^{2+}	0.705	$MgSO_4^0$	0.80	$MgCO_3^0$	0.58
Ca^{2+}	0.793	$CaSO_4^0$	0.65	$CaCO_3^0$	0.60
Na^+	1.33	$NaSO_4^-$	1.23	$NaCO_3^-$	0.97
K^+	1.96	KSO_4^-	1.14		
SO_4^{2-}	1.07				
Cl^-	2.03	$NaCl^0$	1.99	KCl^0	2.16
CO_3^{2-}	0.955	$MgCl^+$	1.18	$CaCl^+$	1.09
HCO_3^-	1.18	$MgHCO_3^+$	0.85	$CaHCO_3^+$	0.88

bulk sediment (including pore water) and an irrigation coefficient (D_I) for the flushing of the pores in Zone I. He then added the two fluxes in Fickian formulation:

$$F_{iBI} = - D_B (\partial \phi c_i / \partial z)_t - \phi D_I (\partial c_i / \partial z)_t \quad (2\text{-}36)$$

where:

F_{iBI} = flux of dissolved component due to bioturbation of bulk sediment plus irrigation of burrows;
D_B, D_I = bioturbation and irrigation coefficients, respectively, in terms of area of total sediment squared per unit of time; and
c_i = mass of component i per unit volume of pore water.

The reasons why the depth dependency of porosity shows up in the first partial flux and not in the second are as follows: (1) the first flux takes into account the change in concentration in terms of mass per unit volume of total sediment (c_{bi} in Eqs. 2-28a and b) which is then written in terms of the concentration in pore water alone by multiplying by the porosity, and (2) only the irrigation coefficient in the second term has to be corrected from its bulk to its pore-water value. This contribution of biodiffusion can, in turn, be added to the molecular diffusion term in Eq. 2-28a, to give the general diagenetic equation for the bioturbated Zone I:

$$(\partial c_{bi} / \partial t)_z = (\partial \phi c_i / \partial t)_z =$$
$$\{\partial [D_B (\partial \phi c_i / \partial z) + \phi (D_I + D_i) (\partial c_i / \partial z)] / \partial z\}_t - (\partial \phi v c_i / \partial z)_t + \Sigma R_i \quad (2\text{-}37a)$$

Berner (1980) pointed out that, in principle, a gradient in porosity in the bioturbated layer will contribute to the biodiffusional flux, because this follows from the incorporation of porosity in the first derivative between brackets on the right-hand side of the equation. Bioturbation tends to diminish differences in properties, including porosity, within the bioturbated layer. In a well-established and active zone of bioturbation, therefore, the coefficients in Eq. 2-37a may be lumped together in an apparent diffusion coefficient, D_a, (in terms of area squared of total sediment per unit of time) for I, in that:

$$D_a = \phi (D_B + D_I + D_i) \quad (2\text{-}37b)$$

When this is substituted into Eq. 2-37a, and when advection is ignored ($v \approx 0$), one arrives at the equations that describe processes in the bioturbated layer (e.g., see Schink and Guinasso, 1978; Aller, 1978, 1980a,b; and Aller and Yingst, 1985):

$$(\partial c_{bi} / \partial t)_z = \partial [D_a (\partial c_{bi} / \partial z)]_t / \partial z + \Sigma R_i \quad (2\text{-}37c)$$

It is obvious that only in the case of high benthic activity the difference between Eqs. 2-37a and 2-28a becomes substantial.

Berner (1980) compared published D_B values and found that they are usually three orders of magnitude (or even less) lower than D_s values. Aller (1982) concluded that there is a decrease in biological reworking from shallow ($D_B \approx 10^{-6}$ cm^2 s^{-1}) to deep-sea ($D_B \approx 10^{-8}$ cm^2 s^{-1}) environments. In near-shore organic-rich muds D_B values may become higher, up to the same order of magnitude as D_s at most; however, the apparent diffusion coefficient may be 10 – 100 times higher than the D_s values for the same deposit (Aller, 1982), apparently due to the effect of irrigation. An extensive mathematical analysis of the role of biodiffusion coefficients in modeling of concentration profiles in sediments is given by Boudreau (1986a,b) and Boudreau and Imboden (1987).

When not bioturbation and related irrigation, but mechanical disturbance of the topmost layer in a sediment column takes place, one can, by the same token, use an apparent diffusion coefficient for this top layer, as was done by Vanderborght et al. (1977a). Because the use of the apparent coefficients does not fall in the framework of the conventional physical model for diffusion (random motion), these coefficients are often indicated as mass transfer coefficients or transport coefficients.

For the description of the overall processes in the sediments buried in Zone II, molecular diffusion takes over completely. This means that the boundary conditions defined by the model have to take into account continuity of concentrations and fluxes for the horizon where Zone I changes into Zone II (at $z = L$):

$$(c_{bi})_{\text{zone I}} = (c_{bi})_{\text{zone II}} \tag{2-38a}$$

$$D_a\ (\partial c_{bi}/\partial z)_t = D_i\ (\partial c_{bi}/\partial z)_t \tag{2-38b}$$

Solutions of these equations with the appropriate boundary conditions and assumptions (e.g., steady state) can be found in the cited papers.

The second approach is put forward by Aller (1978, 1980a,b, 1982, 1983, 1984) and Aller and Yingst (1985). The layer burrowed by deposit-feeders is considered as an array of hollow cylinders in closest packing, as illustrated in Fig. 2-16. It is assumed that burrows are flushed fast enough to assume the water composition therein to be virtually the same as that of overlying bottom water. Fluxes of pore-water components by the process of diffusion have, apart from the vertical direction, also a lateral direction, i.e., from the sediment cylinder towards the burrow in the center. In the latter case, it is more appropriate to use radial rather than orthogonal coordinates to formulate the processes. In the case of cylindrical symmetry, this gives the equation for Fick's second law:

$$(\partial c_{bi}/\partial t)_r = (1/r)\ \partial[rD_i\ (\partial c_{bi}/\partial r)]_t/\partial r \tag{2-39}$$

where: r is the radius of the cylinder.

Because diffusion now occurs in three dimensions, Eqs. 2-28a and 2-39 must be combined to give (ignoring advection and assuming a constant bulk diffusion coefficient for this layer):

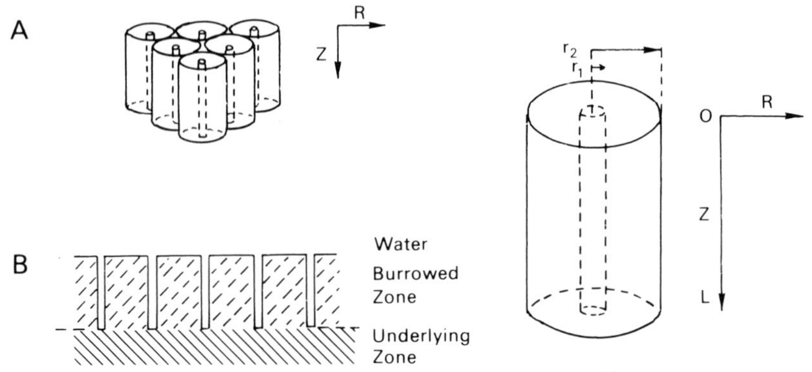

Fig. 2-16. (A) Sketch of the uppermost region of a deposit visualized ideally as packed cylinders filled with sediment and with a hollow in each one. (B) Vertical cross-section of a deposit having the idealized diffusion geometry of (A). (C) The simple cylinder of sediment with a hollow in the center represents an average microenvironment within the bioturbated zone. The radius of the hollows is r_1, distance between two neighbouring hollows is $2r_2$, depth of the hollow is L. (Modified after Aller, 1980a.)

$$\partial c_{bi}/\partial t = D_i\,(\partial^2 c_{bi}/\partial z^2) + (D_i/r)\,\{\partial[r(\partial c_{bi}/\partial r)]/\partial r\} + \Sigma R_i \quad (2\text{-}40)$$

where the boundary conditions are:

c_{bi} $= c_{sw}$ for $z = 0$ and for $r = r_i$;

$\partial c_{bi}/\partial r$ $= 0$ for $r = r_2$ (pore-water solutions go through maximum or minimum half-way between any two adjacent burrows); and

$(\partial c_{bi}/\partial r)_{\text{zone I}}$ $= (\partial c_{bi}/\partial z)_{\text{zone II}}$, for $z = L$

Analytical solutions for these differential equations for steady-state conditions ($\partial c_{bi}/\partial t = 0$) can be found in Aller's work (e.g., 1980a, 1982). Still, the mixture of two coordinate systems in Eq. 2-40 makes the model more complicated to handle. Boudreau (1984) showed that it is possible to apply a one-dimensional diffusion model that is equivalent to the radial diffusion model. The equation equivalent to Eq. 2-40 is:

$$(\partial c_{bi}/\partial t)_z = D_s\,(\partial^2 c_{bi}/\partial z^2)_t - p(c_{bi} - c_{sw}) + \Sigma R_i \quad (2\text{-}41)$$

where $p = p\,(z,t)$ = fraction of constituent exchanged per unit time. The condition for the equivalence of Eqs. 2-41 and 2-40 is that the concentration gradient at $r = r_i$ can be approximated by assuming linear gradients between burrow concentrations, c_{sw}, and the concentration in the sediment, $c_{bi}\,(z,t)$. The term p, in units of reciprocal time, which is a measure of the transport between non-adjacent points in the sediment overlying water system, was introduced by Imboden (1981) as a "nonlocal" source or sink term. The model was advantageously used by Emerson et al. (1984) and was also adopted by Aller and Yingst (1985) as superior in general to apparent diffusion models if irrigated burrows are present.

The reaction term: production or consumption

Under the ΣR_i term are comprised processes such as mineralization of organic matter, oxidation or reduction of sediment constituents, precipitation or dissolution of mineral phases, radioactive decay and concomitant production of radionuclides, and adsorption/desorption, c.q. ion-exchange. The adsorption/desorption processes are usually treated separately from the other processes, as done in this chapter.

Berner's treatment of these processes is based on his 1976 paper.

There are also many so-called sorption isotherms that describe the equilibrium concentrations of a component between solution and solids at constant temperature. The most frequently used isotherms are:

The Freundlich isotherm:

$$c_{is} = a\, c_i^{(1/n)} \tag{2-42}$$

The Langmuir isotherm:

$$c_{is} = a\, c_i/(b + c_i) \tag{2-43}$$

where:

c_{is}, c_i = concentrations of adsorbed and dissolved i in terms of moles per unit mass of total sediment solids and per unit volume of solution, respectively; and

a, b, n = fitting parameters.

The Langmuir isotherm becomes linear when $b >> c_i$;

$$c_{is} \approx K\, c_i \tag{2-44}$$

where: $K = a/b$.

Adsorption can also be considered as an ion-exchange process of the form:

$$A_1^{n+} + B_s^{n+} = A_s^{n+} + B_1^{n+} \tag{2-45}$$

for the case of exchange of equivalent ions. Ion-exchange as a chemical equilibrium reaction can be formulated as:

$$K = (A_s)\,(B_1)/(A_1)\,(B_s) \tag{2-46}$$

Working on a scale where ion activities are equal to ion molarities in solution and also assuming that (1) activities and molarities of each of the adsorbed species are about equal, (2) no great preference exists for one ion over the other for sites at the solid surface, and (3) the concentration of one ion (e.g., A) is much lower than that of the other, then Eq. 2-46 can be simplified to:

$$(A_s) = K' \ (A_1) \tag{2-47}$$

where: $K' = K\,(B_s)/(B_1)$.

Under the above-mentioned assumptions, adsorption as a physicochemical process can be approximated by a linear isotherm, which is the usual approach taken to incorporate explicitly adsorption in the diagenetic equations. One has to be aware of the underlying assumptions, however, when applying this in diagenetic studies that take sorption into account.

The R_i term can be split into a term describing adsorption of i and a term representing all other reactions affecting i:

$$\Sigma R_i = R_{i\,(\mathrm{ads})} + \Sigma R_j \tag{2-48}$$

where:

$R_{i(\mathrm{ads})}$ = rate of change of dissolved i, due to equilibrium adsorption in mass per unit volume of pore water per unit of time; and
ΣR_j = all other, slow, reactions affecting the concentration of i.

In order to transform concentrations in pore water into concentrations in bulk sediment, the left-hand side and right-hand side of Eq. (2-48) must be multiplied by porosity, ϕ. Because the enhanced transport of the solid phase takes place by bioturbation only, the analogous expression for the change in time of the concentration of adsorbed i in the solids, is:

$$\partial[(1-\phi)\,d_s c_{is}/\mathrm{d}t]_z = \partial\{D_B[\partial(1-\phi)_s d_s c_{is}/\partial z]_t/\partial z\}\,\partial z - [\partial(1-\phi)\,d_s w c_{is}/\partial z]_t + (1-\phi)\,d_s R_{is} + (1-\phi)\,d_s\,\Sigma R_{js} \tag{2-49}$$

where:

R_{is} = rate of change of adsorbed i due to equilibrium sorption per unit mass of total solids per unit of time; and
ΣR_{js} = rates of other reactions affecting adsorbed i.

The enhanced change of porosity and total solids due to bioturbation can be added to the change due to burial of pore water and sediment, as given in Eqs. 2-25a and b:

$$(\partial\phi/\partial t)_z = -(\partial\phi v/\partial z)_t + \partial[D_B(\partial\phi/\partial z)/\partial z]_t \tag{2-50a}$$

$$[\partial(1-\phi)d_s/\partial t]_z = -[\partial(1-\phi)\,d_s w/\partial z]_t + \{\partial D_B[\partial(1-\phi)d_s/\partial z]/\partial z\}_t \tag{2-50b}$$

Mass balance considerations require that:

$$R_{is} = -\phi R_{i(\text{ads})}/(1 - \phi)d_s \quad (2\text{-}51)$$

Assuming linear adsorption, formulated as:

$$c_{is} = K' c_i \quad (2\text{-}52a)$$

and taking K' as constant with depth and time, the following equations are derived for the depth and time derivates of the adsorbed concentration of i:

$$(\partial c_{is}/\partial t)_z = K' (\partial c_i/\partial t)_z \quad (2\text{-}52b)$$

and:

$$(\partial c_{is}/\partial z)_t = K' (\partial c_i/\partial z)_t \quad (2\text{-}52c)$$

These assumptions can only be made for a uniform sediment type, constant surface per unit mass of sediment, and a uniform temperature throughout the sediment column under consideration.

Biodiffusionally-transported adsorbed constituent i re-equilibrates with the pore-water solution and, therefore, constitutes an additional transport mechanism of i. This can be expressed by adding Eqs. 2-40 and 2-49, and substituting Eq. 2-48. The immediate result is that the R_i and R_{is} terms cancel out due to Eq. 2-45. Subsequently, derivatives of the products can be written in extended form as sums of the derivatives of the single variables. In order to eliminate implicit equations in the summed equation, Eqs. 2-50a and b can be used by multiplying both the left-hand and right-hand side by the same concentrations:

$$c_i (\partial\phi/\partial t)_z = -c_i \{(\partial\phi v/\partial z)_t + [\partial D_B (\partial\phi/\partial z)/\partial z]_t\} \quad (2\text{-}53a)$$

$$c_{is} [\partial(1 - \phi) d_s/\partial t]_z =$$
$$- c_{is} \{\partial[(1 - \phi) d_s w/\partial z]_t + D_B [\partial(1 - \phi) d_s/\partial z]_t\} \quad (2\text{-}53b)$$

After subtraction of these latter equations from the combined equation, and furthermore introducing:

$$\phi K = (1 - \phi) d_s K' \quad (2\text{-}54)$$

Berner (1980) arrived at the following equation, which takes into account adsorption of constituent i on sediment particles:

$$(\partial c_i/\partial t)_z = 1/[\phi(1 + K)] \{\partial[\phi (1 + K)D_B + D_I + D_s] (c_i/c)/\partial z\}_t/\partial z -$$
$$\{[\phi (v + Kw) - (1 + K)D_B (\partial\phi/\partial z)_t - \phi D_B (\partial K/\partial z)_t]/[\phi (1 + K)]\} \times$$
$$(\partial c_i/\partial z)_t + \Sigma R_j/(1 + K) + [(1 - \phi)d_s \Sigma R_{js}]/[\phi (1 + K)] \quad (2\text{-}55)$$

Whereas K' is assumed to be constant, K is not necessarily constant, being dependent on porosity. If porosity and density of solids are constant with depth, so are K and w. The latter follows from the combined assumptions of constant K' and ϕ. When compaction is constant and impressed flow is absent, then $v = w$. Under these conditions, Eq. 2-55 reduces to:

$$(\partial c_i/\partial t)_z = [1/(1 + K)] \{\partial[(1 + K)D_B + D_I + D_s (\partial c_i/\partial z)]/\partial z\}_t - w(\partial c_i/\partial z)_t + 1/(1 + K) [\Sigma R_j + (1 - \phi)d_s\Sigma R_{js}/\phi] \quad (2\text{-}56)$$

Equation 2-56 shows that the processes of biodiffusion and burial become dominant with increasing values of K. Below the zone of bioturbation, where $D_B = D_I = 0$ and D_s = constant (from earlier assumptions), Eq. 2-56 reduces to:

$$(\partial c_i/\partial t)_z = [D_s/(1 + K)] (\partial^2 c_i/\partial z^2) - w(\partial c_i/\partial z) + [1/(1 + K)] [\Sigma R_j + (1 - \phi)d_s \Sigma R_{js}/\phi] \quad (2\text{-}57)$$

This equation is often used to describe the combined effects of molecular diffusion, burial, and production/consumption within the sediment column for a constituent i subject to equilibrium sorption; however, many implicit assumptions go with this equation. As pointed out by Berner (1980), the relative importance of burial becomes greater with increasing K's, that reduce the role of molecular diffusion. For deep-sea sediments the role of burial cannot be ignored when $K > 100$.

The remaining R_j terms in Eqs. 2-55 to 2-57 stand for slow, non-equilibrium reactions producing or consuming component i. Among such reactions are biological degradation and mineralization of buried organic matter with related redox reactions, and dissolution or precipitation of inorganic phases in response to super- or undersaturation of pore waters. The kinetics of the biological degradation depend on the type of organic matter. This was experimentally studied by Henrichs and Doyle (1986), who found decomposition rates that differed by orders of magnitude between the very labile and almost refractory organic matter. The degradation is highest in the upper part of the sediment and decreases rapidly with depth. This can be attributed to a rapid breakdown of the most decomposable groups in the bulk organic matter, leaving less labile material for burial, and/or the change in redox regime from aerobic to anaerobic at some depth in the sediment. Infaunal biota act in a way to carry more labile organic matter down and less metabolizable organic matter up in the bioturbated layer. Berner (1980) and Westrich and Berner (1984) proposed to rank, within total organic matter (G_T), the most metabolizable to the very slowly metabolizable fractions (G_i), as follows:

$$G_T = \Sigma G_i \quad (2\text{-}58a)$$

$$- dG_T/dt = \Sigma k_{c_i} G_i \quad (2\text{-}58b)$$

where:

k_{c_i} = first-order decay "constant" for decomposition of metabolizable fraction;
G_i = molar amount of individual metabolizable fraction; and
n = number of individual metabolizable fractions;

with boundary conditions:

$G_i = G_{i(0)}$ at $t = 0$;
$G_i = 0$ at $t = \infty$.

This gives, for each individual fraction, the time dependency of the inventory of metabolizable organic fraction as follows:

$$G_{i\,(t)} = G_{i\,(0)}\,[\exp(-k_{c_i}t)] \tag{2-58c}$$

The depth-dependent equations were derived by Billen (1982a):

$$\text{for } z \leq z_B\text{:}\quad (\partial G_i/\partial t) = D_{Bs}\,(\partial^2 G_i/\partial z^2) - w(\partial G_i/\partial z) - k_{c_i}G_i \tag{2-59a}$$

and:

$$\text{for } z > z_B\text{:}\quad (\partial G_i/\partial t) = -\,w(\partial G_i/\partial z) - k_{c_i}G_i \tag{2-59b}$$

with the following boundary conditions:

$G_i = G_i^0$, for $z = 0$,
G_i = finite, for $z = \infty$,
continuity for $z = z_B$,

where:

D_{Bs} = diffusion constant for solid particles in biologically and/or physically perturbated layers of bulk sediment; and
z_B = depth of perturbated layer.

The solutions for steady-state conditions are:

$$\text{for } z \leq z_B\text{: } G_i = G_i^0 \exp\{[w - (w^2 + 4D_{Bs}k_{c_i})^{1/2}]/2D_{Bs}\}z \tag{2-59c}$$

$$\text{for } z > z_B\text{: } G_i = G_i^{z_B} \exp\,[-\,k_{c_i}(z - z_B)/w] \tag{2-59d}$$

For $w << 2\,(k_{c_i}D_B)^{1/2}$, Eq. 2-59c becomes:

$$G_i = G_i^0 \exp\,[-z(k_{c_i}/D_{Bs})^{1/2}] \tag{2-59e}$$

The vertical profile for total organic matter, being equal to the sum of all G_i's, is characterized by a gradual exponential decrease. Middelburg (1989) found that the C-org profiles can be modeled by a simple power function. Boudreau and Ruddick (1991) used a more elaborate model for the decay of organic matter.

The rate of decomposition of organic matter is coupled with the rate of consumption of electron acceptors (oxidants: O_2, NO_3^-, SO_4^{2-}, etc.). The relation follows from the stoichiometry of the oxidation reaction of organic material. For sulphate reduction (Eq. 2-12b) the relation is:

$$R_{SO_4} = 1/2\ \Sigma k_i\ G_{i\,(0)}\ [\exp(-\ k_{c_i} t)] \qquad (2\text{-}60)$$

Because $G_{i\,(t)}$ decreases exponentially with time and, consequently, with depth, so do consumption of O_2 and the reduction of SO_4^{2-}, and concomitant production of CO_2 and HS^-. The degradation of organic matter produces soluble inorganic species such as NH_4^+, and $H_2PO_4^-/HPO_4^{2-}$. The stoichiometry of these reactions, in relation to the metabolized C-org, depends strongly on the burial rate, because N-org are P-org are preferably mineralized. Aller (1980a, 1980b, 1982, 1984) generalized the reaction term in the following equation:

$$R = R_0 \exp(-\beta z) + R_i + k_i\,(c_{i\,(eq)} - c_i) \qquad (2\text{-}61)$$

where:

R_0, R_i = constant;
k_i = first-order rate constant for production or consumption of inorganic constituent *i*;
$c_{i\,(eq)}$ = steady-state equilibrium concentration of constituent *i*; and
c_i = actual concentration of constituent *i*.

These terms are generally representative of the rate of production or consumption of inorganic constituents during decomposition of organic matter, and are limited by substrate (G_i's) reactivity rather than solute buildup or depletion. In the first-order term on the right-hand side, k_i is the first-order rate constant, c_i and $c_{i(eq)}$ are the actual and the steady-state equilibrium concentrations, respectively. For the release of dissolved Si from biogenic silica, this equation has proven to be applicable. But, as discussed by Berner (1980), in cases where the surface of the precipitating or dissolving mineral phase is poisoned by adsorption of other ions or is coated by, for instance, organic matter, the rate may become power-dependent with respect to the difference between actual and equilibrium concentrations.

PART III: EXAMPLES OF MODELING OF EARLY DIAGENETIC PROCESSES

General oxidant consumption and redox potential

The relation between measured Eh values and the actual oxidation state of sediments is often negated, because of intrinsic imperfections in the electrode system that is used and the mixture of multicomponent (sometimes multiphase), reversible or irreversible (in an electrochemical sense) redox couples involved. Still, redox measurements in sediments are often used as a diagnostic, practical means of putting a number on the oxidation state in different layers, sometimes with good agreement between the observations of oxidation and reduction processes in these layers. This is particularly true for the MnO_2/Mn^{2+} and the SO_4^{2-}/HS^- couples. An interesting redox model, with restricted use under specified assumptions and applicable to the top of marine sediments, was elaborated by Billen (1982b). He took into account the following redox couples: O_2/H_2O, $Mn(IV)/Mn^{2+}$, $NO_3^-/N_2/NH_4^+$, $Fe(III)/Fe^{2+}$, SO_4^{2-}/HS^-, CO_2/CH_4, where the Roman numerals indicate the oxidation state in solid phases. Redox reactions are already given in Eqs. 2-26a to 2-35. In his model, he considered the following solid phases: MnO_2, $MnCO_3$ (rhodochrosite), $Fe(OH)_{3\ (amorphous)}$, $FeCO_3$ (siderite), FeS (mackinawite). Throughout the sediment, he assumed a temperature of 20°C and a pH of 7.5. He described the solid particle mixing in the upper layer due to bioturbation, physical mixing by a dispersion coefficient D_B, and the pore water mixing due to the same processes by a dispersion coefficient D_I. For O_2 he preferred the reversible O_2/H_2O_2 couple instead of the irreversible O_2/H_2O couple. He then calculated the O_2 profile in pore water as follows:

$$(O_2) = \phi \exp^{10}[(Eh - 0.658)/0.03] \quad (2\text{-}62)$$

where: Eh = redox potential (in volts) and (O_2) = concentration of dissolved oxygen (in moles per kg pore water).

The boundary between the stability fields of MnO_2 and $MnCO_3$ (rhodochrosite) is calculated as:

$$Eh_{MnO_2/MnCO_3} = 0.572 + 0.031 \log(HCO_3) \quad (2\text{-}63)$$

Further, taking into account mass conservation for the three species involved (i.e., MnO_2, $MnCO_3$, and Mn^{2+}) he calculated that:

$$\{[MnO_2] + [MnCO_3]\} + (D_{Mn}/D_s)(Mn^{2+}) = [Mn^0] \quad (2\text{-}64)$$

where:

[ ] = concentration of *i* (moles per unit volume of sediment);
$[Mn^0]$ = manganese content (moles per unit volumes of sediment) of uppermost surface layer; and
D_s = dispersion coefficient for solid phase.

The profile of MnO_2 within the stability field of MnO_2 is then:

$$[MnO_2] = [Mn^0] - [(D_{Mn}/D_s) - 1]\,\phi \exp^{10}[-(Eh - 0.364)/0.03] \quad (2\text{-}65)$$

For the Fe couples, the boundary between the solid phases $Fe(OH)_{3(amorph)}$ and $FeCO_3$ was calculated to be represented by:

$$Eh_{Fe(OH)_3/FeCO_3} = 0.176 + 0.059 \log(HCO_3) \quad (2\text{-}66)$$

where: (HCO_3) = concentration of bicarbonate ion (in moles per kg solvent).
Consideration of conservation of mass gives the following equation:

$$[Fe(OH)_3] + [FeCO_3] + [FeS] + (D_{Fe}/D_s)\,[Fe^{2+}] = [Fe^0] \quad (2\text{-}67)$$

where:

$[Fe^0]$ = iron content (moles per unit volume of sediment) in the uppermost sediment layer.

Within the stability field of $Fe(OH)_3$, its distribution can be formulated as follows:

$$[Fe(OH)_3] = [Fe^0] - [(D_{Fe}/D_s) - 1]\phi\, \exp^{10}[-(Eh - 0.229)/0.059] \quad (2\text{-}68)$$

The boundary between the stability fields of siderite and mackinawite is given by:

$$Eh_{FeCo_3/FeS} = -0.219 - 0.007\,[(HCO_3^-) - (SO_4^{2-})] \quad (2\text{-}69)$$

where: (SO_4^{2-}) = sulphate concentration (in moles per kg solvent).
Mass conservation considerations imply that:

$$(SO_4^{2-}) + (HS^-) + (D_s/D_{ssp})\,[FeS] = [S^0] \quad (2\text{-}70)$$

where:

$[S^0]$ = sulphate concentration in bulk sediment in uppermost layer; and
D_{ssp} = diffusion coefficient for dissolved sulphur species (HS^- and SO_4^{2-}).

This then leads to the following expression for the redox potential at which all available Fe is converted into mackinawite and below which free hydrogen sulphide can accumulate:

$$Eh_{FeS} = -0.219 + 0.007 \log\{(-[1/\phi\,(HCO_3^-)]\,[(D_s/D_{ssp})\,[Fe^0] - [S^0] + 10^{-4.15}\,(HCO_3^-)]\} \quad (2\text{-}71)$$

Also, the SO_4^{2-} profiles can be formulated as follows:

$$(SO_4^{2-}) = [S^0] \quad (2\text{-}72a)$$

within the stability field of reactive $Fe(OH)_3$, and as:

$$(SO_4^{2-}) = (HCO_3^-) \exp^{10} [(Eh + 0.219)/0.007] \quad (2\text{-}72b)$$

for intermediate Eh, between stability fields of reactive $Fe(OH)_3$ and mackinawite, and as:

$$(SO_4^{2-}) = \left[\{[S^0] - (D_s/D_{ssp})[Fe^0]\} \exp^{10} [(Eh + 0.252)/0.007]\right] \Big/ \{1 + \exp^{10} [(Eh + 0.252)/0.007]\} \quad (2\text{-}72c)$$

within the stability field of mackinawite.

The reduction of bicarbonate to methane leads to the following equations, expressing the conservation of mass and assuming the absence of precipitation or dissolution of calcium carbonate and no gas phase (formation of bubbles of methane):

$$[HCO_3^-] + [CH_4] = [C^0] \quad (2\text{-}73)$$

where $[C^0]$ = dissolved bicarbonate content of bulk sediment in uppermost layer.

The HCO_3^- profile will then be described by:

$$[HCO_3^-] = \left(\{[C^0] + (k_c[\text{C-org}]^0) [1 - \exp(-\alpha_c z)]/\alpha_c^2 D_{HCO_3}\} \times \times \exp^{10}[(Eh + 0.293)/0.007]\right) / \{1 + \exp^{10}[(Eh + 0.293)/0.007]\} \quad (2\text{-}74)$$

where:

$[\text{C-org}]^0$ = labile organic carbon content in bulk sediment in uppermost layer;
α_c = $(k_c/D_s)^{1/2}$; and
k_c = first-order rate constant for degradation of labile organic matter.

In the foregoing equations, the numbers 0.007, 0.30, and 0.059 are the appropriate Nernst slopes for the redox couples under consideration. The other numbers are the standard potentials of these couples for a pH of 7.5 and a temperature of 20°C. Billen (1982a) did not present a redox model for the nitrate profile because he did not consider nitrate to be in equilibrium with other oxidants. The NO_3^- profile is discussed in the following section. The combined redox reactions should, for each horizon, meet the electroneutrality criterion. Examples of the thermodynamic equilibrium model are given in Billen (1982a) and shown in Fig. 2-17. In the following paragraphs the subsequent redox couples and the related models for profiles of pore water constituents will be discussed.

O_2 consumption

The diffusion of O_2 into sediments from bottom water is a rapid process. Estimated times for readjustment of the profile in the top of the sediment column are of the order of magnitude of hours at maximum. In particular for deep-sea sediments, this means that molecular diffusion is more important than bioturbation for realistically describing O_2 profiles. Bouldin (1968) discussed solutions of the differential equations of the type Eq. 2-28a ignoring the second term that takes into account advection, for a number of steady-state and transient-state cases, which describe the profiles in the diffusional fluxes across the sediment – water interface.

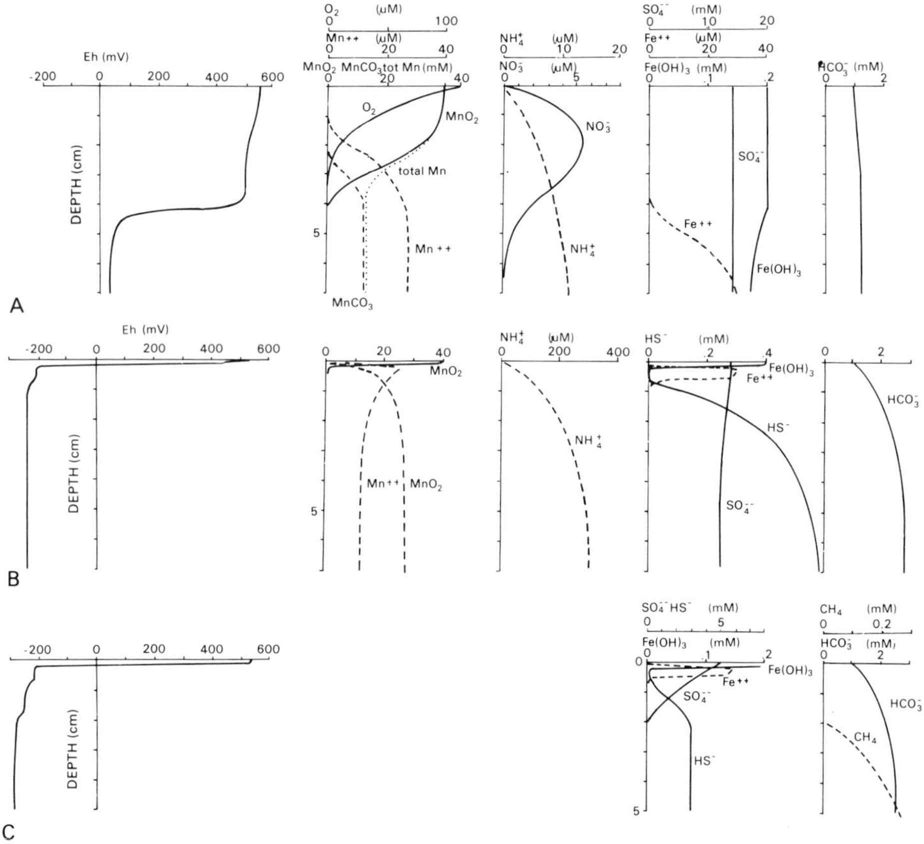

Fig. 2-17. Calculated vertical profiles of mineral redox species in sediments according to a thermodynamic equilibrium model for two different values of the flux of organic matter depositing on the sediment surface. The dispersion coefficients for the interstitial phase (pore-water mixing) $D_i = 10^{-4}$ $cm^2\ s^{-1}$ and for the solid phase (solid-particle mixing) $D_s = 10^{-7}\ cm^2\ s^{-1}$. The rate constant for degradation of labile C-org, $k_C = 3 \times 10^{-8}\ s^{-1}$.

(A) Suboxic diagenesis: case of a marine sediment with a flux of organic matter of 8×10^{-9} mM cm^{-2} s^{-1} (30 g C $m^{-2}\ a^{-1}$); overlying water contains 230 mM O_2, no NO_3^-, 28 mM SO_4^{2-} and 2 mM HCO_3^-; upper sediment layer contains 40 mM cm^{-3} MnO_2 and 200 μM cm^{-3} reactive Fe(III) oxide; porosity = 0.5.

(B) Anoxic diagenesis: case of a marine sediment with a flux of organic matter of 10^{-7} mM $cm^{-2}\ s^{-1}$ (378 g C $m^{-2}\ y^{-1}$) with same composition of overlying water and upper sediment layer as in (A).

(C) Anoxic diagenesis with CH_4 production: case of a fresh-water sediment with a flux of organic matter of 10^{-7} mM $cm^{-2}\ s^{-1}$, with only 1 mM SO_4^{2-} in the overlying water. (Modified after Billen, 1982a.)

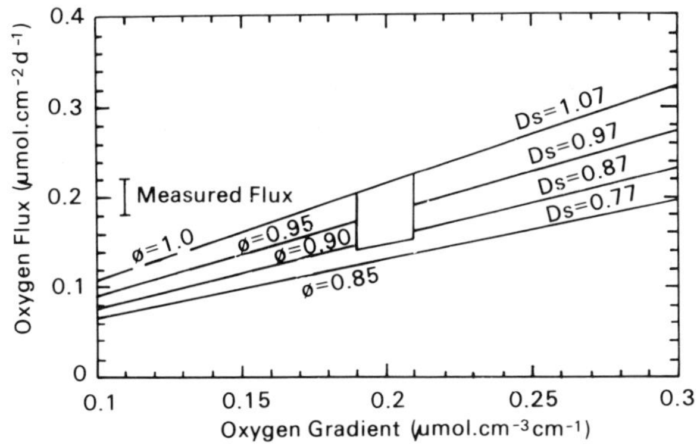

Fig. 2-18. Benthic O_2 flux versus the O_2 gradient at the sediment – water interface. Different lines represent the linear relationships predicted by Fick's law of diffusion with different boundary conditions limiting the porosity (ϕ) and whole sediment diffusion coefficient (D_s) for oxygen. The parallelogram encloses the range of oxygen fluxes derived from the data collected as near as possible to the sediment – water interface of the studied sediments from the E. North Pacific at a depth of about 3750 m. (Modified after Reimers and Smith, 1986.)

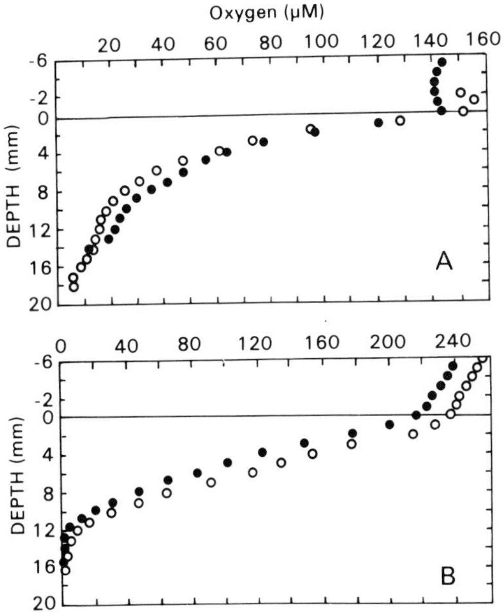

Fig. 2-19. The O_2 gradients measured with microelectrode in sediments from cores obtained in the E. North Pacific Ocean at a depth of about 3750 m using box corer (A) and a "free vehicle grab respirometer" (B). The symbols designate two separate profile determinations, with closed circles preceding open circles in time. The 0-cm depth horizon marks the sediment – water interface. (Modified after Reimers and Smith, 1986.)

The simple model for steady state, in which the profile stays constant with respect to the sediment – water interface, implies that O_2 consumption within the sediment equals the flux of O_2 across this interface:

$$F_{O_2} = -\phi^0 D_{O_2} [\partial(O_2)/\partial z]_{z=0} \quad (2\text{-}75)$$

where:

ϕ^0 = porosity at the uppermost layer of sediment; and
D_{O_2} = diffusion coefficient for dissolved oxygen.

All other parameters have their usual meaning as defined in the previous section.

Reimers and Smith (1986) calculated a family of curves, representing the relationship between the flux of O_2 across and its gradient at the interface, for a number of sediment porosities and related bulk diffusion coefficients, as shown in Fig. 2-18. An example of the O_2 profile obtained in shipboard measurements in box cores, is given in Fig. 2-19.

An interesting case of O_2 consumption is described by Wilson et al. (1985). They studied a sediment core from the Madeira A.P., having a normal pelagic sequence with intercalation of organic-rich turbidites. The pore-water profiles of O_2 and NO_3^- show that this organic matter, with still a high labile C-org content relative to the pelagic sediments in the area, is being progressively oxidized below the sediment – water interface. The rate of oxidation is supposed to be first-order in the C-org content. The stoichiometry of O_2 consumption and C-org combustion is given in Eq. 2-4b ($\Delta O_2/\Delta$C-org = 138/106). Assuming a steady-state profile, constant porosity, absence of other reducing constituents, and ignoring advection, the general diagenetic Eq. 2-22a can be written for both the C-org and O_2:

$$-D_B(\partial^2[\text{C-org}]/\partial z^2) + k_c[\text{C-org}] = 0 \quad (2\text{-}76a)$$

$$\phi D_{O_2}[\partial^2(O_2)/\partial z^2] - (138/106)k_c[\text{C-org}] = 0 \quad (2\text{-}76b)$$

The general solution of Eq. 2-76a for the boundary condition:

$[\text{C-org}] = 0$ for $z = \infty$

is:

$$[\text{C-org}] = [\text{C-org}]^0 \exp(-\alpha_c z) \quad (2\text{-}76c)$$

where [C-org] is the concentration of labile C-org in bulk sediment (superscript zero shows concentration at sediment – water interface).

Equation 2-76c can be substituted into Eq. 2-76b. Taking into account that the slope of the O_2 concentration becomes constant with increasing depth, Wilson et al. (1985) arrived at the following equation describing the O_2 profile:

$$(O_2) = (138/106)(D_B/\phi D_{O_2})[\text{C-org}]^0 \exp(-\alpha_c z) + (z - z_{O_2=0})S \quad (2\text{-}76d)$$

where:

$z_{O_2} = 0$ = depth at which all oxygen is consumed; and
S = slope of the linear part of the oxygen profile.

An equivalent equation can be arrived at for the NO_3^- profile. The results of their models fit the data points quite well, as shown in Fig. 2-20.

A similar model that takes into account advection was used by Grundmanis and Murray (1982). An interesting application of models to O_2 profiles is given by Revsbech et al. (1986). They simulated the profiles for sediments in which light penetration can produce photosynthesis by microalgae in the sediments. They used

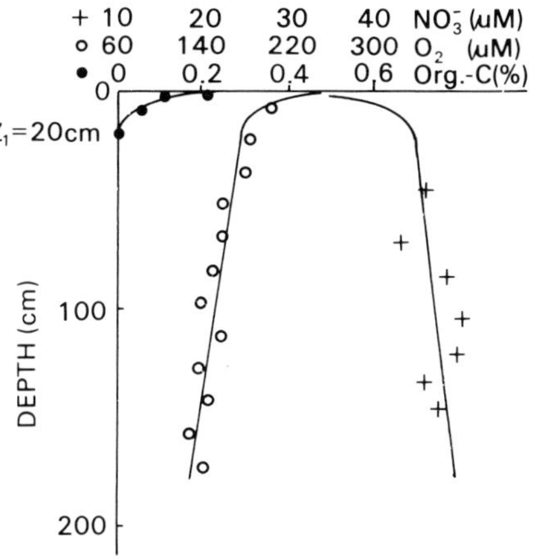

Fig. 2-20. Variation of NO_3^-, O_2 and labile C-org contents in interstitial water for station 10552 on Cape Verde Abyssal Plain with a slow pelagic accumulation (ca. 0.4 cm ka^{-1}). z_1 = depth below which the profiles of oxygen and nitrate are linear. (Modified after Wilson et al., 1985.)

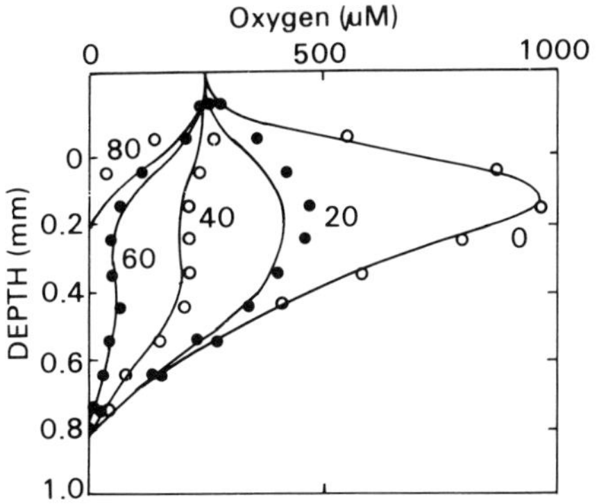

Fig. 2-21. Experimental data (circles) and computer simulation (curves) of O_2 profiles at various times (in seconds) after darkness. Time 0 represents the steady-state profile in the light. (Modified after Revsbech et al., 1986.)

Eq. 2-28a, ignoring advection, and for ΣR_i distinguishing between consumption and production of O_2 at depth z. A numerical solution, suited for computer handling, was used to fit the curves under illumination with photosynthesis ($t = 0$) until darkness with respiration only ($t > 0$). The very quick response (t in seconds) of the O_2 profiles in the uppermost sediment layers, as shown in Fig. 2-21, is remarkable. Similar profiles and quick response times were reported in earlier work of the same group (Revsbech et al., 1980b).

The O_2 consumption occurs also by oxidation of reduced constituents in the sediment that migrate upward from where they are produced. The most extensively studied case is the consumption of O_2 by nitrifying bacteria.

Nitrification and denitrification

Nitrate in pore waters is initially supplied by the entrapped bottom water in sediments. When organic matter is mineralized under oxic conditions, the following sequence of reactions takes place (Suess et al., 1980):

Oxidative deamination: (c.q. ammonification)	$R_1-CH(NH_2)-R_2 + 0.5\ O_2 + H^+ \rightarrow NH_4^+ + R_1-CO-R_2$	(2-77a)
Oxidation of ammonium:	$NH_4^+ + 0.5\ O_2 \rightarrow NH_2OH + H^+$	(2-77b)
	$NH_2OH + O_2 \rightarrow NO_2^- + H_2O + H^+$	(2-77c)
Oxidation of nitrite:	$NO_2^- + 0.5\ O_2 \rightarrow NO_3^-$	(2-77d)

These ammonification and nitrification reactions may be spatially separated, giving rise to measurable ammonium, and nitrite peaks in the oxic zone. But when reaction rates are high enough as compared to the time scale applicable to the sedimentation regime, nitrate will be the dominant end product in this sequence at any depth. This means that nitrate, derived from metabolized organic matter by nitrifying bacteria, is added to the nitrate pool derived from the bottom water. This sets up a concentration gradient. When O_2 is depleted, NO_3^- will become the next electron acceptor, with stoichiometry as given in Eqs. 2-6a, b, c. Many studies are devoted to modeling measured combined N profiles, too many to present or even mention them all; therefore, a selection is made. First, the approach of Goloway and Bender (1982), that was already briefly mentioned (cf. Fig. 2-3) is followed.

Their first model (Fig. 2-3, top), showing pore water NO_3^- concentrations that asymptotically rise toward an upper limit at depth, will apply when only O_2 reduction and nitrification occur in the sediment column. This means that not all O_2 is consumed or, in other words, that all labile organic matter is consumed before O_2 is. These authors assumed that (1) the rate of O_2 consumption (R_{O_2}) and NO_3^- production (R_{NO_3}) are first order with respect to the concentration of C-org solid (i.e., $R_{\text{c-org}} = k_c$ [C-org]), (2) that organic matter is randomly mixed downcore by bioturbation (D_B), (3) that advection can be ignored, and (4) that steady state exists. In that case, Eq. 2-28a as applied to organic matter, can be written as:

$$D_B\ (\partial^2[\text{C-org}]/\partial z^2) + R_{\text{C-org}} = 0 \qquad (2\text{-}78)$$

where the boundary conditions are:

$$[\text{C-org}] = [\text{C-org}]^0 \qquad \text{at } z = 0$$
$$[\text{C-org}] = 0 \qquad \text{at } z = \infty$$

The solution of Eq. 2-78 is:

$$[\text{C-org}] = [\text{C-org}]^0 \exp(-\alpha_c z) \tag{2-79}$$

By the same token, the equation describing the NO_3^- profile is:

$$D_{NO_3^-}\,[\partial^2(NO_3^-)/\partial z^2] + R_{NO_3^-} = 0 \tag{2-80}$$

with boundary conditions:

$$(NO_3^-) = (NO_3^-)^0 \qquad \text{at } z = 0$$
$$(NO_3^-) = (NO_3^-) \qquad \text{at } z = \infty$$

Introducing the stoichiometric relation:

$$\gamma_3 = d(NO_3^-)/d[\text{C-org}]$$

or, in integrated form:

$$R_{NO_3^-} = -\gamma_3\, R_{\text{C-org}}$$

and combining with Eq. 2-79 leads to a differential equation for which the solution is:

$$\ln\left(1 - [(NO_3^-) - (NO_3^-)^0]/[(NO_3^-)^\infty - (NO_3^-)^0]\right) = -\alpha_c z \tag{2-81}$$

The fluxes of NO_3^- out and of O_2 into the sediment can be calculated from:

$$F_{NO_3^-} = D_{NO_3^-}\,[\partial(NO_3^-)/\partial z]_{z=0} = D_{NO_3^-}\,\alpha_c\,[(NO_3^-)^\infty - (NO_3^-)^0] \tag{2-82a}$$
$$F_{O_2} = -F_{NO_3^-}/\gamma_2 \tag{2-82b}$$

where γ_2 = stoichiometric constant = $d(NO_3^-)/d(O_2)$; and $F_{NO_3^-}$ is flux of NO_3^- in mass per unit area of total sediment per unit of time; and $D_{NO_3^-}$ is whole sediment diffusion coefficient of nitrate.

The second model of Goloway and Bender (1982) (Fig. 2-3, middle) takes into account depletion of O_2 at a certain horizon and subsequent denitrification below that horizon. Assumptions made are that (1) the rate of nitrification is proportional to the O_2 concentration in the pore water (in the first model it was assumed that O_2 consumption rate was first order with respect to C-org concentration), i.e.,

$R_{O_2} = k_{O_2} \times [O_2]$, (2) no advection occurs, and (3) steady state exists. This gives the
following set of equations:

$$D_{O_2}\,[\partial^2(O_2)/\partial z^2] - k_{O_2}(O_2) = 0 \tag{2-83}$$

where k_{O_2} = rate constant for oxygen consumption, with boundary conditions:

$(O_2) = (O_2)^0$ at $z = 0$

$(O_2) = 0$ at $z = \infty$

for which the solution is:

$$(O_2) = (O_2)^0 \exp[-(k_{O_2}/D_{O_2})^{1/2}\,z] = 0 \tag{2-84}$$

The relation with nitrate in the oxic zone is then:

$$D_{NO_3^-}\,[\partial^2(NO_3^-)/\partial z^2] - \gamma_2\,k_{O_2}(O_2)^0 \exp(-k_{O_2}/D_{O_2})^{1/2}\,z = 0 \tag{2-85}$$

The boundary conditions are:

$(NO_3^-) = (NO_3^-)^0$ at $z = 0$

$(NO_3^-) = (NO_3^-)_{z'}$ at $z = z'$

where z' = depth at which O_2 is depleted.
The solution is:

$$(NO_3^-) = (NO_3^-)^0 + \gamma_2\,(O_2)^0\,(D_{O_2}/D_{NO_3^-})\,[1 - \exp(-k_{O_2}/D_{O_2})^{1/2}\,z] + Cz \tag{2-86}$$

where C = integration constant.

For the sediment below the z' horizon, Goloway and Bender (1982) assumed first-order denitrification rates, i.e., $R_{NO_3^-} = -k_{N_2}\,(NO_3^-]$, where k_{N_2} = first-order rate constant for denitrification. They neglected any small amount of NO_3^- produced by residual O_2 in this zone. The equation then becomes:

$$D_{NO_3^-}\,[\partial^2(NO_3^-)/\partial^2 z] - k_{NO_3^-}(NO_3^-) = 0 \tag{2-87}$$

with boundary conditions:

$(NO_3^-) = (NO_3^-)_{z'}$ at $z = z'$

$(NO_3^-) = 0$ at $z = \infty$

which has as solution:

$$(NO_3^-) = (NO_3^-)_{z'} \exp[-(k_{N_2}/D_{NO_3^-})^{1/2}(z - z')] \tag{2-88}$$

This is the same expression as derived by Billen (1982b). The continuity of fluxes and concentrations at $z = z'$ requires that:

$$[\partial(NO_3^-)/\partial z)_{eq.\ 2\text{-}85}] = [\partial(NO_3^-/\partial z)_{eq.\ 2\text{-}88}]$$

which means that the integration constant C in Eq. 2-86 can now be calculated, after substitution and rearrangement of this equation for the interval $0 \leq z \leq z'$, leading to:

$$\begin{aligned}(NO_3^-) = {} & (NO_3^-)^0 + \gamma_2 (O_2)^0 (D_{O_2}/D_{NO_3^-}) [1 - \exp(-k_{O_2}/D_{O_2})^{1/2} z] + \\ & \{z/[z' + (D_{NO_3^-}/k_{N_2})^{1/2}]\} \times \\ & \times \{F_{O_2}\gamma_2 \exp[(-k_{O_2}/D_{O_2})^{1/2} z/D_{NO_3^-}(k_{N_2}/D_{NO_3^-})^{1/2}] - \\ & (NO_3^-)^0 - \gamma_2 (O_2)^0 (D_{O_2}/D_{NO_3^-}) [1 - \exp(-k_{O_2}/D_{O_2})^{1/2} z']\}\end{aligned} \tag{2-89}$$

This equation, as discussed by Goloway and Bender (1982) has some flaws that they consider, however, to have only minor consequences. The benthic fluxes of NO_3^- can be calculated by the product of the diffusion coefficients and concentration gradients at the boundaries $z = 0$ and $z = z'$, respectively.

The third model of Goloway and Bender (1982) (Fig. 2-3, bottom), as already stipulated, describes a profile very similar to the second one, but in addition it exhibits a linear decrease in nitrate concentration between the zones of O_2 reduction and NO_3^- reduction. The profile in the upper zone can be described by Eq. 2-86 and the linear zone by Eq. 2-74, with the boundary conditions:

$(NO_3^-) = (NO_3^-)_{z'}$ at $z = z'$

$(NO_3^-) = (NO_3^-)_M$ at $z = z_M$ (= horizon, where the linear decrease of the nitrate concentration ends).

The solutions for these equations are:

For $0 \leq z \leq z'$:

$$\begin{aligned}(NO_3^-) = {} & (NO_3^-)^0 + \gamma_2 (O_2)^0 (D_{O_2}/D_{NO_3^-}) [1 - \exp(-k_{O_2}/D_{O_2})^{1/2} z] + \\ & [z/(z_M - z')] \{(NO_3^-)_M - (NO_3^-)_{z'} + (z_M - z')(k_{O_2}/D_{O_2})^{1/2} \times \\ & \times (O_2)^0 D_{NO_3^-} D_{O_2} \gamma_2 \exp[-(k_{O_2}/D_{O_2})^{1/2} z']\}\end{aligned} \tag{2-90a}$$

For $z > z'$:

$$(NO_3^-) = z/(z_M - z')\,[(NO_3^-)_M - (NO_3^-)_{z'}] + [1/(z_M - z')]\,[z_M\,(NO_3^-)_{z'} - z'\,(NO_3^-)_M] \tag{2-90b}$$

All these equations can be solved by iterative least-squares fitting of the observed profiles. Goloway and Bender (1982) solved the equations for a number of cores, achieving reasonable to excellent fits of the observed data points by the derived models. The results are demonstrated in Fig. 2-22.

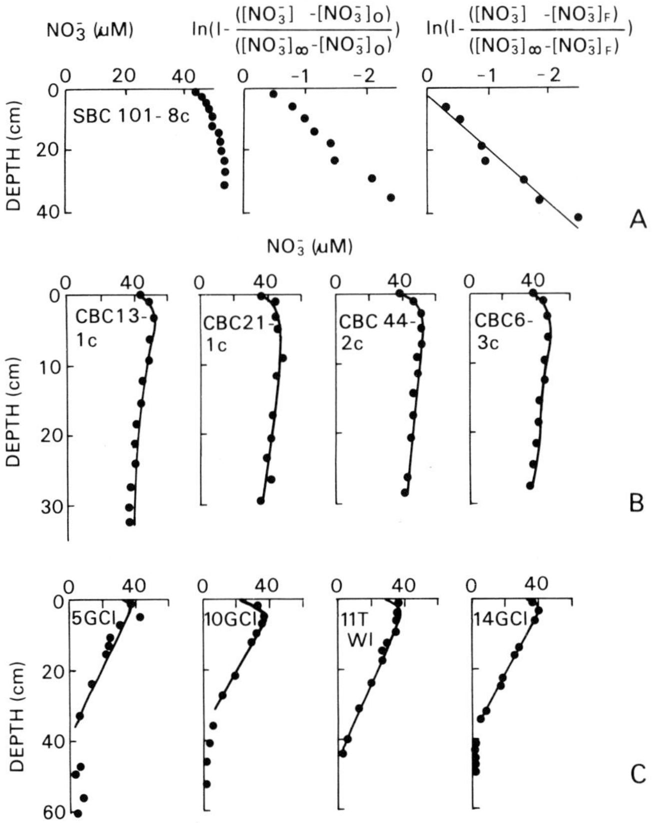

Fig. 2-22. Hypothetical pore water NO_3^- profiles. In the top figure asymptotic NO_3^- content is defined as $(NO_3^-)^\infty$. (A) Observed and derived model profiles of group 1 (top of Fig. 2-3). The first column shows observed profiles; the second column shows model profiles derived using sediment–water interface at zero depth; and the third column shows model profiles calculated by taking zero depth as that of the first pore water sample. (Sta.SBC 101-8C from MANOP site S).
(B) Observed and derived model NO_3^- profiles of group 2 (middle of Fig. 2-3). Dots represent observed concentration, whereas solid lines represent derived profiles. All cores were obtained at stations at MANOP site C.
(C) Observed and derived model NO_3^- profiles of group 3 (bottom of Fig. 2-3). All cores were obtained from Eastern Equatorial Atlantic sediments. (Modified after Goloway and Bender, 1982.)

Billen (1982b) arrived at basically the same equation for the NO_3^- profile in the zone of nitrification ($z \leq z'$), but he expressed nitrification by a rate constant k_{NO_3} for the production of nitrate rather than by relating this process to O_2 consumption. His model results for measured profiles of N-org and N-inorg are shown in Fig. 2-23. Similar results and discussions were earlier presented by Billen (1978).

Jahnke et al. (1982a,b) used a combination of stoichiometric and diffusional models to fit the observed pore-water data. In their latter paper, they also performed a sensitivity analysis of the models to the different parameters. Their model equations, based on Eq. 2-22a, are as follows:

Oxic zone:

$$\partial\{(1 - \phi)\,[\text{C-org}]\}/\partial t = \partial\,\{(1 - \phi)\,D_B\,(\partial\,[\text{C-org}]/\partial z)\}/\partial z - \partial\,\{(1 - \phi)\,w\,[\text{C-org}]\}/\partial z - k_{O_2}\,(1 - \phi)\,[\text{C-org}] \quad (2\text{-}91a)$$

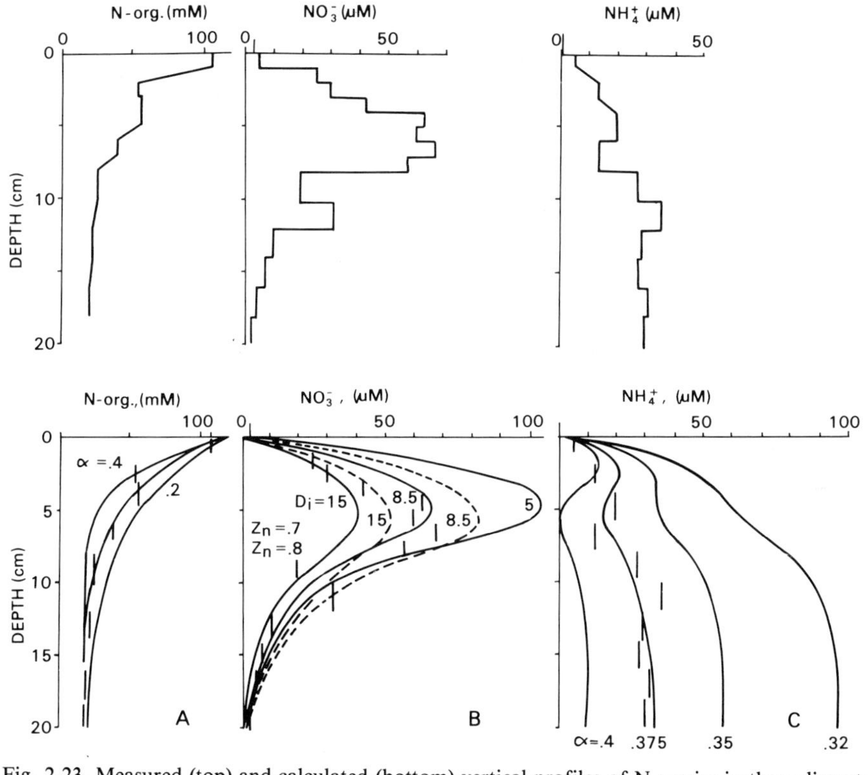

Fig. 2-23. Measured (top) and calculated (bottom) vertical profiles of N species in the sediments of the Southern Bight of the North Sea.
(Bottom left) $\alpha = (k_C/D_s)^{1/2} = 0.2$, 0.3 and 0.4 (cm^{-1}), where D_s is the dispersion coefficient for solid phase.
(Bottom middle) $k_{NO_3^-} = 0.5 \times 10^{-6}$ μmoles cm^{-3} s^{-1}; $z_{NO_3^-}$ (depth of zone of nitrification) is equal to 7 cm (solid curve) and 8 cm (dashed curve); $k_{N_2} = 5.3 \times 10^{-6}$ s^{-1}; D_i (dispersion coefficient of interstitial phase) = 15, 8.5 and 5 × 10^{-5} cm^2 s^{-1}.
(Bottom right) $k_{NH_4^+} = 2.1 \times 10^{-6}$ μmoles cm^{-3} s^{-1}; $\alpha = 0.4$, 0.375, and 0.32 (cm^{-1}). (Modified after Billen, 1982b.)

$$\partial[\phi(O_2)]/\partial t = \partial\{[\gtrapprox D^0_{O_2}/F_f]\,\partial(O_2)/\partial z\}/\partial z - \gamma_1 k_{O_2}(1-\phi)\,[\text{C-org}] \qquad (2\text{-}91b)$$

$$\partial[\phi(NO_3^-)]/\partial t = \partial\{[D^0_{NO_3^-}/F_f]\,\partial(NO_3^-)/\partial z\}/\partial z + \gamma_3 k_{O_2}(1-\phi)\,[\text{C-org}] \qquad (2\text{-}91c)$$

Zone of denitrification:

$$\partial\{(1-\phi)\,[\text{C-org}]\}/\partial t = \partial\{(1-\phi)\,D_B\,(\partial[\text{C-org}]/\partial z)\}/\partial z - \partial\{(1-\phi)\,w\,[\text{C-org}]\}/\partial z - k_{N_2}(1-\phi)\,\phi\,[\text{C-org}]\,(NO_3^-) \qquad (2\text{-}91d)$$

$$\partial[\phi(O_2)]/\partial t = \partial\{[\gtrapprox D_{O_2}/F_f]\,\partial(O_2)/\partial z\}/\partial z - \gamma_4 k_{O_2}(1-\phi)\,[\text{C-org}] \qquad (2\text{-}91e)$$

$$\partial[\phi(NO_3^-)]/\partial t = \partial\{[\gtrapprox D_{NO_3^-}/F_f]\,\partial(NO_3^-)/\partial z\}/\partial z - \gamma_3 k_{N_2}(1-\phi)\phi\,[\text{C-org}]\,(NO_3^-) \qquad (2\text{-}91f)$$

with the following boundary conditions:
At $z = 0$:

$(O_2) = (O_2)^0$ (bottom water)

$(NO_3^-) = (NO_3^-)^0$ (bottom water)

C-org flux as specified.

at $z = z_{max}$:

$\partial(O_2)/\partial z = 0$

$\partial(NO_3^-)/\partial z = 0$

C-org flux $= (1-\phi)\,w\,[\text{C-org}]_{max}$

Where:

D_{O_2}	= molecular diffusion coefficient of O_2 (10.6×10^{-6} cm^2 s^{-1});
$D_{NO_3^-}$	= ditto of nitrate (8.8×10^{-6} cm^2 s^{-1});
F_f	= formation factor $= d_s E = d_s\,\phi\,(1-\phi)$
γ_1	= O_2/C-org ratio during oxic respiration (138/106);
γ_3	= NO_3^-/C-org ratio during oxic respiration (16/106);
γ_4	= NO_3^-/C-org ratio during denitrification (94.4/106);
z_{max}	= maximum depth modeled; and
$[\text{C-org}]_{max}$	= organic carbon content at z_{max} (mol cm$^{-3}{}_{sed}$).

The model results for site C (calcareous ooze) and site S (siliceous ooze) in the

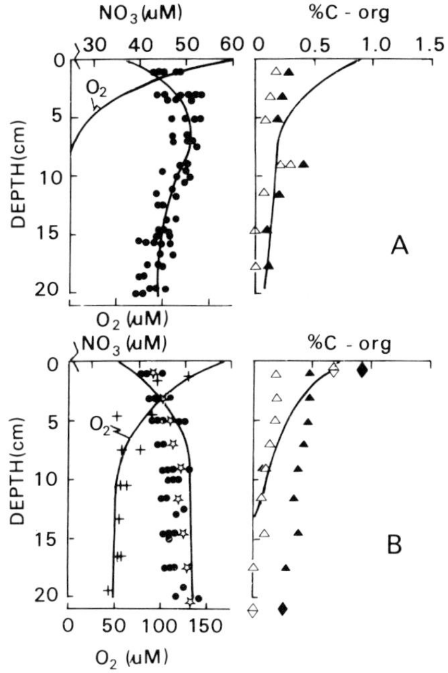

Fig. 2-24. Comparison between the measured data and model results at site C (A) and site S (B). The solid triangles used for the C-org content represent measured values; the open triangles represent estimates of labile C-org metabolized in the upper 20 cm, calculated by subtracting the C-org content from the deepest interval sampled from the other values. (Modified after Jahnke et al., 1982a.)

Central Pacific Ocean (MANOP sites) are shown in Fig. 2-24 using the following model parameters:

	Site C	Site S
D_B (cm^2 per 1000 y)	30	30
w (mm per 1000 y)	10	0.4 – 1
C-org flux (μmole cm^{-2} y^{-1})	4.2	2.5
k_{O_2}(oxic) (10^{-11} s^{-1})	7.3	4.2
k_{N_2}(denitr.) (10^{-4} cm^2 mol^{-1} s^{-1})	1.1	

The O_2 profiles are strongly dependent on the flux of C-org. The NO_3^- and C-org profiles are only slightly dependent on the mixing and accumulation rates of the sediments and strongly on the O_2 respiration constant and the flux of C-org.

Manganese oxide reduction

Murray et al. (1984) made an interesting observation that the apparent oxidation state of Mn-oxide decreases from high values in the top to lower values at and below

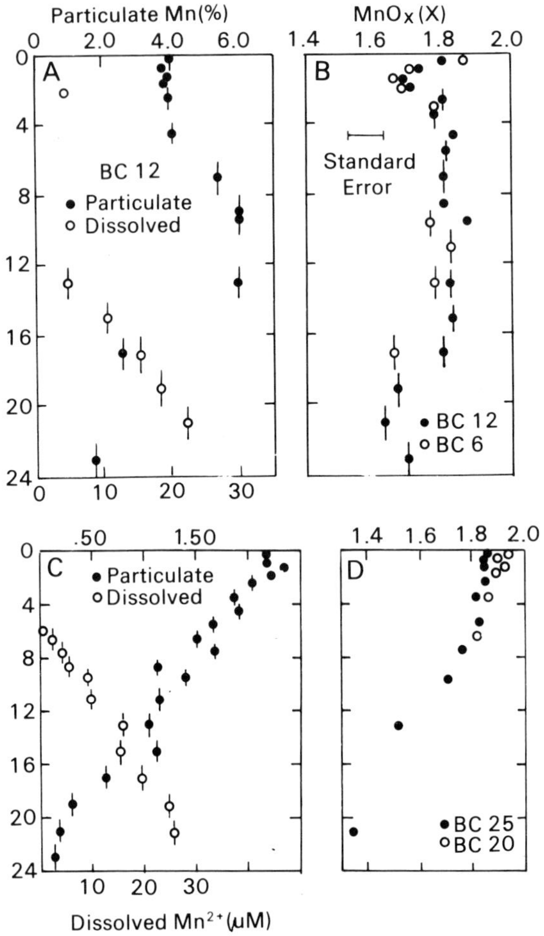

Fig. 2-25 (A) Solid-phase and pore-water Mn concentrations; and (B) Mn oxidation state of the solid phase, MnO_x *(X)*, at MANOP site H; (C) The Mn concentrations in the solid phase and pore water; and (D) oxidation state Mn for solid phase, MnO_x *(X)*, for MANOP site M for four different box cores as indicated. (Modified after Kalhorn and Emerson, 1984.)

TABLE 2-8

Some pertinent data on the sedimentary regime at MANOP sites H and M

	site H	site M
Mn flux to sediments (mg cm^{-2} per 1000 y)	0.46	1.65
Sedimentation rate (cm per 1000 y)	0.66	1.0
Particulate C-org flux (mM cm^{-2} per 1000 y)	12	13.5
Bioturbation coefficient (cm^2 per 1000 y)	24 – 90	160
Area coverage Mn nodules (%)	≈ 20	

the horizon where reduction of Mn-oxide occurs. Kalhorn and Emerson (1984) elaborated that in a simple mass balance approach. The actual profiles for MANOP sites H and M are shown in Fig. 2-25. The oxidation state at a depth of 0.5 – 2 cm at site H is minimum. Suboxic reduction commences at a shallower depth at site M than at site H. Some data on input and mixing parameters at the different sites are given in Table 2-8 (Kalhorn and Emerson, 1984; Emerson et al., 1985). The slightly higher C-org rain in combination with the higher sedimentation rate causes the onset of Mn-oxide reduction at site M at a shallower depth than at site H, as well as a higher C-org content (factor $\approx$ 2) in the upper part of the sediment. Surprisingly, the accumulation of particulate Mn is lower at site M than at site H. These authors offered two explanations for that. Firstly, it cannot be ruled out that the zone of suboxic diagenesis was raised to the very top of the sediment column in the past. This would also have allowed Mn to escape from the sediment into the bottom water. Secondly, Mn may escape by reduction in the top of the sediment column because of the higher C-org content, when, locally (microenvironments), the rate of O_2 supply cannot keep up with its demand and suboxic diagenesis takes over.

A simple mass balance, applied to the shallow depth minimum (0.5 – 2 cm) as observed at site H, assuming steady state, is as follows:

$$f^0 F - s f^{0-2} c^{0-2} + D_B(f^{2-4} c^{2-4} - f^{0-2} c^{0-2})/\Delta z + R = 0 \qquad (2\text{-}92)$$

where:

f = fraction of particulate Mn(II) relative to total Mn at levels or intervals as indicated by superscripts (in cm);
F = particulate flux of Mn(II);
s = sedimentation rate;
D_B = bioturbation rate (tuned to the total particulate Mn profile for reasons of consistency);
c = content of particulate Mn at depth intervals given as superscripts;
Δz = distance between the considered depth intervals (i.e., 2 cm);
R = rate of MnO_x reduction ($1 < x \leq 2$)

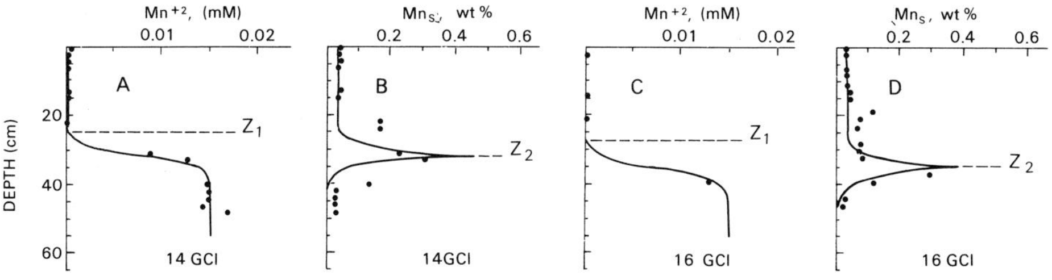

Fig. 2-26. The Mn concentrations and predicted profiles for core 14, using k_{OX} = 12.0 a^{-1} and k_{red} = 2.1 × 10^{-3} a^{-1}. The Mn concentrations in the pore water the solid phase and predicted profiles for core 16, using k_{OX} = 10.7 y^{-1} and k_{red} = 1.5 × 10^{-3} y^{-1}. (Modified after Burdige and Gieskes, 1983.)

For the profiles at site H (Fig. 2-25), the following average values apply:

$f^0(MnO_{1.8}) = 0.2$; $f^{0-2}(MnO_{1.7}) = 0.3$; $f^{2-4}(MnO_{1.8}) = 0.2$;

$c^{0-2} = 0.0147$ g Mn cm^{-3} (3.9% Mn); and

$c^{2-4} = 0.015$ g Mn cm^{-3} (4.0% Mn).

Equation 2-92 can now be solved to obtain the reduction rate (R) required to maintain the oxidation state minimum: $R = 19$ mg cm^{-2} (10^{3}y)$^{-1}$ for $D_B = 24$ cm^2 (10^3)$^{-1}$; and 77 mg cm^{-2} (10^{3}y)$^{-1}$ for $D_B = 90$ cm^2 (10^{3}y)$^{-1}$. This is stoichiometrically equivalent to the oxidation of 1 – 6% of the C-org rain at this site. The reduced Mn, assumed that about 1 – 10% is being able to escape to the bottom water, can support the nodule growth as measured at this site. The rate of Mn-oxide reduction required to deplete the sediments of particulate Mn at site M would be at least an order of magnitude lower than that calculated at site H. This is probably the reason why no subsurface minimum of the Mn oxidation state shows up here.

Burdige and Gieskes (1983) presented a model for the redox processes in marine sediments and related pore waters. Their model takes into account an oxidation zone ($Mn^{2+} \rightarrow MnO_x$) separated by a redox boundary from a reduction zone ($MnO_x \rightarrow Mn^{2+}$). The redox boundary is that depth in the sediments below which Mn^{2+} is favored over solid MnO_x phases ($z \approx 2$ in Fig. 2-26A). This boundary can be determined by comparison of the Mn concentrations in pore waters and the solid phase with an idealized profile (Fig. 2-26A), and/or by comparison with the pore-water NO_3^- profiles, because NO_3^- usually is significantly depleted at a depth at which Mn oxide reduction becomes important. The general diagenetic Eq. 2-28a can be used for pore water and solid phase Mn profiles, assuming steady state:

$$\text{for } Mn^{2+}:\ D_{Mn}\,[\partial^2(Mn)_1/\partial z^2] - w\,[\partial(Mn)_1/\partial z] + R_z = 0 \qquad (2\text{-}93a)$$

$$\text{for } MnO_x:\ - w\{\partial[Mn]_s/\partial z\} - [\phi/(1-\phi)]\,R_z = 0 \qquad (2\text{-}93b)$$

where D_{Mn} = whole sediment diffusion coefficient of dissolved Mn. Further implicit assumptions are (1) that the porosity (ϕ) and diffusion coefficient (D_{Mn}) are constant over the considered depth interval, (2) that advection (w) is constant and equal to the sedimentation rate, (3) that solid-phase diffusion (bioturbation) can be ignored, and (4) that the rain of particulate Mn-oxide is constant in time. The model assumes first-order kinetics for the reduction and oxidation reactions:

$$R_{z(red)} = k_{red}\,[Mn]_{s(red)}$$

$$R_{z(ox)} = k_{ox}\,(Mn)_{1(ox)}$$

where:
$[Mn]_{s(red)}$ = content of solid MnO_x below redox zone; and
$(Mn)_{1(ox)}$ = concentration of dissolved Mn above redox zone.
These kinetic equations can be substituted into Eqs. 2-93a, b giving two sets of equa-

tions, one for the oxidizing zone, and one for the reducing zone.

The appropriate boundary conditions to solve these equations are:
At depth z_1 (cf., Fig. 2-26A):

$$(\mathrm{Mn})_{1\,(\mathrm{ox})}\,(z_1) = 0$$

$$[\mathrm{Mn}]_{\mathrm{s}\,(\mathrm{ox})}\,(z_1) = [\mathrm{Mn}]_{\mathrm{s}}^{0}$$

At depth z_2 (cf., Fig. 2-26B):

$$(\mathrm{Mn})_{1\,(\mathrm{ox})}\,(z_2) = (\mathrm{Mn})_{1\,(\mathrm{red})}\,(z_2)$$

$$[\mathrm{Mn}]_{\mathrm{s}\,(\mathrm{ox})}\,(z_2) = [\mathrm{Mn}]_{\mathrm{s}\,(\mathrm{red})}\,(z_2)$$

Continuity of fluxes: $\{\partial\,[\mathrm{Mn}]_{1\,(\mathrm{ox})}/\partial z\}_{z_2} = \{\partial\,[\mathrm{Mn}]_{1\,(\mathrm{red})}/\partial z\}_{z_2}$

At depth $z = \infty$: $(\mathrm{Mn})_1$ remains finite.

Assuming that advection in pore waters is negligible compared to diffusion, i.e.:

$$D_{\mathrm{Mn}}\,k_{\mathrm{ox}} >> w^2, \text{ and } D_{\mathrm{Mn}}\,k_{\mathrm{red}} >> w^2$$

the solutions for the sets of equations for the oxidizing and reducing zones become:

$$(\mathrm{Mn})_{1\,(\mathrm{ox})} = A \sinh\,[(k_{\mathrm{ox}}/D_{\mathrm{Mn}})^{1/2}\,(z - z_1)] \tag{2-94a}$$

$$(\mathrm{Mn})_{1\,(\mathrm{red})} = G - (1 - \phi)/\phi\,(E\,w^2\,D_{\mathrm{Mn}}/k_{\mathrm{red}})\exp[-(k_{\mathrm{red}}/w)\,(z - z_2)] \tag{2-94b}$$

$$[\mathrm{Mn}]_{\mathrm{s}\,(\mathrm{ox})} = [\mathrm{Mn}]_{\mathrm{s}}^{0} \cosh\,[(k_{\mathrm{ox}}/D_{\mathrm{Mn}})^{1/2}\,(z - z_1)] \tag{2-94c}$$

$$[\mathrm{Mn}]_{\mathrm{s}\,(\mathrm{red})} = E \exp[-(k_{\mathrm{red}}/w)\,(z - z_2)] \tag{2-94d}$$

where:

$A = [\mathrm{Mn}]_{\mathrm{s}}^{0}\,(1 - \phi)\,w/\phi D_{\mathrm{Mn}}\,(k_{\mathrm{ox}}/D_{\mathrm{Mn}})^{1/2}$

$E = [\mathrm{Mn}]_{\mathrm{s}}^{0} \cosh\,(k_{\mathrm{ox}}/D_{\mathrm{Mn}})^{1/2}\,L_{\mathrm{ox}}$

$G = A\,[\sinh\,(k_{\mathrm{ox}}/D_{\mathrm{Mn}})^{1/2}\,L_{\mathrm{ox}} + (k_{\mathrm{ox}}/D_{\mathrm{Mn}})^{1/2}/(k_{\mathrm{red}}/w)\cosh\,(k_{\mathrm{ox}}/D_{\mathrm{Mn}})^{1/2}\,L_{\mathrm{ox}}]$

$L_{\mathrm{ox}} = z_2 - z_1$

All parameters can be determined from the measured or estimated porosity, sedimentation rate, and concentration profiles. Only the rate constants for oxidation and reduction are unknown and can be estimated by curve fitting. Examples of those fits are given in Fig. 2-26 for sediments from the Eastern Equatorial Atlantic (data from Froelich et al., 1979) at depths of 4170 m (14GC1) and 3310 m (16GC1).

The rate constants thus determined are: $k_{ox} = 12.8\ (\pm\ 4.7)\ y^{-1}$ and $k_{red} = 1.99\ (\pm\ 0.31) \times 10^{-3}\ y^{-1}$, respectively. The values of k_{ox} compare reasonably well with other reported values obtained in experiments for heterogeneous oxidation rates, i.e., in the presence of reactive solid surface sites provided by mineral phases, and also with the results for microbial oxidation as determined by Tebo and Emerson (1985). The values for k_{red} are lower by one to several orders of magnitude than reported for shallow-water sediments (Holdren et al., 1975; Aller, 1980c). This can be attributed to much lower labile organic matter contents in deep-sea sediments. It should be noted that these curve-fits were applied on profiles that do not exhibit the ideal steady-state shape (Fig. 2-26). The profiles of solid Mn-oxide with a tendency for two peaks might well be indicative of non steady-state profiles.

Boudreau and Scott (1978), using a simplified form of Eq. 2-93a by ignoring advection (second-term cancels), modeled the dissolved Mn profile assuming that the apex of the profile is situated at the sediment – water interface. Their solution is:

$$(\mathrm{Mn})_{1\,(\mathrm{ox})} = [(\mathrm{Mn})_{1\,(z_2)} \sinh\,(k_{ox}/D_{\mathrm{Mn}})^{1/2} z]/\sinh\,(k_{ox}/D_{\mathrm{Mn}})^{1/2} z_2 \qquad (2\text{-}95a)$$

The flux arriving at the sediment – water interface is given by:

$$F_{\mathrm{Mn}\,(z_0)} = [\phi D_{\mathrm{Mn}}\,(k_{ox}/D_{\mathrm{Mn}})^{1/2}\,(\mathrm{Mn})_{1\,(z_2)}]/\sinh\,(k_{ox}/D_{\mathrm{Mn}})^{1/2} z_2 \qquad (2\text{-}95b)$$

For some realistic values of the bulk diffusion coefficient, pore-water concentrations below the redox boundary, and thicknesses of the oxidizing layers, Boudreau and Scott compared graphically the maximum possible fluxes with the range of observed accumulation rates of Mn nodules. Their conclusion was that, under the most favorable conditions, a thickness of the oxidizing zone not exceeding 40 cm is required to match the flux with the lowest observed accretion rate. But for more moderate and realistic estimates of the parameters, this thickness has to be 20 cm or, more likely, even lower. In some deep-sea sediments, the thickness of the oxidizing layer is greater and, therefore, these authors demonstrated that accretion of manganese nodules from seawater can account for their growth.

An example of the possible contribution of diagenetically-mobilized Mn to the growth of Mn nodules in the Peru Basin is given by Marchig and Reyss (1984). As was discussed for the profile in Fig. 2-25A, B (Manop site H), the occurrence of reducing microenvironments close to the sediment – water interface, but still within the oxidizing layer, may in particular cases contribute to nodule growth. An attempt to model the role of such environments in their formation and the effect on pore-water profiles was made by Jahnke (1985). His conclusion was that the formation of suboxic zones within the particles, that are embedded in oxygenated pore waters, is favored by large particle size, rapid rates of O_2 consumption, and slow internal diffusivity. Although the model was used for oxygen and nitrogen profiles, it could be used as well for Mn profiles.

Sundby and Silverberg (1985) modeled manganese profiles and, based thereon, calculated Mn fluxes in the benthic boundary layer in the lower St. Lawrence estuary. They showed the highly dynamic nature of Mn cycling in these sediments, with turn-over times in the order of $10^{0.3}-10^2$ days. Some of the fluxes can illustrate this, as shown in Fig. 2-27.

Aller (1980c) applied his two-dimensional model (cf., Eq. 2-40), which takes into account the influence of burrowing in the top of the sediments, to the dissolved Mn profile in sediments in Long Island Sound. A slight improvement over the one-dimensional model (i.e., only vertical molecular diffusion) was obtained, as shown in Fig. 2-28. The shape of the curve is determined by a shallow MnO_x redox zone and a sink for Mn^{2+} at depth.

Sulphate reduction

Berner (1980) applied Eq. 2-57 to describe the profile of sulphate in marine pore waters for the case of steady-state diagenesis:

$$D_{SO_4}\,[\partial^2(SO_4^{2-})/\partial z^2] \;-\; w\partial(SO_4^{2-})/\partial z \;-\; \gamma_5 F_f k_{SO_4^{2-}} G_T \;=\; 0 \qquad \text{(2-96a)}$$

where:

$D_{SO_4^{2-}}$ = whole sediment diffusion coefficient of sulphate ion;
γ_5 = stoichiometric coefficient relating the number of moles of sulphate reduced per mole of C-org oxidized to CO_2;
$k_{SO_4^{2-}}$ = sulphate reduction rate; and
G_T = content of metabolizable organic matter (moles/unit mass of total solids).

The underlying assumptions are that (1) this applies to the zone below the zone of bioturbation, (2) the diffusion rate and porosity are constant, (3) no adsorption of sulphate occurs, (4) compaction and externally-induced water flow are absent, and (5) the reduction of sulphate is first order with respect to the organic matter content. The steady-state distribution of C-org can be formulated as:

$$-\;w\,(\partial G_T/\partial z) \;-\; k_{SO_4} G_T \;=\; 0 \qquad \text{(2-96b)}$$

To solve these equations, the following boundary conditions are used:

$z = 0$, for depth at the base of the bioturbated zone!;
$(SO_4^{2-})^0$ = sulphate concentration at $z = 0$ as measured;
$z = \infty$, $G_T = 0$, and $(SO_4^{2-}) = (SO_4^{2-})^\infty$.

The latter condition means that metabolizable C-org is supposed to be all used up, even if there is not enough sulphate to support that, which means that (SO_4^{2-}) may reach an imaginary negative value. If negative, this (SO_4^{2-}) content must be determined by extrapolation from exponential curve fitting of the concentration data. The solutions for Eqs. 2-96a and b are:

$$(SO_4^{2-}) = (SO_4^{2-})^{\infty} + [(SO_4^{2-})^0 - (SO_4^{2-})^{\infty}] \exp[(-k_{SO_4^{2-}}/w)\,z] \quad (2\text{-}97a)$$

$$G_T = G_T^0 \exp[-(k_{SO_4^{2-}}/w)\,z] \quad (2\text{-}97b)$$

where:

$$G_T^0 = G_{z=0}$$

with:

$$(SO_4^{2-})^0 - (SO_4^{2-})^{\infty} = (w^2 F \gamma_5 G_T{}^0)/(w^2 + k_{SO_4^{2-}} D_{SO_4^{2-}}) \quad (2\text{-}97c)$$

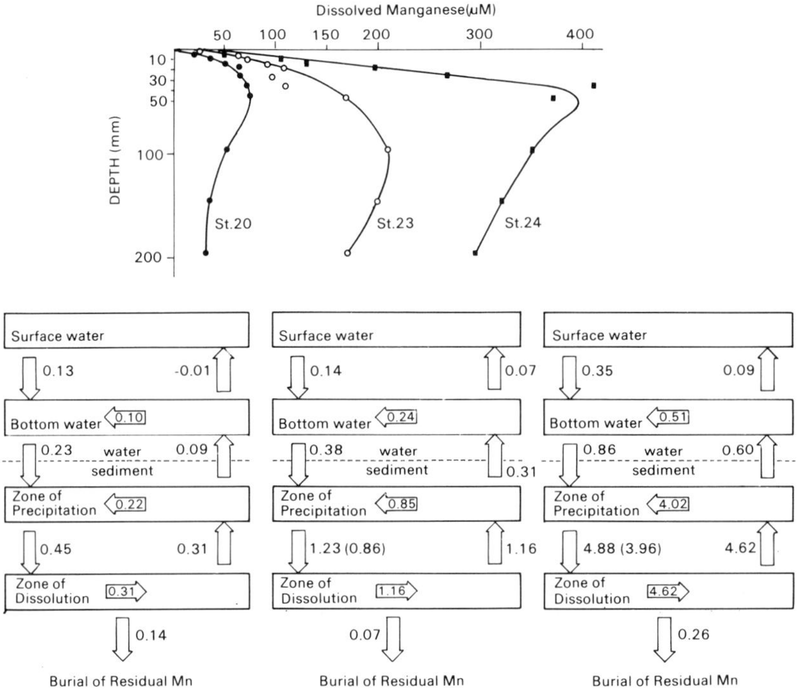

Fig. 2-27. (Top) Distribution of dissolved Mn in the pore water from the Laurentian Trough in sediments at a depth of about 350 m. Fluxes were calculated using the following parameters for the three stations 20, 23 and 24, respectively: sedimentation rate $= 0.99 \times 10^{-2}$, 0.47×10^{-2}, and 1.2×10^{-2} kg m^{-2} day^{-1}; particle mixing coefficient $= 1.4 \times 10^{-8}$, 8.6×10^{-7}, 3.2×10^{-6} m^2 day^{-1}; whole sediment diffusion coefficient for dissolved Mn $= 2.4 \times 10^{-5}$, 2.5×10^{-5}, and 2.5×10^{-5} m^2 day^{-1}; Mn content of particles collected in the sediment traps = 13.6, 29.1, and 29.1 mmole kg^{-1}; Mn content in the zone above the dissolution zone = 44.4, 79.1, and 70.6 mM kg^{-1}; Mn content of sediment below dissolution zone = 14.4, 15.5 and 21.6 mmole kg^{-1}; gradients of particulate Mn across the dissolution zone = ?, -9.9×10^5, and -12.6×10^5 mmole m^{-4}; gradients of dissolved Mn across sediment – water interface $= 3.7 \times 10^3$, 12.5×10^3, and 21.8×10^3 mmole m^{-4}.
(Bottom) Calculated fluxes of Mn in sediments at stations 20, 23 and 24 (from left to right). The numbers in parentheses are the fractions of the downward flux which is due to diffusional transport of particulate Mn within the sediment. (Modified after Sundby and Silverberg, 1985.)

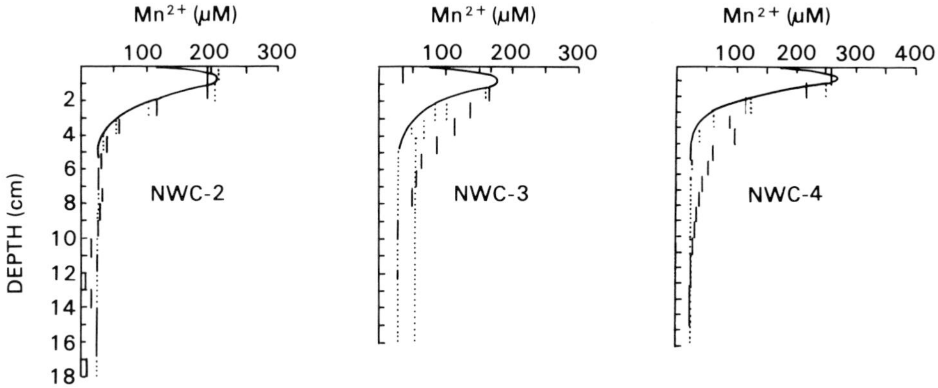

Fig. 2-28. Pore water Mn^{2+} profiles predicted by the two-dimensional cylinder model (dashed or dotted bars) and one-dimensional model (continuous profile lines), compared with the actual measured profiles in three cores from Long Island Sound at a depth of about 15 m. (Modified after Aller, 1980b.)

Both the SO_4^{2-} and C-org profiles indicate an exponential decrease with depth. This equation is commonly used to model SO_4^{2-} profiles in the sediment column not subject to bioturbation. Because of the coupling of ammonium and phosphate production with sulphate reduction, similar equations can be used for the profiles of these species assuming that only adsorption takes place with no precipitation (Berner, 1977):

$$(NH_4^+) = (NH_4^+)^0 + (N/C)_{org}\gamma_5 G_T{}^0 w^2 \times$$
$$\times\ [1 - \exp(-k_{NH_4^+}/w)\,z]/[D_{NH_4^+}k_{NH_4^+} + (1 + K_{NH_4^+})\,w^2] \qquad (2\text{-}97d)$$

$$(PO_4^T) = (PO_4^T)^0 + (P/C)_{org}\gamma_5 G_T{}^0 w^2 \times$$
$$\times\ [1 - \exp(-k_{PO_4^T}/w)\,z]/[D_{PO_4^T}k_{PO_4^T} + (1 + K_{PO_4^T})\,w^2] \qquad (2\text{-}97e)$$

where:

$(N/C)_{org}$, $(P/C)_{org}$ = average ratio of organic nitrogen and organic phosphorus to organic carbon in organic matter undergoing mineralization;

$k_{NH_4^+}$, $k_{PO_4^T}$ = first-order rate constants for mineralization of N-org and P-org;

$D_{NH_4^+}$, $D_{PO_4^T}$ = whole-sediment diffusion constants for dissolved ammonium and phosphate;

$(NH_4^+)^0$, $(PO_4^T)^0$ = concentrations of dissolved ammonium and phosphate at $z = 0$;

$K_{NH_4^+}$, $K_{PO_4^T}$ = linear adsorption constants relating to the same units (per unit volume or per unit mass), so that K's are dimensionless.

If plots of dissolved ammonium or phosphate versus sulphate prove to be straight lines, an overall stoichiometric decomposition reaction for organic matter is suggested; non-linearity indicates that stoichiometric modeling is impossible. The reaction may be caused by preferential breakdown of certain organic compounds or by precipitation reactions. If decomposition is stoichiometric, then the rate constants $k_{SO_4^{2-}} = k_{NH_4^+} = k_{PO_4^T} = k$, and G_T^0 stand for the same organic matter in Eqs. 2-97a – e. The adsorption constants K_{NH_4} and $K_{PO_4^T}$ in these equations are for linear isotherms (cf., Eq. 2-98a). Other sorption isotherms may be more appropriate (Berner, 1980; Boatman and Murray, 1982); however, low concentrations, the use of a linear isotherm constitutes a good description of the sorption behavior. Under the assumption of stoichiometric decomposition and for linear plots of dissolved ammonium or phosphate concentrations versus dissolved sulphate concentration, the slopes of these lines are given by:

$$\partial (NH_4^+)/\partial (SO_4^{2-}) = -(N/C)_{org} (w^2 + k_S D_{SO_4^{2-}}) \times \times \{\gamma_5 [k_N D_{NH_4^+} + w^2 (1 + K_{NH_4^+})]\}^{-1} \quad (2\text{-}98a)$$

$$\partial (PO_4^T)/\partial (SO_4^{2-}) = -(P/C)_{org} [k_{SO_4^{2-}} D_{SO_4^{2-}} + w^2 (1 + K_{SO_4^{2-}})] \times \times \{\gamma_5 [k_P D_{PO_4^T} + w^2 (1 + K_{PO_4^T})]\}^{-1} \quad (2\text{-}98b)$$

where:

$K_{SO_4^{2-}}$ = linear adsorption constant (dimensionless); and

$k_{SO_4^{2-}}$ = first-order rate constant for sulphate reduction.

These equations show that the C:N:P ratios of the decomposing organic matter can be derived from these slopes when the diffusion coefficients, sedimentation rates, and sorption constants are known, whereas k is found by curve fitting. Examples of such model calculations can be found in Rosenfeld (1981) for Long Island Sound (C/N = 6.7 and 13.3 for Sachem and Foam sites, respectively) and Elderfield et al. (1981b) for Narragansett Bay (C/N = 6.4 and 6.0 for Sabin Point and Jamestown North stations, respectively). These results show that the C/N ratio of the decomposing matter does not have to be equal to the Redfield ratio (6.6), a fact that was discussed earlier. As can be seen in Eq. 2-98, the adsorption can only be neglected if $kD/w^2 >> (1 + K)$. It was already mentioned that the rate of degradation is related to the rate of sedimentation (Eq. 2-14), which means that the inequality can be recasted as $D >> 25(1 + K)$ cm^2 y^{-1}. The values of $D_{NH_4^+}$ are in the order of 2.3×10^2 cm^2 y^{-1} and the values of $K_{NH_4^+}$ in the order of 1.5 (Berner, 1980). This means that the difference in the inequality relation is by about a factor of three, which implies that adsorption cannot be completely ignored.

The zone of bioturbation was considered in the models of Aller (e.g., 1980a,b, 1982). The transport of dissolved constituents can be formulated as was done in Eq. 2-40. The R-terms for sulphate reduction and ammonium production are described as a function of depth:

$$R = R^0 \exp(-\beta z) + R' \qquad (2\text{-}99)$$

where: R^0, R', and β are constants. Equation 2-99 can be substituted into Eq. 2-40. The boundary conditions are:

$r = r_1$ and $c = c^0$ (concentration in bottom water) for $z = 0$
(= sediment – water interface);
$\partial c/\partial r = 0$ for $r = r_2$; and
$\partial c/\partial z = B$ for $z = L$ (L = depth of bioturbation).

The physical meaning of r_1 and r_2 was shown in Fig. 2-16. The first condition specifies a constant concentration of solute along the sediment – water interface and within the burrow core. The second condition specifies that the concentration reaches a maximum or minimum halfway between any two burrows. The third condition follows from the notion of continuity of solute fluxes between the bioturbated and underlying zones, the observed concentration gradient being equal to B. The solution of the combined Eqs. 2-40 and 2-99 with these boundary conditions is rather complex and can be found in Aller (1980b, 1982). Aller (1982) applied the model to concentration profiles in pore water from Mud Bay, S. Carolina. The consumption and production functions for SO_4^{2-} and NH_4^+ were estimated by incubation of sediment. The dimensions of the burrows and their intervals were estimated from the type and number of infaunal individuals per unit of sediment surface in combination with their average burrow radii. The burrow depths were estimated from X-radiographs. The model concentration profile fits the observed data points quite well, as demonstrated in Fig. 2-29. The model also enables to demonstrate the influence of burrow spacings (approximate number of infaunal individuals per unit area) and burrow size (type of individuals) on the profiles, as shown in Fig. 2-30 for NH_4^+. This approach has a disadvantage because of mathematical complexity, but offers useful insight into the effects of bioturbation in a variety of sedimentary regimes. It was already mentioned that less complex models can be used (cf. Aller and Yingst, 1985).

Jørgensen (1979) presented an interesting variant by breaking Eq. 2-99a into one for the $^{32}SO_4^{2-}$ and one for the $^{34}SO_4^{2-}$ profiles. He further took into account the fact that the sulphate reduction decreases exponentially with depth, by substituting $\gamma z^{-\beta}$ for the last term in Eq. 2-99a (γ and β are constants below the zone of bioturbation). He then arrived at the following set of equations:
(1) Total SO_4^{2-} profile:

$$D_{SO_4^{2-}} [\partial^2 (SO_4^{2-})/\partial z^2] - w [\partial (SO_4^{2-}]/\partial z] - \gamma z^{-\beta} = 0 \qquad (2\text{-}100a)$$

(2) $^{32}SO_4^{2-}$ profile:

$$D_{SO_4^{2-}} [\partial^2 (^{32}SO_4^{2-})/\partial z^2] - w [\partial (^{32}SO_4^{2-})/\partial z] - \alpha\gamma z^{-\beta} [(^{32}SO_4^{2-})]/[(SO_4^{2-}) + (\alpha - 1)(^{32}SO_4^{2-})] = 0 \qquad (2\text{-}100b)$$

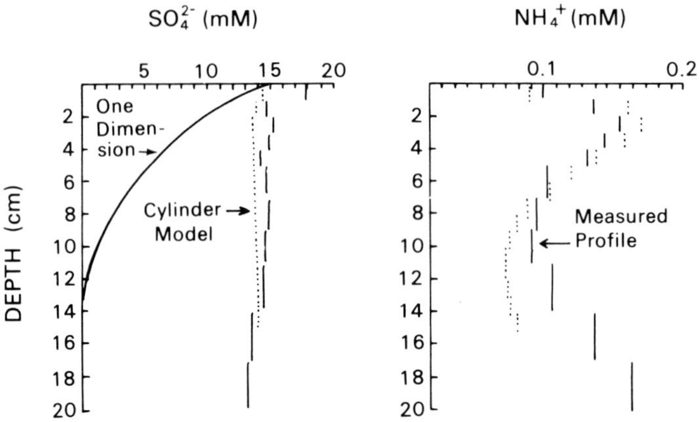

Fig. 2-29. SO_4^{2-} and NH_4^+ concentration profiles in pore water from Mud Bay sediments, S. Carolina, illustrating the behavior of the cylindrical microenvironments model. Solid vertical bars: measured concentrations; dashed vertical bars: cylinder model profiles; solid continuous curves: one-dimensional model profiles having same diffusion, reaction, and boundary constants as used in the cylinder model. The NH_4^+ profile predicted by the one-dimensional model is off-scale and not plotted. (Redrawn from Aller, 1982).

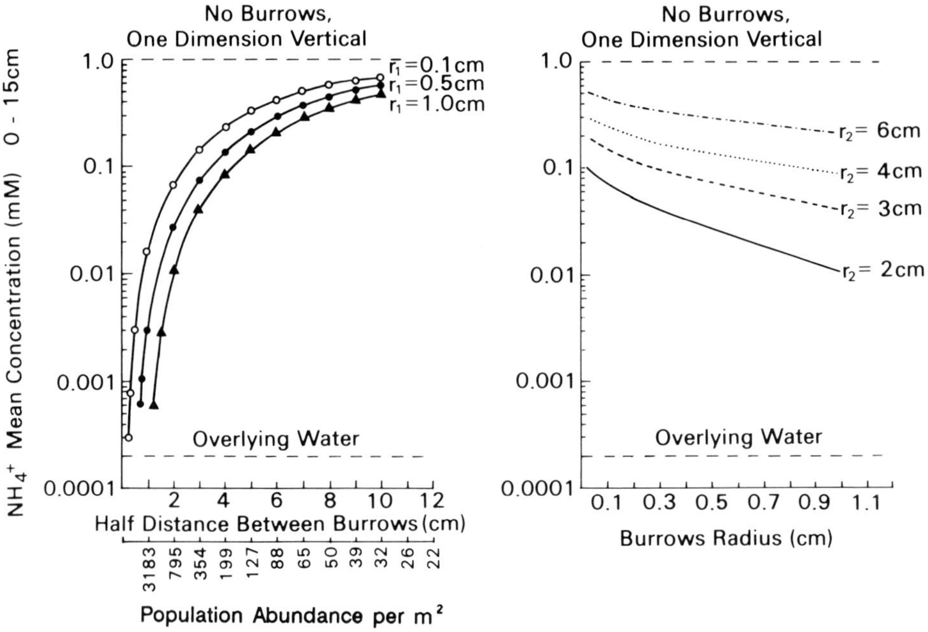

Fig. 2-30. (Left) Expected average concentration of NH_4^+ in 0–15 cm interval as a function of burrow spacing or abundance (r_2), with burrow size being fixed (r_1). Equivalent population abundances (N) per m^2 are indicated beneath r_2 values. Possible concentrations are bounded above by the concentration predicted by the one-dimensional model and below by the assumed overlying water concentration. (Right) Expected average concentration of NH_4^+ in 0–15 cm interval as a function of burrow size (r_1) with fixed spacing (r_2) between burrow axes. (Modified after Aller, 1980a.)

(3) $^{34}SO_4^{2-}$ profile:

$$D_{SO_4^{2-}}\,[\partial^2(^{34}SO_4^{2-})/\partial z^2] - w\,[\partial(^{34}SO_4^{2-})/\partial z] - \gamma z^{-\beta}\,(^{34}SO_4^{2-})/[\alpha(SO_4^{2-}) - (\alpha - 1)\,(^{34}SO_4^{2-})] = 0 \quad \text{(2-100c)}$$

where α = isotopic fractionation coefficient.

It is assumed that the diffusion coefficient for sulphate, $D_{SO_4^{2-}}$, does not depend on the isotopic composition. Consumption of SO_4^{2-} produces HS^-, which is partly converted into transient Fe(II)-sulphides and ultimately into pyrite. When it is assumed that pyrite is formed in a constant proportion (= f) to produced HS^- and that no isotopic fractionation is involved in this conversion, the following set of equations can be used to describe the profiles of dissolved HS^- and of FeS_2:
(1) Total H_2S profile:

$$D_{\Sigma H_2S}\,[\partial^2\,(\Sigma H_2S)/\partial z^2] - w\,[\partial(\Sigma H_2S)/\partial z] + (1 - f)\gamma z^{-\beta} = 0 \quad \text{(2-101a)}$$

(2) $H_2{}^{32}S$ profile:

$$D_{\Sigma H_2S}\,[\partial^2(H_2{}^{32}S)/\partial z^2] - w\,[\partial(H_2{}^{32}S)/\partial z] + \gamma z^{-\beta}\,\{\alpha(^{32}SO_4^{2-})/[(SO_4^{2-}) + (\alpha - 1)\,(^{32}SO_4^{2-})] - f(H_2{}^{32}S)/(\Sigma H_2S)\} = 0 \quad \text{(2-101b)}$$

(3) $H_2{}^{34}S$ profile:

$$D_{\Sigma H_2S}\,[\partial^2(H_2{}^{34}S)/\partial z^2] - w\,[\partial(H_2{}^{34}S)/\partial z] + \gamma z^{-\beta}\,\{(^{34}SO_4^{2-})/[\alpha(SO_4^{2-}) - (\alpha - 1)\,(^{34}SO_4^{2-})] - f\,(H_2{}^{34}S)/(\Sigma H_2S)\} = 0 \quad \text{(2-101c)}$$

Below the bioturbated zone, pyrite is not displaced from the horizon in which it is formed, so:
(1) Total FeS_2 profile:

$$- w\,\{\partial[FeS_2]/\partial z\} + f\gamma z^{-\beta} = 0 \quad \text{(2-102a)}$$

(2) $Fe^{32}S_2$ profile:

$$- w\,\{\partial[Fe^{32}S_2]/\partial z\} + f\gamma z^{-\beta}\,[(H_2{}^{32}S)/(\Sigma H_2S)] = 0 \quad \text{(2-102b)}$$

(3) $Fe^{34}S_2$ profile:

$$- w\,\{\partial[Fe^{34}S_2]/\partial z\} + f\gamma z^{-\beta}\,[(H_2{}^{34}S)/(\Sigma H_2S)] = 0 \quad \text{(2-102c)}$$

These equations were used to model the observations in sediments in Limfjorden, consisting mainly of soft silt and clay, with an organic matter content of 8–12% (on dry weight basis), deposited in water depths between 5 and 10 m, in a salinity range from 23 to 29 units. Modeling starts below the zone of bioturbation ($z > 15$ cm), which means that only the contribution of diagenesis below this zone is considered.

The input parameters that were used are as follows:

$D_{SO_4^{2-}} = 4 \times 10^{-6}$ cm^2 s^{-1};

$D_{H_2S} = 7 \times 10^{-6}$ cm^2 s^{-1};

$\gamma z^{-\beta} = 0.5 \times 10^{-10} z^{-2.15}$ mole SO_4^{2-} cm^{-2} s^{-1};

$w = 0.6 \times 10^{-8}$ cm s^{-1} (= 2 mm y^{-1});

$f \approx 0.1$ and 0.95; and

$\alpha = 1.025$ and 1.05.

The boundary conditions at $z = 15$ cm are:

(1) (SO_4^{2-}) as inferred from the actual salinity and isotopic composition as in normal seawater;

(2) (ΣH_2S) = 0 and [FeS_2] = 0, which means that the model describes only additional diagenetically-produced sulphide compounds; and

(3) gradients of sulphate and sulphide are chosen in such a manner that their total concentrations stay finite at all depths.

The results obtained using this model are shown in Fig. 2-31.

Jørgensen (1979) commented on the openness of the system for both sulphate and hydrogen sulphide. One can take the influx of SO_4^{2-} through the boundary layer of the system as a measure of its openness. One can also compare the reduction rate with the supply rate at a certain depth; supply is by diffusion plus sedimentation, i.e., by enclosed pore water. The system is then open with respect to sulphate (diffusion supply rate divided by sulphate reduction rate) × 100%. By the same token, the production rate of HS^- (= reduction rate of SO_4^{2-}) can be compared to the rate of all losses (diffusion plus sedimentation plus precipitation) at a certain depth. The openness is then defined as: (diffusion loss rate divided by sedimentation plus precipitation rates) × 100%. The sedimentation rate is usually negligible as compared to the precipitation rate.

The diffusion of the species with different isotopic composition is proportional to their gradient according to Eq. 2-33, and the isotopic ratio of the flux is:

$$F_{^{32}S}/F_{^{34}S} = [\partial(^{32}S)/\partial z]/[\partial(^{34}S)/\partial z] \qquad (2\text{-}103)$$

where: (^{32}S), (^{34}S) = fractional concentrations of SO_4^{2-} and ΣH_2S with respect to the isotopes indicated.

This can be used to demonstrate that the up and down fluxes of the different isotopes of the S-species do not have the same isotopic composition as has the *S*-pool in the layers. This is shown in Fig. 2-32. Goldhaber and Kaplan (1980) used the same notion to reconcile their observations in sediments of the Gulf of California. They argued that the openness of sediments for diffusion of SO_4^{2-} decreases from shallow water with high sedimentation rates and high organic matter content, to deep water with low sedimentation rates and low organic matter content, e.g., Santa Barbara Basin with 54 – 70% of SO_4^{2-} by diffusion and Joides Site 34 with 21 – 35%.

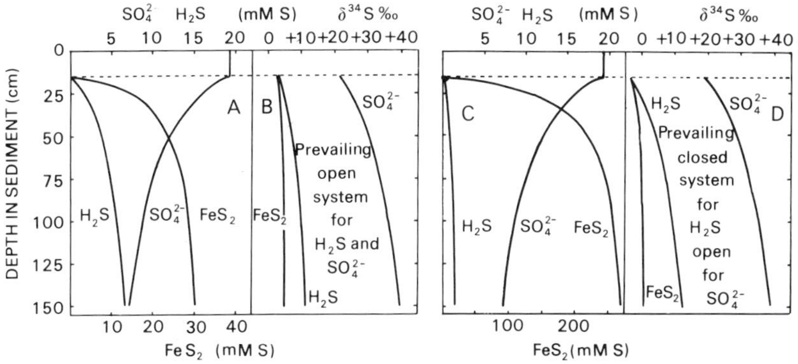

Fig. 2-31. Calculated concentrations (A,C) and isotopic compositions (B,D) of SO_4^{2-}, H_2S and FeS in a model sediment with open system conditions prevailing for sulphide (f = 0.1) and sulphate (A,B), and closed system conditions for sulphide (f = 0.95) but open for sulphate (C,D) (f = depth-dependent bacterial sulphate reduction rate). The bacterial isotope fractionation factor is held constant in all cases (α = 1.025). (After Jørgensen, 1979.)

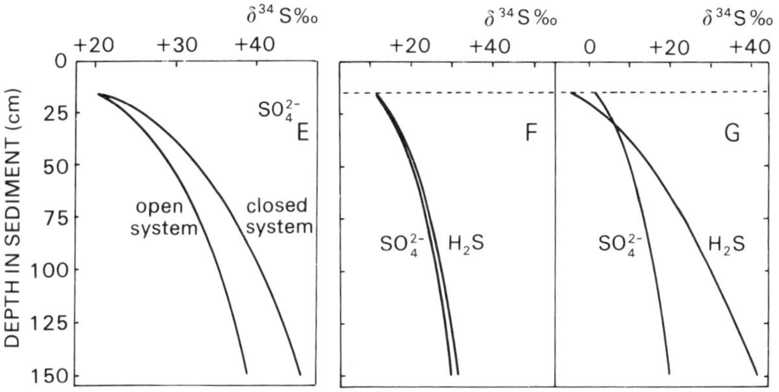

Fig. 2-32. (E) Isotopic composition of SO_4^{2-} in a sediment under open and closed system conditions; the two curves are calculated from the same SO_4^{2-} gradient. (F) Isotopic composition of diffusing SO_4^{2-} and H_2S in a model sediment under prevailing open and closed system conditions (G) for sulphide. The two S-species do not diffuse with the same isotopic composition as they occur in the pore water (cf. Fig. 2-31B,D). Both systems are considered to be open for sulphate. (Modified after Jørgensen, 1979.)

Methane production

Devol et al. (1984) presented a model for coupled sulphate reduction and methane oxidation. They considered the following reactions:

$$2\ CH_2O + SO_4^{2-} \leftrightharpoons HS^- + 2\ HCO_3^- \qquad (2\text{-}104a)$$

$$CH_4 + SO_4^{2-} \leftrightharpoons HS^- + HCO_3^- + H_2O \qquad (2\text{-}104b)$$

Dividing the sediment column into two layers, they assumed that in the upper layer SO_4^{2-} is abundant and the reaction rates can be taken to be first order with respect to the particulate organic matter and to methane, whereas in the lower layer SO_4^{2-} concentrations are so low that they are rate limiting and, consequently, the reaction rates are first order with respect to sulphate. The general diagenetic equation, Eq. 2-22a, can be applied to the SO_4^{2-} and CH_4 distributions, which for steady-state conditions, ignoring advection, and assuming that no significant CH_4 production takes place until SO_4^{2-} has been depleted, leads to the following set of equations:
Upper layer:

$$D_{SO_4^{2-}}\ [\partial^2\ (SO_4^{2-})^u/\partial z^2] - \gamma_6 k_{CH_4(u)}\ (CH_4)^u - R^0 \exp(-\beta z) = 0 \qquad (2\text{-}105a)$$

$$D_{CH_4}\ [\partial^2 (CH_4)^u/\partial z^2] - k_{CH_4(u)}(CH_4)^u = 0 \qquad (2\text{-}105b)$$

Lower layer:

$$D_{SO_4^{2-}}\ [\partial^2 (SO_4^{2-})^l/\partial z^2] - [(k_{CH_4(l)}/\alpha)^l + k_{SO_4^{2-}(l)}]\ (SO_4^{2-})^l = 0 \qquad (2\text{-}105c)$$

$$D_{CH_4}\ [\partial^2 (CH_4)^l/\partial z^2] - k_{CH_4(l)}\ (SO_4^{2-})^l/\alpha = 0 \qquad (2\text{-}105d)$$

where:

D_{CH_4} = bulk sediment diffusion coefficients for dissolved CH_4;
u, l = designation of upper and lower layer, respectively;
$k_{CH_4(u)}$ = first-order rate constant for methane oxidation for upper layer;
$k_{CH_4(l)}$ = first-order rate constant for methane oxidation for lower layer;
$k_{SO_4^{2-}(l)}$ = first-order rate constant for sulphate reduction due to organic matter oxidation when sulphate is limiting;
γ_6 = stoichiometric coefficient (number of moles of sulphate reduced per mole of methane oxidized, according to Eq. 2-104b equal to 1); and
R^0 = sulphate reduction rate due to degradation of particulate organic matter at depth $z = 0$.

The coordinate system is chosen in such a manner that the transition zone between the upper and lower layer is situated at $z = 0$, and the sediment–water interface, at $z = -h$.

The appropriate boundary conditions are:

(1) at $z = 0$: $(CH_4)^u = (CH_4)^0$; $(SO_4^{2-})^l = (SO_4^{2-})^0$; $(CH_4)^l = (CH_4)^0$;
$[\partial (CH_4)^u/\partial z] = [\partial (CH_4)^l/\partial z]$;

(2) at $z = -h$: $(CH_4)^u = 0$;

(3) at $z = \infty$: $(SO_4^{2-})^l = 0$; and

(4) moreover, the flux of sulphate through the transition zone ($z = 0$) has to be equal to the integrated rate of sulphate reduction in the lower zone, formulated as:

$$D_{SO_4^{2-}} [(SO_4^{2-})^u/z] = (k_{CH_4(l)} + k_{SO_4^{2-}(l)}) (SO_4^{2-})^0 \exp(-rz)\, dz$$

where:

$$r = [(k_{CH_4(l)} + k_{SO_4^{2-}(l)})/D_{SO_4^{2-}}]^{1/2}$$

With these boundary conditions, the solutions of Eqs. 2-105b, c, d, are as follows:

$$(CH_4)^u = A\,(CH_4)^0 \exp(nz) + B\,(CH_4)^0 \exp(-nz) \quad \text{(2-106a)}$$

$$(SO_4^{2-})^l = (SO_4^{2-})^0 \exp(-rz) \quad \text{(2-106b)}$$

$$(CH_4)^l = [k_{CH_4(l)}/(D_{CH_4}\, r^2)]\,(SO_4^{2-})^0 \exp(-rz) + n\,(CH_4)^0 (A-B)\, z + [k_{CH_4(l)}/(D_{CH_4}\, r)]\,(SO_4^{2-})^0 z - [k_{CH_4(l)}/(D_{CH_4}\, r^2)]\,(SO_4^{2-})^0 + (CH_4)^0 \quad \text{(2-106c)}$$

Finally, the solution of Eq. 2-99a becomes:

$$(SO_4^{2-})^u = (D_{CH_4}/D_{SO_4^{2-}})\,[A(CH_4)^0 \exp(nz) + B(CH_4)^0 \exp(-nz)] + (R^0/\beta^2 D_{SO_4^{2-}}) \exp(-\beta z) + [(k_{CH_4(l)} + k_{SO_4^{2-}(l)})/(D_{SO_4^{2-}}\, r)]\,(SO_4^{2-})^0 z - [k_{CH_4(u)}/(n\, D_{SO_4^{2-}}]\,(CH_4)^0 (A-B)\, z + [R^0/(\beta D_{SO_4^{2-}})]\, z - (D_{CH_4}/D_{SO_4^{2-}})\,(CH_4)^0 - R^0/(\beta^2 D_{SO_4^{2-}}) + (SO_4^{2-})^0 \quad \text{(2-106d)}$$

where:

$n = (k_{CH_4(u)}/D_{CH_4})^{1/2}$;

$A = -\exp(-2nh)/[1 - \exp(-2nh)]$; and

$B = 1/[1 - \exp(-2nh)]$.

Equations 2-106 a, b, c, d are coupled. Devol et al. (1984) applied their model to sediments in Saanich Inlet and in Skan Bay and estimated that the bulk diffusion coefficients in these sediments for sulphate are 5×10^{-6} and 3.8×10^{-6} cm^2 s^{-1}, respectively, and for methane 10×10^{-6} and 6.1×10^{-6} cm^2 s^{-1}, respectively. The concentrations of SO_4^{2-} and CH_4 at depth $z = 0$ can be estimated from the observed profiles, as is the case for the value of h, by assuming that the changes from organic matter or methane limitation to sulphate limitation will be reflected as slope changes in the SO_4^{2-} and CH_4 profiles. Alternatively, the SO_4^{2-} and CH_4 concentrations can be estimated by pinpointing the depth of the CH_4 oxidation and SO_4^{2-} reduction maxima, respectively (cf., Fig. 2-33). The rate of CH_4 oxidation from Eqs. 2-105b and c must be equal at $z = 0$, i.e., $k_{CH_4(u)}\ (CH_4)^0 = k_{CH_4(l)}\ (SO_4^{2-})^0$. Also the rate of sulphate reduction by organic matter at $z = 0$ must be equal when using Eqs. 2-105a and c, i.e., $R^0 = k_{SO_4^{2-}(l)}\ (SO_4^{2-})^0$. This eliminates two more unknowns and leaves three unknowns, i.e., the rate constants $k_{CH_4(l)}$ and $k_{SO_4^{2-}(l)}$ as well as the attenuation constant, β, that can be determined on the basis of three actual profiles. Devol et al. (1984) applied this model to profiles in sediments of the Saanich Inlet and Skan Bay. The best fits were obtained with the following values for the unknown parameters:

	Saanich Inlet	Skan Bay
$k_{CH_4(l)}$ (in 10^{-6} s^{-1})	1.4	1.9
$k_{SO_4^{2-}(l)}$ (in 10^{-7} s^{-1})	5.3	1.0
β	0.16	0.19

The relationships between the depth and SO_4^{2-} and CH_4 concentrations are shown in Fig. 2-33. The model is sensitive to the input parameters. Devol et al. (1984) stated that the diffusion coefficients control the absolute magnitude of the reaction rates, whereas the choice of the boundary conditions regulates the shape of the model profiles, i.e., the depth of peaks and slope changes. The choice of the diffusion coefficients has direct influence on the inferred reaction rates and on the estimation of the relative amount of SO_4^{2-} consumed by CH_4 oxidation.

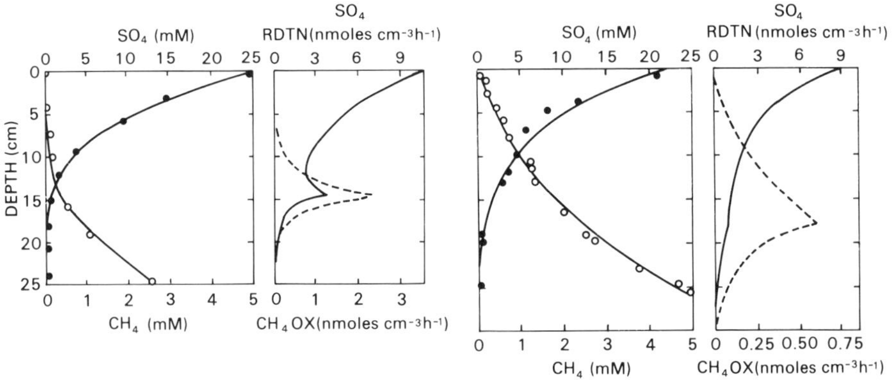

Fig. 2-33. Model fits to observed profiles for SO_4^{2-} and CH_4 in the Saanich Inlet (left two) and Skan Bay (right two). Sulphate: solid circles; methane: open circles; sulphate reduction (SO_4^{2-} RDTN) rate: solid lines; methane oxidation (CH_4 Ox) rate: dashed lines. Average values (± 1 standard deviation) of the measured sulphate reaction are indicated by bars. (Modified after Devol et al., 1984.)

Production of carbon dioxide and alkalinity

Emerson et al. (1980, 1982a) presented a stoichiometric model for the alkalinity profile in pore waters. They defined the "potential alkalinity increase" (*PAI*), and "potential total CO_2 increase" (*PCI*), as the increase in alkalinity or total CO_2 predicted for a pore-water system open to molecular diffusion, assuming initial calcite saturation and an organic matter stoichiometry as given in Eq. 2-2. For *PAI* the following relation should hold:

$$\begin{aligned} PAI = \Delta TA_{inorg} &+ \partial TA/\partial(O_2) \quad (D_{O_2}/D_{HCO_3^-}) \quad \Delta(CO_2) \; + \\ &\partial TA/\partial(NO_3^-) \quad (D_{NO_3^-}/D_{HCO_3^-}) \quad \Delta(NO_3^-) \; + \\ &\partial TA/\partial(Mn^{2+}) \quad (D_{Mn}/D_{HCO_3^-}) \quad \Delta(Mn^{2+}) \; + \\ &\partial TA/\partial(Fe^{2+}) \quad (D_{F_e}/D_{HCO_3^-}) \quad \Delta(Fe^{2+}) \; + \\ &\partial TA/\partial(SO_4^{2-}) \quad (D_{SO_4^{2-}}/D_{HCO_3^-}) \quad \Delta(SO_4^{2-}) \end{aligned} \qquad (2\text{-}107)$$

where: $\Delta\ TA_{inorg}$ = increase in titration alkalinity (*TA*) due to purely inorganic dissolution of $CaCO_3$ required to resaturate the pore waters in sediments, that are below the saturation horizon in the ocean.

A similar equation for *PCI* is obtained by substituting CO_2 for *TA* in Eq. 2-107. The partial derivatives are given in Table 2-5, with a preference for model-derived values when $CaCO_3$ is present (only given for the first three derivatives in the last two columns). Diffusion coefficients are taken from Li and Gregor (1974), without correction for porosity and tortuosity, because these effects are supposed to cancel out in the ratios of the diffusivities. The Δ concentration terms are the differences between bottom-water and pore-water concentrations of the various species.

Emerson et al. (1980, 1982a) compared the results for measured changes in *TA* and total CO_2 as a function of O_2 consumption at MANOP sites C and S in the Equatorial Pacific with the theoretical ratios of solid biogenic $CaCO_3$ to C-org in the particulate flux to the sediments. The correlations shown in Fig. 2-34 are not perfect. The (O_2) values were not measured in-situ, but inferred from the NO_3^- profile by using the following relation:

$$\Delta(O_2) = (138/106)\ (D_{NO_3^-}/D_{O_2})\ \Delta(NO_3^-) \qquad (2\text{-}108)$$

According to these authors, this may give somewhat erroneous results and, therefore, explain the outlying data points for site C. The data points for site S seem to fall within the $CaCO_3$/C-org ratio range of 0.4 – 1.2. This ratio is about 0.9 in sediment trap samples at this site. These authors showed the relation between the increase of (Ca^{2+}) and alkalinity for both sites, which leads them to believe that the ΔTA is due to dissolution of $CaCO_3$ (Fig. 2-34).

Emerson and Bender (1981) developed a model for the effect of degradation of organic matter on the preservation of $CaCO_3$ in the oxic top of marine sediments. The budget and the reactions for $CaCO_3$ and C-org are presented in Fig. 2-35. For a steady-state profile, ignoring advection, the diagenetic equation for the carbonate ion can be formulated (cf. Eq. 2-28a) as:

$$D_{CO_3^{2-}} \, [\partial^2(CO_3^{2-})/\partial z^2] + [\partial(CO_3^{2-})/\partial(\Sigma CO_2)]_{Ca^{2+}} \, R_{\Delta CO_2(ox)} + [\partial(CO_3^{2-})/\partial(\Sigma CO_2)]_{O_2} \, R_{\Delta CO_2(dis)} = 0 \quad (2\text{-}109)$$

where:

$[\partial(CO_3^{2-})/\partial(\Sigma CO_2)]_{Ca^{2+}}$ = equilibrium ratio of carbonate ion concentration to the ΣCO_2 concentration as a result of CO_2 increase from C-org oxidation alone;

$[\partial(CO_3^{2-})/\partial(\Sigma CO_2)]_{O_2}$ = ratio of carbonate ion concentration to the ΣCO_2 concentration as a result of dissolution of $CaCO_3$ alone;

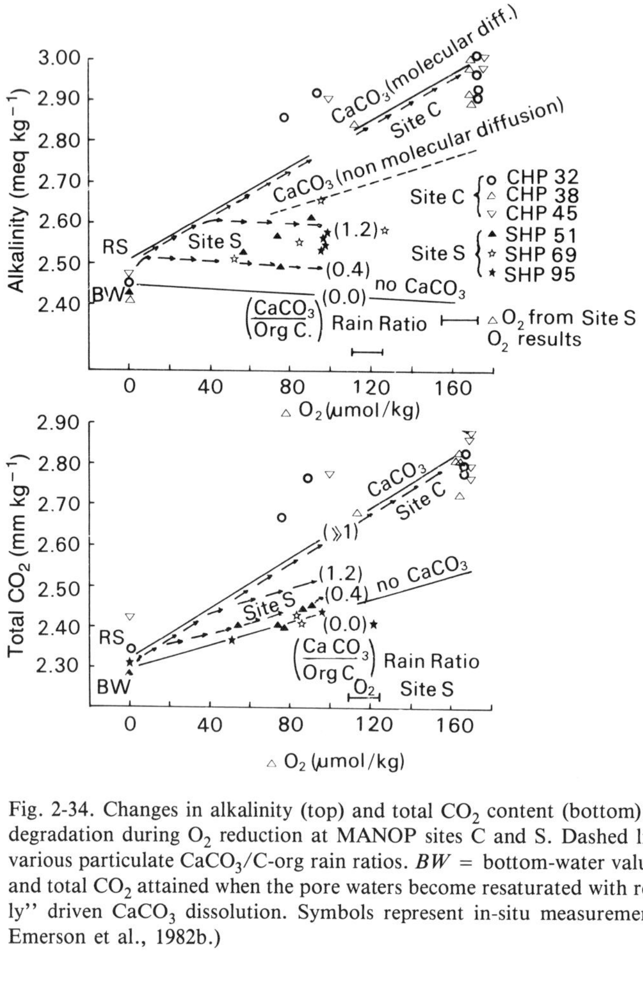

Fig. 2-34. Changes in alkalinity (top) and total CO_2 content (bottom) as a function of organic matter degradation during O_2 reduction at MANOP sites C and S. Dashed lines are the predicted values for various particulate $CaCO_3$/C-org rain ratios. *BW* = bottom-water values; *RS* = the value of alkalinity and total CO_2 attained when the pore waters become resaturated with respect to calcite by "inorganically" driven $CaCO_3$ dissolution. Symbols represent in-situ measurements at the sites. (Modified after Emerson et al., 1982b.)

$R_{\Delta CO_2(ox)}$ = production rate of total dissolved inorganic carbon as a result of degradation of organic matter; and

$R_{\Delta CO_2(dis)}$ = rate of increase of total dissolved inorganic carbon as a result of dissolution of $CaCO_3$.

The production of CO_2 due to degradation of organic matter is assumed to be first order in labile organic matter (G_T), with a rate constant of k_c:

$$R_{CO_2(ox)} = k_c[\text{C-org}] \qquad (2\text{-}110)$$

A steady-state C-org profile can be formulated as follows:

$$D_B(\partial^2[\text{C-org}]/\partial z^2) - k_c[\text{C-org}] = 0 \qquad (2\text{-}111)$$

for which the solution is:

$$[\text{C-org}] = [\text{C-org}]^0 \exp -\alpha_c z \qquad (2\text{-}112)$$

The flux of labile C-org to the sediment (F_c) equals the depth-integrated rate of its degradation:

$$[\text{C-org}]^0 = F_c/\alpha_c \qquad (2\text{-}113)$$

The rate of production of CO_2 due to dissolution of $CaCO_3$ is a function of the degree of undersaturation of $CaCO_3$ in pore waters:

$$R_{CO_2(dis)} = k_{cc}\,[(CO_3^{2-})_s - (CO_3^{2-})]^n \qquad (2\text{-}114)$$

where:

$(CO_3^{2-})_s > (CO_3^{2-})$;

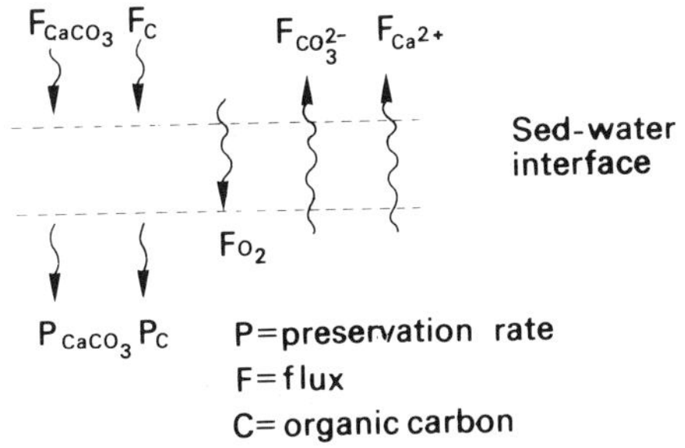

Fig. 2-35. Schematic representation of fluxes and preservation rates in C-org-rich deep-sea sediments. Fluxes of particulates and dissolved species to or across the sediment – water interface are designated by F's, whereas burial fluxes are designated by P's. The Ca mass balance is: $F_{CaCO_3} - P_{CaCO_3} = F_{Ca^{2+}}$. (Modified after Emerson and Bender, 1981.)

subscript s = at saturation;
k_{cc} = rate constant for dissolution of $CaCO_3$; and
n = experimentally-determined exponent for dissolution behavior ($n > 1$).

Although the dissolution rate is in fact not linear, these authors choose to simplify their model by assuming that $n = 1$.

The rate constant k_{cc} is related to the $CaCO_3$ content of the sediment. But for the depth interval of interest it is assumed to be constant.

The following equation is obtained upon substitution of Eqs. 2-110, 2-112 and 2-114 (with $n = 1$) into Eq. 2-109:

$$D_{CO_3^{2-}}\,[\partial^2(CO_3^{2-})/\partial z^2] + f_c k_c[\text{C-org}]^0 \exp\{-\alpha_c[\text{C-org}]\} + f_{cc}k_{cc}\,[(CO_3^{2-})_s - (CO_3^{2-})]\,(F_\partial) = 0 \qquad (2\text{-}115)$$

where:

f_c = $[\partial\,(CO_3^{2-})/(\Sigma CO_2)]_{Ca}$ (assumed to be constant over the depth interval of calculation);

f_{cc} = $[\partial(CO_3^{2-})/\partial(\Sigma CO_2)]_{O_2}$ (assumed to be constant over the depth interval of calculation);

$\Delta(CO_3^{2-})$ = $(CO_3^{2-}) - [CO_3]_s$;

(F_∂) = delta function ($=1$ for $\Delta(CO_3^{2-}) > 0$; $= 0$ for $\Delta(CO_3^{2-}) \leq 0$).

The boundary conditions are as follows:

$(CO_3^{2-}) = (CO_3^{2-})^0$ (bottom water) at $z = 0$;

$(CO_3^{2-}) = 0$ at $z = \infty$.

On introducing an inverse depth scale for the change in carbonate concentration due to $CaCO_3$ dissolution:

$$\alpha_{cc} = (k_{cc}f_{cc}/D_{CO_3^{2-}})^{1/2}$$

the following cases and their solutions can be distinguished:

(a) $\Delta(CO_3^{2-})^0 \leq 0$
i.e., bottom water is (under)saturated with respect to $CaCO_3$, which means that the pore waters will not become supersaturated.
The solution to Eq. 2-115 is then:

$$\Delta(CO_3^{2-}) = \Delta(CO_3^{2-})^0 \exp(-\alpha_{cc}z) + f_{cc}\,\alpha_c F_c\,[\exp(-\alpha_{cc}z) - \exp(-\alpha_c z)]/D_{CO_3^{2-}}(\alpha_c^2 - \alpha_{cc}^2) \qquad (2\text{-}116a)$$

(b) $\Delta(CO_3^{2-})^0 > 0$
i.e., bottom water is supersaturated with respect to $CaCO_3$.
The solution to Eq. 2-115 is then:

$$\Delta(CO_3^{2-}) = \Delta(CO_3^{2-})^0 + \{f_c F_c [1 - \exp(-\alpha_c z)]/\alpha_c D_{CO_3^{2-}}\} + Az \quad (2\text{-}116b)$$

where: A = integration constant.
Two cases can be distinguished:

(b$_1$) $\Delta(CO_3^{2-})^0 \geq f_c F_c/\alpha_c D_{CO_3^{2-}}$

Then $\Delta(CO_3^{2-})$ will never become equal to zero, $A = 0$, and the second condition will never be fulfilled for the time scale of the described processes.

(b$_2$) $\Delta(CO_3^{2-})^0 < f_c F_c/\alpha_c D_{CO_3^{2-}}$

In this case, there exists a depth, Z, where $\Delta(CO_3^{2-})$ changes sign:
For $z > Z$, $\Delta(CO_3^{2-}) = 0$, and A can be expressed as a function of Z:

$$\Delta(CO_3^{2-}) = [f_c F_c \alpha_c / D_{CO_3^{2-}} (\alpha_c^2 - \alpha_{cc}^2)] \times$$
$$\times \exp(-\alpha_c Z)\{\exp[-(z - Z)\alpha_{cc}] - \exp[-(z - Z)\alpha_c]\} \quad (2\text{-}116c)$$

For $0 \leq z \leq Z$:

$$\Delta(CO_3^{2-}) = \Delta(CO_3^{2-})^0 (1 - z/Z) + (f_c F_c/\alpha_c D_{CO_3^{2-}}) \times$$
$$\{[1 - \exp(-\alpha_c z)] - z[1 - \exp(-\alpha_c Z)]/Z\} \quad (2\text{-}116d)$$

Continuity of concentrations at depth $z = Z$ requires that the depth derivatives of (ΔCO_3^{2-}) in Eqs. 2-116c and d are equal, which leads to the relationship:

$$f_c F_c (Z\alpha_c \alpha_{cc} + \alpha_c + \alpha_{cc}) \exp(-\alpha_c z) =$$
$$(\alpha_c + \alpha_{cc}) \{f_c F_c + D_{CO_3^{2-}} \alpha_c \Delta(CO_3^{2-})^0\} \quad (2\text{-}116e)$$

Because addition of CO_2 from degradation of organic matter, at constant Ca^{2+} concentration (i.e., no reaction with $CaCO_3$), diminishes the dissolved carbonate concentration due to its reaction with CO_2 giving HCO_3^-, the value of f_c will be negative. All other terms in Eq. 2-116e being positive, therefore, the following inequality must exist in order for this equation to have a solution:

$$f_c F_c + D_{CO_3^{2-}} \alpha_c \Delta(CO_3^{2-})^0 < 0 \quad (2\text{-}117a)$$

or:

$$\Delta(CO_3^{2-})^0 < -f_c F_c/\alpha_c D_{CO_3^{2-}} \quad (2\text{-}117b)$$

Above the boundary Z, no dissolution of $CaCO_3$ occurs, which means that the flux of Ca^{2+} at $z = Z$ must be equal to the flux at $z = 0$. The flux of Ca^{2+} is then:

$$F_{Ca^{2+}} = -k_{cc} \int_{Z}^{\infty} \Delta(CO_3^{2-})\,dz \qquad (2\text{-}118)$$

This gives the following set of equations:
For $\Delta(CO_3^{2-})^0 < 0$ (undersaturated bottom water):

$$F_{Ca^{2+}} = -k_{cc} [\Delta(CO_3^{2-})^0/\alpha_{cc} + (f_c F_c)/D_{CO_3^{2-}} (\alpha_c + \alpha_{cc})\,\alpha_{cc}] > 0 \qquad (2\text{-}119a)$$

For $\Delta(CO_3^{2-}) \geq 0$ and $\Delta(CO_3^{2-})^0 < -f_c F_c/\alpha_c D_{CO_3^{2-}}$:

$$F_{Ca^{2+}} = -k_{cc} [f_c F_c \exp(-\alpha_c z)]/D_{CO_3^{2-}} (\alpha_c + \alpha_{cc})\alpha_{cc} \qquad (2\text{-}119b)$$

For $\Delta(CO_3^{2-})^0 >> 0$ and $\Delta(CO_3^{2-})^0 \geq f_c F_c/\alpha_c D_{CO_3^{2-}}$:

$$F_{Ca^{2+}} = 0 \qquad (2\text{-}119c)$$

These equations can be used to calculate the per cent $CaCO_3$ (X) in the sediments as a function of C-org fallout (rain) (F_c), the reverse depth scale for degradation of C-org (α_c), and the dissolution rate of $CaCO_3$ (k_{cc}). The F_{cc} and X can be determined by solving a set of nonlinear equations obtained by combination of Eq. 2-119a or Eq. 2-119b with equations for (a) the Ca mass balance (cf., Fig. 2-35), (b) the rate constant for $CaCO_3$ dissolution in sediments ($k = k^*X$, where k^* is the rate constant in sediments entirely composed of $CaCO_3$) and (c) the fraction X of the sediment which is $CaCO_3$ $\{X = P_{cc}/[(F_{cc}/X^0) - F_{cc} + P_{cc}]\}$, where X^0 is the $CaCO_3$ fraction at $z = 0$. Emerson and Bender (1981) then attempted to quantify the key parameters in their model:

(a) F_c/F_{cc}: based on sediment trap data, a value of $0.5-1.0$ was adopted (ratio of flux of C-org to flux of C-inorg in rain of particulates towards sediment).

(b) $1/(k_{cc}/D_{CO_3^{2-}})^{1/2}$ (scale depth for $CaCO_3$ dissolution): based on an analysis of in-situ and laboratory experiments for the dissolution rate of $CaCO_3$, in combination with the best estimate of the total sediment diffusion coefficient for carbonate ions, they adopted a value of $0.1-1.0$ mm.

(c) $1/(k_c/D_B)^{1/2}$ (scale depth for degradation of organic matter): based on a mass balance for O_2 used in the degradation of organic matter, they adopted a value of $2-20$ mm.

(d) $F_{CO_3^{2-}}/F_{Ca}$ (measure of the source of the carbonate ion which neutralizes metabolic CO_2). The F_{CO_3} is the flux of carbonate in bottom water across the sediment – water interface and F_{Ca} is the flux of calcium generated by dissolution of $CaCO_3$. When this ratio is > 1, dissolution of $CaCO_3$ is slow as compared to degradation of C-org; when it is < 1, then the $CaCO_3$ dissolution can keep up with this degradation. When the depth of organic matter degradation is small, i.e., situated very close to the sediment – water interface, then diffusion of CO_3^{2-} from bottom water is more likely to neutralize the produced CO_2.

The model results are shown in Fig. 2-36 for different values of the parameters (a – d). Emerson and Bender (1981) visualized that dissolution of $CaCO_3$ can occur in sediments deposited above the saturation depth. This is also inferred from the data obtained in the North Equatorial Atlantic (Fig. 2-37) where it is strongly suggested that corrosion of $CaCO_3$ takes place in sediments well above the defined saturation depth in the water column.

McCorkle et al. (1985) modeled the effect of organically-derived CO_2 on the ^{13}C profile of pore waters. They started out with stoichiometric equations. For the change in total dissolved inorganic CO_2 they used:

$$\Delta(\Sigma CO_2) = -(1 + \gamma_7)/\gamma_1 \text{ (oxic consumption) or}$$
$$= -(1 + \gamma_2)/\gamma_3 \text{ (nitrate consumption),} \quad (2\text{-}120)$$

where: $\gamma_7 = \Delta CaCO_3/\Delta C\text{-org}$ = stoichiometric ratio of $CaCO_3$ dissolution to C-org oxidation.

This equation can be used for the stable carbon isotopes as well. Using an average value of – 20‰ for ^{13}C of C-org and +1‰ for biogenic C-inorg, the equations become:

$$\Delta(\Sigma^{12}CO_2) = -(S_1 + \gamma_7 S_3)/\gamma_1$$

$$\Delta(\Sigma^{13}CO_2) = -(S_2 + \gamma_7 S_4)/\gamma_1$$

where: S_1 (=0.9891075) and S_3 (0.9888767) are the fractions of ^{12}C, whereas S_2 (=0.0108925) and S_4 (=0.0111233) are the fractions of ^{13}C in organic matter and calcium carbonate (relative to PDB), respectively.

For an open pore-water system, the diffusion of the species has to be considered. The diffusion coefficient for O_2 is about twice that of HCO_3^- and the maintenance of $CaCO_3$ dissolution in pore waters requires reaction of produced CO_2 with $CaCO_3$ in roughly equal amounts ($\alpha \approx 1$). This means that Eq. 2-120 becomes:

$$\Delta(\Sigma CO_2) = -[(1 + \gamma_7)/\gamma_1]\,(D_{O_2}/D_{HCO_3^-}) \approx -3\Delta(O_2) \quad (2\text{-}121a)$$

Assuming that the differences in diffusivities of species containing ^{12}C or ^{13}C are negligible, this equation can again be used to describe the changes in the pools of the carbon isotopes:

$$\Delta(\Sigma^{12}CO_2) = [(S_1 + \gamma_7 S_2)/\gamma_1]\,(D_{O_2}/D_{HCO_3^-}) \quad (2\text{-}121b)$$

$$\Delta(\Sigma^{13}CO_2) = [(S_2 + \gamma_7 S_4)/\gamma_1]\,(D_{O_2}/D_{HCO_3^-}) \quad (2\text{-}121c)$$

When the concentrations of O_2 and total dissolved inorganic carbon in bottom water as well as their isotopic compositions are known, the total dissolved inorganic

carbon and its isotopic composition in pore water can be calculated as a function of the O_2 concentration in pore water. A test of this stoichiometric model for MANOP site C revealed a reasonable agreement, whereas for the MANOP site S this agreement was not as good. McCorkle et al. (1985) then used the diagenetic Eq. 2-22a in the following form (steady state):
Pore water:

$$\partial[\phi D_i\,(\partial c_i/\partial z)]/\partial z \;-\; \partial\,(\phi v c_i)/\partial z \;-\; \gamma_i\,(1-\phi)\,k_c[\text{C-org}] = 0 \qquad (2\text{-}122a)$$

Sediment:

$$\partial\,\{(1-\phi)D_s\,\partial[\text{C-org}]/\partial z\}/\partial z \;-\; \partial\{(1-\phi)w[\text{C-org}]\}/\partial z \;-\; (1-\phi)\,k_c[\text{C-org}] = 0 \qquad (2\text{-}122b)$$

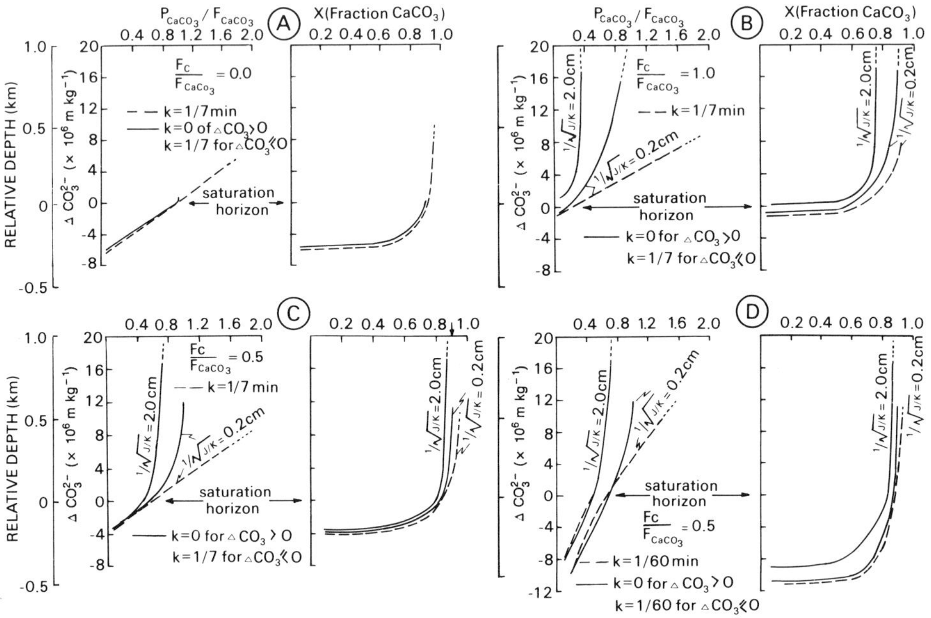

Fig. 2-36. The preservation/rain ratio (P_{CaCO_3}/F_{CaCO_3}) and the equivalent fraction *(X)* of $CaCO_3$ in sediments as a function of depth for three different molar ratios of the particulate carbon to particulate carbonate rain rate (F_C/F_{CaCO_3}). The relationship between X and P_{CaCO_3}/F_{CaCO_3} ratio is given by: $X = P_{CaCO_3}/[(F_{CaCO_3}/X^0) - F_{CaCO_3} + P_{CaCO_3}]$, where the superscript (0) indicates the fraction of $CaCO_3$ in the region of no dissolution. The relative depth scale is calculated using the relationship between the (CO_3^{2-}) content in the seawater and depth presented by Broecker and Takahashi (1978). (A) $F_C/F_{CaCO_3} = 0$, (B) $F_C/F_{CaCO_3} = 0.5$, (C) $F_C/F_{CaCO_3} = 1$, (D) as (B), except that the rate constant for $CaCO_3$ dissolution *(k)* is 1/(60 min). The solid lines are for different organic matter degradation rates $1/(j/K)^{1/2}$ (where j = first-order rate constant of C-org degradation, and K = effective mixing rate). The broken lines show the results for precipitation of $CaCO_3$ above Z (= depth of change between undersaturation and supersaturation with respect to $CaCO_3$). (Modified after Emerson and Bender, 1981.)

where:

D_s = particle mixing rate (function of z; set equal to 4.8×10^{-9} cm^2 s^{-1} for $z \leq 5$ cm and linear decrease to 1×10^{-20} cm^2 s^{-1} at z = 12 and below);
γ_i = reaction stoichiometry given by Eqs. 2-4b and 2-6b; and
c_i = dissolved species concentration (O_2, NO_3^-, $^{12}CO_2$, $^{13}CO_2$).

For denitrification, when (O_2) < 6 μM, the last term of Eq. 2-116a becomes: $-\gamma_i(1-\phi)k_c C_W$[C-org].

Boundary conditions (bottom water) used for the two MANOP sites C and S were:

$$(O_2) = 167\ \mu M;\ (\Sigma CO_2) = 2.351\ mM;\ (NO_3^-) = 35\ \mu M;\ \partial^{13}C = -0.14‰$$

The flux of O_2 (= $\phi D_{O_2}\, \partial O_2/\partial z$) is related to the rate of organic rain, F_c, by the following equation (Emerson and Bender, 1981):

$$F_{O_2} = \gamma_i\, r_c\, F_c$$

where: r_c = respiratory coefficient of sedimentary C-org ($\approx$ 0.9).

The results of this model (McCorkle et al., 1985) are shown in Fig. 2-38. The results show that the gradient in $\partial^{13}C$ in the top of the sediment depends highly on the C-org rain. Because infaunal benthos probably will use (in part) carbonate of the pore water for the formation of protective tests, these tests, when the biological effect may be ignored, will be accordingly lighter than is the case for epifaunal

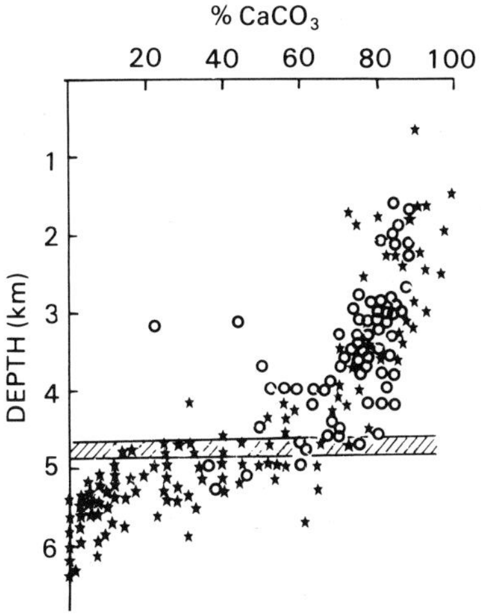

Fig. 2-37. The $CaCO_3$ content (% by weight) in sediments of the North Equatorial Atlantic Ocean as a function of depth. The dashed region indicates the depth of the "critical carbonate ion concentration" from Broecker and Takahashi (1981). (Modified after Emerson and Bender, 1981.)

species. The isotopic composition of these tests in the sedimentary record, therefore, does not solely depend on the deep-water $\partial^{13}C$ signal.

Production of dissolved phosphate and silica

It is already mentioned that profiles of dissolved phosphate and silica are often not smooth, but have irregular shapes with high or low values for several intervals. This can be caused not only by sampling or analytical artifacts, but also by differences in the amount of labile P-org or by adsorption or precipitation reactions. The appropriate diagenetic equation describing such processes in the bioturbated or physically perturbated zone was given in Eq. 2-41. The net production rate of phosphate can be formulated (Berner, 1980; Billen, 1982) as follows:

$$R_{PO_4^{3-}} = (1/\gamma_B)\, k_c G_T - k_m\, [(PO_4^{3-}) - (PO_4^{3-})_{eq}] \qquad (2\text{-}123)$$

where:

γ_8 = stoichiometric C/P ratio of labile organic matter being degraded;
k_c = rate constant for degradation of labile C-org;
k_m = linear rate constant for precipitation or dissolution of phosphate; and
$(PO_4^{3-})_{eq}$ = phosphate concentration at equilibrium with solubility-controlling solid phase.

Berner (1980) showed phosphate profiles to be expected for different values of k_m. Multiple extrema can be caused by alternating layers of sluggish and rapid formation of authigenic phosphate minerals, respectively. Nucleation of such minerals,

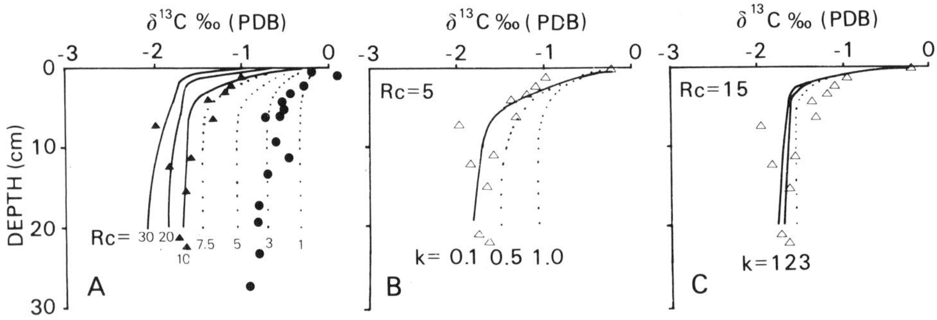

Fig. 2-38. (A) Pore water $\delta^{13}C$ data from MANOP sites C (triangles) and S (circles) plotted with a set of model $\delta^{13}C$ profiles. All model profiles have $k_C = 10^{-9}\ s^{-1}$. Carbon rain rates (left to right) are 30, 20, 10, 7.5, 5, 3, and 1 mmol C $cm^{-2}\ a^{-1}$. Solid curves indicate runs where pore water O_2 decreases to zero; dotted curves are runs where O_2 is preserved.
(B) Site C data and model $\delta^{13}C$ profiles for R_C = 5; and
(C) R_C = 15 μmol C $cm^{-2}\ a^{-1}$, with values of k_C as indicated ($\times\ 10^{-9}\ s^{-1}$). Dotted curves = O_2 preserved in pore waters; solid curves = runs where O_2 goes to zero (this is observed at site C, with an estimate of $k = 1-2 \times 10^{-9}\ s^{-1}$). (Modified after McCorkle et al., 1985.)

like apatite, can be favored by the presence of $CaCO_3$ surfaces or hindered by dissolved Mg. For nucleation of mineral phases like apatite and vivianite, the solution has to become supersaturated with respect to these phases. Once nuclei with the critical radii are present, precipitation will proceed.

Krom and Berner (1981) modeled the P distribution in Long Island Sound sediments (FOAM site). The profiles of P-org in the solid phase and of PO_4^{3-} in the dissolved phase are shown in Fig. 2-39. They chose their coordinate origin (z = 0) at the depth where practically no more bioturbation takes place (± 15 cm). Their empirical curve fits are then given by:

$$[\text{P-org}] = [\text{P-org}]^0 \exp(-0.015\,z) \tag{2-124a}$$

$$(PO_4^{3-}) = (PO_4^{3-})^0 + [(PO_4^{3-})^\infty - (PO_4^{3-})^0\}\,[1 - \exp(-0.012z)] \tag{2-124b}$$

where:

$[\text{P-org}]$ = content of organic P at depth z (in cm);
$[\text{P-org}]^0$ = content of organic P at depth $z = 0$;
(PO_4^{3-}) = concentration of dissolved phosphate species at depth z (in cm); and
$(PO_4^{3-})^0$ = concentration of dissolved phosphate species at depth $z = 0$.

These authors then used Eq. 2-55, ignoring the R_{js} term, substituted Eq. 2-124b, and solved for $R_{PO_4^{3-}}(z)$ to obtain:

$$R_{PO_4^{3-}}(z) = D_{PO_4^{3-}}(0.012)^2\,[(PO_4^{3-})^\infty - (PO_4^{3-})^0] + (1 + K_P)\,0.012\,w\,[(PO_4^{3-})^\infty - (PO_4^{3-})^0]\exp(-0.012z) \tag{2-125}$$

where: K_P = adsorption constant for phosphate.

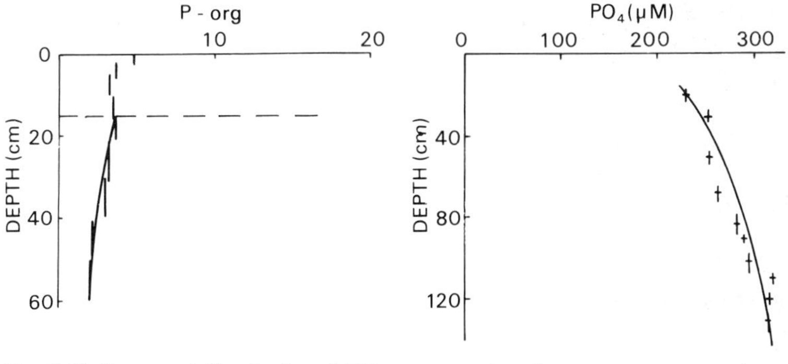

Fig. 2-39. P-org and dissolved total PO_4 concentrations in pore water versus depth in a sediment from Long Island Sound (FOAM site). The curve for P-org represents the best fit curves to (1) $[\text{P-org}] = [\text{P-org}]^0 \exp(-bz)$, and (2) $(PO_4^{3-}) = [(PO_4^{3-})^\infty - (PO_4^{3-})^0][1 - \exp(-\beta x)] + (PO_4^{3-})^0$, where β = attenuation factor. The dashed horizontal line represents the base of the zone affected by temperature fluctuations and bioturbation. (Modified after Krom and Berner, 1981.)

In order to relate the production of phosphate to the mineralization of P-org, they assumed steady-state diagenesis over the depth interval under consideration:

$$\Delta[\text{P-org}] = 1/w \int_{z_1}^{z_2} R(z)\,dz \qquad (2\text{-}126a)$$

which, after substitution of Eq. 2-125, gives:

$$\Delta[\text{P-org}] = -D_{PO_4^{3-}}(0.012)^2\,[(PO_4^{3-})^{\infty} - (PO_4^{3-})^0)] +$$

$$(1 + K_P)(0.012w)\,[(PO_4^{3-})^{\infty} - (PO_4^{3-})^0]\,\exp(-0.012z)/(0.012w)\Big|_{z_1}^{z_2}$$

(2-126b)

Based on experimental results, Krom and Berner (1981) used a value of 76 $cm^2\,y^{-1}$ for the total sediment diffusion coefficient for phosphate D_{PO_4}, and a value of 1.8 for the phosphate adsorption constant K_P. They estimated that the rate of burial of sediment particles $w = 0.05-0.1$ cm a^{-1}. Using Eq. 2-126 b, they converted the change in P-org content from μmol l^{-1} to μmol g^{-1} of dry sediment for the $15-55$ cm (true depth) interval, and found that Δ[P-org] $= 0.8$ μmol g^{-1} for $w = 0.05$ cm y^{-1} and 0.4 μmol g^{-1} for $w = 0.1$ cm y^{-1}. Considering all assumptions that are involved in this calculation, the authors concluded that these results compare reasonably well with the value of 1.4 μmol g^{-1} that can be inferred from the P-org profile in Fig. 2-39.

Vanderborght et al. (1977a) presented a model for the dissolved Si profile in North Sea muds off the coast of Belgium. The profiles based on actual measurements show that the upper 3.5 cm of these sediments is highly disturbed and there is an almost homogeneous concentration of dissolved Si in this top layer. In the underlying, more consolidated sediment, Si concentration increases with a pronounced concave-down profile. Assuming a steady-state profile, Eq. 2-28a can be applied, using a linear constant for the dissolution rate of solid amorphous silica:

$$D_{Si}\,[\partial^2(Si)/\partial z^2] - w\,\partial(Si)/\partial z] + k_{Si}\,[(Si)^{\infty} - (Si)\} = 0 \qquad (2\text{-}127)$$

where: (Si) = concentration of dissolved silicium, and k_{Si} = rate constant for dissolution of biogenous silica. Using a two-layer model for the upper (u < 3.5 cm) and the lower (l ≥ 3.5 cm) part of the sediment, the following boundary conditions were formulated:

$(Si) = (Si)^0$ at $z = 0$;

$D_{Si(u)}\,[\partial(Si)/\partial z]_{(u)} = D_{Si(l)}\,[\partial(Si)/\partial z]_l$ (continuity of fluxes at $z = 3.5$ cm); and:

$(Si)^{\infty}$ = finite for $z = \infty$

Solutions to Eq. 2-127 with these boundary conditions are:

$$(\mathrm{Si})^{\mathrm{u}} = (\mathrm{Si})^{\infty} - [(\mathrm{Si})^{\infty} - (\mathrm{Si})^{0}] \exp(wz/2D_{\mathrm{Si(u)}}) \times$$
$$\times \{[\cosh \alpha_{\mathrm{u}} (3.5 - z) + r_{D_{\mathrm{Si}}} \sinh \alpha_{\mathrm{u}} (3.5 - z)]/$$
$$[\cosh (3.5z) + r_{D_{\mathrm{Si}}} \sinh(3.5z)]\} \qquad (2\text{-}128a)$$

$$(\mathrm{Si})^{\mathrm{l}} = (\mathrm{Si})^{\infty} - [(\mathrm{Si})^{\infty} - (\mathrm{Si})^{0}] \exp \{[(w/2D_{\mathrm{Si(l)}}) - d_{\mathrm{l}}] (z - 3.5)\} \times$$
$$\{[\exp(3.5w)/2D_{\mathrm{Si(l)}}]/[\cosh(3.5d_{\mathrm{u}})] + r_{D_{\mathrm{Si}}} \sinh(3.5\alpha_{\mathrm{u}})\} \qquad (2\text{-}128b)$$

where:

$$\alpha_{\mathrm{u}} = \{[w^2/(2D_{\mathrm{Si(u)}})^2] + (k_{\mathrm{Si}}/D_{\mathrm{Si(u)}})\}^{1/2}$$
$$\alpha_{\mathrm{l}} = \{[w^2/(2D_{\mathrm{Si(l)}})^2] + (k_{\mathrm{Si}}/D_{\mathrm{Si(l)}})\}^{1/2}$$
$$r_{D_{\mathrm{Si}}} = (D_{\mathrm{Si(l)}}\,\alpha_{\mathrm{l}})/(D_{\mathrm{Si(u)}}\,\alpha_{\mathrm{u}})$$

For $w = 0.03$ cm y^{-1}, $D_{\mathrm{Si(l)}} = 31.5$ cm^2 y^{-1}, and estimating $(\mathrm{Si})^{\infty}$ from the observed profile, the unknown parameters are k_{Si} and $D_{\mathrm{Si(u)}}$. The best fit to the observed profile is given by $k_{\mathrm{Si}} = 15.8$ y^{-1} and $D_{\mathrm{Si(u)}} = 3150$ cm^2 y^{-1}. This means that the effective diffusion in the upper, disturbed part of the sediment is greater than molecular diffusion by two orders of magnitude. The curve fit of the data points is given in Fig. 2-40.

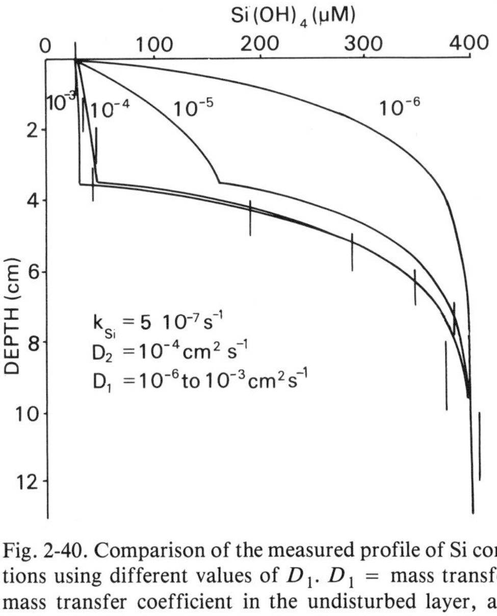

Fig. 2-40. Comparison of the measured profile of Si concentration in interstitial water and model calculations using different values of D_1. D_1 = mass transfer coefficient in the upper disturbed layer, D_2 = mass transfer coefficient in the undisturbed layer, and k_{Si} = rate constant for dissolution of opal. (Modified after Vanderborght et al., 1977a.)

Emerson et al. (1984) used a model for dissolved Si profiles in Puget Sound sediments, in which the enhanced exchange between the upper part of the sediment and bottom water by bioturbation is formulated by incorporation of a nonlocal term, as given in Eq. 2-35. For steady state and ignoring advection, the equation is:

$$D_{Si}\,[\partial^2(\mathrm{Si})/\partial z^2] - \pi\,[(\mathrm{Si}) - (\mathrm{Si})^0/\phi] + k_{Si}\,[(\mathrm{Si})^\infty - (\mathrm{Si})] = 0 \qquad (2\text{-}129)$$

where: π = depth-dependent nonlocal source parameter.
The boundary conditions are:

$(\mathrm{Si}) = (\mathrm{Si})^0/\phi$ at $z = 0$ ($(\mathrm{Si})^0$ is the Si-concentration in overlying bottom water);

$[\partial(\mathrm{Si})/\partial z]_b = G$ at $z = b$ (= gradient in (Si) at $z = b$).

The solution to this equation is:

$$(\mathrm{Si}) = \{[k_{Si}(\mathrm{Si})^\infty + \pi(\mathrm{Si})^0/\phi]/k'\}\,[1 - \exp(\alpha_{Si}z)] +$$
$$\{[(\mathrm{Si})^0/p]\exp\alpha_{Si}\,z\} + G\sinh(\alpha_{Si}\,z)/\cosh(\alpha_{Si}\,b) +$$

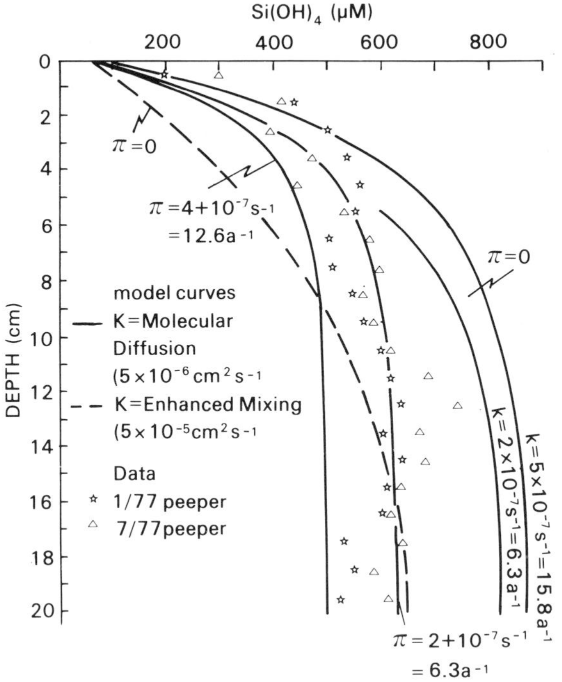

Fig. 2-41. Model solution for the depth distribution of dissolved Si for enhanced mixing ($D = 5 \times 10^{-5}$ cm^2 s^{-1}) with no nonlocal transport ($\pi = 0$) and molecular diffusion ($D = 5 \times 10^{-6}$ cm s^{-1}), with three different values of the nonlocal term ($\pi = 0, 2, 4, \times 10^{-7}$ s^{-1}). The opal dissolution rate, k, is assumed to be 5×10^{-7} s^{-1}, except where indicated. Dissolved Si concentrations in peeper samples are shown for comparison. π = nonlocal source parameter. (Modified after Emerson et al., 1984.)

$$\{[\exp\,(z + b)\,\alpha_{Si} - \exp - (z - b)]/\cosh\,(\alpha_{Si}\,b)\} \times$$
$$\times \{[k_{Si}(Si)^{\infty} + \pi(Si)^{0}/\phi]/k' - (Si)^{0}/\phi\} \qquad (2\text{-}130)$$

where: $\alpha_{Si} = (r_{DSi}/D_{Si})^{1/2}$ and $k' = (k_{Si} + \pi)$.

Values for k_{Si} were obtained from experiments and found to be between 6.3 y^{-1} and 15.8 y^{-1} at 10°C; $(Si)^0$ and $(Si)^{\infty}$ were equal to 70 and 850 μmol l^{-1}, respectively. The D_{Si} was taken as 158 $cm^2\ y^{-1}$ at 10°C. The fits of the data points for various values of the parameters are shown in Fig. 2-41. This model shows that π does not vary much in the top of the sediments and that the use of this nonlocal source parameter can be combined with the use of conventional total sediment diffusion coefficients in a one-dimensional model to obtain good fits of the observed data. More recent examples of modeling of silica diagenesis can be found in Boudreau (1990a,b).

ACKNOWLEDGEMENTS

This chapter was largely prepared during a six months stay in 1986 at the Scripps Institution of Oceanography, La Jolla, California. The writer is grateful to Dr. Joris Gieskes for his sponsorship and hospitality, and for the pleasant and fruitful discussions on many subjects in marine chemistry. Financial support was received from the Senior Scientists Awards Program (No. 259/85) of NATO Scientific Affairs Division. Comments on earlier versions of this chapter were received from Dr. Jeffrey S. Hanor, Dr. A. Lerman, Dr. Kenneth H. Nealson and Dr. J.J. Middelburg. Their suggestions were seriously considered and partly followed. The author is also grateful to Professor George V. Chilingarian and Dr. Karl H. Wolf for their valuable suggestions.

LIST OF SYMBOLS

Unless otherwise stated, the following symbols are used in model equations. The values have to be chosen in a consistent set of units.

$c_{bd} = c_{bi}$	concentration of dissolved constituent i [in mass per unit volume of total (bulk) sediment]
c_{bs}	concentration of solid constituent [in mass per unit volume of total (bulk) sediment]
$c_d = c_i$	concentration of dissolved constituent i (in mass per unit volume of pore water)
c_{is}	concentration of adsorbed constituent i (in mass per unit mass of total sediment solids)
c_{sw}	concentration of constituent i in seawater (in mass per unit volume)
d_s	mean density of total sediment solids (in mass per unit volume)
D	water depth

D_a	apparent diffusion coefficient (in area of sediment per unit time)
D_B	biodiffusion coefficient (in area of sediment per unit time)
D_{Bs}	diffusion coefficient for solid particles by biological and/or physically perturbated layers of total sediment (in area of total sediment per unit time)
D_i	diffusion coefficient of constituent *i* (in area of total sediment per unit time)
D_i^0	molecular diffusion coefficient of constituent *i* in water (in area per unit time)
D_I	irrigation coefficient (in area of total sediment per unit time)
D_s	dispersion coefficient for the solid phase
$E = \phi/(1-\phi)$	relative pore volume (dimensionless)
E^0	relative pore volume at 1 bar (dimensionless)
Eh	redox potential
f	function
f_c	change in CO_3^{2-} concentration relative to change in total CO_2 concentration due to oxidation of organic carbon alone (constant Ca^{2+} concentration)
f_{cc}	change in CO_3^{2-} concentration relative to change in total CO_2 concentration due to dissolution of $CaCO_3$ alone (no O_2 consumption)
f_i	driving force acting on ion *i*
$F_f = d_s\,\phi(1-\phi)$	formation factor
F_i	flux of constituent *i* (in mass per unit area of total sediment per unit of time; positive upward)
F_δ	delta function
g	acceleration of gravity
G_i	amount of individual metabolizable fractions of labile organic carbon
G_T	total amount of labile organic carbon
h	considered length of sediment column (length)
H	considered sediment layer or horizon
k	Boltzmann constant
k_2	rate constant for degradation of not so labile (rather refractory) organic carbon ($time^{-1}$)
$k_c = k_{c_i}$	rate constant for degradation of labile organic carbon ($time^{-1}$)
k_{cc}	rate constant for dissolution of $CaCO_3$ ($time^{-1}$)
k_{CH_4}	rate constant for oxidation of methane ($time^{-1}$)
k_i	rate constant for production or consumption of inorganic constituent *i*
k_m	rate constant for precipitation or dissolution ($time^{-1}$)
k_{N_2}	rate constant of denitrification ($time^{-1}$)

$k_{NH_4^+}$ rate constant of ammonification of organic nitrogen (time^{-1})
$k_{NO_3^{2-}}$ rate constant of nitrification in oxic zone (time^{-1})
$k_{O_2} = k_{ox}$ rate constant of oxygen respiration/consumption (time^{-1})
$k_{PO_4^{3-}}$ rate constant of mineralization of organic phosphorus (time^{-1})
k_{red} rate of oxygen respiration (consumption) in reduced zone
k_{Si} rate constant for dissolution of biogenous silica (time^{-1})
$k_{SO_4^{2-}}$ rate of sulphate reduction
$K_{NH_4^+}$ linear adsorption constant for ammonium (dimensionless)
$K_P = K_{PO_4^{3-}}$ phosphate adsorption constant (dimensionless)
L stoichiometric coefficient relating the number of moles of sulphate reduced per mole of organic carbon oxidized to CO_2
m sedimentation rate of solids (mass per unit time)
n $(k_{CH_4}/D_{CH_4})^{1/2}$
$M(z)$ mass of overlying sediment (dependent of depth z)
p fraction of constituent i exchanged between pore water and water in burrow per unit time
P^0 initial pressure (in bar)
r radius of burrow (length^{-1})
r_c respiratory coefficient for sedimentary C-org
R gas constant (= 8.3147 joules deg^{-1} mole^{-1})
R_c primary production in surface water
R_i production or consumption reaction within the sediment column affecting the concentration of constituent i (in mass per unit volume of total sediment per unit time)
$R_{i\,(ads)}$ rate of change of dissolved constituent i due to sorption (in mass per unit volume of pore water per unit time)
R_{is} rate of change of adsorbed constituent i due to sorption (in mass of total solids per unit time)
R_{js} rates of slow reactions affecting the adsorption of constituent i (in mass of total solids per unit time)
s sedimentation rate (in length per unit time)
T absolute temperature in Kelvin
t time
u_i mobility of ion i
v velocity of flow relative to the sediment – water interface
v_d velocity of flow at depth d below the sediment – water interface
v_H velocity of flow relative to a fixed horizon in the sediment
w burial rate of layer below the sediment – water interface
w_d burial rate of layer at depth d below which the porosity

	remains constant ($\phi = \phi_d$)
w_H	burial rate of horizon H
z	depth below the sediment – water interface (positive downward)
z_B	depth of bioturbated or physically perturbated layer
z_m, z_n	charge of cation and anion
α	isotopic fractionation constant
$\alpha_c \sqrt{(k_c / D_B)}$	reciprocal length parameter for degradation of organic carbon
$\alpha_{cc} = \sqrt{(k_{cc} f_{cc} / D_{CO_3^{2-}})}$	ditto for change in carbonate concentration due to $CaCO_3$ dissolution
β	attenuation factor
γ_1	$\delta(O_2)/\delta[\text{C-org}]$
γ_2	$\delta(NO_3^-)/\delta(O_2)$ in the zone of nitrification
γ_3	$\delta(NO_3^-)/\delta[\text{C-org}]$ in the oxic zone
γ_4	$\delta(NO_3^-)/\delta[\text{C-org}]$ in the zone of denitrification
γ_5	$\delta(SO_4^{2-})/\delta[\text{C-org}]$
γ_6	$\delta(SO_4^{2-})/\delta(CH_4)$
γ_7	$\delta[CaCO_3]/\delta[\text{C-org}]$
γ_8	$\delta(\Sigma PO_4^{3-})/\delta[\text{C-org}]$
ϵ	dielectric constant
η	viscosity
μ_i	chemical potential
π	nonlocal source parameter
ϕ	porosity (in volume of interconnected pore water per unit volume of total sediment)
ϕ_d	constant porosity at and below a certain depth d
ϕ_H	porosity at a certain horizon H
ϕ^0	initial porosity
Φ	electrical potential
τ	tortuosity
Ω	electrical resistivity of bulk sediment
Ω_0	electrical resistivity of pore water

REFERENCES

Aller, R.C., 1978. Experimental studies of changes produced by deposit feeders on pore water, sediment, and overlying water chemistry. *Am. J. Sci.*, 278: 1185 – 1234.

Aller, R.C., 1980a. Quantifying solute distributions in the bioturbated zone of marine sediments by defining an average microenvironment. *Geochim. Cosmochim. Acta*, 44: 1955 – 1965.

Aller, R.C., 1980b. Diagenetic processes near the sediment – water interface of Long Island Sound I. Decomposition and nutrient element geochemistry. *Adv. Geophys.*, 22: 237 – 350.

Aller, R.C., 1980c. Diagenetic processes near the sediment – water interface of Long Island Sound sediments II. Fe and Mn. *Adv. Geophys.*, 22: 351 – 415.

Aller, R.C., 1982. The effects of macrobenthos on chemical properties of marine sediments and overlying water. In: P.J. McCab and M.J.S. Tevesz (Editors), *Animal – Sediment Interactions.* Plenum Press, New York, N.Y., pp. 53 – 102.

Aller, R.C., 1983. The importance of the diffusive permeability of animal burrow linings in determining marine sediment chemistry. *J. Mar. Res.*, 41: 299 – 322.

Aller, R.C., 1984. The importance of relict burrow structures and burrow irrigation in controlling sedimentary solute distributions. *Geochim. Cosmochim. Acta*, 48: 1929 – 1934.

Aller, R.C. and Yingst, J.Y., 1985. Effects of the marine deposit-feeders *Heteromastus filiformis* (Polychaeta), *Macoma balthica* (Bivalvia), and *Tellina texana* (Bivalvia) on averaged sedimentary solute transport, reaction rates, and microbial distributions. *J. Mar. Res.*, 43: 615 – 645.

Aller, R.C., Yingst, J.Y. and Ullman , W.J., 1983. Comparative biogeochemistry of water in intertidal *Onuphis* (polychaeta) and *Upogebia* (crustacea) burrows: temporal patterns and causes. *J. Mar. Res.*, 41: 571 – 604.

Aller, R.C., Mackin, J.E., Ullman, W.J., Chen-Hou, W., Shing-Min, T., Jian-Cai, J., Yong-Nian, S. and Jia-Zhen, H., 1985. Early chemical diagenesis, sediment – water solute exchange, and storage of reactive organic matter near the mouth of the Changjiang, East China Sea. *Cont. Shelf Res.*, 4: 227 – 251.

Alperin, M.J. and Reeburgh, W.S., 1984. Geochemical observations supporting anaerobic methane oxidation. In: L. Crawford and R.S. Hanson (Editors), *Microbial Growth on C-1 Compounds.* Am. Soc. Microbiol., Washington D.C., pp. 282 – 289.

Alperin, M.J. and Reeburgh, W.S., 1985. Inhibition experiments on anaerobic methane oxidation. *Appl. Environ. Microbiol.*, 50: 940 – 945.

Anderson, L.G., Hall, P.O.J., Iverfeldt, A., Rutgers van der Loeff, M.M., Sundby, B. and Westerlund, S.F.G., 1986. Benthic respiration measured by total carbonate production. *Limnol. Oceanogr.*, 31: 319 – 329.

Applin, K.R. and Lasaga, A.C., 1984. The determination of SO_4^{2-}, $NaSO_4^-$, and $MgSO_4^0$ tracer diffussion coefficients and their application to diagenetic flux calculations. *Geochim. Cosmochim. Acta*, 48: 2151 – 2162.

Archie, G.E., 1942. The electrical resistivity log as an aid in determining some reservoir characteristics. *Trans. Am. Inst. Min. Metall. Pet. Eng.*, 146: 54 – 62.

Barnes, R.D. and Goldberg, E.D., 1976. Methane production and consumption in anoxic marine sediments. *Geology*, 4: 297 – 300.

Bender, M.L. and Heggie, D.T., 1984. Fate of organic carbon reaching the deep sea floor: a status report. *Geochim. Cosmochim. Acta*, 48: 977 – 986.

Bender, M.L., Fanning, K.A., Froelich, P.N., Heath, G.R. and Maynard, V., 1977. Interstitial nitrate profiles and oxidation of sedimentary organic matter in the eastern equatorial Atlantic. *Science*, 198: 605 – 608.

Bender, M., Martin, W., Hess, J., Sayles, F., Ball, L. and Lambert, C., 1987. A whole-core squeezer for interfacial pore-water sampling. *Limnol. Oceanogr.*, 32: 1214 – 1225.

Berelson, W.M., Hammond, D.E. and Johnson, K.S., 1987. Benthic fluxes and the cycling of biogenic silica and carbon in two southern California borderland basins. *Geochim. Cosmochim. Acta*, 51: 1345 – 1364.

Berger, W.H., Finkel, R.C., Killingley, J.S. and Marchig, C., 1983. Glacial – Holocene transition in deep-sea sediments: manganese spike in the east-equatorial Pacific. *Nature*, 303: 231 – 233.

Berner, R.A., 1976. Inclusion of adsorption in the modelling of early diagenesis. *Earth Planet. Sci. Lett.*, 29: 333 – 340.

Berner, R.A., 1977. Stoichiometric models for nutrient regeneration in anoxic sediments. *Limnol. Oceanogr.*, 22: 781 – 786.

Berner, R.A., 1980. *Early diagenesis.* Princeton Univ. Press, Princeton, N.J., 241 pp.

Berner, R.A., 1982. Burial of organic carbon and pyrite sulfur in the modern ocean: its geochemical and environmental significance. *Am. J. Sci.*, 282: 451 – 453.

Berner, R.A., 1984. Sedimentary pyrite formation: an update. *Geochim. Cosmochim. Acta*, 48: 605 – 615.

Berner, R.A., 1985. Sulphate reduction, organic matter decomposition and pyrite formation. *Philos. Trans. R. Soc. London*, A 315: 25 – 38.

Berner, R.A. and Raiswell, R., 1983. Burial of organic carbon and pyrite sulfur in sediments over Phanerozoic time: a new theory. *Geochim. Cosmochim. Acta*, 47: 855 – 862.

Berner, R.A. and Raiswell, R., 1984. C/S method for distinguishing freshwater from marine sedimentary rocks. *Geology*, 12: 365 – 368.

Berner, R.A. and Westrich, J.T., 1985. Bioturbation and the early diagenesis of carbon and sulfur. *Am. J. Sci.,* 284: 193 – 206.

Billen, G., 1978. A budget of nitrogen recycling in North Sea sediments off the Belgian coast. *Estuarine Coastal Mar. Sci.,* 7: 127 – 146.

Billen, G., 1982a. Modelling the processes of organic matter degradation and nutrients recycling in sedimentary systems. In: D.B. Nedwell and C.M. Brown (Editors), *Sediment Microbiology.* Academic Press, London, 243 pp.

Billen, G., 1982b. An idealized model of nitrogen recycling in marine sediments. *Am. J. Sci.,* 282: 512 – 541.

Boatman, C.D. and Murray, J.W., 1982. Modelling exchangeable NH_4^+ adsorption in marine sediments: Process and controls of adsorption. *Limnol. Oceanogr.,* 27: 99 – 110.

Boesen, C. and Postma, D., 1988. Pyrite formation in anoxic environments of the Baltic. *Am. J. Sci.,* 288: 575 – 603.

Bogdanov, Y., Gurvich, Y.G. and Lisitsyn, A.P., 1980a. Model for the accumulation of calcium carbonate in bottom sediments of the Pacific Ocean. *Geochem. Int.,* 17: 125 – 132.

Bogdanov, Y., Gurvich, Y.G. and Lisitsyn, A.P., 1980b. A model of organic carbon accumulation in bottom sediments of the Pacific Ocean. *Geochem. Int.,* 17: 151 – 159.

Borel-Curial, F. and Rio, M., 1988. Distribution de la silice dans les eaux interstitielles des sédiments océaniques à microfossiles silicieux. *C.R. Acad. Sci. Paris,* Sér. II, 306: 1271 – 1276.

Boudreau, B.P., 1984. On the equivalence of nonlocal and radial diffusion models for porewater irrigation. *J. Mar. Res.,* 42: 731 – 735.

Boudreau, B.P., 1986a. Mathematics of tracer mixing in sediments: I. Spatially-dependent, diffusive mixing. *Am. J. Sci.,* 286: 161 – 198.

Boudreau, B.P., 1986b. Mathematics of tracer mixing in sediments: II. Nonlocal mixing and biological conveyor-belt phenomena. *Am. J. Sci.,* 286: 199 – 238.

Boudreau, B.P., 1987. A steady-state diagenetic model for dissolved carbonate species and pH in the porewaters of oxic and suboxic sediments. *Geochim. Cosmochim. Acta,* 51: 1985 – 1996.

Boudreau, B.P., 1990a. Modelling early diagenesis of silica in non-mixed sediments. *Deep Sea Res.,* 37: 1543 – 1567.

Boudreau, B.P., 1990b. Asymptotic forms and solutions of the model for silica-opal diagenesis in bioturbated sediments. *J. Geophys. Res.,* 95C: 7367 – 7379.

Boudreau, B.P. and Scott, M.R., 1978. A model for the diffusion-controlled growth of deep-sea manganese nodules. *Am. J. Sci.,* 278: 903 – 929.

Boudreau, B.P. and Imboden, D.M., 1987. Mathematics of tracer mixing in sediments: III. The theory of nonlocal mixing within sediments. *Am. J. Sci.,* 287: 693 – 719.

Boudreau, B.P. and Ruddick, B.R., 1991. On a reactive continuum representation of organic matter diagenesis. *Am. J. Sci.:* 507 – 538.

Bouldin, D.R., 1968. Models for describing the diffusion of oxygen and other mobile constituents across the mud – water interface. *J. Ecol.,* 56: 77 – 87.

Broecker, W.S. and Peng, T.-H., 1982. *Tracers in the Sea.* Eldigio Press, Columbia Univ., Palisades, New York, N.Y., 690 pp.

Broecker, W.S. and Takahashi, T., 1978. The relationship between lysocline depth and in situ carbonate concentration. *Deep-Sea Res.,* 25A: 65 – 95.

Burdige, D.J. and Gieskes, J.M., 1983. A pore water/solid phase diagenetic model for manganese in marine sediments. *Am. J. Sci.,* 283: 29 – 47.

Burdige, D.J. and Kepkay, P.E., 1983. Determination of bacterial manganese oxidation rates in sediments using an in-situ dialysis technique. I. Laboratory studies. *Geochim. Cosmochim. Acta,* 47: 1907 – 1916.

Calvert, S.E. and Piper, D.Z., 1984. Geochemistry of ferromanganese nodules from DOMES Site A, Northern equatorial Pacific: Multiple diagenetic metal sources in the deep sea. *Geochim. Cosmochim. Acta,* 48: 1913 – 1928.

Chanton, J.P., Martens, C.S. and Goldhaber, M.B., 1987a. Biogeochemical cycling in an organic-rich coastal marine basin. 7. Sulfur mass balance, oxygen uptake and sulfide retention. *Geochim. Cosmochim. Acta,* 51: 1187 – 1200.

Chanton, J.P., Martens, C.S. and Goldhaber, M.B., 1987b. Biogeochemical cycling in an organic-rich coastal marine basin. 8. A sulfur isotopic budget balanced by differential diffusion across the sediment – water interface. *Geochim. Cosmochim. Acta,* 51: 1201 – 1208.

Christensen, J.P. and Rowe, G.T., 1984. Nitrification and oxygen consumption in northwest Atlantic deep-sea sediments. *J. Mar. Res.*, 42: 1099 – 1116.

Christensen, J.P., Devol, A.H. and Smethie, W.M., 1984. Biological enhancement of solute exchange between sediments and bottom water on the Washington continental shelf. *Cont. Shelf Res.*, 3: 9 – 23.

Christensen, J.P., Smethie, W.M. and Devol, A.H., 1987a. Benthic nutrient regeneration and denitrification on the Washington continental shelf. *Deef-Sea Res.*, 34A: 1027 – 1047.

Christensen, J.P., Murray, J.W., Devol, A.H. and Codispoti, L.A., 1987b. Denitrification in continental shelf sediments has major impact on the oceanic nitrogen budget. *Global Biogeochem. Cycles,* 1: 97 – 116.

Claypool, G.E. and Kaplan, I.R., 1974. The origin and distribution of methane in marine sediments. In: I.R. Kaplan (Editor), *Natural Gases in Marine Sediments.* Plenum Press, New York, N.Y., 324 pp.

Comans, R.N. and Middelburg, J.J., 1987. Sorption of trace metals on calcite: Applicability of the surface precipitation model. *Geochim. Cosmochim. Acta,* 51: 2587 – 2591.

Corliss, B.H., 1985. Microhabitat of benthic foraminifera within deep-sea sediments. *Nature,* 314: 435 – 438.

Crank, J., 1975. *The Mathematics of Diffusion.* Clarendon Press, Oxford. 414 pp.

Crill, P.M. and Martens, C.S., 1986. Methane production from bicarbonate and acetate in an anoxic marine sediment. *Geochim. Cosmochim. Acta,* 50: 2087 – 2097.

Crill, P.M. and Martens, C.S., 1987. Biogeochemical cycling in an organic-rich coastal marine basin. 6. Temporal and spatial variations in sulfate reduction rates. *Geochim. Cosmochim. Acta,* 51: 1175 – 1186.

Degens, E.T. and Mopper, K., 1975. Early diagenesis of organic matter in marine soils. *Soil Sci.,* 119: 65 – 72.

Degens, E.T. and Mopper, K., 1976. Factors controlling the distribution and early diagenesis of organic material in marine sediments. In: J.P. Riley and R. Chester (Editors), *Chemical Oceanography, Vol. 6,* 2nd Ed. Academic Press, London, pp. 59 – 113.

De Lange, G.J., 1984. Chemical composition of interstitial water in cores from the Madeira Abyssal Plain/eastern North-Atlantic. *Meded. Rijks Geol. Dienst,* 38: 199 – 207.

De Lange, G.J., 1986. Early diagenetic reactions in interbedded pelagic and turbiditic sediments in the Nares Abyssal Plain (western North Atlantic): Consequences for the composition of sediment and interstitial water. *Geochim. Cosmochim. Acta,* 50: 2543 – 2562.

De Lange, G.J. and Rispens, F.B., 1986. Indication of a diagenetically induced precipitate of an Fe – Si mineral in sediment from the Nares Abyssal Plain, western North Atlantic. *Mar. Geol.,* 73: 85 – 97.

Devol, A.H., Anderson, J.J., Kuivila, K. and Murray, J.W., 1984. A model for coupled sulfate reduction and methane oxidation in the sediments of Saanich Inlet. *Geochim. Cosmochim. Acta,* 48: 973 – 1004.

De Vooys, C.G.N., 1979. Primary production in aquatic environments. In: B. Bolin, E.T. Degens, S. Kempe and P. Ketner (Editors), *The Global Carbon Cycle.* Wiley, Chichester, pp. 259 – 292.

Dickson, A.G., 1981. An exact definition of total alkalinity and a procedure for the estimation of alkalinity and total inorganic carbon from titration data. *Deep-Sea Res.,* 28A: 609 – 623.

Eaton, A., 1979. The impact of anoxia on Mn fluxes in the Chesapeake Bay. *Geochim. Cosmochim. Acta,* 43: 429 – 432.

Ehrlich, H.L., 1981. *Geomicrobiology.* Marcel Dekker, New York, N.Y., 393 pp.

Einsele, G., 1977. Range, velocity, and material flux of compaction flow in growing sedimentary sequences. *Sedimentology,* 24, 639 – 655.

Elderfield, H., Luedtke, N., McCaffrey, R.J. and Bender, M., 1981a. Benthic flux studies in Narragansett Bay. *Am. J. Sci.,* 281: 768 – 787.

Elderfield, H., McCaffrey, R.J., Luedtke, N., Bender, M. and Truesdale, V.W., 1981b. Chemical diagenesis in Narragansett Bay sediments. *Am. J. Sci.,* 281: 1021 – 1055.

Emerson, S., 1985. Organic matter preservation in marine sediments. In: E.T. Sundquist and W.S. Broecker (Editors), *The Carbon Cycle and Atmospheric* CO_2*: Natural Variations Archean to Present. Geophys. Monogr.,* 32. Am. Geophys. Union, Washington, D.C., pp. 78 – 87.

Emerson, S. and Bender, M., 1981. Carbon fluxes at the sediment – water interface of the deep-sea: calcium carbonate preservation. *J. Mar. Res.,* 39: 139 – 162.

Emerson, S., Jahnke, R., Bender, M., Froelich, P., Klinkhammer, G., Bowser, C. and Setlock, G., 1980. Early diagenesis in sediments from the eastern equatorial Pacific. I. Pore water nutrient and carbonate results. *Earth Planet. Sci. Lett.,* 49: 57 – 80.

Emerson, S., Grundmanis, V. and Graham, D., 1982a. Carbonate chemistry in marine pore waters: MANOP sites C and S. *Earth Planet. Sci. Lett.*, 61: 220 – 232.

Emerson, S., Kalhorn, S., Jacobs, L., Tebo, B.M., Nealson, K.N. and Rosson, R.A., 1982b. Environmental oxidation rate of manganese (II): bacterial catalysis. *Geochim. Cosmochim. Acta,* 46: 1073 – 1079.

Emerson, S., Jahnke, R. and Heggie, D., 1984. Sediment – water exchange in shallow water estuarine sediments. *J. Mar. Res.*, 42: 709 – 730.

Emerson, S., Fisher, K., Reimers, C. and Heggie, D., 1985. Organic carbon dynamics and preservation in deep-sea sediments. *Deep-Sea Res.*, 32: 1 – 21.

Feijtel, T.C., Delaune, R.D. and Patrick, W.H., 1988. Seasonal pore water dynamics in marshes of Barataria Basin, Louisiana. *Soil Sci. Soc. Am. J.*, 32: 59 – 67.

Fenchel, T.M. and Jørgensen, B.B., 1977. Detritus food chains of aquatic ecosystems: the role of bacteria. *Adv. Microb. Ecol.*, 1: 1 – 58.

Filipek, L.H. and Owen, R.M., 1981. Diagenetic controls of phosphorus in outer continental-shelf sediments from the Gulf of Mexico. *Chem. Geol.*, 33: 181 – 204.

Fisher, T.R., Carlson, P.R. and Barber, R.T., 1982. Sediment nutrient regeneration in three North Carolina Estuaries. *Estuarine Coastal Shelf Sci.*, 14: 101 – 116.

Froelich, P.N., Klinkhammer, G.P., Bender, M.L., Luedtke, N.A., Heath, G.R., Cullen, D., Dauphin, P., Hammond, D., Hartmann, B. and Maynard, V., 1979. Early oxidation of organic matter in pelagic sediments of the eastern equatorial Atlantic: suboxic diagenesis. *Geochim. Cosmochim. Acta,* 48: 1075 – 1090.

Froelich, P.N., Bender, M.L., Luedtke, N.A., Heath, G.R. and De Vries, T., 1982. The marine phosphorus cycle. *Am. J. Sci.*, 282: 474 – 511.

Giblin, A.E. and Howarth, R.W., 1984. Porewater evidence for the dynamic sedimentary iron cycle in salt marshes. *Limnol. Oceanogr.*, 29: 47 – 63.

Gieskes, J.M., 1975. Chemistry of interstitial waters of marine sediments. *Annu. Rev. Earth Planet. Sci.*, 3: 433 – 453.

Gieskes, J.M., 1983. The chemistry of interstitial water of deep sea sediments: interpretation of Deep Sea Drilling Data. In: J.P. Riley and R. Chester (Editors), *Chemical Oceanography, Vol. 8,* 2nd Ed. Academic Press, London, pp. 222 – 269.

Goldhaber, M.B. and Kaplan, I.R., 1980. Mechanisms of sulfur incorporation and isotope fractionation during early diagenesis in sediments of the Gulf of California. *Mar. Chem.*, 9: 95 – 143.

Goloway, F. and Bender, M., 1982. Diagenetic models of interstitial nitrate profiles in deep sea suboxic sediments. *Limnol. Oceanogr.*, 27: 624 – 638.

Graybeal, A.L. and Heath, G.R., 1984. Remobilization of transition metals in surficial pelagic sediments from the eastern Pacific. *Geochim. Cosmochim. Acta,* 48: 965 – 975.

Grundmanis, V. and Murray, J.W., 1982. Aerobic respiration in pelagic marine sediments. *Geochim. Cosmochim. Acta,* 46: 1101 – 1120.

Hale, S., 1975. The role of benthic communities in the nitrogen remineralization in coastal waters. *Limnol. Oceanogr.*, 23: 684 – 694.

Hartmann, M., 1979. Evidence for early diagenetic mobilization of trace metals from discolorations of pelagic sediments. *Chem. Geol.*, 26: 277 – 293.

Hartmann, M., Müller, P.J., Suess, E. and Van Der Weijden, C.H., 1973. Oxidation of organic matter in recent marine sediments. *"Meteor" Forschungs-Ergeb., Reihe C,* 12: 74 – 86.

Hartmann, M., Müller, P.J., Suess, E. and Van Der Weijden, C.H., 1976. Chemistry of Late Quaternary sediments and their interstitial waters from the NW African continental margin. *"Meteor" Forschungs-Ergeb., Reihe C,* 24: 1 – 67.

Hartwig, E.D., 1978. Factors affecting respiration and photosynthesis by the benthic community of a subtidal siliceous sediment. *Mar. Biol.*, 46: 283 – 293.

Heath, G.R., Moore, T.C. and Dauphin, J.P., 1976. Organic carbon in deep-sea sediments. In: N.R. Andersen and A. Malahoff (Editors), *The Fate of Fossil Fuel CO_2 in the Oceans.* Plenum Press, New York, N.Y., pp. 605 – 625.

Henrichs, S.M. and Doyle, A.P., 1986. Decomposition of ^{14}C-labeled organic substances in marine sediments. *Limnol. Oceanogr.*, 31: 765 – 778.

Henrichs, S.M. and Farrington, J.W., 1984. Peru upwelling region sediments near 15°S. 1. Remineralization and accumulation of organic matter. *Limnol. Oceanogr.*, 29: 1 – 19.

Henrichs, S.M. and Reeburgh, W.S., 1987. Anaerobic mineralization of marine sediment organic matter: rates and the role of anaerobic processes in the oceanic carbon economy. *Geomicrobiol. J.,* 5: 191 – 237.

Henriksen, K., Rasmussen, M.B. and Jensen, A., 1983. Effect of bioturbation on microbial nitrogen transformations in the sediment and fluxes of ammonium and nitrate to the overlying water. In: R. Hallberg (Editor), *Environmental Biogeochemistry. Ecol. Bull. (Stockholm),* 35: 193 – 205.

Hesse, R., 1986. Diagenesis #11. Early diagenetic pore water water/sediment interaction: Modern offshore basins. *Geosci. Canada,* 13: 165 – 193.

Hinga, K.R., Sieburth, J. McN. and Heath, G.R., 1979. The supply and use of organic material at the deep-sea floor. *J. Mar. Res.,* 37: 557 – 579.

Holdren, G.R., Bricker, O.P. and Matisoff, G., 1975. A model for the control of dissolved manganese in the interstitial waters of Chesapeake Bay. In: T.M. Church (Editor), *Marine Chemistry in the Coastal Environment.* Am. Chem. Soc. Symp. Ser. 18, Washington, D.C, 710 pp.

Holland, H.D., 1978. *The Chemistry of the Atmosphere and the Oceans.* Wiley, New York, N.Y., 351 pp.

Horrigan, S.G. and Capone, D.G., 1985. Rates of nitrification and nitrate reduction in nearshore marine sediments at near ambient substrate concentrations. *Mar. Chem.,* 16: 317 – 327.

Howarth, R.W. and Jørgensen, B.B., 1984. Formation of ^{35}S-labelled elemental sulfur and pyrite in coastal marine sediments (Limfjorden and Kysing Fjord, Denmark) during short-term $^{35}SO_4^{2-}$ reduction measurements. *Geochim. Cosmochim. Acta,* 48: 1807 – 1818.

Howes, B.L., Dacey, J.W.H. and King, G.M., 1984. Carbon flow through oxygen and sulfate reduction pathways in salt marsh sediments. *Limnol. Oceanogr.,* 29: 1037 – 1051.

Imboden, D.M., 1975. Interstitial transport of solutes in non-steady state accumulating and compacting sediments. *Earth Planet. Sci. Lett.,* 27: 221 – 228.

Imboden, D.M., 1981. *Tracers and Mixing in the Aquatic Environment.* Habilitations Thesis. Swiss Federal Inst. Technol., Dübendorf, Switzerland, 137 pp.

Iversen, N. and Blackburn, T.N., 1981. Seasonal rates of methane oxidation in anoxic marine sediments. *Appl. Environ. Microbiol.,* 41: 1295 – 1300.

Iversen, N. and Jørgensen, B.B., 1985. Anaerobic methane oxidation rates at the sulfate – methane transition in marine sediments from Kattegat and Skagerrak (Denmark). *Limnol. Oceanogr.,* 30: 944 – 955.

Jahnke, R.A., 1985. A model of microenvironments in deep-sea sediments: formation and effects on porewater profiles. *Limnol. Oceanogr.,* 30: 956 – 965.

Jahnke, R.A., Heggie, D., Emerson, S. and Grundmanis, V., 1982a. Pore waters of the central Pacific Ocean: nutrient results. *Earth Planet. Sci. Lett.,* 61: 233 – 256.

Jahnke, R.A., Emerson, S.R. and Murray, J.W., 1982b. A model of oxygen reduction, denitrification, and organic matter mineralization in marine sediments. *Limnol. Oceanogr.,* 27: 610 – 623.

Jeffrey, A.LW., Pflaum, R.C., Brooks, J.M. and Sackett, W.M., 1983. Vertical trends in particulate organic carbon $^{13}C/^{12}C$ ratios in the upper water column. *Deep-Sea Res.,* 30A: 971 – 983.

Jenkins, M.C. and Kemp, W.M., 1984. The coupling of nitrification and denitrification in two estuarine sediments. *Limnol. Oceanogr.,* 29: 609 – 619.

Johnson, K.S., 1981. The calculation of ion pair diffusion coefficients: a comment. *Mar. Chem.,* 10: 195 – 208.

Jørgensen, B.B., 1977. The sulfur cycle of a coastal marine sediment (Limfjorden, Denmark). *Limnol. Oceanogr.,* 22: 814 – 832.

Jørgensen, B.B., 1979. A theoretical model of the stable sulfur isotope distribution in marine sediments. *Geochim. Cosmochim. Acta,* 43: 363 – 374.

Jørgensen, B.B., 1982. Mineralization of organic matter in the sea bed – the role of sulphate reduction. *Nature,* 296: 643 – 645.

Jørgensen, B.B., 1983. Processes at the sediment – water interface. In: B. Bolin and R.B. Cook (Editors), *The Major Biogeochemical Cycles and Their Interactions.* SCOPE 21, Wiley, Chichester, pp. 477 – 575.

Jørgensen, B.B. and Revsbech, N.P., 1985. Diffusive boundary layers and the oxygen uptake of sediments and detritus. *Limnol. Oceanogr.,* 30: 111 – 122.

Jørgensen, B.B. and Sørensen, J., 1985. Seasonal cycles of O_2, NO_3^- and SO_4^{2-} reduction in estuarine sediments: the significance of an NO_3^- reduction maximum in spring. *Mar. Ecol. Prog. Ser.,* 24: 65 – 74.

Kalhorn, S. and Emerson, S., 1984. The oxidation state of manganese in surface sediments of the deep sea. *Geochim. Cosmochim. Acta,* 48: 897 – 902.
Katz, A. and Ben-Yaakov, S., 1980. Diffusion of seawater ions. Part II. The role of activity coefficients and ion pairing. *Mar. Chem.,* 8: 263 – 280.
Kipphut, G.W. and Martens, G.S., 1982. Biogeochemical cycling in an organic-rich coastal marine basin – 3. Dissolved gas transport in methane-saturated sediments. *Geochim. Cosmochim. Acta,* 46: 2049 – 2060.
Klinkhammer, G.P., 1980. Early diagenesis in sediments from the eastern equatorial Pacific, II. Pore water model results. *Earth Planet. Sci. Lett.,* 49: 81 – 101.
Klinkhammer, G.P. and Bender, M.L., 1980. The distribution of manganese in the Pacific Ocean. *Earth Planet. Sci. Lett.,* 46: 361 – 384.
Klinkhammer, G.P., Heggie, D.T. and Graham, D.W., 1982. Metal diagenesis in oxic marine sediments. *Earth Planet. Sci. Lett.,* 61: 211 – 219.
Klump, J.V. and Martens, C.S., 1981. Biogeochemical cycling in an organic rich coastal marine basin – II. Nutrient sediment – water exchange processes. *Geochim. Cosmochim. Acta,* 45: 101 – 121.
Knowles, R., 1982. Denitrification. *Microbiol. Rev.,* 46: 43 – 70.
Krom, M.D. and Berner, R.A., 1981. The diagenesis of phosphorus in a nearshore marine sediment. *Geochim. Cosmochim. Acta,* 45: 207 – 216.
Kvenfolden, K.A., 1988. Methane hydrate – A major reservoir of carbon in the shallow geosphere? *Chem. Geol.,* 71: 41 – 51.
Lasaga, A.C., 1979. The treatment of multi-component diffusion and ion pairs in diagenetic fluxes. *Am. J. Sci.,* 279: 324 – 346.
Lasaga, A.C. and Holland, H.D., 1976. Mathematical aspects of non-steady state diagenesis. *Geochim. Cosmochim. Acta,* 40: 257 – 266.
Lasaga, A.C., Berner, R.A. and Garrels, R.M., 1985. An improved geochemical model of atmospheric CO_2 fluctuations over the past 100 million years. In: E.T. Sundquist and W.S. Broecker (Editors), *The Carbon Cycle and Atmospheric CO_2: Natural Variations Archean to Present. Geophys. Monogr.,* 32. Am. Geophys. Union, Washington, D.C., pp. 397 – 411.
Lebel, J., Silverberg, N. and Sundby, B., 1982. Gravity core shortening and pore water chemical gradients. *Deep-Sea Res.,* 29: 1365 – 1372.
Lerman, A., 1975. Maintenance of steady state in oceanic sediments. *Am. J. Sci.,* 275: 609 – 635.
Lerman, A., 1979. *Geochemical Processes.* John Wiley, New York, N.Y., 481 pp.
Lerman, A. and Lietzke, T.A., 1977. Fluxes in a growing sediment layer. *Am. J. Sci.,* 277: 25 – 37.
Leventhal, J.S., 1983. An interpretation of carbon and sulfur relationships in Black Sea sediments as indicators of environments of deposition. *Geochim. Cosmochim. Acta,* 47: 133 – 137.
Li, Y.-H. and Gregory, S., 1974. Diffusion of ions in sea water and in deep-sea sediments. *Geochim. Cosmochim. Acta,* 38: 703 – 714.
Liu, K.K. and Kaplan, I.R., 1984. Denitrification rates and availability of organic matter in marine environments. *Earth Planet. Sci. Lett.,* 68: 88 – 100.
Lovley, D.R. and Phillips, E.J.P., 1986. Organic matter mineralization with reduction of ferric iron in anaerobic sediments. *Appl. Environ. Microbiol.,* 51: 683 – 689.
Lyle, M., 1983. The brown-green color transition in marine sediments: A marker of the Fe(III) – Fe(II) redox boundary. *Limnol. Oceanogr.,* 28: 1026 – 1033.
Mach, D.L., Ramirez, A. and Holland, H.D., 1986. Organic phosphorus and carbon in marine sediments. *Am. J. Sci.,* 278: 429 – 441.
Mackin, J.E., 1987. Boron and silica behavior in salt-marsh sediments: implications for paleo-boron distributions and the early diagenesis of silica. *Am. J. Sci.,* 287: 197 – 241.
Mackin, J.E. and Aller, R.C., 1984a. Dissolved Al in sediments and waters of the east China Sea: Implications for authigenic mineral formation. *Geochim. Cosmochim. Acta,* 48: 281 – 297.
Mackin, J.E. and Aller, R.C., 1984b. Diagenesis of dissolved aluminum in organic-rich estuarine sediments. *Geochim. Cosmochim. Acta,* 48: 299 – 313.
Manheim, F.T., 1970. The diffusion of ions in unconsolidated sediments. *Earth Planet. Sci. Lett.,* 9: 307 – 309.
Manheim, F.T., 1976. Interstitial waters of marine sediments. In: J.P. Riley and R. Chester (Editors), *Chemical Oceanography, Vol. 6,* 2nd Ed. Academic Press, London, pp. 115 – 186.
Marchig, V. and Reyss, J.L., 1984. Diagenetic mobilization of manganese in the Peru Basin sediments. *Geochim. Cosmochim. Acta,* 48: 1349 – 1352.

Martens, C.S., 1984. Recycling of organic carbon near the sediment – water interface in coastal environments. *Bull. Mar. Sci.,* 35: 566 – 575.
Martens, C.S. and Berner, R.A., 1977. Interstitial water chemistry of anoxic Long Island Sound sediments. I. Dissolved gases. *Limnol. Oceanogr.,* 22: 10 – 25.
Martin, W.R. and Bender, M.L., 1988. The variability of benthic fluxes and sedimentary remineralization ratio in response to seasonally variable organic carbon rain rates in the deep sea: A modeling study. *Am. J. Sci.,* 288: 561 – 574.
McCorkle, D.C. and Emerson, S.R., 1988. The relationship between pore water carbon isotopic composition and bottom water oxygen concentration. *Geochim. Cosmochim. Acta,* 52: 1169 – 1178.
McCorkle, D.C., Emerson, S.R. and Quay, P.D., 1985. Stable carbon isotopes in marine porewaters. *Earth Planet. Sci. Lett.,* 74: 13 – 26.
McDuff, R.E., 1978. Conservative behavior of calcium and magnesium in the interstitial waters of marine sediments: identification and interpretation. Ph.D. thesis, Univ. California, San Diego, Calif., 183 pp.
McDuff, R.E. and Ellis R.A., 1979. Determining diffusion coefficients in marine sediments: a laboratory study of the validity of resistivity techniques. *Am. J. Sci.,* 279: 666 – 675.
Middelburg, J.J., 1989. A simple rate model for organic decomposition in marine sediments. *Geochim. Cosmochim. Acta,* 53: 1577 – 1581.
Middelburg, J.J., De Lange, G.J. and Van Der Weijden, C.H., 1987. Manganese solubility control in marine pore waters. *Geochim. Cosmochim. Acta,* 51: 759 – 763.
Miller, L.G., 1980. Dissolved inorganic carbon isotope ratios in reducing marine sediments. M.Sc. thesis, Univ. South. Calif., Los Angeles, Calif., 101 pp.
Müller, P.J., 1975. Zur Diagenese stickstoffhaltiger Substanzen in marinen Sedimenten unter oxydierende und reduzierende Bedingungen. Diss. Christian – Albrechts Universität, Kiel, 179 pp.
Müller, P.J. and Mangini, A., 1980. Organic carbon decomposition rates in sediments of the Pacific manganese-nodule belt dated by ^{230}Th and ^{231}Pa. *Earth Planet. Sci. Lett.,* 51: 94 – 114.
Müller, P.J. and Suess, E., 1979. Productivity, sedimentation rate, and sedimentary organic matter in the oceans. I. Organic carbon preservation. *Deep-Sea Res.,* 26A: 1347 – 1362.
Murray, J.W., Grundmanis, V. and Smethie, W.M., 1978. Interstitial water chemistry in the sediments of Saanich Inlet. *Geochim. Cosmochim. Acta,* 42: 1011 – 1026.
Murray, J.W., Emerson, S. and Jahnke, R., 1980. Carbonate saturation and the effect of pressure on the alkalinity of interstitial waters from the Guatemala Basin. *Geochim. Cosmochim. Acta,* 44: 962 – 972.
Murray, J.W., Balistrieri, L.S. and Paul, B., 1984. The oxidation state of manganese in marine sediments and ferromanganese nodules. *Geochim. Cosmochim. Acta,* 48: 1237 – 1247.
Nealson, K.H., 1986. The microbial manganese cycle. In: W.E. Krumbein (Editor), *Microbial Geochemistry.* Blackwell, Oxford, pp. 191 – 222.
Nedwell, D.B., 1984. The input and mineralization of organic carbon in anaerobic aquatic sediments. *Adv. Microb. Ecol.,* 7: 93 – 131.
Oenema, O., 1988. Early diagenesis in recent fine-grained sediments in the Eastern Scheldt. Ph.D. thesis, Univ. Utrecht, 222 pp.
Oremland, R.S. and Taylor, B.F., 1978. Sulfate reduction and methanogenesis in marine sediments. *Geochim. Cosmochim. Acta,* 42: 209 – 214.
Palmer, M.R., 1985. Rare earth elements in foraminifera tests. *Earth Planet. Sci. Lett.,* 73: 285 – 298.
Pedersen, T.F. and Calvert, S.E., 1990. Anoxia vs. productivity: what controls the formation of organic-carbon-rich sediments and sedimentary rocks. *AAPG Bull.,* 74: 454 – 466.
Pedersen, T.F. and Price, N.B., 1982. The geochemistry of manganese carbonate in Panama Basin sediments. *Geochim. Cosmochim. Acta,* 46: 59 – 68.
Pelet, R., 1981. Preservation and alteration of present-day sedimentary organic matter. *Adv. Org. Geochem.,* pp. 241 – 250.
Peng, T.-H. and Broecker, W.S., 1987. C/P ratios in marine detritus. *Global Biogeochem. Cycles,* 1: 155 – 161.
Pikal, M.J., 1971. Ion-pair formation and the theory of mutual diffusion in a binary electrolyte. *J. Phys. Chem.,* 75: 663 – 675.
Price, N.B., 1976. Chemical diagenesis in sediments. In: J.P. Riley and R. Chester (Editors), *Chemical Oceanography, Vol. 6.* 2nd ed. Academic Press, London, pp. 1 – 58.

Raiswell, R. and Berner, R.A., 1985. Pyrite formation in euxinic and semi-euxinic sediments. *Am. J. Sci.,* 285: 710–724.
Rau, G.H., Sweeney, R.E. and Kaplan, I.R., 1982. Plankton $^{13}C/^{12}C$ ratio changes with latitude: differences between northern and southern oceans. *Deep-Sea Res.,* 29A: 1035–1039.
Redfield, A.C., Ketchum, B.H. and Richards, F.A., 1963. The influence of organisms on the composition of seawater. In: M.H. Hill (Editor), *The Sea, Vol. 2.* Wiley, New York, N.Y., pp. 26–77.
Reeburgh, W.S., 1982. A major sink and flux control for methane in marine sediments: anaerobic consumption. In: K.A. Fanning and F.T. Manheim (Editors), *The Dynamic Environment of the Ocean Floor.* Lexington Books, Lexington, Mass., pp. 203–217.
Reeburgh, W.S., 1983. Rates of biogeochemical processes in anoxic sediments. *Annu. Rev. Earth Planet. Sci.,* 11: 269–298.
Reimers, C.E. and Suess, E., 1983. The partitioning of organic carbon fluxes and sedimentary organic matter decomposition rates in the ocean. *Mar. Chem.,* 13: 141–168.
Reimers, C.E., Kalhorn, S., Emerson, S.R. and Nealson, K.H., 1984. Oxygen consumption rates in pelagic sediments from the Central Pacific: First estimates from microelectrode profiles. *Geochim. Cosmochim. Acta,* 48: 903–911.
Reimers, C.E. and Smith, K.L., 1986. Reconciling measured and predicted fluxes of oxygen across the deep sea sediment–water interface. *Limnol. Oceanogr.,* 31: 305–318.
Reimers, C.E., Fischer, K.M., Merewether, R., Smith, K.L. and Jahnke, R.A., 1986. Oxygen microprofiles measured in situ in deep ocean sediments. *Nature,* 320: 741–744.
Revsbech, N.P., Jørgensen, B.B. and Blackburn, T.H., 1980a. Oxygen in the sea bottom measured with a microelectrode. *Science,* 207: 1355–1356.
Revsbech, N.P., Sørensen, J., Blackburn, T.H. and Lomholt, J.P., 1980b. Distribution of oxygen in marine sediments measured with microelectrodes. *Limnol. Oceanogr.,* 25: 403–411.
Revsbech, N.P., Jørgensen, B.B. and Brix, O., 1981. Primary production of micro algae in sediments measured by oxygen microprofile, $H^{14}CO_3$-fixation, and oxygen exchange methods. *Limnol. Oceanogr.,* 26: 717–730.
Revsbech, N.P., Madsen, B. and Jørgensen, B.B., 1986. Oxygen production and consumption in sediments determined at high spatial resolution by computer simulation of oxygen microelectrode data. *Limnol. Oceanogr.,* 31: 293–304.
Rieke, H.H. and Chilingarian, 1974. *Compaction of Argillaceous Sediments.* Elsevier, Amsterdam, 424 pp.
Ritger, S., Carson, B. and Suess, E., 1987. Methane-derived authigenic cabonates formed by subduction-induced pore-water expulsion along the Oregon/Washington margin. *Geol. Soc. Am. Bull.,* 98: 147–156.
Romankevich, Y.A., 1984. *Geochemistry of Organic Matter in the Ocean.* Springer, Berlin etc., 334 pp.
Rosenfeld, J.K., 1981. Nitrogen diagenesis in Long Island Sound sediments. *Am. J. Sci.,* 281: 436–462.
Rowe, G.T. and Denning, J.W., 1985. The role of bacteria in the turnover of organic carbon in deep-sea sediments. *J. Mar. Res.,* 43: 925–950.
Rowe, G.T. and Howarth, R., 1985. Early diagenesis of organic matter in sediments off the coast of Peru. *Deep-Sea Res.,* 32: 43–55.
Rudd, J.W.M. and Taylor, C.D., 1980. Methane cycling in aquatic environments. In: M.R. Droop and H.W. Jannasch (Editors), *Advances in Aquatic Microbiology.* Academic Press, London, pp. 77–150.
Rutgers Van Der Loeff, M.M., 1980. Time variation in interstitial nutrient concentrations at an exposed subtidal station in the Dutch Wadden Sea. *Neth. J. Sea Res.,* 14: 123–143.
Rutgers Van Der Loeff, M.M., Van Es, F.B., Helder, W. and De Vries, R.T.P., 1981. Sediment–water exchanges of nutrients and oxygen on tidal flats in the Ems–Dollard estuary. *Neth. J. Sea Res.,* 15: 113–129.
Rutgers Van Der Loeff, M.M., Andersen, L.G., Hall, P.O.J., Iverfeldt, A., Josefson, A.B., Sundby, B. and Westerlund, S.F.G., 1984. The asphyxiation technique: An approach to distinguishing between molecular diffusion and biologically mediated transport at the sediment–water interface. *Limnol. Oceanogr.,* 29: 675–686.
Sansone, F.J. and Martens, C.S., 1981. Methane production from acetate and associated methane fluxes from anoxic coastal sediments. *Science,* 211: 707–709.
Sansone, F.J. and Martens, C.S., 1982. Volatile fatty acid cycling in organic-rich marine sediments. *Geochim. Cosmochim. Acta,* 46: 1575–1589.

Sayles, F.L., 1981. The composition and diagenesis of interstitial solutions – II. Fluxes and diagenesis at the water – sediment interface in the high latitude North and South Atlantic. *Geochim. Cosmochim. Acta,* 45: 1961 – 1986.

Schink, D.R. and Guinasso, N.L., 1978. Redistribution of dissolved and adsorbed materials in abyssal marine sediments undergoing biological stirring. *Am. J. Sci.,* 278: 687 – 702.

Schink, D.R., Guinasso, N.L. and Fanning, K.A., 1975. Processes affecting the concentration of silica at the sediment – water interface of the Atlantic Ocean. *J. Geophys. Res.,* 80: 3013 – 3031.

Seitzinger, S.P., Nixon, S.W. and Pilson, M.E.Q., 1984. Denitrification and nitrous oxide production in a coastal marine ecosystem. *Limnol. Oceanogr.,* 29: 73 – 83.

Shaw, D.G., Alperin, M.J., Reeburgh, W.S. and McIntosh, D.J., 1984. Biogeochemistry of acetate in anoxic sediments of Skan Bay, Alaska. *Geochim. Cosmochim. Acta,* 48: 1819 – 1825.

Sherwood, B.A., Sager, S.L. and Holland, H.D., 1987. Phosphorus in foraminiferal sediments from the North Atlantic Ridge cores and in pure limestones. *Geochim. Cosmochim. Acta,* 51: 1861 – 1866.

Sheu, D.-D. and Presley, B.J., 1986. Variations of calcium carbonate, organic carbon and iron sulfide in anoxic sediments from the ORCA basin, Gulf of Mexico. *Mar. Geol.,* 70: 103 – 118.

Silverberg, N., Bakker, J., Edenborn, H.M. and Sundby, B., 1987. Oxygen profiles and organic carbon fluxes in Larentian Trough sediments. *Neth. J. Sea Res.,* 21: 95 – 105.

Skyring, G.W., 1987. Sulfate reduction in coastal ecosystems. *Geomicrobiol. J.,* 5: 295 – 374.

Smith, K.L. and Hinga, K.R., 1983. Sediment community respiration in the deep sea. In: G.T. Rowe (Editor), *The Sea. Vol. 8: Deep-sea Biology.* Wiley, New York, N.Y., pp. 331 – 370.

Sørensen, J., 1987. Nitrate reduction in marine sediment: pathways and interactions with iron and sulfur cycling. *Geomicrobiol. J.,* 5: 401 – 421.

Sørensen, J. and Jørgensen, B.B., 1987. Early diagenesis in sediments from Danish coastal water: microbial activity and Mn – Fe – S geochemistry. *Geochim. Cosmochim. Acta,* 51: 1583 – 1590.

Suess, E., 1979. Mineral phases formed in anoxic sediments by microbial decomposition of organic matter. *Geochim. Cosmochim. Acta,* 43: 339 – 352.

Suess, E. and Müller, P.J., 1980. Productivity, sedimentation rate and sedimentary organic matter in the oceans. In: *Proc. C.N.R.S. Symp. Benthic Boundary Layer,* Marseille, France, pp. 17 – 26.

Suess, E., Müller, P.J., Powell, H.S. and Reimers, C.E., 1980. A closer look at nitrification in pelagic sediments. *Geochem. J.,* 14: 129 – 137.

Suess, E., Carson, B., Ritger, S., Moore, J.C., Jones, M.L., Kulm, L.D. and Cochrane, G.R., 1985. Biological communities at vent sites along the subduction zone off Oregon. In: M.L. Jones (Editor), The Hydrothermal Vents of the Eastern Pacific: An Overview. *Biol. Soc. Wash. Bull.,* 6: 474 – 484.

Sundby, B. and Silverberg, N., 1985. Manganese fluxes in the benthic boundary layer. *Limnol. Oceanogr.,* 30: 372 – 381.

Sundby, B., Anderson, L.G.., Hall, P.O.J., Iverfeldt, A., Rutgers Van Der Loeff, M. and Westerlund, S.F.G., 1986. The effect of oxygen on release and uptake of cobalt, manganese, iron and phosphate at the sediment – water interface. *Geochim. Cosmochim. Acta,* 50: 1281 – 1288.

Sweeney, R.E. and Kaplan, I.R., 1980. Diagenetic sulfate reduction in marine sediments. *Mar. Chem.,* 9: 165 – 174.

Tebo, B.M., 1983. The ecology and ultrastructure of marine manganese oxidizing bacteria. Ph.D. thesis, Univ. California, San Diego, Calif., 220 pp.

Tebo, B.M. and Emerson, S., 1985. Effect of oxygen tension, Mn(II) concentration and temperature on the microbially catalyzed Mn(II) oxidation rate in a marine fjord. *Appl. Environ. Microbiol.,* 50: 1268 – 1273.

Thomson, J., Wilson, T.R.S., Culkin, F. and Hydes, D.J., 1984. Non-steady state diagenetic record in eastern equatorial Atlantic sediments. *Earth Planet. Sci. Lett.,* 71: 23 – 30.

Ullman, W.J. and Aller, R.C., 1982. Diffusion coefficients in nearshore marine sediments. *Limnol. Oceanogr.,* 27: 552 – 556.

Vanderborght, J.-P., Wollast, R. and Billen, G., 1977a. Kinetic models of diagenesis in disturbed sediments. Part 1. Mass transfer properties and silica diagenesis. *Limnol. Oceanogr.,* 22: 787 – 793.

Vanderborght, J.-P., Wollast, R. and Billen, G., 1977b. Kinetic models of diagenesis in disturbed sediments. Part 2. Nitrogen diagenesis. *Limnol. Oceanogr.,* 22: 794 – 803.

Van Cappellen, P. and Berner, R.A., 1988. A mathematical model for the early diagenesis of phosphorus and fluorine in marine sediments: apatite precipitation. *Am. J. Sci.,* 288: 289 – 333.

Van Genuchten, M.T. and Alves, W.J., 1982. Analytical solutions of the one-dimensional convective – dispersive solute transport equation. Agricultural Res. Serv. U.S. Dep. Agriculture, Washington D.C., *Tech. Bull.,* 1661, 149 pp.

Walsh, J.J., Premuzic, E.T., Gaffney, J.S., Rowe, G.T., Harbottle, G., Stoenner, R.W., Balsam, W.L., Betzer, P.R. and Macko, S.A., 1985. Organic storage of CO_2 on the continental slope off the mid-Atlantic bight, the southeastern Bering Sea, and the Peru coast. *Deep-Sea Res.,* 32A: 853 – 883.

Warford, A.L., Kosiur, D.R. and Doese, P.R., 1979. Methane production in Santa Barbara Basin sediments. *Geomicrobiol. J.,* 1: 117 – 137.

Westrich, J.T. and Berner, R.A., 1984. The role of sedimentary organic matter in bacterial sulfate reduction: The G model tested. *Limnol. Oceanogr.,* 29: 236 – 249.

Whiticar, M.J., 1978. Relationships of interstitial gases and fluids during early diagenesis in some marine sediments. Diss., Christian – Albrechts Univ., Kiel, 152 pp.

Whiticar, M.J., 1982. The presence of methane bubbles in the acoustically turbid sediments of Eckenfördner Bay, Baltic Sea. In: K.A. Fanning and F.T. Manheim (Editors), *The Dynamic Environment of the Ocean Floor.* Lexington, Mass., pp. 219 – 235.

Whiticar, M.J. and Faber, E., 1985. Methane oxidation in sediment and water column environments – Isotopic evidence. *Org. Geochem.,* 10: 759 – 768.

Whiticar, M.J., Faber, E. and Schoell, M., 1986. Biogenic methane formation in marine and freshwater environments: CO_2 reduction vs. acetate fermentation – isotopic evidence. *Geochim. Cosmochim. Acta,* 50: 693 – 709.

Willey, J.D., 1974. The effect of pressure on the solubility of amorphous silica in seawater at 0°C. *Mar. Chem.,* 2: 239 – 250.

Williams, L.A. and Crerar, D.A., 1985. Silica diagenesis. II. General mechanisms. *J. Sediment. Petrol.,* 55: 312 – 321.

Williams, P.J. leB, 1975. Biological and chemical aspects of dissolved organic material in seawater. In: J.P. Riley and G. Skirrow (Editors), *Chemical Oceanography, Vol. 2,* 2nd ed. Academic Press, London, pp. 301 – 363.

Wilson, T.R.S., 1978. Evidence for denitrification in aerobic pelagic sediments. *Nature,* 274: 354 – 356.

Wilson, T.R.S., Thomson, J., Colley, S., Hydes, D.J. and Higgs, N.C., 1985. Early organic diagenesis: the significance of progressive subsurface oxidation fronts in pelagic sediments. *Geochim. Cosmochim. Acta,* 49: 811 – 822.

Wollast, R., 1974. The silica problem. In: E.D. Goldberg (Editor), *The Sea. Vol. 5: Marine Chemistry.* Wiley, New York, N.Y., pp. 359 – 392.

Yayanos, A.A., Dietz, A.S. and Van Boxtel, R., 1979. Isolation of a deep-sea barophilic bacterium and some of its growth characteristics. *Science,* 205: 808 – 809.

Yayanos, A.A., Dietz, A.S. and Van Boxtel, R., 1981. Obligately barophilic bacterium from the Mariana Trench. *Proc. Natl. Acad. Sci. USA,* 78: 5212 – 5215.

Chapter 3

THE RECOGNITION OF SOFT-SEDIMENT DEFORMATIONS AS EARLY-DIAGENETIC FEATURES – A LITERATURE REVIEW

A.J. VAN LOON

INTRODUCTION

Geological records show that sediments (Table 3-1), in so far as they are preserved (Fig. 3-1; Van Loon, 1989), commonly become buried, then consolidated, lithified and – sometimes – metamorphosed. The early-diagenetic stage, i.e., the period between sedimentation and lithification, includes some important changes within most types of sediments; the most obvious is the common reduction of thickness by compaction, a process which includes reduction of pore size, particularly in fine-grained sediments (cf. Rieke and Chilingarian, 1974), and, most commonly, expulsion of pore water. Compaction, a process which is due to the vertical force exerted by gravity upon the overlying sediment layers, changes the original geometry of the sediment (see, e.g., Chilingarian and Wolf, 1975, 1976; Kraus, 1988) and might, therefore, be considered as a deformational process (on a microscale) affecting the grain-to-grain contact relations of huge masses of sediments. The internal structure of a sediment which is being compacted, however, does not really change except for a certain loss of volume due to the decrease of pore space and to a reorientation of grains sometimes with a well-developed *a*/*b*-plane. This results most commonly in a thinning of the sedimentary unit involved. Units with lateral transitions between sediment types that react differently to compaction (e.g., due to a difference in the original pore space) will thus show a different degree of thinning in the course of time, a phenomenon known as "differential compaction" (Fig. 3-2). As compaction forms part of the natural development of most sediment types, it is, however, not generally considered as a form of (early-diagenetic) deformation, even though it meets the most important criterion, i.e., a change in the original grain-to-grain relationship. Compaction is generally considered as a diagenetic process (cf. Wolf and Chilingarian, 1976).

There are, however, many more processes that do affect the internal structure of the sediment during early diagenesis (Table 3-2). The structures produced by these processes (Table 3-2) range from very simple to extremely complicated. Several types of structures may be formed that have their counterpart in lithified rocks, where such structures are generally due to endogenic forces (Table 3-3) and, consequently, used for interpretation of the structural history of the area. It is, therefore, important to distinguish between such "tectonic" structures on one hand and early-diagenetic deformations on the other (cf. Meier and Thomas, 1969). Such a distinction, of course, requires that early-diagenetic structures be recognized as such.

Diagenetic deformation structures that are formed after lithification do exist (e.g., solution breccias, creep). Such structures only rarely show similarities with soft-rock deformations and are, therefore, not considered here.

TABLE 3-1

General characteristics of the five categories of sediments and sedimentary rocks* (after Braakman, 1974)

Category	Schematic genesis	Examples
Clastic sediment	Deposition of particles due to the presence (or absence) of action by water, wind, or ice	sand/sandstone* clay/shale* diamict/diamictite* mud/mudstone*
Chemical sediment	Precipitation from a saturated solution	salt/rock salt* oölite/oölite* lime/limestone*
Organogenic sediment	Gradual accumulation (either or not in-situ) of inorganic material formed originally by an organism	coral colony*/reef* diatomooze/diatomite* algae/algal mat* shell layer/coquina*
Organic sediment	Organic material, composed of faeces of organisms, (parts of) dead organisms, or conversion products of dead organisms	guano/? wood/coal* bacteria, etc./oil plants/natural gas resin/amber*
Pyroclastic sediment	Fragments transported through the air during volcanic eruptions	volcanic ash/tuff* bombs/lapillistone*

* The terms "sediment" and "sedimentary rock" are used by some authors for unconsolidated and for lithified material, respectively. Other authors use the terms as synonyms. The right-hand column presents both unconsolidated and lithified examples; the latter group is marked with an asterisk.

Terminology

It should be emphasized that the term "early-diagenetic deformation" is frequently used as a synonym for "soft-sediment deformation". This is not entirely correct as the latter category also includes synsedimentary deformation structures (formed during the depositional process and, thus, before diagenesis in a strict sense). It should be noticed, however, that there is a gradual transition from synsedimentary into early-diagenetic structures. A well-known example of a structure that may start as a synsedimentary deformation but that may continue to be deformed during a more or less extended early-diagenetic period, is the load cast.

A short time span may be present between the deposition of a sediment (the synsedimentary stage) and the moment that the sediment becomes covered by new layers. An example of such a situation is found in tidal flats, where some parts undergo sedimentation in cycles, e.g., during each flood tide (upper tidal flats) or during each spring tide (salt marshes). If this sedimentation pattern results in a succession that, in the geological record, does not show real hiatuses or other signs of important interruptions in sedimentation, the period between deposition of a specific layer and its coverage by a new one represents a period during which the

TABLE 3-2

Early-diagenetic deformational processes

Type	Deformational agent	Examples of structures
Bioturbation	Organisms	Root-induced fissures Burrows
Cryoturbation	Freezing/melting alternation	Dead-ice faults Pingos
Glaciturbation	Moving ice	Glacial folds Ice-push induced kinks
Thermoturbation	Temperature-induced stress	Heat-induced cracks Cold-induced cracks
Graviturbation	Gravity-induced processes	Slump Gravifossum
Hydroturbation	Water	Desiccation cracks Wave-induced breccias
Chemoturbation	Chemical reaction(s)	Crystal-growth imprints Solution infilling
Atmoturbation	Weather	Rain-drop imprints Imprints made by wind-blown fragments
Endoturbation	Endogenic activity	Fault breccias Convolutions
Astroturbation	Processes in the universe	Meteorite crater Meteorite imprint

sediment may be affected by deformational agents that affect the (temporary) sedimentary surface. This stage was termed the "metadepositional" stage by Nagtegaal (1963) and has also been termed "metasedimentary" stage by later authors. It is obvious that more fortunate terms would have been possible, because confusion with a stage of metamorphosis is not unlikely. Anyway, "metadepositional" structures (Fig. 3-3) belong to the wider category of early-diagenetic structures. A well-described deformational structure that is formed at least partly during the period between sedimentation and coverage by new sediment, is the gravifossum. This structure starts as a load cast but continues loading in the highly unstable, thixotropic, water-saturated substratum, which finally results in subsidence of a block along fault planes. The gravifossum, therefore, might be considered as a sedimentary (micro)graben.

Another term commonly used in the literature is "penecontemporaneous deformation". This term is used in a very loose sense. In the author's opinion, the prefix "pene" does exclude synsedimentary structures: the Greek prefix "pene" means "almost"; "penecontemporaneous", therefore, means "more or less at the same time" (as sedimentation). Another restriction is that the term was never applied to deformations of lithified sediments. This inevitably leads to the conclusion that the term "penecontemporaneous" is, at least in this context, synonymous with "early diagenetic". The latter term is older, more generally used, better known, applicable to more situations, and – most importantly – well-defined. Use of the term "penecontemporaneous deformations" should, therefore, be avoided.

STUDIES ON EARLY-DIAGENETIC DEFORMATIONS

Although geology is now relatively well established, certain problems have not yet been solved, partly because geologists were interested more in other topics. Diagenetic features, such as sedimentary deformations, were certainly not the first aspects that drew the attention of geologists, implying that research into early-

Fig. 3-1A.

Fig. 3-1. Deformation structures with typically an extremely small and an extremely large preservational potential (Figs. 3-1A and 1B, respectively).
(A) Resting mark of a fish on the surface of an intertidal sand body at Genêts, Mont-Saint-Michel Bay, France.
(B) Tight, asymmetric fold in ice-pushed glaciofluvial sediments of Weichselian age (sand pit near Andebølle, Denmark).

TABLE 3-3

Examples of deformation structures occurring in unconsolidated sediments and in lithified sedimentary rocks, which show similar aspects in spite of genetically different origins

Structure	Most common origin in unconsolidated sediments	Most common origin in lithified material
Fault	Dead-ice melting Differential compaction Extreme loading	Tectonic activity
Fold	Glacial push Mass movement Thixotropic behavior	Tectonic activity
Breccia	Wave action Mass movement Differential rigidity	Faulting Partial solution
Kinking	Glacial push Compaction-induced lateral pressure	Tectonic activity

Fig. 3-2.

diagenetic deformations is a fairly recent development. Even when it had become clear that investigation of these structures could contribute to a better understanding of geological history, there were periods of less and of more intensive research (most disciplines face comparable oscillations of interest in specific topics). These periods of more and of less interest in early-diagenetic deformations can be reconstructed from a survey of the literature. The following periods (with, of course, gradual transitions) can be distinguished:

(1) From the beginning of the nineteenth to the middle of the twentieth century (with only little – though some – interest in early-diagenetic structures);
(2) from about 1950 to about 1960 (when sedimentology developed an interest in diagenetic processes was raised);
(3) from about 1960 to about 1970 (when sedimentology came of age, as expressed in the more routine-like approach of diagenetic problems, and when insight into early-diagenetic deformational processes was deepened by thorough investigations rather than enriched by innovative ideas);
(4) from about 1970 to about 1980 (when the problems related to basin analysis brought renewed interest in diagenetic processes about, particularly in relation to facies distribution); and
(5) from about 1980 to the present day (as far as can be reliably characterized now, a period of consolidation and inventory, rather than a period of fresh ideas regarding early-diagenetic deformations. During this period the sedimentological results concerning these deformations were applied in other earth-science disciplines, especially structural geology and stratigraphy).

The five periods mentioned above are dealt with separately in the following sections. It must be pointed out, however, that this approach has some disadvantages:

(a) The various periods have, as mentioned above, no boundaries but only gradual transitions (the approach followed here may therefore – unavoidably but incorrectly – deal with work of specific authors in two sections, although this work is part of an uninterrupted project);
(b) it can only be determined afterwards how long a specific period lasted and what were its main characteristics (which implies that the "central theme" need not necessarily be expressed by the titles of the papers referred to in this review); and
(c) a specific period can have a "central theme" only if high-standard research in the earlier period has paved the way for a more general new approach.

Fig. 3-2. Compaction reduces the thickness of unconsolidated sediments by vertical pressure. Differences inside the deposit with respect to resistance to pressure result in differential compaction, as shown in the subrecent sediments exposed in the reclaimed area in the central Netherlands.

(A) Strongly compacted peat layer (dark, left) and less compacted clastics (right). Both are overlain by a unit with a laterally changing thickness, reflecting the differential compaction of the material underneath. Subvertical section near Emmeloord in the Noordoostpolder.

(B) Overview of the surface in the Wieringermeerpolder. Two thin peat layers (dark) separate three layers of clay formed in tidal flats and tidal marshes. The sediments have been partly "decapitated", following a pattern that results from differential compaction in the subsoil. (Photograph courtesy of Rijksdienst voor de IJsselmeerpolders.)

Fig. 3-3. Bude Sandstone (lower Westphalian) near Bude, England, with metasedimentary faults and other deformations in graded layers (vertical section).

THE PERIOD BEFORE 1950

The first period comprises the timespan when geology developed, but with only occasional sedimentological observations (including early-diagenetic deformations), commonly "by-products" of regional studies. This period of "pre-sedimentology" geology was characterized by field work that was often a masterpiece of observational skill and analysis. It was nevertheless still not yet generally recognized that many deformation structures in hard-rock sediments had been formed before lithification.

Literature from the 18th century

Irregular structures, that would now be recognized as soft-sediment deformations, had already been described by numerous investigators in the 19th century, among them Strangeways (1821), Lyell (1841, 1851; Fig. 3-4), Vanuxem (1842), Dana (1849), Darwin (1851), Sorby (1859), Oldham and Mallet (1872), Salisbury (1885), Kavanaugh (1889), Diller (1890), Gosselet (1890), Weston (1891), Hay (1892), Walther (1893/1894), Cross (1894), Case (1895), Pavlow (1896), Todd

Fig. 3-4. Picture of Charles Lyell, who described several types of soft-sediment deformations in 1851, e.g., raindrop imprints.

(1896), Crosby (1897), Salisbury and Atwood (1897), Gressley (1898), and Whitten (1898).

This listing of authors is interesting from a historical point of view, not only because several famous pioneer geologists apparently noticed soft-sediment deformations, but also because the frequency of descriptions of these structures increased with the course of time. This may be due partly to the relative inaccessibility of the older literature (the famous role of "historical perspective"), but is certainly also a sign of the exponential growth in the number of scientific (here: geological) publications. This trend has remained the same in the present century (although there are now signs that the steepest part of the "S-curve" has been passed), so that it has become impossible to provide a complete survey, even of specific early-diagenetic deformation structures.

The descriptions from the past century often refer to "concretion-like structures" (Fig. 3-5), the type of deformations that would, after having been correctly interpreted by Macar (1948) in the middle of the present century, form the first real impulse for systematic research into early-diagenetic deformation structures.

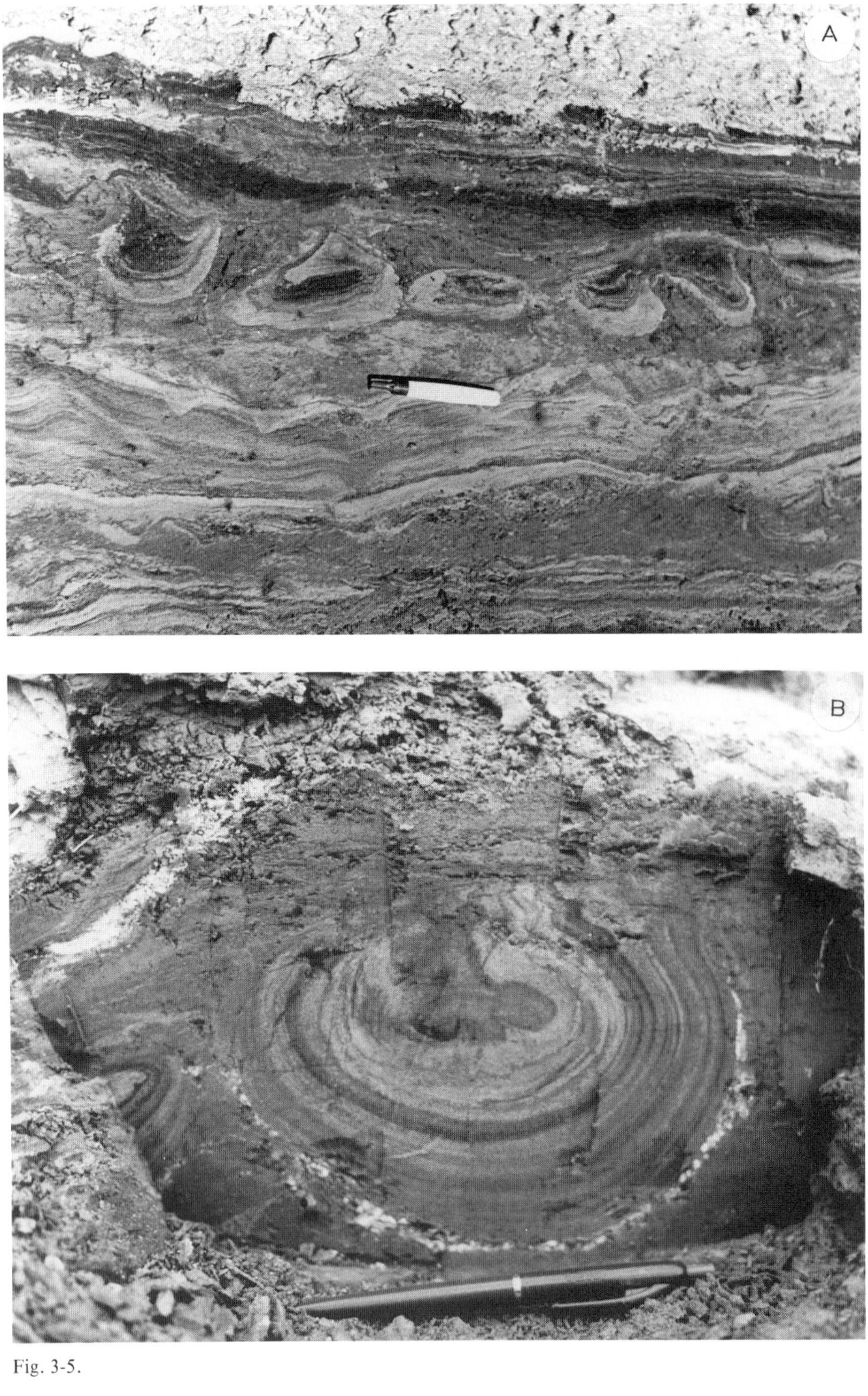

Fig. 3-5.

Literature from 1900 to 1950

Some "concretion-like structures" were recognized already in the last century as interesting deformation phenomena, but their precise genesis remained unsolved for a long time: it was only some forty years ago that these structures, now known as "pseudonodules", were correctly interpreted as a result of load casting in such a way that a sandy layer is (almost) completely reshaped into isolated, rounded balls in which the originally lower boundary forms the outer surface, whereas the originally uppermost material is found in the core. The relatively long duration of the period of insufficient insight into the genesis of pseudonodules might lead to the conclusion that soft-sediment deformations did not constitute a "hot item" of geological research during the first half of the present century. This conclusion is fully justified, although an increasing number of papers were published that were – usually in part – devoted to this topic (e.g., Sorby, 1908; Deeley, 1916). Articles concerned primarily with such deformations were, among others, those by Bailey and Weir (1932) and Kent (1945).

Many early-diagenetic structures were only mentioned in regional studies, however, as phenomena of little interest (a.o. Grabau, 1900; Ells, 1902, 1903; Walther, 1904; Cushing et al., 1910; Baker, 1916; Wilson, 1918; Ward, 1922; Wanless, 1922, 1923; Collins, 1925; Earp, 1937; Gripp, 1944; Beets, 1946; Boswell, 1949). The frequent descriptions of soft-sediment deformations in regional studies must be attributed partly to the role played by geographers, who were those carrying out many of the field studies. These investigators were also greatly interested in glacigenic deposits. It is thus not unexpected that comparatively many deformations were described from such glacigenic sediments by, among others, Bretz (1913), Lahee (1914), Haughton et al. (1925), Dreimanis (1935), Denny (1936), Caldenius (1938), Goldthwait and Kruger (1938), Kruger (1938), Carruthers (1939), Anderson (1940), Rice (1940), Sharp (1942), Steeger (1944), Troll (1944), Bryan (1946), and Schafer (1949).

Among the several types of nonglacigenic deformations that were recognized within soft sediments during these early stages, two types received relatively most attention. The first type consists of clastic dikes (Fig. 3-6) and related structures. These have been described and analyzed by Greenly (1900), Ransome (1900), McCallie (1903), Newsom (1903), Campbell (1904), Lawler (1923), Jenkins (1925), Russell (1927), Falcon (1929), S.K. Roy (1929), Hawley and Hart (1934), Miser

---

Fig. 3-5. More or less isolated "balls" in layers of different lithology, which have attracted the attention of earth scientists in the nineteenth century. They described "balls" of this type, frequently found in hard-rock deposits, commonly as "concretion-like structures".

(A) Pseudonodules (subvertical section) in the subrecent Almere Member of the Groningen Formation, central Netherlands. They are remnants of a layer that was almost completely broken up by load casting of a sandy layer into the underlying, water-saturated silty/humic material.

(B) Detail of a load cast (vertical section), showing the almost concentric structure (oldest laminae outside; youngest material in the center). This concentric buildup explains why early investigators compared these structures with concretions. Kleszczów Graben, central Poland.

(1935), Simpson (1935), Kruger (1938), Lupher (1944), Moret (1945), C.J. Roy (1946), Gulinck (1949), Waterston (1950) and others.

The second type of deformation structures that also received much attention was the type due to mass movements (Fig. 3-7), particularly of sediments deposited on slopes (e.g., of subaerial hills, of lakes, of sea coasts, and of continents). Examples

Fig. 3-6. Bude Sandstone (lower Westphalian) near Bude, England, with a sandstone dike intruded in shales (oblique section).

Fig. 3-7. Slump (vertical section) in a sandy turbidite succession of the Poumanous Formation (Cretaceous) near Pobla de Segur, Spain.

were described and interpreted by, for example, Miller (1908, 1922), Hahn (1913), E.W. Shaw (1914), Arkhanguelski (1930), Hadding (1931), Henderson (1935), O.T. Jones (1937, 1939), Lippert (1937), Goguel (1938), Lamont (1938), Klinger (1939), Rice (1939), Cooper (1943), Fairbridge (1946, 1947), Gulinck (1948), Kuenen (1949), Macar and Antun (1949), Van Straaten (1949), and Migliorini (1950).

Many other structures, found in layers deposited in a wide variety of environments, were only mentioned occasionally (e.g., Ells, 1903; Hobbs, 1907; Kindle, 1914, 1916, 1917; B. Smith, 1916; Day, 1928; Quirke, 1930; Chadwick, 1931; W.H. Monroe, 1932; Häntzschel, 1935, 1939, 1941; Dineley, 1936; McKee, 1938, 1945; Maxon, 1940; Cope, 1945; and Shrock, 1948).

This period also witnessed the first laboratory experiments aimed at the formation of soft-sediment deformations. Interesting experiments of this type were carried out by Schofield and Keen (1929), Rettger (1935), and Dobrin (1941). Researchers, with Boswell (1949, 1950) as a most important example, who would now be considered as specialists in material science, also contributed to a better understanding of the deformational processes.

THE EARLY AGE OF SEDIMENTOLOGY

The development of sedimentology as a separate geological discipline was gradual and occurred in the fifties. Increasing interest in the conditions and processes related to sedimentation led to more and more research on sedimentary structures. It is no wonder that the often complex sedimentary deformations (Fig. 3-8) also attracted much attention.

A description by the Belgian geographer (!) Paul Macar (1948, 1950, 1958) of pseudonodules from the Devonian in the Belgian Ardennes was the first, to the author's knowledge, that regarded the conditions of the sedimentary environment and of the sediment itself as essential for the interpretation of the deformational structure and for the reconstruction of its genesis.

Several of the earth scientists who became involved in sedimentology in these pioneer years became aware of comparable phenomena. There was a rapidly growing flow of papers on soft-sediment deformations (e.g., Prentice, 1960), partly still as details appended to broader research topics, but also partly as truly sedimentological analyses of depositional and early-diagenetic conditions. This was reflected in, for example, the title of a paper by Cloud (1960): "Gas as a sedimentary and diagenetic agent".

It was not unexpected that the gradual development of sedimentology, and the interest of geographers in sedimentary structures, continued to result in a large number of descriptions of soft-sediment deformations in regional studies. There were such studies by, among others, Kiersch (1950), Bump (1951), McKee et al.

Fig. 3-8. Complex deformations (subvertical section) in the subrecent Almere Member of the Groningen Formation, central Netherlands. The central layer shows partly rigid, partly plastic deformation and even some fluidization. Note the lateral grain-size variation in the layers underneath, which may have triggered the above deformations by differential compaction that induced slopes.

(1953), Newell et al. (1953), McKee (1954), Pepper et al. (1954), Fuller (1955), Wiggers (1955), Greensmith (1956), Kuenen and Sanders (1956), Robson (1957), Kingma (1958), MacKay (1958), Smith and Rast (1958), Marchant and Black (1959), Wood and Smith (1959), and Perry and Dickens (1960).

The contributions by physical geographers were still numerous and the attention paid to deformations in glacigenic sediments (Fig. 3-9) reflects this interest on the part of geographers (e.g., Horberg, 1951; Gripp, 1952; Schwarzbach, 1952; Carruthers, 1953; Wolfe, 1953; Morell and Hilly, 1956; Michalska, 1957; Johnsson, 1959; Péwé, 1959; A.J. Smith, 1959; Virkkala, 1959, 1960; and Matisto, 1960).

Soft-sediment deformation was, however, also recognized as a problem in itself: this subject became more and more considered as a research topic that deserved detailed analysis. Various papers were published, on this topic in general, on specific aspects, such as thixotropic behavior (Boswell, 1952) – a common characteristic of silt-rich material (Fig. 3-10) – or on specific structures (Rich, 1951; Broadhurst, 1954; H.B. Stewart, 1956; Moore and Scruton, 1957; Conaster, 1958; Macar, 1958; Neruchev and Il'im, 1959; Aurola, 1960; and Carozzi, 1960). There also appeared some of the first papers on the classification of, and related terminology for, these structures (Packham, 1954; Sullwold, 1959, 1960; and Holland, 1960).

Fig. 3-9. A common deformation structure in glacigenic sediments: a small "graben", probably due to collapse after melting of a buried lense of dead-ice, with slump-like structures filling the depression that was formed on top of the "graben". Glaciolimnic sediments (vertical section), Żary area, western Poland.

Fig. 3-10. Typical deformation of silt-rich material with thixotropic characteristics. There is a gradual change from liquefaction in the center of the structure (here: the left part of the photograph) via plastic deformation (middle) to more brittle deformation (right part). Subvertical section in the subrecent Almere Member, Groningen Formation, near Emmeloord, central Netherlands.

Clastic dikes continued to be of much interest (although numerous descriptions of dikes had already been published and the reconstruction of their genesis had become common knowledge), and there appeared reports by, for example, J.N. Monroe (1950), Birham (1952), K.G. Smith (1952), Allison (1953), Gottis (1953), Vintanage (1954), and Dzulynski and Radomski (1956).

Structures related to mass transport

The other "inheritance" from the previous period, an interest in structures formed as a result of reworking and resedimentation, received a great impetus due to the research into turbidity currents, a phenomenon that would remain the "hot item" of sedimentology for some ten years after the publication of the famous article by Kuenen and Migliorini (1950), "Turbidity currents as a cause of graded bedding". Deformation structures in – or otherwise related to – turbidites were described and analyzed by Emery (1950), Książkiewicz (1951), Natland and Kuenen (1951), Kuenen and Ménard (1952), Kuenen (1953a,b, 1956), Kuenen and Carozzi (1953), Crowell (1955), Sullwold (1958), Colacicchi (1959), Dzulynski et al. (1959), Hardy and Williams (1959), Seilacher (1959), Ten Haaf (1959), and Halicki (1960). It is most probable that the analysis of turbidites led to the interest in convolute lamina-

Fig. 3-11. Pleistocene Lissan Formation (vertical section) with well-developed deformation horizon of slumped, slightly consolidated limestone. Northern Arava, Israel.

tion, a structure that was also described frequently by Ten Haaf (1956), Holland (1959), Sanders (1960), Sullwold (1960), E. Williams (1960), and others.

Other forms of mass transport, especially slumping and sliding (Fig. 3-11), were also the object of much interest. There were interesting reports by Boswell (1953), O.T. Jones (1953), Destombes and Jeanette (1955), Kühn-Velten (1955), Crowell (1955), Książkiewicz (1958), Rigby (1958), Williams and Prentice (1958), and Nichols (1960).

Other approaches

Even though sedimentology was in its pioneer stages in the fifties, it is clear from the literature that there were new approaches, also with regard to early-diagenetic deformations. This was reflected by, for example, the increase in field inventories and analyses of specific deformation structures, but also by the far more commonly followed experimental approach (see, among others, Skempton and Northey, 1952; Richardson and Zaki, 1954; Dzulynski and Slaczka, 1958; Kuenen, 1958; Metzner and Whitlock, 1958; Sanford, 1959).

Some studies were devoted to specific deformation structures, such as load casts (Prentice, 1956, 1958; Kelling and Walton, 1957; Kuenen and Prentice, 1957; Hills,

1958; McCrossan, 1958; and Holland, 1960); escape structures (Gill and Kuenen, 1958; Oldershaw, 1960); so-called cylindrical structures (Gableman, 1955; Berthois, 1958; Phoenix, 1958; and Arai, 1959); and various types of sole marks (Rich, 1950; Kaye and Power, 1954; Prentice, 1956; Kuenen, 1957; Kuenen and Prentice, 1957; and Lamont, 1957).

Finally, there was a first attempt to investigate possible relationships between structures and depositional environments, for example by Van Straaten (1954) for tidal flats. His paper ("Sedimentology of Recent tidal flat deposits and the Psammites du Condroz (Devonian)") became famous, not for his inventory of deformation structures in tidal-flat deposits, but because of the comparison made by Van Straaten between recent and ancient tidal flats, their deposits, and their sedimentary structures (including deformations).

THE 1960-1970 PERIOD: EMPHASIS ON ENVIRONMENTAL ANALYSIS

The sixties were a period during which sedimentology came of age as a scientific discipline. Methods and techniques developed during the previous decade were

Fig. 3-12. Example of a deformation structure that is preserved only in exceptional cases. Moisture on a muddy substratum forms ice crystals when the temperature drops below 0°C. The ice crystals (bright white on the photograph) deform the mud during their growth, thus leaving ice-crystal imprints after melting ("flowers" or needles, generally less than 1 mm thick). A period of one hour with a temperature above 0°C is generally sufficient to destroy the imprints by fluidization of the mud. (View from above.) Ginzling, Austria. (Black circle is appr. 5 cm.)

refined and were used for environmental analysis of fossil sediments; there was a simultaneous, rapid growth of the interest in recent environments. Consequently, it was observed that the preservational potential of structures encountered in recent sediments (Fig. 3-12) may be fairly divergent, so that an inventory of (deformational) structures in recent sediments formed in a particular environment may yield quite different data than a similar inventory in (well exposed) ancient rocks formed under exactly the same conditions. The study of recent environments also clarified many depositional and deformational processes that were responsible for the present state of fossil sediments.

Environmental analyses required *regional* (instead of local) inventories of sedimentological features, including structures that gave insight into the direction of the paleoslopes (sole marks, slump heads, etc.). These inventory-type studies yielded a mass of data on soft-sediment deformations (set in the regional context that had been largely ignored earlier), both in (sub)recent unconsolidated sediments and in older, consolidated and lithified rocks (e.g., Voight, 1962). Regional soft-rock and hard-rock studies that provided data on soft-sediment deformation were carried out by McIver (1961), Crook (1961), G.P. Jones (1961, 1962), De Vries Klein (1962), Murphy and Schlanger (1962), Allen (1963), Harms et al. (1963), Selley et al. (1963), Ballance (1964a,b), Coleman et al. (1964), Kunert (1964), Crowell et al. (1966), Hubert (1966), Laming (1966), Middlemost (1967), Tada (1968), Selley (1969), Sevon (1969), Weaver (1969), P.F. Williams (1969), Gradziński (1970) and others.

The structures acquired greater importance for regional analyses and environmental interpretation as a result of more fundamental studies on soft-sediment deformations (e.g., Boswell, 1961; Kuenen, 1961; Dimitrieva et al., 1962; Doeglas, 1962; Arogyaswamy, 1963; Sutton, 1963; Black, 1964; Mountain, 1964; Artyushkov, 1965; Davies, 1965; Macar, 1965; Kelling and Williams, 1966; Middlemost, 1967; Mikadze, 1967; Conybeare and Cook, 1968; B.R. Rust, 1968; Howard and Lahrengel, 1969; Meier and Thomas, 1969; P.F. Williams, 1969; Allen, 1970; Kirkland and Anderson, 1970; Wigley and Sergeant, 1970, and G.E. Williams, 1970).

Continuing investigations were carried out, both theoretical and based on laboratory experiments, regarding engineering – geological properties of suspensions and "fresh" deposits of various grain-size compositions (White, 1961; Rosenquist, 1966; Mandl and Luque, 1970). Moreover, many experiments were performed to produce deformation structures (McKee et al., 1962a,b; Selley and Shearman, 1962; Dzulynski and Walton, 1963; Kuenen, 1963, 1965; Dzulynski, 1965a,b, 1966; Dzulynski and Radomski, 1966; and McKee and Goldberg, 1969). These studies contributed to a growing insight into the sediment properties that favor soft-sediment deformation.

The steady effort to improve nomenclature and terminology in this field greatly aided communication between researchers of different disciplines (McKee, 1964; Pettijohn and Potter, 1964; Seilacher, 1964b; Elliot, 1965; Nagtegaal, 1965; Gubler et al., 1966; Conybeare and Crook, 1968; Allen, 1970).

Age of consolidation

As already mentioned, it was in this decade that sedimentology became a well-developed discipline. Some of the youthful enthusiasm disappeared, to be replaced by maturity, in the form of a consolidation phase. This became particularly evident where the study of graded beds is concerned. The vivid interest in turbidites, raised for a large part by the extremely interesting flume experiments performed by Kuenen in particular, had some unfortunate consequences: many geologists who were not involved themselves in turbidite research, got the – incorrect – impression that "graded beds" and"turbidites" were synonyms. Consequently, several reconstructions of basin paleogeography had been based on this false idea. When this became evident, the confidence in sedimentology in general – and in the significance of sedimentary structures in particular – diminished, particularly from the side of structural geologists. This mistrust, which would soon appear to be unjustified, was possibly a major reason why the sixties were not the most inventive period with regard to new approaches in the study of sedimentary structures and sedimentary deformations.

Relatively few new pathways were followed and the approach remained mainly traditional. This was reflected by, for instance, the choice of topics studied touching soft-sediment deformations; the topics remained almost unchanged.

Environmental analyses did not yet take full advantage of ongoing research into soft-sediment deformations, although it was recognized that deformation structures might be of environmental significance (for instance, because information is provided about continuity of sedimentation, paleoslope, salinity, frequency of changes in depositional conditions, etc.). This is illustrated, for example, by the title of a paper by Oomkens (1966): "Environmental significance of sand dikes"; and one by Burne (1970): "The origin and significance of sand volcanoes in the Bude Formation (Cornwall)". Some studies in specific environments were carried out in deserts (Peacock, 1966; Glennie, 1970), coastal dunes (Bigarella et al., 1969), tidal flats (Evans, 1965), rivers (McKee et al., 1967; Pękala, 1967; Coleman, 1969), deltas (Coleman and Gagliano, 1965), shallow-marine environments (Weimer and Hoyt, 1964) and deep-sea fans (Piper and Marshall, 1969). Considerable attention, however, was still devoted to the deformations in the much-studied glacigenic environment (Fig. 3-13) (Dylik, 1961; Galloway, 1961; Hansen et al., 1961; Makovska, 1961; Mojski, 1961; Johnsson, 1962; Lachenbruch, 1962; Butrym et al., 1964; Jahn and Czerwiński, 1965; Rasmussen, 1965; Theakstone, 1965, 1970; Watson, 1965; Banerjee, 1966; Dionne, 1966, 1969, 1975; Dahl, 1967, 1968; Dylik and Maarleveld, 1967; Eissmann, 1967; Lundqvist, 1967; Okko, 1967; Tricart, 1967; Dreimanis, 1969; Gangloff, 1970; and McArthur and Onesti, 1970).

Mass-transported sediments, with their frequent deformations, also remained a prime research object. This included both the turbidites (Dewey, 1962; Houtz and Wellman, 1962; Parea, 1962; A.D. Stewart, 1962; Ballance, 1964a; Banerjee, 1966; Morgenstern, 1967; and Wentworth, 1967) and the slumps (and related structures) (Matthews, 1961; Morgan, 1961; Dott and Howard, 1962, 1963; McCall, 1962; Dott, 1963; Nagtegaal, 1963; A.D. Stewart, 1963; Chamberlin, 1964; Dill, 1964; Grant-Mackie and Lowry, 1964; Jardine, 1965; Johnson and Heron, 1965; Sanders,

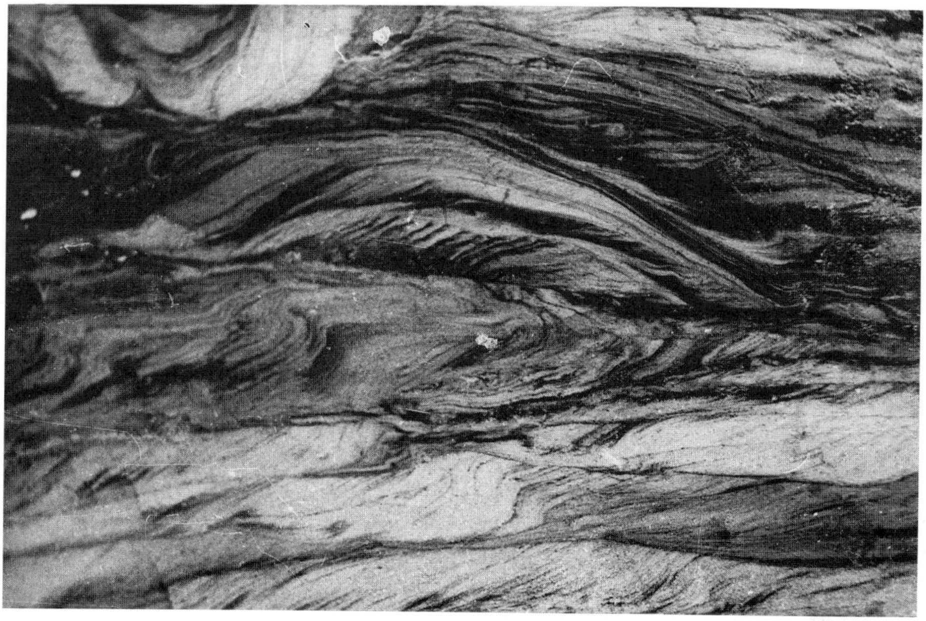

Fig. 3-13. Deformed current ripples (vertical section) in glaciolimnic sediments. The deformation is mainly due to glacial push, but additional, smaller, deformations result from intraformational movements along inclined interfaces. Sand pit near Włostów, central Poland.

1965; Scheidegger and Potter, 1965; Andresen and Bjerrum, 1967; Mikulenko, 1967; Morgenstern, 1967; Misik, 1968; Gregory, 1969; J.N. Monroe, 1969; Lajoie, 1970; and Van Loon, 1970).

Even the various types of deformation structures that received most attention were almost identical: (1) the turbidite-related or other convolutions (Holland, 1961; Dott and Howard, 1962, 1963; Dzulynski and Smith, 1963; Nagtegaal, 1963; E. Williams, 1963; Davies, 1965; Dzulynski and Slaczka, 1965; Ghent and Henderson, 1965; Sutton and Lewis, 1966; Okko, 1967; and Anketell and Dzulynski, 1968a); (2) the ''classical'' clastic dikes (Michel, 1962; Newcomb, 1962; Duncan, 1964; Harms, 1965; Peterson, 1965, 1966; Hayashi, 1966; Lambrecht and Thorez, 1966; Oomkens, 1966; and Andreux, 1967); (3) load casts (Dzulynski and Kotlarczyk, 1962; Jardine, 1965; Macar, 1965; Pękala, 1967; Anketell and Dzulynski, 1968a,b; Gry, 1968; and Anketell et al., 1970); (4) escape structures (Nichols and Yehle, 1961; Bondesen, 1966; Burne, 1970; and Ridd, 1970); (5) cylindrical structures (Schlee, 1963); (6) sole marks (Plessman, 1961); and (7) structures that were interpreted (on the basis of comparison with recent structures of known origin; particularly widespread deformations – mainly load casts and convolutions in units with layers of alternating grain size; brecciation of more rigid layers; fluidization of thixotropic layers – in levels that may intersect layers) as triggered by earthquakes (Verzilin, 1961; Olausson and Uusitalo, 1963; Reimnitz and Marshall, 1965; Barret, 1966; Coulter and Migliaccio, 1966; Tuthill and Laird, 1966; Foster and Karlstrom, 1967; Ambrasseys and Sharma, 1969; and Seilacher, 1969).

Bioturbation as an early-diagenetic phenomenon

All the literature mentioned above gives, correctly, an impression of consolidation, but there was one important new development: the finding that, in many environments, living organisms play an important role by reworking, obliterating, destroying or reshaping structures (and textures) in the unconsolidated sediment (Fig. 3-14), and as agents producing completely new structures. Seilacher (1964a,b) contributed much to the research in this field, but other investigators published material on this subject more or less simultaneously (Goldring, 1964; Weimer and Hoyt, 1964; Ghent and Henderson, 1965; and Piper and Marshall, 1969).

Bioturbation could develop in this period as a major field of study, because this research topic fitted well in the more general interest of sedimentologists for recent environments. It was soon found that the rate of bioturbation (possibly, finally resulting in mottled or even completely homogeneous sediments) gives some information about the rate of sedimentation rather than about the number of burrowing organisms within a specific environment. It was also found that recent environments, with their characteristic fauna, show differences in the type and frequency of bioturbations that are preserved. Particularly studies carried out by Reineck (e.g., Reineck et al., 1968) in recent tidal flats contributed greatly to the

Fig. 3-14. Oligocene sands at Kessel-Lo (Belgium) completely restructured by burrows made by *Ophiomorpha* (vertical section).

recognition of bioturbations as important diagenetic features that may give indications about the (palaeo)environment.

1970–1980: BASIN ANALYSIS AND PALAEOGEOGRAPHIC RECONSTRUCTIONS

Environmental analysis in sedimentology, developed in the sixties, cannot be considered an aim in itself. Analyses of this kind must obviously be fitted into a much broader framework. It is thus not surprising that the seventies can be characterized as a period of integration in sedimentology: the results of several regional investigations, when combined with studies from other geological disciplines, were used as input data for basin analyses and palaeogeographical reconstructions. Some major work on these topics had admittedly already been done; a splendid example is the book by Potter and Pettijohn (1963; 2nd ed. 1977) entitled "Palaeocurrents and basin analysis". Research into palaeocurrents as a major tool in basin analysis nevertheless only became part of basic sedimentological studies in the seventies. This was reflected by the sudden, increased attention to sole marks and other palaeocurrent indicators, and to their use in reconstructions of basin development (e.g., B.J. Smits, 1971; Van Loon, 1972; Nilsen and Simoni, 1973; and Begin, 1975).

The fact that early-diagenetic deformations had been recognized as features that might yield information about environmental conditions (Fig. 3-15) during and (relatively) shortly after sedimentation, becomes apparent from an overwhelmingly large number of environmental studies concerning these structures. The environments studied, also for their soft-sediment deformations, include:

(1) typically terrestrial environments, such as deserts (e.g., B.G. Jones, 1972), glacigenic areas (Aario, 1971; Jahn, 1971, 1977; Mörner, 1972, 1973; Banerjee, 1973; Kowalczyk, 1974; Ashwell, 1975; Jersak, 1975; Kostyaew, 1975; Michel, 1975; Rymer and Sims, 1976; Sugden and John, 1976; Daniel, 1977; Königsson and Linde, 1977; J. Shaw, 1977; Vandenberghe and Gullentops, 1977; Berthelsen, 1979; Brodzikowski and Van Loon, 1979; Schwan and Van Loon, 1979; and Schwan et al., 1980a,b), and even volcanic areas (Pederson and Surlyk, 1977; R.J. Stewart, 1978);

(2) aquatic continental environments, such as rivers (Fraser and Cobb, 1974; Leeder, 1975) and lakes (Reineck, 1974; Sims, 1975; Hesse, 1976; Rymer and Sims, 1976; Sims and Rymer, 1976; Stone, 1976; Theakstone, 1976; and Shaw, 1977);

(3) coastal environments above the shoreline, such as coastal dunes (McKee et al., 1971; McKee and Bigarella, 1972), more or less at sea level (tidal flats: Dionne, 1976; De Vries Klein, 1977; also estuaries: De Boer, 1979), and just below sea level (e.g., lagoons (Fig. 3-16): Van Loon and Wiggers, 1975a,b,c, 1976a,b,c; Brodzikowski and Van Loon, 1979); and

(4) marine environments with their soft-sedimentary deformations ranging from shallow-marine (Schwars, 1975) and deep-sea clastics (Nilson and Simoni, 1973) to folds in salt deposits (Wardlaw, 1972).

The increased interest in the relationship between the characteristics of early-diagenetic features and environment or basin configuration was one reason why soft-sediment deformations were described so frequently. A second reason is that regional studies were more and more considered complete only if sufficient attention

Fig. 3-15. A well-developed gravifossum (subvertical section) in the Almere Member of the Groningen Formation, central Netherlands. This deformation structure, representing an extreme form of syn- and metadepositional loading, was described so far only from sediments deposited in shallow, brackish waters with relatively limited current activity.

Fig. 3-16. Lagoons in moderate climates comprise commonly locations where deposits with a relatively high silt content are deposited. The frequent occurrence of deformation structures in sediments with such a composition have resulted in many descriptions of these structures. A well studied example is the Almere Member of the Groningen Formation, which became accessible when areas in the central Netherlands became reclaimed.

was devoted to sedimentary structures, including deformations (Dionne, 1971a,b,c, 1975; Dionne and Laverdière, 1972; Tyler, 1972; Dionne and Gangloff, 1975; Albers, 1976; Button and Vos, 1977; Unrug, 1977; and Shabica, 1978).

Specific types of soft-sediment deformations retained much appeal. Rascoe (1975), for instance, studied tectonically deformed unconsolidated rocks (tectonically induced early-diagenetic deformations). Most attention, however, was, of course, paid to the "classical" types of deformations, such as: (1) load casts (Aario, 1971; Kivinen, 1971; Weaver, 1976; Reinhardt and Cleaves, 1978); (2) convolute lamination (Lowe, 1975a,b; Allen, 1977; De Boer, 1979); (3) clastic dikes (!) (e.g., Daley, 1971; Dionne, 1971a; Heron et al., 1971; Kerns, 1971; Lindström, 1971; Balynskiy, 1972; Marschalko, 1972; Morrow, 1972; Setty and Wagle, 1972; Yanushevich, 1972; Chandler, 1973; Tada, 1973; Dionne and Shilts, 1974; Pierce and Peterson, 1974; Zupon and Abbot, 1975); (4) structures in turbidites (Chipping, 1972; Mutti and Ricci Lucchi, 1972; Negendank, 1972; Middleton and Hampton, 1973; Hirayama and Nakajima, 1977; Montenat and Seilacher, 1978), and related features, such as slumps (Naganuma, 1973; Hampton, 1975, 1979; Stone, 1976; Pickering, 1979); (5) escape structures (Fig. 3-17) (Lovell, 1974; Rautman and Dott, 1977); (6) desiccation cracks (Donovan and Foster, 1972); and (7) bioturbations (Frey, 1975; Ahlbrandt et al., 1978).

Interest in the significance of sedimentary deformations for environmental inter-

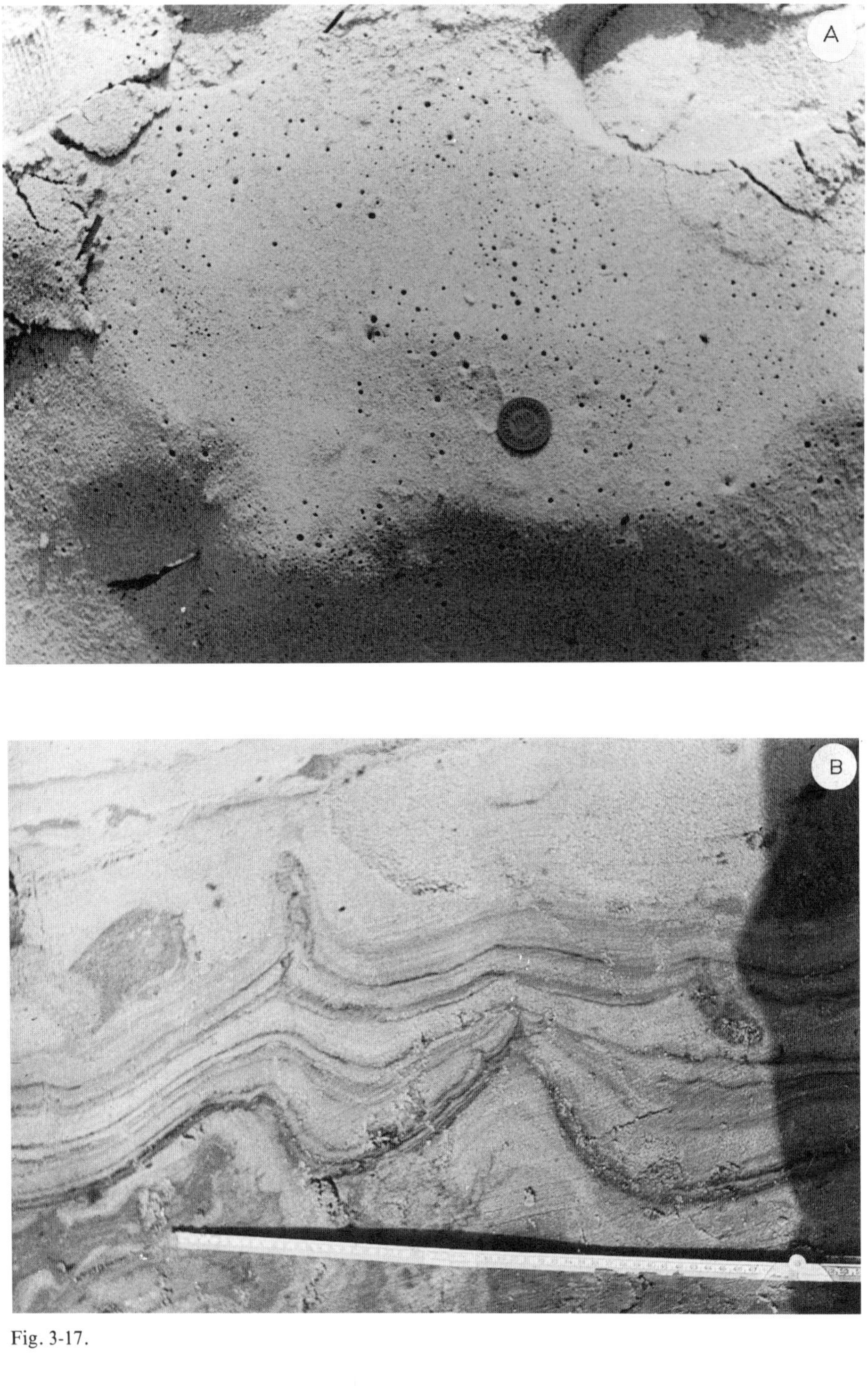

Fig. 3-17.

pretations was not the only reason for studying soft-sediment deformations. These themselves remained a research topic (Allen and Banks, 1972; Danilov, 1973; Pettijohn et al., 1973; Lowe and LoPiccolo, 1974; Pettijohn, 1975; and Hendry and Stauffer, 1977). Even very rare deformations, such as structures formed under the influence of a nearby sill, were studied (Bacon and Duffield, 1978). Research was even extended out to the universe, as expressed in a study by Carr (1977) on Martian impact craters, with signs of ejecta emplacement by surface flow. Not only such deformations, triggered by "foreign" forces, were examined but structures formed as a result of earthquakes also received much attention (e.g., Francis, 1971; Hayashi, 1971; Reimnitz, 1972; Damberger, 1973; Sims, 1973, 1975; Yeleyeva, 1974; Coates, 1975a,b; Hesse, 1976; Krinitsky and Bonis, 1976; Rymer and Sims, 1976; Sims and Rymer, 1976; Weaver, 1976; and Montenat, 1980).

Finally, work was continuing on engineering–geological aspects of early-diagenetic deformations (e.g., Martheiades, 1971), through experiments (Fig. 3-18) (e.g., Mandl et al., 1977), and with respect to terminology (e.g., Roe, 1972).

THE POST-1980 PERIOD

Far from being a period of lessening interest in early-diagenetic deformations, the eighties have seen even more publications than ever devoted – directly or indirectly – to these features. Deformation structures are now considered as phenomena that deserve no less attention than do, for example, primary sedimentary structures, if the development of the geological processes responsible for the present state of the rocks is to be reconstructed. It is interesting in this context that a major chapter of Allen's (1982) well-known work on sedimentary structures was completely devoted to soft-sediment deformations. General recognition of the importance of early-diagenetic deformations is also reflected by the titles of several papers, among them those by Mills (1983: "Genesis and diagnostic value of soft-sediment deformation structures – a review"), Maltman (1984: "On the term 'soft-sediment deformation' "), Brodzikowski and Van Loon (1985a: "Inventory of deformational structures as a tool for unravelling the Quaternary geology of glaciated areas") and Van Loon and Brodzikowski (1987: "Problems and progress in the research on soft-sediment deformations").

It is beyond the scope of this chapter even to review briefly the various types of studies or the study techniques that are now being applied; even less to mention all different types of deformational structures (but see Tables 3-2 and 3-3) and the (sub)environments in which they occur most frequently. Moreover, there are now several databases that can provide all the necessary data about these topics, at least

---

Fig. 3-17. Escape structures are formed when a fluid (commonly water) or a gas (commonly air) is pressed out of the pores of a sediment.

(A) Recent beach sand (Djerba, Tunisia) with escape structures made by air, pressed out of the pores during swash downbeach (vertical view from above).

(B) Recent sand body in the Noordoostpolder, The Netherlands, where water was pressed out of the pores when the sand underwent compaction due to the weight of the overlying material (vertical section).

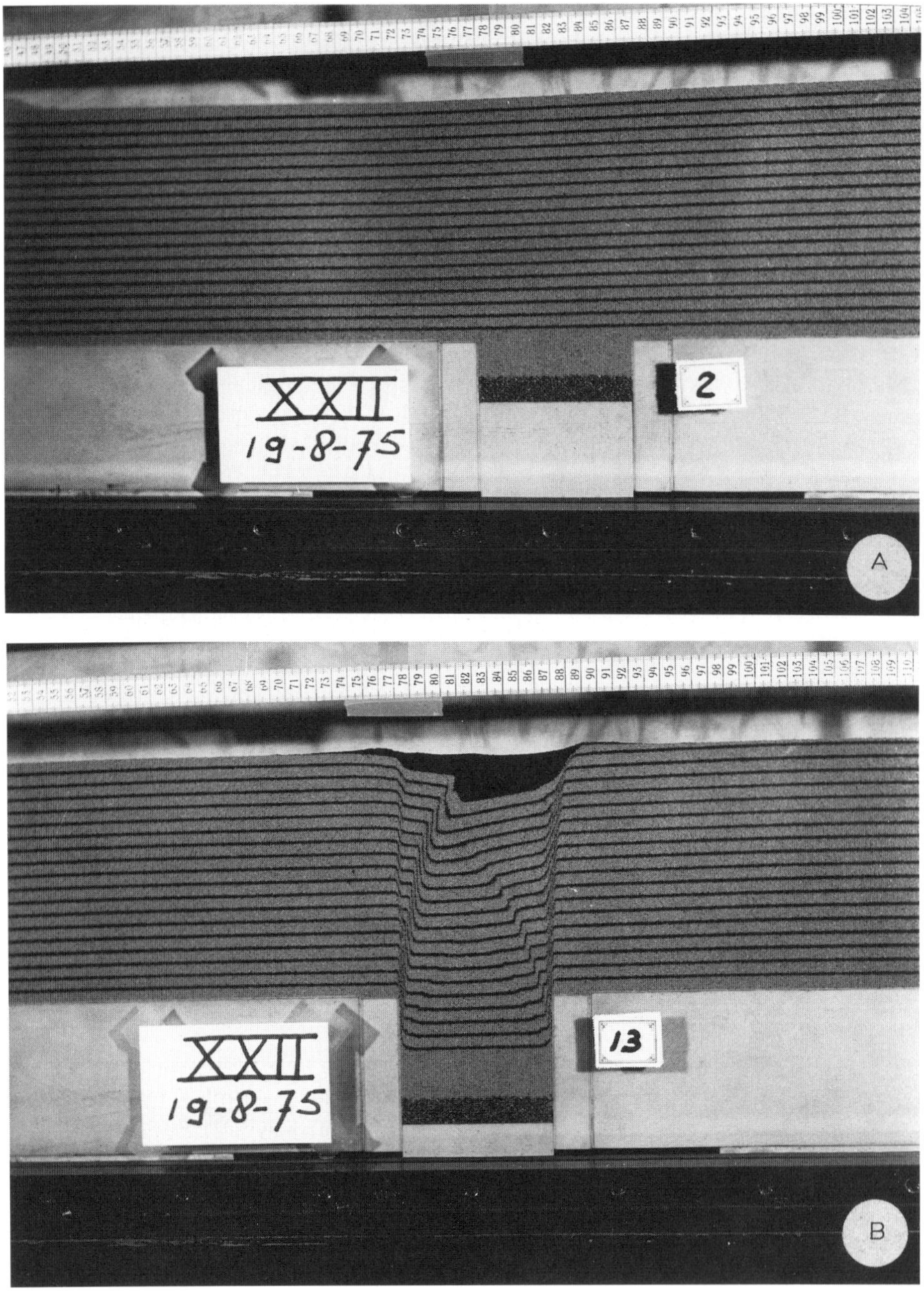

Fig. 3-18. Experiments at Koninklijke/Shell Exploratie en Produktie Laboratorium in The Hague, The Netherlands, with cohesionless sand. (Courtesy Shell Research B.V.).
(A) Original material.
(B) Result of lateral pressure, showing "grabens" with curved faults and upthrusts, resembling the gravifossums shown in Fig. 3-15.

from the current decade. It will, therefore, suffice to mention here only a selection of the more interesting publications, to give an impression of the work being done.

The eighties have become, as concerns sedimentology, the period in which models were a basic tool. Models were developed for a wide variety of topics, ranging from glacitectonism in general (Aber, 1982; Aber et al., 1989) to kink structures in unconsolidated sands (Van Loon et al., 1984; Fig. 3-19) and in fine-grained sediments

Fig. 3-19. Saalian glaciotectonically pushed pure sands near Balderhaar, Federal Republic of Germany, with a well-developed kink zone (subvertical section).

(Van Loon et al., 1985). There was also much interest in the processes underlying the deformational activity, such as: (1) dilatancy (Brodzikowski, 1981); (2) liquefaction and fluidization (Horowitz, 1982; Plint, 1983; Van de Poll and Plint, 1983; Chough and Chun, 1987); (3) rheoplasis (Plint et al., 1983); (4) stress systems during mass transport (Van Loon, 1983); (5) quick-clay behavior (Torrance, 1983; Smalley et al., 1984); and (6) nontectonic brecciation (Fig. 3-20; Brodzikowski and Van Loon, 1985b). Part of this work was based on experiments (a.o. Plint et al., 1983).

There was, of course, still much interest in "classical" topics such as glacigenic deformations (Bell, 1981; Carvalho, 1981; Eissmann, 1981, 1982, 1984, 1985; Rust, 1981; Schwan and Van Loon, 1981; French and Gilbert, 1982; Reinson and Rosen, 1982; Swart and Hiller, 1982; Brodzikowski, 1983; Brodzikowski and Van Loon, 1983; Postma et al., 1983; Brodzikowski et al., 1984; Thomas and Connell, 1984; Visser et al., 1984, 1987; Drozdowski, 1985, Eissmann et al., 1985; Eyles and Clark, 1985; Thomas and Connell, 1985; De Groot et al., 1987; Van der Meulen, 1988; Eyles et al., 1989; and O'Brien, 1989) and mass-transport deformations (Elliott and Lapido, 1981; Naylor, 1981; Postma et al., 1983; Walker, 1984; Alvarez et al., 1985; Broster and Hicock, 1985; Eyles and Clark, 1985; and Schwab and Lee, 1987). Moreover, the various types of structures themselves were still being studied, including desiccation cracks (Fig. 3-21) (Plummer and Gostin, 1981; Allen, 1984, 1986; Stear, 1985; Van der Westhuizen et al., 1989), sole marks (Van de Poll and Patel, 1981; G.A. Smith, 1984), and other palaeocurrent indicators (Fritz and Harrison, 1985), escape structures (Glennie and Buller, 1983; Postma, 1983; Nocita,

Fig. 3-20. Brecciated varved deposit (oblique section) of glaciolimnic origin. The breccias were formed by rigid reaction to stresses due to glacial push. Żary area, western Poland.

Fig. 3-21. Recent tidal flat in the Baie Mont-Saint-Michel, France, with three generations of well-developed desiccation cracks (oblique view).

1988), load casts (Puziewicz and Wojewoda, 1984; Allen, 1985), convolutions (Visher and Cunningham, 1981), and, of course, clastic dikes (Kumar and Singh, 1982; Von Brunn and Talbot, 1986). Even raindrop imprints, the deformation structures described already by Lyell (1851), still received attention (Van der Westhuizen et al., 1989).

DISCUSSION

The fact that geology could develop as a science of crucial importance to mankind is due to careful analyses of rock features and subsequent derivation of "geological laws". One of the first principles recognized was the law of superposition: layers gradually become younger in an upward direction, at least if the succession is undisturbed. It is, therefore, not surprising that features that appear at first sight to be in contradiction with the superposition principle, received (and still receive) much attention. Clastic dikes and related phenomena, such as diapirs (Fig. 3-22), are such features. If present, they are easily detected, and it is consequently only logical that descriptions of such phenomena are some of the very first descriptions of these features that would now be termed "early diagenetic" (e.g., Diller, 1890).

Layering of sedimentary rocks has long been the clue for unravelling the stratigraphy of rocks. It can thus be understood that, in addition to the layer-intersecting dikes, other types of irregular layering also attracted attention. This is reflected by the interest in structures, such as convolutions and slumps.

Interest in a more systematic explanation of deformations that had apparently occurred before lithification, only developed in mid-twentieth century, when sedimentology emerged as a separate discipline. The rapidly growing concern with sedimen-

Fig. 3-22. Gravel pit on the island of Funen, Denmark. The apparent pile in the center does not represent a dumping or storage site, but a clay diapir that intruded the glaciofluvial sands and gravels. The diapir was left "in-situ" when the surrounding material was excavated. The diapir intruded the overlying sands and gravels during the Weichselian, probably when the margin of the land ice-cover induced local pressure gradients.

tology, without doubt favored by the importance of facies analysis for the petroleum industry, soon led to a fairly good understanding of most early-diagenetic deformation processes, the conditions that determined them, and the resulting structures (cf. Wolf and Chilingarian, 1988). Thus, soft-sediment deformations gradually became a basic topic for study during field work.

The "normal" inventory of early-diagenetic deformations, made in order to unravel the history of the sediment before lithification, has resulted in an overwhelming number of descriptions and analyses, sometimes only as "by-products", sometimes as primary research topics. Many of these descriptions add no new information other than from a regional or stratigraphic point of view, so that one might question whether such data really contribute to knowledge. Good examples are the ongoing publications of clastic dikes and load-casts (Fig. 3-14). This never-ending interest can possibly be explained by the fact that individual researchers become intrigued and fascinated by some of the – commonly complicated – deformation structures and consider them interesting features that should be communicated to colleagues (even though one may be aware that comparable structures have been dealt with extensively by earlier workers).

Part of integrated research

The above is no plea for terminating research in this field: more data are still required to get a better insight into the distribution of the various types of deformations over the sedimentary facies. Research, including experiments, regarding relationships between grain-size distribution, anisotropy, pressure gradients, pore volume and pore size, etc., will remain of even greater importance.

Moreover, the study of early-diagenetic features, including soft-sediment deformations, is an important part of the much wider research in the field of earth sciences. It cannot be emphasized enough that only combination of data from all possible subdisciplines will generate both a better understanding of the Earth as an object in itself, and a more efficient – and responsible – use of the resources that Earth offers to Mankind.

It is quite interesting in this framework that some analyses based on the approach mentioned above for the study of soft-sediment deformations, but carried out to unravel the genesis of plutonic rocks, have yielded most interesting results (Elliston, 1984, 1985). In a paper on orbicular granites, Elliston (1984) stated that ". . . that orbicular granites crystallized from a hydrosilicate system. These hydromagmas must have contained sufficient water to enable them to behave as gelatinous colloid systems. Alternate dynamic and static conditions in such systems would account for all the observations. (. . .) In addition to thixotropy, accretion, concretion, syneresis and diffusion, other properties of gels, such as cohesion, differing gel densities, plasticity, fracturability, change in phase boundary conditions due to syneresis, gel condensation, dehydration and crystallization of hydrolysates, are specific characteristic of the macromolecular system. (. . .) The physicochemical conditions required for the genesis of granitic orbicules are those which occur in other natural gelatinous systems, such as fine-grained wet sediments and gelatinous accumulations of newly precipitated mineral matter under hydrothermal conditions''.

In a paper on the genesis of the texture of the famous Rapakivi granite, Elliston (1985) also stated: ''The main features of a macromolecular system in which surface energy and particle interactions are predominant in the bulk behavior of the material, are: 1. gel plasticity; 2. gel cohesion and fracturability; 3. gel diffusability; 4. gel enhancement of the crystal growth; 5. reversible hydrolysis; 6 concretion; 7. thixotropy; 8. accretion; 9. rheopexy; 10. syneresis; 11. adsorption; 12. desorption''.

These statements made by Elliston for granites are also true, at least for a considerable part, for soft-sediment deformations, thus stressing that the study of processes has a higher fundamental value than the description of the final results of these processes. This is an additional stimulus for subsequent studies of early-diagenetic conditions (cf. Borst, 1982). Analysis of man-made deformations (Fig. 3-23) might be of much use if the precise deformational conditions are known. Much more information could thus be gathered touching the relationship between early-diagenetic deformations and other processes.

Fig. 3-23. An anthropogenic deformation structure: heaps of peaty clay embedded in intertidal clays off Ostend, Belgium. This ''breccia'' was formed in Medieval times during peat digging.

ACKNOWLEDGEMENTS

The author is indebted to Dr. K. Brodzikowski (University of Lodz, Poland), who greatly helped in literature search, particularly regarding works published in Eastern Europe. Marie-Louise Schönbaum-Desbarats kindly corrected the English text. The help extended by the editors, Drs. Karl H. Wolf and George V. Chilingarian is gratefully acknowledged.

REFERENCES

Aario, R., 1971. Kuormituksen aiheuttamia deformaatiarakenteita Kempeleen harjussa. *Geologi,* 23: 6–7.

Aber, J.S., 1982. Model for glacitectonism. *Bull. Geol.Soc. Den.,* 30: 79–90.

Aber, J.S., Croot, D.G. and Fenton, M.M., 1989. *Glaciotectonic Landforms and Structures.* Kluwer, Dordrecht, 157 pp.

Ahlbrandt, T.S., Andrews, S. and Gwynne, D.T., 1978. Bioturbation in eolian deposits. *J. Sediment. Petrol.,* 48: 839–848.

Albers, H.J., 1976. Feinstratigraphie, Faziesanalyse und Zyklen des Untercampans (Vaalser Grünsand – Hervian) von Aachen und dem niederländisch-belgischen Limburg. *Geol. Jahrb.,* 34: 3–68.

Allen, J.R.L., 1963. Depositional features of Dittonian rocks: Pembrokeshire compared with the Welsh borderland. *Geol. Mag.,* 100: 375–389.

Allen, J.R.L., 1970. *Physical Processes of Sedimentation.* Elsevier, New York, N.Y., 248 pp.

Allen, J.R.L., 1977. The possible mechanics of convolute lamination in graded sand beds. *Q. J. Geol. Soc. London,* 134: 19–31.

Allen, J.R.L., 1982. *Sedimentary Structures, their Character and Physical Basis, Vol. 2.* Developments in Sedimentology, 30B, Elsevier, Amsterdam, 663 pp.

Allen, J.R.L., 1984. Truncated fossil contraction polygons (?Devensian) in the Mercia Mudstone Formation (Trias), Oldbury upon Severn, Gloucestershire. *Proc. Geol. Assoc.,* 95: 263–273.

Allen, J.R.L., 1985. Wrinkle marks: an intertidal sedimentary structure due to aseismic soft-sediment loading. *Sediment. Geol.,* 41: 75–95.

Allen, J.R.L., 1986. On the curl of desiccation polygons. *Sediment. Geol.,* 46: 23–31.

Allen, J.R.L. and Banks, N.L., 1972. An interpretation and analysis of recumbent-folded deformed cross-bedding. *Sedimentology,* 19: 257–283.

Allison, I.S., 1953. Clastic dikes in Quaternary lake sediments in Oregon. *Geol. Soc. Am. Bull.,* 64: 1499.

Alvarez, W., Colacicchi, R. and Montanari, A., 1985. Synsedimentary slides and bedding formation in Apennine pelagic limestones. *J. Sediment. Petrol.,* 55: 720–734.

Ambrasseys, N.N. and Sharma, S.K., 1969. Liquefaction of soils induced by earthquakes. *Seismol. Soc. Am. Bull.,* 59: 651–664.

Anderson, J.G.C., 1940. Glacial drifts near Roslin, Midlothian. *Geol. Mag.,* 77: 470–473.

Andresen, A. and Bjerrum, L., 1967. Slides in subaqueous slopes in loose sand and silt. In A.F. Richards (Editor), *Marine Geotechnique.* Univ. Illinois Press, Urbana, Ill., pp. 221–229.

Andrieux, J., 1967. Etude de quelques filons clastiques intraformationels du flysch Albo-Aptien des zones externes du Rif (Maroc). *Bull. Soc. Géol. Fr.,* 7e Série, 9: 844–849.

Anketell, J.M., Cęgla, J. and Dzulynski, S., 1970. On the deformational structures in systems with reversed density gradients. *Rocz. Polsk. Towarz. Geol.,* 40: 3–30.

Anketell, J.M. and Dzulynski, S., 1968a. Patterns of density-controlled convolutions involving statistically homogeneous and heterogeneous layers. *Rocz. Polsk. Towarz. Geol.,* 38: 401–409.

Anketell, J.M. and Dzulynski, S., 1968b. Transverse deformational patterns in unstable sediments. *Rocz. Polsk. Towarz. Geol.,* 38: 411–416.

Arai, J., 1959. Cylindrical structures in the Tertiary sediments of the Chichibu Basin, Saitama Prefecture, Japan. *Bull. Chichibu Mus. Nat. Hist.,* 9: 61–68.

Arkhangel'skiy, A.D., 1930. Slides of sediments on the Black Sea bottom and the importance of this

phenomenon for geology. *Bull. Moscow Soc. Natur., Geol. Sec, N.S.,* 38: 38 – 80.
Arogyaswamy, R.N.P., 1963. Some secondary structures in the upper Gondwana sediments near Ongur. *India Geol. Surv. Res.,* 92: 231 – 234.
Artyushkov, E.V., 1965. Convective deformations developed in feebly lithified sedimentary rocks. *Dokl. Akad. Nauk S.S.S.R. Izv. Ser. Geol.,* 12: 79 – 101.
Ashwell, I.Y., 1975. Glacial and late glacial processes in western Iceland. *Geogr. Ann.,* 57: 225 – 245.
Aurola, E., 1960. Folded structures in the wall of the gravel pit at Piispanristi near Turku. *Geologi,* 12: 34 – 37.
Bacon, C.R. and Duffield, W.A., 1978. Soft-sediment deformation near the margin of a basalt sill in the Pliocene Coso Formation, Inyo County, California. *Geol. Soc. Am., Abstr. Progr.,* 10: 94.
Bailey, E.B. and Weir, J., 1932. Submarine faulting in Kimmeridgian times, east Sutherland. *Trans. R. Soc. Edinburgh,* 57: 429 – 467.
Baker, M.B., 1916. The geology of Kingston and vicinity. *Ont. Bur. Mines, Annu. Rep.,* 25(3): 1 – 36.
Ballance, P.F., 1964a. Streaked-out mud ripples below Miocene turbidites, Puriri Formation, New Zealand. *J. Sediment. Petrol.,* 34: 91 – 101.
Ballance, P.F., 1964b. The sedimentology of the Waitemata Group in the Takapuna section, Auckland. *N. Z. J. Geol. Geophys.,* 7: 466 – 499.
Balynskiy, N.A., 1972. Klasticheskiye dayki Donbassa i otlichiye ikh ot klasticheskikh inyektsiy drugikh ygolnykh mestorozhdeniy. *Litol. Polezn. Iskop.,* 7: 136 – 141.
Banerjee, I., 1966. Turbidites in a glacial sequence: a study from the Talchir Formation, Ranigani coalfield, India. *J. Geol.,* 74: 593 – 606.
Banerjee, I., 1973. Sedimentology of Pleistocene glacial varves in Ontario, Canada. *Can. Geol. Surv. Bull.,* 226, 60 pp.
Barrett, P.J., 1966. Effects of the 1964 Alaskan earthquake on some shallow-water sediments in Prince William Sound, Southeast Alaska. *J. Sediment. Petrol.,* 36: 992 – 1006.
Beets, C., 1946. Miocene submarine disturbances of strata in northern Italy. *J. Geol.,* 54: 229 – 245.
Begin, Z.B., 1975. Paleocurrents in the Plio-Pleistocene Samra Formation (Jericho region, Israel). *Sediment. Geol.,* 14: 191 – 218.
Bell, C.M., 1981. Soft-sediment deformation in sandstone related to the Dwyka glaciation in South Africa. *Sedimentology,* 28: 321 – 329.
Berthelsen, A., 1979. Recumbent folds and boudinage structures formed by subglacial shear: an example of gravity tectonics. In: W.J.M. van der Linden (Editor), *Van Bemmelen and His Search for Harmony. Geol. Mijnbouw,* 58: 253 – 260.
Berthois, L., 1958. Note sur la formation de structures cylindriques dans le grès. *Bull. Soc. Géol. Fr.,* 6: 315 – 324.
Bigarella, J.J., Becker, R.D. and Duarte, G.M., 1969. Coastal dunes from Parana (Brazil). *Mar. Geol.,* 7: 5 – 55.
Birman, J.H., 1952. Pleistocene clastic dykes in weathered granite-gneiss, Rhode Island. *Am. J. Sci.,* 250: 721 – 734.
Black, F.B., 1964. Cryptoexplosive structure near Versailles, Kentucky. *Geol. Surv. Res. Pap.,* 501-B.
Bondesen, E., 1966. Observations on Recent sand volcanoes. *Dansk Geol. Foren. Medd.,* 16: 195 – 198.
Borst, R.L., 1982. Some effects of compaction and geological time on the pore parameters of argillaceous rocks. *Sedimentology,* 29: 291 – 298.
Boswell, P.H.G., 1949. A preliminary examination of the thixotropy of some sedimentary rocks. *Q. J. Geol. Soc. London,* 104: 499 – 526.
Boswell, P.H.G., 1950. Thixotropic and allied phenomena in geological deposits. *Proc. Liverpool Geol. Soc.,* 20: 86 – 105.
Boswell, P.H.G., 1952. The determination of the thixotropic limits of sediments. *Liverpool Manchester Geol. J.,* 1: 1 – 22.
Boswell, P.H.G., 1953. The alleged subaqueous sliding of large sheets of sediment in the Silurian rocks of north Wales. *Liverpool Manchester Geol. J.,* 1: 148 – 152.
Boswell, P.H.G., 1961. *Muddy Sediments.* W. Heffer and Sons, Cambridge, 140 pp.
Braakman, J.H., 1974. Afzettingsgesteenten. *Grote Spectrum Encyclopedie,* 1: 202 – 208.
Bretz, J.H., 1913. Glaciation of the Puget Sound region. *Wash. Geol. Surv. Bull.,* 8: 244 pp.
Broadhurst, F.M., 1954. A note on contorted gritstones in the Rowarth area of north Derbyshire. *Liverpool Manchester Geol. J.,* 1: 240 – 245.

Brodzikowski, K., 1981. Dilatancy and the course of the deformational process in unconsolidated sediments. *Ann. Soc. Geol. Poloniae,* 51/52: 83 – 98.

Brodzikowski, K., 1983. Deformacje metasedymentacyjne w osadach czwartorzędu okolic Jaroszowa. *Acta Univ. Wratislaviensis,* 655; *Prace Inst. Geogr., Seria A:* 15 – 55.

Brodzikowski, K., Burdukiewicz, J.M. and Van Loon, A.J., 1984. Deformational processes and environment of Late Vistulian fluvial sedimentation in Kopanica Valley (Late Palaeolithic settlement area). In: J.K. Kozlowski and S.K. Kozlowski (Editors), *Advances in Palaeolithic and Mesolithic Archaeology. Archaeol. Interreg.,* 5: 79 – 94.

Brodzikowski, K. van Van Loon, A.J., 1979. Comparison of metasedimentary structures and their genesis in some Holocene lagoonal sediments of the Netherlands and Pleistocene (Mindel) glaci-fluvial sediments of Poland. *Bull. Acad. Polon. Sci., Série Sci. Terre,* 27: 95 – 105.

Brodzikowski, K. and Van Loon, A.J., 1983. Sedimentology and deformational history of unconsolidated Quaternary sediments in the Jaroszów Zone (Sudetic Foreland). *Geol. Sudetica,* 18: 121 – 196 (+ 20 plates).

Brodzikowski, K. and Van Loon, A.J., 1985a. Inventory of deformational structures as a tool for unravelling the Quaternary geology of glaciated areas. *Boreas,* 14: 175 – 188.

Brodzikowski, K. and Van Loon, A.J., 1985b. Penecontemporaneous non-tectonic brecciation of unconsolidated silts and muds. *Sediment. Geol.,* 41: 269 – 282.

Broster, B.E. and Hicock, S.R., 1985. Multiple flow and support mechanisms and the development of inverse grading in a subaquatic glacigenic debris flow. *Sedimentology,* 32: 645 – 657.

Bryan, K., 1946. Cryopedology – the study of frozen ground and intensive frost-action with suggestions on nomenclature. *Am. J. Sci.,* 244: 622 – 642.

Bump, J.D., 1951. White River badlands in South Dakota. *Guidebook Field Conf. Western South Dakota Soc. Vertebrate Paleont.,* pp. 35 – 46.

Burne, R.V., 1970. The origin and significance of sand volcanoes in the Bude Formation (Cornwall). *Sedimentology,* 15: 211 – 218.

Butrym, J., Cęgla, J., Dzulynski, S. and Nakonieczny, S., 1964. New interpretation of periglacial structures. *Folia Quat.,* 17: 1 – 34.

Button, A. and Vos, R.G., 1977. Subtidal and intertidal clastic and carbonate sedimentation in a microtidal environment: an example from the lower Proterozoic of South Africa. *Sediment. Geol.,* 18: 175 – 200.

Caldenius, C., 1938. Carboniferous varves, measured at Paterson, New South Wales. *Geol. Fören. Stockholm Förh.,* 60: 349 – 364.

Campbell, M.R., 1904. Conglomerate dikes in southern Arizona. *Am. Geol.,* 33: 135 – 138.

Carozzi, A.V., 1960. Microscopic arched flow structures and spiral structures in sedimentary rocks. *Bull. Inst. Nat. Genevois,* 60: 1 – 23.

Carr, M.H., 1977. Martian impact craters and emplacement of ejecta by surface flow. *J. Geophys. Res.,* 82: 4055 – 4066.

Carruthers, R.G., 1939. On northern glacial drifts: some peculiarities and their significance. *Q. J. Geol. Soc. London,* 95: 299 – 333.

Carruthers, R.G., 1953. *Glacial Drifts and the Undermelt Theory.* Harold Hill and Son, Newcastle upon Tyne, 38 pp.

Carvalho, G.S., 1981. Gelistruturas nos depositos de um terraço no vale do Rio Cavado (Penida, Minho, Portugal). *Mem. Not. Publ. Mus. Lab. Mineral. Geol. Univ. Coimbra,* 91/92: 153 – 164.

Case, E.C., 1895. On the mud and sand dikes of the White River, Miocene. *Am. Geol.,* 15: 248 – 254.

Chadwick, G.H., 1931. Storm rollers. *Geol. Soc. Am. Bull.,* 42: 242.

Chamberlin, T.K., 1964. Mass transport of sediment in the heads of Scripps submarine canyon, California. In: R.L. Miller (Editor), *Papers in Marine Geology.* MacMillan Co, New York, N.Y.

Chandler, F.W., 1973. Clastic dykes at Whitefish Falls, Ontario and the base of the Huronian Gowganda Formation. *Geol. Assoc. Canada, Spec. Pap.,* 12: 199 – 209.

Chilingarian, G.V. and Wolf, K.H. (Editors), 1975. *Compaction of Coarse-Grained Sediments, I.* Developments in Sedimentology, 18A. Elsevier, Amsterdam, 552 pp.

Chilingarian, G.V. and Wolf, K.H. (Editors), 1976. *Compaction of Coarse-Grained Sediments, II.* Developments in Sedimentology, 18B. Elsevier, Amsterdam, 808 pp.

Chilingarian, G.V. and Wolf, K.H. (Editors), 1988. *Diagenesis, I.* Developments in Sedimentology, 41. Elsevier, Amsterdam, 592 pp.

Chilingarian, G.V. and Wolf, K.H. (Editors), 1988. *Diagenesis, II.* Developments in Sedimentology, 43. Elsevier, Amsterdam, 268 pp.

Chipping, D.H., 1972. Sedimentary structure and environment of some thick sandstone beds of turbidite type. *J. Sediment. Petrol.,* 42: 587 – 595.

Chough, S.K. and Chun, S.S., 1987. Intrastratal rip-down clasts, Late Cretaceous Uhangri Formation, Southwest Korea. *J. Sediment. Petrol.,* 58: 530 – 533.

Cloud, Jr., P.E., 1960. Gas as a sedimentary and diagenetic agent. *Am. J. Sci.,* 258A: 35 – 45.

Coates, D.R., 1975a. Quaternary sediment deformation as a seismic indicator in the St. Lawrence Lowland, New York. *Geol. Soc. Am., Abstr. Progr.,* 7: 1030 – 1031.

Coates, D.R. (Editor), 1975b. *Identification of Late Quaternary Sediment Deformation and its Relation to Seismicity in the St. Lawrence Lowland, New York.* New York State Res. Dev., Author Rep., 30, 268 pp.

Colacicchi, R., 1959. Dichi sedimentari deo flysch oligomiocenico della Sicilia nord-orientale. *Eclogae Geol. Helv.,* 51: 901 – 916.

Coleman, J.M., 1969. Brahmaputra river: channel processes and sedimentation. *Sediment. Geol.,* 3: 129 – 239.

Coleman, J.M. and Gagliano, S.M., 1965. Sedimentary structures: Mississippi deltaic plain. In: G.V. Middleton (Editor), *Primary Sedimentary Structures and Their Hydrodynamic Interpretation. Soc. Econ. Paleontol. Mineral., Spec. Pap.,* 12: 133 – 148.

Coleman, J.M., Gagliano, S.M. and Webb, J.E., 1964. Minor sedimentary structures in a prograding distributary. *Mar. Geol.,* 1: 240 – 258.

Collins, W.H., 1925. North shore of Lake Huron. *Geol. Surv. Can., Mem.,* 143, 160 pp.

Conaster, W.E., 1958. The contorted strata of the Cynthiana Limestone. M.Sc. thesis (unpubl.), Univ. Cincinnati, 112 pp.

Conybeare, C.E.B. and Crook, K.A.W., 1968. Manual of sedimentary structures. *Canberra Bur. Min. Res. Geol. Geophys. Bull.,* 112, 327 pp.

Cooper, J.R., 1943. Flow structure in the Berea Sandstone and Bedford Shale of central Ohio. *J. Geol.,* 51: 190 – 203.

Cope, F.W., 1945. Intraformational contorted rocks in the Upper Carboniferous of the southern Pennines. *Q. J. Geol. Soc. London,* 101: 139 – 176.

Coulter, H.W. and Migliaccio, R.R., 1966. Effects of the earthquake of March 27, 1964, at Valdez, Alaska. *U. S. Geol. Surv., Prof. Pap.,* 542-C, 36 pp.

Crook, K.A.W., 1961. Stratigraphy of the Patry Group (Upper Devonian – Lower Carboniferous), Tamworth-Nundle district, N.S.W. *J. Proc. R. Soc. New South Wales,* 94: 189 – 207.

Crosby, W.O., 1897. The great fault and accompanying sandstone dikes of Ute Pass, Colorado. *Science,* 5: 604 – 607.

Cross, C.W., 1894. Intrusive sandstone dikes in granite. *Geol. Soc. Am. Bull.,* 5: 225 – 230.

Crowell, J.C., 1955. Directional-current structures from the Prealpine flysch, Switzerland. *Geol. Soc. Am. Bull.,* 66: 1351 – 1384.

Crowell, J.C., 1957. Origin of pebble mudstones. *Geol. Soc. Am. Bull.,* 63: 993 – 1010.

Crowell, J.C., Hope, R.A., Kehle, J.E., Ovenshine, A.T. and Sams, R.H., 1966. Deep-water sedimentary structures, Pliocene Pico Formation, Santa Paula Creek, Ventura Basin, California. *Calif. Div. Mines Geol., Spec. Rep.,* 89, 40 pp.

Cushing, H.P., Fairchild, H.L., Ruedeman, R. and Smyth, C.H., 1910. Geology of the Thousand Islands region, Alexandria Bay, Cape Vincent, Clayton, Grindstone, and Theresa quadrangles, N.Y. *N. Y. State Mus., Bull.,* 145: 194 pp.

Dahl, R., 1967. Late-glacial moraine forms in the Narvik district, Norway. *Norsk Geogr. Tidsskr.,* 21: 157 – 241.

Dahl, R., 1968. Late-glacial accumulation, drainage, and ice recession in the Narvik-Skjomen district, Norway. *Norsk Geogr. Tidsskr.,* 22: 101 – 165.

Daley, B., 1971. Diapiric and other deformational structures in an argillaceous Oligocene limestone. *Sediment. Geol.,* 6: 29 – 51.

Damberger, H.H., 1973. Physical properties of the Illinois Herin (No. 6) coal before burial as inferred from earthquake-induced disturbances. *C. R. 7e Congr. Int. Stratigr. Géol. Carbonif.,* 2: 341 – 350.

Dana, J.D., 1849. *Geology, United States Exploring Expedition during the Years 1838, 1839, 1840, 1841, 1842, under the Command of Charles Wilkes.* U.S.N., Vol. 10, Philadelphia, 756 pp.

Daniel, E., 1977. Fossil ice wedge at Penstrop, Skane. *Geol. Fören. Stockholm Förh.*, 99: 420–423.
Danilov, I.D., 1973. Plastic and ruptural deformations in basin sediments. *Biul. Peryglacj.*, 22: 329–337.
Darwin, C., 1851. *Geological Observations on Coral Reefs, Volcanic Islands and on South America, Part III.* Smith, Elder and Co., London.
Davies, H.G., 1965. Convolute lamination and other structures from the lower coal measures of Yorkshire. *Sedimentology,* 5: 305–325.
Day, A.E., 1928. Pipes in the coast sandstone of Syria. *Geol. Mag.,* 65: 412–415.
De Boer, P.L., 1979. Convolute lamination in modern sands of the estuary of the Oosterschelde, The Netherlands, formed as the result of entrapped air. *Sedimentology,* 26: 283–294.
Deeley, R.M., 1916. Trail and underplight. *Geol. Mag.,* 53: 2–5.
De Groot, Th., Cleveringa, P. and Klijnstra, B., 1987. Frost-mound scars and the evolution of a Late Dryas environment (northern Netherlands). *Geol. Mijnbouw,* 66: 239–250.
Denny, C.S., 1936. Periglacial phenomena in southern Connecticut. *Am. J. Sci.,* 232: 322–342.
Destombes, J. and Jeannette, A., 1955. Etude pétrographique et sédimentologique de la série Acadienne de Casablanca – présence de glissements sous-marins (slumpings). *Notes Serv. Géol. Maroc,* 11: 75–98.
De Vries Klein, G., 1962. Sedimentary structures in the Keuper marl (Upper Triassic). *Geol. Mag.,* 99: 137–144.
De Vries-Klein, G., 1977. *Clastic Tidal Facies.* Continuing Education Publ. Co., Champaign, Ill. 149 pp.
Dewey, J.F., 1962. The provenance and emplacement of Upper Arenigian turbidites in Co. Mayo, Eire. *Geol. Mag.,* 99: 238–252.
Dill, R.F., 1964. Sedimentation and erosion in Scripps submarine canyon head. In: R.L. Miller (Editor), *Papers in Marine Geology.* MacMillan Co., New York, N.Y., pp. 23–41.
Diller, J.S., 1890. Sandstone dikes. *Geol. Soc. Am. Bull.,* 1: 411–442.
Dineley, D.L., 1936. Contortions in the Bovey Beds (Oligocene), southwest England. *Biul. Peryglacj.,* 12: 151–160.
Dionne, J.-C., 1966. Formes de cryoturbation fossiles dans le sud-est du Québec. *Cah. Géogr. Québec,* 10: 89–100.
Dionne, J.-C., 1969. Nouvelles observations de fentres de gel fossiles sur la côte sud du Lac-Saint-Laurent. *Rev. Géogr. Montréal,* 23: 307–316.
Dionne, J.-C., 1970. Structures sédimentaires dans du fluvio-glaciaire, Lac-Saint-Laurent, Québec. *Rev. Géogr. Montréal,* 24: 255–263.
Dionne, J.-C., 1971a. Dyke de till dans sables glacio-lacustres, haute vallée de la Chaudière. *Ann. ACFAS,* 38: 72.
Dionne, J.-C., 1971b. Structure cylindrique verticale dans un dépôt quaternaire non consolidé. *Int. Assoc. Sedimentol. Congr., Progr. Abstr.:* 22–23.
Dionne, J.-C., 1971c. Contorted structures in unconsolidated Quaternary deposits, Lake Saint-Jean and Saguenay regions, Quebec. *Rev. Géogr. Montréal,* 25: 5–34.
Dionne, J.-C., 1975. Paleoclimatic significance of late Pleistocene ice-wedge casts in southern Quebec. *Palaeogeogr., Palaeoclimatol., Palaeoecol.,* 17: 65–76.
Dionne, J.-C., 1976. Miniature mud volcanoes and other injection features in tidal flats, James Bay, Quebec. *Can. J. Earth. Sci.,* 13: 422–428.
Dionne, J.-C. and Gangloff, P., 1975. Vertical cylindrical structures in Quaternary deposits, southern Quebec. *Geol. Soc. Am., Abstr. Progr.,* 7: 49.
Dionne, J.-C. and Laverdière, C., 1972. Structure cylindrique verticale dans un dépôt meuble Quaternaire, au nord de Montréal, Québec. *Can. J. Earth Sci.,* 9: 528–543.
Dionne, J.-C. and Shilts, W., 1974. A Pleistocene clastic dyke, upper Chaudière valley, Quebec. *Can. J. Earth Sci.,* 11: 1594–1605.
Dmitrieva, E.V., Erchova, G.I. and Oreshnikova, E.I., 1962. *Structures and Textures of Sedimentary Rocks: Part I, Clastic and Argillaceous Rocks.* Moscow, Gosgeoltekhizdat.
Dobrin, M.B., 1941. Some quantitative experiments on a fluid saltdome model and their geological implications. *Trans. Am. Geophys. Union,* 22: 528–542.
Doeglas, D.J., 1962. The structure of sedimentary deposits of braided rivers. *Sedimentology,* 1: 167–190.

Donovan, R.N. and Foster, R.J., 1972. Subaqueous shrinkage cracks from the Caithness flagstone series (Middle Devonian) of northeast Scotland. *J. Sediment. Petrol.*, 42: 309 – 317.
Dott, Jr., R.H., 1963. Dynamics of subaqueous gravity depositional processes. *Am. Assoc. Pet. Geol. Bull.*, 47: 104 – 128.
Dott, Jr., R.H. and Howard, J.K., 1962. Convolute lamination in non-graded sequences. *J. Geol.*, 70: 114 – 120.
Dott, Jr., R.H. and Howard, J.K., 1963. Convolute lamination: a reply. *J. Geol.*, 71: 659 – 660.
Dreimanis, A., 1935. The rock-deformations caused by inland-ice on the left bank of the Daugava at Dole Island, near Riga in Latvia. Izdevis A. Gulbis, Riga: 12 – 18.
Dreimanis, A., 1969. Till wedges as indicators of direction of glacial movement. *Geol. Soc. Am. Abstr. Progr.*, 7: 52 – 53.
Drozdowski, E., 1985. On the effects of bedrock protuberances upon the depositional and relief-forming processes in different marginal environments of Spitsbergen glaciers. *Palaeogeogr., Palaeoclimatol., Palaeoecol.*, 51: 397 – 413.
Duncan, Jr., J.R., 1964. Structural significance of clastic dykes in a selected exposure of the Modelo Formation, Santa Monica Mountains, California. *South. Calif. Acad. Sci. Bull.*, 63: 157 – 163.
Dylik, J., 1961. Analyse sédimentologique des formations de versant remplissant les dépressions fermées aux environs de Lodz. *Biul. Peryglacj.*, 10: 57 – 74.
Dylik, J. and Maarleveld, G.C., 1967. Frost cracks, frost fissures and related polygons. *Meded. Geol. Stichting*, 18: 7 – 22.
Dzulynski, S., 1965a. Experiments on clastic wedges. *Bull. Acad. Polon. Sci., Série Sci. Géol. Géogr.*, 13: 301 – 303.
Dzulynski, S., 1965b. New data on experimental production of sedimentary structures. *J. Sediment. Petrol.*, 35: 196 – 212.
Dzulynski, S., 1966. Sedimentary structures resulting from convection-like pattern of motion. *Rocz. Polsk. Tow. Geol.*, 36: 3 – 21.
Dzulynski, S. and Kotlarczyk, J., 1962. On load-casted ripples. *Rocz. Polsk. Tow. Geol.*, 32: 148 – 159.
Dzulynski, S., Książkiewicz, M. and Kuenen, Ph.H., 1959. Turbidites in flysch of Polish Carpathian mountains. *Geol. Soc. Am. Bull.*, 70: 1089 – 1118.
Dzulynski, S. and Radomski, A., 1956. Clastic dykes in the Carpathian flysch. *Rocz. Polsk. Tow. Geol.*, 26: 225 – 264.
Dzulynski, S. and Radomski, A., 1966. Experiments on bedding disturbances produced by the impact of heavy suspensions upon horizontal sedimentary layers. *Bull. Acad. Polon. Sci., Série Sci. Géol. Géogr.*, 14: 227 – 230.
Dzulynski, S. and Slączka, A., 1958. Directional structures and sedimentation of the Krosno Beds. *Bull. Acad. Polon. Sci., Série Sci. Géol. Géogr.*, 6: 205 – 260.
Dzulynski, S. and Slączka, A., 1965. On ripple-load convolution. *Bull. Acad. Polon. Sci., Série Sci. Géol. Géogr.*, 13: 135 – 139.
Dzulynski, S. and Smith, A.J., 1963. Convolute lamination, its origin, preservation and directional significance. *J. Sediment. Petrol.*, 33: 616 – 627.
Dzulynski, S. and Walton, E.K., 1963. Experimental production of sole markings. *Trans. Edinburgh Geol. Soc.*, 19: 279 – 305.
Dzulynski, S. and Walton, E.K., 1965. *Sedimentary Features of Flysch and Greywackes.* Developments in Sedimentology, 7. Elsevier, Amsterdam, 274 pp.
Earp, J.R., 1937. The higher Silurian rocks of the Kerry districts, Montgomeryshire. *Q. J. Geol. Soc. London*, 94: 125 – 160.
Edelman, C.H., Florschütz, F. and Jeswiet, J., 1936. Über Spätpleistozäne und frühholozäne kryoturbate Ablagerungen in den östlichen Niederlanden. *Geol. Mijnbouw*, 11: 301 – 336.
Eissmann, L., 1967. Glaziäre Destruktionszonen (Rinnen, Becken) im Altmoränengebiet des norddeutschen Tieflandes. *Geologie*, 16: 804 – 832.
Eissmann, L., 1981. Periglaziäre Prozesse und Permafroststrukturen aus sechs Kaltzeiten des Quartärs. Ein Beitrag zur Periglazialgeologie aus der Sicht des Saale – Elbe – Gebietes. *Altenburger Naturwiss. Forsch.*, 1: 3 – 171.
Eissmann, L., 1982. Zum Ablauf der Elstereiszeit in der leipziger Tieflandsbucht unter besonderer Berücksichtigung geschiebeanalytischer Befunde. *Z. Geol. Wiss., Berlin*, 10: 771 – 781.
Eissmann, L., 1984. Beiträge zur Periglazialgeologie des Saale – Elbe – Gebietes I. Der Tropfboden von

Süptitz, Kreis Torgau. *Abh. Ber. Nat. Mus. Mauritianum Altenburg,* 11: 106 – 113.
Eissmann, L., 1985. 50 Millionen Jahre Subrosion. Über Persistenz und Zyklizität von Auslaugungsprozessen im Weiszelsterbecken. *Geol. Geophys. Veröff. KMU Leipzig,* 3(2): 31 – 65.
Eissmann, L., Priese, O. and Richter, E., 1985. Die Geologie des Naherholungsgebietes Kulkwitz – Miltitz bei Mankranstädt. Ein Leitprofil des Glaziärs under Periglaziärs in Sachsen. *Abh. Ber. Nat. Mus. Mauritianum Altenburg,* 11: 217 – 248.
Elliot, R.E., 1965. A classification of subaqueous sedimentary structures based on rheological and kinematical parameters. *Sedimentology,* 5: 193 – 209.
Elliott, T. and Lapido, K.O., 1981. Syn-sedimentary gravity slides (growth faults) in the coal measures of south Wales. *Nature,* 291(5812): 220 – 222.
Elliston, J.N., 1984. Orbicules: an indication of the crystallisation of hydrosilicates, I. *Earth Sci. Rev.,* 20: 265 – 344.
Elliston, J.N., 1985. Rapakivi texture: an indication of the crystallisation of hydrosilicates, II. *Earth. Sci. Rev.,* 22: 1 – 92.
Ells, R.W., 1902. The district around Kingston, Ontario. *Annu. Rep. Geol. Surv. Can.,* 1901-A: 176.
Ells, R.W., 1903. Notes on some interesting rock contacts in the Kingston district, Ontario. *Trans. R. Soc. Can.,* 9: 103.
Emery, K.O., 1950. Contorted strata at Newport Beach, California. *J. Sediment. Petrol.,* 20: 111 – 115.
Evans, G., 1965. Intertidal flat sediments and their environments of deposition in the Wash. *Q. J. Geol. Soc. London,* 121: 209 – 245.
Eyles, N. and Clark, B.M., 1985. Gravity-induced soft-sediment deformation in glaciomarine sequences of the Upper Proterozoic Port Askaig Formation, Scotland. *Sedimentology,* 32: 789 – 814.
Eyles, N., Eyles, C.H. and McCabe, A.M., 1989. Sedimentation in an ice-contact subaqueous setting: the mid-Pleistocene "North Sea Drifts" of Norfolk, U.K. *Quat. Sci. Rev.,* 8: 57 – 74.
Fairbridge, R.W., 1946. Submarine slumping and the location of oil bodies. *Am. Assoc. Pet. Geol. Bull.,* 30: 84 – 92.
Fairbridge, R.W., 1947. Possible causes of intraformational disturbances in Carboniferous varve rocks of Australia. *J. Proc. R. Soc. N. S. W.,* 81: 99 – 121.
Falcon, N.L., 1929. Pipes in coast sandstones. *Geol. Mag.,* 66: 48.
Foster, H.L. and Karlstrom, T.N.V., 1967. Ground breakage and associated effects in the Cook Inlet area, Alaska, resulting from the March 27, 1964, earthquake. *U.S. Geol. Surv. Prof. Pap.,* 543-F: 28 pp.
Francis, T.J.G., 1971. Effects of earthquakes on deep-sea sediments. *Nature,* 233: 98 – 101.
Fraser, G.S. and Cobb, J.C., 1974. Deformed sediments associated with outwash deposits in the Woodfordian (Pleistocene) of northeastern Illinois. *Geol. Soc. Am., Abstr. Progr.,* 6: 508 – 509.
French, H.M. and Gilbert, R., 1982. Periglacial phenomena near Churchill, Manitoba. *Nat. Can.,* 109: 433 – 444.
Frey, R.W., 1975. *The Study of Trace Fossils.* Springer, New York, N.Y., 562 pp.
Fritz, W.J. and Harrison, S., 1985. Transported trees from the 1982 Mount St. Helens sediment flows: their use as paleocurrent indicators. *Sediment. Geol.,* 42: 49 – 64.
Fuller, J.O., 1955. Source of Sharon conglomerate of northeastern Ohio. *Geol. Soc. Am. Bull.,* 66: 159 – 176.
Gableman, J.W., 1955. Cylindrical structures in Permian(?) siltstone, Eagle County, Colorado. *J. Geol.,* 63: 214 – 227.
Galloway, R.W., 1961. Ice wedges and involutions in Scotland. *Biul. Peryglacj.,* 10: 169 – 193.
Gangloff, P., 1970. Structures de gélisols réliques dans la région de Montréal. *Rev. Géogr. Montréal,* 24: 241 – 253.
Ghendt, E.D. and Henderson, R.A., 1965. Significance of burrowing structures in the origin of convolute laminae. *Nature,* 207: 1286 – 1287.
Gill, W.D. and Kuenen, Ph.H., 1958. Sand volcanoes on slumps in the Carboniferous of County Clare, Ireland. *Q. J. Geol. Soc. London,* 113: 441 – 460.
Glennie, K.W., 1970. *Desert Sedimentary Environments.* Developments in Sedimentology, 14, Elsevier, Amsterdam, 222 pp.
Glennie, K.W. and Buller, A.T., 1983. The Permian Weissliegend of NW Europe: the partial deformation of aeolian dune sands caused by the Zechstein transgression. *Sediment. Geol.,* 35: 43 – 81.
Goguel, J., 1938. Glissements sous-marins dans le Crétacé inférieur. *Bull. Soc. Géol. Fr.,* 5 – 8:

251 – 256.
Goldring, R., 1964. Trace fossils and the sedimentary surface in shallow-water marine sediments. In: L.M.J.U. van Straaten (Editor), *Deltaic and Shallow Marine Deposits.* Developments in Sedimentology, 1. Elsevier, Amsterdam, pp. 136 – 143.
Goldthwait, J.W. and Kruger, F.C., 1938. Weathered rock in and under the drift in New Hampshire. *Geol. Soc. Am. Bull.,* 49: 1183 – 1197.
Gosselet, J., 1890. Les "demoiselles" de Lihus. *Ann. Soc. Géol. Nord,* 17: 35 – 45.
Gottis, C., 1953. Les filons clastiques "intraformationnels" du "flysch" Numidien tunisien. *Bull. Soc. Géol. Fr.,* 6: 775 – 784.
Grabau, A.H., 1900. Siluro – Devonic contact on Eroe County, New York. *Geol. Soc. Am. Bull.,* 11: 347 – 376.
Gradziński, R., 1970. Sedimentation of dinosaur-bearing Upper Cretaceous deposits of the Nemegt Basin, Gobi Desert. *Paleontol. Pol.,* 21: 147 – 229.
Grant-MacKie, J.A. and Lowry, D.C., 1964. Upper Triassic rocks of Kiritehere, southwest Auckland, New Zealand, part 1: submarine slumping of Norian strata. *Sedimentology,* 3: 296 – 317.
Greenly, E., 1900. On sandstone pipes in the Carboniferous limestone at Dwlban Point, East Anglesey. *Geol. Mag.,* 7: 20 – 24.
Greensmith, J.T., 1956. Sedimentary structures in Upper Carboniferous of north and central Derbyshire, England. *J. Sediment. Petrol.,* 26: 343 – 355.
Gregory, M.R., 1969. Sedimentary features and penecontemporaneous slumping in the Waitemata Group, Whangapareoa Peninsula, North Auckland, New Zealand. *N. Z. J. Geol. Geophys.,* 12: 248 – 282.
Gresley, W.S., 1898. Clay-veins vertically intersecting coal measures. *Geol. Soc. Am. Bull.,* 9: 35 – 58.
Gripp, K., 1944. Enstehung und künftige Entwicklung der Deutschen Bucht. *Arch. Dtsch. Seewarte Marinbos,* 63: 1 – 44.
Gripp, K., 1952. Zwei Beiträge zur Frage der periglazialen Vorgänge. *Meyniania,* 1: 2 – 118.
Gry, H., 1968. Belastningsstrukturer (load casts) fra nexösandstenen. *Dan. Geol. Foren. Medd.,* 18: 101 – 106.
Gubler, Y., Bugnicourt, D., Faber, J., Kubler, B. and Nyssen, R., 1966. Essai de nomenclature et charactérisation des principales structures sédimentaires. *Chambre Synd. Rech. Prod. Pétrole Gaz Natur,* 291 pp.
Gulinck, M., 1948. Sur des phénomènes de glissement sous-aquatique et quelques structures particulières dans les sables Landéniens. *Bull. Soc. Belge Géol.,* 57: 12 – 30.
Gulinck, M., 1949. Poches et pipes de sable dans le Landénien, près de Havay. *Bull. Soc. Belge Géol.,* 58: 403 – 413.
Hadding, A., 1931. On subaqueous slides. *Geol. Fören. Stockholm Förh.,* 53: 377 – 393.
Hahn, F.F., 1913. Untermeerische Gleitung bei Trenton Falls (Nordamerika) und ihr Verhältnis zu ähnlichen Störungsbildern. *Neues Jahrb. Mineral. Geol. Paläontol.,* 36: 1 – 41.
Halicki, B., 1960. O roznej genezie strukturalnych deformacji osadów w srodowisku hydroplastycznym. *Biul. Peryglacj.,* 7: 165 – 167.
Hampton, M.A., 1975. Competence of fine-grained debris flows. *J. Sediment. Petrol.,* 45: 834 – 844.
Hampton, M.A., 1979. Buoyancy in debris flows. *J. Sediment. Petrol.,* 49: 753 – 758.
Hansen, E., Porter, S.C., Hall, B.A. and Hills, A., 1961. Decollement structures in glacial-lake sediments. *Geol. Soc. Am. Bull.,* 72: 1415 – 1418.
Häntzschel, W., 1935. Erhaltungsfähige Schliefspuren von Gischt am Nordseestrand. *Nat. Volk.,* 65: 461 – 465.
Häntzschel, W., 1939. Brandungswalle, Rippeln und Fleisfiguren am Strande von Wangeroog. *Nat. Volk,* 69: 40 – 48.
Häntzschel, W., 1941. Entgasungs-Krater im Wattenschlick. *Nat. Volk,* 71: 312 – 314.
Hardy, C.T. and Williams, J.S., 1959. Columnar contemporaneous deformation. *J. Sediment. Petrol.,* 29: 281 – 287.
Harms, J.C., 1965. Sandstone dykes in relation to Laramide faults and stress distibution in the southern Front Range, Colorado. *Geol. Soc. Am. Bull.,* 76: 981 – 1002.
Harms, J.C., MacKenzie, D.B. and MacCubbin, D.G., 1963. Stratification in modern sands of the Red River, Louisiana. *J. Geol.,* 71: 566 – 580.
Harris, C. and Ellis, S., 1980. Micromorphology of soils in soliflucted materials, Okstindan, Northern

Norway. *Geoderma,* 23: 11 – 29.

Haughton, S.H., Krige, L.J. and Krige, A.V., 1925. On intraformational folding connected with glacial beds in the Table Mountain Sandstone. *Proc. Geol. Soc. S. Afr.,* 28: 19 – 25.

Hawley, J.E. and Hart, R.C., 1934. Cylindrical structures in sandstone. *Geol. Soc. Am. Bull.,* 45: 1017 – 1034.

Hay, R., 1892. Sandstone dikes in northwestern Nebraska. *Geol. Soc. Am. Bull.,* 3: 50 – 55.

Hayashi, T., 1966. Clastic dykes in Japan. *Jpn. J. Geol. Geogr.,* 37: 1 – 20.

Hayashi, T., 1971. "Quake sheet" in the Miocene series of Saku-shima, Aichi prefecture, Japan. *Aichi Univ. Educ. Bull.,* 20: 193 – 201.

Henderson, S.M.K., 1935. Ordovician submarine disturbances in the Girvan district. *Trans. R. Soc. Edinburgh,* 58: 487 – 507.

Hendry, H.E. and Stauffer, M.R., 1977. Penecontemporaneous folds in cross-bedding: inversion of facin criteria and mimicry of tectonic folds. *Geol. Soc. Am. Bull.,* 88: 809 – 812.

Heron, S.D., Judd, J.B. and Johnson, H.S., 1971. Clastic dykes associated with soil horizons in the North and South Carolina coastal plain. *Geol. Soc. Am. Bull.,* 82: 1801 – 1810.

Hesse, R., 1976. Einige ungewöhnliche sekundäre Sedimentstrukturen im lakustrinen Unterkarbon Neuschottlands (Kanada) und ihre Deutung als Erdbebenzeiger. *Eclogae Geol. Helv.,* 69: 196 – 200.

Hills, E.S., 1958. Load-casts and flame structures. *Geol. Mag.,* 95: 171 – 172.

Hirayama, J. and Nakajima, T., 1977. Analytical study of turbidites, Otadai Formation, Baso Peninsula, Japan. *Sedimentology,* 24: 747 – 779.

Hobbs, W.H., 1907. *Earthquakes – An Introduction to Seismic Geology,* Appleton-Century, New York, N.Y., 336 pp.

Holland, C.H., 1959. On convolute bedding in the Lower Ludlovian rocks of northeast Radnorshire. *Geol. Mag.,* 96: 230 – 236.

Holland, C.H., 1960. Load-cast terminology and origin of convolute bedding. *Geol. Soc. Am. Bull.,* 71: 633 – 634.

Holland, C.H., 1961. Origin of convolute lamination. *Geol. Mag.,* 98: 168.

Horberg, L., 1951. Intersecting minor ridges and periglacial features in the Lake Agassiz Basin, North Dakota. *J. Geol.,* 59: 1 – 18.

Horowitz, D.H., 1982. Geometry and origin of large-scale deformation structures in some ancient wind-blown sand deposits. *Sedimentology,* 29: 155 – 180.

Houtz, R.E. and Wellman, H.W., 1962. Turbidity current at Kadavu Passage, Fiji. *Geol. Mag.,* 99: 57 – 62.

Howard, J.D. and Lahrengel, C.F., II, 1969. Large nontectonic deformational structures from Upper Cretaceous rocks of Utah. *J. Sediment. Petrol.,* 39: 1032 – 1039.

Hubert, J.F., 1966. Sedimentary history of Upper Ordovician geosynclinal rocks, Girvan, Scotland. *J. Sediment. Petrol.,* 36: 677 – 699.

Jahn, A., 1971. Plejstocenskie struktury glacitektoniczne w swietle obserwacji z obzarów wspolczesnie zlodowacenonych. *Poznan. Tow. Przyj. Nauk, Wydzial Matemat.-Przyr. Prace Kom. Geogr.-Geol.,* 13: 131 – 141.

Jahn, A., 1977. Struktury zwiazane z klinami lodowyni w osadach plejstocenskich. *Stud. Geol. Polonica,* 52: 177 – 194.

Jahn, A. and Czerwiński, J., 1965. The role of impulses in the process of periglacial soil structure formation. *Acta Univ. Wratislaviensis,* 44; *Stud. Geogr.,* 7: 3 – 24.

Jardine, W.G., 1965. Note on a temporary exposure in central Glasgow of Quaternary sediments with slump and load structures. *Scott. J. Geol.,* 1: 221 – 224.

Jenkins, O.P., 1925. Clastic dikes of eastern Washington and their geologic significance. *Am. J. Sci.* (5th Ser.), 10: 234 – 246.

Jersak, J., 1975. Frost fissures in loess. *Biul. Peryglacj.,* 24: 245 – 258.

Johnson, H.S. and Heron, D., Jr., 1965. Slump features in the McBean Formation and younger beds, Riley Cut, Calhoun County, South Carolina. *Geol. Notes,* 9: 37 – 44.

Johnsson, G., 1959. True and false ice-wedges in southern Sweden. *Geogr. Ann.,* 41: 15 – 33.

Johnsson, G., 1962. Periglacial phenomena in southern Sweden. *Geogr. Ann.,* 44: 378 – 404.

Jones, B.G., 1972. Deformational structures in siltstone resulting from the migration of an Upper Devonian aeolian dune. *J. Sediment. Petrol.,* 42: 935 – 940.

Jones, G.P., 1961. Sedimentary structures in the Bima Sandstone. *Niger. Geol. Surv.,* 6: 14 pp.

Jones, G.P., 1962. Deformed cross-stratification in Cretaceous Bima Sandstone, Nigeria. *J. Sediment. Petrol.,* 32: 231 – 239.
Jones, O.T., 1937. On the sliding or slumping of submarine sediments in Derbyshire, north Wales, during the Ludlow period. *Q. J. Geol. Soc. London,* 93: 241 – 283.
Jones, O.T., 1939. The geology of the Colwyn Bay district; a study of submarine slumping during the Salopian period. *Q. J. Geol. Soc. London,* 95: 335 – 376.
Jones, O.T., 1953. On submarine slumping in the Lower Ludlow rocks of north Wales. *Geol. Mag.,*, 90: 220 – 221.
Kavanaugh, S.J., 1889. On modern concretions from the St. Lawrence. With remarks (by J. W. Dawson) on cylinders found in Potsdam Sandstone. *Can. Rev. Sci.,* 3: 292 – 294.
Kaye, C.A. and Power, W.R., 1954. A flow cast of very recent date from northeastern Washington. *Am. J. Sci.,* 252: 309 – 310.
Kelling, G. and Walton, E.K., 1957. Load cast structures: their relationship to upper surface structures and their mode of formation. *Geol. Mag.,* 94: 481 – 491.
Kelling, G. and Williams, B.P.J., 1966. Deformation structure of sedimentary origin in the Lower Limestone Shales (basal Carboniferous) of South Pembrokeshire, Wales. *J. Sediment. Petrol.,* 36: 927 – 939.
Kent, P.E., 1945. Contemporaneous disturbance in lacustrine beds in Kenya. *Geol. Mag.,* 82: 130 – 135.
Kerns, E., 1971. Clay dikes in the Pittsburgh coal (Pennsylvanian) of southwestern Greene County, Pennsylvania. M.Sc. thesis (unpubl.), Univ. of West Virginia, Morgantown, W. Va.
Kiersch, G.A., 1950. Small-scale structures and other features of Navajo Sandstone, northern part of San Rafael Swell, Utah. *Am. Assoc. Pet. Geol. Bull.,* 34: 923 – 942.
Kindle, E.M., 1914. Columnar structure in limestone. *Bull. Mus. Can.,* 2: 35 – 44.
Kindle, E.M., 1916. Small pit and mound structures developed during sedimentation. *Geol. Mag.,* 3: 443 – 444.
Kindle, E.M., 1917. Deformation of unconsolidated beds in Nova Scotia and southern Ontario. *Geol. Soc. Am. Bull.,* 28: 323 – 334.
Kingma, J.T., 1958. The Tangaporutuan sedimentation in central Hawke's Bay, New Zealand. *N. Z. J. Geol. Geophys.,* 1: 1 – 30.
Kirkland, D.W. and Anderson, R.Y., 1970. Microfolding in the Castile and Todilto evaporites, Texas and New Mexico. *Geol. Soc. Am. Bull.,* 81: 3259 – 3282.
Kivinen, E., 1971. Pari volokuvaa kuormituksen aiheuttamista muutoksista heikkamassa. *Geologi,* 23: 65.
Klinger, F.E., 1939. Sediment-Rollen (unterwasser Gleitung) in Muschelsandstein bei Saarlantern. *Senckenbergiana Lethaea,* 21: 311 – 314.
Königsson, L.-K. and Linde, L.A., 1977. Glaciotectonically disturbed sediments at Ronnerum on the island of Oland. *Geol. Fören. Stockholm Förh.,* 99: 68 – 72.
Kostyaew, A.G., 1975. The periglacial zone of western Eurasian plains. *Biul. Peryglacj.,* 24: 55 – 68.
Kowalczyk, G., 1974. Kryoturbationsartige Sedimentstrukturen im Pliozän und Altquartär der südlichen Niederrheinischen Bucht. *Eiszeitalter Ggw.,* 25: 141 – 156.
Kraus, M.J., 1988. Nodular remains of Early Tertiary forests, Bighorn Basin, Wyoming. *J. Sediment. Petrol.,* 58: 888 – 893.
Krinitsky, E.L. and Bonis, S.B., 1976. Notes on earthquake shaking in soils – Guatemala earthquake of 4 February 1976. *U.S. Army Chief Eng. Inf. Rep.,* 33 pp.
Kruger, F.C., 1938. A clastic dike of glacial origin. *Am. J. Sci.,* 35: 305 – 307.
Książkiewicz, M., 1951. Slip bedding in the Carpathian flysch. *Rocz. Pol. Tow. Geol.,* 19: 493 – 501.
Książkiewicz, M., 1958. Osuwiska podmorskie we fliszu karpeckim. *Rocz. Pol. Tow. Geol.,* 28: 123 – 150.
Kuenen, Ph.H., 1949. Slumping in the Carboniferous rocks of Pembrokeshire. *Q. J. Geol. Soc. London,* 104: 365 – 385.
Kuenen, Ph.H., 1953a. Graded bedding with observations on Paleozoic rocks of Britain. *Verh. K. Ned. Akad. Wet., Amsterdam, Afd. Natuurk.,* 20: 1 – 47.
Kuenen, Ph.H., 1953b. Significant features of graded bedding. *Am. Assoc. Pet. Geol. Bull.,* 37: 1044 – 1066.
Kuenen, Ph.H., 1956. The difference between sliding and turbidity flow. *Deep-Sea Res.,* 3: 134 – 139.
Kuenen, Ph.H., 1957. Sole markings of graded graywacke beds. *J. Geol.,* 65: 231 – 258.

Kuenen, Ph.H., 1958. Experiments in geology. *Trans. Geol. Soc. Glasgow,* 23: 1 – 28.
Kuenen, Ph.H., 1961. Some arched and spiral structures in sediments. *Geol. Mijnbouw,* 40: 71 – 74.
Kuenen, Ph.H., 1963. Experimentele sedimentstructuren. *Verh. K. Ned. Akad. Wet., Amsterdam, Afd. Natuurk.,* 72: 65 – 66.
Kuenen, Ph.H., 1965. Value of experiments in geology. *Geol. Mijnbouw,* 44: 22 – 36.
Kuenen, Ph.H. and Carozzi, A.V., 1953. Turbidity currents and sliding in geosynclinal basins in the Alps. *J. Geol.,* 61: 363 – 372.
Kuenen, Ph.H. and Menard, H.W., 1952. Turbidity currents, graded and non-graded deposits. *J. Sediment. Petrol.,* 22: 83 – 96.
Kuenen, Ph.H. and Migliorini, C.I., 1950. Turbidity currents as a cause of graded bedding. *J. Geol.,* 58: 91 – 127.
Kuenen, Ph.H. and Prentice, J.W., 1957. Flow-markings and load-casts. *Geol. Mag.,* 94: 173 – 174.
Kuenen, Ph.H. and Sanders, J.E., 1956. Kulm and Flözleeres graywackes, Sauerland and Oberharz, Germany. *Am. J. Sci.,* 254: 649 – 671.
Kühn-Velten, H., 1955. Subaquatische Rutschungen in höherem Oberdevon des Sauerlandes. *Geol. Rundsch.,* 44: 3 – 25.
Kumar, S. and Singh, T., 1982. Sandstone dykes in Siwalik Sandstone – sedimentology and basin analysis – Subansiri district (NEFA), Eastern Himalaya. *Sediment. Geol.,* 33: 217 – 236.
Kunert, R., 1964. Kleinzyklen in der oberen Schiefertonzone SU 5 des nordöstlichen Harzvorlandes. *Ber. Geol. Ges. DDR,* 9: 35 – 40.
Lachenbruch, A.H., 1962. Mechanics of thermal contraction cracks and ice-wedge polygons in permafrost. *Geol. Soc. Am., Spec. Pap.,* 70: 69 pp.
Lahee, F.H., 1914. Contemporaneous deformation: a criterion for aqueo-glacial sedimentation. *J. Geol.,* 22: 786 – 790.
Lajoie, J. (Editor), 1970. Flysch Sedimentology in North America. *Geol. Assoc. Can., Spec. Pap.,* 7: 272 pp.
Lambrecht, L. and Thorez, J., 1966. Filons clastiques intraformationnels dans le Namurien de Belgique. *C. R. Acad. Sci.,* 263: 1556 – 1559.
Laming, D.J.C., 1966. Imbrication, paleocurrents, and other sedimentary features in the lower New Red Sandstone, Devonshire, England. *J. Sediment. Petrol.,* 36: 940 – 959.
Lamont, A., 1938. Contemporaneous slumping and other problems at Bray Series, Ordovician and Lower Carboniferous horizons, in County Dublin. *Proc. R. Irish Acad.,* 45: 1 – 32.
Lamont, A., 1957. Slow anti-dunes and flow marks. *Geol. Mag.,* 94: 472 – 480.
Lawler, T.B., 1923. On the occurrence of sandstone dikes and chalcedony veins in the White River Oligocene. *Am. J. Sci.,* 5: 160 – 172.
Leeder, M.R., 1975. Pedogenic carbonates and flood sediment accretion rates: a quantitative model for alluvial arid-zone lithofacies. *Geol. Mag.,* 112: 257 – 270.
Lindström, M., 1971. Small scale domes and piercing-structures in the Lower Ordovician limestones of Oeland (S.E. Sweden). Comment on a paper by W.A. van Wamel. *Proc. K. Ned. Akad. Wet.,* B-74: 93 – 95.
Lippert, H., 1937. Gleit-Faltung in subaquatischen und subaerischen Gestein. *Senckenbergiana Lethaea,* 19: 355 – 375.
Lovell, J.P.B., 1974. Sand volcanoes in the Silurian rocks of Kirkcudbrightshire. *Scott. J. Geol.,* 10: 161 – 162.
Lowe, D.R., 1975a. The origin of convolute lamination. *Am. Assoc. Pet. Geol., Soc. Econ. Paleontol. Mineral., Abstr. Annu. Meet.,* 2: 92.
Lowe, D.R., 1975b. Water escape structures in coarse-grained sediments. *Sedimentology,* 22: 157 – 204.
Lowe, D.R. and LoPiccolo, R.D., 1974. The characteristics and origins of dish and pillar structure. *J. Sediment. Petrol.,* 44: 484 – 501.
Lundqvist, J., 1967. Submoräna sediment i Jämtlands Län. *Sver. Geol. Unders.,* C 618, 267 pp.
Lupher, R.L., 1944. Clastic dikes of the Columbia Basin region, Washington and Idaho. *Geol. Soc. Am. Bull.,* 55: 1431 – 1461.
Lyell, C., 1841. *Elements of Geology.* London, 2nd ed.
Lyell, C., 1851. On fossil rain marks of the Recent, Triassic, and Carboniferous periods. *Q. J. Geol. Soc. London,* 7: 238 – 247.
Macar, P., 1948. Les pseudonodules du Famennien et leur origine. *Ann. Soc. Géol. Belg.,* 72: 47 – 74.

Macar, P., 1950. Pseudo-nodules en terrains meubles. *Ann. Soc. Géol. Belg.,* 75: 111 – 115.
Macar, P., 1958. Les déformations péné-contemporaines de la sédimentation. *Rev. Quest. Sci.,* 19: 5 – 33.
Macar, P., 1965. Les déformations non-tectoniques des roches sédimentaires. *Rev. Univers. Mines,* 4: 141 – 150.
Macar, P. and Antun, P., 1949. Pseudo-nodules et glissement sousaquatique dans l'Emsien inférieur de l'Oesling. *Ann. Soc. Géol. Belg.,* 73-B: 121 – 150.
MacKay, J.R., 1958. The valley of the lower Anderson river, N.W.T. *Geogr. Bull.,* 11: 36 – 56.
Makowska, A., 1961. Diapiry ilow warwowych w Baniosze pod Warszawa. In: S.Z. Różycki (Editor), Prace o Plejstocenie Polski srodkowej. *Pol. Akad. Nauk, Warsawa:* 159 – 176.
Maltman, A., 1984. On the term "soft-sediment deformation". *J. Struct. Geol.,* 6: 589 – 592.
Mandl, G., De Jong, L.N.J. and Maltha, A., 1977. Shear zones in granular material. An experimental study of their structure and mechanical genesis. *Rock Mech.,* 9: 95 – 144.
Mandl, G. and Luque, F., 1970. Fully developed plastic shear flow of granular materials. *Géotechnique,* 20: 277 – 307.
Marchant, S. and Black, C.D.G., 1959. The nature of the clay pebble-beds and associated rocks of southwest Ecuador. *Q. J. Geol. Soc. London,* 95: 317 – 338.
Marschalko, R., 1972. Termin klasticke zily. *Geol. Pr.,* 2 pr., 58: 231 – 238.
Martheiades, E., 1971. Erosion and deposition of cohesive materials. In: H. W. Shen (Editor), *River Mechanics.* Private publication, pp. 25 – 91.
Matisto, A., 1960. Fossilista routamaat a Espoon Tapiolassa. *Geologi,* 12: 18 – 19.
Matthews, P.E., 1961. Slump structures in the Table Mountain Series of Natal. *Trans. Proc. Geol. Soc. S. Afr.,* 64: 55 – 71.
Maxon, J.H., 1940. Gas pits in non-marine sediments. *J. Sediment. Petrol.,* 10: 142 – 145.
McArthur, D.S. and Onesti, L.J., 1970. Contorted structures in Pleistocene sediments near Lansing, Michigan. *Geogr. Ann.,* 52: 186 – 193.
McCall, G.J.H., 1962. Reversed graded bedding. *N. Z. J. Geol. Geophys.,* 5: 666.
McCallie, S.W., 1903. Sandstone dikes near Columbus, Georgia. *Am. Geol.,* 32: 199 – 202.
McCrossan, R.G., 1958. Sedimentary "boudinage" structure in the Upper Devonian Ireton Formation of Alberta. *J. Sediment. Petrol.,* 28: 316 – 320.
McIver, N.L., 1961. Upper Devonian marine sedimentation in the central Appalachians. Ph.D. thesis, Johns Hopkins Univ., Maryland, Md., 346 pp.
McKee, E.D., 1938. Original structures in the Colorado river flood deposits. *J. Sediment. Petrol.,* 8: 77 – 83.
McKee, E.D., 1945. Small-scale structures in the Coconino Sandstone of northern Arizona. *J. Geol.,* 53: 313 – 325.
McKee, E.D., 1954. Stratigraphy and history of the Moenkopi Formation of Triassic age. *Geol. Soc. Am. Mem.,* 61: 133 pp.
McKee, E.D., 1964. Inorganic sedimentary structures. In: J. Imbrie and N.D. Newell (Editors), *Approaches to Paleoecology.* Wiley, New York, N. Y., pp. 275 – 295.
McKee, E.D. and Bigarella, J.J., 1972. Deformational structures in Brazilian coastal dunes. *J. Sediment. Petrol.,* 42: 670 – 681.
McKee, E.D., Crosby, E.J. and Berryhill, H.L., Jr., 1967. Flood deposits, Bijou Creek, Colorado, June 1965. *J. Sediment. Petrol.,* 37: 829 – 851.
McKee, E.D., Douglas, J.R. and Rittenhouse, S., 1971. Deformation of lee-side laminae in eolian dunes. *Geol. Soc. Am. Bull.,* 82: 359 – 378.
McKee, E.D., Eveson, C.G. and Grundy, W.D., 1953. Studies in sedimentology of the Shinarump Conglomerate of northeastern Arizona. *U.S. A. En. C.,* RME-3089: 48 pp.
McKee, E.D. and Goldberg, M., 1969. Experiments on formation of contorted structures in mud. *Geol. Soc. Am. Bull.,* 80: 231 – 244.
McKee, E.D., Reynolds, M.A. and Baker, C.H., 1962a. Laboratory studies on deformation in unconsolidated sediment. *U.S. Geol. Surv., Prof. Pap.,* 450-D: 151 – 155.
McKee, E.D., Reynolds, M.A. and Baker, C.H., 1962b. Experiments on intraformational recumbent folds in cross-bedded sand. *U.S. Geol. Surv., Prof. Pap.,* 450-D: 155 – 160.
Meier, R. and Thomas, U., 1969. Differences between synsedimentary and tectonic structures of the Tanne Graywacke; Guntersberge – Alexisbad – Gernrode, Harz. *Geologie,* 18: 334 – 343.

Metzner, A.B. and Whitlock, M., 1958. Flow behavior of concentrated (dilatant) suspensions. *Trans. Soc. Rheol.*, 2: 239 – 254.
Michalska, Z., 1957. Struktury peryglacjalne w osadach zbiornika interstadialnego w Gaskach kolo Ciechanowa. *Biul. Peryglacj.*, 5: 91 – 103.
Michel, J.-P., 1962. Description de déformations quaternaires semblables à des diapirs dans les alluvions anciennes de la Seine et de la Marne près de Paris. *Bull. Soc. Géol. Fr.*, Sér. 7, 4: 795 – 799.
Michel, J.-P., 1975. Periglaciaire des environs de Paris. *Biul. Peryglacj.*, 24: 259 – 352.
Middlemost, E., 1967. Note on the soft-sediment deformational structures in the basal beds of the Table Mountain Series, Cape Town. *Trans. Proc. Geol. Soc. S. Afr.*, 71: 119 – 120.
Middleton, G.V. and Hampton, M.A., 1973. Sediment gravity flows: mechanics of flow and deposition. In: *Turbidites and Deep Water Sedimentation. Pacific Sect., Soc. Econ. Paleontol. Mineral.*, Los Angeles, Calif., pp. 1 – 38.
Migliorini, C.I., 1950. Dati a conferma della risedimentazione delle arenarie del macigno. *Mem. Soc. Toscana Sci. Nat.*, 57: 1 – 15.
Mikadze, G.A., 1967. Contorted bedding in the lower Eocene in the vicinity of Borzhom. *Akad. Nauk Gruz. S.S.R. Soobshch.*, 47: 351 – 356.
Mikulenko, K.N., 1967. Submarine slump structures in Paleocene and Eocene deposits. *Int. Geol. Rev.*, 9: 1353 – 1364.
Miller, W.J., 1908. Highly folded between non-folded strata of Trenton Falls, N.Y. *J. Geol.*, 16: 428 – 433.
Miller, W.J., 1922. Intraformational compacted rocks. *J. Geol.*, 30: 587 – 610.
Mills, P.C., 1983. Genesis and diagnostic value of soft-sediment deformation structures – a review. *Sediment. Geol.*, 35: 83 – 104.
Miser, H.D., 1935. Cylindrical structures in sandstone (discussion). *Geol. Soc. Am. Bull.*, 46: 2008 – 2011.
Misik, M., 1968. Traces of submarine slumping and evidence of hypersaline environment in the middle Triassic of the west Carpathian Mountains. *Slov. Akad. Vied Geol. Sbornik*, 19: 205 – 224.
Mojski, J.E., 1961. Stratigraphy of cryoturbate structures in the Würm-age deposits in the southern part of the Dorohucza Basin (Lublin upland). *Biul. Peryglacj.*, 10: 235 – 256.
Monroe, J.N., 1950. Origin of the clastic dikes of the Rockwell area, Texas. *Field Lab.*, 18: 133 – 143.
Monroe, J.N., 1969. Slumping structures caused by organically derived gases in sediments. *Science*, 164: 1394 – 1395.
Monroe, W.H., 1932. Earth cracks in Mississippi. *Am. Assoc. Pet. Geol. Bull.*, 16: 214 – 215.
Montenat, C., 1980. Rélation entre déformations synsédimentaires et paléoséismicité dans le Messinien de San Miguel de Salinas (Cordillères Bétiques orientales, Espagne). *Bull. Soc. Géol. Fr.*, 7 – 22: 501 – 509.
Montenat, C. and Seilacher, A., 1978. Les turbidites messiniennes à *Helminthoides* et *Palaeodictyon* du bassin de Vera (Cordillères Bétiques orientales). Indications paléobathymétriques. *Bull. Soc. Géol. Fr.*, 7 – 20: 319 – 322.
Moore, D.G. and Scruton, P.C., 1957. Minor internal structures of some recent unconsolidated sediments. *Am. Assoc. Pet. Geol. Bull.*, 41: 2722 – 2751.
Morell, J. and Hilly, J., 1956. Nouvelles observations sur les formations quaternaires dans le département de Bône et particulièrement dans le Massif du Cap de Fer et de l'Edough. *Quaternaria*, 3: 179 – 201.
Moret, L., 1945. A propos du mode de formation des "filons clastiques". *Ann. Univ. Grenoble*, 21: 99 – 101.
Morgan, J.P., 1961. Genesis and paleontology of Mississippi mud lumps, part 1. *La. Geol. Surv., Geol. Bull.*, 35, 116 pp.
Morgenstern, N.R., 1967. Submarine slumping and the initiation of turbidity currents. In: *Marine Geotechnique.* Proc. Int. Res. Conf. (Monticello, Ill.). Univ. Illinois Press, Urbana, Ill., pp. 189 – 220.
Mörner, N.A., 1972. The first report on till wedges in Europe and late-Weichselian ice flows over southern Sweden. *Geol. Fören. Förh.*, 94: 581 – 587.
Mörner, N.A., 1973. New find of till wedges in Nova Scotia, Canada. *Geol. Fören. Förh.*, 95: 272 – 273.
Morrow, D.W., 1972. An injection structure in a Permian limestone, northern British Columbia. *J. Sediment. Petrol.*, 42: 230 – 235.

Mountain, E.D., 1964. Overturned cross-bedding in Wittenberg Quartzite. *Trans. Proc. Geol. Soc. S. Afr.,* 67: 203 – 209.
Murphy, M.A. and Schlanger, S.O., 1962. Sedimentary structures in Ilhad and Sao Sebastio Formations (Cretaceous), Reconcavo Basin, Brazil. *Am. Assoc. Pet. Geol.,* 46: 457 – 477.
Mutti, E. and Ricci Lucchi, F., 1972. Le torbiditi dell' Appennino settentrionali: introduzione all' analysi di facies. *Mem. Soc. Geol. Ital.,* 2: 161 – 169.
Naganuma, Y., 1973. The discovery and geological meaning of minor slump structure from the Koshiba Formation of the Kazusa Group, in southern area of Yokohama City, Kanagaira prefecture, Japan. *J. Geol. Soc. Jpn.,* 79: 311 – 313.
Nagtegaal, P.J.C., 1963. Convolute lamination, metadepositional ruptures and slumping in an exposure near Pobla de Segur (Spain). *Geol. Mijnbouw,* 42: 363 – 374.
Nagtegaal, P.J.C., 1965. An approximation to the genetic classification of non-organic sedimentary structures. *Geol. Mijnbouw,* 44: 347 – 352.
Natland, M.L. and Kuenen, Ph.H., 1951. Sedimentary history of the Ventura Basin, California, and the action of turbidity currents. *Soc. Econ. Paleontol. Mineral., Spec. Publ.,* 2: 76 – 107.
Naylor, M.A., 1981. Debris flow (olistostromes) and slumping on a distal passive continental margin: the Palombini limestone – shale sequence of the northern Apennines. *Sedimentology,* 28: 837 – 852.
Negendank, J.F.W., 1972. Turbidite aus dem Unterrotliegenden des Saar – Nake-Gebietes (ein Beitrag zur Sedimentologie limnischer Ablagerungen). *Neues Jahrb. Geol. Paläontol. Monatsh.,* 5: 561 – 572.
Neruchev, S.G. and Il'Im, A.F., 1959. About syngenetic deformation of layers in the Sinisi deposits in the river Mai in relation to evidence of seismic shock. Vses. Nauchno-Issled. *Geol. Razved. Inst. Trudy,* 130: 75 – 80.
Newcomb, R.C., 1962. Hydraulic injection of clastic dykes in the Touchet Beds, Washington, Oregon and Idaho. *Geol. Soc. Oreg. Cty Geol. News Lett.,* 28: 70.
Newell, N.D., Rigby, J.K., Fischer, A.G., Whiteman, A.J., Hickox, J.E. and Bradley, J.S., 1953. *The Permian Reef Complex of the Guadaloupe Mountains Region, Texas and New Mexico.* Freeman, San Francisco, Calif. 236 pp.
Newsom, J.F., 1903. Clastic dikes. *Geol. Soc. Am. Bull.,* 14: 227 – 268.
Nichols, D.R., 1960. Slump structures in Pleistocene lake sediments of the Copper River Basin, Alaska. *U.S. Geol. Surv. Prof. Pap.,* 400-B: 353 – 354.
Nichols, D.R. and Yehle, L.A., 1961. Mud volcanoes in the Copper River Basin, Alaska. In: *Geology of the Arctic.* Univ. Toronto Press, pp. 1063 – 1087.
Nilsen, T.H. and Simoni, Jr., T.R., 1973. Deep-sea fan paleocurrent patterns of the Eocene Butano Sandstone, Santa Cruz Mountains, California. *U.S. Geol. Surv. J. Res.,* 1: 439 – 452.
Nocita, B.W., 1988. Soft-sediment deformation (fluid escape) features in a coarse-grained pyroclastic surge deposit, north-central New Mexico. *Sedimentology,* 35: 275 – 285.
O'Brien, P.E., 1989. Subglacial sedimentary features in Late Palaeozoic sedimentary rocks, central Victoria, Australia. *Sediment. Geol.,* 61: 1 – 15.
Okko, M., 1967. Convolute lamination in a Late Pleistocene deposit at Pannujarvi, Tuu-os, Finland. *Finland Comm. Geol. Bull.,* 229: 123 – 131.
Olausson, E. and Uusitalo, S., 1963. On the influence of seismic vibrations on sediments. *C.R. Finland Soc. Geol.,* 35: 101 – 114.
Oldershaw, W., 1960. Probable sand volcanoes in the lower Proterozoic at Tennant Creek, N.T. *J. Geol. Soc. Austr.,* 6: 197 – 199.
Oldham, D.R. and Mallet, R., 1872. Notice on some of secondary effects of the earthquake of 10th January, 1869, in Cachar. *Q. J. Geol. Soc. London,* 28: 255 – 270.
Oomkens, E., 1966. Environmental significance of sand dikes. *Sedimentology,* 7: 145 – 148.
Packham, G.H., 1954. Sedimentary structures as an important factor in the classification of sandstones. *Am. J. Sci.,* 252: 466 – 476.
Parea, G.C., 1962. Il flysch ad elmintoidi di Serra Mazzoni (Appennino modense); strutture sedimentarie e modo di deposizione. *Bull. Geol. Soc. Ital.,* 80: 159 – 174.
Pavlow, A.D., 1896. On dikes of Oligocene sandstone in Russia. *Geol. Mag.,* N. S. 3: 49 – 53.
Peacock, J.D., 1966. Contorted beds in the Permo-Triassic aeolian sandstones of Morayshire. *Bull. G. B. Geol. Surv.,* 24: 157 – 162.
Pederson, G.K. and Surlyk, F., 1977. Dish structure in Eocene volcanic ash layers, Denmark. *Sedimentology,* 24: 581 – 590.

Pękala, K., 1967. Load structures in the polygenetic terrace of the San River at the mouth of the Wisznia river. *Ann. Univ. Mariae Curie-Sklodowska,* B-19: 177 – 190.
Pepper, J.F., De Witt, Jr., W. and Demarest, D.F., 1954. Geology of the Bedford Shale and Berea Sandstone in the Appalachian Basin. *U.S. Geol. Surv. Prof. Pap.,* 259: 111 pp.
Perry, W.J. and Dickens, J.M., 1960. The geology of the Badgeradda area, western Australia. *Austr. Bur. Miner. Resour. Geol. Geophys. Rep.,* 46: 38 pp.
Peterson, G.L., 1965. Emplacement of the northwest Sacramento Valley sandstone dikes. *Geol. Soc. Am. Spec. Pap.,* 87: 222 – 223.
Peterson, G.L., 1966. Structural interpretation of sandstone dikes, northwest Sacramento Valley, California. *Geol. Soc. Am. Bull.,* 77: 833 – 842.
Pettijohn, F.J., 1975. *Sedimentary Rocks,* 3rd ed. Harper and Row, New York, N.Y., 628 pp.
Pettijohn, F.J. and Potter, P.E., 1964. *Atlas and Glossary of Primary Sedimentary Structures.* Springer, New York, N.Y., 370 pp.
Pettijohn, F.J., Potter, P.E. and Siever, R., 1973. *Sand and Sandstones.* Springer, New York, N.Y., 618 pp.
Péwé, T.L., 1959. Sand-wedge polygons (tesselations) in the McMurdo Sound region, Antarctica. *Am. J. Sci.,* 257: 545 – 552.
Phoenix, D.A., 1958. Sandstone cylinders as possible guides to paleomovement of ground water. *Guidebook 9th Field Conf., N.M. Geol. Soc.,* pp. 194 – 196.
Pickering, K.T., 1979. Possible retrogressive flow slide deposits from the Kongsfjord Formation: a Precambrian submarine fan, Finnmark, N. Norway. *Sedimentology,* 26: 295 – 306.
Pierce, S.E. and Peterson, G.L., 1974. Some mineral and elemental variations in the Bodden Canyon Formation and their bearing on the origin of sandstone dikes, northeast Sacramento Valley, California. *Geol. Soc. Am. Abstr.,* 6: 235 – 236.
Piper, D.J.W. and Marshall, N.F., 1969. Bioturbation of Holocene sediments on La Jolla deep sea fan, California. *J. Sediment. Petrol.,* 39: 601 – 606.
Plessman, W., 1961. Stromungsmarken in klastischen Sedimenten und ihre geologische Auswertung. *Geol. Jahrb.,* 78: 503 – 566.
Plint, A.G., 1983. Liquefaction, fluidization and erosional structures associated with bituminous sands of the Bracklesham Formation (Middle Eocene) of Dorset, England. *Sedimentology,* 30: 525 – 535.
Plint, A.G., Van de Poll, H.W. and Patel, I.W., 1983. Experiments in rheoplasis during sediment intrusion. *Marit. Sediments Atl. Geol.,* 19: 11 – 19.
Plummer, P.S. and Gostin, A.V., 1981. Shrinkage cracks: desiccation or synaeresis? *J. Sediment. Petrol.,* 51: 1147 – 1156.
Postma, G., 1983. Water escape structures in the context of a depositional model of a mass flow dominated conglomeratic fan delta (Abrioja Formation, Pliocene, Almería Basin, SE Spain). *Sedimentology,* 30: 91 – 103.
Postma, G., Roep, T.B. and Ruegg, H.J., 1983. Sandy-gravelly mass-flow deposits in an ice-marginal lake (Saalian, Leuvenumsche Beek Valley, Veluwe, The Netherlands), with emphasis on plug-flow deposits. *Sediment. Geol.,* 34: 59 – 82.
Potter, P.E. and Pettijohn, F.J., 1963. *Paleocurrents and Basin Analysis.* Springer, Berlin, 296 pp.
Potter, P.E. and Pettijohn, F.J., 1977. *Paleocurrents and Basin Analysis,* 2nd ed. Springer, New York, N.Y., 425 pp.
Prentice, J.E., 1956. The interpretation of flow markings and load casts. *Geol. Mag.,* 93: 393 – 400.
Prentice, J.E., 1958. Anti-dunes and flame structures. *Geol. Mag.,* 95: 169 – 171.
Prentice, J.E., 1960. Flow structures in sedimentary rocks. *J. Geol.,* 68: 217 – 224.
Puziewicz, J. and Wojewoda, J., 1984. Origin of load structures and graded bedding of dark minerals in the Strzegom granites (SW Poland) – a sedimentological and petrological approach. *Neues Jahrb. Geol. Mineral., Monatsh.,* 8: 353 – 364.
Quirke, T.T., 1930. Spring pits, sedimentation phenomena. *J. Geol.,* 38: 88 – 91.
Ransome, F.L., 1900. A peculiar clastic dike near Ouray, Colorado, and its deposit of silver ore. *Trans. Min. Eng. Inst.,* 30: 227 – 236.
Rascoe, Jr., B., 1975. Tectonic origin of pre-consolidation deformation in Late Pennsylvanian rocks, Bartlesville, Oklahoma. *Am. Assoc. Pet. Geol. Bull.,* 59: 1626 – 1638.
Rasmussen, H.W., 1965. Strukturer dannet ved jordflydning udglidning og iss-tapning i Kvartaere smeltevandsflejringer. *Dan. Geol. Foren. Medd.,* 15: 470 – 485.

Rautman, C.A. and Dott, Jr., R.H., 1977. Dish structures formed by fluid escape in Jurassic shallow marine sandstones. *J. Sediment. Petrol.,* 47: 101 – 106.
Reimnitz, E., 1972. Effects in the Copper River delta. In: *The great Alaska Earthquake of 1964: Oceanography and Coastal Engineering. Natl. Acad. Sci., Washington,* pp. 290 – 302.
Reimnitz, E. and Marshall, N.F., 1965. Effects of the Alaska earthquake and tsunami on the recent deltaic sediments. *J. Geophys. Res.,* 70: 2363 – 2376.
Reinhardt, J. and Cleaves, E.T., 1978. Load structures at the sediment – saprolite boundary, Fall Line, Maryland. *Geol. Soc. Am. Bull.,* 89: 307 – 313.
Reineck, H.-E., 1974. Schichtgefüge der Ablagerungen im tieferen Seebecken des Bodensees. *Senckenbergiana Maritima,* 6: 47 – 63.
Reineck, H.-E., Dörjes, J., Gadow, S. and Hertweck, G., 1968. Sedimentologie, Faunenzonierung und Faziesabfolge vor der Ostküste der inneren Deutschen Bucht. *Senckenbergiana Lethaea,* 49: 261 – 309.
Reinson, G.E. and Rosen, P.S., 1982. Preservation of ice-formed features in a subarctic sandy beach sequence: geologic implications. *J. Sediment. Petrol.,* 52: 463 – 471.
Rettger, R.E., 1935. Experiments on soft-rock deformation. *Am. Assoc. Pet. Geol. Bull.,* 19: 271 – 292.
Ricci Lucchi, F., 1970. *Sedimentografia Atlante Fotografico della Strutture Primari de Sedimenti.* Zanichelli, Bologna, 288 pp.
Rice, R.C., 1939. Contorted beds in the Trias of the northwest Wirral. *Proc. Liverpool Geol. Soc.,* 17: 361 – 370.
Rice, R.C., 1940. Contorted drift with carbonaceous banding, at Grange, Wirral. *Proc. Liverpool Geol. Soc.,* 18: 9 – 13.
Rich, J.L., 1950. Flow markings, groovings, and interstratal crumplings as criteria for the recognition of slope deposits, with illustrations from Silurian rocks of Wales. *Am. Assoc. Pet. Geol. Bull.,* 34: 717 – 741.
Rich, J.L., 1951. Three critical environments of deformation and criteria for recognizing rocks deposited in each of them. *Geol. Soc. Am. Bull.,* 62: 1 – 20.
Richardson, J.F. and Zaki, W.N., 1954. Sedimentation and fluidization. *Trans. Inst. Chem. Eng.,* 32: 35 – 53.
Ridd, M.F., 1970. Mud volcanoes in New Zealand. *Am. Assoc. Pet. Geol. Bull.,* 54: 601 – 616.
Rieke, III, H.H. and Chilingarian, G.V., 1974. *Compaction of Argillaceous Sediments,* Developments in Sedimentology, 16, Elsevier, Amsterdam, 424 pp.
Rigby, J.K., 1958. Mass movements in Permian rocks of Trans-Pecos Texas. *J. Sediment. Geol.,* 28: 298 – 315.
Robertson, P.B., 1983. Haughton impact structure: structural and morphological aspects. *Can. J. Earth Sci.,* 20.
Robertson, W.A., 1962. Umbrella-shaped fossils(?) from the Lower Proterozoic of the Northern Territory of Australia. *J. Geol. Soc. Aust.,* 9: 87 – 89.
Robson, D.A., 1957. A sedimentary study of the Fell Sandstones of the Coquet Valley, Northumberland. *Q. J. Geol. Soc. London,* 112: 241 – 262.
Roe, L.M., 1972. The origin and classification of pillow structures at Irondequoit Bay, Rochester, New York. M.Sc. thesis, Univ. Rochester, Rochester, N.Y.
Rosenquist, I.T., 1966. Norwegian research into the properties of quick clay. *Eng. Geol. Int. J.,* 1: 445 – 450.
Roy, C.J., 1946. Clastic dikes of the Pikes Peak region. *Geol. Soc. Am. Bull.,* 57: 1226 – 1227.
Roy, S.K., 1929. Columnar structure in limestone. *Science,* 70: 140 – 141.
Russell, W.L., 1927. The origin of the sandstone dikes of the Black Hills region. *Am. J. Sci.,* 14: 402 – 408.
Rust, B.R., 1968. Deformed cross-bedding in Tertiary – Cretaceous sandstone. Arctic Canada. *J. Sediment. Petrol.,* 38: 87 – 91.
Rust, I.C., 1981. Early Paleozoic Pakhuis Tillite, South Africa. In: M.J. Hambrey and W.B. Harland (Editors), *Earth's pre-Pleistocene Glacial Record.* Cambridge University Press, Cambridge, pp. 113 – 117.
Rymer, M.J. and Sims, J.D., 1976. Preliminary survey of modern glaciolacustrine sediments for earthquake-induced deformational structures, south-central Alaska. *U.S. Geol. Surv. Open File Rep.,* 76 – 373: 29 pp.
Salisbury, R.D., 1885. Columnar structures in subaqueous clay. *Science,* 5: 287.

Salisbury, R.D. and Atwood, W.W., 1897. Drift phenomena in the vicinity of Devils Lake and Baraboo, Wisconsin. *J. Geol.,* 5: 131 – 147.
Sanders, J.E., 1960. Origin of convoluted laminae. *Geol. Mag.,* 97: 409 – 421.
Sanders, J.E., 1965. Primary sedimentary structures formed by turbidity currents and related sedimentation mechanisms. In: G.V. Middleton (Editor), *Primary Sedimentary Structures and their Hydrodynamic Interpretation.* Soc. Econ. Paleontol. Mineral. Spec. Publ., 12: 192 – 219.
Sanford, A.R., 1959. Analytical and experimental study of simple geologic structures. *Geol. Soc. Am. Bull.,* 70: 19 – 52.
Schafer, J.P., 1949. Some periglacial features in central Montana. *J. Geol.,* 57: 154 – 174.
Scheidegger, A.E. and Potter, P.E., 1965. Textural studies of graded bedding, observation and theory. *Sedimentology,* 5: 225 – 228.
Schlee, J.S., 1963. Sandstone pipes of the Laguna area, New Mexico. *J. Sediment. Petrol.,* 33: 112 – 123.
Schofield, R.K. and Keen, B.A., 1929. Rigidity in weak clay suspensions. *Nature,* 123: 492 – 493.
Schwab, W.C. and Lee, H.J., 1987. Causes of two-slope failure types in continental-shelf sediment, northeastern Gulf of Alaska. *J. Sediment. Petrol.,* 58: 1 – 11.
Schwan, J. and Van Loon, A.J., 1979. Structural and sedimentological characteristics of a Weichselian kame terrace at Sönderby Klint, Funen, Denmark. *Geol. Mijnbouw,* 58: 305 – 319.
Schwan, J. and Van Loon, A.J., 1981. Structure and genesis of a buried ice-pushed zone near Rold (Funen, Denmark). In: A.J. van Loon (Editor), *Quaternary Geology: a Farewell to A.J. Wiggers. Geol. Mijnbouw,* 60: 385 – 394.
Schwan, J., Van Loon, A.J., Steenbeek, R. and Van der Gaauw, P., 1980. Intraformational clay diapirism and extrusion in Weichselian sediments at Ormehøj (Funen, Denmark). *Geol. Mijnbouw,* 59: 241 – 250.
Schwan, J., Van Loon, A.J., Van der Gaauw, P.G. and Steenbeek, R., 1980. The sedimentary sequence of a Weichselian intraglacial lake at Ormehøj (Funen, Denmark). *Geol. Mijnbouw,* 59: 129 – 138.
Schwars, H.-U., 1975. Sedimentary structures and facies analysis of shallow marine carbonates (Lower Muschelkalk, Middle Trias, southwestern Germany). *Contrib. Sedimentol.,* 3: 100 pp.
Schwarzbach, M., 1952. Ein Pseudo-Eiskeil aus den Albaner Bergen bei Rom. *Geol. Rundsch.,* 40: 56 – 57.
Seilacher, A., 1959. Zur ökologischen Characteristik von Flysch und Molasse. *Eclogae Geol. Helv.,* 51: 1062 – 1078.
Seilacher, A., 1964a. Biogenic sedimentary structures. In: J. Imbrie and N.D. Newell (Editors), *Approaches to Paleoecology.* John Wiley, New York, N.Y., pp. 296 – 316.
Seilacher, A., 1964b. Sedimentological classification and nomenclature of trace fossils. *Sedimentology,* 3: 253 – 256.
Seilacher, A., 1969. Fault-graded beds interpreted as seismites. *Sedimentology,* 13: 155 – 159.
Selley, R.C., 1969. Torridonian alluvium and quicksands. *Scott. J. Geol.,* 5: 328 – 346.
Selley, R.C. and Shearman, D.J., 1962. The experimental production of sedimentary structures in quicksands. *Proc. Geol. Soc. London,* 159: 101 – 102.
Selley, R.C., Shearman, D.J., Sutton, J. and Watson, J.V., 1963. Some underwater disturbances in the Torridonian of Skye and Raasay. *Geol. Mag.,* 100: 224 – 243.
Setty, M.G.A.P. and Wagle, B.G., 1972. Clastic dike from Baga, Goa. *Curr. Sci.,* 41: 731 – 732.
Sevon, W.D., 1969. Stratigraphy and sedimentology of the Tertiary rocks of the Mandamus – Dove River area, north Canterbury, New Zealand. *N. Z. J. Geol. Geophys.,* 12: 283 – 309.
Shabica, C.W., 1978. Sedimentary structures from the Carbondale Formation (Middle Pennsylvanian) of northern Illinois. *Geology,* 33: 541 – 568.
Sharp, R.P., 1942. Periglacial involutions in northeastern Illinois. *J. Geol.,* 50: 113 – 133.
Shaw, E.W., 1914. The mud lumps at the mouths of the Mississippi. *U.S. Geol. Surv. Prof. Pap.,* 85: 11 – 28.
Shaw, J., 1977. Sedimentation in an Alpine lake during deglaciation, Okanagan Valley, British Columbia, Canada. *Geogr. Ann.,* 59: 221 – 240.
Shrock, R.R., 1948. *Sequence in Layered Rocks.* McGraw-Hill, New York, N.Y., 507 pp.
Simpson, G.G., 1935. Cylindrical structures in sandstone (discussion). *Geol. Soc. Am. Bull.,* 46: 2011 – 2014.
Sims, J.D., 1973. Earthquake-induced structures in sediments of Van Norman Lake, San Fernando, California. *Science,* 182: 161 – 163.

Sims, J.D., 1975. Determining earthquake recurrence intervals from deformational structures in young lacustrine sediments. *Tectonophysics,* 29: 141 – 152.
Sims, J.D. and Rymer, M.J., 1976. Correlation of deformed glaciolacustrine sediments and historic earthquakes, Skilak Lake, Kenai Peninsula, Alaska. *Geol. Soc. Am. Abstr. Progr.,* 8: 410.
Skempton, A.W. and Northey, R.D., 1952. The sensitivity of clays. *Geotechnique,* 3: 30 – 35.
Smalley, I.J., Fordham, C.J. and Callander, P.F., 1984. Towards a general model of quick clay development. *Sedimentology,* 31: 595 – 598.
Smith, A.J., 1959. Structures in the stratified late-glacial clays of Windermere, England. *J. Sediment. Petrol.,* 29: 447 – 453.
Smith, A.J. and Rast, N., 1958. Sedimentary dykes in the Dalradian of Scotland. *Geol. Mag.,* 95: 234 – 240.
Smith, B., 1916. Ball- or pillow-form structure in sandstone. *Geol. Mag.,* N.S. 3: 146 – 156.
Smith, G.A., 1984. Setulflike scour-remnant features in a Neogene volcaniclastic flood deposit, central Oregon. *J. Sediment. Petrol.,* 55: 53 – 56.
Smith, K.G., 1952. Structure plan of clastic dikes. *Trans. Am. Geophys. Union,* 33: 889 – 892.
Smits, B.J., 1971. Paleocurrent directions and deformed cross-bedding in the Molteno stage (Triassic), Stromberg Mountains, northeast Cape province (South Africa). *Palaeogeogr., Palaeoclimatol., Palaeoecol.,* 9: 123 – 131.
Sorauf, J.E., 1965. Flow rolls of upper Devonian rocks of south-central New York state. *J. Sediment. Petrol.,* 35: 553 – 563.
Sorby, H.C., 1859. On the contorted stratification of the drift of the coast of Yorkshire. *Proc. Geol. Polytech. Soc. West Riding Yorks.,* 1849 – 1859: 220 – 224.
Sorby, H.C., 1908. On the application of quantitative methods to the study of the structure and history of rocks. *Q. J. Geol. Soc. London,* 64: 171 – 233.
Stear, W.M., 1985. Comparison of the bedform distribution and dynamics of modern and ancient sandy ephemeral flood deposits in the southwestern Karoo region, South Africa. *Sediment. Geol.,* 45: 209 – 230.
Steeger, A., 1944. Diluviale Bodenfrosterscheinungen am Niederrhein. *Geol. Rundsch.,* 34: 520 – 538.
Stewart, A.D., 1962. Greywacke sedimentation in the Torridonian of Colonsay and Oronsay. *Geol. Mag.,* 99: 399 – 419.
Stewart, A.D., 1963. On certain slump structures in the Torridonian sandstone of Applecross. *Geol. Mag.,* 100: 205 – 218.
Stewart, Jr., H.B., 1956. Contorted sediments in modern coastal lagoon explained by laboratory experiments. *Am. Assoc. Pet. Geol. Bull.,* 40: 153 – 179.
Stewart, R.J., 1978. Neogene volcaniclastic sediments from Atka Basin, Aleutian Ridge. *Am. Assoc. Pet. Geol. Bull.,* 62: 87 – 97.
Stone, B.D., 1976. Analysis of slump slip lines and deformation fabric in slumped Pleistocene lake beds. *J. Sediment. Petrol.,* 46: 313 – 325.
Strangeways, T.H.P., 1821. Description of strata in the Brook Pulcovea near the village of Great Pulcovea in the neighbourhood of St. Petersberg. *Trans. Geol. Soc. London,* 5: 382 – 458.
Sugden, D.E. and John, B.S., 1976. *Glaciers and Landscape.* Wiley, New York, N.Y., 376 pp.
Sullwold, Jr., H.H., 1958. Turbidity currents in the Modelo Formation of the Santa Monica Mountains. In: *A Guide to the Geology and Oil Fields of the Los Angeles and Ventura Regions. Am. Assoc. Pet. Geol., Pac. Sect.,* 204 pp.
Sullwold, Jr., H.H., 1959. Nomenclature of load deformation in turbidites. *Geol. Soc. Am. Bull.,* 70: 1247 – 1248.
Sullwold, Jr., H.H., 1960. Load cast terminology and origin of convolute bedding, further comments. *Geol. Soc. Am. Bull.,* 71: 635 – 636.
Sutton, R.F., 1963. Involutions in surficial deposits, northwestern Ontario. *Geol. Soc. Am. Bull.,* 74: 789 – 794.
Sutton, R.G. and Lewis, T.L., 1966. Regional patterns of cross-laminae and convolutions in a single bed. *J. Sediment. Petrol.,* 36: 225 – 229.
Swart, R. and Hiller, N., 1982. Soft-sediment deformation of sandstone related to the Dwyka glaciation in South Africa – discussion. *Sedimentology,* 29: 749 – 751.
Tada, M., 1968. On the sedimentary structure observed in the Miocene Musazawa Formation, Shizukuishi Basin, Iwate-gun, Iwate prefecture, northeast Japan. *Iwate Univ. Technol. Rep.,* 3: 29 – 42.

Tada, M., 1973. On the injection structures observed in the late Miocene Musazawa Formation, Shizukuishi Basin, Iwate prefecture, northeast Japan. *Tohoku Univ. Sci. Rep.*, 2 – 6: 423 – 428.
Ten Haaf, E., 1956. Significance of convolute lamination. *Geol. Mijnbouw,* 18: 188 – 194.
Ten Haaf, E., 1959. Graded beds of the northern Apennines, Ph.D. thesis, Univ. Groningen, Groningen, 102 pp.
Theakstone, W.H., 1965. Contorted glacial lake sediments and ice blocks in outwash deposits at Osterdalsisen, Norway. *Geogr. Ann.,* 47: 39 – 44.
Theakstone, W.H., 1970. Sediments, structures and processes: studies at the Osterdalsisen glacier-dammed lake. *Aarhus Univ. Geogr. Inst. Skr.,* 2, 11 pp.
Theakstone, W.H., 1976. Glacial lake sedimentation, Austerdalsisen, Norway. *Sedimentology,* 23: 671 – 688.
Thomas, G.S.P. and Connell, R.J., 1985. Iceberg drop, dump, and grounding structures from Pleistocene glacio-lacustrine sediments. Scotland. *J. Sediment. Petrol.,* 55: 243 – 249.
Todd, J.E., 1896. Log like concretions and fossil shores. *Am. Geol.,* 17: 347 – 349.
Torrance, J.K., 1983. Towards a general model of quick clay development. *Sedimentology,* 30: 547 – 555.
Tricart, J., 1967. Traité de géomorphologie: le modèle des régions périglaciaires. *Soc. Ed. Enseign. Sup., Paris,* 512 pp.
Troll, C., 1944. Strukturboden, Solifluktion und Frostklima der Erde. *Geol. Rundsch.,* 34: 545 – 694.
Tuthill, S.J. and Laird, W.M., 1966. Geomorphic effects of the earthquake of March 27, 1964, in the Martin-Bering Rivers area, Alaska. *U.S. Geol. Surv. Prof. Pap.,* 543-B: 27 pp.
Tyler, J.H., 1972. Pigeon Point Formation: an Upper Cretaceous shoreline succession, central California coast. *J. Sediment. Petrol.,* 42: 537 – 557.
Unrug, R., 1977. Ancient deep-sea traction currents deposits in the Lgota Beds (Albian) of the Carpathian flysch. *Rocz. Pol. Tow. Geol.,* 47: 355 – 370.
Vandenberghe, J. and Gullentops, F., 1977. Contribution to the stratigraphy of the Weichsel Pleniglacial in the Belgian coversand area. *Geol. Mijnbouw,* 56: 123 – 128.
Van de Poll, H.W. and Patel, I.M., 1981. Flute casts and related structures on moulded silt injection surfaces in continental sandstone of the Boss Point Formation: southeastern New Brunswick, Canada. *Marit. Sed. Atl. Geol.,* 17: 1 – 22.
Van de Poll, H.W. and Plint, A.G., 1983. Secondary sedimentary structures associated with fluidization zones in Permo – Carboniferous redbeds of Prince Edward Island, Canada. *Marit. Sed. Atl. Geol.,* 19: 49 – 58.
Van der Meulen, S., 1988. The spatial facies of a group of pingo remnants on the southeast Frisian till plateau (The Netherlands). *Geol. Mijnbouw,* 67: 61 – 74.
Van der Westhuizen, W.A., Grobler, N.J., Loock, J.C. and Tordiffe, E.A.W., 1989. Raindrop imprints in the Late Archaean – Early Proterozoic Ventersdorf Supergroup, South Africa. *Sediment. Geol.,* 61: 303 – 309.
Van Loon, A.J., 1970. Grading of matrix and pebble characteristics in syntectonic pebbly mudstones and associated conglomerates, with examples from the Carboniferous of northern Spain. *Geol. Mijnbouw,* 49: 41 – 55.
Van Loon, A.J., 1972. A prograding deltaic complex in the Upper Carboniferous of the Cantabrian Mountains (Spain): the Prioro – Tejerina Basin. *Leidse Geol. Meded.,* 48: 1 – 81.
Van Loon, A.J., 1983. The stress system in mudflows during deposition, as revealed by the fabric of some Carboniferous pebbly mudstones in Spain. *Geol. Mijnbouw,* 62: 493 – 498.
Van Loon, A.J., 1990. Geologie rond het vriespunt. *Grondboor Hamer,* 44: 1 – 3.
Van Loon, A.J. and Brodzikowski, K., 1987. Problems and progress in the research on soft-sediment deformations. *Sediment. Geol.,* 50: 167 – 193.
Van Loon, A.J., Brodzikowski, K. and Gotowała, R., 1984. Structural analysis of kink bands in unconsolidated sands. *Tectonophysics,* 104: 351 – 374.
Van Loon, A.J., Brodzikowski, K. and Gotowała, R., 1985. Kink structures in unconsolidated fine-grained sediments. In: R. Hesse (Editor): *Sedimentology of Siltstone and Mudstone. Sediment. Geol.,* 41: (2/4): 283 – 300.
Van Loon, A.J. and Wiggers, A.J., 1975a. Holocene lagoonal silts (formerly called "sloef") from the Zuiderzee. *Sediment. Geol.,* 13: 47 – 55.
Van Loon, A.J. and Wiggers, A.J., 1975b. Composition and grain-size distribution of the Holocene

Dutch "sloef" (Almere Member of the Groningen Formation). *Sediment. Geol.,* 13: 237 – 251.
Van Loon, A.J. and Wiggers, A.J., 1975c. Erosional features in the lagoonal Almere Member ("sloef") of the Groningen Formation (Holocene, central Netherlands). *Sediment. Geol.,* 13: 253 – 265.
Van Loon, A.J. and Wiggers, A.J., 1976a. Abrasion as an agent for sand supply in a Holocene lagoon (Almere and Zuiderzee Members, Groningen Formation) in The Netherlands. *Sediment. Geol.,* 15: 293 – 307.
Van Loon, A.J. and Wiggers, A.J., 1976b. Primary and secondary synsedimentary structures in the lagoonal Almere Member (Groningen Formation, Holocene, The Netherlands). *Sediment. Geol.,* 16: 89 – 97.
Van Loon, A.J. and Wiggers, A.J., 1976c. Metasedimentary "graben" and associated structures in the lagoonal Almere Member (Groningen Formation, The Netherlands). *Sediment. Geol.,* 16: 237 – 254.
Van Straaten, L.M.J.U., 1949. Occurrence in Finland of structures due to subaqueous sliding of sediments. *Finland Comm. Geol. Bull.,* 144: 9 – 18.
Van Straaten, L.M.J.U., 1954. Sedimentology of recent tidal flat deposits and the Psammites du Condroz (Devonian). *Geol. Mijnbouw,* 16: 25 – 47.
Vanuxem, L., 1842. *Geology of New York,* Pt. III. Surv. 3rd District, 306 pp.
Verzilin, N.N., 1961. Diversity of evidence of ancient earthquakes in Lower Cretaceous deposits on northern Fergania. *Leningrad Univ. Vestnik, Ser. Geol. Geogr.,* 24.
Vintanage, P.W., 1954. Sandstone dikes in the South Platte area, Colorado. *J. Geol.,* 62: 493 – 500.
Virkkala, K., 1959. On the Late Glacial frost phenomena in southern Finland. *Finland Comm. Geol. Bull.,* 184: 21 – 39.
Virkkala, K., 1960. Fossilisesta routamaasta. *Geologi,* 12: 30 – 31.
Visher, G.S. and Cunningham, R.D., 1981. Convolute laminations – a theoretical analysis: example of a Pennsylvanian sandstone. *Sediment. Geol.,* 28: 175 – 188.
Visser, J.N.J., Colliston, W.P. and Terblanche, J.C., 1984. The origin of soft-sediment deformation structures in Permo – Carboniferous glacial and proglacial beds, South Africa. *J. Sediment. Petrol.,* 54: 1183 – 1196.
Visser, J.N.J., Loock, J.C. and Colliston, W.P., 1987. Subaqueous outwash fan and esker sandstones in the Permo – Carboniferous Dwyka Formation of South Africa. *J. Sediment. Petrol.,* 57: 467 – 478.
Voight, E., 1962. Frühdiagenetische Deformation der turonen Plänerkalke bei Halle/ Westf. *Geol. Mitt. Staatsinst.,* 33: 147 – 275.
Von Brunn, V. and Talbot, C.J., 1986. Formation and deformation of subglacial intrusive clastic sheets in the Dwyka Formation of northern Natal, South Africa. *J. Sediment. Petrol.,* 56: 35 – 44.
Walker, R.G., 1963. Distinctive types of ripple-drift cross lamination. *Sedimentology,* 2: 173 – 188.
Walker, R.G., 1985. Mudstones and thin-bedded turbidites associated with the Upper Cretaceous Wheeler Gorge conglomerates, California: a possible channel-levee complex. *J. Sediment. Petrol.,* 55: 279 – 290.
Walther, J., 1893/1894. *Einleitung in die Geologie als historische Wissenschaft – Beobachtungen über die Bildung der Gesteine und ihre organischen Einschlüsse,* 1. G. Fischer, Jena.
Walther, J., 1904. Die Fauna der Solnhofener Plattenkalke bionomisch betrachtet. *Festschr. 70. Geburtstage Ernst Haeckel,* Jena.
Wanless, H.R., 1922. The lithology and stratigraphy of the White River beds of South Dakota. *Trans. Am. Philos. Soc.,* 61: 184 – 203.
Wanless, H.R., 1923. The lithology and stratigraphy of the White River beds of South Dakota. *Trans. Am. Philos. Soc.,* 62: 190 – 269.
Ward, F., 1922. The geology of a portion of the Badlands. *S.D. Geol. Nat. Hist. Surv. Bull.,* 11.
Wardlaw, N.C., 1972. Syn-sedimentary folds and associated structures in Cretaceous salt deposits of Sergipe, Brazil. *J. Sediment. Petrol.,* 42: 572 – 577.
Waterston, C.D., 1950. Note on the sandstone injections of Eathie Haven, Cromarty. *Geol. Mag.,* 87: 133 – 139.
Watson, E., 1965. Periglacial structures in the Aberystwyth region of central Wales. *Proc. Geol. Assoc.,* 76: 443 – 462.
Weaver, D.W. (Editor), 1969. Geology of the northern Channel Islands. *Am. Assoc. Pet. Geol., Soc. Econ. Paleontol. Mineral. Pac. Sect., Spec. Publ.,* 200 pp.
Weaver, J.D., 1976. Seismically-induced load structures in the basal coal measures, south Wales. *Geol. Mag.,* 113: 535 – 543.

Weimer, R.J. and Hoyt, J.H., 1964. Burrows of *Calianassa major* Say, geologic indicators of littoral and shallow neritic environments. *J. Paleontol.,* 38: 761 – 767.

Wentworth, C.M., 1967. Dish structure, a primary sedimentary structure in coarse turbidites. *Am. Assoc. Pet. Geol. Bull.,* 51: 485.

Weston, T.C., 1891. Notes on concretionary structure in various rock formations in Canada. *Trans. Nova Scotia Inst. Sci.,* 8: 137 – 142.

White, W.A., 1961. Colloid phenomena in sedimentation of argillaceous rocks. *J. Sediment. Petrol.,* 31: 560 – 570.

Whitten, W.M., 1898. "Quicksand pockets" in "blue clay" of South Bend. *Proc. Indiana Acad. Sci.,* (1897): 234 – 240.

Wiggers, A.J., 1955. De wording van het Noordoostpoldergebied. Ph.D. thesis, Univ. Amsterdam, Amsterdam, 214 pp.

Wigley, P.B. and Sergeant, R.E., 1970. Penecontemporaneous sedimentary structures in the Ravenna facies of the New Albany Shale. *Geol. Soc. Am. Abstr.,* 2: 248 – 249.

Williams, B.J. and Prentice. J.E., 1958. Slump structures in the Ludlovian rocks of north Herfordshire. *Proc. Geol. Assoc.,* 68: 286 – 293.

Williams, E., 1960. Intra-stratal flow and convolute folding. *Geol. Mag.,* 97: 208 – 214.

Williams, E., 1963. Convolute folds and movement in water-logged sediments. In: *Syntaphryl Tectonics and Diagenesis – A Symposium.* Univ. Tasmania Press, Hobart, pp. 11 – 16.

Williams, G.E., 1970. Origin of disturbed bedding in Torridon Group sandstones. *Scott. J. Geol.,* 6: 409 – 411.

Williams, P.F, 1969. Note on some deformation structures of sedimentary origin in the Little Haven – Amroth coal field, Pembrokeshire. *Geol. Mag.,* 106: 395 – 411.

Wilson, M.E., 1918. Timiskaming County, Quebec. *Can. Geol. Surv. Mem.,* 103: 197 pp.

Wolf, K.H. and Chilingarian, G.V., 1976. Diagenesis of sandstones and compaction. In: G.V. Chilingarian and K.H. Wolf (Editors), *Compaction of Coarse-Grained Sediments, II.* Developments in Sedimentology, 18B. Elsevier, Amsterdam, pp. 69 – 444.

Wolf, K.H. and Chilingarian, G.V., 1988. Ore-related diagenesis – an encyclopedic review. In: G.V. Chilingarian and K.H. Wolf (Editors), *Diagenesis, I.* Developments in Sedimentology, 41. Elsevier, Amsterdam, pp. 25 – 553.

Wolfe, P.E, 1953. Periglacial frost-thaw basins in New Jersey. *J. Geol.,* 61: 133 – 141.

Wood, A. and Smith, A.J., 1959. The sedimentation and sedimentary history of the Aberystwyth grits (upper Llandoverian). *Q. J. Geol. Soc. London,* 114: 163 – 195.

Yanushevich, Yu. D., 1972. Clastic dikes in deposits of the northwestern Caucasus. *Lithol. Miner. Resour.,* 7: 391 – 392.

Yeleyeva, I.V., 1974. Traces of ancient earthquakes in the Beleya Graben, eastern Transbaikalia. *Vyssh. Uchebn. Zav. Izv., Geol. Rezved.,* 5: 39 – 45.

Zupan, A.J. and Abbott, W.H., 1975. Clastic dikes: evidence for post-Eocene(?) tectonics in the upper coastal plain of South Carolina. *S. C. Div. Geol., Geol. Notes,* 19: 16 – 23.

Reprinted from: *Diagenesis, III. Developments in Sedimentology, 47.*
Edited by K.H. Wolf and G.V. Chilingarian.

Chapter 4

CLIMATIC INFLUENCE ON DIAGENESIS OF FLUVIAL SANDSTONES

PRODIP K. DUTTA

INTRODUCTION

Driven by economic realities to develop an efficient tool for exploration and exploitation of petroleum resources in siliciclastic reservoirs, sandstone diagenesis received considerable attention during the last two decades. During this time, clastic diagenesis has evolved from its documentation and descriptive approach into a process-oriented discipline in sedimentary petrology. As result, a significant progress has been made in the understanding of various aspects of sandstone diagenesis. Some of the accomplishments made in this respect include: (1) recognition of authigenic nature of clay minerals; (2) widespread development of secondary porosity; (3) involvement of meteoric water in cementation; (4) importance of hydrologic regime in mass-transfer; (5) use of isotope geochemistry in reconstructing the thermal history and geochemical evolution of pore-fluid chemistry through time; (6) radiometric dating of diagenetic episodes; (7) role of organic maturation in hydrocarbon generation; and (8) an overall understanding of chemical diagenesis through application of thermodynamic principles. In spite of such progress, scientists are yet to have a clear understanding regarding the source of cement, a fundamental question in sand cementation. The problem of mass-transfer is another critical area that needs attention. The question of mass-transfer is interlinked with the "source" problem. How far is the source from the site of authigenesis? Without knowing the source location, no mass-transfer mechanism can be formulated. Isotope geochemistry had been helpful in offering a partial answer to this question of cement source. But a better picture may emerge through mass-balance calculations. For this reason one needs to treat diagenesis of the entire sedimentary package in a basin, rather than to deal with "sandstone diagenesis" as a single entity in isolation. Mass-balance approach will also need a better assessment of the nature of the starting material. What controlled the nature of the initial materials? How did pre- and syndepositional processes shape these initial materials? How important is the link between pre-, syn-, and postdepositional processes that make the final product?

Diagenesis may be defined as the combination of physical, biological, and chemical processes that bring an overall textural, chemical and mineralogical change subsequent to deposition. All these changes are mostly accomplished during burial. Textural changes are brought about mostly through porosity – permeability reduction, whereas chemical and mineralogical changes are attained through alteration, dissolution and precipitation. Diagenesis, by convention, is considered to be outside the domain of surficial weathering and metamorphism (see Larsen and Chilingar, 1979, 1983).

Porosity reduction due to compaction is the most important *physical* process in diagenesis (see Chilinger and Wolf, 1976). Berner (1980) observed that in the upper few hundred meters, sands undergo only minor particle reorientation and, as a result, decrease of porosity with depth is minimal. Tickell et al. (1933) compacted loose sands at a pressure of 4150 psi (lb inch^{-2}) and found a reduction in porosity of only 1 – 2%. This pressure corresponds to a burial depth of 1100 m. Inasmuch as most sandbodies are, at least, partially lithified at this depth, porosity reduction by compaction may not be an important process. At very great burial depths, however, sands undergo compaction by deformation, breakage, and interpenetration (Berner, 1980). Appreciable loss of porosity due to compaction in sands is accomplished only in the presence of ductile grains (Rittenhouse, 1971).

A *biological* process, such as burrowing activities of organisms, is common only within the top one meter of sediment (Berner, 1980). Such burrowing activities are widespread mostly in marine environments and have very little importance in most continental sediments. The bulk of diagenetic change, consequently, is related to *chemical processes* designated as "chemical diagenesis". Chemical diagenesis involves a series of reactions between solids and migrating pore fluids during burial in the subsurface. These reactions form a suite of authigenic minerals and textural modification in sediments. Through the study of these products (i.e., the neoformed mineral suites) and textural modifications, physicochemical conditions of the diagenetic environment can be reconstructed.

Diagenesis, in a classical sense, is considered to be a postdepositional event and most diagenetic studies tend to focus only on postdepositional processes/factors that turn "loose sands" into "lithified sandstones". The predepositional processes through their control on mineralogical composition and texture, however, seem to influence burial diagenesis. These predepositional processes largely remained outside the domain on diagenesis. There seems to be an aspect in clastic diagenesis which has been overlooked as to how the pre- and syndepositional processes and factors that shaped the character of starting materials (the detritus) have influenced the postdepositional diagenetic processes. Moreover, no major attempt has been made to understand how surface processes influence pore-water chemistry at shallow depths and its influence on diagenesis, particularly during the early diagenetic stage. A complete understanding of diagenesis will not be possible unless one considers the processes and/or factors that control interstitial water chemistry, the mineralogical composition of sand, and the texture of clastic sediments.

Both mineralogical composition and texture of sediments are controlled by processes and/or factors such as source area characteristics, tectonic setting, rigor of transport, environment of deposition, and climate. There have been very few attempts to relate or connect diagenesis with climatically-induced factors or processes. The importance of tectonic setting (in source areas) in diagenesis of siliciclastic sediments (in depositional milieu) have received some attention (Siever, 1979; Dickinson and Suczek, 1979). Hayes (1979) in a general way attempted to relate most predepositional and syndepositional factors/processes to chemical diagenesis. He concluded that the "initial mineralogy and texture profoundly influence its diagenetic history . . . To understand diagenetic history of a sandstone one must know what the starting materials were."

To understand the diagenetic history it is possibly necessary to go one step further "back" to evaluate the factors/processes that shaped the character of the starting materials. Climate in the source area is such a factor that has an important bearing on chemical diagenesis in siliciclastic sediments in the depositional area. The objective of this chapter is to review and analyze the climatic control on the initial materials: i.e., groundwater chemistry during shallow burial and mineralogy of detrital sand. These, in turn, demonstrate the importance of predepositional and synburial climate on postdepositional diagenetic processes. It is relatively easier to monitor climatic control on groundwater chemistry and detrital mineralogy in a cratonic setting. This chapter, therefore, will focus on the chemical diagenesis of siliciclastic sediments in block-faulted basins within a craton.

CLIMATIC CONTROL ON GROUNDWATER CHEMISTRY AND SOIL MINERALOGY

Introduction

The ultimate source rock (parent) of all sedimentary particles (daughter) is the crystalline lithosphere. This lithosphere crystallized at a high temperature and, except for the extrusive rocks, formed under high pressure deep within the crust. This was a chemical environment with no or very little liquid water and free oxygen. As these rocks are exhumed and exposed to (1) low temperature and pressure, (2) unlimited supply of free oxygen, (3) liquid water in most environments, and (4) carbon dioxide, the rocks try to reestablish a new equilibrium through both mechanical and chemical disintegration. This is the basic premise that will guide this discussion to demonstrate how climate, through chemical weathering/pedogenesis*, controls: (1) groundwater chemistry in siliciclastic sediments at shallow burial depths; and (2) an overall mineralogical change between the source rock and the final detritus in the depocenter.

Volumetrically, mechanical weathering is an insignificant process compared to chemical weathering in rock decay. Fragmentation of rocks begins as they are exposed to the Earth's surface due to the release of superincumbent pressure; initially, the rocks break into large blocks. Frost wedging further disintegrates rocks into smaller fractions. Frost wedging is important in areas where either the daily (night and day) or the seasonal (summer and winter) temperature varies around the freezing point, 0°C (Konishchev and Rogov, 1983). Such environments are present in the high mountains in the tropics and at lower elevations in temperate and subarctic regions. Little moisture, which is present even in most arid environments, helps chemical decay along grain/crystal boundaries and facilitates mechanical separation of grains into sand-size particles. Organisms also cause mechanical disintegration. Lichens hyphae (roots) disintegrate rocks as they expand and contract during the wetting and drying process. Chewing and grinding actions of burrowing animals

---

* In this chapter pedogenesis and chemical weathering are used as synonymous terms and so are soil and weathering profiles. See Lelong et al. (1976) and Wopfner and Schwarzbach (1976) for additional information.

further disintegrate mineral matter into finer fractions (Blatt et al., 1980). One of the most important aspects of mechanical fragmentation is to accelerate chemical weathering by creating more surface area. In a way, mechanical and chemical weathering aid each other and go hand in hand. Mechanical weathering is most common in arid and/or mountainous regions. Otherwise, rock decay is essentially a chemical process on the Earth's surface.

*Factors controlling chemical weathering in the source area**

Throughout the humid and semi-arid regions of the globe, which constitute nearly 75% of the Earth's surface (Tarbuck and Lutgens, 1988), chemical weathering appears to be the most important process in rock disintegration (Strakhov, 1967; Garrels and Mackenzie, 1971). Through such processes an overall mineralogical change takes place between parent and weathered mantle. These mantle materials are mostly made up of unaltered and partially altered primary minerals, neoformed secondary minerals, and soil moisture/pore water with dissolved solids. Through chemical weathering groundwaters also acquire their mineral matter. The total extent of chemical weathering determines the volume of primary minerals which have undergone dissolution, alteration and decay to produce secondary minerals and the amount of ions released through chemical reactions. The amount of ions thus released and the amount of rain water percolating through the weathering profile will ultimately control groundwater chemistry at shallow depths. A general scheme of chemical weathering at the Earth's surface, a reference point where the lithosphere, hydrosphere, biosphere, and atmosphere interact, is as follows:

Primary minerals (*lithospheric materials*) + Chemical reagents (*materials of hydrospheric, atmospheric and biospheric origin*) →

→ Stable primary minerals + Secondary minerals (*precursors of clastic sediments*)

+ Dissolved solids in water (*precursors of chemical sediments and authigenic minerals in sediment and soil*)

The primary minerals are mostly of igneous and/or metamorphic origin, whereas the chemical reagents are made up of rain water, atmospheric carbon dioxide, and oxygen. Additional carbon dioxide is available in abundance in most soil horizons. There are also few chemical constituents, such as HCl, H_4SiO_4, etc., which may be present in the environment and may take part in chemical reactions. But such reagents are very local in nature and discussions on these reagents and related reac-

* The reader should also consult the work of R. L. Folk (1974) entitled "Petrology of Sedimentary Rocks", Hemphills, Austin, Texas, especially his ideas on mineralogical versus textural maturity of sediments, which is important in diagenesis (editorial comment).

tions are beyond the scope of this chapter. Unstable primary feldspars and ferromagnesian minerals are prone to chemical destruction, whereas quartz, chemically the most stable mineral, remains relatively unaffected even under intense chemical weathering conditions. Most primary aluminosilicates are completely or partially destroyed, producing secondary clay minerals and/or oxides and hydroxides of aluminum and iron. Both unaltered primary minerals and newly-formed secondary minerals remain at the site of weathering and form the precursors of clastic sediments. The ionic species released during chemical weathering are dissolved in rain water and move away from the site of weathering by surface runoff and groundwater. These dissolved chemical constituents are the source of chemical sediments and early authigenic cements in siliciclastic sediments during their shallow burial. In the chemical reaction shown above, the total chemical and mineralogical changes between parent materials and the products through interaction of chemical reagents depend upon three independent factors:

(1) The nature of primary minerals, a thermodynamic entity which determines the chemical stability of a mineral suite;
(2) Relief, which controls the rate of flushing and, in turn, determines the total time allowed to interact between solids and chemical reagents; and
(3) Climate, which is a factor that may be defined here by parameters like precipitation, seasonality, prevailing temperature of the environment, and amount of carbon dioxide in the soil/weathering profile.

Except for the availability of oxygen and carbon dioxide in the atmosphere, other factors like rainfall, carbon dioxide in soil horizons, and atmospheric temperature are extremely variable. This causes the rate of decay to be a variable component. Consequently, the weathering products differ widely.

Nature of primary minerals

The chemical stability of different minerals to chemical weathering on the Earth's surface is extremely variable. This stability is ultimately related to the bond strength between oxygen and the various cations present in the mineral (Keller, 1957; Nicholls, 1963). The strongest bond which is formed between oxygen and silicon is covalent in nature. The order of bond strengths *nearly* duplicates the order of increasing amount of covalent bond character. The bond strength decreases among the common cations in the Earth's crust in the following order: Al, Fe^{3+}, Mg, Fe^{2+}, Mn, Ca, Na, and K (Table 4-1).

The stability of the minerals, therefore, depends on the number of strong bonds present in the mineral. The silicates with high Si/O ratios are the most stable ones, and silicate minerals become less stable either because of an increased substitution of aluminum for silicon or a small number of Si – O – Si bonds in the mineral. In the "mineral stability" series, quartz is the most stable because it is completely made up of covalent Si – O – Si bonds. Chemically, calcic plagioclase and olivine are least stable among silicate minerals because of increased substitution of aluminum for silicon in the case of calcic plagioclase and the absence of Si – O – Si bonds in olivine (Loughnan, 1969). This explains the relative chemical stability of common rock-forming minerals in nature. Goldich (1938) studied soil profiles and showed that the

TABLE 4-1

Relative strengths of common cation – oxygen bonds in silicate minerals (after Nicholls, 1963)

Bond	Relative strengths
Si – O	2.40
Al – O	1.65
Fe^{3+} – O	1.40
Mg – O	0.90
Fe^{2+} – O	0.85
Mn – O	0.80
Ca – O	0.70
Na – O	0.35
K – O	0.25

TABLE 4-2

Bowen's reaction series and stability of common rock-forming minerals in weathering profiles

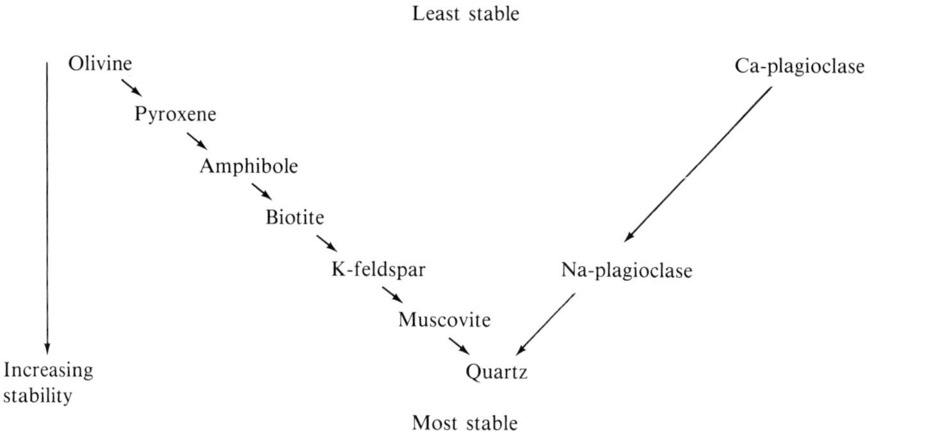

common rock-forming minerals could be arranged in an order dependent on the degree of weathering. This order is the same as Bowen's reaction series reversed. The minerals formed at the highest temperatures and pressures were found to be the least stable under surface conditions (Table 4-2).

It follows then that the rocks with minerals with weaker bonds, such as ferromagnesian minerals and calcic-plagioclase (high in the Bowen's reaction series; ultrabasic and basic composition) are less stable than the rocks having minerals with stronger cation – oxygen bonds, such as quartz, alkali feldspar, and muscovite (acidic composition). In addition to the chemical buildup of minerals, the grain size is also important in chemical weathering. In fine-grained materials, because of higher surface to volume ratio, the chemical reactions proceed faster. Thus, a fine-grained basalt is chemically more reactive and weathers faster compared to a coarse-grained granite.

The role of relief in chemical weathering in the source area

The relief of an area has a tremendous effect on the rate of chemical weathering and, consequently, on the nature of the weathered products. Relief exerts its influence in several ways by controlling: (a) the rate of surface runoff of rain water and, hence, the rate of moisture intake by the weathered materials; (b) the rate of infiltration and, therefore, the rate of leaching of the soluble constituents; and (c) the rate of erosion of the weathered products and, thereby, the duration of weathering and the rate of exposure of fresh rock (Loughnan, 1969).

In mountainous terrains relief is high, the surface slope is steep, and most of the rain water is lost through surface runoff with little or no infiltration. Here, chemical weathering is a very slow process even under conditions of a very moist climate (Strakhov, 1967). In areas with very little relief, as in poorly drained swamps, chemical reactions proceed until equilibrium is reached. Once equilibrium is established between the reactants and products, no further reaction takes place unless the saturated solution is removed from the system. In such a situation chemical weathering is restricted. For the continuity of the chemical reaction, continuous flushing is necessary. Such conditions are attained in well-drained areas of low relief as in the shield and craton and/or in areas with moderate relief as in the cratonized part of a continent or areas of a dissected magmatic arc provenance in the process of evolving into a craton. In a craton, mechanical removal of surface debris takes place more slowly than chemical decomposition of the rock. This results in a very thick weathered zone in a warm humid climate. A relatively thin but intensely weathered zone may develop even in hilly areas if a dynamic equilibrium between the rate of weathering and the rate of erosion is established. Velbel (1985) observed such an equilibrium condition in the moist climate of the southern Appalachian Mountains. The intensity of chemical weathering seems to be inversely related to topographic slope.

Climate

Climate controls the temperature of weathering environments and the nature and abundance of chemical reagents. The most important chemical reagent on the Earth's surface is meteoric water with dissolved carbon dioxide. All geochemical reactions require the presence of water in liquid form. In areas where the temperature is below 0°C, liquid water is absent. Geochemical reactions and, consequently, chemical weathering is virtually nonexistent in such an extreme cold climate. Even in some cold regions, however, if liquid water is present, at least during a part of the year, chemical weathering seems to be common. Paucity of water in liquid form, low temperature, and lack of biological activity and vegetation make chemical weathering an extremely slow process under such extreme cold conditions (Ugolini, 1986). The rates of chemical reactions and, in turn, the intensity of chemical weathering are greatly enhanced by increased temperature. An increase of 10°C in temperature accelerates all chemical reactions, 2 to 2.5 times (Strakhov, 1967).

As the rain water precipitates, it dissolves atmospheric CO_2 and generates carbonic acid, the main reagent in chemical weathering. Rain water has a pH value of approximately 5.7. In vegetated areas, physical, chemical and microbiological breakdown of vegetal matter generates abundant CO_2. In soils, the partial pressure of CO_2 may vary from 10 to 400 times higher than the atmospheric CO_2 (Merkle, 1955; Holland, 1978). So the acidity of meteoric water is maximum in humid forested areas where plant materials are rapidly oxidized, as in the humid tropics, and least in cold and arid regions. The total extent of chemical weathering, therefore, is largely a function of precipitation. It is possibly more reasonable to assume that it is the volume of infiltrated water rather than the volume of total precipitation that is important in chemical weathering. Volume of infiltration largely depends on total precipitation and relief. The water involved in chemical weathering may be divided into two parts: (1) static water, i.e., water that remains in contact with solids for the duration of the chemical reaction; and (2) moving water that flushes the system and removes the chemical constituents that have been released through interaction between static water and solids. Precise quantitative estimation of the amount of water involved in these processes is difficult. As an approximation, Grantham and Velbel (1988) used a parameter, "effective precipitation" (stream discharge per watershed unit area), as a quantity which, along with the "relief factor", determines the intensity of chemical weathering.

Climate and chemical reaction mechanisms in source areas

The mechanisms of chemical reactions and, consequently, their products depend on the amount of water available for chemical reactions and the prevailing atmospheric temperature. Depending on climatic factors, therefore, rocks are altered chemically, in a number of weathering mechanisms: acidolysis*, alkalinolysis*, salinolysis*, and hydrolysis (Pedro and Sieffermann, 1979). Acidolysis reactions are common in very harsh, cold, but humid climates. In such a climatic milieu, vegetal matter, characterized mostly by conifers, decomposes very slowly. The rate of reaction between reagents and minerals is also very slow. In warm arid environments, due to lack of rain, the minerals are chemically altered by salinolysis and/or alkalinolysis. In such extreme climatic regimes the resultant change in composition between the parent and soil mineralogy is small, and groundwater chemistry is characterized by a high concentration of chemical constituents as the result of excessive evaporation. The climatic control on groundwater chemistry is shown in Table 4-3. Water from the arid region is many times more concentrated than water

* These terms are used by French geochemists and soil scientists to define the processes of breaking down of silicates due to different climatic conditions.

TABLE 4-3

General chemical character of spring and well water in granitic country from different climatic environments (after Feth et al., 1964)

Source	Approximate average climate of the area	Total dissolved solids (ppm)
Springs – Mojave Desert	Warm arid	579
Springs – California coast	Temperate semi-arid	134
Springs – Idaho Batholith	Temperate semi-humid	100
Wells – North Carolina	Temperate humid	141

from the humid country. Chemical weathering on the Earth's surface is mainly accomplished by hydrolysis, meaning literally "break up by means of water". This process operates under the influence of dilute water (in areas of appreciable rainfall) and CO_2 in soils with the formation of a weak acid, H_2CO_3. Hydrolysis destroys silicates, extracting the soluble cation and neutral silicic acid as shown by the dissolution reaction of forsterite:

$$Mg_2SiO_4 + 4\ H_2CO_3 \rightarrow 2\ Mg^{2+} + 4\ HCO_3^- + H_4SiO_4 \qquad (4\text{-}1)$$

This is a case of complete dissolution of the silicate mineral without leaving behind any solid product. Here, all the products of chemical reaction are in a dissolved state. When an aluminosilicate is involved, an additional product, i.e., an insoluble aluminum compound, is formed as shown by the following equation:

$$NaAlSi_3O_8\text{(albite)} + H_2CO_3 + 7H_2O \rightarrow$$
$$Na^+ + 3\ H_4SiO_4 + \underset{\text{(gibbsite)}}{Al(OH)_3} + HCO_3^- \qquad (4\text{-}2)$$

This is a case of complete hydrolysis where the flow is such that the system is in equilibrium with the products and the reactants. At moderate flow rates, albite is changed to kaolinite according to the following equation:

$$2NaAlSi_3O_8 + 2H_2CO_3 + 9H_2O \rightarrow$$
$$2Na^+ + 4H_4SiO_4 + \underset{\text{(kaolinite)}}{Al_2Si_2O_5(OH)_4} + 2HCO_3^- \qquad (4\text{-}3)$$

When the flow rates are slower, materials are removed at a relatively slow rate from the weathering site and, if magnesium-bearing minerals (like biotite, amphibole, olivine, etc.) are present, montmorillonite forms in place of kaolinite or gibbsite as follows:

$$3.33\ NaAlSi_3O_8 + 5.32\ H_2O + 1.32\ H^+ + 0.67\ Mg^{2+} \rightarrow$$
$$2.66\ Na^+ + \underset{\text{(montmorillonite)}}{Na_{0.67}Al_{3.33}Mg_{0.67}Si_8O_{20}(OH)_4} + 1.99\ H_4SiO_4 \qquad (4\text{-}4)$$

The above equations demonstrate how hydrolysis under different water supply conditions (flushing rates) produce different products implying the importance of rainfall in chemical weathering.

Modern evidence of climatic control on groundwater chemistry and soil mineralogy in the source area

The preceding analysis infers that the intensity of chemical weathering in controlling soil mineralogy and groundwater chemistry is the result of an interplay of factors like source-rock composition, relief and climate. Documentation of chemical weathering from varied climatic regions and in different source-rock terrains under different relief conditions support this inference (Feth et al., 1964; Ruxton, 1970; Tardy, 1971; Darnell, 1974; Velbel, 1985).

In a tropical climate in northeast Papua, covering an area of nearly 7000 km^2, Ruxton (1970) demonstrated the role of climate as well as relief on soil mineralogy. The area under investigation represents a rapidly uplifted block of mostly Tertiary basalt – andesite and pre-Tertiary phyllites and metabasalts. Deep weathering has taken place on the ridge crest (having gentle slopes) on all rock types and produced kaolinitic clay and abundant quartz and opaque minerals. On hill slopes, with slope angles of 35° – 40°, weakly-weathered profiles have developed. The mineralogy of the slope soils is dominated by lithic fragments with subordinate feldspar and very little quartz. Composition of sand in the fine-grained fraction from hill crests and slopes shows wide variations (Fig. 4-1). Ruxton's work shows that tropical weathering can generate a mature quartz sand even in the highly unstable magmatic arc provenance. At the same time, the maturity trend, as shown by the arrows in Fig. 4-1,

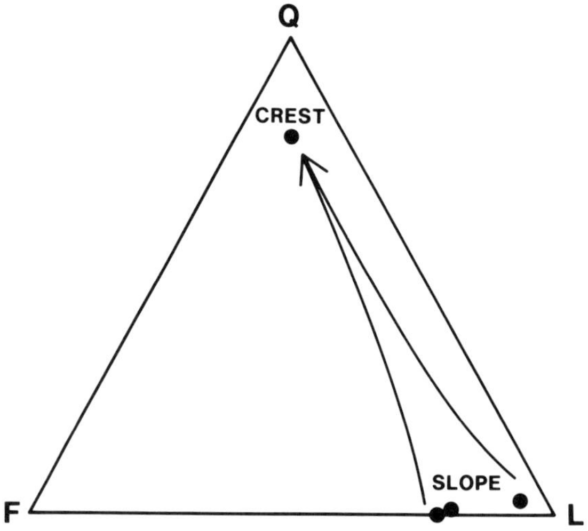

Fig. 4-1. Dual control of climate and slope angle on soil profiles. Mineralogically immature soils are developed on the hill slope and mature soils on the hill crest. Both soil profiles are developed on similar rock type characterized by basic igneous, meta-graywacke, and meta-volcanics within a magmatic arc setting in a warm humid climate. In an arid climate such differences in soil mineralogy due to slope differences will not be observed. (Data from Ruxton, 1970.)

focusses on the importance of combined effects of slope (relief) and climate in chemical weathering and, in turn, its control on sand composition. In an attempt to understand the role of climate on soil mineralogy, Darnell (1974) documented the mineralogy of sand-size particles of soils developed on plutonic and metamorphic rocks on slopes under tree cover and grass cover in a semi-arid climate. Even under semi-arid conditions the mineralogical differences between source rock and soil, due to chemical weathering, are appreciable. Quartz tends to be concentrated in the coarse-grained fraction because of its high resistance to weathering. On the other hand, feldspars accumulate in the fine-sand fraction due to their succeptibility to chemical weathering. Chemical weathering seems to be more intense in a tree-covered area compared to the grass-covered section.

In an investigation dealing with the weathering rates of rock-forming silicates in natural forested watersheds in the southern Blue Ridge Mountains in the Appalachian region, Velbel (1985) observed appreciable mineralogical differences between parent rock and soil mineralogy. In a temperate (the mean annual average temperature is 12.8°C) and humid climate (annual rainfall ranges from 1700 mm at lower elevation to 2500 mm on the upper slopes), saprolitization on the hill slopes (27%) completely depleted sodic plagioclase, biotite, and garnet from a parent metamorphic schist. Broadly the soil profile is in dynamic equilibrium with the rate of saprolitization of 3.8 cm per 1000 years, where the rate of denudation is about 4 cm per 1000 years. The sands derived from this highly-weathered metamorphic source terrain will have a very different composition than they would if there had not been any chemical weathering. All these examples from different source terrains under different relief conditions from varied climatic regions, i.e., tropical, temperate semi-arid, and temperate humid, show that chemical weathering can make an appreciable difference between soil and source rock mineralogy.

Climatic influences on groundwater chemistry at shallow depths have been documented from different climatic regions of the world by many workers. The works of Feth et al. (1964) and Tardy (1971) are considered pioneering in this respect. In a classic study, Feth et al. (1964) attempted to answer how, why, and from what sources groundwater acquires its mineral content. In their study on groundwater chemistry from granitic rocks in Sierra Nevada, California and Nevada, they observed that the groundwater acquires mineral matter from chemical weathering of the lithosphere. Starting with snow, the source of virtually all recharge in the region, mineral content increases on the average 7.5 times, as melt water comes in contact with soil and saprolite (ephemeral spring), and then doubles again during deeper penetration (perennial spring) of the water (Table 4-3). A schematic diagram (Fig. 4-2) shows how, why and from where the groundwater acquires mineral species. Feth et al. also observed that spring and well waters from granitic country – but from different climatic regions – differ significantly, confirming the control of climate on groundwater chemistry (Table 4-4). Most concentrated water is observed in arid regions of Mojave deserts and dilute waters from humid regions of Idaho and North Carolina. Mojave desert water shows about four times the mineral content of the average North Carolina well water. In their study it was also observed that, under moderate climatic conditions, the lithosphere also influences the chemistry of groundwater. They noted that the groundwater of perennial springs in volcanic terrains of the Sierra Nevada and southern Cascade Mountains differ from the groundwater in granitic terrains from the same area by more than 1.7 times the mineral content (Table 4-5) and having a significantly higher percentage of magnesium. The mean content of dissolved solids in the granitic

TABLE 4-4

Changes in average concentration of selected constituents in snow and ephemeral and perennial springs in Nevada (data from Feth et al., 1964)

Constituents	Snow (42 samples) (ppm)	Ephemeral springs (15 samples) (ppm)	Perennial springs (56 samples) (ppm)
SiO_2	0.16	16.40	24.60
Al	–	0.03	0.01
Fe	–	0.03	0.03
Ca	0.40	3.11	10.40
Mg	0.17	0.70	1.70
Na	0.46	3.30	5.95
K	0.32	1.09	1.57
HCO_3	2.88	20.00	54.60
SO_4	0.95	1.00	2.38
Cl	0.50	0.50	1.06
F	–	0.07	0.09
NO_3	0.07	0.02	0.28
Total dissolved solids	5.91	46.15	102.67

TABLE 4-5

Comparison of chemistry of groundwater in granitic and volcanic terrains in similar climatic environments of the Sierra Nevada Mountains and Cascade Mountains (data from Feth et al., 1964)

Constituents	Perennial springs in granitic terrain (ppm)	Perennial springs in volcanic terrain (ppm)
SiO_2	24.60	40.60
Al	0.01	0.04
Fe	0.03	0.01
Ca	10.40	15.32
Mg	1.70	6.63
Na	5.95	8.41
K	1.57	2.18
HCO_3	54.60	99.00
SO_4	2.38	2.26
Cl	1.06	1.60
F	0.09	0.07
NO_3	0.28	0.60
Total	102.67	176.18

spring water is 103 ppm, whereas in the volcanic springs it is 176 ppm. The greater content of magnesium in the volcanic terrain is caused presumably by the higher proportion of Mg-bearing minerals in volcanic rocks.

Tardy (1971), on a much wider scale (ranging from the cold climatic regions in northern Europe to the warm tropical belt in the Ivory Coast and Malagasy, through the warm arid regions of Africa), demonstrated that the nature and relative abundance of weathered products in soils and the chemistry of water are largely controlled by the prevailing climate. Even seasonal changes are also reflected in the products of weathering. Tardy documented clay mineral assemblages and the co-existing water in soils from the granitic provenance from Norway in the north to Malagasy in the south, covering most of the climatic belts of the world (Table 4-6). Though

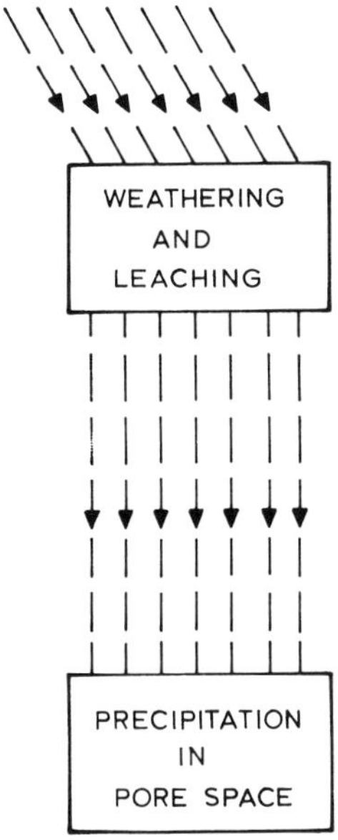

COMPOSITION OF RAIN AND SNOW[*]

CONSTITUENT	CONCENTRATION (ppm)
SiO_2	0.16
Ca	0.40
Mg	0.17
Na	0.46
K	0.32
HCO_3	2.88
SO_4	0.95
Cl	0.50
NO_3	0.07

AVERAGE COMPOSITION OF GROUNDWATER[+]

CONSTITUENT	CONCENTRATION (ppm)
SiO_2	31.00
Al	0.57
Fc	0.42
Ca	75.00
Mg	27.00
Na	8.00
K	4.20
HCO_3	319.00
Cl	10.30
SO_4	39.30
NO_3	9.30

AVERAGE GROUNDWATER IS OVERSATURATED WITH RESPECT TO THE FOLLOWING MINERALS

MINERAL	LOG SATURATION INDEX
Hematite	18.87
Smectite	10.00
Kaolinite	6.91
Chlorite	4.55
Quartz	0.96
Calcite	0.55

SOURCES: [*]Feth et al.,1964; [+]White et al.,1963

Fig. 4-2. Evolution of groundwater chemistry showing the early control of groundwater at shallow depths and the potential for early authigenic cement precipitation. (After Dutta, 1983.)

TABLE 4-6

Mean chemical composition of water, coexisting neoformed mineral assemblages in soil and shallow sediments underlain by granite, and overall climate of the area (after Tardy, 1971)

Location	SiO_2	Ca	Mg	Na	K	HCO_3	SO_4	Cl	Total dissolved solids (ppm)	Mean annual rainfall (mm)	Mean annual tempera-ture (°C)	Number of dry months in a year	Overall climate	Neoformed clays in arenes (sand) and soils
Norway	3.0	1.7	0.6	2.6	0.4	4.9	4.6	5.0	22.8	1500	< 10	0	cold humid	Common: vermiculite, montmorillonite, chlorite Rare: kaolinite
Vosges	11.5	5.8	2.4	3.3	1.2	15.9	10.9	3.4	54.2	1200	8	2	cold humid	Common: vermiculite, montmorillonite
Brittany	15.0	4.4	2.6	13.3	1.3	13.4	3.9	16.2	70.1	800	11	3	temperate humid	
Central massif	15.1	5.8	2.4	4.2	1.2	12.2	3.7	2.6	44.9	1000	10	2	temperate humid	Common: montmorillonite kaolinite, gibbsite
Corsica	13.2	8.1	4.0	16.5	1.4	40.3	8.6	22.0	114.1	1000	15	4	temperate humid	
Chad	85.0	8.0	2.5	15.7	3.4	54.4	1.4	3.0	181.3	16*	28*	12*	warm semi-arid	Common: montmorillonite, (dominant), kaolinite (subordinate)
Sahara	9.0	40.0	–	30.0	1.8	30.4	20.0	4.0	135.2^+	< 10	30	12	warm arid	Data not available
Ivory Coast	8.0	< 1.0	< 0.1	0.2	0.6	6.1	0.5	< 3.0	< 19.5	1400	25	5	warm humid	Common: kaolinite (dominant), gibbsite (subordinate)
Malagasy (high plateaus)	10.6	0.4	0.1	0.9	0.6	6.1	0.7	1.0	20.4	1200	18	6	warm humid	Common: gibbsite
Malagasy** (Eastern coast)	16.2	2.8	1.4	4.3	0.4	18	2.4	5.5	51	2200	24	0	warm humid	Common: kaolinite

* Climatic data of Northern Chad from Rudloff, 1981.
** Water sample collected from basaltic country.
$^+$ Water sample collected immediately after rain.

the mechanism of chemical decay was hydrolysis, the variations in water chemistry and clay mineral assemblages were largely related to climatic variations in terms of precipitation and atmospheric temperature of the area. The waters in humid regions are dilute, whereas in arid regions they are relatively more concentrated. In case of humid regions, such as Norway, the Vosges (France), Ivory Coast and Malagasy, the waters are dilute and fall into the kaolinite stability field (Fig. 4-3). Some water samples plot within the kaolinite field, but close to the kaolinite/montmorillonite phase boundary shown by the cross-hatched area. This is observed in samples collected in Corsica, Brittany, and Central Massif. These are humid regions, but sampling was done during a long dry season. The third group of samples correspond to waters collected in arid countries (Chad, Sahara). These waters, which are most concentrated and are characteristic of an arid climate, fall within the montmorillonite field. Tardy's work also demonstrated the control of the source-rock composition on water chemistry. Water samples collected from the same region having a similar climatic pattern, but from different underlying source rocks, show different geochemical properties. The groundwater in the underlying basic rock terrain shows a higher mineral content compared to the waters collected from a granitic country, an observation also made by Feth et al. (1964; see Table 4-5).

The neoformed clay mineral assemblages in different countries seem to be in

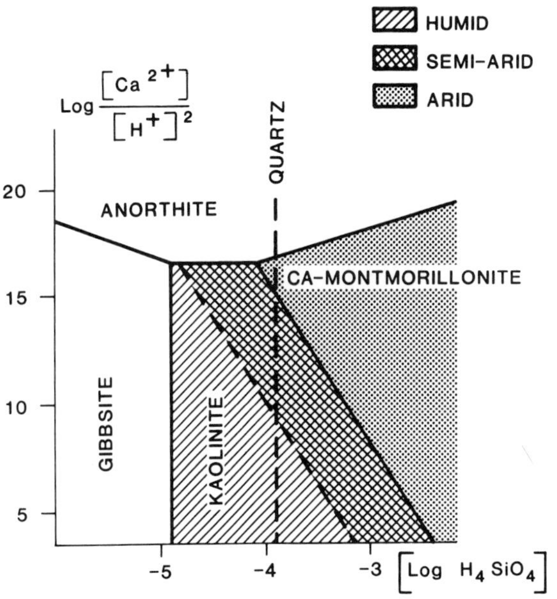

Fig. 4-3. Stability relation of anorthite, Ca-montmorillonite, kaolinite and gibbsite, at 25°C and 1 atmosphere as a function of [Ca^{2+}], pH, and [H_4SiO_4]. The waters from humid regions of Norway, the Vosges, the Ivory Coast, and Malagasy plot well within the kaolinite stability field (striped area). Water from Brittany, Central Massif, and Corsica plot within the kaolinite stability field but close to the montmorillonite phase boundary (cross-hatched area). Though these regions are rather humid, sampling was done during a long dry season. Water from arid countries like Chad and Sahara plot within the montmorillonite stability field.

equilibrium with the water except in the case of Norway and the Vosges. The water is in thermodynamic equilibrium with kaolinite, while there is very little kaolinite in the clay mineral assemblages in these countries (Table 4-6). This inconsistency is also related to climate. In northern Europe, where the climate is cold but humid, there are two geochemical environments present. The surface and the microfractures within the crystals form one environment and the other is outside the crystals in circulating dilute water. Within the crystal, hydrolysis is mild. Basic cations are released but silica is retained in part. Through such mild hydrolysis, feldspars and micas give rise to vermiculite and montmorillonite. Outside the crystal, the waters are dilute because they are renewed by drainage. The amount of ions present in this water is too low to precipitate kaolinite. In contrast, in arid to semi-arid countries evaporation leads to the concentration of solutions and neoformation of montmorillonite. In these examples it is seen that in warm humid regions water is dilute where kaolinite and gibbsite characterize the soil mineralogy, whereas in both cold humid and warm arid regions the clay minerals are typified by cation-rich clays (see Tardy, 1971).

Synthesis

The examples of modern geochemical processes demonstrate the importance of three variables, i.e., source-rock composition, relief, and climate that control chemical weathering and their products. On a large (global) scale a close relationship between relief and source-rock composition can be established. These two parameters may be combined and designated as plate tectonic setting (Dickinson and Suczek, 1979). For example, young mountains with high relief are mostly made up of volcanic suites, whereas the shield areas with low relief are characterized by coarse-grained acid igneous rocks. By combining relief and source-rock composition, the number of variables that influence chemical weathering are reduced to two, i.e., plate tectonic setting and climate. The fact that chemical weathering is mainly controlled by these two variables is apparent in the soil map of the world (Fig. 4-4). Mountain soil, which is immature and thin, develops on steep hill slopes that cut across all the climatic zones of the world. Such soils are mostly controlled by tectonic regimes acting through relief where chemical weathering is minimal. Except for the mountain soil, the formation of most other soil types is influenced by climate. Because of extreme climatic variations across the globe, however, the thickness of soil profiles as well as the nature of weathered products are extremely variable between the equator and the poles. Such a hypothetical weathered profile between the equator and the north pole, developed on tectonically stable areas, is shown in Fig. 4-5. In this figure, two intense weathering zones are observed. The first corresponds to podzols and podzolic soils of relatively humid and temperate zones of North America, Europe, and western Siberia. Further north of this zone, there is a development of a thin immature tundra soil with little chemical weathering. South of the temperate zone, the soil formation is dominated by the lack of moisture and, thereby, lack of much chemical weathering. The soil is thin with little organic matter and is characterized by a light-colored surface horizon overlying a hardpan. The most intense chemical weathering is observed in areas covered by

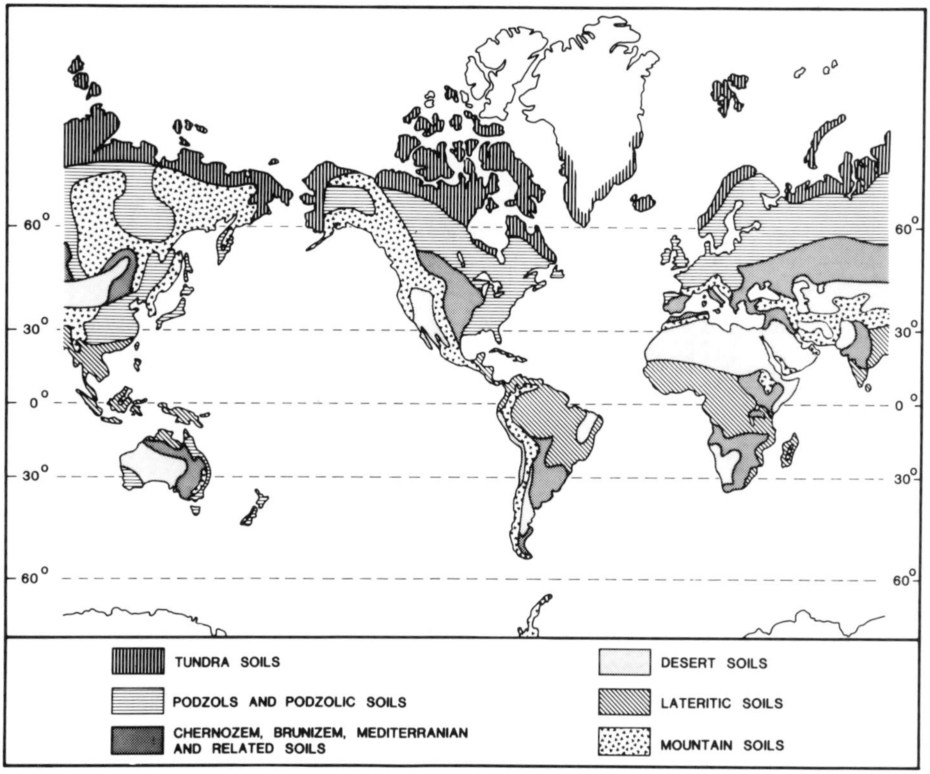

Fig. 4-4. A simplified soil map of the world showing the climatic control on pedogenesis except in the case of mountain soil, the genesis of which is mainly controlled by relief. The immature mountain soil belt cuts across all climatic zones, whereas most other soil zones are climatically influenced. (Modified after McIntyre, 1980.)

tropical forests where a very thick lateritic soil is developed. The thickness of the weathering profile may range up to 100 to 120 m (Strakhov, 1967). In addition, there are two narrow belts of soils developed in the steppes and in the savannas. The soils in the steppes may be considered as a transition between temperate podzolic and desert soils, whereas savanna soils are transitional between desert and lateritic soils.

The tectonic and climatic control on the nature of weathered products are summarized in Fig. 4-6. Within the craton, climate plays the most dominant role in influencing the products. On the other hand, climate has relatively less influence in shaping the nature of the products in tectonically active regions, such as collision margins and block-faulted basins within the craton. In stable cratonic settings within the tropics, the weathering profiles are dominated by primary quartz and kaolinite and/or gibbsite as secondary products. In cool to temperate humid and in warm, relatively arid conditions, the soil mineralogy is characterized by quartz and some altered and/or unaltered feldspars, whereas the secondary minerals are mostly

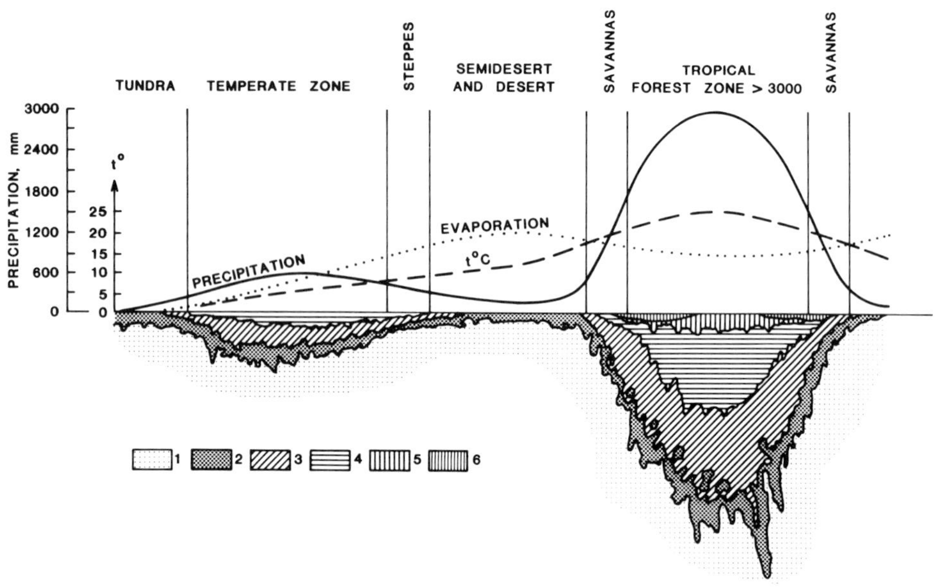

Fig. 4-5. Schematic diagram showing the formation of weathering mantle in tectonically stable areas. Both the thickness and the nature of weathered materials are influenced by climate. 1: Fresh rock; 2: little altered chemically; 3: hydromica – montmorillonite – beidelite zone; 4: kaolinite zone; 5: ocher – Al_2O_3 zone; 6: soil armour, Fe_2O_3 + Al_2O_3. (After Strakhov, 1967.)

kaolinite and smectite. In very cold and warm arid regions, lack of chemical weathering does not result in much change between soil mineralogy and the mineralogy of the parent rock. Secondary minerals, if present, are dominated by cation-rich silicates. In regions with high relief, mechanical weathering predominates and the mineralogy of the weathered products and the parent material are very similar. Soil mineralogy dominated by unstable primary and cation-rich secondary minerals may be related to climate (arid), tectonic setting (with high relief), or a combination of both. The mature soil minerals are characteristic of humid climate only.

The chemical evolution of groundwater at shallow depths has been shown earlier in Fig. 4-2. As the rain or meltwater percolates through weathered zones, the chemical constituents released through chemical reactions are dissolved in the percolating water and make their downward journey, becoming a part of the groundwater system. Due to the lack of chemical weathering along the young mountainous regions, surficial processes have little control on groundwater chemistry. Initially, the percolating water is dilute. Subsequently, the amount of dissolved solids in porewater is mostly influenced by detrital mineralogy of sediments. Even at shallow depths, dissolution of highly-reactive detritus that characterizes sediments in such a setting makes the groundwater highly concentrated. Similar concentrated water is also present in arid regions due to excessive evaporation. In humid regions intense chemical weathering releases abundant chemical constituents, but high precipitation

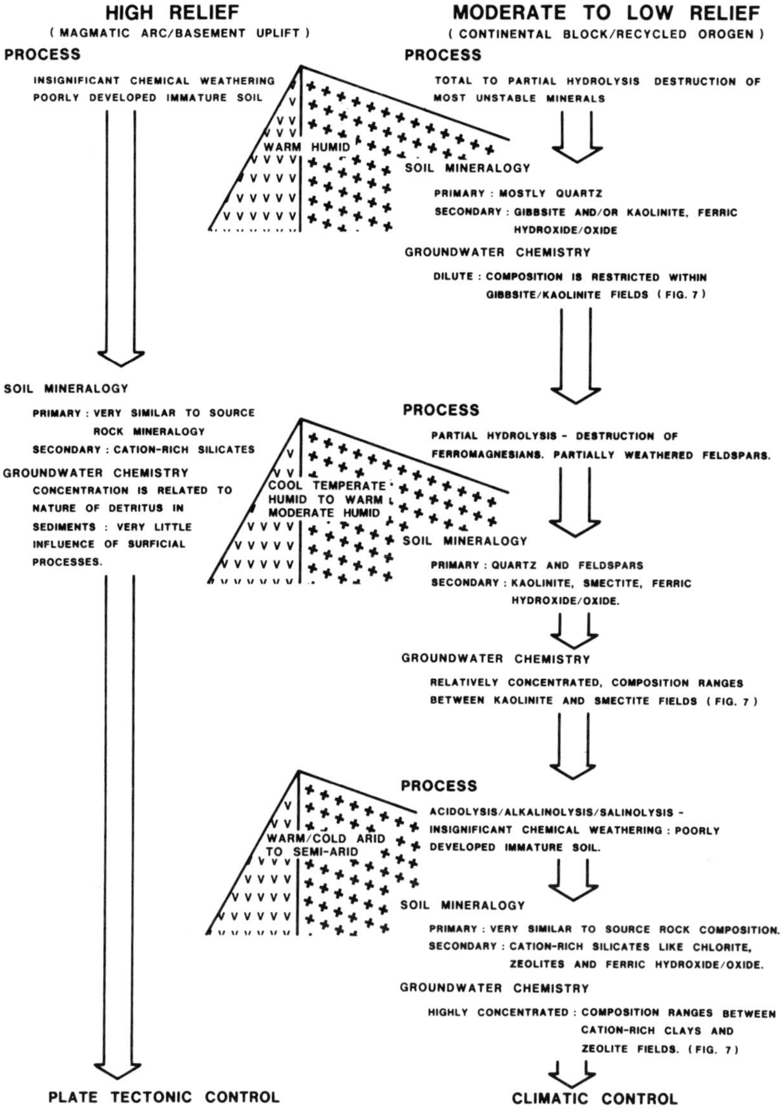

Fig. 4-6. Schematic diagram showing the plate-tectonic control and climatic control on soil mineralogy and groundwater chemistry. Climatic control is mostly significant in cratonic settings and in many cratonized mountain systems. Tectonic control is more pronounced in young mountain systems including block-faulted mountains within craton.

makes the groundwater dilute. The interrelationships among climate, water chemistry at shallow depths, and coexisting neoformed silicate minerals are schematically shown in Fig. 4-7. In arid regions the soil and groundwater are highly concentrated. Such water will plot farthest from the origin of the diagram. Depending on the concentration of various cations in water in such harsh climate, cation-

TABLE 4-7

Framework composition of selected sands* and sandstones

	Composition			Stratigraphic Name/Environment	Age	Location	Reference
	O	F	L				
1	3	25	72	North Range beds	Triassic	New Zealand	Boles, 1974
2	7	34	59	Turbidite sand	Neogene	Alka Basin, Alaska	Stewart, 1978
3	8	38	54	Shoalhaven Group	Late Permian	New South Wales, Australia	Dutta and Wheat (in review)
4	9	21	70	Moehau Fm.	Mesozoic	New Zealand	Skinner, 1972
5	13	15	72	Marginal sea floor	Neogene	Japan and Phillipines seas	Harrold and Moore, 1975
6	15	69	16	Purari graywacke	Cretaceous	Papua	Edwards, 1950
7	21	28	51	Turbidite sand	Neogene	Komardorskiy Basin, Bering Sea	Stewart, 1977
8	24	9	67	Sierra Nevada foothills belt	Upper Jurassic	California, USA	Behrman, 1978
9*	28	38	34	Low-order stream sand	Holocene	Rocky Mountains, USA	Basu, 1975
10	30	28	42	Bear Lake Fm.	Miocene	Bristol Basin, Alaska	Galloway, 1974
11	35	30	35	Eugeosynclinal sediment	Eocene	Oregon, USA	Dott, 1965
12*	41	16	43	Big-river sand	Holocene	Rhine River, France	Potter, 1978
13	44	12	44	Paskapoo Fm.	Paleocene	Alberta, Canada	Carrigy, 1971
14	46	5	49	Trenchard Group	Carboniferous	England	Jones, 1972
15	49	44	7	Cutler Fm.	Permian	Colorado, USA	Suttner and Dutta, 1986
16	50	13	37	Vester Fm.	Upper Triassic	Oregon, USA	Dickinson et al., 1979
17*	53	24	23	Big-river sand	Holocene	Brahmaputra River Bangladesh	Potter, 1978
18	54	42	4	Talchir Fm.	Lower Permian	Raniganj Basin, India	Suttner and Dutta, 1986

TABLE 4-7 (continued)

Framework composition of selected sands* and sandstones

	Composition			Stratigraphic Name/Environment	Age	Location	Reference
	O	F	L				
19	54	36	10	Fountain Fm.	Permian Pennsylvanian	Colorado, USA	Suttner and Dutta, 1986
20	58	19	23	Aberystwyth Grit	Silurian	Wales	Okada, 1967
21*	60	28	12	Low-order stream sand	Holocene	Appalachian Mountains, USA	Basu, 1975
22	61	3	36	Bradford sand	Devonian	Pennsylvania, USA	Krynine, 1940
23*	68	10	22	Big-river sand	Holocene	Nile River, Egypt	Potter, 1978
24*	70	5	24	Big-river sand	Holocene	Mekong River, Vietnam	Potter, 1978
25*	74	3	23	Low-order stream	Holocene	Appalachian Mountains, USA	Young, 1975
26	78	3	19	Taconic molasse	Lower Paleozoic	Appalachian Mountains, USA	Pettijohn et al., 1973
27	81	3	16	Deese Fm.	Pennsylvanian	Oklahoma, USA	Jackobsen, 1959
28	85	7	8	Yellow sands	Permian	North Sea	Pryor, 1971
29	91	8	1	Panchet Fm.	Lower Triassic	Raniganj Basin, India	Dutta, 1983
30	94	6	0	Miogeoclinal sandstone	Precambrian/Cambrian	Mojave Desert, USA	Lobo and Osborne, 1976
31	98	0	2	Hawkesbury sandstone	Middle Triassic	New South Wales, Australia	Dutta and Wheat (in review)
32	99	1	0	Mahadeva Fm.	Upper Triassic/Lower Jurassic	Raniganj Basin, India	Suttner and Dutta, 1986
33*	100	0	0	Big-river sand	Holocene	Orinoco River, Venezuela	Johnsson et al., 1988.

rich silicate minerals, like zeolite, palygorskite, chlorite, smectite, sepiolite, etc., will form. On the other end of the spectrum gibbsite forms in extreme warm humid conditions, as in the tropics. In between these two extreme climatic conditions relatively cation-poor silicates, such as kaolinite and silica, precipitate. (Editorial note: Many publications are available on the climatic control of zeolites in sandstones/conglomerates.)

CLIMATIC CONTROL ON DETRITAL MINERALOGY OF FLUVIAL SAND

Introduction

A fundamental question the sedimentary petrologists are trying to answer is: Why are sandstones compositionally so varied (Table 4-7) though they are ultimately derived from a few, rather restricted suites of igneous and metamorphic rocks? What happens during the transformation from the starting material, i.e., the source rock, to the final product, the sandstone? Attempts have been made to understand the entire spectrum of changes from lithospheric material to sand generation and, subsequently, the transformation of sand into sandstone. In the preceding section on the climatic control on the mineralogy of soils, the precursors of clastic sediments have been discussed. In this section the discussion focuses on the climatic control on detrital mineralogy of modern fluvial sands. In a way this discussion is an extension of the review of the climatic control on soil mineralogy. Here, the analysis is

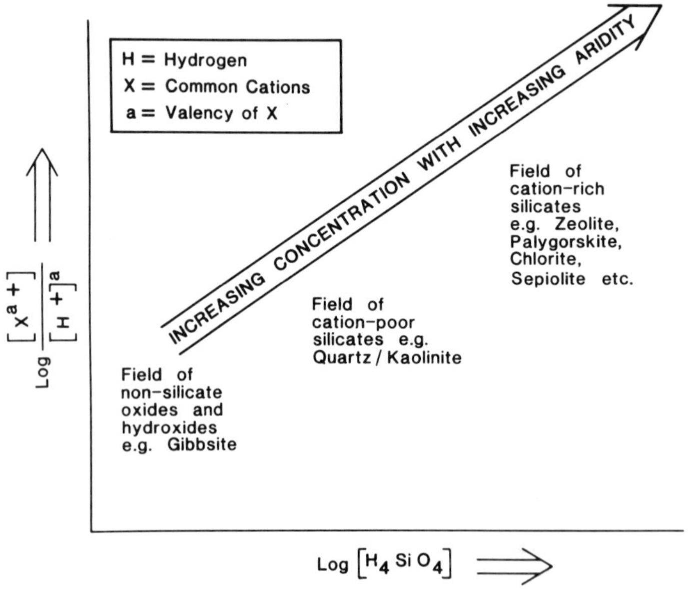

Fig. 4-7. Schematic diagram showing the relationship among pore-water chemistry, climate, and the nature of silicate minerals in sediment.

based on modern fluvial sands from different parts of the world, having different climates, source rocks, and relief conditions.

In a sequential order of change from parent rock to sediment through soil formation or through saprolitization, the study of Holocene sands under controlled conditions (with known variables) have identified climate as one of the most important factors in controlling sand/sandstone composition. These studies have been done on two very widely different scales. On a regional and global scale, Garner (1959), Potter (1978, 1986), Franzinelli and Potter (1983), and Johnsson et al. (1988) have studied the texture and composition of Holocene sands from the big rivers of the world. On a much smaller scale, Mann and Cavarock (1973), Basu (1975), Young (1975), Suttner et al. (1981), and Grantham and Velbel (1988) have documented the effects of climate on controlling the first-cycle fluvial sand composition in lower-order* streams.

Evidence of climatic control on mineralogy of modern sand in low-order streams

Based on the analysis of their own data and those from literature on soil and sand mineralogy, Suttner et al. (1981) observed a strong correlation between composition of stream sand and the composition of the sand fraction of soils, which serve as the immediate source material for stream sands (Fig. 4-8). Because of these close compositional similarities between soil and low-order stream sand, they argued that climate primarily controls sand mineralogy in sediments. The dual influence of grain size and climate on composition are well demonstrated in the same size-compositional diagram in Fig. 4-8. In spite of an overall good correlation, anomalies are observed in the fine-grained fraction with respect to the rock fragments and monocrystalline quartz derived from metamorphic source rocks. Both depletion of rock fragments and concentration of monocrystalline quartz in fine-grained fluvial sand from metamorphic sources are attributed to the mechanical destruction of metamorphic rock fragments associated with fluvial transport. The same process would explain the relative abundance of monocrystalline quartz. The destruction of rock fragments as observed by Suttner et al. (1981) is consistent with Cameron and Blatt's (1971) conclusion that schistose rock fragments are highly susceptible to destruction during fluvial transport.

Keeping all variables (except climate) controlling the sand composition constant, Basu (1975) and Young (1975) were able to document the effect of climate on sand composition. Basu collected sand from first- and second-order streams draining plutonic rocks from temperate humid climate in the southern Appalachians and a temperate semi-arid region in the Rocky Mountains. Young (1975) also collected sands from a temperate humid climate in the Appalachians and temperate semi-arid climate in the Rockies from first- and second-order streams, but draining high-rank

* Stream order is a quantitative classification of river channel segments according to their hierarchical position in a drainage network. Fingertip tributaries are first-order streams. Successively higher-orders are formed by the junction of two stream segments of the same order, i.e., two first orders → second order, two second orders → third order, etc.

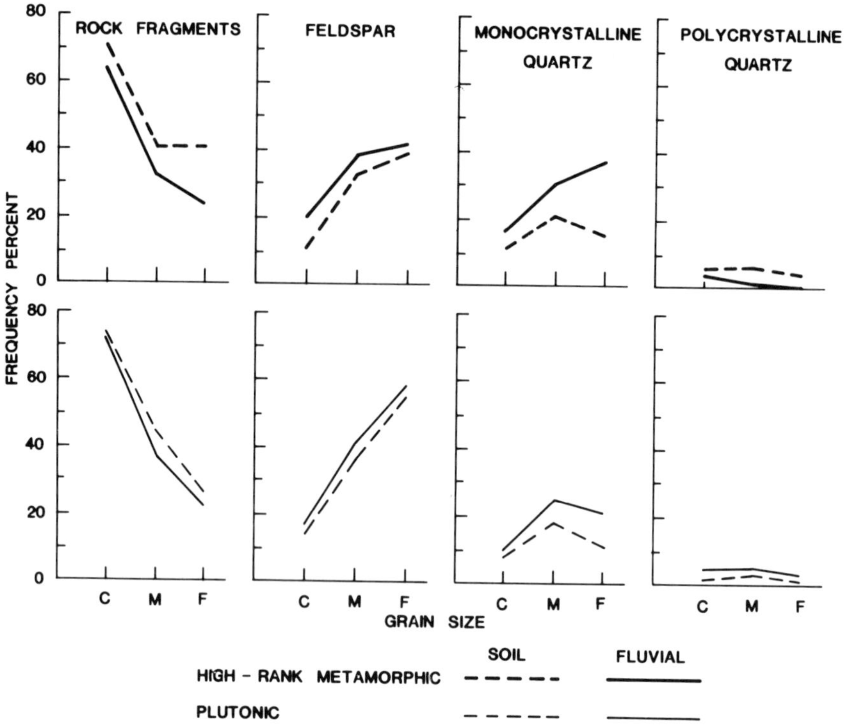

Fig. 4-8. Size-compositional plot of soil and sand in first- and second-order streams draining the same soil horizons. These plots show the initial control of sand composition as related to pedogenesis. The plots also show the composition as a function of size. (After Suttner et al., 1981.)

metamorphic rocks. In both studies the mineralogical maturity in the humid climate were found to be greater. Their data in Fig. 4-9 highlight climatically-induced compositional maturity in fluvial sand under temperate humid condition relative to a temperate semi-arid climate. Mann and Cavarock (1973) documented the effect of climate on three source-rock types: (1) plutonic, (2) metamorphic, and (3) first-cycle sedimentary rock. Under warm temperate (average annual temperature of 15°C), humid (average annual precipitation of 1140 mm), and low-relief (slope varies between 10 and 20 m km^{-1}) settings, weathering conditions in the southern Appalachians nearly completely destroyed ferromagnesian minerals, plagioclase, and micaceous rock fragments from both granitic and high-rank metamorphic rocks. Their study also indicated the size-dependent compositional variations as subsequently observed by Darnell (1974), Suttner et al. (1981), and Grantham and Velbel (1988). A significant conclusion of their work is that chemical weathering of first-cycle sediment did not produce marked variation in composition of second-cycle sand.

Grantham and Velbel (1988) observed that in the southern Blue Ridge Mountains,

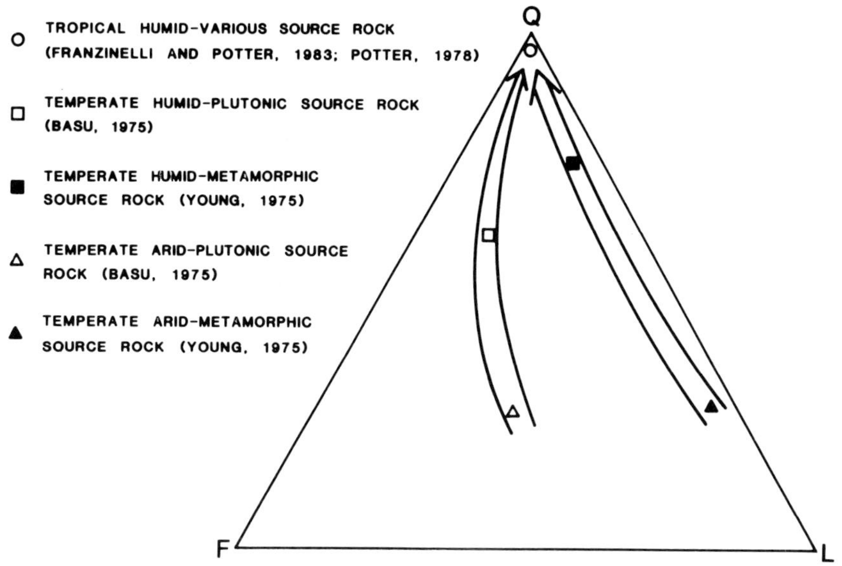

Fig. 4-9. Compositional maturity trends observed in Holocene sands from different climatic belts indicating the climatic control on sand composition. (Data from Basu, 1975; Young, 1975; Franzinelli and Potter, 1983.)

North Carolina (U.S.A.), chemical weathering in a small drainage basin underlying similar parent rock type is controlled by relief and precipitation. Combining relief ratio and intensity of weathering, they approximated the total extent of chemical weathering over a period of time which they defined as Cumulative Chemical Weathering Index (*CCWI*):

$$CCWI = \text{Effective Precipitation} \times 1/\text{Relief Ratio}$$

The effective precipitation is equal to the stream discharge per watershed unit area and the relief ratio is equal to the maximum relief divided by the maximum length of the watershed. This equation can be used to assess or predict the extent of chemical weathering in a given setting where the temperature of the environment and the source-rock composition are known. In this study, they observed that the rock fragment content in sand is related to the Cumulative Chemical Weathering Index (*CCWI*), a measure of total extent of chemical weathering. Plots of modal abundance of rock fragments against *CCWI* for different grain sizes (i.e., coarse, medium, and fine) show that as the *CCWI* increases, the content of rock fragments systematically decreases linearly (Fig. 4-10). The most significant mineralogical maturity, however, is observed due to the combined effect of relief and tropical climate. In the tropical Barro Colorado Island (Panama) the chemical weathering in low-relief terrain is so intense that quartz-rich sands are produced from different source rocks including volcanic-type rocks (Johnsson and Stallard, 1989).

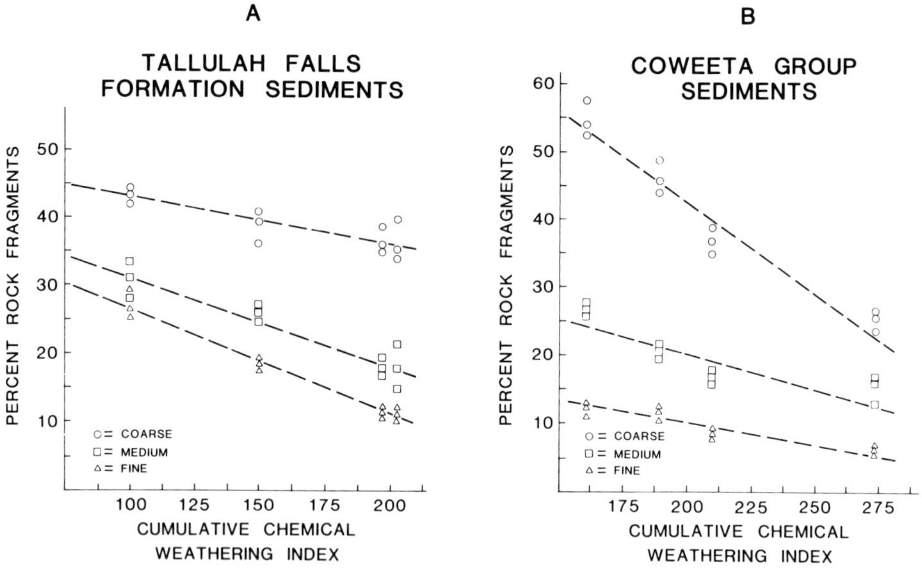

Fig. 4-10. Percentage of rock fragments versus the Cumulative Chemical Weathering Index (*CCWI*) for watersheds draining (A) the Tallulah Falls Formation and (B) the Coweeta Group bedrock. Rock fragments decrease systematically as a result of increasing *CCWI*. (After Grantham and Velbel, 1988.)

Evidence of climatic control on sand mineralogy on a global scale

Present-day understanding of the detrital mineralogy of fluvial sands on a global scale and their probable genesis is based on the findings of Potter (1978, 1986), Franzinelli and Potter (1983), and Johnsson et al. (1988). In his study of modern big river sands, Potter (1978) showed that the total feldspar and rock fragments in sands from tropical low-relief rivers varies between 1 and 13% with an average of 5%. In contrast, the same parameter in sands from rivers in both high- and low-relief settings, but in either temperate or arctic regions, range from 20 to 87%, with an average of 47% (Fig. 4-11). Moderate- to low-relief rivers, draining through the humid tropics, generate supermature quartz sand (Sao Francisco, Parana, Congo and Niger). Low-relief rivers, draining either temperate/arctic humid, or cold/warm dry areas, produce immature sands (Rio Grande, Orange, Moose, Mackenzie, Shatt-al-Arab; Fig. 4-12). Franzinelli and Potter (1983) studied the composition, texture and chemistry of fluvial sands of the Amazon River system in order to relate the texture and composition of the sand to source rock and climate. They noted that the low-relief rivers, draining the Precambrian Guyana and Brazilian shields within the humid tropics, generate first-cycle mature sand including quartz arenite. First-cycle quartz arenite sands have also been reported from the Orinoco River basin, Colombia and Venezuela (Johnsson et al., 1988). In this study, the authors observed that first-cycle quartz arenite can be generated in diverse tectonic settings as a result of intense chemical weathering in a tropical climate. Exceedingly pure quartz arenite

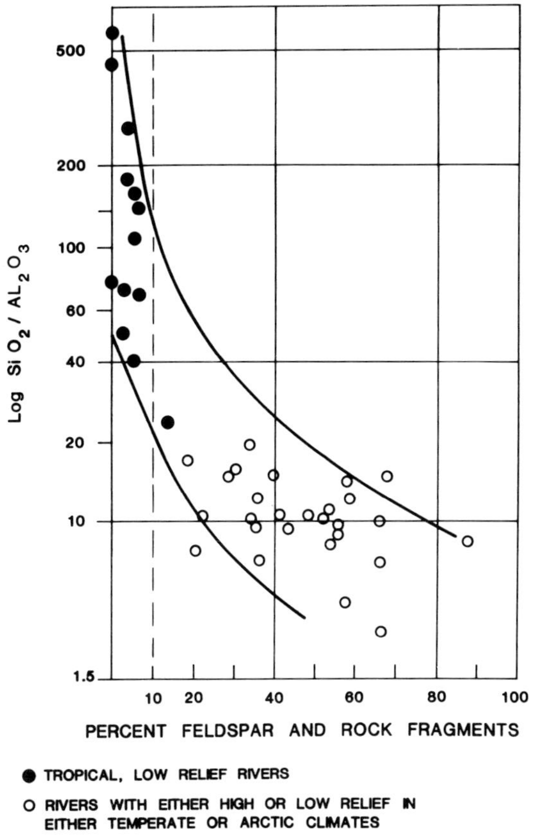

Fig. 4-11. Climatic influence on sand composition and sand chemistry. Sands with high quartz and high SiO_2/Al_2O_3 ratios (solid circles) are mostly from low-relief tropical river basins. Open circles represent low-relief rivers draining either humid temperate/arctic or cold/warm dry areas. (After Potter, 1978.)

sands are forming today within continental block provenance of the lowland Guyana shield of granitic composition. Similar quartz-rich sands are also being generated in the Andean foreland basin as a result of chemical weathering over an extended period. Inasmuch as the first-cycle quartz arenites can be produced in very different tectonic settings, the authors concluded that the climate has the capacity to obliterate the tectonic signature.

Garner (1959) made one of the first attempts to understand the role of climate in controlling the texture of detritus. He made a detailed study of the grain size of sediments in the provenance area and the river systems draining the Andean Mountains in South America. He observed that, in spite of high relief and proximity to the elevated mountains, the river sediments are characterized by a high percentage of clay- and silt-size particles. He concluded that the nature of detrital materials

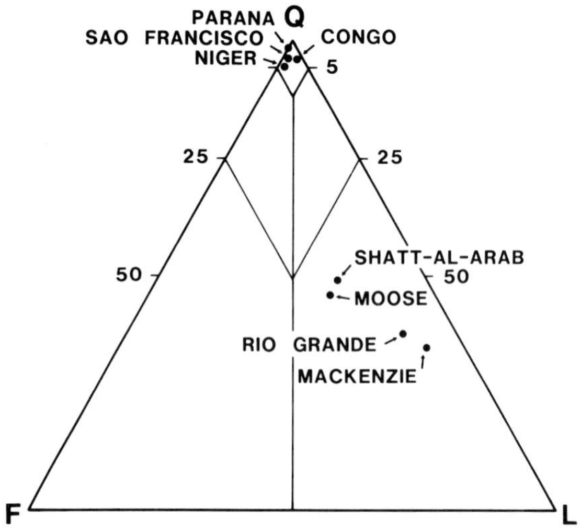

Fig. 4-12. Big-river sand composition. Supermature quartz arenite sands are observed in rivers draining low-relief tropical rivers (Congo, Niger, Sao Francisco, Parana). Immature sands are observed in rivers in moderate- to low-relief terrains draining either cold or warm arid regions. (After Potter, 1978.)

were primarily controlled by climate, even masking the effects of high relief, source-rock composition, and distance of transport.

Synthesis

Table 4-8 and Fig. 4-9 show the compositions of first-cycle fluvial sand derived from plutonic and metamorphic source rocks (Basu, 1975; Young, 1975; Franzinelli and Potter, 1983; Velbel, 1988; Johnsson et al., 1988). The data come from three different climatic belts, i.e., temperate semi-arid (Rocky Mountains), temperate humid (southern Appalachians), and tropical humid (South America). The compositional trends shown by arrows in Fig. 4-9 demonstrate the effects of climate on fluvial sand composition. The degree of sand maturation in a granitic source rock from a temperate semi-arid (annual average: precipitation – 250–600 mm; temperature – 5° – 10°C; QFL ratio – 28:38:34) through temperate humid (annual average: precipitation – 1000–1500 mm; temperature – 13° – 16°C; QFL ratio – 60:28:12) and finally to humid tropics (annual average: precipitation – > 2000 mm; temperature – > 27°C; QFL ratio – 100:0:0) is large. Roughly a 15°C increase in average temperature and an average increase of 500 mm in rainfall can change the sand composition from a first-cycle arkose to a first-cycle quartz arenite. The maturity of big-river sands from mixed provenances (Figs. 4-11 and 4-12) is also strongly influenced by climate. The data from modern fluvial environments both from low-order streams and big rivers of the world suggest that climate has a first-order control on sand composition in moderate- to low-relief terrains.

TABLE 4-8

Composition of first-cycle fluvial sand from moderate-to-low-relief terrains in different climatic regions of the world.

Source of data	Temperate semi-arid			Temperate humid			Warm humid		
	Q	F	L	Q	F	L	Q	F	L
Basu (1975)*	28	38	34	60	28	12		–	
Young (1975)**	29	3	68	74	3	23		–	
Franzinelli and Potter (1983)*		–			–		100	0	0
Grantham and Velbel (1988)**		–		74	0	26		–	
Johnsson et al. (1988)*		–			–		100	0	0

* Granitic source rock.
** High-rank metamorphic source rock.

CLIMATIC CONTROL ON EARLY DIAGENESIS

Introduction

Reduction and modification of primary porosity and permeability, and accompanying lithification due to precipitation of authigenic cement are the end results of chemical diagenesis in sandstones. Most siliciclastic sands have an initial porosity in the range of 35 – 40% and a permeability of several darcys, whereas most oil or gas reservoir sandstones have porosities ranging from 10 to 25%, and permeabilities of a few to a few hundred millidarcys (Hayes, 1979). Thus, porosity reduction may range from 10 to 30%, with a permeability reduction of several darcys. The bulk of this porosity reduction is mainly due to the chemical diagenesis. Many alteration reactions of detrital minerals with addition of water cause volume expansion and, in turn, reduction in porosity. The alteration of feldspars and micas to clay minerals, and plagioclase to laumontite are such examples. Porosity reduction could be brought about by pressure solution involving mostly quartz. As mentioned earlier, porosity reduction can also be accomplished by compaction. A literature survey (e.g., Chilingarian and Wolf, 1976) on diagenesis of siliciclastic sandstones reveals that even all three processes together (compaction, pressure solution, and volume expansion due to alteration reactions) cannot account for such a large volume of porosity reduction. In practice, such processes can reduce porosity only marginally. Porosity reduction, therefore, is largely due to precipitation of cement from pore solution. But "where do the solutions and their dissolved components come from?"

There are only a few possibilities regarding the source of cement in siliciclastic sediments. These sources are: (1) externally-derived chemical constituents from weathering profiles transported by circulating meteoric water, (2) internally-derived solutes from the dissolution of minerals, (3) solutes in pore water derived from dewatering of shale, (4) ions released during clay diagenesis, and (5) waters of hydrothermal and metamorphic origins. In order to understand the role of climate in diagenesis, it is essential to know the nature of various sources of cements during different stages of diagenesis and specifically identify the sources influenced by climate. In addition, it is also necessary to evaluate the depth range where the meteoric water dominates the early authigenic process. Finally, one needs to know the "time frame" of different diagenetic episodes: when did it happen?

Source of cement in sandstone

In the past, most researchers on diagenesis made no serious attempt to identify the source of cements. In many studies an internal source from dissolution of detrital minerals has been implied. Probing questions on this aspect are being raised in recent years. Chemically-rich interstitial water in mud beds, expelled during compaction has been advocated as a possible source of cement (Rieke and Chilingarian, 1974; Land and Dutton, 1978; Boles and Franks, 1979; Hayes, 1979). Considering even a very high mud/sand ratio of 100:1 at the time of deposition, the volume of expelled water from mud to the volume of sand would be nearly in the proportion of 40:1. In this calculation, the assumption has been made that compaction in sand is zero and the porosity reduction in mud is about 40%, from an average initial porosity of 60 – 20% during burial, up to a depth of about 1000 m (Von Engelhardt, 1977). This means that for each unit volume of sand body there are 40 unit volumes of water expelled from the mud. If one assumes that only 10% porosity is reduced by precipitation of cement, then for each unit volume of cement precipitated in pore space there are 400 unit volumes of pore solution available from compacted shale. Considering the average chemical composition of a pore fluid, it is estimated that at least 10^5 volumes of water is required to cement each cm^3 of pore space in sandstone (Land and Dutton, 1978; Bjorlykke, 1979, 1988; Blatt, 1979; Dutta and Suttner, 1986). This means that only 1/250 part ($400/10^5$) of cement may be available from dewatering of shale.

Diagenesis of clay has also been cited as a possible source of cement in sandstone (Boles and Frank, 1979). At a burial temperature in the range of 75° – 100°C and beyond, kaolinite and illite/smectite (I/S) interlayer clays are converted to illite (Hower et al., 1976; Boles and Frank, 1979; Dutta, 1983; Srodon and Ebert, 1984; Dutta and Suttner, 1986; Suter, 1986). Illitization of I/S interlayer clays releases silica, iron, and magnesium (Hower et al., 1976). These authors have also observed that the bulk composition of shale does not change as a function of mineralogical change with depth except for Ca. Chemically, the shale acted as a closed system except for Ca cation. Dutta (1983) demonstrated that the transformation of kaolinite to illite, unlike the illitization of I/S interlayer clays, needs silica, K, and Mg. He observed that all the chemical constituents needed for illitization of kaolinite were internally derived from the dissolution of detrital minerals. It seems, therefore, that illitization in a shale bed, where both kaolinite and smectite are present, will involve two different types of reactions, such as:

Smectite + Al + K = Illite + Si + Mg + Fe (Hower et al., 1976)*
Kaolinite + Si + K + Mg = Illite + Al (Dutta, 1983)

Illitization of interlayer I/S releases silica and cations like iron and magnesium (Hower et al., 1976; Boles and Frank, 1979), whereas transformation of kaolinite to illite releases Al, but will need silica and cations like potassium and magnesium (Dutta, 1983). These reactions are complementary to each other: one releases and the other absorbs silica and other cations involved in the formation of illite. In addition, in both reactions, potassium is necessary and is available through dissolution of K-feldspar and mica (Hower et al., 1976; Dutta, 1983). It has only been inferred, but has not yet been documented through mass-balance calculations that clay diagenesis, which involves mainly illitization of I/S interlayer clays and kaolinite, had been a reasonable source of cement. It is possible that the processes of compaction, pressure solution, alteration reaction involving volume expansion, dewatering of shale, and clay diagenesis, which individually may contribute only marginally, cumulatively may be significant in reducing primary porosity. Yet, taken even cumulatively, these processes seem to leave a considerable gap between the amount of porosity reduction observed in most sandstones and the porosity reduction caused by these processes. Dissolution of detrital minerals, a common process during late diagenesis, seems to be the most important source of cement. Such a process, however, is, in a way, only a chemical readjustment within the sediment system without contributing much towards reduction of primary porosity. In sandstone diagenesis, therefore, an external source of cement becomes imperative. The external source of cement, excluding hydrothermal and metamorphic water**, has to be of meteoric origin in continental setting.

Based on textural evidences and mass-balance calculations, Dutta (1983) and Dutta and Suttner (1986) noted that a considerable part of authigenic cement in quartzo-feldspathic sandstones has formed early during shallow burial. They observed no dissolution, alteration, or replacement relationship among a set of authigenic minerals, which they identified as "neoformed early cement" and the detrital components. From these observations, these authors concluded that the source of the cement in quartzo-feldspathic sandstones, during early diagenesis, was externally derived. An external source of meteoric origin during early diagenesis has also been supported by isotopic data (Longstaffe, 1984; Dutta, 1985; Dutton and Land, 1985; Dutta and Suttner, 1986; Ayalon and Longstaffe, 1988). Longstaffe (1984) observed kaolinite, smectite, and calcite as pore-lining and pore-filling cements in the Milk River aquifer sandstone of Upper Cretaceous age. Oxygen isotopic composition of these authigenic minerals are in isotopic equilibrium with a pore fluid of meteoric origin. An internal source through dissolution of detrital minerals at this stage would have made the pore fluid heavier in ^{18}O compared to meteoric water. Though Longstaffe did not deal with the aspect of the source of cement, the isotopic evidence implied that the source of cement must have been an external one of

* The reactions are not written in stoichiometric proportion.

** Hydrothermal and metamorphic waters have deep-seated origin and are relatively infrequent in cratonic setting. These sources, therefore, are not discussed in this chapter.

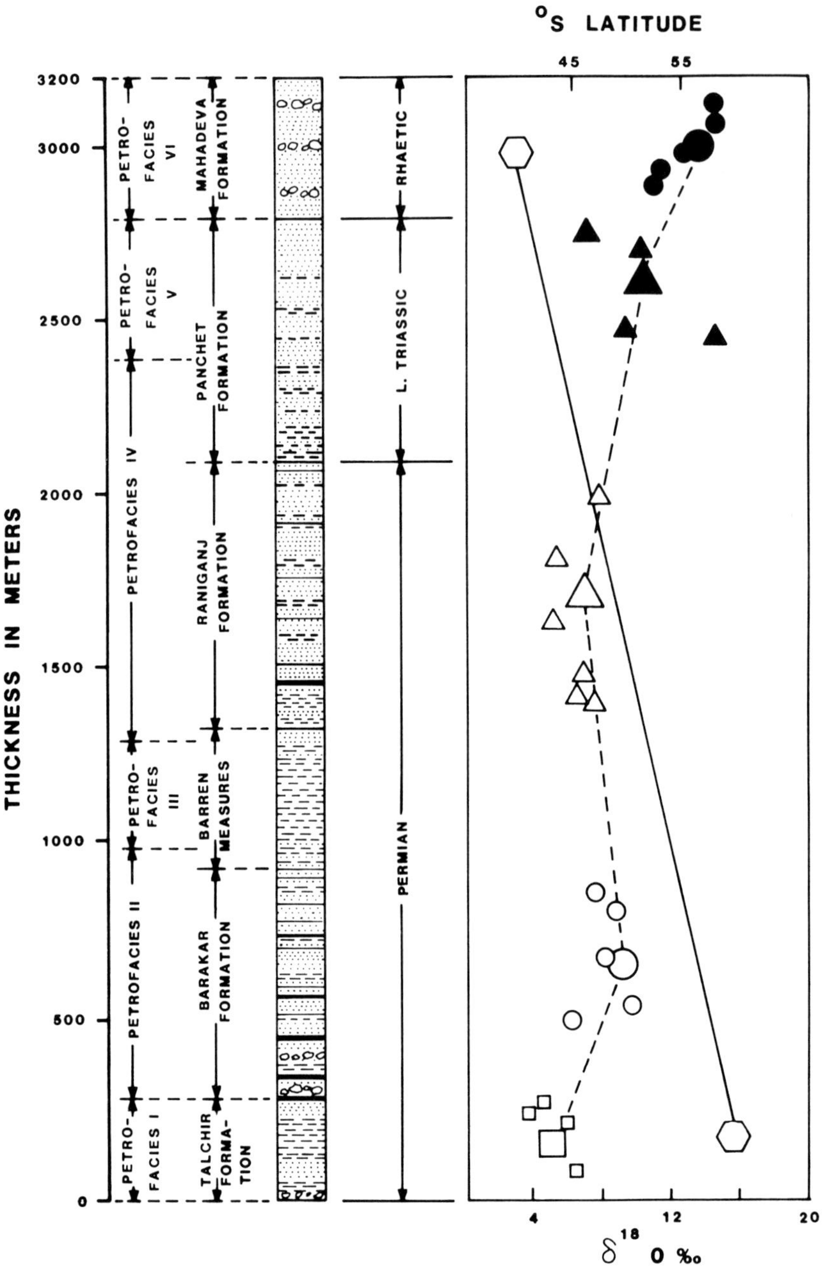

Fig. 4-13. Oxygen isotope composition of early authigenic clay in different petrofacies of Gondwana Supergroup shows gradual change with the absolute age of the sediments. Roman numerals represent Gondwana petrofacies. Small symbols represent oxygen isotopic composition of individual samples and the large symbols represent the average value for the same parameter for each petrofacies. (After Dutta and Suttner, 1986.)

meteoric origin. Using oxygen isotopic composition of authigenic clay minerals in Permo-Triassic Gondwana sandstones of India, Dutta (1985) identified the source of cement during early diagenesis. He observed that the oxygen isotopic compositions of early authigenic clay cements in sandstones showed a gradual increase from +5.00‰ in Sakmarian time, to +13.20‰ in Rhaetic time (Fig. 4-13). This gradual increase of $\delta^{18}O$ values with decreasing age shows a strong correlation with the changing latitudinal location of the sample site from a 60°S during Sakmarian time to a position around 38°S during Rhaetic time, respectively (Fig. 4-14). The changing pattern of $\delta^{18}O$ values of authigenic clays has been interpreted as a result of corresponding change in isotopic composition of coexisting meteoric water due to a northerly shift of the basin during early diagenesis. From these isotopic and paleogeographic data, Dutta concluded that the aqueous solution, involved in early diagenesis, was externally derived and was of meteoric origin. Oxygen isotopic compositions of early authigenic clay cements in Permo-Pennsylvanian sandstones of the Cutler and Fountain formations of Colorado (U.S.A.) also suggest a meteoric origin for early cements (Dutta and Suttner, 1986). Dutton and Land (1985) also

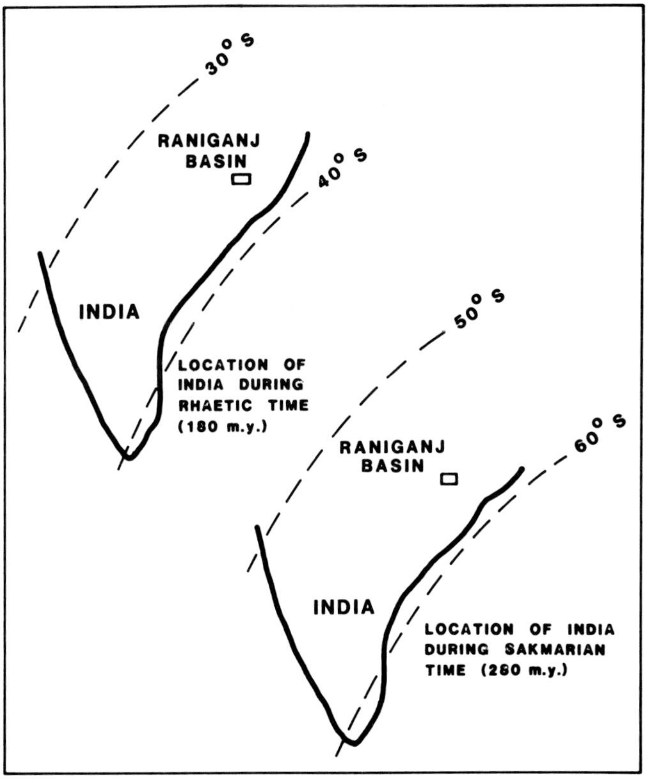

Fig. 4-14. Schematic diagram showing the changing latitudinal locations of the Raniganj Basin during Gondwana sedimentation from a higher latitudinal position to mid-latitudinal position. This migration accentuated the climatic change from a frigid condition during Lower Permian (Sakmarian) time to relatively warm humid conditions during Upper Triassic (Rhaetic) time.

observed involvement of meteoric water during early diagenesis of Pennsylvanian arkosic sandstones in the Anadarko Basin in Texas (U.S.A.). Based on oxygen isotope studies, Ayalon and Longstaffe (1988) concluded that an external source of meteoric or brackish water played an important role in early cementation of chlorite and, possibly, calcite in the Upper Cretaceous Basal Belly River sandstones in Alberta (Canada).

Considering an open system for fluid flow at shallow depths, an external source of cement driven by meteoric water seems to be an important factor in early diagenesis. At shallow burial depths, the coarse-grained sediments behave as highly porous and permeable media where fluids can move fast and freely. The chemical system within such media can also be considered as an "open system" involving transfer of chemical constituents over long distances (Wood and Surdam, 1979). Inasmuch as the pore water moves rapidly, the residence time is too short to initiate any dissolution reaction, because the kinetics of the dissolution reaction is slow (Kramer, 1968). In spite of the relatively fast movement of pore water, however, neoformation of minerals can take place because, in this case, the kinetics of precipitation is fast (Kramer, 1968). Thus, at shallow burial depths, the pore water imposes its chemistry on the sediment system through authigenesis (Merino and Ortoleva, 1981). A dynamic "open system" such as this, can precipitate large amounts of material per unit time, suggesting that such a system will be more effective in cementing a porous body (Wood and Surdam, 1978). This implies that a significant part of authigenic cement, derived from an external source such as soil-weathering profile, may form at shallow burial depths when the sediments still behave as a part of an open dynamic system. Documentation to support this model of an "open system" during shallow burial has been made in modern soils and sands at shallow depths. Neoformation of clay and hydroxides in soils and shallow sands have been observed in different parts of the world. With few exceptions, such neoformed minerals seem to be in chemical and isotopic equilibrium with the coexisting water of meteoric origin (Feth et al., 1964; Lawrence and Taylor, 1971; Tardy, 1971).

Petrographic, geochemical, and isotopic evidences presented by authors cited in this discussion suggest that a part of the authigenic cement forms early during shallow burial. During this stage, the source of the cement was externally derived where meteoric water acted as a carrier of solutes from weathering profiles to the sites of authigenesis (Fig. 4-2). Inasmuch as the pore-water chemistry is controlled by climate, it should be possible to relate the early authigenic minerals to climate through pore-water chemistry. At this point, it is necessary to point out that the validity and the conclusions drawn from the above analyses is true for predominantly quartzo-feldspathic source rocks. In case of highly reactive unstable minerals or volcanic glass, dissolution is common even at shallow burial depths (Davies et al., 1979; Mathisen, 1984). In such a case, the groundwater chemistry will be controlled both by detrital mineralogy and climate.

Time of early cementation – An early diagenetic model

There have been various attempts to answer the question of timing of early authigenesis in terms of absolute age. A direct answer to this question is possible only by radiometric dating of early authigenic minerals. Common methods of dating authigenesis are by K/Ar (Potassium/Argon) and Rb/Sr (Rubidium/Strontium) techniques using illite and glauconite. Inasmuch as illite is a late diagenetic mineral, dating illite will yield only late diagenetic event. Glauconite typically forms in a marine environment at the sediment – water interface or within the top few centimeters of muddy sediment and the top few meters of coarse sandy sediment (Odin and Dodson, 1982). Any radiometric age of glauconite, therefore, will indicate the age of the bed or at best the initiation of the early authigenic process.

Early authigenic smectite, chlorite, zeolite, etc., may also be used for radiometric dating; however, none of these minerals so far have been proven to yield reliable age (Bogg, 1986). Dutta and Suttner (1986) have tried to make a qualitative estimate to answer the question on timing of early diagenesis. Their method is based on calculations on the average chemical composition of groundwater in siliciclastic sediments at shallow depths, on the volume of such water necessary to lithify sandstone, on the type of groundwater flow system, and on the time necessary to flow the required volume through the sediment.

Volume of groundwater involved in early cementation

In general, groundwater within a depth of a few hundred meters is a dilute aqueous solution, particularly in noncarbonate terrains. The groundwater analyses shown in Fig. 4-2 represent the average values up to a depth of 400 m in siliciclastic sediments (average of 20 samples; White et al., 1963). At this depth, water contains, on an average, 31 ppm aqueous silica. The equilibrium solubility of quartz at 25°C and 1 atmosphere is 6 ppm. This leaves 25 ppm of aqueous silica available for the formation of any silica-bearing mineral. In order to precipitate one cubic centimeter of quartz, 10^5 cm^3 of pore solution are necessary*. To lithify a hypothetical sand body 100 km long × 20 km wide × 20 m thick (Fig. 4-15), by reducing 10% of its original pore space by silica cement, 4×10^{15} cm^3** of pore space have to be cemented. Because each cubic centimeter of cement needs 10^5 cm^3 of pore water, the volume of water (V) necessary to move through the sand body per year, through a cross-section 20 km wide and 20 m thick, will be 4×10^{20} cm^3 [4×10^{15} cm^3 (total volume of cement) × 10^5 cm^3 (volume of pore solution necessary to precipitate each cm^3 of cement)].

Timing of early cementation

Groundwater may move longitudinally within a regional groundwater flow system as shown in Fig. 4-15. Considering the velocity of groundwater flow through

---

* 25 ppm of silica in groundwater is available (in this specific case) for the formation of silica cement. Considering the density of groundwater to be 1 g cm^{-3}, 25 g of silica will be present in 10^6 g of groundwater. Therefore, 2.65 g of silica (1 cm^3 of silica) will be present in $10^6 \times 2.65/25$ g of water (= 1.06×10^5 g of water or approximately 10^5 cm^3 of groundwater).

** Volume of sand body = 100 km × 20 km × 20 m

= (100 × 1000 × 100) cm × (20 × 1000 × 100) cm × (20 × 100) cm

= 4×10^{16} cm^3

Ten percent of the above sand body:

= $1/10 \times 4 \times 10^{16}$ cm^3 = 4×10^{15} cm^3.

a sediment to be 200 cm per year (Fetter, 1980), the volume of water, Q, flowing through the sand body per year through a cross-section 20 km wide and 20 m thick (Fig. 4-15) will be:

$$\begin{aligned} Q &= 200 \text{ cm y}^{-1} \times (20 \text{ km} \times 20 \text{ m}) \\ &= 200 \text{ cm y}^{-1} \times (20 \times 1000 \times 100)\text{cm} \times (20 \times 100) \text{ cm} \\ &= 8 \times 10^{11} \text{ cm}^3 \text{ y}^{-1} \end{aligned}$$

Thus, the time t necessary for the flow of the required volume of water, V, is equal to:

$$\begin{aligned} t &= V/Q = 4 \times 10^{20} \text{ cm}^3/8 \times 10^{11} \text{ cm}^3 \text{ y}^{-1} \\ &= 500 \text{ million years.} \end{aligned}$$

This figure of 500 million years for lithification of sand is untenable in most geological settings. Geological evidences suggest that sandstones, much more extensive than that used in this example, are lithified within a few tens of million years (Blatt, 1979). But, if one assumes that cementation is caused by vertically-circulating groundwater as shown in Fig. 4-16 while the sand unit is at a relatively shallow depth, then the water flowing vertically through the sand body per year through a cross-section (plan view of the sand body) 100 km long and 20 km wide will be:

$$Q' = 200 \text{ cm y}^{-1} \times (100 \text{ km} \times 20 \text{ km}) = 4 \times 10^{15} \text{ cm}^3 \text{ y}^{-1}$$

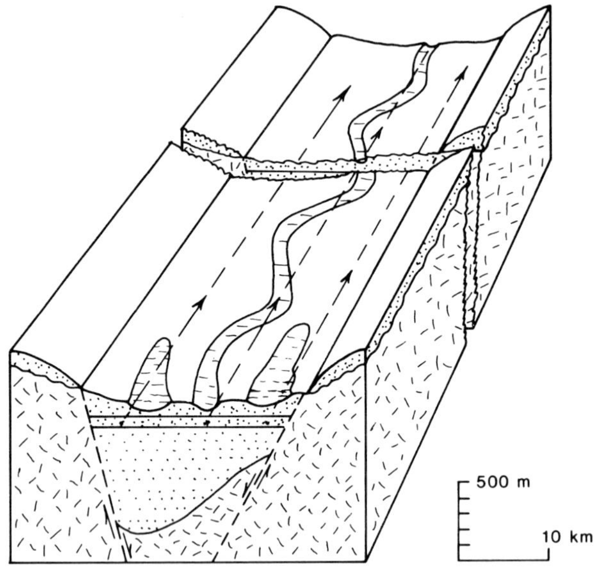

Fig. 4-15. A hypothetical sedimentary succession within a block-faulted basin. The thin strip represents a sandbody having the following dimensions: 20 m thick, 20 km wide, and 100 km long. The dashed lines with arrows show the regional groundwater flow through the sandbody along the regional slope of the valley floor with very low hydraulic gradient.

Thus, the time necessary to flow the required volume V will be:

$$t = V/Q' = 4 \times 10^{20}\ \text{cm}^3 / 4 \times 10^{15}\ \text{cm}^3\ \text{y}^{-1}$$
$$= 100{,}000\ \text{years}$$

This is the minimum time necessary to cement the hypothetical sand body. In this calculation, an absolute vertical flow has been assumed for convenience of calculation. But in practice it is obvious from the flow path shown in Fig. 4-16 that the time needed to cement the sand body will be more than the calculated value of 100,000 years because the water will take a longer path. This approximation for partial cementation of the sand body seems to be more reasonable compared to the value obtained by assuming longitudinal flow. It is necessary, however, to examine the validity of the assumption of vertical flow by examining the groundwater flow system.

Groundwater flow mechanism

In order to understand the nature of groundwater movement in intracratonic basins, the mechanism of groundwater flow in the sediment needs to be evaluated. Such evaluation can best be made by following Toth's (1963) theoretical analysis of groundwater flow in small drainage basins. Intracratonic basins are often characterized by block-faulted troughs occupied by sediments. The essential elements in such sedimentary basins are that they are bounded by uplands on both sides, whereas the sediments themselves underlie an area with little relief at the valley floor (Figs. 4-15 and 4-16).

Following Hubert (1940), Darcy's law may be written as follows:

$$q = -k \frac{\varrho g}{\mu} \cdot \frac{\mathrm{d}h}{\mathrm{d}l}$$

where: q = the rate of fluid flow per unit time per unit cross-sectional area, k = permeability, ϱ = fluid density, μ = fluid viscosity, g = acceleration due to gravity, and $\mathrm{d}h/\mathrm{d}l$ = hydraulic gradient.

This equation indicates that the quantity of groundwater flow in a particular sediment body is a function of hydraulic gradient, because the other parameters may

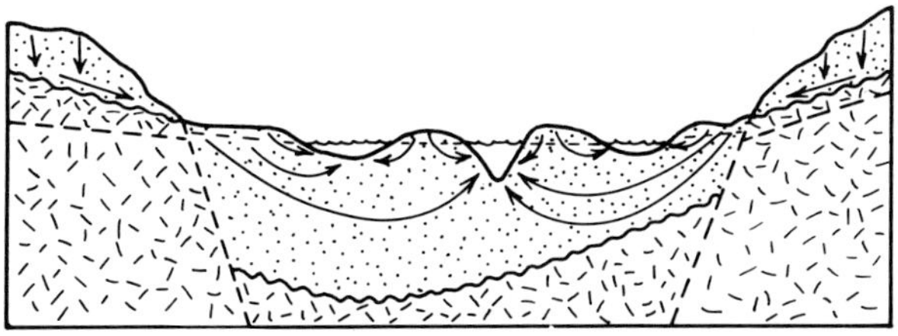

Fig. 4-16. Cross-section along the hypothetical sandbody shown in Fig. 4-15 illustrating the vertical flow of water with a high hydraulic gradient. Local flow systems indicated by small arrows are restricted to shallow depths and have a relatively large vertical flow component. Intermediate flow systems shown with long arrows are dominated by a more horizontal flow.

be assumed constant for a particular setting. In intracratonic block-faulted basins, the slopes of the valley walls greatly exceed the longitudinal slopes of the valley floor (Fig. 4-15). The differences in hydraulic gradients in these two perpendicular directions as a response to the differences in slopes cause the longitudinal component to be negligible compared to the transverse component. Based on this assumption, the groundwater flow in such basins can be treated as a two-dimensional flow system (Toth, 1963). Figure 4-16 is a schematic cross-section of the valley, reflecting the importance of vertical components in the flow system as compared to the horizontal longitudinal flow system shown in Fig. 4-15. This model postulates that 90% of the recharge water does not penetrate deeper than 76 – 91 m (250 – 300 ft; Toth, 1963). A similar view was also expressed by Ubell (1962). He concluded that below a certain depth in loose sediments, water does not move in voids until the state of stress is disturbed by boring. This implies that groundwater movement is extremely slow at greater burial depths.

Toth's (1963) theoretical analysis and the observed nature of groundwater movement in sediment (Plotnikov and Bogomolov, 1958; Ubell, 1962) support the contention of the formation of early authigenic minerals by a vertical flow system during shallow burial of sediments. If 91 m (300 ft) is the approximate cut-off depth of major recharge by groundwater flow activity in a basin having the shape and dimensions shown in Fig. 4-15, then, at least, formation of a part of authigenic cement and, in turn, porosity reduction must take place within shallow burial depths during early diagenesis. The amount of pore reduction by early authigenic cement, however, will be mainly controlled by water chemistry, the amount of recharge, hydrologic parameters, and the subsidence rate of the basin. Considering the rate of burial of 80 to 20 m Ma^{-1} in cratonic basins (Miall, 1981), the time necessary to be buried to a depth of 91 m by a particular sediment layer is about 4.5 m.y. to a little over a million years. Thus, the various approaches, based on the volume of water necessary for early cementation, rate of burial, and the nature of groundwater flow, yield an age of the same order of magnitude for the time of early cementation.

Synthesis

Combining petrographic data, early authigenic mineralogical assemblage, and their oxygen isotopic composition, it has been possible to infer about the source of cement during early and shallow-burial diagenetic stages. Based on the early diagenetic model, a qualitative estimate about the time of early cementation can also be made. From these indirect evidences it is reasonable to estimate that early cementation takes place within a "few" million years and within a depth of a "few" hundred meters. During this stage the source of cement is derived primarily from the weathering profile, driven by groundwater flow of meteoric origin. It is still very speculative to put a numerical value on the term "few". But based on the assessment made in this discussion, it is expected that partially lithified sand should be present at least at a burial depth of 100 – 200 m, ranging in age from 2.5 to 5 million years. It is observed in nature that even at a depth of a few hundred meters some quartzo-feldspathic sediments are not cemented well. There are innumerable examples of carbonate-cemented sandstones at the surface and at shallow depths.

James (1985) observed a carbonate-cemented glacial outwash sublitharenite of Wisconsinian age. Dutta (unpublished work) also made similar observations in Pleistocene quartzo-feldspathic glacial sediments in northern Indiana (U.S.A.). This is possibly because the kinetics of dissolution and precipitation of carbonate minerals are fast and, therefore, a relatively large volume of carbonate cement may form early. Cementation of Oligocene Frio Sandstone in the Gulf Coast (U.S.A.) possibly has taken place within 6 million years after deposition (Land, 1984). In the Frio Sandstone calcite forms 5.3% and quartz forms 2.5% cement by volume. Early cementation in this case is largely dominated by calcite. Unlike carbonate-cemented sediments, there are no extensive developments of silicate-cemented sands at shallow depths. This may again be related to the kinetics of dissolution and precipitation of silicate minerals. In this case, the rates are much slower compared to carbonate minerals. In spite of the presence of authigenic silicate cements in Milk River aquifer sandstone, Longstaffe (1984) observed that the sandstones are rather loosely cemented. One does not have a clear answer to this question at this stage. This is a new area in sandstone diagenesis and there is a great need to answer the following question: "How early is early diagenesis?"

EARLY DIAGENETIC MINERAL ASSEMBLAGE AND CLIMATE

Introduction

Early diagenesis during shallow burial demands an either continuous or at least intermittent flux of aqueous solution from an external source. This enables replenishment of mineral matter consumed in precipitation of cement. Hence, a balance is necessary between the supply of chemical constituents from weathering profiles and consumption of these materials at the site of authigenesis. A balance between the rate of supply and the rate of consumption is rarely attained because the rate of kinetics of dissolution and neoformation are different (Kramer, 1968). Except in extreme conditions, the Earth's climate is seasonal. The rate of chemical reaction will also vary, therefore, causing non-steady state conditions. Seasonal variations in climate will also cause fluctuations in concentration of various ionic species in groundwater. An overall humid condition may be punctuated by a dry spell, where cation-rich silicates may precipitate as a minor constituent amid a predominance of cation-poor silicates. Similarly, in many arid regions flash floods may bring a brief period of relatively humid conditions and temporarily cause the groundwater to become dilute. Comparatively more stable cation-poor silicates may precipitate from such relatively dilute pore solutions and the authigenic mineral assemblage may contain a minor fraction of stable silicates. Climatic interpretation and the evaluation of climatic control on early diagenesis, therefore, must be based on relative abundances of neoformed mineral species as well as on the overall assemblage in order to determine the range of variability in terms of water chemistry and, in turn, climate. If an assemblage is characterized by 90% cation-rich minerals, such as chlorite/sepiolite/smectite, zeolite, etc., and 10% cation-poor silicates, such as kaolinite/quartz, the climate was possibly an overall arid climate with a short seasonal humid condition.

Although there had not been much effort to relate authigenic minerals to climate, literature search reveals that there is a close relationship between the nature and abundance of authigenic minerals and the prevailing climate of the provenance – basin area during sedimentation and early diagenesis. Silica, both as quartz or chalcedony, along with kaolinite seem to dominate the cement type in sandstones associated with coal beds, which are products of temperate to warm humid climate. A similar authigenic cement association is also common in quartz-rich sandstones, many of which are related to intense chemical weathering in relatively temperate to warm humid conditions (Tallman, 1948; Friedman, 1954; Greensmith, 1957; Standard, 1969). Clarke and Keller (1984) observed a gibbsite-cemented Pliocene sandstone in Florida (U.S.A.). Here, the climate during deposition and shortly after, was mainly dominated by warm humid conditions, except for a cold spell during Pleistocene glaciation. On the other hand, cation-rich silicate cements are common in many arkosic sandstones formed in arid regions. Walker et al. (1978) documented the occurrence of cation-rich authigenic silicates, i.e., montmorillonite, potassium feldspar, and zeolite in first-cycle arkosic alluvium of Cenozoic age deposited within a warm desert environment. This causal relationship between the nature and abundance of authigenic cements in sandstones and the prevailing climate during early diagenesis has not received the attention it deserves. The following discussion is an attempt to demonstrate the close relationship between climate and the nature and abundance of early authigenic minerals in sandstone.

Early authigenic minerals in quartzo-feldspathic sandstones and their climatic significance

Dutta (1981, 1983) and Dutta and Suttner (1986) attempted to establish explicitly the relationship between early diagenesis and climate through its control on pore-water chemistry. Based on early authigenic mineral assemblages in sandstones of the Permo-Triassic Gondwana Supergroup in India and the Permo-Pennsylvanian Cutler and Fountain formations of Colorado (U.S.A.), they demonstrated the climatic influence on early diagenesis, which commenced immediately after deposition. Their work is possibly the first attempt to focus on the role of climate as one of the most important factors that influences diagenesis in siliciclastic sediments.

Gondwana Supergroup, India

In the Raniganj Basin, India, the Gondwana sediments of fluvial origin range in age from Lower Permian to possibly Lower Jurassic. These sediments, deposited within block-faulted intracratonic basins, were derived mostly from high-grade metamorphic and plutonic rocks of granitic composition. Because of their derivation from a coarse crystalline source, they are considered to be of first-cycle origin. Based on sandstone composition, the Gondwana succession has been subdivided into six petrofacies. A cyclical compositional trend through time [arkose (petrofacies I) → subarkose to quartz arenite (petrofacies II) → arkose to subarkose (petrofacies

IV)* → subarkose (petrofacies V) → quartz arenite (petrofacies VI)] is observed in the Gondwana succession (Fig. 4-17). The early authigenic mineral assemblages in different petrofacies are also shown in the same figure. Similar to the detrital composition, the early authigenic minerals in sandstones also show a systematic cyclical variation with respect to their nature and abundances. The early authigenic cements are typified by kaolinite, smectite, chlorite and quartz. All these minerals occur in various combinations and various proportions. Kaolinite and quartz characterize petrofacies II and petrofacies VI. Petrofacies I and IV show the presence of chlorite, smectite, kaolinite and quartz. Even the relative abundances of each mineral species in different petrofacies show considerable variations (Fig. 4-17).

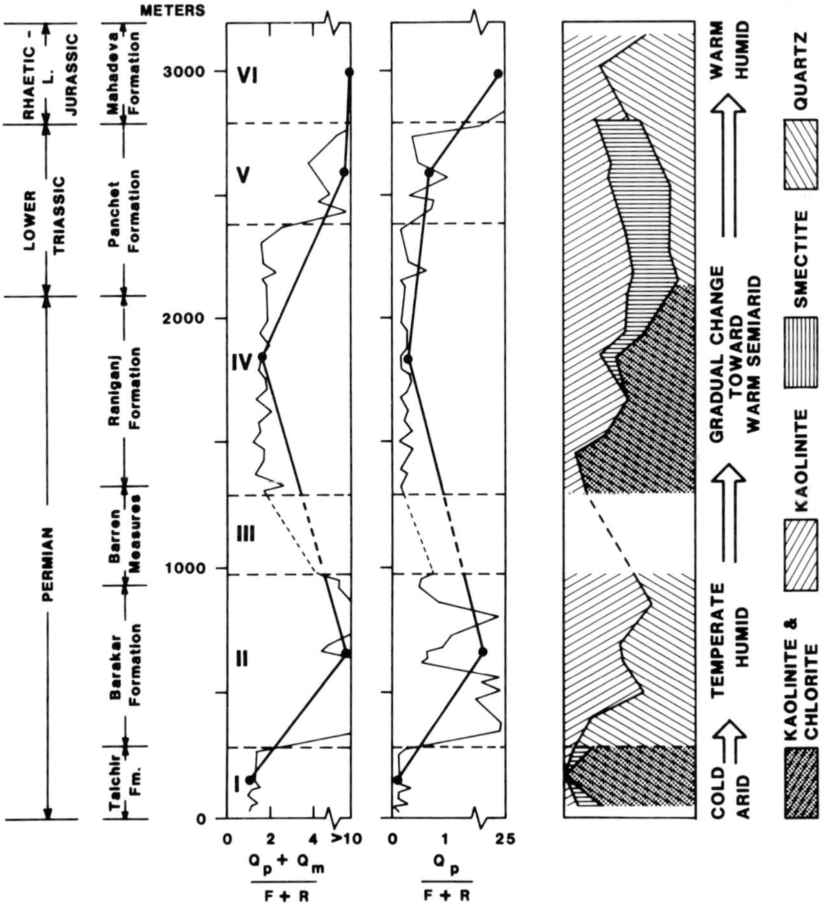

Fig. 4-17. Stratigraphic variations in framework composition and early authigenic mineral assemblages in different petrofacies of Gondwana Supergroup, Raniganj Basin, India.

* Petrofacies III is an argillaceous unit and, therefore, is not a data source.

Cutler and Fountain Formations, Colorado (U.S.A.)

Like the Gondwana succession in India, the Permo-Pennsylvanian Cutler and Fountain sediments have a similar geological history. These sediments are of fluvial and first-cycle origin and were deposited within block-faulted cratonic basins. In spite of these similarities, the detrital as well as the early authigenic minerals in the Cutler and Fountain sandstones show a very different pattern in their stratigraphic distribution (Figs. 4-18 and 4-19). The Cutler sandstones are very immature and are classified as arkose. Compositional variations in sandstones are not observed in the stratigraphic column. The early authigenic mineral assemblage is characterized by the presence of chlorite, smectite, laumontite and kaolinite in various proportions. Authigenic quartz has not been observed. Among the clay minerals, chlorite and smectite dominate the assemblage with minor kaolinite (Fig. 4-18). Based on facies characteristics and sandstone composition the Fountain Formation has, informally, been divided into upper and lower parts. Compositionally, although the Fountain

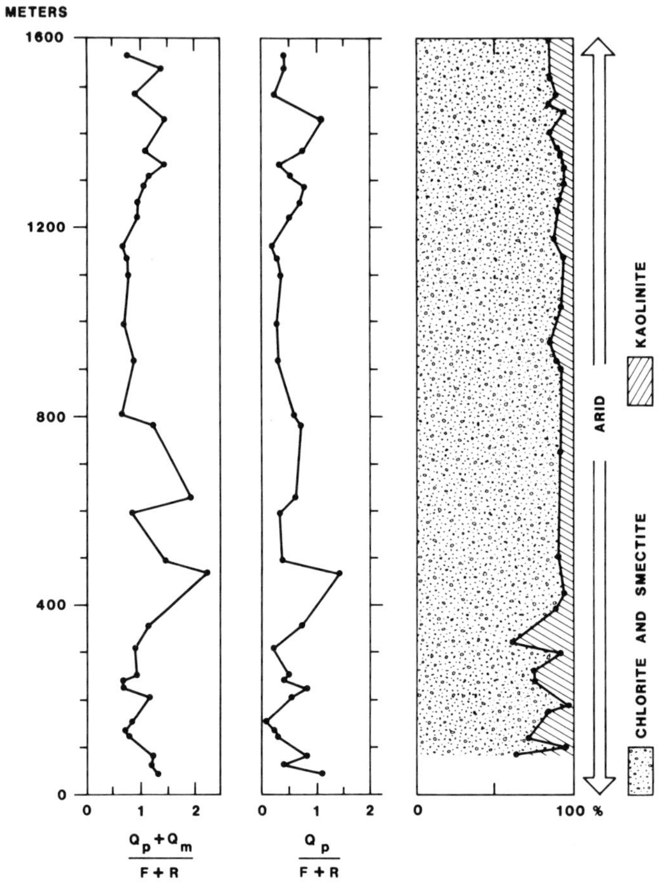

Fig. 4-18. Stratigraphic variations in framework composition and early authigenic mineral assemblages in the Cutler Formation of Permian age, Colorado (U.S.A.).

sandstones are immature, they are relatively more mature compared to the Cutler sandstones and range in composition from that of an arkose to that of a subarkose. A slight compositional variation is observed between upper and lower sections, the lower one being slightly more mature. Early authigenic minerals in the Fountain sandstones, like in the Gondwana sandstones, are characterized by chlorite, smectite, kaolinite, and quartz. Kaolinite is the most abundant, whereas chlorite and smectite form the minor constituents (Fig. 4-19).

Synthesis

The nature of the early authigenic minerals in sandstones in all of the three sedimentary successions discussed above has been interpreted to be related to

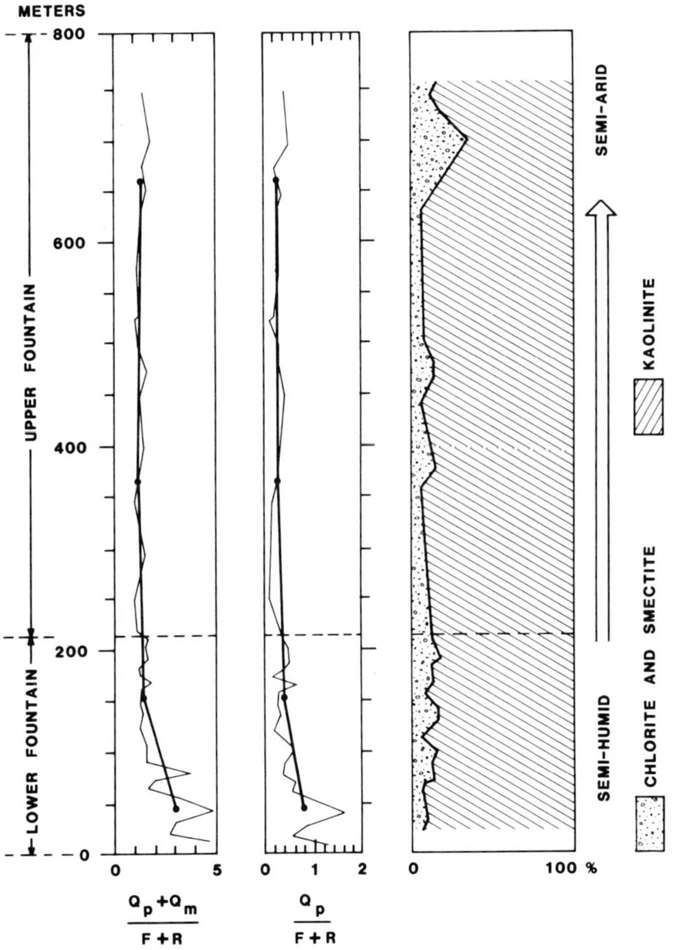

Fig. 4-19. Stratigraphic variations in framework composition and early authigenic mineral assemblages in the Fountain Formation of Pennsylvanian age, Colorado (U.S.A.).

climate through its control on pore-water chemistry during early diagenesis. In this interpretation the stability diagrams representing coexisting mineral phases have been used to reconstruct the water chemistry during precipitation of these minerals. Such an approach readily relates climate and early diagenesis because, at shallow depths during early diagenesis, pore-water chemistry is largely controlled by climate. Microprobe analysis as well as wet chemical analyses data of chlorite and smectite from all three suites (i.e., Gondwana, Cutler, and Fountain), indicate that both of these minerals are Mg-rich or contain appreciable amount of Mg (unpublished data of Dutta). Mg-activity, therefore, has been used as one of the chemical parameters to draw the stability diagrams in order to reconstruct the paleo-hydrogeochemistry during early diagenesis.

In the Gondwana sequence petrofacies I is dominated by abundant chlorite and subordinate amounts of smectite, kaolinite, and very little quartz. This implies that during early diagenesis of petrofacies I, the pore water was highly concentrated and such water would plot mostly within the stability field of chlorite (Fig. 4-20). Presence of minor amounts of smectite, kaolinite, and quartz indicate a relatively dilute water during their formation. This fluctuation in water chemistry possibly is related to the seasonal increase in water supply to the groundwater system, possibly

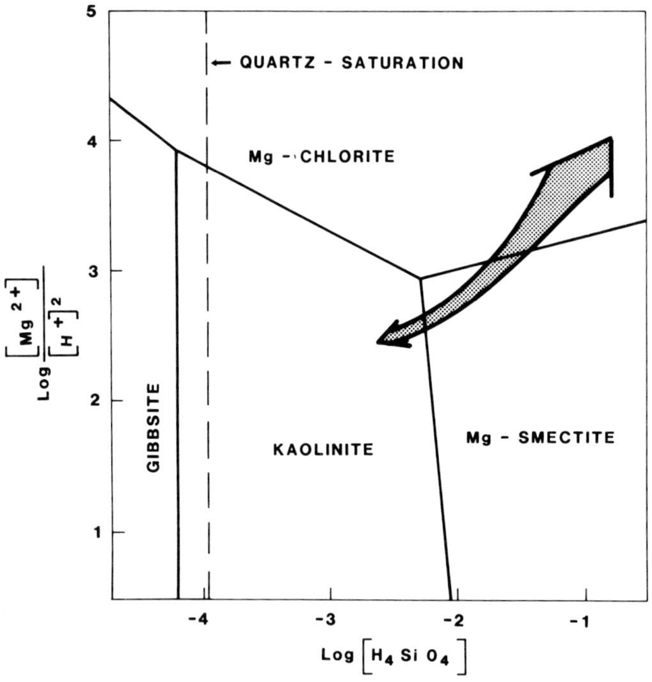

Fig. 4-20. Stability relations of Mg-chlorite, Mg-smectite, kaolinite, gibbsite, and quartz at 25°C and one atmosphere pressure as a function of [Mg^{2+}], pH, and [H_4SiO_4]. The arrow indicates the range of fluctuation in pore-water chemistry during early diagenesis of petrofacies I. The width of the arrow is a relative measure of the abundance of any particular mineral species in the assemblage. In petrofacies I, chlorite dominates the assemblage, whereas smectite and kaolinite are minor constituents.

as a result of spring thaw or a seasonal change in the precipitation pattern. Such a seasonal change will effectively make the environment more humid within a broad cold/frigid climate, which prevailed during sedimentation and early diagenesis of petrofacies I. Nearly equal proportions of kaolinite and quartz in petrofacies II and VI (Fig. 4-17) indicate a sharp change in pore-water chemistry, which plot within a rather restricted kaolinite – quartz field (Fig. 4-21). The uniformly dilute nature of pore-water chemistry indicates a uniformly humid condition during the early diagenesis of petrofacies II and VI.

The early authigenic mineral assemblage in petrofacies IV is mostly dominated by chlorite with subordinate kaolinite and quartz. The relative abundances of both quartz and kaolinite, however, are higher than in petrofacies I. Towards the upper part of the petrofacies, smectite reappears as a minor constituent (Fig. 4-17). Like petrofacies I, a concentrated aqueous solution dominated the pore-water chemistry with occasional dilution as indicated by the water chemistry, which moved back and forth between the chlorite and kaolinite/quartz fields and, subsequently, through the chlorite – smectite – kaolinite/quartz fields (Fig. 4-22). Like petrofacies I, the pore-water chemistry during early diagenesis of petrofacies IV was mainly influenced by aridity punctuated by relatively more humid conditions, which may be at-

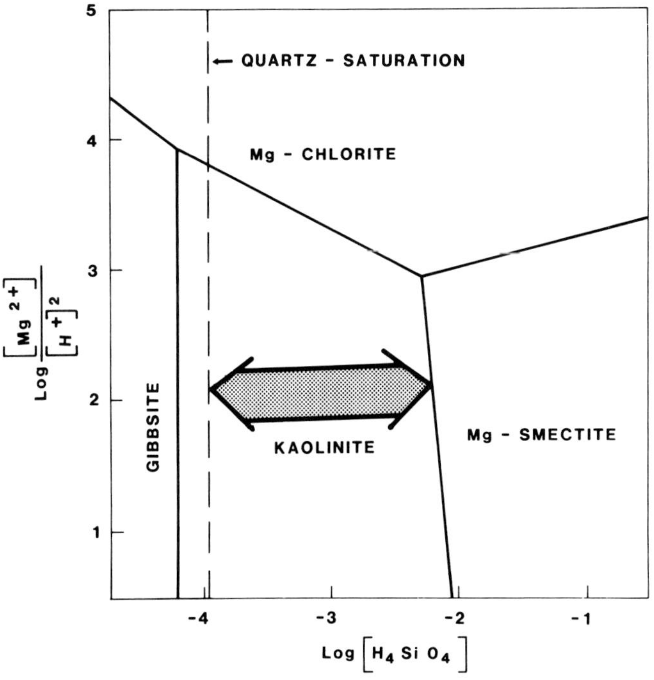

Fig. 4-21. Stability relations of Mg-chlorite, Mg-smectite, kaolinite, gibbsite, and quartz at 25°C and one atmosphere pressure as a function of $[Mg^{2+}]$, pH, and $[H_4SiO_4]$. The arrow indicates that the pore-water chemistry was uniformly restricted within the stability field of kaolinite up to the quartz saturation line during early diagenesis of petrofacies II and VI. In these petrofacies both kaolinite and quartz dominate the assemblage.

tributed to seasonal change in climate. Relative abundances of kaolinite and quartz in this petrofacies compared to petrofacies I possibly indicate that seasonal variations with increased humid condition became more pronounced. Climatic change towards more humid conditions with time is further indicated by the early authigenic mineral assemblage in petrofacies V (Fig. 4-17). In this petrofacies, chlorite is totally absent, whereas smectite appears in more abundance than in any other petrofacies, and the early authigenic mineral assemblage is characterized by almost equal proportions of kaolinite, smectite, and quartz. This assemblage would indicate a pore solution, the field of concentration of which lies in between petrofacies II/VI and petrofacies I/IV as shown in Fig. 4-23, characteristic of a semi-humid condition. Culmination of this humid trend is seen in petrofacies VI, dominated by quartz and kaolinite in equal proportion with complete elimination of smectite. Here, the pore-water chemistry was restricted within the kaolinite field up to the quartz saturation line (Fig. 4-21), indicating a very humid condition throughout.

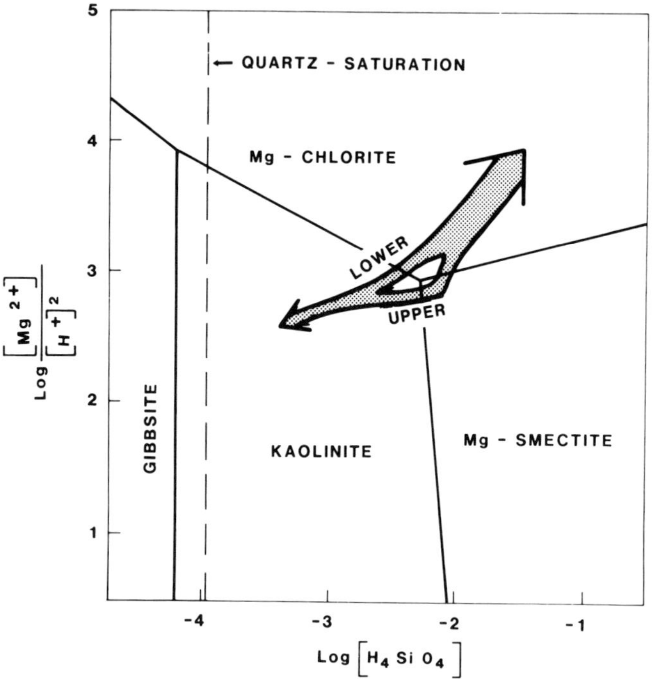

Fig. 4-22. Stability relations of Mg-chlorite, Mg-smectite, kaolinite, gibbsite, and quartz at 25°C and one atmosphere pressure as a function of $[Mg^{2+}]$, pH, and $[H_4SiO_4]$. The arrow indicates the range of fluctuation of water chemistry during early diagenesis of petrofacies IV. The width of the arrow is a relative measure of the abundance of any particular mineral species in the assemblage. Note the fluctuation path: in the lower part of the section, the pore-water chemistry changed from the chlorite field to the kaolinite field indicating mainly lowering of silica activity, whereas the upper part indicates that activities of both silica and Mg were lowered. The width of the arrow is a measure of the relative abundance of a mineral species in the assemblage.

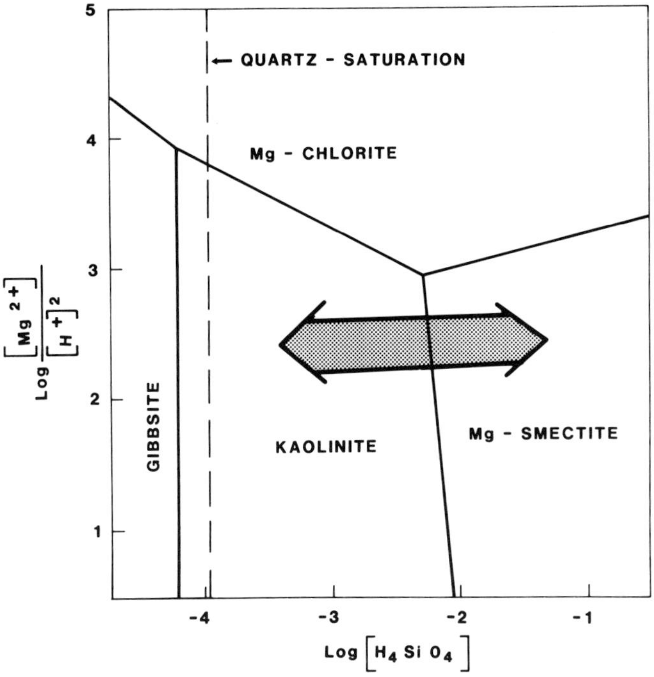

Fig. 4-23. Stability relations of Mg-chlorite, Mg-smectite, kaolinite, gibbsite, and quartz at 25°C and one atmosphere pressure as a function of [Mg^{2+}], pH, and [H_4SiO_4]. The arrow indicates that the pore-water chemistry was uniformly restricted within the stability fields of kaolinite and smectite during early diagenesis of petrofacies V. Within this range quartz is also stable.

The climatic vicissitude during early diagenesis of Gondwana sandstones deduced from the early authigenic mineral assemblage reconstructed through paleo-hydrogeochemistry indicates that the climate changed from arid (petrofacies I) → humid (petrofacies II) → semi arid (petrofacies IV) → semi-humid (petrofacies V) → humid (petrofacies VI). Climatic interpretation based on paleo-groundwater chemistry is unable to infer about the temperature parameter. The overall climatic interpretation in terms of aridity/humidity, however, is supported by independent evidence based on detrital mineralogy (Suttner and Dutta, 1986) and paleontological evidence (Lele, 1976; Shah, 1976) during Gondwana sedimentation*. This climatic change during Gondwana sedimentation/early diagenesis is attributed to an overall global change in climate from a cold condition during Lower Permian time to a relatively warm humid condition during an Upper Triassic to Lower Jurassic period (Fischer, 1982). The climatic change on the Indian subcontinent was further accentuated by the northerly movement of the Indian plate from a higher latitudinal position to a more mid-latitudinal location (Fig. 4-14).

* The early diagenetic episode is interpreted to have succeeded immediately after sedimentation and, therefore, both of the processes are considered to be geologically contemporaneous.

Unlike the Gondwana sandstones, the early authigenic minerals in Cutler sandstones – like their detrital counterparts – do not show much compositional variation with time. The assemblage is dominated by chlorite and smectite. Laumontite occurs sporadically throughout the whole section. Kaolinite occurs as a very minor constituent, whereas authigenic quartz has not been observed. The predominance of chlorite and smectite suggests that the coexisting pore water was highly concentrated with points plotting within chlorite – smectite fields (Fig. 4-24). In the absence of any volcanic source rocks, sporadic occurrences of laumontite throughout the entire stratigraphic section indicate an unusual pore-water chemistry characteristic of an extreme arid climate (Figs. 4-7 and 4-24). Such aridity appears to have persisted throughout the early diagenetic episode of the Cutler Formation. Minor kaolinite possibly indicates occassional rain/cloudbursts as observed in many arid regions of the world. A similar climatic interpretation has also been deduced independently, based on a detrital mineral assemblage during deposition of Cutler sediments (Mack, 1977; Suttner and Dutta, 1986).

The early authigenic mineral assemblage in the Fountain sandstones unlike the Cutler sandstones is dominated by kaolinite; chlorite and smectite are less abundant in the Fountain sediments compared to the Cutler sandstones (Fig. 4-25). Quartz is also present throughout most of the succession. Laumontite is totally absent. Abun-

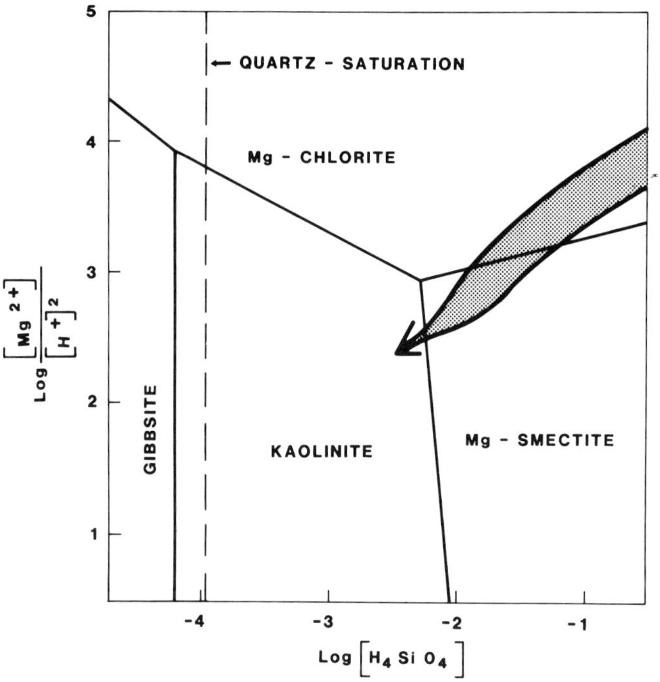

Fig. 4-24. Stability relations of Mg-chlorite, Mg-smectite, kaolinite, gibbsite, and quartz at 25°C and 1 atmosphere pressure as a function of $[Mg^{2+}]$, pH, and $[H_4SiO_4]$. The arrow shows the range of water-chemistry fluctuations during early diagenesis of the Cutler sandstones. The width of the arrow indicates the relative abundance of different mineral species in the assemblage.

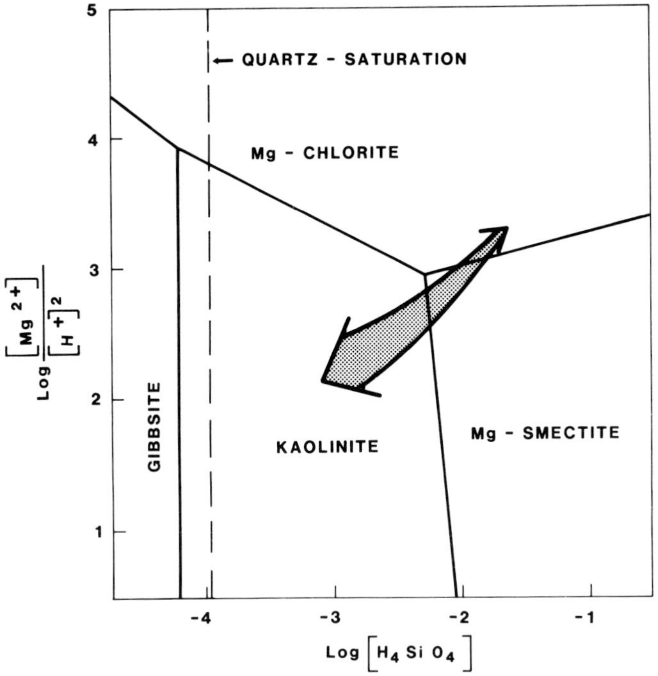

Fig. 4-25. Stability fields of Mg-chlorite, Mg-smectite, kaolinite, gibbsite, and quartz at 25°C and one atmosphere pressure as a function of [Mg^{2+}], pH, and [H_4SiO_4]. The arrow indicates the range of fluctuations in pore-water chemistry during early diagenesis of the Fountain sandstones. The width of the arrow is a relative measure of the abundance of any mineral in the assemblage. Kaolinite dominates the assemblage, whereas smectite and chlorite form minor constituents.

dance of kaolinite possibly represents a relatively humid condition, whereas small contents of chlorite and smectite may be related to seasonal aridity. Presence of thin coal seams in the lower part of the Fountain sandstones (Mack, 1977) confirm this interpretation of a relatively humid condition. A slight increase in the chlorite and smectite contents in the upper part of the Fountain Formation (Fig. 4-19) possibly indicates a slight shift in climate towards aridity during early diagenesis of the upper Fountain sediments. The detrital mineral assemblages also indicate a semi-humid condition for lower Fountain sandstones, whereas the upper part has been interpreted as having been formed in a semi-arid climate (Suttner and Dutta, 1986).

Climatic changes through time was the primary cause of the fluctuations in water chemistry, which resulted in the characteristic distribution of early authigenic cements observed in all the three units that have been discussed. Perhaps the strongest evidence that the early authigenic minerals in the Gondwana, Cutler, and Fountain formations are mainly products of open groundwater systems, having a chemistry controlled by climate, is the vertical zonation of the cement assemblages. A strong correlation exists between framework mineralogy and cement throughout the entire stratigraphic column as seen in Figs. 4-17 to 4-19. Cation-rich silicates of early authigenic origin are abundant in immature sandstones, whereas the most stable authigenic cements are common in mature sandstones. This correlation could suggest that framework mineralogy, acting independently of climate, controlled cement mineralogy. Inasmuch as textural evidences indicate that during the early diagenetic stage the detrital minerals did not experience any dissolution, replacement, and alteration in relationship with any of the early authigenic cement, the detrital minerals could not have influenced this cement. Instead, Dutta and Suttner (1986) argued that variations in framework mineralogy were themselves the result of changes in climate. Certainly, unstable ferromagnesian minerals were destroyed very early in diagenesis. But evidence of extensive total dissolution of other silicate minerals is not present. Cations released through early destruction of the ferromagnesian minerals during times of aridity were not readily flushed out of the pore spaces and were soon precipitated as the early cation-rich cements. During periods of high rainfall, these cations were more effectively removed from the system. But, beyond this probability, there is little evidence of significant framework-mineral control on early cement genesis.

CLIMATIC CONTROL ON DEEP BURIAL DIAGENESIS

Introduction

Deep burial diagenesis, in this chapter, is defined as a stage characterized by dissolution/replacement/alteration of minerals, generation of secondary porosity, and precipitation of neoformed minerals. In an earlier section, it has been argued that at greater burial depths the sediments are cut-off from the bulk of the groundwater flow of local and intermediate flow systems. This implies that at greater burial depths the supply of chemical constituents that are brought from outside the sediment system is restricted. It is imperative, therefore, that during the late diagenetic stage the pore-water chemistry and, consequently, the authigenic minerals should mainly be controlled by dissolution/alteration reactions. Innumerable petrographic data based on textural evidences (Schmidt and McDonald, 1979; Walker, 1984; McBride, 1985; Dutta and Suttner, 1986) and oxygen isotope data from deep basinal brines (Clayton et al., 1966) also confirm the above conclusion. Inasmuch as the detrital minerals and early authigenic cements in sand/sandstone, in most provenances, are largely controlled by climate, the pore-water chemistry during late diagenesis will also be related to climate.

Mineral dissolution during burial

The threshold conditions that initiate dissolution/alteration reactions during burial diagenesis are possibly related to the combination of various factors like pore-fluid chemistry, velocity of pore fluid, chemical stability of minerals, and depth/temperature relationship. Though dissolution of highly-reactive solids has been observed at shallow depths, in most cratonic settings the process of dissolution/alteration/replacement seems to initiate at a considerable depth. It is a common observation that different minerals experience dissolution progressively with depth depending on the chemical stability of mineral species within a specific physicochemical environment. Except for a few ultrastable heavy minerals, almost all minerals are susceptible to dissolution during diagenesis (McBride, 1985). Table 4-9 shows the stability of common detrital grains in sandstone in diagenetic environment.

Mineralogically, sandstones in cratonic settings are mostly made up of quartz, feldspars and rock fragments. Accessary minerals other than micas are, volumetrically, insignificant in common sandstones. Most coarse crystalline rock fragments contain quartz, feldspars, and micas. In cratonic settings, the mineralogical composition of sandstones, therefore, may be considered to be made up mostly of quartz, feldspars, and micas. Common authigenic minerals during early stage are quartz, carbonates, iron oxides, and clay minerals, like kaolinite, smectite, chlorite, and interlayered clays. Less commonly, feldspars and laumontite may also be present. The above-mentioned primary and authigenic mineralogic association broadly defines a partially-cemented sandbody, a product of early diagenesis.

Dissolution during late diagenesis is common, even in the case of one of the most stable minerals like quartz. Quartz dissolution enriches pore solution in silica. Micas, generally, do not show much dissolution, but they are subjected to alteration. Such alteration releases silica and K, Fe, Mg, Ca, etc., leaving behind an iron-stained non-descript mass of aluminosilicates (unpublished micro-probe data of the author). Micas also alter to clays and in the process also change the pore-fluid chemistry. Dissolution of feldspars is one of the most common dissolution features in craton-derived sediments. Plagioclase dissolution makes Ca, Na, Al, and Si available to the pore fluid, whereas dissolution of K-feldspar enriches the pore fluid with K, Si, and Al. Dissolution of early-formed authigenic minerals at this stage also

TABLE 4-9

Detrital grains that dissolve in diagenetic environment

The grains are listed in approximate sequence of greatest susceptibility (top) to lowest susceptibility to dissolution (bottom) (after McBride, 1985)

Light minerals/grains	Heavy minerals
Carbonate rock fragments	Olivine, pyroxene
Volcanic glass and rock fragments	Andalusite, sillimanite
Plagioclase (> 10% An)	Amphibole
Perthite	Epidote/zoisite
K-feldspar	Sphene
Chert	Kyanite
Metamorphic rock fragments	Staurolite
Quartz	Garnet
Micas	Apatite, chloritoid
Biotite	Spinel
Chlorite	Rutile, tourmaline
Muscovite	

influences pore-water chemistry. Because of the relative abundance and relative chemical instability of carbonates in diagenetic environments, however, their dissolution is one of the most important factors that influence the chemistry of pore water.

Late diagenetic reactions and mass-balance

Many investigations on sandstone diagenesis demonstrate a causal relationship between dissolution of detrital minerals on one hand, and alteration and neoformation of minerals, on the other. Close relationship between the illitization of early-formed kaolinite and I/S interlayer clay and dissolution of K-feldspar and alteration of micas is common in many arkosic sandstones and argillaceous sediments buried at, and above, a depth of 2000 m (Hower et al., 1976; Boles and Frank, 1979; Dutta, 1983; Dutta and Suttner, 1986; Suter, 1986). The mechanism of illitization has been documented in a detailed study of burial metamorphism of Miocene–Oligocene argillaceous sediments of the Gulf Coast (U.S.A.) (Hower et al., 1976). These authors observed that mixed-layer I/S clay undergoes a conversion from less than 20% to about 80% illite layers over a depth interval of 2000 m to 3700 m, according to the equation:

$$\text{Smectite} + \text{Al} + \text{K} = \text{Illite} + \text{Si}$$

In addition, iron and magnesium are also released from smectite in this reaction. Both K and Al necessary for illitization of smectite were derived from dissolution of K-feldspar and mica. It was also noted that K-feldspar disappears below a depth of 3700 m, after which no more conversion of the remaining 20% smectite layers takes place. Presence of smectite layers within the thermodynamic stability field of illite possibly indicates that the chemical constituents necessary for the conversion of mixed I/S layers to illite were not available in the environment. In the Gulf Coast, the presence or absence of detrital K-feldspar and micas determined the clay mineral assemblage during burial diagenesis.

Dutta (1983) observed a close chemical relationship between authigenic and detrital minerals, as well as among authigenic minerals, in the Gondwana sandstones of India. Illitization of early authigenic kaolinite and dissolution of feldspars and mica is one such relationship. Based on textural evidences and mass-balance calculations, he concluded that illitization of kaolinite was volume per volume rather than mole per mole, according to the reaction:

$$1.41\ \underset{\text{(Kaolinite)}}{Al_{1.96}Si_{2.03}O_5(OH)_4} + 0.4477\ H_4SiO_4 + 0.75\ K^+ + 0.12\ Mg^{2+} =$$

$$\underset{\text{(Illite)}}{K_{0.75}Mg_{0.12}Al_{2.59}Si_{3.31}O_{10}(OH)_2} + 2.4808\ H_2O + 0.1736\ Al^{3+} + 0.4692\ H^+$$

This reaction shows that each mole of illite was derived from 1.41 moles of kaolinite and, in addition, 0.75 mole of K, 0.12 mole of Mg, and 0.4477 mole of

Si were necessary. Mass-balance calculations showed that all chemical constituents necessary for conversion of kaolinite to illite, were derived from dissolution of K-feldspar and micas. As in the Gulf Coast, illite formation in Gondwana sandstones is also depth/temperature dependent. The first appearance of illitization of kaolinite is observed at a present-day depth of 1700 m. Considering erosion of at least 600 m since the exhumation of the Gondwana sediments (Dutta, 1983), the initiation of illitization roughly corresponds to the depth – temperature range observed by Hower et al. (1976) in the Gulf Coast. Illitization of kaolinite in the Gondwana sandstones was also, as in the Gulf Coast, a function of detrital mineralogy.

Volumetrically, carbonates form one of the most important groups of authigenic cement in sandstones and are common in both early- and late-diagenetic stages. Carbonates are possibly the most pervasive of all authigenic minerals in sandstones. Early carbonate cements in siliciclastic sediments are frequent in arid regions because of highly-concentrated groundwater. Early carbonate cements are also frequent in humid regions when the sediments are closely associated with carbonate rocks. Late-diagenetic carbonates are common mostly in arkosic rocks as replacements of detrital and early authigenic cements. Multiple generations of carbonate cements due to frequent dissolution and precipitation are also common in most siliciclastic sediments. This is mainly due to high solubility of carbonate minerals and fluctuations of partial pressure of CO_2 that controls the precipitation and dissolution of carbonate minerals. A change of partial pressure of CO_2 during burial is the result of organic reactions, such as maturation of organic matter and thermal degradation of kerogen (Schmidt and McDonald, 1979; Franks and Forester, 1984; Shanmugam, 1985). Because of multiple generation and remobilization of carbonate cements from an earlier episode, it is difficult to trace the primary source of carbonate cements. This makes mass-balance calculations difficult at times.

Based on textural evidences, Dutta (1986b) made mass-balance calculations in order to find the source of late carbonate cements in Gondwana sandstones of India (Table 4-10). A considerable gap was observed between the amounts of cations, like Ca, Mg, and Fe, derived internally and the respective amounts of these cations needed to form the various carbonate cements at this stage. The source of these cations is not clear. Remobilization of early carbonate cement, originally formed from an external source during the early stage, may explain the gap in the mass-balance calculations. Though remobilization does explain the gap, a considerable part of late carbonate cement was derived mainly from dissolution of plagioclase (source of Ca) and micas (source of Fe, and Mg) (Table 4-10). Mass-balance calculations can account for most other late-diagenetic minerals in the Gondwana sandstones. Dissolution of detrital minerals like quartz, feldspars and micas, as well as the early-formed silicate and carbonate cements, supply the most common cations that are necessary to form the most frequent late-diagenetic minerals (Table 4-10). Mass-balance calculations support the conclusion that the nature and the amount of authigenic cements of late diagenetic origin are largely controlled by the mineralogical composition of Gondwana sandstones.

TABLE 4-10

Mass-balance between detrital/early-authigenic and late-diagenetic minerals in different petrofacies of Gondwana Supergroup, India (after Dutta, 1983)

Mineral	Weight of oxide and cations (in grams) released by dissolution per 100 grams of detrital and early diagenetic minerals									Weight of oxide and cations (in grams) consumed to form late-diagenetic minerals in a volume occupied by 100 grams of detrital and early-diagenetic minerals								
	SiO_2	Al	K	Mg	Fe	Ca	Ti	Mn	Na	SiO_2	Al	K	Mg	Fe	Ca	Ti	Mn	Na
Petrofacies I																		
K-feldspar	7.77	1.17	1.68															
Plagioclase	12.46	4.51				0.66			1.24									
Quartz	1.00																	
Muscovite	0.47	0.17	0.04	0.01	0.05													
Biotite	0.79	0.23	0.07	0.13	0.23	0.01	0.03											
Illite										2.22	0.45	0.82	0.08					
Carbonates															11.54		0.33	
Iron oxide														6.37				
Total	22.49	6.08	1.81	0.14	0.28	0.67	0.03		1.24	2.22	0.45	0.82	0.08	6.37	11.54		0.33	
Petrofacies II																		
K-feldspar	3.98	0.61	0.84						0.03									
Plagioclase	2.70	0.54				0.14			0.27									
Quartz	1.21																	
Muscovite	3.58	1.29	0.50	0.05	0.39													
Biotite	0.24	0.07	0.02	0.04	0.07		0.01											
Kaolinite	0.15	0.07																
Diagenetic quartz										0.57								
Illite										2.22	0.45	0.82	0.08					
Carbonates													2.02	13.29	0.69		0.19	
Iron oxide														0.28				
Total	11.86	2.58	1.34	0.09	0.46	0.14	0.01		0.30	2.79	0.45	0.82	2.10	13.57	0.69		0.19	
Petrofacies IV																		
K-feldspar	1.80	0.28	0.38						0.01									
Plagioclase	4.49	0.90				0.24			0.44									
Quartz	0.40																	
Muscovite	0.90	0.32	0.12	0.01	0.10													
Biotite	1.34	0.42	0.10	0.20	0.43	0.01	0.04											
Garnet	0.10	0.08			0.06	0.10	0.01	0.01										
!llite										0.50	0.10	0.19	0.02					
Carbonates													1.41	9.30	0.48		0.13	
Iron oxide														3.99				
Total	9.03	2.00	0.60	0.21	0.59	0.35	0.05	0.01	0.45	0.50	0.10	0.19	1.43	13.29	0.48		0.13	

TABLE 4-10 (continued)

Mineral	Weight of oxide and cations (in grams) left in pore solution after formation of late-diagenetic minerals, derived from dissolution per 100 grams of detrital and early-diagenetic minerals									Weight of oxide and cations (in grams) derived from unknown source(s) necessary to form late-diagenetic minerals in a volume occupied by 100 grams of detrital and early-diagenetic minerals								
	SiO_2	Al	K	Mg	Fe	Ca	Ti	Mn	Na	SiO_2	Al	K	Mg	Fe	Ca	Ti	Mn	Na
Petrofacies I																		
K-feldspar																		
Plagioclase																		
Quartz																		
Muscovite																		
Biotite																		
Illite																		
Carbonates																		
Iron oxide																		
Total	20.27	5.65	0.99	0.06	−6.09	−10.87	0.03	−0.33	1.24					6.09	10.87		0.33	
Petrofacies II																		
K-feldspar																		
Plagioclase																		
Quartz																		
Muscovite																		
Biotite																		
Kaolinite																		
Diagenetic quartz																		
Illite																		
Carbonates																		
Iron oxide																		
Total	9.07	2.13	0.52	−2.01	−13.11	−0.55	0.01	−0.19	0.30				2.01	13.11	0.55		0.19	
Petrofacies IV																		
K-feldspar																		
Plagioclase																		
Quartz																		
Muscovite																		
Biotite																		
Garnet																		
Illite																		
Carbonates																		
Iron oxide																		
Total	8.53	1.90	0.41	−1.22	−12.70	−0.13	0.05	−0.12	0.45				1.22	12.70	0.13		0.12	

Synthesis

In the Gulf Coast a progressive authigenic mineralogical change is observed in the argillaceous sediments of Oligocene – Miocene age. This change, in a way, is mostly a chemical readjustment within a relatively closed chemical system. The formation, as well as the abundance of authigenic cement at this stage, is closely related to the dissolution of detrital components in sediments. The same conclusion was drawn by Dutta (1983, 1986a, 1986b), i.e., that during the late-diagenetic stage the continental sediments behaved mostly as a closed system, where the late-diagenetic mineral assemblage is mainly controlled by detrital and early-authigenic minerals. Inasmuch as, in a cratonic setting, the composition of sandstone is largely controlled by climate, the nature and abundance of late authigenic minerals are also related to climate through its influence on detrital minerals.

CONCLUSIONS

Chemical diagenesis has been assumed, by many workers, to be largely some complicated function of: (1) framework mineralogy, (2) pore-water chemistry, and (3) pressure and temperature conditions during burial. Each one of these factors, however, is intricately related to many more dependent and independent variables that are to be ascertained in order to understand clastic diagenesis. Many of these factors that relate to pre- and syndepositional clastic diagenesis also influence the postdepositional diagenetic process. Climate is such a predepositional and syn-early burial factor that influences chemical diagenesis profoundly. In addition, plate-tectonic setting also plays a major role in chemical diagenesis. Most dependent and independent variables that influence chemical diagenesis may broadly be categorized under two major independent variables, i.e., climate and plate-tectonic setting as shown in Fig. 4-26.

Earlier, climatic control on diagenesis has largely been ignored or considered to be a relatively insignificant or undeterminable factor in chemical diagenesis. In this chapter an attempt has been made to focus on the importance of climate in diagenesis of siliciclastic sandstones of continental origin. Climate influences the chemistry of groundwater at shallow depth through its control on the intensity of chemical weathering and the amount of subsurface infiltration. Groundwater chemistry at this stage has great influence on early authigenesis and, in turn, controls subsequent diagenetic episodes and reservoir quality. For example, if smectite forms very early before fluid migration into the reservoir rock, the fluid may not be able to migrate into the reservoir because smectite may choke the pore throats upon swelling as soon as it comes in contact with water. On the other hand, early carbonates tend to generate secondary porosity during fluid migration into the reservoir rocks. Formation of early carbonates, smectite and other silicate cements is largely a function of climatically-controlled pore-fluid chemistry. As a predepositional factor, climate also influences chemical diagenesis through its control on detrital minerals. During deep burial, the chemistry of the pore fluid is primarily controlled by dissolution of detrital minerals. Pore-water chemistry at this stage is

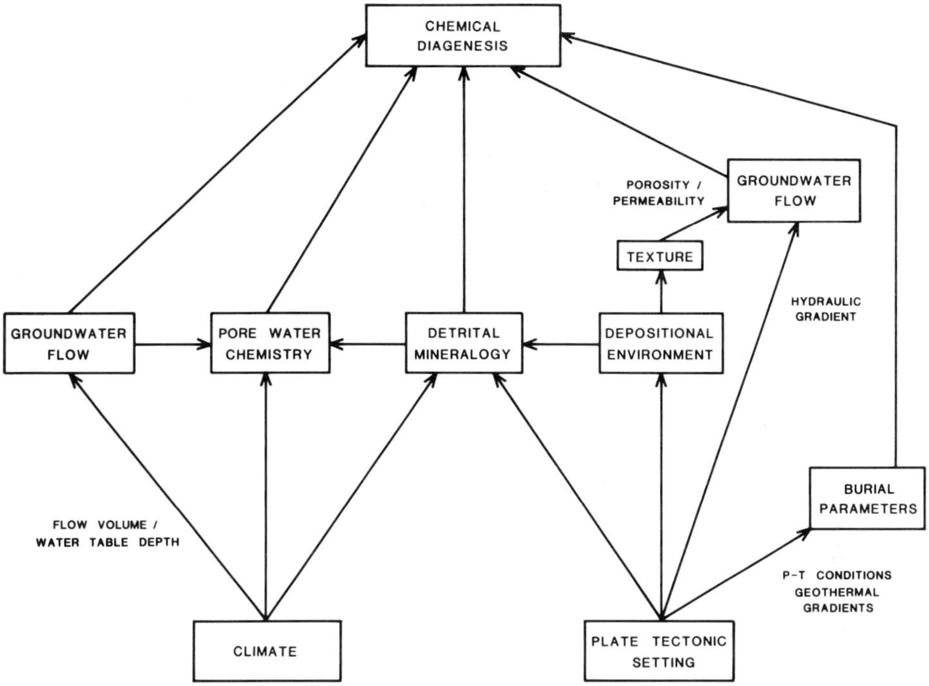

Fig. 4-26. Flow chart showing first-order control of both climate and plate-tectonic setting on chemical diagenesis.

also influenced by dissolution of early authigenic cement. In this chapter, a close link between the nature and abundance of cement in sandstones and climate has been established. In the future, by integrating climatic parameters in chemical diagenetic model, one will have a better understanding of diagenetic processes. This will enable geologists to better assess the potential and quality of siliciclastic petroleum and groundwater reservoirs.

ACKNOWLEDGEMENTS

An earlier version of this manuscript was reviewed by Deba Bhattacharayya and Calvin James. The manuscript in its present form was reviewed by H. Michael Velbel, Dr. Karl H. Wolf and Dr. George V. Chilingarian. The author is grateful to all the reviewers for their constructive criticism. The author, however, accepts all responsibility for the ideas and views expressed in this chapter. Drafting of the figures by Norman Cooprider is greatly appreciated.

REFERENCES

Ayalon, A. and Longstaffe, F.J., 1988. Oxygen isotope studies and pore-water evolution in the western Canada sedimentary basin: Evidence from the Upper Cretaceous Basal Belly River Sandstone, Alberta. *J. Sediment. Petrol.*, 58: 489 – 505.

Basu, A., 1975. Petrology of Holocene fluvial sand derived from plutonic source rocks. Unpubl. Ph.D. diss., Indiana University, Bloomington, Ind., 138 pp.

Behrman, P.G., 1978. Paleogeography and structural evolution of a Middle to Late Jurassic volcanic arc, Sierra Nevada foothills, California. Unpubl. Ph.D. diss., Univ. California, Berkeley, Calif., 301 pp.

Berner, R.K., 1980. *Early Diagenesis: A Theoretical Approach.* Princeton Univ. Press, Princeton, N.J., 241 pp.

Bjorlykke, K., 1979. Cementation of sandstone – discussion. *J. Sediment. Petrol.*, 49: 1358 – 1359.

Bjorlykke, K., 1988. Sandstone diagenesis in relation to preservation, destruction and creation of porosity. In: G.V. Chilingarian and K.H. Wolf (Editors), *Diagenesis, I.* Developments in Sedimentology, 41. Elsevier, Amsterdam, pp. 555 – 588.

Blatt, H., 1979. Diagenetic processes in sandstone. In: P.A. Scholle and P.R. Schluger (Editors), *Aspects of Diagenesis.* Soc. Econ. Paleontol. Mineral., Spec. Publ., 26: 141 – 157.

Blatt, H., Middleton, G. and Murray, R., 1980. *Origin of Sedimentary Rocks* (2nd ed.). Prentice Hall, Englewood Cliffs, N.J., 782 pp.

Bogg, S., Jr., 1986. *Principles of Sedimentology and Stratigraphy.* Merrill Publ. Co., Columbus, Ohio, 784 pp.

Boles, J.R., 1974. Structure, stratigraphy, and petrology of mainly Triassic rocks, Hokonui Hills, Southland, New Zealand. *N. Z. J. Geol. Geophys.*, 17: 337 – 374.

Boles, J.R. and Franks, S.G., 1979. Cementation of sandstones – Reply. *J. Sediment. Petrol.*, 49: 1362.

Cameron, K.L. and Blatt, H., 1971. Durabilities of sand-sized schist and volcanic rock fragments during fluvial transport, Elk Creek, Black Hills, South Dakota. *J. Sediment. Petrol.*, 41: 565 – 575.

Carrigy, M.A., 1971. Lithostratigraphy of the uppermost Cretaceous (Lance) and Paleocene strata of the Alberta plains. *Alta Res. Counc. Bull.*, 27: 161 pp.

Chilingarian, G.V. and Wolf, K.H., 1976. *Compaction of Coarse-Grained Sediments, II.* Developments in Sedimentology, 18B. Elsevier, Amsterdam, 808 pp.

Clarke, O.M., Jr. and Keller, W.D., 1984. A gibbsite cemented quartz-sandstone. *J. Sediment. Petrol.*, 54: 154 – 148.

Clayton, R.N., Friedman, I., Graff, D.L. and Mayeda, T.K., 1966. The origin of saline formation waters, I. Isotopic composition. *J. Geophys. Res.*, 71: 3869 – 3882.

Darnell, N., 1974. A comparison of surficial, in situ sediments overlying plutonic rock of Boulder Batholith and gneissic rocks of the southern Tobacco Root Mountains. Unpubl. M.Sc. thesis, Indiana Univ. Bloomington, Ind., 126 pp.

Davies, D.K., Almon, W.R., Bonis, S.B. and Hunter, B.E., 1979. Deposition and diagenesis of Tertiary – Holocene volcaniclastics, Guatemala. In: P.A. Scholle and D.R. Schluger (Editors), *Aspects of Diagenesis.* Soc. Econ. Paleontol. Mineral., Spec. Publ., 26: 281 – 306.

Dickinson, W.R. and Suczek, C.A., 1979. Plate tectonics and sandstone composition. *Bull. Am. Assoc. Pet. Geol.*, 63: 2164 – 2182.

Dickinson, W.R., Helmold, K.P. and Stein, J.A., 1979. Mesozoic lithic sandstones in central Oregon. *J. Sediment. Petrol.*, 49: 501 – 516.

Dott, R.H., 1965. Mesozoic – Cenozoic tectonic history of the southwestern Oregon Coast in relation to Cordilleran orogenesis. *J. Geophys. Res.*, 70: 4687 – 4707.

Dutta, P.K., 1981. Early authigenic clay minerals in sandstones as paleoclimatic indicator (abstract). *Geol. Soc. Am., Abstr. Progr.*, 13(7): 443.

Dutta, P.K., 1983. The role of climate in the evolution of detrital and authigenic mineralogy in sandstone from the Gondwana Supergroup, India. Unpubl. Ph.D. diss., Indiana Univ., Bloomington, Ind., 169 pp.

Dutta, P.K., 1985. In search of the origin of cement in siliciclastic sandstones: An isotopic approach. *Chem. Geol. (Isotope Geosci. Sect.)*, 52: 337 – 348.

Dutta, P.K., 1986a. Role of feldspar in determining the nature of authigenic minerals in burial diagenesis (abstract). *Geol. Soc. Am., Abstr. Progr.*, 18(4): 287.

Dutta, P.K., 1986b. Source of silicate and carbonate cements during deep burial diagenesis (abstract). *Abstr., Am. Assoc. Pet. Geol.,* pp. 590 – 591.

Dutta, P.K. and Suttner, L.J., 1986. Alluvial sandstone composition and paleoclimate, II. Authigenic mineralogy. *J. Sediment. Petrol.,* 56: 346 – 358.

Dutta, P.K. and Wheat, R.W. (in review). Tectonic and climatic signatures in detrital grains in a foreland setting in the Sydney Basin, Australia.

Dutton, S.F. and Land, L.S., 1985. Meteoric burial diagenesis of Pennsylvania arkosic sandstones, southwestern Anandarko Basin, Texas. *Bull. Am. Assoc. Pet. Geol.,* 69: 22 – 38.

Edward, A.B., 1950. The petrology of Cretaceous graywackes of the Purari Valley, Papua. *Proc. R. Soc. Victoria,* 60: 163 – 171.

Feth, J.H., Robertson, G.E. and Poltzer, W.L., 1964. Source of mineral constituent in water from granitic rocks, Sierra Nevada, California and Nevada. *U.S. Geol. Surv. Water Supply Pap.* 15351: 70 pp.

Fetter, C.W., Jr., 1980. *Applied Hydrogeology.* Charles E. Merrill, Columbus, Ohio, 488 pp.

Fischer, A.G., 1982. Long-term climatic oscillation recorded in stratigraphy. In: *Climate in Earth History.* National Academy Press, Washington, D.C., pp. 97 – 104.

Franks, S.G. and Forester, R.W., 1984. Relationships among secondary porosity, pore-fluid chemistry and carbon dioxide, Texas Gulf Coast. In: D.A. McDonald and R.C. Surdam (Editors), *Clastic Diagenesis. Mem. Am. Assoc. Pet. Geol.,* 37: 63 – 79.

Franzinelli, E. and Potter, P.E., 1983. Petrology, chemistry and texture of modern river sands, Amazon River system. *J. Geol.,* 91: 23 – 39.

Friedmann, M., 1954. Miocene orthoquartzite from New Jersey. *J. Sediment. Petrol.,* 24: 235 – 241.

Galloway, W.E., 1974. Deposition and diagenetic alteration of sandstone in northeast Pacific arc-related basins: Implications for graywacke genesis. *Bull. Geol. Soc. Am.,* 85: 379 – 390.

Garner, H.F., 1959. Stratigraphic-sedimentary significance of contemporary climate and relief in four regions of the Andes Mountains. *Bull. Geol. Soc. Am.,* 70: 1327 – 1368.

Garrels, R.M. and Mackenzie, F.T., 1971. *Evolution of Sedimentary Rocks.* W.W. Norton, New York, N.Y., 397 pp.

Goldich, S.S., 1938. A study in rock weathering. *J. Geol.,* 46: 17 – 58.

Grantham, J.H. and Velbel, M.A., 1988. The influence of climate and topography on rock-fragment abundance in modern fluvial sands of the southern Blue Ridge Mountains, North Carolina. *J. Sediment. Petrol.,* 58: 219 – 227.

Greensmith, J.T., 1957. Lithology, with particular reference to cementation of Upper Carboniferous sandstones in northern Derbyshire, England. *J. Sediment. Petrol.,* 27: 405 – 416.

Harrold, P.J. and Moore, J.C., 1973. Composition of deep-sea sands from marginal basins of the northwestern Pacific. In: *Initial Rep. Deep Sea Drilling Project,* 18: 915 – 924.

Hayes, J.B., 1979. Sandstone diagenesis – the hole truth. In: P.A. Scholle and D.R. Schluger (Editors), *Aspects of Diagenesis. Soc. Econ. Paleontol. Mineral., Spec. Publ.,* 26: 127 – 140.

Holland, H.D., 1978. *The Chemistry of the Atmosphere and Oceans.* Wiley, New York, N.Y., 351 pp.

Hower, J., Eslinger, E., Hower, M.E. and Perry, E.A., 1976. Mechanism of burial metamorphism of argillaceous sediments: 1. Mineralogical and chemical evidence. *Bull. Geol. Soc. Am.,* 87: 725 – 737.

Hubert, M.K., 1940. The theory of groundwater motion. *J. Geol.,* 48: 785 – 944.

Jackson, T.A. and Keller, W.D., 1970. A comparative study of the role of lichens and "inorganic" processes in the chemical weathering of recent Hawaiian lava flows. *Am. J. Sci.,* 269: 446 – 466.

Jacobsen, L., 1959. Petrology of Pennsylvanian sandstones and conglomerates of the Ardmore Basin. *Okla. Geol. Surv. Bull.,* 79: 144.

James, W.C., 1985. Early authigenesis, Atheton Formation (Quaternary): A guide for understanding early cement distribution and grain modification in nonmarine deposits. *J. Sediment. Petrol.,* 55: 135 – 146.

Johnsson, M.J. and Stallard, R.F., 1989. Physiographic control on the composition of sediments derived from volcanic and sedimentary terrains on Barro Colorado Island, Panama. *J. Sediment. Petrol.,* 59: 769 – 781.

Johnsson, M.J., Stallard, R.F. and Meade, R.H., 1988. First-cycle quartz arenite in the Orinoco River basin, Venezuela and Colombia. *J. Geol.,* 96: 263 – 277.

Jones, P.C., 1972. Quartzarenite and litharenite facies in the fluvial foreland deposits of the Trenchand Group (Westphalian), Forest of Dean, England. *Sediment. Geol.,* 8: 177 – 198.

Keller, W.D., 1957. *The Principles of Chemical Weathering.* Lucas Brothers Publ., Columbia, Mo., 112 pp.
Konishchev, V.N. and Rogov, V.V., 1983. The cryogenic evolution of mineral matter (an experimental model). *Proc. 4th Int. Conf. Permafrost,* pp. 656 – 659.
Kramer, J.R., 1968. Mineral – water equilibria in silicate weathering. *Proc. 23rd Int. Geol. Congr., Sect. 6,* Prague, pp. 149 – 160.
Krynine, P.D., 1940. Petrology and genesis of the Third Bradford Sand. *Bull. Pennsylvania State College,* 29, 134 pp.
Land, L.S., 1984. Frio sandstone diagenesis, Texas Gulf Coast: A regional isotopic study. In: D.A. McDonald and R.C. Surdam (Editors), *Clastic Diagenesis. Mem. Am. Assoc. Pet. Geol.,* 37: 47 – 62.
Land, L.S. and Dutton, S.P., 1978. Cementation of a Pennsylvanian deltaic sandstone: Isotopic data. *J. Sediment. Petrol.,* 48: 1167 – 1176.
Larsen, G. and Chilingar, G., 1979. *Diagenesis in Sediments and Sedimentary Rocks, 1.* Developments in Sedimentology, 25A, Elsevier, Amsterdam, 579 pp.
Larsen, G. and Chilingar, G., 1983. *Diagenesis in Sediments and Sedimentary Rocks, 2.* Developments in Sedimentology, 25B, Elsevier, Amsterdam, 572 pp.
Lawrence, J.R. and Taylor, H.P., 1971. Deuterium and O-18 correlation: clay minerals and hydroxides in Quaternary soils compared to meteoric waters. *Geochim. Cosmochim. Acta,* 35: 993 – 1003.
Lele, K.M., 1976. Paleoclimatic implications of Gondwana flora. *Geophytology,* 6: 207 – 229.
Lelong, F., Tardy, Y., Grandin, G., Trescades, J.J. and Boulage, B., 1976. Pedogenesis, chemical weathering and processes of formation of some supergene ore deposits. In: K.H. Wolf (Editor), *Handbook of Stratabound and Stratiform Ore Deposits,* Vol. 3. Elsevier, Amsterdam, pp. 93 – 174.
Lobo, C.F. and Osborne, R.H., 1976. Petrology of the Precambrian – Cambrian quartzose sandstones in the eastern Mojave Desert, southeastern California. *J. Sediment. Petrol.,* 46: 829 – 846.
Longstaffe, F.J., 1984. The role of meteoric water in diagenesis of shallow sandstones stable isotope studies of the Milk River Aquifer and Gas Pool, Southeastern Alberta. In: D.A. McDonald and R.C. Surdam (Editors), *Clastic Diagenesis. Mem. Am. Assoc. Pet. Geol.,* 37: 81 – 98.
Loughnan, F.C., 1969. *Chemical Weathering of the Silicate Minerals.* Elsevier, New York, N.Y., 154 pp.
Mack, G.H., 1977. The effects of depositional environment on detrital mineralogy: The Permian Cutler – Cedar Mesa facies transition near Moab, Utah. Unpublished Ph.D. diss., Indiana University, Bloomington, Ind., 152 pp.
Mann, W.R. and Cavarock, V.W., 1973. Composition of sand released from three source areas under humid, low relief weathering in the North Carolina Piedmont. *J. Sediment. Petrol.,* 43: 870 – 881.
Mathisen, M.E., 1984. Diagenesis of Plio – Pleistocene nonmarine sandstones, Cagayon Basin, Philippines: Early development of secondary porosity in volcanic sandstones. In: D.A. McDonald and R.C. Surdam (Editors), *Clastic Diagenesis. Mem. Am. Assoc. Pet. Geol.,* 37: 177 – 193.
McBride, E.F., 1985. Diagenetic processes that affect provenance determinations in sandstone. In: G.G. Zuffa (Editor), *Provenance of Arenites.* NATO ASI Ser. C, D. Reidel, Dordrecht, 148: 95 – 113.
McIntyre, M.P., 1980. *Physical Geography,* 3rd ed. Wiley, New York, 507 pp.
Merino, E. and Ortoleva, P., 1981. Chemical kinetic competition between sandstone and porefluids – A quantitative application to the genesis of redox transition fronts (abstract). *Geol. Assoc. Can. Annu. Meet.,* 6: A-39.
Merkle, F.G., 1955. Oxidation-reduction processes in soils. In: F.E. Bear (Editor), *Chemistry of the Soil.* Reinhold, New York, N.Y., pp. 200 – 218.
Miall, A.D., 1981. Alluvial sedimentary basins: Tectonic setting and basin architecture. In: A.D. Miall (Editor), *Sedimentation and Tectonics in Alluvial Basins. Geol. Assoc. Can., Spec. Pap.,* 23: 1 – 34.
Nicholls, G.D., 1963. Environmental studies in sedimentary geochemistry. *Sci. Progr.,* 51: 12 – 31.
Odin, G.S. and Dodson, M.H., 1982. Zero isotopic ages of glauconite. In: G. S. Odin (Editor), *Numerical Dating in Stratigraphy.* Wiley, New York, N.Y., pp. 277 – 306.
Okada, H., 1967. Composition and cementation of some Lower Paleozoic grits in Wales. *Kyushu Univ. Mem. Fac. Sci., Ser. D, Geol.,* 18: 261 – 276.
Pedro, G. and Sieffermann, 1979. Weathering of rocks and formation of soils. In: F.R. Siegal (Editor), *Review of Research in Modern Problems in Geochemistry.* UNESCO: pp. 39 – 55.
Pettijohn, F.J., Potter, P.E. and Siever, R., 1973. *Sand and Sandstone.* Springer, New York, N.Y., 618 pp.

Plotnikov, N.A. and Bogomolov, G.B., 1958. Classification of underground water resources and their reflections on maps. *Int. Assoc. Sci. Hydrol., Gen. Assem. Toronto,* 2: 525 pp.
Potter, P.E., 1978. Petrology and chemistry of modern big river sands. *J. Geol.,* 86: 423 – 449.
Potter, P.E., 1986. South America and a few grains of sand: Part I – Beach sands. *J. Geol.,* 94: 301 – 319.
Pryor, W.A., 1971. Petrology of the Permian Yellow Sands of northeastern England and their North Sea basin equivalents. *Sediment. Geol.,* 6: 221 – 254.
Rieke, H.H. and Chilingarian, G.V., 1974. *Compaction of Argillaceous Sediments.* Developments in Sedimentology, 16. Elsevier, Amsterdam, 424 pp.
Rittenhouse, G., 1971. Mechanical compaction of sands containing different percentages of ductile grains: A theoretical approach. *Bull. Am. Assoc. Pet. Geol.,* 55: 92 – 96.
Rudloff, W., 1981. *World Climates.* Wissenschaftliche Verlagsgesellschaft, Stuttgart, 632 pp.
Ruxton, P.B., 1970. Labile quartz-poor sediments from young mountain ranges in northwest Papua. *J. Sediment. Petrol.,* 40: 1262 – 1270.
Schmidt, V. and McDonald, D.A., 1979. Texture and recognition of secondary porosity in sandstones. In: P.A. Scholle and D.R. Schluger (Editors), *Aspects of Diagenesis. Soc. Econ. Paleontol. Mineral., Spec. Publ.,* 26: 175 – 208.
Shah, S.C., 1976. Climates during Gondwana Era in peninsular India: Faunal evidence. *Geophytology,* 6: 186 – 206.
Shanmugam, G., 1985. Types of porosity in sandstones and their significance in interpreting provenance. In: G.G. Zuffa (Editor), *Provenance of Arenites.* NATO ASI Ser. C, Reidel, Dordrecht, 148: 115 – 137.
Siever, R., 1979. Plate-tectonic control on diagenesis. *J. Geol.,* 87: 127 – 155.
Skinner, D.N.B., 1972. Subdivision and petrology of the Mesozoic rocks of the Coromandel (Manaia Hill Group). *N. Z. J. Geol. Geophys.,* 15: 203 – 227.
Srodon, J. and Ebert, D.D., 1984. Illite. *Rev. Mineral.,* 13: 495 – 544.
Standard, J.C., 1969. Hawkesbury sandstone, Sydney basin, Australia. In: G.H. Packham (Editor), *J. Geol. Soc. Austr.,* 16: 407 – 417.
Stewart, R.J., 1977. Neogene turbidite sedimentation in Komandorskiy basin, western Bering Sea. *Bull. Am. Assoc. Pet. Geol.,* 61: 192 – 206.
Stewart, R.J., 1978. Neogene volcaniclastic sediments from Atka basin, Aleutian Ridge. *Bull. Am. Assoc. Pet. Geol.,* 62: 87 – 97.
Strakhov, N.M., 1967. *Principles of Lithogenesis,* Vol. 1. Consultants Bureau, New York, N.Y., 245 pp.
Suter, T.D., 1986. Evidence of the diagenetic formation of illite from smectite in the Huron Member of the Ohio Shale (abstract). *Geol. Soc. Am. Abstr. Progr.,* 18(4): 326.
Suttner, L.J. and Dutta, P.K., 1986. Alluvial sandstone composition and paleoclimate, I. Framework mineralogy. *J. Sediment. Petrol.,* 56: 329 – 345.
Suttner, L.J., Basu, A. and Mack, G.H., 1981. Climate and the origin of quartz arenite. *J. Sediment. Petrol.,* 51: 1235 – 1246.
Tallmann, S.L., 1949. Sandstone types. Their abundance and cementing agents. *J. Geol.,* 57: 582 – 591.
Tarbuck, E.J. and Lutgens, F.K., 1988. *Earth Science.* Merill, Columbus, Ohio., 612 pp.
Tardy, Y., 1971. Characterization of the principal weathering types by the geochemistry of waters from some European and African crystalline massifs. *Chem. Geol.,* 7: 258 – 271.
Tickell, F.G., Mechem, O.E and McCurdy, R.C., 1933. Some studies on the porosity and permeability of rocks. *Trans. Min. Metall. Eng.,* 103: 250 – 260.
Tallman, S.L., 1949. Sandstone types: Their abundance and cementing agents. *J. Geol.,* 57: 582 – 591.
Toth, J., 1963. A theoretical analysis of groundwater flow in small drainage basins. *J. Geophys. Res.,* 68: 4795 – 4811.
Ubell, K., 1962. A felszin alatti vizkeszlet. *Hydrol. Kozl. Budapest, Hungary,* 42: 94 – 104 (English summary).
Ugolini, F.C., 1986. Processes and rates of weathering in cold and polar desert environments. In: S.M. Colman and D.P. Dethier (Editors), *Rates of Chemical Weathering of Rocks and Minerals.* Academic Press, London, pp. 193 – 235.
Velbel, M.A., 1985. Geochemical mass balances and weathering rates in forested watersheds in the southern Blue Ridge. *Am. J. Sci.,* 285: 904 – 930.
Von Engelhardt, W., 1977. *The Origin of Sediments and Sedimentary Rocks.* Wiley, New York, N.Y.

Walker, T.R., 1984. Diagenetic alteration of potassium feldspar in arkosic sandstone. *J. Sediment. Petrol.,* 54: 1 – 16.

Walker, T.R., Waugh, B. and Crone, A.J., 1978. Diagenesis of first-cycle desert alluvium of Cenozoic age, southeastern United States and northwestern Mexico. *Bull. Geol. Soc. Am.,* 89: 19 – 32.

White, D.E., Hem, J.D. and Waring, G.A., 1963. Chemical composition of subsurface waters. *U.S. Geol. Surv. Prof. Pap.,* 440-D: 67 pp.

Wood, J.R. and Surdam, R.C., 1979. Application of convective diffusion models to diagenetic processes. In: P.A. Scholle and P.R. Schluger (Editors), *Aspects of Diagenesis. Soc. Econ. Paleontol. Mineral., Spec. Publ.,* 26: 243 – 250.

Wopfner, H. and Schwarzbach, M., 1976. Ore deposits in the light of palaeoclimatology. In: K.H. Wolf (Editor), *Handbook of Stratabound and Stratiform Ore Deposits,* Vol. 3. Elsevier, Amsterdam, pp. 43 – 92.

Young, S.W., 1975. Petrography of Holocene fluvial sand derived from regionally metamorphosed source rocks. Unpubl. Ph.D. Diss., Indiana University, Bloomington, Ind., 144 pp.

Chapter 5

DIAGENESIS OF DEEP-SEA VOLCANICLASTIC SANDSTONES

YONG IL LEE

INTRODUCTION

The study of sandstone diagenesis has undergone rapid expansion over the past decades. It has been demonstrated that the original composition of sandstone controls in part the nature of diagenetic changes (Carrigy and Mellon, 1964; Blatt, 1979; Hayes, 1979; Vavra, 1983). During lithification, the burial depth, temperature and pore-water chemistry also play significant roles (Bjorlykke, 1983; Saigal et al., 1988). Sandstone diagenesis proceeds through several systematic steps, starting with pore-space reduction by compaction, and later with rim cementation, pore-fill cementation, and alteration and transformation of mineral phases in more-deeply buried sandstone (Wilson and Pittman, 1977). It is for this reason that porosity and permeability decrease downhole in an oil well. Continued solution at depth, however, causes the development of secondary porosity within sandstone (Schmidt and McDonald, 1979; Burley and Kantorowicz, 1986).

The diagenetic study of deep-sea sediments has centered on the transformation of biogenic silica, alteration of volcanic material and formation of oxides, such as manganese nodules (Ernst and Calvert, 1969; Heath and Moberly, 1971; Mizutani, 1977; Hein et al., 1978; Kastner, 1981), and the results are well known. At present, however, a generalization of deep-sea sandstone diagenesis is not possible because in different ocean basins there is different geodynamic influence upon sandstone composition and diagenesis. Also, lack of extensive studies on such subjects hampers its generalization. The main contribution of this chapter, therefore, is to discuss some specific aspects of the diagenetic processes and products of deep-sea sandstones and to point out their controlling factors.

Certain sandstone components, such as volcanic fragments, undergo rapid alteration under low temperatures and pressures. Thus, volcaniclastic sands are more susceptible to diagenetic alteration because of the chemical instability and reactivity of their framework grains (Whetten and Hawkins, 1970; Davies et al., 1979; Dickinson and Suczek, 1979). The diagenetic reactions occurring in volcaniclastic sandstones have been reviewed by Surdam and Boles (1979). They divided diagenetic processes into early and late diagenetic changes. Early diagenesis is characterized by hydration and carbonitization reactions. The most significant hydration reactions are glass to zeolite and plagioclase to zeolite. During the late stages of diagenesis, dehydration reactions become dominant. Such factors as fluid flow and fluid composition are suggested to be as significant as depth of burial in controlling the distribution of diagenetic mineral phases in volcaniclastic sandstones.

The present discussion is limited to one sandstone clan, i.e., volcaniclastic sandstones. The volcaniclastic sandstones in deep-sea environments occur in specific ocean basins, near volcanic sources. In the tectonic sense, they belong to the active plate margins. Only a few studies have been done on sandstone diagenesis in active plate margins (Galloway, 1974, 1979; Davies et al., 1979; Lee and Klein, 1986; Lee, 1987, 1988).

DISTRIBUTION OF DEEP-SEA SANDSTONES AND TECTONIC SETTING

The distribution of the deep-sea sands has been studied to infer the tectonic influence on sandstone composition with the advent of plate tectonics (Dickinson and Suczek, 1979; Ingersoll and Suczek, 1979; Dickinson and Valloni, 1980; Valloni and Maynard, 1981; Maynard et al., 1982; Valloni and Mezzadri, 1984; Gergen and Ingersoll, 1986; Packer and Ingersoll, 1986). Valloni and Maynard (1981) subdivided the ocean basins into four types, namely: trail edge, leading edge, back-arc, and fore-arc basins. The average composition of the deep-sea sands in different tectonic settings is shown in Table 5-1. Other studies provide similar results, with some refinements, to the petrologic models of Table 5-1.

As shown in Table 5-1, the abundant occurrence of volcaniclastic grains is restricted mostly to the back-arc and fore-arc basins. Sands deposited in these basins are derived mostly from andesitic island-arc sources (Karig, 1975; Klein, 1975a,b, 1985; Klein and Lee, 1984).

The characteristics of back-arc basins have been investigated during the last 30 years. The mechanisms proposed for their development are complex, as are many of the characteristics of convergent plate boundaries. It is generally accepted, however, that back-arc basins are extensional in origin and are characterized by large heat flow (Karig, 1970, 1971, 1975; Watts and Weissel, 1975; Hilde et al., 1977; Watts et al., 1977; Kobayashi and Nakada, 1978; Anderson, 1980; Klein et al., 1980; Hsui and Toksoz, 1981; Brooks et al., 1984, amongst others).

TABLE 5-1

Average composition of deep-sea sands from various tectonic settings (after Valloni and Maynard, 1981)

Basin	Number of samples	Q	F	L	C/Q	P/F	Lv/L
Trailing-edge							
(a) Atlantic type	29	66	23	12	0.11	0.30	0.13
(b) Collision (Indian Ocean)	26	58	29	14	0.04	0.63	0.04
Leading-edge							
(a) Subduction	8	16	53	31	–	0.72	0.98
(b) Strike-slip	50	34	39	27	0.29	0.65	0.33
Back-arc	53	20	29	51	0.48	0.61	0.84
Fore-arc	20	8	17	75	–	0.87	0.99
	186						

Q: quartzose; F: feldspathic; L: lithic; C: chert; Lv: volcanic lithic fragments; P: plagioclase.

Most of the back-arc basins in the western Pacific Ocean are not undergoing active extension now (Toksoz and Bird, 1977), but substantial evidence such as magnetic anomaly maps indicates that they were spreading actively in the past (Weissel, 1981). Some inactive basins occur directly behind andesitic island arcs, but many occur behind remnant arcs. Karig (1971) proposed that inactive basins located farther from the present-day volcanic chain are older.

Fore-arc basins are relatively large topographic depressions lying between the volcanic arc and the structural high (trench-slope break) at the top of the accretionary prism formed by the subduction complexes (Reading, 1982). The configuration of modern fore-arc regions has been classified into four types, such as shelved, sloped, terraced, and ridged fore-arcs (Dickinson and Seeley, 1979).

Diagenesis of volcaniclastic sandstones in back-arc basins are discussed extensively first, followed by diagenesis of the fore-arc basins. By comparison between diagenesis of sandstones in both basins, general conclusions in active plate margins have been drawn.

DIAGENESIS OF BACK-ARC BASIN SANDSTONES

Sandstones in the back-arc basin considered here are from seven Deep Sea Drilling Project cores (Sites 299, 297, 445, 446, 453, 286 and 285) in the western Pacific Ocean (Fig. 5-1). Table 5-2 provides location data, basement ages and DSDP site report references.

TABLE 5-2

Location of DSDP sites discussed in this chapter, basement ages and depth interval of sandstones (after Lee and Klein, 1986)

Site	Latitude; longitude	Location	Depth interval of sandstones (below subbottom depth) (m)	Basement age (m.y. BP)	Site report reference
299	39° 29.69′ N 137° 39.72′ E	Toyama Fan, Sea of Japan	0 – 532	0 – 5	Karig (1975)
297	30° 52.36′ N 134° 09.89′ E	Shikoku Basin	330 – 570	23 – 25	Karig (1975)
445	25° 29.69′ N 133° 12.49′ E	Daito Basin	647 – 892	44 – 48	Klein et al. (1980)
446	24° 42.04′ N 132° 46.49′ E	Daito Basin	172 – 628	45 – 51	Klein et al. (1980)
453	17° 54.42′ N 143° 40.95′ E	Western Mariana Trough	303 – 443	4.7 – 5	Hussong et al. (1981)
286	26° 32.92′ S 166° 22.18′ E	New Hebrides Basin	206 – 649	37 – 45	Andrews et al. (1975)
285	26° 49.16′ S 175° 48.24′ E	South Fiji Basin	453 – 564	12 – 15	Andrews et al. (1975)

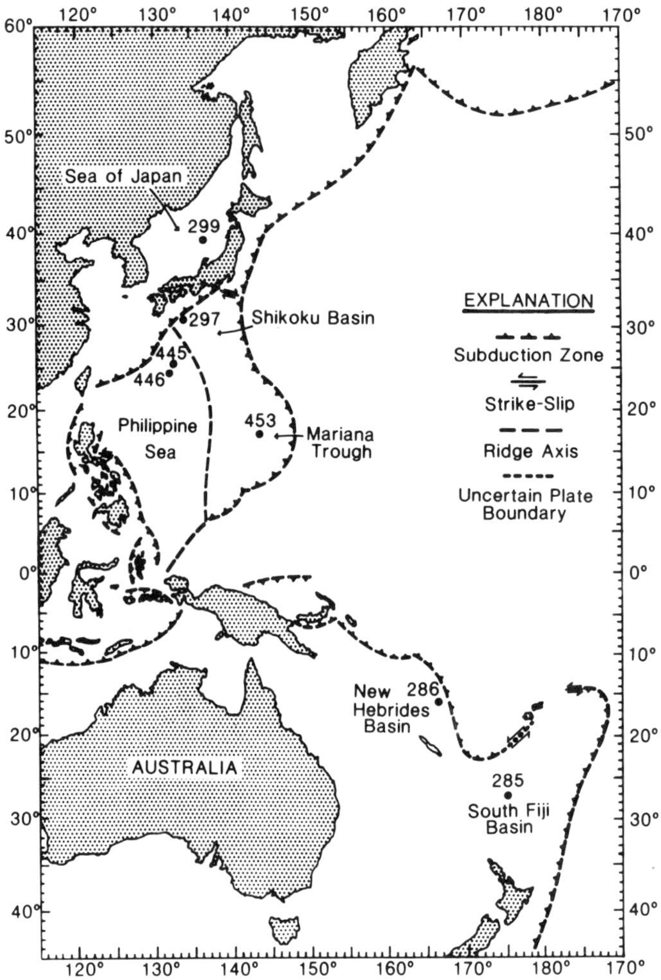

Fig. 5-1. Map of the western Pacific Ocean showing back-arc basin sites containing sandstones discussed in this chapter. (After Lee and Klein, 1986; courtesy of *Sedimentology*.)

Sandstone petrology

Sandstone of the back-arc basin setting ranges in particle size from fine silt to coarse sand. Most show poor sorting and grains are subrounded to angular in most samples, except at Site 299 and 297 where rounded grains are common. Detrital mineralogy is relatively simple, consisting mostly of plagioclase and recognizable volcanic rock and glass fragments with accessory quartz (Table 5-3). At Site 297, quartz and sedimentary rock fragments are more common. The relative proportions of these constituents, especially quartz and volcanic glass and rock fragments, vary between sites and downhole. Sandstones were classified according to Folk (1968;

ite		clay	cristo-balite	Other authigenic			Recrystallization		Remarks
-andite	anal-cite			zeo-lite	cal-cite	chert	fossil	sedimentary rock frag-ments	
			tr			0.6	7.3	15.9	
		tr		tr	tr	0.6	6.1	17.4	
		0.3			1.2	0.8	0.8	15.4	
		0.3			0.9	0.3	4.4	13.3	clinop.
	tr	8.6			1.1	0.9			heulan.
	tr	0.5			1.6	0.8	6.7	9.3	
		1.1			1.7	0.3	7.7	33.3	
		1.4		0.9	1.2		12.6	33.3	
	tr	0.6		0.6	1.4	0.3	7.1	17.8	authi. Ab, K-sp
	tr	1.1	tr	tr	1.1	0.9	12.3	18.9	authi. Ab, K-sp
	tr	1.4	tr		1.1	1.9			
		0.3			0.8	1.4			
	1.5	0.3	tr	1.1	2.9	2.0			illite
	tr	0.5		0.9	2.0				
	10.8	0.6		1.2	1.4	1.1			authi. Ab
	9.0	2.5		0.6	1.5	2.4			authi. Ab, K-sp
	6.6	0.3		1.1	2.6	0.8			authi. Ab
	8.7	1.1		0.5	0.3				
	2.3	0.3		2.4	0.6				
	3.2	1.4		0.6	0.3	tr			authi. Ab
	3.6	1.6		0.9	1.4	1.2			
	1.7	1.3		tr	2.2	0.9			
					0.5				
		0.3	tr						authi. K-sp
		tr				0.4			
		7.5	tr						authi. K-sp
						0.5	tr		
	14.3	1.3	4.3						
	0.9	0.9			0.4		tr		
							8.2	24.0	
		0.6			0.4		11.8	8.0	
				1.1	tr		17.7	7.9	
		1.1			1.5		18.5	9.1	
	tr	16.4	tr	8.0			18.1		
		tr			tr				
					tr	tr			
					tr				
					tr				
		2.1			tr				
		tr			tr				

TABLE 5-3

Modal analysis of sandstones from DSDP sites 286, 297, 299, 445, 446 and 453 in back-arc basins of the western P

Sample (interval in cm)	Sub-bottom depth (m)	Detrital grains						
		quartz	plagio-clase	K-feld-spar	volcanic rock frag-ments	volcanic glass frag-ments	tary rock frag-ments	meta-morph rock frag-ments
Site 286 (% based on 230 points)								
12-2(50 – 55)	208.53	3.7	13.4		21.7	15.2		
13-2(101 – 109)	229.05	4.0	6.7		42.9	19.4	14.3	
14-2(92 – 96)	246.94	2.0	6.1		22.2	8.9		
19-1(105 – 110)	340.57	0.5	2.8		23.1	61.3	0.5	
21-2(87 – 89)	379.88	tr	8.1		45.5	10.6	0.4	
26-3(30 – 33)	475.82	0.5	22.6	0.5	32.1	21.3	0.5	
29-3(124 – 127)	533.75	0.8	3.6		8.7	80.9		
31-1(98 – 102)	658.50	tr	3.7		1.9	55.6		
32-2(78 – 82)	588.80	3.4	13.8	0.4	10.4	53.8		
Site 297 (% based on 230 points)								
15-4(146 – 148)	339.47	9.2	2.3		0.5		56.8	
16-2(49 – 53)	354.51	25.4	5.6	0.8	tr		22.0	
17-2(74 – 79)	397.76	45.9	14.3	1.0	1.0		27.5	1.1
18-2(10 – 15)	430.13	42.7	7.5	0.4	tr		12.9	
19-1(119 – 125)	447.72	44.0	6.0	0.4	1.1		6.8	tr
22-2(139 – 141)	554.90	27.6	tr		12.6		36.2	
Site 299 (% based on 230 points)								
9-3(124 – 130)	80.27	22.1	4.4	0.6	1.7	43.1	2.8	
12-4(48 – 54)	109.51	10.3	0.6		tr	60.3		
14-5(30 – 32)	129.81	34.2	3.5	0.9	tr	39.0	0.4	
15-5(63 – 69)	139.66	7.3	4.8	1.5	7.7	25.8	tr	
16-2(42 – 44)	144.43	2.9	4.8	0.7	14.4	17.3	tr	
16-4(10 – 24)	147.07	13.0	8.4	2.0	23.3	18.3	0.5	
17-3(48 – 53)	155.51	22.1	12.4	1.8	22.1	15.5	0.8	
35-1(43 – 47)	475.45	3.3	3.8	0.4	26.3	37.1		
38-5(135 – 137)	529.86	16.4	3.2		50.5	tr		
Site 445(% based on 350 points)								
69-2(70 – 72)	647.20	tr	0.5		46.3		2.0	
69-2(143 – 149)	647.93	0.9	2.8	0.6	67.5		1.1	
71-1(90 – 91)	664.90	tr	0.9		19.8		15.1	
71-2(37 – 40)	665.87	0.3	3.6	0.3	33.4		2.6	
74-1(93 – 97)	693.47	0.3	2.6		60.8		6.1	
75-3(23 – 25)	705.23	1.2	11.0	1.2	44.0		27.4	
75-6(97 – 100)	710.47	0.3	1.6	0.3	42.8		22.3	

cific Ocean (modified after Lee and Klein, 1986)

							Matrix	Cements					
c	horn blende	pyro-xene	chert	musco-vite	glauco-nite	fossil (foram %)		anhy-drite	calcite early	calcite late	zeolite clino-heulandite	zeolite anal-cite	clay
		2.7				12.0(0)	42.4						
	tr	1.2				2.8(tr)	7.9						
		tr		0.4	tr	0.8(tr)	56.1						
		tr		0.9		0.9(50)	7.6						
		2.0		0.4		0.8(50)	31.4						
	tr	1.8		0.5		tr	19.9						
		0.8				1.2(0)	3.9						
							38.3						
		3.3					14.2						
				tr	tr		30.9						
		0.4					43.3						
			5.1		tr		4.1						
			2.2				33.4	1.0					
		tr	0.8	tr	tr		35.6	0.5					4.9
	1.6	0.8		2.4			18.9						
						8.8(50)	13.7	1.2					0.5
		0.6			tr	7.7(17)	19.7	0.8					
		0.4					17.9	0.5					3.1
			1.0		tr	6.7(0)	45.0	0.9					tr
		0.5	0.5				59.6						
	tr	1.5	4.5		0.5	0.5(0)	25.7						
	tr	1.8	4.4	0.4	tr	1.3(0)	21.2						
	0.8	0.4	tr			1.3(0)	25.8						
	0.9		tr	0.9		tr	28.2						
	tr					30.6(95)	14.1		1.6		5.6		0.8
	tr		0.8			1.7(95)	5.1				10.7		7.6
	tr	tr	0.3			42.6(94)	18.2		1.5		0.9		0.6
	0.3					2.3(100)	53.4			tr	tr		4.3
	0.3					15.7(95)	12.5		2.7		3.8		0.6
	0.6	0.9	0.6			0.3(100)	8.9				tr		tr
	1.6	0.5	0.5			59.0(83)	9.6		3.9		tr		0.3

sedimentary rock fragments	metamorphic rock fragments	hornblende	pyroxene	chert	muscovite	glauconite	fossil (foram %)	Matrix	Cements			
									anhydrite	calcite		zeo
										early	late	clin heu
23.9		0.3	0.6				3.9(57)	8.6		1.4		1.
26.2		0.6	0.3	0.3			4.5(71)			18.7		t
16.3							34.9(99)			17.3		
22.1			0.3				15.3(98)			17.0		
2.4							6.2(100)	tr			3.9	8.
29.9							13.6(80)	8.0		9.2		t
5.4		0.3					4.3(82)	15.0		8.5		
6.9			0.3				4.8(85)	12.0		7.5		0.
6.5							1.7(82)			18.4		0.
6.8							3.0(81)			1.9	0.3	11.
							tr	4.6				t
1.4							0.5(83)	6.1		1.1		1.
				1.2			tr	1.9		0.5	1.8	6.
											7.0	10.
		0.6					tr				3.7	
				0.6			tr				2.7	
tr							tr				8.0	
2.7							1.1(90)				4.6	
0.9							0.5(8)	2.3				
tr											tr	
tr												
										0.3		
7.3			0.3				69.1(100)			5.6		3.
0.9		0.6	3.1	0.3			0.3(6.2)	30.4				9.
		4.1	6.9					23.0				
0.5		1.1	4.3					21.5				t
1.3		3.6	3.1				0.5(100)	25.9				
0.4		2.6	2.2	tr			tr	9.1		1.3		
		2.2	5.2				0.4(100)	35.9				
17.2		0.3	3.6				48.6(98)			17.5		
21.1		tr	0.4				27.3(82)	11.7		2.8	0.8	5.
46.8			0.5				32.6(65)			11.1	1.1	2.
27.6			tr				26.2(98)			8.4		17.
			0.5				9.4(100)	9.9			0.5	3.
			0.5									
0.9			0.4					61.9				
7.6			tr				tr	17.2		tr		
2.3			0.5				tr	11.7				
0.5			1.4					42.8				
1.4			0.4					23.5				
			1.7					30.9		tr		
			1.5					48.5				
								34.6				

	Other authigenic			Recrystallization		Remarks
cristobalite	zeolite	calcite	chert	fossil	sedimentary rock fragments	
1.4			0.9			
0.8		tr				
3.7		tr				
2.4		tr				
0.8			tr			
0.5						
tr						
0.5						
0.8						
		1.0				
		tr				
			0.5			
			1.9			
		0.8				
	tr					
tr				2.4	tr	
tr						
tr		tr		4.6		
			0.6		22.8	
		0.3		9.1	17.2	authi. Ab

TABLE 5-3 (continued)

Sample (interval in cm)	Sub-bottom depth (m)	Detrital grains				
		quartz	plagio-clase	K-feld-spar	volcanic rock frag-ments	volcanic glass frag-ments
Site 445 (continued)						
76-1(23 – 28)	711.73	0.8	11.7		48.4	
76-1(122 – 125)	712.72		6.8		41.9	
77-4(99 – 104)	726.49	0.3	0.8		28.1	
78-3(58 – 63)	734.08	0.3	3.7		38.9	
78-5(0 – 4)	736.50	tr	tr		68.7	
79-2(143 – 148)	743.93		2.4		33.0	
82-4(1 – 5)	771.50	0.3	3.4		59.0	
83-4(9 – 13)	779.59	0.3	1.5		62.7	
84-2(124 – 126)	790.24	tr	2.8		67.4	
85-1(18 – 20)	797.18	tr	2.1		71.9	
86-1(92 – 94)	797.92		3.0		88.1	
86-4(73 – 75)	811.73	0.3	2.2		84.8	
88-2(24 – 26)	827.24	0.3	1.7		78.0	
89-1(131 – 142)	836.31	tr	0.8		77.7	
90-2(11 – 14)	846.11	tr	2.3		78.5	
90-2(17 – 21)	846.17	0.3	2.2		78.2	
90-2(22 – 26)	846.22	0.8	3.6		76.3	
91-2(64 – 70)	856.14	0.3	2.7		77.9	
91-4(84 – 88)	859.34	0.3	0.3		90.3	
92-3(40 – 48)	866.90	0.6	10.0		84.0	
93-4(127 – 132)	878.77	0.9	4.3		86.1	
94-2(78 – 82)	884.78	1.8	9.9		80.8	
94-4(132 – 134)	888.32		0.3		13.6	
Site 446 (% based on 350 points)						
20-1(84 – 88)	137.34	1.5	13.4	0.6	39.1	
23-2(24 – 27)	201.76		4.6		60.8	
24-1(146 – 148)	211.97		0.5		64.5	
27-1(80 – 82)	239.81	1.0	4.9		59.4	
29-1(13 – 15)	258.13	4.3	12.2		49.1	
30-4(97 – 102)	273.00	0.3	3.5		49.8	
34-4(31 – 35)	307.31		3.0		9.7	
38-cc(33 – 38)	353.00	1.2	1.2		28.9	
39-2(84 – 88)	355.34				4.7	
40-1(81 – 86)	363.31		tr		17.9	
41-1(71 – 75)	372.73	6.2	0.9		44.6	
Site 453(% based on 230 points)						
29-1(53 – 58)	266.05	3.6	1.4		26.2	6.3
33-2(63 – 69)	305.66	0.9	1.3	tr	3.9	75.4
34-2(30 – 35)	314.83	0.5	0.5		0.5	79.2
39-3(108 – 110)	364.59	2.3	1.9	tr	6.1	44.2
41-3(48 – 52)	383.00	3.3	4.2		6.1	61.0
42-1(116 – 122)	390.19	2.3	1.4		5.9	57.7
42-2(24 – 26)	390.75	2.1	5.1		22.8	17.7
45-3(92 – 97)	421.45	2.4	3.9		30.9	26.6

Fig. 5-2); most are litharenites with accessory feldspathic litharenites. Sandstones at Sites 297 and 299 are the most mineralogically mature.

Almost all sandstones occurred on the tie-line between sedimentary and volcanic rock fragments in the rock fragment daughter triangle. These sandstones are classified as volcanic arenite, except at Site 297 and three samples at Site 446, which are sedarenites.

The principal framework component in these sandstones consists of volcanic rock fragments at nearly all of the sites (Table 5-3). Andesitic rock fragments are by far the most common volcanic rock fragments. Detrital volcanic glass is a major constituent in sandstones at most sites occurring both as framework grains and as matrix, especially at Sites 453 and 286. At Site 297, recognizable sedimentary rock fragments comprise the major detrital component. Sandstone matrix at Sites 299 and 297 consists mostly of terrigeneous clay and volcanic glass.

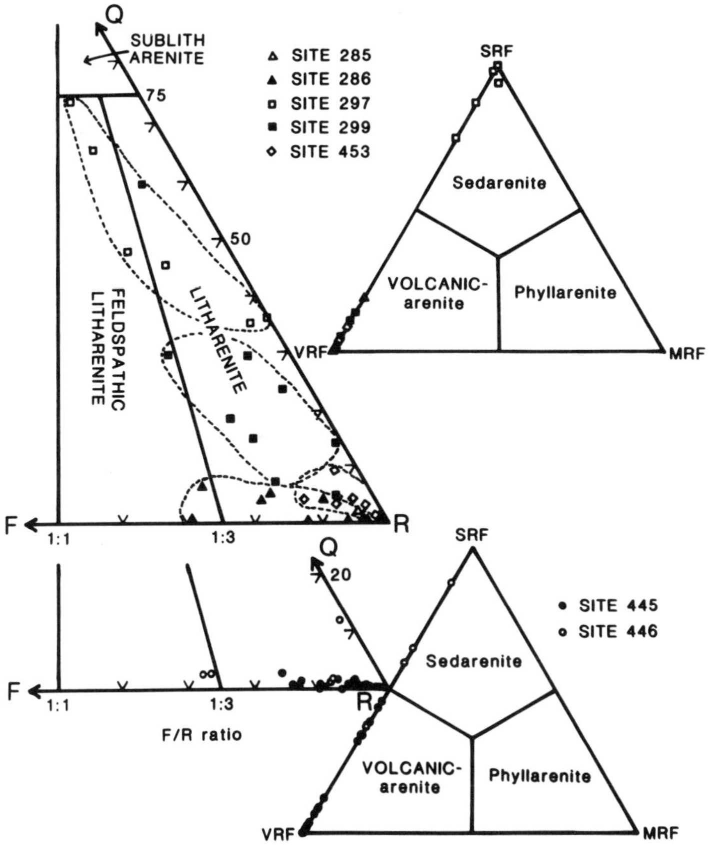

Fig. 5-2. Classification of sandstones from Pacific Ocean back-arc basins. (After Lee and Klein, 1986; courtesy of *Sedimentology*.)

Diagenetic phenomena

Although a large variety of diagenetic phenomena was observed at each site, most represent the end-product of cementation, complex alteration and replacement, and dissolution. Authigenic cements include anhydrite cement (Fig. 5-3), micritic and sparry calcite rim and pore fills, clay rims oriented normal to detrital grain surfaces (Figs. 5-4 and 5-5), well-crystallized pore-filling zeolites including erionite, clinoptilolite, heulandite and analcite (Figs. 5-6, 5-7, 5-8 and 5-9), and minor quantities of syntaxial feldspar (Plate 5-9) and fossil overgrowths. Mechanically-compacted softer grains, such as limestone fragments and fossils, also occur. Alteration and replacement phenomena include volcanic glass matrix altered to clay minerals and zeolites, cristobalite layers around volcanic rock fragments and altered and chloritized volcanic rock fragments. These rock fragments also show replacement by chlorite, zeolite, chert and calcite, whereas plagioclase is replaced by calcite and sericite. Alteration of components, such as limestone fragments and calcareous fossils, is common. Cristobalite, mica, and quartz overgrowths were also observed.

Pore-lining and pore-filling cementation

Pore spaces in these sandstones contain a variety of cements ranging from clay coats on detrital grain surfaces to complex sequences of zeolite and calcite.

Anhydrite pore-fill cement: Anhydrite pore-fill cement, at Site 299 (Fig. 5-3), shows an oval to elongate shape and develops independently of grain composition.

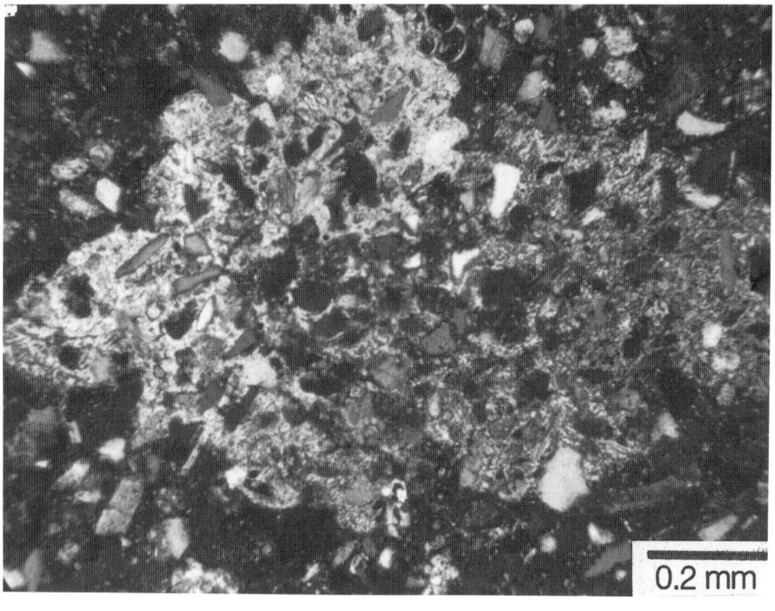

Fig. 5-3. Photomicrograph showing a white patch of anhydrite cement. Detrital grains are floating in anhydrite cement. Site 299, sample 299-9-3 (124 – 130), subbottom depth of 80.27 m. Crossed nicols.

Anhydrite pore-fill cement was observed also at Site 297 showing radiating clusters of elongate, oval-shaped crystals.

Clay minerals: Clay mineral rims are common on volcaniclastic grains. These rims are 30 – 50 μm thick. At Site 297, no diagenetic smectite was observed. At Site 445, smectite (Figs. 5-4 and 5-5), chlorite and illite are oriented normal to detrital grain surfaces, and are absent at grain contacts suggesting an authigenic origin. X-ray dif-

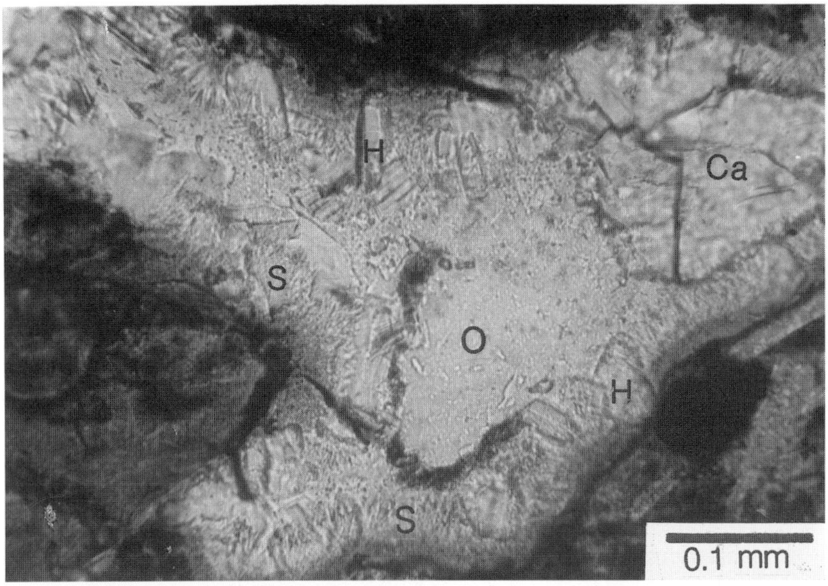

Fig. 5-4. Photomicrograph showing sequence of cementation of smectite (*S*) rim cement, followed by heulandite (*H*) pore-fill cement, which in turn is followed by late sparry calcite (*Ca*) cement. *O*: open space. Site 445, sample 445-78-5 (0 – 4), subbottom depth of 736.5 m. Plane light. (After Lee and Klein, 1986; courtesy of *Sedimentology*.)

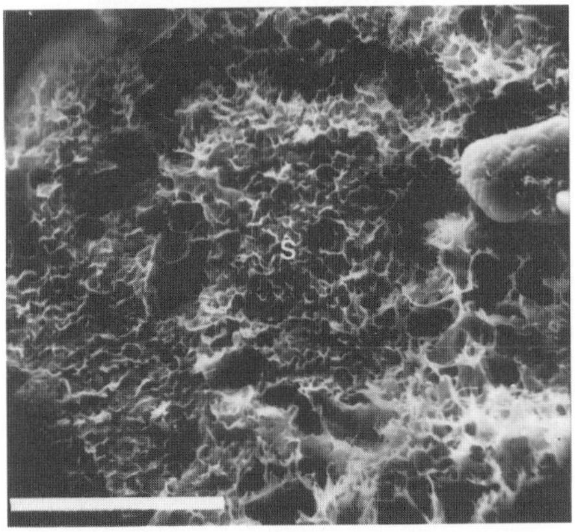

Fig. 5-5. SEM micrograph showing smectite (*S*) rim cement developed normal to volcaniclastic grain surface. Site 445, sample 445-91-2 (64 – 70), subbottom depth of 856.14 m. Bar scale is 10 μm.

fraction (XRD) analysis indicated that smectite was mixed-layered, with 20% illite layers, and chlorite is iron-rich. Minor quantities of discrete illite appear as a pore-bridging cement in the deep borehole (Fig. 5-9). Smectite and chlorite are observed also at Sites 446, 453 and 286.

Erionite: Erionite cement was observed at Site 286, showing bean-shaped bundles of needles (Fig. 5-6). Each bundle is about 8 – 10 μm in length, about 5 μm thick

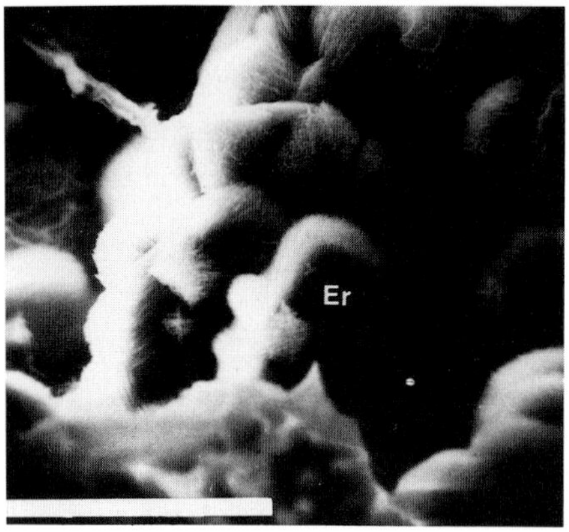

Fig. 5-6. SEM micrograph showing erionite (*Er*) pore-fill cement consisting of bundles of needles which enclose one another. Site 286, sample 286-21-2 (87 – 89), subbottom depth of 379.88 m. Bar scale is 10 μm.

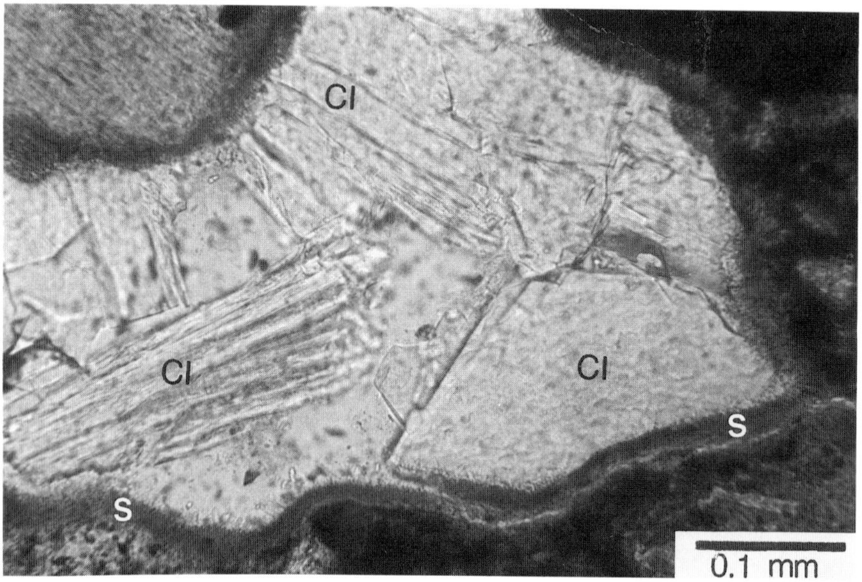

Fig. 5-7. Photomicrograph showing clinoptilolite (*Cl*) in different orientations. Smectite (*S*) rim cement developed earlier along the volcaniclastic grain. Site 445, sample 445-69-2 (70 – 72), subbottom depth of 647.20 m. (After Lee and Klein, 1986; courtesy of *Sedimentology*.)

and consists of hundreds of individual needles, each slightly less than a micron thick. The sheaf of erionite needles encloses poikilitically another erionite sheaf, both with pore spaces lined with smectite.

Heulandite group minerals: Both clinoptilolite and heulandite were identified and classified according to Boles (1972).

Clinoptilolite: Clinoptilolite occurs at Site 445 with heulandite, cristobalite, smectite and altered glass matrix. At Site 446, it replaces both volcanic glass matrix and volcanic rock fragments. In thin section or under the SEM, clinoptilolite occurs as tabular, lath-shaped crystals developing from alteration of volcanic glass (Fig. 5-7). Most crystals display developed cleavage. Some clinoptilolite crystals show evidence of dissolution. The crystal length of clinoptilolite ranges from 0.09 to 0.32 mm with an average of 0.17 mm.

Heulandite: The framework ion ratio of Si to Al (< 4.0) in heulandite differs from this ratio in clinoptilolite, and heulandite is also thermally less stable. It occurs in sandstones at Site 445 with an association identical to clinoptilolite (Fig. 5-8). Probable heulandite was observed at Sites 285 and 286 occurring as euhedral laths and plates. Crystal size is in the range of 0.017 – 0.05 mm in length.

Analcite: Analcite was observed at both Sites 445 and 446. In SEM, analcite crystals show both cubooctahedral and trapezohedral euhedral forms (Fig. 5-9). Some analcites show dissolution features and also form pseudomorphs after clinoptilolite or heulandite. In volcanic rock fragments, analcite replaces plagioclase. The average diameter of analcite is about 0.16 mm.

Calcite: Both early and late calcite cements occur in sandstones at Sites 445 and 446. The distribution between them depends on the relative timing of formation and occurrence. Early cements occur usually as pore-rim and pore-filling cement and are both equant microsparry and sparry. Late calcite cement occurs only in deeper samples as sparry and fills any remaining pore space after the appearance of zeolite (Fig. 5-10). Locally, this later calcite cement replaces zeolite cements, as well as

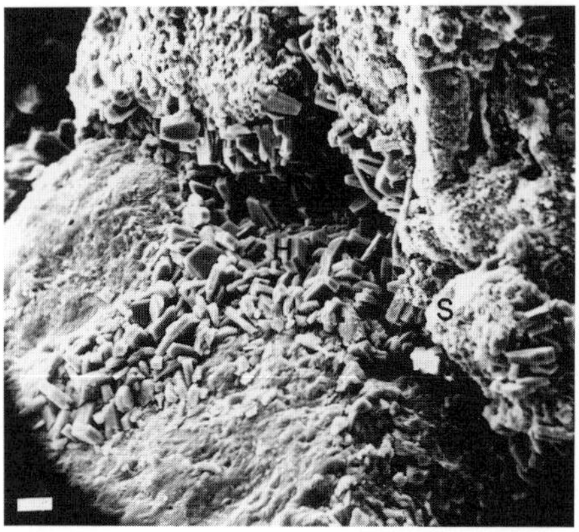

Fig. 5-8. SEM micrograph. The development of heulandite (*H*) pore-fill cement succeeding earlier smectite (*S*) pore rims. Grain contact is shown in middle left and bottom. Site 445, sample 445-85-1 (18 – 20), subbottom depth of 797.18 m. Bar scale is 10 μm.

volcanic fragments and plagioclases. Electron microprobe analyses indicate that both types of calcite are iron-poor (Lee, 1984).

Cristobalite: Cristobalite forms spherulites or lepispheres up to 30 μm in diameter

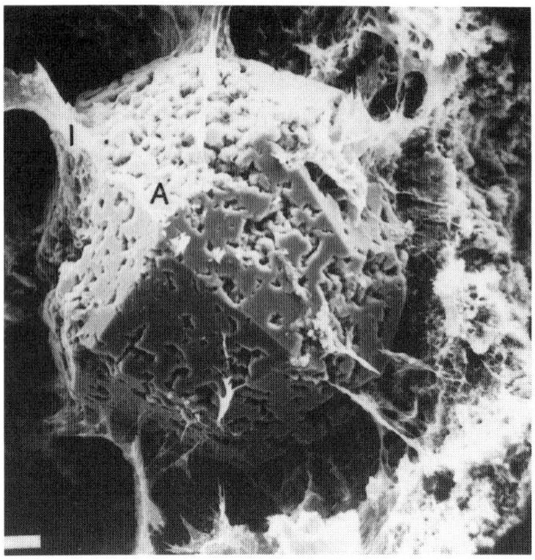

Fig. 5-9. SEM micrograph showing analcite (*A*) cement with partial dissolution and illite (*I*) which bridges between analcite and pore wall. Site 445, sample 445-9-2 (64 – 70), subbottom depth of 856.14 m. Bar scale is 10 μm. (After Lee and Klein, 1986; courtesy of *Sedimentology*.)

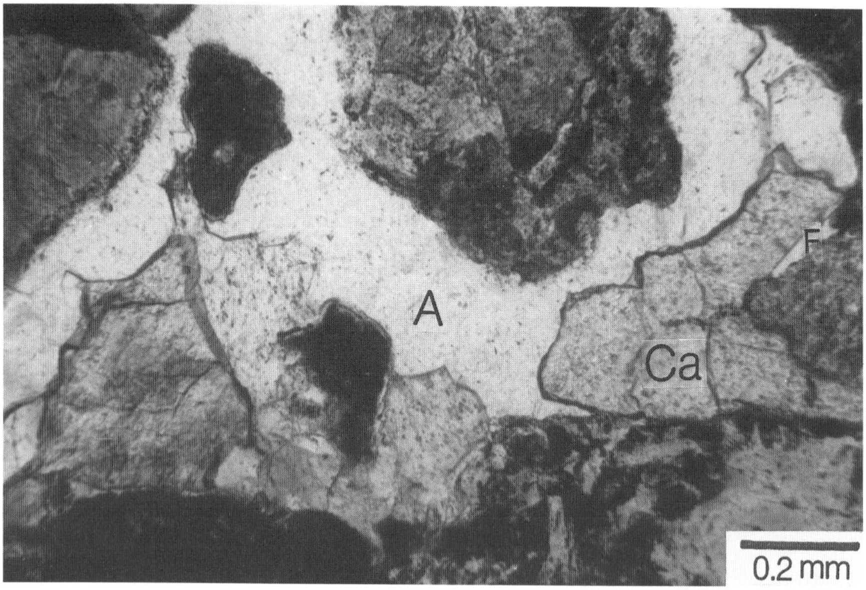

Fig. 5-10. Photomicrograph showing succession of cement. Analcite (*A*) pore-fill cement is followed by late sparry calcite (*Ca*) cement. Calcite cement shows more prominent relief than analcite. Authigenic feldspar (*F*) overgrowth is shown in middle right. Site 445, sample 90-2 (11 – 14), subbottom depth of 846.11 m. Crossed nicols.

at Site 445. It shows flat individual plates with ragged edges and occurs commonly with clinoptilolite (Fig. 5-11). Some cristobalites were precipitated on volcanic rock fragments and are overlain by pore-fill clinoptilolite or heulandite cement.

Replacement and dissolution phenomena

Several replacement, overgrowth and dissolution features occur commonly in these sandstones, including albitization of grains observed in two samples at Site 299, natrolite replacement of volcanic materials, dissolution of grains and cements, and replacement of grains by calcite and chert.

Dissolution of grains: Dissolution of feldspar was observed rarely with plagioclase dissolution occurring along cleavage planes. Some volcanic glass fragments show pitted and etched surfaces under SEM indicating etching by pore solution. Volcanic glass matrix and groundmass of volcanic rock fragments are devitrified and locally volcanic glass matrix is replaced by authigenic calcite. Partial dissolution of volcanic rock fragments was observed in several samples, including total grain dissolution contributing to increased porosity. Augite shows dissolution with a saw-tooth shape.

Authigenic feldspar: Authigenic albite and K-feldspar are present both as overgrowths on detrital feldspar grains (Fig. 5-12) and as rhombic crystals. The authigenic K-feldspar and albite overgrowths are clear and colorless, show uniform extinction, and are in optical continuity with detrital feldspar. Some detrital feldspar underwent albitization (Fig. 5-12).

Chert and natrolite: Both chert and natrolite occur as replacement minerals of volcanic rock fragments. Natrolite was observed only at Site 445, and in thin sec-

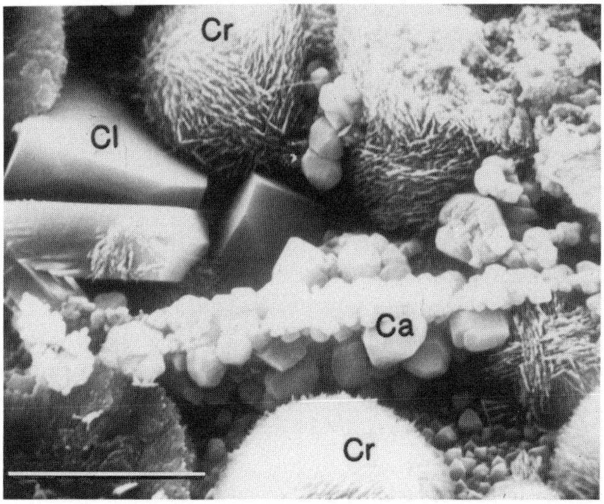

Fig. 5-11. SEM micrograph showing sequential development of cements. First clinoptilolite (*Cl*) pore-fill cement and cristobalite (*Cr*), and then, small crystals of calcite (*Ca*). Site 445, sample 445-71-2 (37 – 40), subbottom depth of 665.87 m. Bar scale is 5 μm.

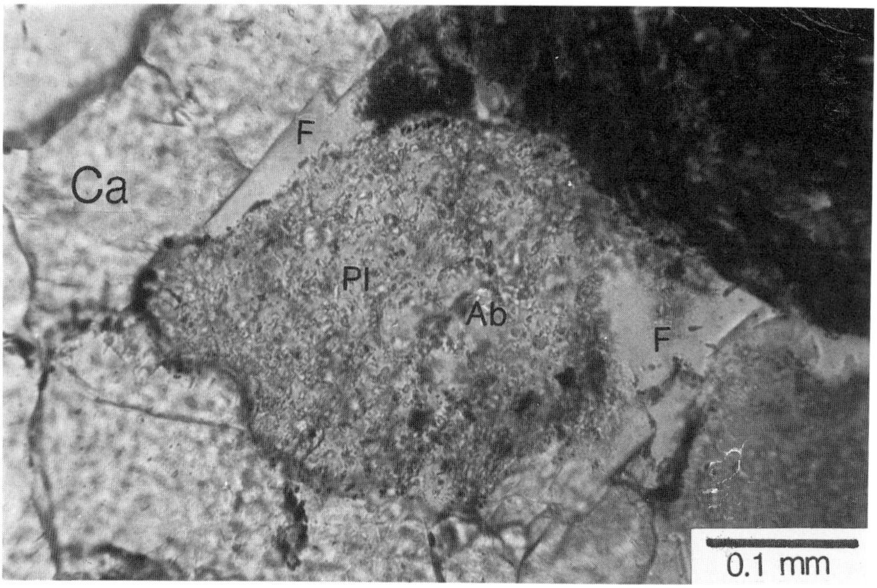

Fig. 5-12. Photomicrograph. Enlarged view of right-hand side of Fig. 5-10 showing feldspar overgrowth (*F*) on the plagioclase (*Pl*) core which is altered and albitized (*Ab*). *Ca*: calcite.

tion, it occurs as a fibrous mineral. In natrolite, sodium is almost the only cation and the Si/Al ratio ranges from 1.47 to 1.48 (Lee, 1984).

Mineral paragenesis

Diagenetic modifications change with depth from moderate to strong, downhole at the DSDP sites in the back-arc basins. Most of the observed changes show a depth-dependency downhole. A mineral paragenetic sequence is established based mainly on observations at Sites 445, 446 and 286, where most diagenetic phenomena were observed (Fig. 5-13).

The first authigenic mineral phases observed downhole are smectite and chlorite rim cements, followed by precipitation of early calcite rim and pore-fill cement. This rim cementation was followed by precipitation of cristobalite, clinoptilolite and heulandite cements filling interstices between smectite, chlorite and early calcite cement. Some zeolites appear to be a direct alteration product of volcanic glass. Most of the clinoptilolite or heulandite, however, show well-developed crystal outlines and were precipitated from solution onto smectite and chlorite rim cements, or on detrital grain surfaces, reducing porosity significantly. Locally, pores are filled completely by zeolites, in which case the contacts between crystals are sutured (Fig. 5-7).

Downhole, clinoptilolites or heulandites are associated with analcite pore-fill cements. This analcite occurs in the pore center, was precipitated late, and shows well-developed euhedral crystals (Fig. 5-14). Late calcite fills any pore space remaining after crystallization of analcite.

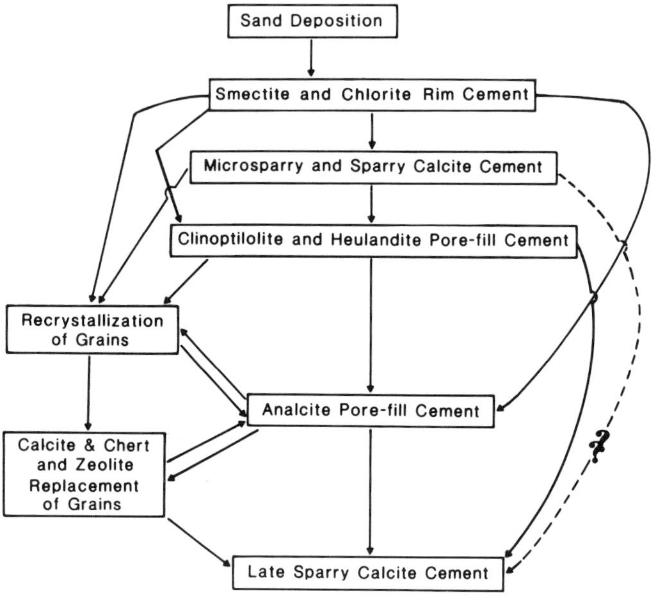

Fig. 5-13. Sandstone mineral paragenetic sequence in the western Pacific back-arc basins. (After Lee and Klein, 1986; courtesy of *Sedimentology*.)

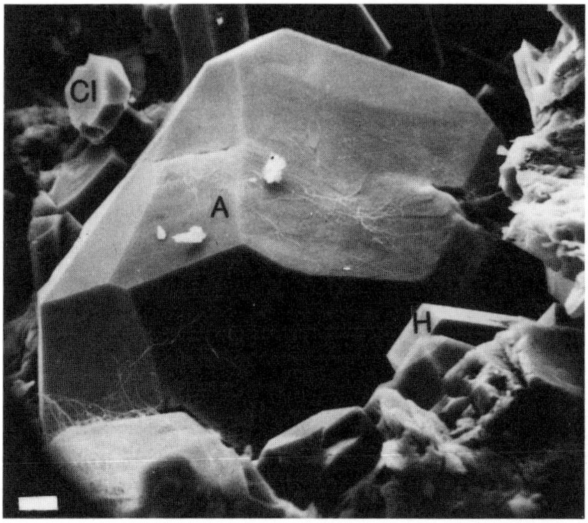

Fig. 5-14. SEM micrograph showing analcite (*A*) crystal occupying center of pore as a pore-filling cement. Dissolution features on clinoptilolite (*Cl*) and heulandite (*H*) suggest analcite developed at the expense of both clinoptilolite and heulandite. Site 445, sample 445-88-2 (24–26), subbottom depth of 827.24 m. Bar scale is 10 μm. (After Lee and Klein, 1986; courtesy of *Sedimentology*.)

Mixtures of volcanic glass and terrigeneous sediment comprise the matrices at Sites 297 and 299. Their diagenetic changes are rare, reaching no further than the earliest stages of the above paragenesis.

A similar diagenetic trend has been observed in ancient back-arc basin sandstones (Bristol Basin; Galloway, 1974, 1979), except that zeolite in the Bristol Basin is laumontite.

Most authigenic minerals are observed within interstices. Reactive volcanic glass matrix appears to be the primary source of these authigenic minerals. The paragenesis of authigenic minerals derived from a volcanic glass precursor is shown in Fig. 5-15. The paragenetic reactions in Fig. 5-15 indicate introduction of some cations into the diagenetic system which are discussed in the next section.

Geochemical aspects

Geochemical study on the authigenic minerals may help to get information about the diagenetic conditions under which these minerals formed. Chemical composition of zeolites and stable isotope composition of carbon and oxygen for zeolites and late sparry calcite pore-fill cements were analyzed using samples taken from Site 445, where the most extensive diagenetic changes were observed.

Isotope composition

The isotopic composition of late sparry calcite and zeolite cements is shown in Tables 5-4 and 5-5 and in Fig. 5-16. Late calcite cements contain $\delta^{18}O$-values in the

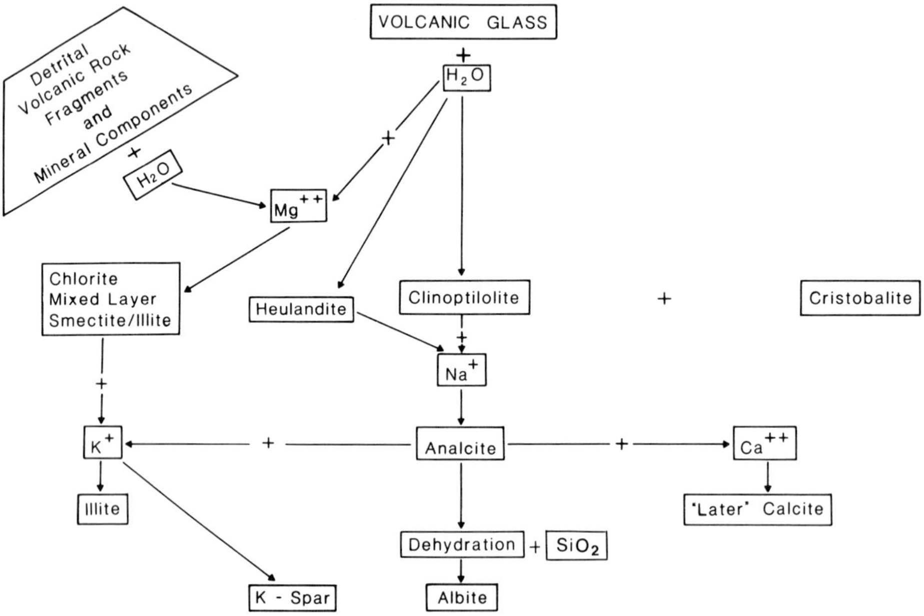

Fig. 5-15. Authigenic mineral paragenesis derived from a volcanic glass precursor. (After Lee, 1988; courtesy of *Chemical Geology*.)

range +25.1 to +27.4‰, which correspond to the isotopic temperature of crystallization ranging from 31° to 44°C (Lee, 1988). These isotopic temperatures are slightly lower than the presumed bottomhole temperatures. This suggests that either the $\delta^{18}O$ of the in-situ pore waters at Site 445 is more positive than assumed, or Site 445 calcite cements seem to have formed in the past when the temperatures were lower than they are today. In either case, the isotopic evidence suggests that calcite cements were formed in the late stage of diagenesis at this site. Table 5-5 shows the oxygen isotopic composition of zeolite pore-fill cements at Site 445. Clinoptilolite and heulandite show oxygen isotope compositions in the range of +30.1 to +22.7‰ and a tendency of decreasing $\delta^{18}O$ with depth. Analcites show much more depletion in ^{18}O than clinoptilolite and heulandite. The overall sequence of oxygen isotopic composition of zeolite indicates progressive depletion of ^{18}O with depth, suggesting that formation of analcite occurred at higher temperatures in an open sediment – water system. The estimated isotopic temperature for analcite formation ranges from 75° to 100°C. The present geothermal gradient cannot account for the oxygen shift in analcites, unless there is a corresponding decrease with depth in the $\delta^{18}O$ of pore water (Fig. 5-16), which is less likely.

These isotopic studies on zeolite and late calcite cements provide important information about diagenetic conditions for authigenic mineral formation. Both zeolite and calcite cements are observed at similar depth horizons. The calculated isotopic temperatures for these minerals, however, are quite different from one another, suggesting that they formed in different thermal regimes during diagenesis. During times of zeolite formation, sandstone experienced more elevated temperatures than those during late calcite formation.

Chemical composition of zeolites

Chemical composition of zeolites at Sites 445 and 446 has been used to infer the chemistry of the pore-water system. Chemical analyses for some representative

TABLE 5-4

Oxygen and carbon isotope data from late calcite pore-fill cements at DSDP Site 445 (after Lee, 1987)

Sample	Subbottom depth (m)	$\delta^{13}C$ (‰ PDB)	$\delta^{18}O$ (‰ SMOW)	Isotopic temperature* ($\delta^{18}O$ −0.3) (°C)	$\delta^{18}O$ of water in equilibrium with solid at 2°C (‰)
445-78-5(9 – 13)D	736.57	+1.04	+27.4	31	−6.5
445-78-5(9 – 13)H	736.57	+0.44	+25.5	42	−8.4
445-82-4(5 – 9)D	771.57	−1.12	+27.4	31	−6.5
445-88-2(27 – 31)	827.28	−1.90	+25.1	44	−8.8
445-89-1(135 – 141)	836.31	−1.41	+26.1	38	−7.8
445-91-2(64 – 70)	846.09	−0.51	+26.6	36	−7.3
445-90-2(8 – 14)	856.14	−0.10	+26.3	37	−7.6

D: drilling; H: hand-picked.
* Friedman and O'Neil (1977).

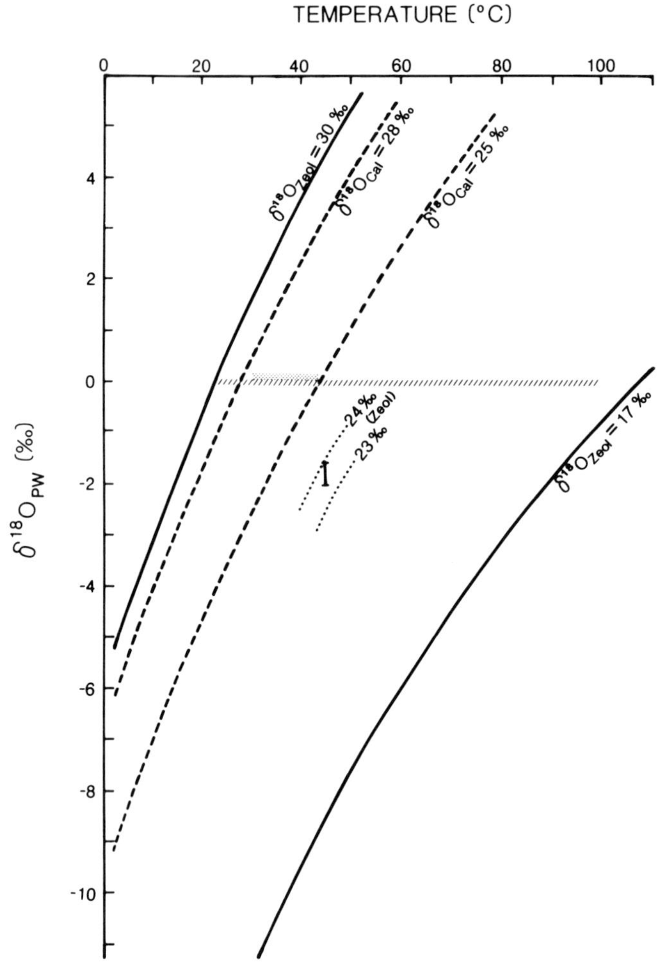

Fig. 5-16. Diagram showing relationship between temperature, $\delta^{18}O$ of pore water, and $\delta^{18}O$ of calcite (dashed curve) and zeolite cements (solid curve) precipitated from pore water at DSDP Site 445. Isotopic temperatures are corrected for Tertiary ocean water. Vertical bar represents range of hypothetical $\delta^{18}O$ value of analcite at a subbottom depth of 850 m. Stippled bar (calcite) and slanted bar (zeolite) represent the calculated isotopic temperature range. (After Lee, 1987; courtesy of *Chemical Geology*.)

clinoptilolite and heulandite are presented in Table 5-6. The average chemical composition of clinoptilolite is $Na_{3.1}\ K_{1.5}\ Ca_{0.8}\ Mg_{0.3}\ Al_{6.8}\ Si_{29.2}\ O_{72} \cdot 21H_2O$ and of heulandite is $Na_{3.1}\ K_{1.2}\ Ca_{1.3}\ Mg_{0.5}\ Al_{7.9}\ Si_{28.1}\ O_{72} \cdot 23H_2O$. Clinoptilolites are predominantly sodic compared to other deep-sea clinoptilolites (Fig. 5-17), which contain potassium as a dominant cation (Boles and Wise, 1978; Stonecipher, 1978). Heulandites also contain more Na than Ca-Mg and K. This difference is accounted for by the pore-water chemistry of the system. If the sediments are relatively impermeable, the system becomes isochemical and preserves K derived from the

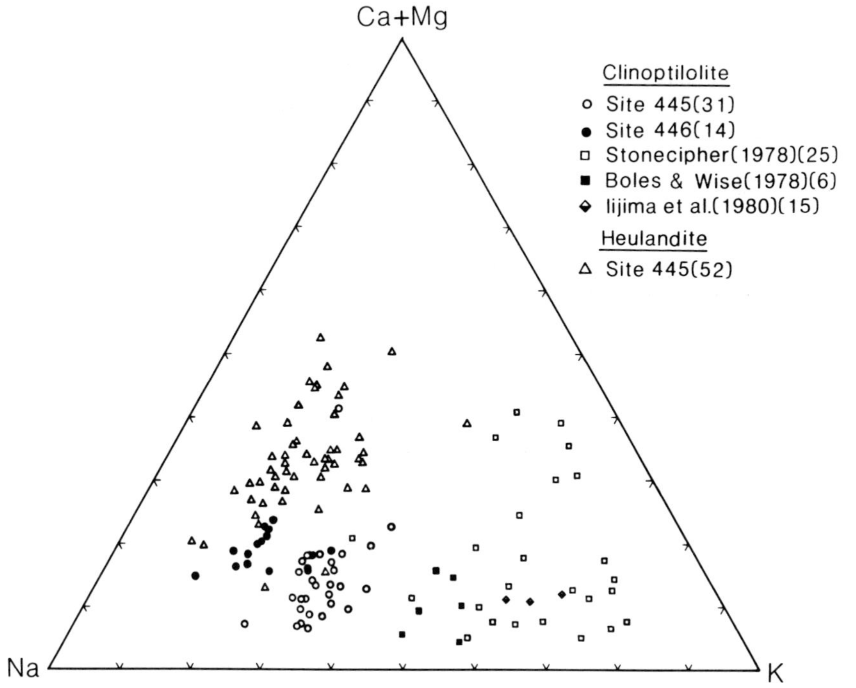

Fig. 5-17. Ternary (Ca + Mg) – Na – K diagram in atomic percent of deep-sea clinoptilolites and heulandites. Numbers in parentheses represent the number of analyses. (After Lee, 1988; courtesy of *Chemical Geology*.)

TABLE 5-5

Oxygen isotope data from zeolite pore-fill cements at DSDP Site 445 (after Lee, 1987)

Sample	Subbottom depth (m)	Mineral	$\delta^{18}O$ (‰ SMOW)	Isotopic temperature* ($\delta^{18}O$ − 0.3) (°C)	$\delta^{18}O$ of water in equilibrium with solid at 2°C (‰)
445-69-2(73 – 78)	647.24	clinoptilolite	+ 30.1	23	− 4.9
445-78-5(9 – 13)	736.57	heulandite	+ 26.7	39	− 8.3
445-88-2(27 – 31)	827.28	heulandite/ clinoptilolite	+ 22.7	62	− 12.3
445-90-2(8 – 10)	846.09	analcite	+ 20.7	77	− 14.4
445-90-2(19 – 21)	846.20	analcite	+ 17.8	100	− 17.3

* Feldspar – water fractionation from O'Neil and Taylor (1967).

dissolution of volcanic material (Kastner and Siever, 1979). In a relatively open pore-water system or in the presence of brines, the relative proportion of K decreases with the increase of other cations. Most host sediments of other deep-sea clinoptilolites are dominantly clays and muds and rare siliceous oozes (Boles and Wise, 1978; Stonecipher, 1978). The dominance of K in clinoptilolite composition may be due to the presence of clay, whereas clinoptilolite and heulandite in sandstones at Sites 445 and 446 may have formed in a relatively more permeable open system, providing Na to the pore waters by advection (Abbott et al., 1983).

Chemical analyses of three representative analcites are presented in Table 5-7. The average chemical composition of analcite is $Na_{15}\,Al_{15}\,Si_{33}\,O_{96}\cdot 16H_2O$. Based on mineral paragenetic sequence (Fig. 5-15), analcite is interpreted to have been formed from clinoptilolite and heulandite mainly by the following reactions:

TABLE 5-6

Chemical analyses and unit cell contents of five representative clinoptilolites and heulandites at Sites 445 and 446 (after Lee, 1988)

	Clinoptilolite			Heulandite	
No.:	445-69-2	445-88-2	446-40-1	445-78-5	445-88-2
Core (cm):	(70 – 72)	(24 – 26)	(81 – 86)	(0 – 4)	(24 – 26)
SiO_2	66.33	68.58	65.71	64.41	65.71
Al_2O_3	13.15	13.31	12.93	15.31	14.99
Na_2O	3.80	2.77	4.23	5.14	2.80
K_2O	2.82	3.34	1.65	1.40	2.80
CaO	1.12	1.48	2.03	1.66	2.40
MgO	0.51	0.50	0.14	0.74	0.72
Fe_2O_3*			0.04	0.18	0.10
H_2O**	12.27	10.02	13.27	11.66	10.48
Total	100	100	100	100	100
Number of cations on the basis of 72 oxygens:					
Si	29.33	29.43	29.24	28.13	28.47
Al	6.83	6.73	6.77	7.88	7.65
Na	3.25	2.30	3.64	4.35	2.36
K	1.59	1.83	0.94	0.79	1.54
Ca	0.53	0.68	0.97	0.78	1.11
Mg	0.34	0.32	0.09	0.48	0.47
Fe			0.01	0.06	0.04
Si/Al	4.28	4.37	4.32	3.57	3.72
Si + Al	36.06	36.16	36.01	36.01	36.12

* Fe_2O_3 is total iron.

** By difference.

clinoptilolite ⟶ analcite

$$Na_{3.1}\,K_{1.5}\,Ca_{0.8}\,Mg_{0.3}\,Al_{6.8}\,Si_{29.2}\,O_{72}\cdot 21\,H_2O + 3.65\,Na^+ \longrightarrow$$
$$0.45\,Na_{15}\,Al_{15}\,Si_{33}\,O_{96}\cdot 16\,H_2O + 14.35\,SiO_2 + 1.5\,K^+ +$$
$$0.8\,Ca^{2+} + 0.3\,Mg^{2+} + 13.9\,H_2O \qquad (5\text{-}1)$$

heulandite ⟶ analcite

$$Na_{3.1}\,K_{1.2}\,Ca_{1.3}\,Mg_{0.5}\,Al_{7.9}\,Si_{28.1}\,O_{92}\cdot 23\,H_2O + 4.85\,Na^+ \longrightarrow$$
$$0.55\,Na_{15}\,Al_{15}\,Si_{33}\,O_{96}\cdot 16\,H_2O + 9.85\,SiO_2 + 1.2\,K^+ +$$
$$1.3\,Ca^{2+} + 0.5\,Mg^{2+} + 14.4\,H_2O \qquad (5\text{-}2)$$

The above reactions require a relatively large activity of sodium, suggesting a continuous supply of sodium from the interstitial water. Continuous sodium supply

TABLE 5-7

Chemical analyses and unit cell contents of three representative analcites at Sites 445 and 446 (after Lee, 1988)

No.: Core (cm):	445-90-2 (11 – 14)	445-90-2 (22 – 26)	446-29-1 (13 – 15)
SiO_2	56.47	55.32	52.22
Al_2O_3	21.40	20.67	19.89
Na_2O	12.70	12.22	11.80
K_2O		0.04	0.04
CaO			0.01
MgO			0.04
Fe_2O_3*		0.15	0.01
H_2O**	9.43	11.60	15.97
Total	100	100	100
Number of cations on the basis of 96 oxygens:			
Si	33.30	33.36	33.18
Al	14.79	14.69	14.87
Na	14.43	14.28	14.51
K		0.04	0.03
Ca			
Mg			0.04
Fe		0.06	0.02
Si/Al	2.25	2.27	2.23
Si + Al	48.09	48.05	48.05

* Fe_2O_3 is total iron.
** By difference.

also explains that both clinoptilolite and heulandite are dominated by sodium over potassium.

The chemical composition and oxygen isotope measurements of zeolites suggest hydrothermal circulation of seawater by advection, which provided thermal conditions during diagenesis to favor the reaction of clinoptilolite – heulandite to

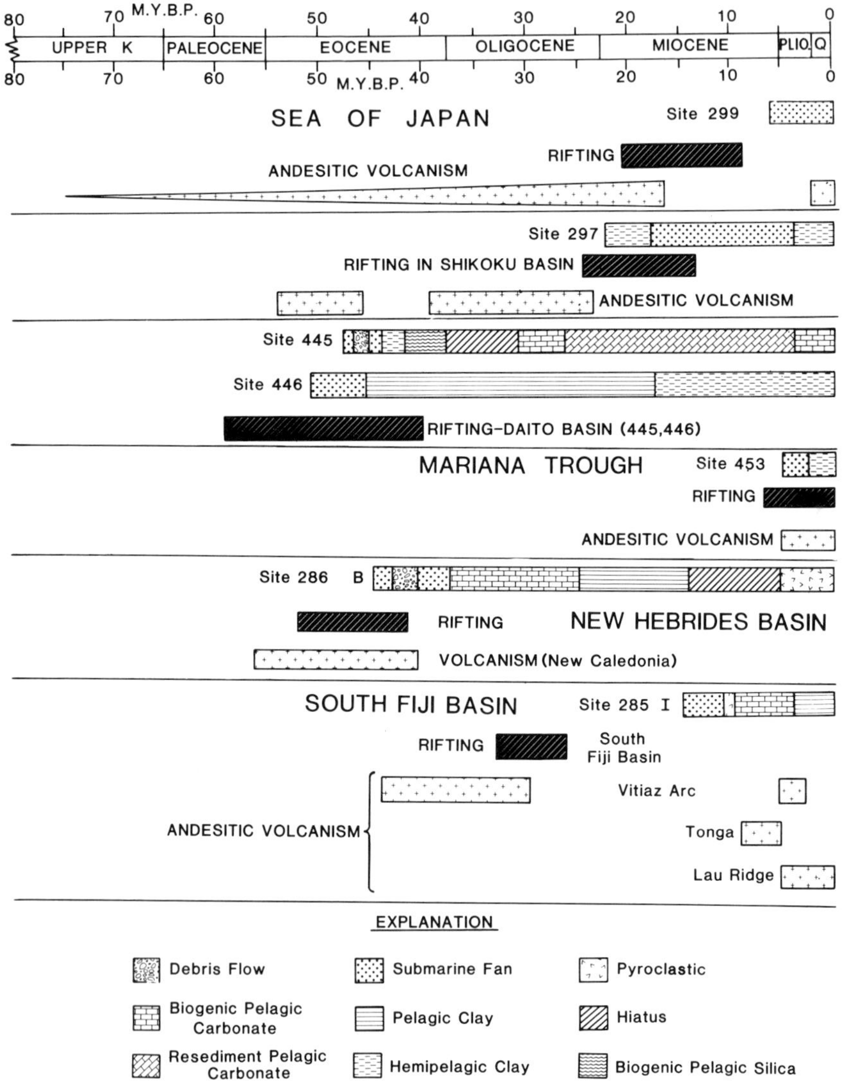

Fig. 5-18. Correlation diagram showing timing of sandstone deposition (as submarine fans and debris flows), other sediment deposition, rifting and andesitic arc volcanism in Pacific back-arc basins considered in this chapter. (After Weissel, 1981; Klein, 1985; Klein and Lee, 1984.)

analcite. This hydrothermal circulation is also implied by the presence of a large number of hiatuses (Klein et al., 1980; Abbott et al., 1981).

Factors for sandstone diagenesis in Pacific back-arc basins

Because major differences in diagenetic changes in sandstones were observed between different sites and different basins, the relevant depositional and geodynamic history of each basin must be considered not only to interpret these diagenetic differences but also to determine the processes controlling diagenesis in an active tectonic domain. A correlation diagram, comparing timing of submarine fan and debris-flow sandstone deposition, basin rifting, and andesitic arc volcanism in each basin (Fig. 5-18), shows the deposition of sandstones to differ with respect to timing of rifting and andesitic volcanic events. Sandstones deposited during active rifting are more likely to be influenced by coeval rates of high heat-flow, but other variables, such as original composition, sandstone age (Fig. 5-19), and burial history (Fig. 5-20), should also be considered.

Original sandstone composition

Original sandstone composition is known to play an important role in diagenesis (Carrigy and Mellon, 1964; Blatt, 1979; Hayes, 1979; Vavra, 1983), because dif-

1. RIFTING HISTORY

Rifting Stage	Early	Late	Post
Site	445,446,453	286,297	285,299

2. ANDESTIC ARC VOLCANISM

Volcanic Stage	Early	Late	Post	
Site	453	286,299	285,297	445(?),446(?)

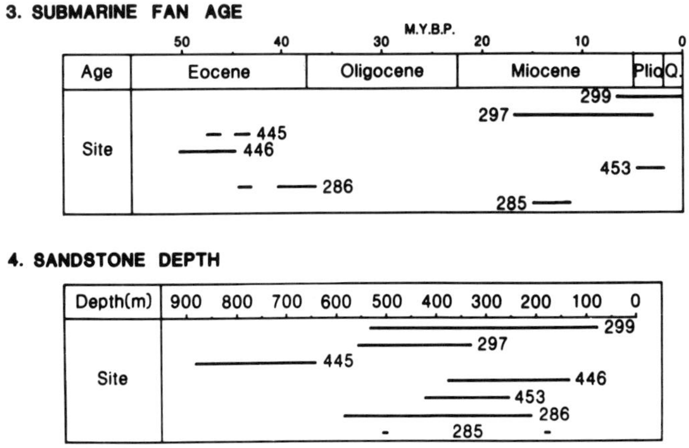

Fig. 5-19. Comparison of sandstone age, depth, rifting, and arc volcanism in back-arc basins, western Pacific. (After Lee and Klein, 1986; courtesy of *Sedimentology*.)

ferent mineral components react differently to post-depositional alteration. Considering the negligible amount of authigenic components, except at Site 445 where it reaches up to 4%, the observed sandstone composition may be used for interpretation.

Sandstones deposited during active basin rifting (Sites 453, 286, 445, 446) contain immature volcanic rock fragments and, therefore, more diagenetic alteration is to be expected in these sandstones. Sandstones at Sites 297 and 299, which were deposited after rifting ceased, contain mineralogically more stable components (Fig. 5-2) and would be, diagenetically, less altered. The greater degree of diagenesis at Sites 445 and 446, the moderate degree of diagenesis at Site 286, and the minimal diagenetic changes at Sites 297 and 299 confirm the interpretation that original composition is a significant variable in the back-arc basin sandstone diagenesis.

Hydrolysis of volcanic glass matrix also contributes to diagenesis by changing the composition of pore solutions (Surdam and Boles, 1979). Sandstones at Sites 445, 446, 286, and 453 contain a considerable volume of volcanic glass matrix (up to 61%), whereas the sandstone matrix at Sites 297 and 299 consists of a mixture of such glass and argillaceous sediment. Alteration of matrix was observed to be greater, therefore, at Sites 445, 446 and 286. At Site 453, few diagenetic changes

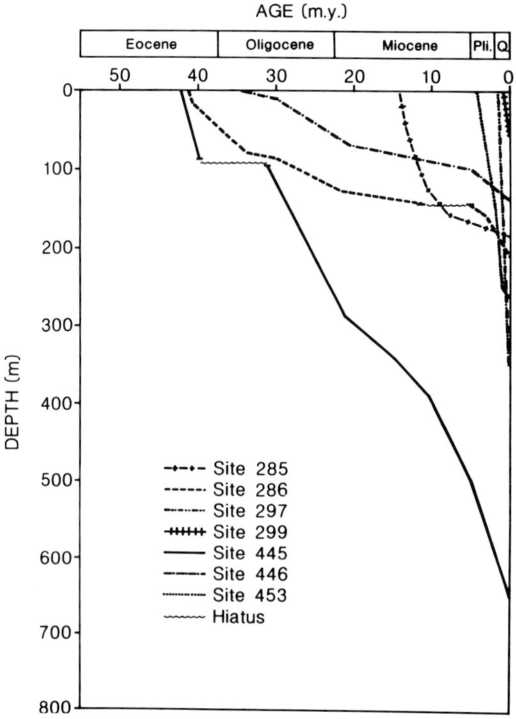

Fig. 5-20. Burial history curves of sandstones in back-arc basins, western Pacific Ocean. (After Lee and Klein, 1986; courtesy of *Sedimentology*.)

were observed because the main component is silicic volcanic glass which reacts more slowly with seawater than mafic glass (Hawkins, 1981).

Rifting history and heat flow

Rifting processes form back-arc basins by seafloor spreading processes similar to those of mid-ocean ridges (Weissel and Watts, 1975; Watts et al., 1977; Weissel, 1981). Back-arc basins are also characterized by high heat-flow (Karig, 1971; Watanabe et al., 1977), which appears to depend on age (Karig, 1970, 1971).

At present, Site 453 occurs in an actively-spreading basin, Sites 297 and 299 occur in mature basins with high heat-flow, and Sites 445, 446, 286 and 285 are located within inactive basins with normal heat-flow (Toksoz and Bird, 1977). When these basins rifted actively, each site was characterized by high heat-flow. If sand was deposited in the basin during rifting, thermal effects from newly-formed basement (Sclater et al., 1980) are to be expected in the overlying sediment. Elevated temperatures during rifting increase reaction rates between sediment and pore water leading to thermal diagenesis (Lee, 1987).

Sandstone deposition occurred during the early stages of basin rifting at Sites 445, 446 and 453, and coeval with late stages of rifting at Sites 286 and 297 (Figs. 5-7 and 5-8). At Sites 299 and 285, sandstone was deposited after rifting terminated. Thermal effects, therefore, are expected to be greater in sandstones at Sites 445, 446 and 453, whereas intermediate thermal diagenesis should be observed at Sites 299 and 285. The observed sandstone diagenesis at Sites 445, 446, 286, and 299 fits this prediction. Oxygen isotope study on zeolite cements at Site 445 also indicates thermal influence from the basement. Sites 453 and 297 are exceptions to this prediction. The original sandstone composition limited diagenesis because of a larger quartz content at Site 297. At Site 453 there is a hydrothermal circulation in the basement (Gieskes, 1981), causing relatively low heat-flow, which may account for the relatively little diagenetic changes.

Andesitic arc volcanism

Most sandstones in back-arc basins are derived from andesitic island arcs (Klein and Lee, 1984; Klein, 1985). These sandstones, however, occur only when andesitic volcanism is associated with, or follows, basin rifting and if sediment yield from a large volcanic drainage area was caused by a large rate of tectonic uplift and associated increased denudation (Klein, 1984, 1985). The influence of andesitic volcanism on diagenesis is in controlling original sandstone composition, and timing of sandy sediment yield into the basins.

Sandstone age and burial depth

Both sandstone age and depth represent the burial factor in sandstone diagenesis. Sandstone age indicates the total time-span during which diagenesis occurred since deposition, whereas burial depth is an independent indicator of the temperature and pressure to which these sediments were subjected. Sandstones at Sites 286, 445, and 446 are both older and experienced more diagenetic alteration, compared with sandstones at the other sites. Burial depths differ between sites, however, suggesting that the observed diagenesis may be caused also by different geothermal gradients.

The most common diagenetic products from volcanic sediments are smectite and zeolites, and their occurrence is partly a function of age (Denoyer de Segonzac, 1970; Boles and Wise, 1978; Kastner and Stonecipher, 1978). Authigenic smectite in marine sediments forms, commonly, during early diagenesis. In deep-sea sediments, phillipsite is most abundant in Miocene or younger sediments, whereas clinoptilolite is most abundant in Eocene or older sediments (Boles and Wise, 1978; Kastner and Stonecipher, 1978), suggesting reaction rates of 10^6 y or more. Moreover, the frequency of occurrence of analcite increases with age (Kastner and Stonecipher, 1978). It should be expected, and it was observed, that sandstones containing clinoptilolite, heulandite, and analcite at Sites 445 and 446, and erionite at Site 286, are older than sandstones at the other sites in which zeolites are absent.

Construction of burial history curves (Fig. 5-20) showed that although Sites 445 and 446 occur in the same basin, their burial histories differ, because Site 445 occurs on the edge of the basin, whereas Site 446 occurs in the center. Burial proceeded at slow to moderate rates at Sites 286, 445, and 446 (Fig. 5-20), whereas burial was rapid at Sites 297, 299, 453, and 285. Sandstones at Sites 445, 446, and 286 were provided more time, therefore, to react with hydrothermally-driven pore waters

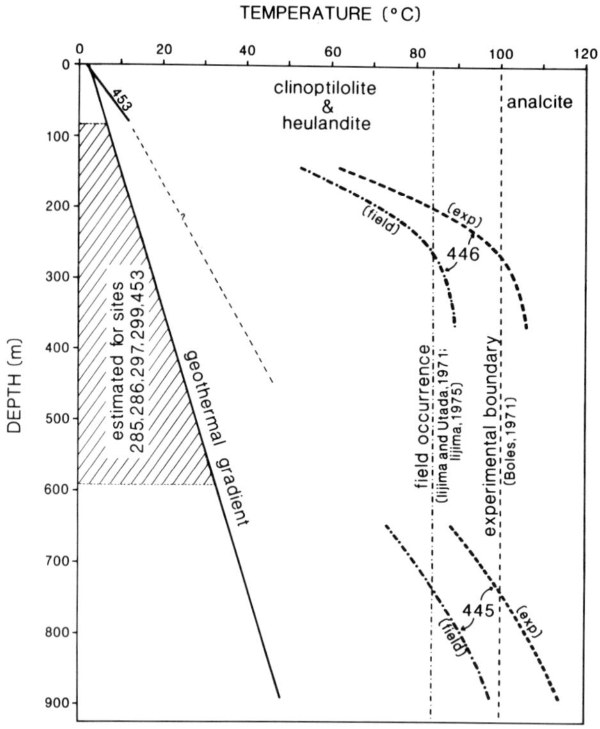

Fig. 5-21. Diagram showing the probable geothermal gradient for each DSDP site, back-arc basin, western Pacific. Curves for Sites 445 and 446 are based on mineral transformation from clinoptilolite and heulandite to analcite according to both field occurrence and experimental observations. (After Lee and Klein, 1986; courtesy of *Sedimentology*.)

because compaction was minimal. At the other sites, rapid burial caused excess compaction, sealed-off hydrothermal circulation, and blanked thermal influences from the basement. Burial rate, therefore, controlled directly the potential thermal diagenesis in these basins.

Paleogeothermal gradient

Based on the diagenetic changes, oxygen isotope composition, and chemical com-

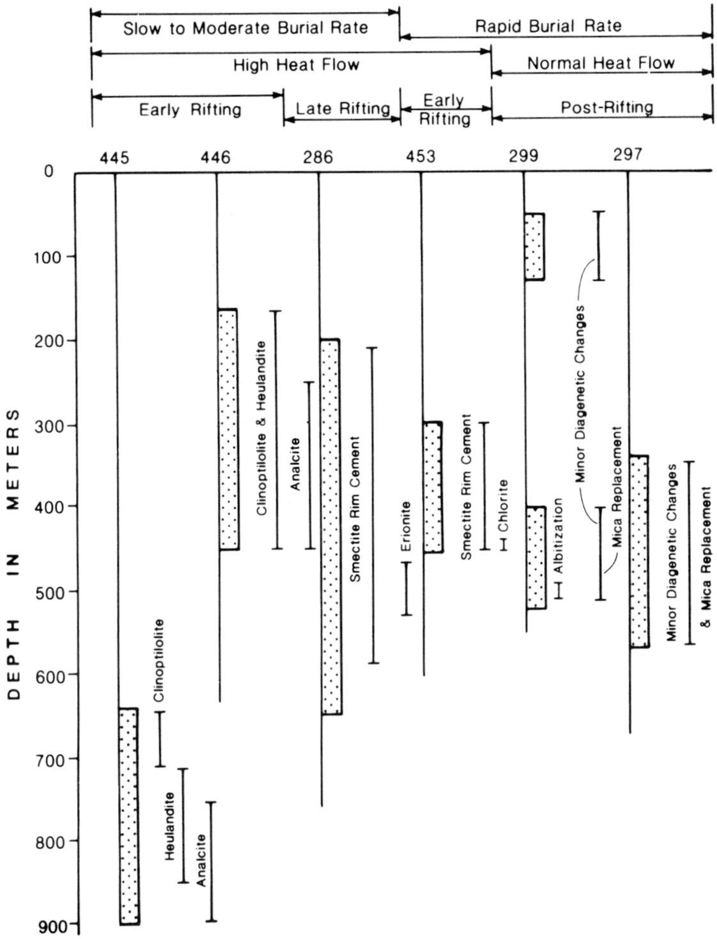

Fig. 5-22. Comparison of major diagenetic changes in each back-arc basin site containing sandstones with different history of burial rate, heat-flow, and rifting history. Vertical line shows total depth of hole. Diagenesis at Sites 445 and 446 was by zeolite mineralization in response to a combination of high heat-flow during deposition and slow to moderate burial rates. Erionite diagenesis and smectite rim cementation at Site 286 was associated with later stages of rifting and was caused by moderate heat-flow and moderate burial rates. Rapid burial rates at Sites 453, 299, and 297 appear to have prevented development of major diagenetic changes, although albitization of basal part of Site 299 indicates initial high heat-flow, which was blanketed later by rapid burial. (After Lee and Klein, 1986; courtesy of *Sedimentology*.)

position of zeolites, paleogeothermal gradients may be established (Fig. 5-21). Crystallization temperatures of zeolites observed at Sites 445 and 446 appear to require a thermal regime in excess of normal geothermal gradients. Elevated temperatures caused by pore-water advection from the basement may account for anomalous zeolite temperatures of crystallization at the shallow subbottom depth. Temperatures of crystallization of later calcite cements may have been influenced by a lower temperature regime probably because of reduction in heat flow. An arbitrary geothermal gradient of 5°C/100 m was calculated using isotope analysis of later calcite cement as a constraint. Sandstones at Sites 285, 286, 297, 299, and 453 would be influenced by a smaller geothermal gradient suggesting less influence from excess temperature. Although Site 286 may be an exception, its deviation from expectation could not be ascertained. Temperature trends for Sites 445 and 446 suggest that these sites experienced elevated temperatures well above average geothermal gradients during basin evolution, sandstone deposition and burial.

In summary, sandstone diagenetic reactions in the back-arc basins depend primarily on changes in heat flow during basinal rifting. Volcaniclastic sandstone deposition occurred during both early stages of rifting when heat flow was high (Sites 445, 446, and 453), later stages of rifting when heat flow was less (Site 286) and well after rifting ceased when heat flow was normal (Sites 285, 297, and 299). The higher degree of diagenesis at Sites 445 and 446 is explained primarily from deposition during early rifting, and by heat-flow modification, associated thermally-driven fluid circulation, older age and deep burial, and a slow rate of

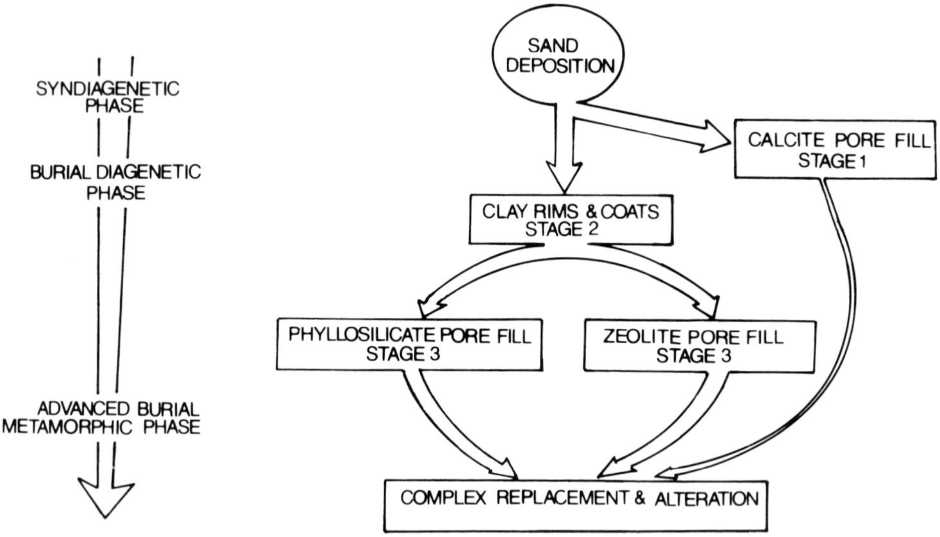

Fig. 5-23. Diagram showing three stages of chemical diagenesis. Formation of early calcite pore-filling cement (stage 1) effectively insulates the sandstone from further diagenetic alteration or compaction until deep burial. Where the fabric is open and cementation has not significantly filled in available pore-space, clay rim and coats (stage 2) and zeolite pore-fill (stage 3) cementation are developed. Physical diagenesis, in the form of grain deformation and mechanical compaction, is especially pronounced in early to intermediate stages of burial. (Simplified after Galloway, 1974.)

burial (Figs. 5-20, 5-21, and 5-22). The moderate diagenetic influences observed at Site 286 are explained in part by a moderate heat-flow regime and slow burial rates (Figs. 5-20 and 5-22). The minimal diagenetic reactions observed at Sites 453, 297, 299, and 285 appear to be influenced by a combination of low heat-flow and rapid burial rates (Figs. 5-20 and 5-22). The mineralogically more mature composition of sandstones at Sites 297 and 299 may account also for the minimal diagenesis observed there.

DIAGENESIS OF FORE-ARC BASIN SANDSTONES

The diagenesis of ancient fore-arc basin sandstones (Chehalis and Grays Harbor basins, northeast Pacific) has been discussed by Galloway (1974, 1979). Most sandstones are feldspathic litharenites rich in volcaniclastic materials, and the depositional environments for these sandstones range from continental to marine shelf setting, although the distinction is not clear in his paper. Compared to sandstones in modern fore-arc and back-arc basins (Table 5-1), sandstones from these ancient fore-arc basins contain more quartz ranging from 0 to 42% with an average of 26% by volume. Plagioclase is dominant also. Plagioclase and volcanic rock fragments comprise the most abundant detrital grains.

Based on the authigenic cementing agents, the sequence of diagenetic changes is divided into three discrete stages (Fig. 5-23): (1) calcite pore-filling cement formation, (2) formation of clay coats and rims, and (3) filling of the remaining open pore spaces either by an authigenic zeolite (typically laumontite) or a well-crystallized phyllosilicate (usually chlorite or montmorillonite). The principal process seems to be burial diagenesis because of great sediment thickness up to 4600 m; therefore, temperature is favored as the primary control of burial diagenetic reactions.

COMPARISON BETWEEN BACK-ARC AND FORE-ARC BASIN SANDSTONE DIAGENESIS

Although both back-arc and fore-arc basins receive sandstones of similar composition from the volcanic arc (Table 5-1 and Fig. 5-13), both may show compositional differences depending on sources, in other words, proximity to the volcanic arc. In fore-arc basins, Dickinson et al. (1979) discussed the sediment sources, such as mélange, volcanic eruption nearby, and mixed provenance. Most extensive diagenetic changes are expected in sandstones derived from volcanic eruptions and mixed provenance because of abundant presence of volcanic materials. The same is true in the back-arc basins. As discussed in the section of original sandstone composition of the back-arc basin sandstone diagenesis, sandstone composition varies depending on the sediment sources. One margin of the back-arc basin is bounded by the volcanic arc and the other by the cratonic continent or a mature island arc. In the back-arc basin, volcanic materials are abundant closer to the volcanic arc and mature sediments occur more towards the landward margin. Sibley and Pentony (1978) showed differences in feldspar types in sandstones in two DSDP sites in the Sea of Japan. Deposition of unstable volcanic materials in the back-arc basins also depends on the rifting and arc volcanism history in the basin. Simple comparison of diagenesis of sandstones between back-arc and fore-arc basins, therefore, cannot be made. It seems clear that in both basins more diagenetic changes may occur closer to the volcanic source.

The main difference in sandstone diagenesis between fore-arc and back-arc basins, however, lies in the thermal regime. In the back-arc basins more heat-flow is measured than in the fore-arc basins (Watanabe et al., 1977). So a relatively high geothermal gradient is expected in the back-arc basins (Galloway, 1974). The large heat-flow in the back-arc basin is derived from the newly-formed basement followed by basin rifting. Such a heat source is absent in the fore-arc basins. Under the same conditions, therefore, the diagenetic changes of sandstone in the back-arc basins may occur at shallower depths of burial compared to those in the fore-arc basins (e.g., Galloway, 1974).

CONCLUSIONS

Volcaniclastic sandstones in deep-sea environments occur in both back-arc and fore-arc basins of the active plate margins. Their main source is the volcanic arc, and sandstones are rich in basaltic to andesitic volcanic rock fragments and glass matrix. These volcanic materials are highly susceptible to alteration under burial conditions.

In back-arc basins, sandstones which were deposited coevally with early stages of basin rifting at a slow rate of burial show greater diagenetic alteration (including pore-space reduction, rim cementation by clay minerals and calcite, pore-fill cementation by zeolites, such as clinoptilolite, heulandite and analcite, and later sparry calcite) than those deposited during later and post-rifting stages and at a more rapid rate of burial. This observation implies that the large heat-flow and fluid circulation associated with rifting strongly affected the sandstone diagenetic history. Similar diagenetic changes may occur in sandstones in fore-arc basins.

Factors controlling sandstone diagenesis in the back-arc basins are sandstone age, sandstone composition, heat flow from the basement, and burial rate. These factors also control diagenetic reactions of fore-arc basin sandstones, although much less influence of heat-flow from the basement is expected. Thus, the main difference in diagenesis of sandstones between fore-arc and back-arc basins lies in the thermal regime during burial. The distribution of thermal regimes among back-arc basins, however, is not homogeneous. Only contemporary deposition of sandstones during early stages of rifting in back-arc basins favors extensive diagenetic alteration because of associated high heat-flow.

Exploration activity for hydrocarbon is expanding into deep-sea areas. In active margins, reservoir heterogeneity could exist because of various geodynamic controls on sandstone diagenesis, which should be carefully studied.

REFERENCES

Abbott, D.H., Menke, W., Hobart, M. and Anderson, R., 1981. Evidence for excess pore pressures in southwest Indian Ocean sediments. *J. Geophys. Res.,* 82: 1813 – 1827.
Abbott, D.H., Menke, W. and Morin, R., 1983. Constraints upon water advection in sediments of the Mariana Trough. *J. Geophys. Res.,* 88: 1075 – 1093.
Anderson, R.N., 1980. Update of heat flow in the east and southeast Asian seas. In: D.E. Hayes (Editor), *Geologic Evolution of Southeast Asian Seas and Islands. Am. Geophys. Union, Geophys. Monogr.,* 23: 318 – 326.
Andrews, J.E., Packam, G.H. et al., 1975. *Initial Reports of the Deep Sea Drilling Project,* Vol. 30. U.S. Government Printing Office, Washington, D.C., 753 pp.
Bjorlykke, K., 1983. Diagenetic reactions in sandstones. In: H. Packham and B.W. Sellwood (Editors), *Sediment Diagenesis.* NATO A.S.I. Ser., Reidel, Dordrecht, pp. 169 – 213.
Blatt, H., 1979. Diagenetic processes in sandstones. In: P.A. Scholle and P.R. Schluger (Editors), *Aspects of Diagenesis. Soc. Econ. Paleontol. Mineral., Spec. Publ.,* 26: 141 – 158.
Boles, J.R., 1971. Synthesis of analcime from natural heulandite and clinoptilolite. *Am. Mineral.,* 56: 1724 – 1734.
Boles, J.R., 1972. Composition, optical properties, cell dimensions and thermal stability of some heulandite group zeolites. *Am. Mineral.,* 57: 1463 – 1493.
Boles, G.R. and Wise, W.S., 1978. Nature and origin of deep sea clinoptilolite. In: L.B. Sand and F.A. Mumpton (Editors), *Natural Zeolites: Occurrence, Properties, Use.* Pergamon, New York, N.Y., pp. 235 – 243.
Brooks, D.A., Carlson, R.L., Harry, D.L., Melia, P.J., Moore, R.P., Rayhorn, J.E. and Tubb, S.G., 1984. Characteristics of back-arc regions. *Tectonophysics,* 102: 1 – 16.
Burley, S.D. and Kantorowicz, J.D., 1986. Thin section and SEM textural criteria for the recognition of cement-dissolution porosity in sandstones. *Sedimentology,* 33: 605 – 614.
Burley, S.D., Kantorowicz, J.D. and Waugh, B., 1985. Clastic diagenesis. In: P.J. Brenchley and B.P. Williams (Editors), *Sedimentology – Recent Developments and Applied Aspects,* Geol. Soc., Blackwell, London, pp. 189 – 228.
Carrigy, M.A. and Mellon, G.B., 1964. Authigenic clay mineral cements in Cretaceous and Tertiary sandstones of Alberta. *J. Sediment. Petrol.,* 34: 461 – 472.
Davies, D.K., Almon, W.R., Bonis, S.B. and Hunter, B.E., 1979. Deposition and diagenesis of Tertiary – Holocene volcaniclastics, Guatemala. In: P.A. Scholle and P.R. Schluger (Editors), *Aspects of Diagenesis. Soc. Econ. Paleontol. Mineral., Spec. Publ.,* 26: 281 – 306.
Denoyer de Segonzac, G., 1970. The transformation of clay minerals during diagenesis and low-grade metamorphism. *Sedimentology,* 15: 281 – 346.
Dickinson, W.R. and Seeley, D.R., 1979. Structure and stratigraphy of forearc regions. *Am. Assoc. Pet. Geol. Bull.,* 63: 2 – 31.
Dickinson, W.R. and Suczek, C.A., 1979. Plate tectonics and sandstone compositions. *Am. Assoc. Pet. Geol. Bull.,* 63: 2164 – 2182.
Dickinson, W.R. and Valloni, R., 1980. Plate tectonics and the provenance of sandstone in modern ocean basins. *Geology,* 8: 82 – 86.
Ernst, W.G. and Calvert, S.E., 1969. Experimental study of the crystallization of porcelanite and its bearing on the origin of some bedded cherts. *Am. J. Sci.,* 267A: 114 – 133.
Folk, R.L., 1968. *Petrology of Sedimentary Rocks.* Hemphill, Austin, Tex., 170 pp.
Friedman, I. and O'Neill, J.R., 1977. Compilation of stable isotope fractionation factors of geochemical interest. *U.S. Geol. Surv., Prof. Pap.* 440-KK: 12 pp.
Galloway, W.E., 1974. Deposition and diagenetic alteration of sandstone in northeast Pacific arc-related basins. *Geol. Soc. Am. Bull.,* 85: 379 – 390.
Galloway, W.E., 1979. Diagenetic control of reservoir quality in arc-derived sandstones: implications for petroleum exploration. In: P.A. Scholle and P.R. Schluger (Editors), *Aspects of Diagenesis. Soc. Econ. Paleontol. Mineral., Spec. Publ.,* 26: 251 – 262.
Gergen, L.D. and Ingersoll, R.V., 1986. Petrology and provenance of the Deep Sea Drilling Project sand and sandstone from the north Pacific Ocean and the Bering Sea. In: W.L. Bilodeau (Editor), *Plate Tectonics and Petrologic Studies. Sediment. Geol.,* 51: 29 – 56.
Gieskes, J.M., 1981. Deep sea drilling interstitial-water studies: implications for chemical alteration of

the ocean crust Layer I and II. In: J.E. Warme, R.G. Douglas and E.L. Winterer (Editors), *The Deep Sea Drilling Project, A Decade of Progress. Soc. Econ. Paleontol. Mineral., Spec. Publ.,* 32: 149 – 167.

Hawkins, D.B., 1981. Kinetics of glass dissolution and zeolite formation under hydrothermal conditions. *Clays Clay Miner.,* 29: 331 – 340.

Hayes, J.B., 1979. Sandstone diagenesis – the hole truth. In: P.A. Scholle and P.R. Schluger (Editors), *Aspects of Diagenesis. Soc. Econ. Paleontol. Mineral., Spec. Publ.,* 26: 127 – 140.

Heath, G.R. and Moberly, R., Jr., 1971. Cherts from the western Pacific, Leg 7, Deep Sea Drilling Project. In: E.L. Winterer, W.R. Riedel et al. (Editors), *Initial Reports of the Deep Sea Drilling Project,* Vol. 7. U.S. Government Printing Office, Washington, D.C., pp. 991 – 1007.

Hein, J.R., Schill, D.W., Barron, J.A., Jones, M.G. and Miller, J., 1978. Diagenesis of late Cenozoic diatomaceous deposits and formation of the bottom simulation reflector in the southern Bering Sea. *Sedimentology,* 25: 155 – 181.

Hilde, T.W.C., Uyeda, S. and Kroenke, L., 1977. Evolution of the western Pacific and its margin. *Tectonophysics,* 38: 145 – 165.

Hsui, A.T. and Toksoz, M.N., 1981. Back-arc spreading: trench migration, continental pull or induced convection? *Tectonophysics,* 74: 89 – 98.

Hussong, D.M., Uyeda, S. et al., 1981. *Initial Reports of the Deep Sea Drilling Project,* Vol. 60. U.S. Government Printing Office, Washington, D.C., 929 pp.

Iijima, A., 1975. Effect of pore water to clinoptilolite – analcime – albite reaction series. *Univ. Tokyo, Faculty Sci. J.,* Sect. 11, 19: 133 – 147.

Iijima, A. and Utada, M., 1971. Present-day zeolite diagenesis of the Neogene geosynclinal deposits in the Niigata oil field, Japan. In: R.F. Gould (Editor), *Molecular Sieve Zeolite. Am. Chem. Soc., Adv. Chem. Ser.,* 101: 342 – 349.

Iijima, A., Matsumoto, R. and Tada, R., 1980. Zeolite and silica diagenesis and sandstone petrography at Sites 438 and 439 off Sanriku, northwest Pacific, Leg 57, Deep Sea Drilling Project. In: R. von Huene, N. Nasu, M. Langseth, H. Okada et al. (Editors), *Initial Reports of the Deep Sea Drilling Project,* Vols. 56 and 57, Part II. U.S. Government Printing Office, Washington, D.C., pp. 1143 – 1158.

Ingersoll, R.V. and Suczek, C.A., 1979. Petrology and provenance of Neogene sand from Nicobar and Bengal fans, DSDP Sites 211 and 218. *J. Sediment. Petrol.,* 49: 1217 – 1228.

Karig, D.E., 1970. Ridges and basins of the Tonga – Kermadec Island arc system. *J. Geophys. Res.,* 75: 238 – 254.

Karig, D.E., 1971. Origin and development of marginal basins in the western Pacific. *J. Geophys. Res.,* 76: 2542 – 2561.

Karig, D.E., 1975. Basin genesis in the Philippine Sea. In: D.E. Karig, J.C. Ingle, Jr. et al. (Editors), *Initial Reports of the Deep Sea Drilling Project,* Vol. 31. U.S. Government Printing Office, Washington, D.C., pp. 875 – 880.

Kastner, M., 1981. Authigenic silicates in deep-sea sediments: formation and diagenesis. In: C. Emiliani (Editor), *The Oceanic Lithosphere, The Sea,* Vol. 7. Wiley, New York, N.Y., pp. 915 – 980.

Kastner, M. and Siever, R., 1979. Low temperature feldspar in sedimentary rocks. *Am. J. Sci.,* 279: 435 – 479.

Kastner, M. and Stonecipher, S.A., 1978. Zeolites in pelagic sediments of the Atlantic, Pacific, and Indian Oceans. In: L.B. Sand and F.A. Mumpton (Editors), *Natural Zeolites: Occurrence, Properties, Use.* Pergamon, New York, N.Y., pp. 199 – 220.

Klein, G. deV., 1975a. Sedimentary tectonics in southwestern Pacific marginal basins based on Leg 30 Deep Sea Drilling Project cores from the South Fiji, Hebrides, and Coral Sea Basins. *Geol. Soc. Am. Bull.,* 86: 1012 – 1018.

Klein, G. deV., 1975b. Depositional facies of Leg 30 Deep Sea Drilling Project cores. In: J.E. Andrews, G.H. Packham et al. (Editors), *Initial Reports of the Deep Sea Drilling Project,* Vol. 30. U.S. Government Printing Office, Washington, D.C., pp. 423 – 449.

Klein, G. deV., 1984. Relative rates of tectonic uplift as determined from episodic turbidite deposition in marine basins. *Geology,* 12: 48 – 50.

Klein, G. deV., 1985. The control of depositional depth, tectonic uplift, and volcanism on sedimentary processes in the back-arc basins of the western Pacific Ocean. *J. Geol.,* 93: 1 – 25.

Klein, G. deV. and Lee, Y.I., 1984. A preliminary assessment of geodynamic controls on depositional systems and sandstone diagenesis in back-arc basins, western Pacific Ocean. *Tectonophysics,* 102:

119 – 152.
Klein, G. deV., Kobayashi, K. et al., 1980. *Initial Reports of the Deep Sea Drilling Project,* Vol. 58. U.S. Government Printing Office, Washington, D.C., 1017 pp.
Kobayashi, K. and Nakada, M., 1978. Magnetic anomalies and tectonic evolution of the Shikoku interarc basin. *J. Phys. Earth,* 26: S391 – S402.
Lee, Y.I., 1984. Petrology and diagenesis of the medium-grained clastic sediments in the back-arc basins of the western Pacific Ocean. Ph.D. diss., University of Illinois at Urbana-Champaign, Urbana, Ill., 208 pp.
Lee, Y.I., 1987. Isotopic aspects of thermal and burial diagenesis of sandstones at DSDP Site 445, Daito Ridge, northwest Pacific Ocean. *Chem. Geol. (Isot. Geosci. Sect.),* 65: 95 – 102.
Lee, Y.I., 1988. Chemistry and origin of zeolites in sandstones at DSDP Sites 445 and 446, Daito Ridge and Basin Province, northwest Pacific. *Chem. Geol.,* 67: 261 – 273.
Lee, Y.I. and Klein, G. deV., 1986. Diagenesis of sandstones in the back-arc basins of the western Pacific Ocean. *Sedimentology,* 33: 651 – 675.
Maynard, J.B., Valloni, R. and Yu, H.-S., 1982. Composition of modern deep-sea sands from arc-related basins. In: J.K. Leggett (Editor), *Trench-Forearc Geology: Sedimentation and Tectonics on Modern and Ancient Active Plate Margins. Geol. Soc. London, Spec. Publ.,* 10: 551 – 561.
Mizutani, S., 1977. Progressive ordering of cristobalitic silica in the early stage of diagenesis. *Contrib. Mineral. Petrol.,* 61: 129 – 140.
O'Neil, J.R. and Taylor, H.P., 1967. The oxygen isotope and cation exchange chemistry of feldspars. *Am. Mineral.,* 52: 1414 – 1437.
Packer, B.M. and Ingersoll, R.V., 1986. Provenance and petrology of Deep Sea Drilling Project sands and sandstones from the Japan and Mariana forearc and backarc regions. In: W.L. Bilodeau (Editor), *Plate Tectonics and Petrologic Studies. Sediment. Geol.,* 51: 5 – 28.
Reading, H.G., 1982. Sedimentary basins and global tectonics. *Proc. Geol. Assoc.,* 93: 321 – 350.
Saigal, G.C., Morad, S., Bjorlykke, K., Egeberg, P.K. and Aagaard, P., 1988. Diagenetic albitization of detrital K-feldspar in Jurassic from offshore, Norway. I. Textures and origin. *J. Sediment. Petrol.,* 58: 1003 – 1013.
Schmidt, V. and McDonald, D.A., 1979. The role of secondary porosity in the course of sandstone diagenesis. In: P.A. Scholle and P.R. Schluger (Editors), *Aspects of Diagenesis. Soc. Econ. Paleontol. Mineral., Spec. Publ.,* 26: 175 – 208.
Sclater, J.G., Jaupart, C. and Galson, D., 1980. The heat flow through oceanic and continental crust and the heat loss of the earth. *Rev. Geophys. Space Phys.,* 18: 269 – 311.
Sibley, D.F. and Pentong, K.J., 1978. Provenance variation in turbidite sediments, Sea of Japan. *J. Sediment. Petrol.,* 48: 1241 – 1248.
Stonecipher, S.A., 1978. Chemistry of deep-sea phillipsite, clinoptilolite, and host sediment. In: L.B. Sand and F.A. Mumpton (Editors), *Natural Zeolites: Occurrence, Properties, Use.* Pergamon, New York, N.Y., pp. 221 – 234.
Surdam, R.C. and Boles, J.R., 1979. Diagenesis of volcanic sandstones. In: P.A. Scholle and P.R. Schluger (Editors), *Aspects of Diagenesis. Soc. Econ. Paleontol. Mineral., Spec. Publ.,* 26: 227 – 242.
Toksoz, M.N. and Bird, P., 1977. Formation and evolution of marginal basins and continental plateaus. In: M. Talwani and W.S. Pittman III (Editors), *Island Arcs, Deep Sea Trenches and Back-Arc Basins.* Am. Geophys. Union, Maurice Ewing Ser., 1: 379 – 393.
Valloni, R. and Maynard, J.B., 1981. Detrital mode of Recent deep-sea sands and their relation to tectonic setting: a first approximation. *Sedimentology,* 28: 75 – 83.
Valloni, R. and Mezzadri, G., 1984. Compositional suites of terrigeneous deep-sea sands of the present continental margins. *Sedimentology,* 31: 353 – 364.
Vavra, C.L., 1983. Mineral reactions and controls on zeolite facies alteration in sandstone in central Transantarctic Mountains, Antarctica. *Am. Assoc. Pet. Geol. Bull.* (Abstract), 67: 563.
Watanabe, T., Langseth, M.G. and Anderson, R.N., 1977. Heat flow in back-arc basins of the western Pacific. In: M. Talwani and W.C. Pittman III (Editors), *Island Arcs, Deep Sea Trenches and Back-Arc Basins.* Am. Geophys. Union, Maurice Ewing Ser., 1: 137 – 161.
Watts, A.B. and Weissel, J.K., 1975. Tectonic history of the Shikoku marginal basin. *Earth Planet. Sci. Lett.,* 25: 239 – 250.
Watts, A.B., Weissel, J.K. and Larson, R.L., 1977. Sea-floor spreading in marginal basins of the western Pacific. *Tectonophysics,* 37: 167 – 181.

Weissel, J.K., 1981. Magnetic lineations in marginal basins of the western Pacific. *Phil. Trans. R. Soc. London, Ser. A,* 300: 223 – 247.
Weissel, J.K. and Watts, A.B., 1975. Tectonic complexities in the South Fiji marginal basin. *Earth Planet. Sci. Lett.,* 28: 121 – 126.
Whetten, J.T. and Hawkins, J.W., 1970. Diagenetic origin of graywacke matrix minerals. *Sedimentology,* 15: 347 – 361.
Wilson, M.D. and Pittman, E.D., 1977. Authigenic clays in sandstones: recognition and influence on reservoir and paleoenvironmental analysis. *J. Sediment. Petrol.,* 47: 3 – 31.

Reprinted from: *Diagenesis, III. Developments in Sedimentology, 47.*
Edited by K.H. Wolf and G.V. Chilingarian.

Chapter 6

A VOLUME AND MASS APPROACH TO CARBONATE DIAGENESIS: THE ROLE OF COMPACTION AND CEMENTATION

WERNER RICKEN

INTRODUCTION

Traditionally, carbonate diagenesis is qualitatively described as changing petrographic textures of calcareous sediments and rocks; this is expressed in various models involving cement types, pore waters, and chemical processes. Diagenesis is treated in various excellent textbooks on carbonates, of which only a few can be mentioned here, such as those by Lippmann (1973), Bathurst (1976), Chilingar et al. (1979), Flügel (1982), Reeder (1983), Schneiderman and Harris (1985), Marshall (1987), Scoffin (1987), Moore (1989), and Tucker and Wright (1990). In this chapter, however, a different approach is taken, insofar as carbonate diagenesis is numerically described in terms of changing volume and carbonate mass. Such an approach is based on the development of basic relationships among the degree of compaction of sediment, porosity, and carbonate content. Thus, a dynamic and quantitative understanding of carbonate diagenesis can be achieved, because the history of compaction, volume reduction, cementation and dissolution can be modeled, and associated important problems can be solved related to the quantity of cement present in a given carbonate sample. Also, it is possible to show the evolution of compaction and porosity, differential compaction, and compactional enrichment of organic matter and minor elements.

This chapter is written in order to demonstrate the principles of the new concept of mass and volume changes in diagenesis. Examples of various diagenetic processes that can be treated and quantified by this approach are given throughout the text. They stem mainly from the author's experience with the fine-grained marl–limestone alternations which were analyzed in several Jurassic to Tertiary sections of Europe and North America. These alternations have recently become a subject of considerable debate, as many authors argued that they originated from orbital cycles, i.e., Milankovitch cycles (Barron et al., 1985; Fischer, 1986; Herbert and Fischer, 1986; Schwarzacher, 1987; Weedon, 1989), whereas others thought that carbonate content variations in these alternations were also enhanced and more pronounced by differential diagenesis (Eder, 1982; Einsele, 1982; Arthur et al., 1984; Simpson, 1985, Ricken, 1985, 1986; Hallam, 1986; Bathurst, 1987).

SOME BASIC CONSIDERATIONS OF COMPACTION IN CALCAREOUS ROCKS

Some preliminary basic considerations of compaction and carbonate content are required first. The original (high) porosities and the degree of compaction in carbonate sediments and rocks are the most important parameters which control their postdiagenetic carbonate contents, because compaction reduces the original pore space and thus determines the amount of carbonate that can be precipitated or removed by dissolution. Compaction, pore space, and carbonate content can be related by a relatively simple numerical expression, providing the basis for further concepts developed herein.

The significance of compaction and its role in pore-space reduction in carbonates has long been a matter of debate. An excellent discussion of this problem appears in a paper by Bathurst (1980). Originally, it was thought that the onset of cementation in carbonates is so early that compaction is either low or even as good as nonexistent (Pray, 1960; Steinen, 1978). Extensive presence of deformed fossils, compactional drapings, etc., however, demonstrate moderate to high degrees of compaction in some calcareous rocks (Carozzi, 1961; Kahle, 1966; Wolfe, 1968; Brown, 1969; Zankl, 1969; Baldwin, 1971; Rieke and Chilingarian, 1974; Chilingarian and Wolf, 1975; Kendall, 1975; Wolf and Chilingarian, 1976; Chilingar et al., 1979; Bathurst, 1983; Meyers and Hill, 1983; Gaillard and Jautee, 1985). Also, porosity reduction data in pelagic carbonate ooze recovered by the Deep Sea and Ocean Drilling Projects (i.e., DSDP/ODP) support the occurrence of compaction (Matter, 1974; Schlanger and Douglas, 1974; Garrison, 1981). Arguments recently made by Bathurst (1987) and Ricken (1986) show that carbonate compaction seems spatially concentrated in relatively narrow zones in bedded carbonate rocks from various environments, characterized by fabrics of mechanical compaction and pressure dissolution.

Derivation of the carbonate compaction equation

The carbonate compaction equation of Ricken (1986, 1987) is a basic and theoretically-founded relationship among the following three sediment or rock parameters: carbonate content, compaction, and porosity. This relationship can be derived by considering a sediment – rock transformation of calcareous sediment containing various proportions of pore space, and carbonate and noncarbonate contents (Fig. 6-1). The noncarbonate fraction is usually composed of clay minerals, quartz, and organic matter (Wedepohl, 1970). During the sediment – rock transformation, the pore volume is reduced due to compaction or cementation and the initial carbonate content is changed because of cementation or carbonate dissolution. Only the noncarbonate fraction remains essentially unaffected (Fig. 6-1). The noncarbonate fraction (NC_d, in vol%), however, is the only constant factor in carbonate diagenesis, when it is standardized to the primary, or decompacted, sediment bulk volume:

$$NC_d = \frac{(100 - K)(100 - n)(100 - C)}{10{,}000} \tag{6-1}$$

where C is the volume of solid carbonates expressed as the percentage of the compacted sediment (bulk) volume; n is the porosity, percent of bulk volume; K is the degree of compaction expressed as the percentage of the original volume of the sedi-

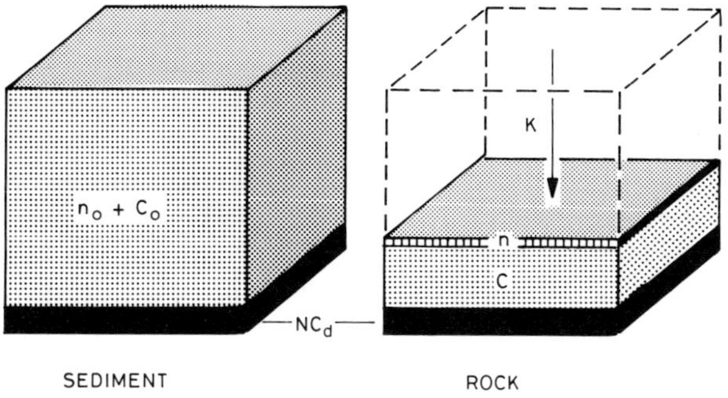

Fig. 6-1. Principles of volume changes during sediment-to-sedimentary-rock transformation for carbonates. Left: Uncompacted sediment with original porosity (n_o) and original carbonate content (C_o). Right: Compacted and lithified calcareous rock volume (K) with diminished porosity (n) and altered carbonate content (C). The non-carbonate fraction (NC_d) remains constant when it is expressed as percentage of the original sediment volume.

ment; and NC_d is the solid noncarbonate fraction expressed as a percentage of the original sediment volume. This NC_d value is standardized to the original sediment volume and will, therefore, be referred to simply as the "standardized noncarbonate content". Equation 6-1 was termed the "carbonate compaction law" by Ricken (1986, 1987), because in most rocks with low porosity, it essentially relates the carbonate volume to the degree of compaction. As the specific grain densities for the carbonate and noncarbonate fractions are very similar, the volume percent of carbonate in Eq. 6-1 is essentially equivalent to the weight percent content of carbonate. The compaction law is valid regardless of whether the diagenetic carbonate system is closed or open.

Carbonate compaction equation for rocks with low porosities

In many lithified calcareous rocks and marls, porosities are below 15%, whereas limestones commonly have porosities below 5% (Bathurst, 1980). Thus, for these low-porosity rocks, the compaction equation (Eq. 6-1) can be simplified to the following:

$$NC_d = \frac{(100 - K_r)(100 - C)}{100} \tag{6-2}$$

where the percentage of compaction (K_r) in low-porosity calcareous rocks is equal to:

$$K_r = 100 - \left(\frac{100\ NC_d}{100 - C}\right) \tag{6-3}$$

Thus compaction can be calculated for porous and lithified sediments with various standardized noncarbonate fractions (i.e., NC_d) by using Eqs. 6-1 and 6-3, respectively. As follows from Eq. 6-3, the degree of compaction in nonporous rocks is non-linearly related to the carbonate content. For a constant value of the standardized noncarbonate fraction content (NC_d), compaction is low at high carbonate contents, large at medium carbonate contents, and very large at low carbonate contents. An illustration for this non-linear relationship is presented in Fig. 6-2. A nonporous limestone sample is subjected to progressive chemical compaction and pressure dissolution. Although compaction increases equally, the relative $CaCO_3$ content (in percent) is non-equally reduced.

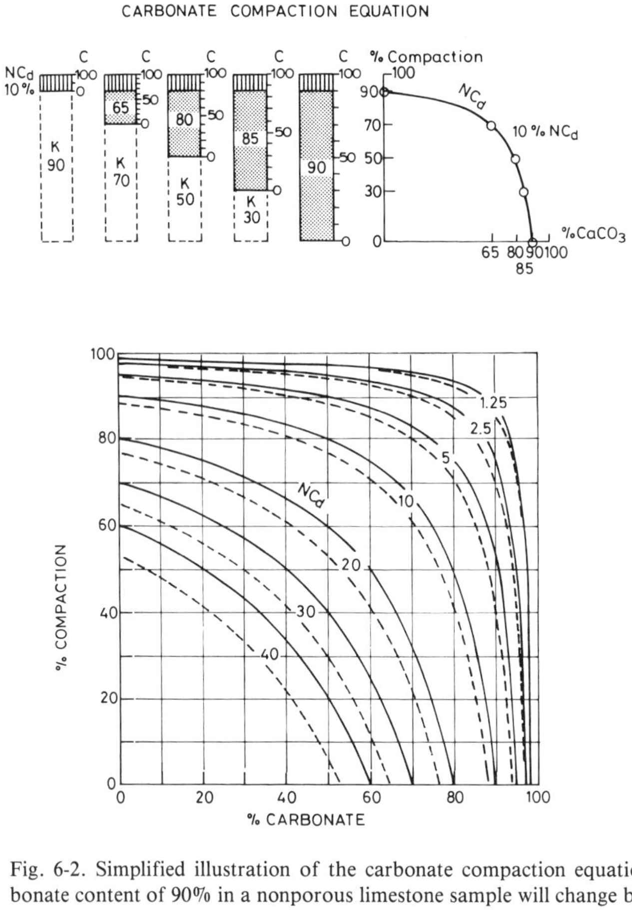

Fig. 6-2. Simplified illustration of the carbonate compaction equation, depicting how the initial carbonate content of 90% in a nonporous limestone sample will change by constantly increasing the degree of compaction (K, in %) and removing carbonate by dissolution. The percentage of carbonate fraction (C) is indicated by small numbers within columns, with a scale on the right-hand side. Note that the same values for carbonate content and compaction will be obtained when a porous sediment having the same standardized percentage of noncarbonate fraction (NC_d) first undergoes mechanical compaction and then cementation. Lower diagram shows the theoretical relationship between degree of compaction (%) and carbonate content (%) for samples with various NC_d contents and porosities ranging from 0 to 15% (solid and dashed curves). (After Ricken, 1986, 1987.)

For most calcareous rocks, however, this curved relationship between the carbonate content and degree of compaction reflects the presence of both less compacted, cemented and more compacted dissolution-affected rock portions. For sites at low degrees of compaction and high carbonate content, the sediment was cemented early in its diagenetic history (after some mechanical compaction) by the precipitation of additional cement in the original pore space, thus inhibiting further compaction. On the other hand, low carbonate content and a higher degree of compaction are usually associated with pressure dissolution of carbonate or chemical compaction. As a consequence, the degree of compaction and the carbonate content in calcareous rocks can be viewed to reflect the diagenetic history, related to mechanical compaction, cementation, and pressure dissolution.

Testing of compaction equation by compaction measurements

In order to test whether the theoretically-derived carbonate-compaction equation is documented in the rock record, compaction, carbonate content, and porosity were measured in the interbedded marl – limestone alternations mentioned above. These values were then compared with the theoretical curves for carbonate content versus degree of compaction. Among these parameters, compaction is the most difficult to determine. Many authors have addressed compaction measurements by using various methods, including grain orientation and deformation (e.g., Wolf and Chilingarian, 1976; Bathurst, 1987), deformation of primary sedimentary structures (e.g., Baldwin, 1971), deformation of vertically-emplaced sedimentary dikes (e.g., Bedaudoin et al., 1984), compactional draping relative to early concretions (e.g., Einsele and Mosebach, 1955; Chanda et al., 1977), density of bioturbation pattern (Gaillard and Jautee, 1985), and experimental studies (Chilingarian and Rieke, 1976; Shinn et al., 1977). Inasmuch as most of these methods can be considered relative and only semiquantitative, the deformation of originally circular bioturbation tubes (parallel to the bedding) was used, because these sedimentary structures deform with the sediment and, therefore, their degree of compaction can be reliably determined (Plessmann, 1966).

Bioturbation tubes are usually formed within the upper meter of the sediment, but below the mixed uppermost sediment layer (Ekdale et al., 1984). Bioturbation structures undergo the same amount of compaction as the surrounding sediment, except burrows with early diagenetic cementation (Fig. 6-3). Burrows suitable for direct compaction measurements must have originally circular tubes, such as Thalassinoides, Teichichnus, Chondrites, and Planolites (Häntzschel, 1975); this can be confirmed by examining the cross-sections of vertically-emplaced burrows. Such burrows are abundant in shelf and pelagic sediments (Kennedy, 1975; Ekdale and Bromley, 1984). Burrows must be parallel to the bedding and only cross-sections perpendicular to the burrow tubes must be used for measuring the burrow axes. Measurements can be performed in the field utilizing suitable rock samples with burrows. During compaction, only the vertical axis (b) is reduced and, thus, the degree of compaction (K) can be expressed as a percentage of the undeformed, horizontal axis (a; see Fig. 6-3B):

$$K = \left(\frac{a - b}{a}\right) 100 = 100 - \left(\frac{100b}{a}\right) \quad (6\text{-}4)$$

Burrows with early cementation can be recognized by their significantly higher carbonate content and lower degree of compaction than the surrounding rock (Fig. 6-3C). Despite this, compaction can be indirectly determined by using the carbonate content of the burrow fill and that of the surrounding rock. From the partial compaction and carbonate content of the burrow, a standardized noncarbonate fraction content (NC_d) can be calculated (Eq. 6-1), which is assumed to be the same for the burrow and the host rock. This will finally allow a calculation of actual rock compaction by using the calculated NC_d value of the burrow and the carbonate content of the host rock. Compaction is obtained either by solving Eq. 6-1 for K or by using Eq. 6-3. Repeated direct and indirect determinations of compaction using burrow deformation show an accuracy of ± 10%. Consequently, only the *means* of several measurements allow a correct determination of compaction. The following example demonstrates this.

Example: What is the degree of compaction in a lithified marl containing cemented burrows (75% $CaCO_3$) with a significantly higher carbonate content than in the surrounding rock (50% $CaCO_3$)? From the degree of shortening of the vertical burrow axes, compaction of the cemented burrow is calculated to be 60% (Eq. 6-4). Because porosities are low enough to be ignored, the NC_d value can be calculated according to Eq. 6-2 using the burrow-tube carbonate content and the degree of compaction, which results in a NC_d of 10%. Under the assumption that the NC_d value (i.e., the noncarbonate fraction of the original bulk sediment volume) is the same for the cemented burrow and the surrounding sediment, the actual degree of compaction can be calculated using the carbonate content of the surrounding rock (50% $CaCO_3$) and the NC_d value of the burrow (Eq. 6-3). Thus, the degree of rock compaction in the rock matrix is calculated to be 80%, which is substantially higher than that indicated by the degree of compaction (60%) determined in the cemented burrows.

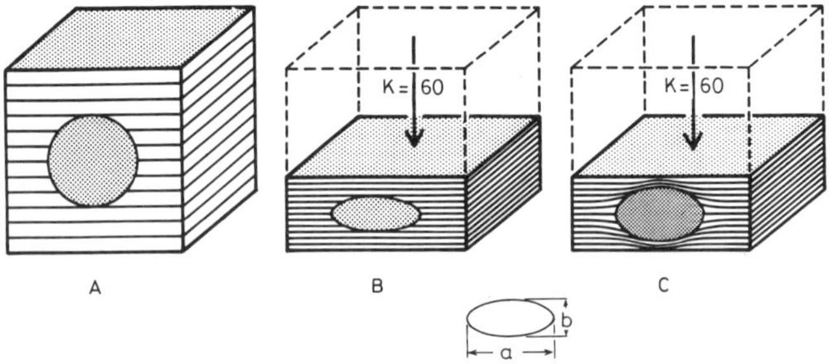

Fig. 6-3. (A). Compaction measurement using the deformation (D) of an originally circular burrow tube. (B). Normally, burrow deformation equals the actual sediment or rock compaction ($D = K = 60\%$). (C). In early-cemented burrows, compaction (K) is higher than the burrow deformation ($D = 40\%$). (From Ricken, 1986.)

In order to determine whether or not there is a predictable relationship between carbonate content and the degree of compaction as suggested by the compaction equation, carbonate rocks with a large variation in the degree of compaction were employed. These are the widely distributed, rhythmically interbedded marl – limestone sequences of shelf to hemipelagic environments which occur throughout the Phanerozoic (Einsele and Ricken, 1991). As previously mentioned, many authors relate the primary deposition to both orbital variations and superimposed depositional noise (e.g., Berger et al., 1984; Fischer, 1986; Weedon, 1989; Fischer et al., 1990). Often, there is also an important diagenetic overprinting, due to alternating cementation and carbonate dissolution in carbonate-rich and carbonate-poor beds, respectively, which is associated with differential compaction and diagenetic enhancement of the initial bedding rhythm (e.g., Hallam, 1986; Ricken, 1986; Ricken and Eder, 1990; Bathurst, 1990).

The interbedded marl – limestone sections utilized for the author's compaction studies represent various formations and environmental settings in Europe and North America. They include epicontinental basins, such as the Upper Jurassic of southern Germany and the Upper Cretaceous of northern Germany, the U.S. Western Interior, the hemipelagic Lower Cretaceous sections of southern France, and the pelagic Upper Cretaceous to Neogene sections of Umbria, Italy. Most of the compaction data were obtained by direct measurements, as explained above. Although there is considerable scattering of measured data points, the degree of compaction and carbonate content exhibit a distinct curvilinear relationship for individual sections, which fits the theoretical curves (Fig. 6-4). Thus, the following two conclusions can be drawn: (1) The carbonate compaction law is well-documented in the rock record and, therefore, can be used for further development of the volume concept of diagenesis as presented herein. (2) For each group of interbedded marls and limestones shown in Fig. 6-4, there is a fairly constant standardized noncarbonate content (NC_d). This indicates that initially interbedded

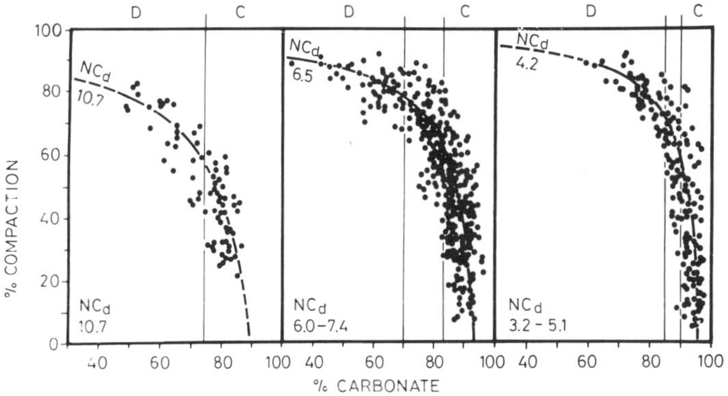

Fig. 6-4. Relationship between measured values of degree of compaction (%) and carbonate content (%) in various European interbedded marl – limestone sections, fitting the curves obtained using the carbonate compaction equation. *D* and *C* denote dissolution-affected and cemented samples, respectively. (After Ricken, 1987.)

marl – limestone sequences were more uniform. Their present carbonate content variation is considerably increased due to differential compaction, which is associated with carbonate dissolution and cementation in alternating beds.

APPLICATION OF THE CARBONATE COMPACTION EQUATION

The application of the compaction equation, gives the possibility to quantify diagenetic processes, i.e., the determination of compaction, decompaction, cement contents, carbonate mass balances, simulation of diagenetic histories, and compactional enrichment of minor elements and organic carbon content.

Determination of compaction

In rocks with porosities below 15%, compaction can be calculated according to Eq. 6-3. Only two parameters must be used, the carbonate content and the standardized content of the noncarbonate fraction (NC_d). For fine-grained carbonate sediments, the NC_d was found to range from 2 to 7% in limestones, from 7 to 15% in marls, and from 15 to 25% in shales. Coarser-grained carbonates have higher NC_d values. According to original porosities observed in carbonate-rich grainstones, packstones, and wackestones (Enos and Sawatzky, 1981), the NC_d is assumed to range from 3 to 6% (90% original carbonate content) and from 6 to 12% (80% original carbonate content). Consequently, compaction in various types of calcareous rocks with various NC_d values can be easily calculated. Further methods of calculations are presented in Ricken (1986).

Simulation of diagenetic histories

The diagenetic history of calcareous sediments and rocks can be graphically shown using a three-dimensional representation of the carbonate compaction equation (Eq. 6-1). Such a representation is expressed as the curvilinear interrelationships among three pairs of parameters, i.e., compaction and carbonate content, compaction and porosity, and carbonate content and porosity (Fig. 6-5). Each one of these relationships is nonlinear, which was already demonstrated for the relationship between carbonate content and compaction (see Fig. 6-2). In a three-dimensional diagram, with compaction, carbonate content, and porosity, as the axes, the parameters span a sphere-like surface for a given standardized noncarbonate fraction content (NC_d). On such a surface, all possible types of calcareous sediment – rock transformations with their various substages are represented (Fig. 6-5). Sediments with high porosities are situated at the bottom edge of the spherical diagram, whereas rocks with low porosities and an altered carbonate content (compared to the original sediment) are represented at the upper edge of the diagram. There are four basic types of sediment – rock transformations (Fig. 6-5):

Mechanical compaction. Mechanical compaction is represented by a curve of increasing compaction, decreasing porosity, and constant carbonate content, i.e., the curve is parallel to the plane containing the K- and n-axes. During burial diagenesis,

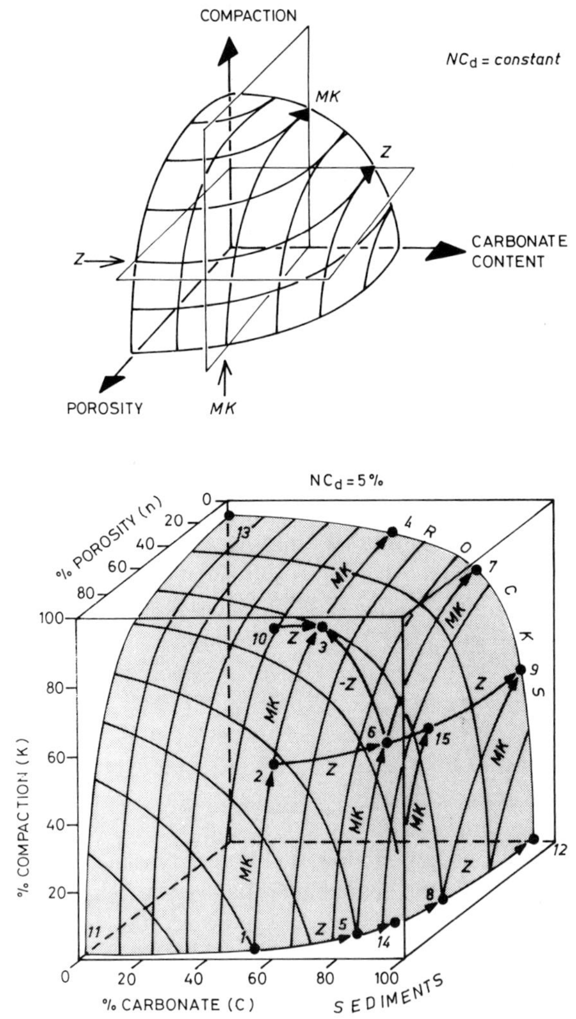

Fig. 6-5. Three-dimensional representation of the carbonate compaction equation. Various types of sediment – rock transformations are denoted as mechanical compaction (*MK*), cementation (*Z*), and chemical compaction ($-Z$). Small numbers indicate examples using various compositions. (From Ricken, 1986, 1987.)

mechanical compaction can change to a combined process of dissolution and in-situ reprecipitation, which is represented by the same curve.

Cementation by additional carbonate. Cement precipitation within the compacted pore space is denoted by increasing carbonate content with decreasing porosity at a constant degree of compaction; the curve is parallel to the plane including the *C*- and *n*-axes. It is thought that additional cementation with solid carbonate inhibits further compaction.

Chemical compaction. The term chemical compaction refers to the various processes of pressure dissolution, expressed by fitted fabrics, dissolution seams, and stylolites (Logan and Semeniuk, 1976; Garrison and Kennedy, 1977; Wanless, 1979; Bathurst, 1987; Füchtbauer and Richter, 1989). An exact definition of chemical compaction on a three-dimensional diagram is not possible because it depends on the size, shape, and mineralogy of grains (e.g., Rittenhouse, 1971). As a result, chemical compaction is described by a somewhat variable curve of increasing compaction, with both carbonate content and porosity decreasing.

Formation of secondary porosity. There are various processes that can generate secondary porosity (Choquette and Pray, 1970). In this context, secondary porosity is created by dissolution of susceptible carbonate grains or cements after the primary porosity was already reduced by other processes. Dissolution forms new pore space; however, the removal of carbonate is not accompanied by increasing compaction as observed for pressure dissolution. Thus, secondary porosity is ideally represented by a curve of increasing porosity with simultaneous reduction in carbonate content while compaction is constant. This curve is essentially that of cementation, but with an opposite sense.

Very often, the diagenetic history cannot be described by one of the simple processes discussed above. Instead, successions and combinations of several of these processes are frequently observed. The diagenetic history can nevertheless be evaluated, as explained by means of the following example involving mechanical compaction followed by cementation (Fig. 6-5).

Example: Consider a fine-grained limestone containing 90% carbonate. Burrow deformation indicates a relatively high degree of compaction of 50%. Because porosity is low enough to be neglected, a standardized content of noncarbonate fraction (NC_d = 5%) is determined using Eq. 6-2; the limestone sample is denoted as point 9 on the three-dimensional representation of the carbonate compaction equation (Fig. 6-5). Point 9 can be explained either by simple mechanical compaction or by a combination of mechanical compaction and cementation. If only mechanical compaction is assumed, the original sediment would be represented by point 8 with a porosity of 50%, which is too low for fine-grained carbonates, for which Hamilton (1976) reported a mean value of 72% porosity. When this porosity is assumed for the original sediment, an original carbonate content (C_o) of 82.1% is determined, either by using Eq. 6-1 (with $NC_d = 5$ and $K = 0$), or by applying Eq. 6-11 (point 14). The diagenetic history can be evaluated by determining the point of intersection between a curve of mechanical compaction, tied to the assumed original sediment (point 14), and a curve of cementation tied to the limestone sample (point 9). This intersection point (point 15) indicates the onset of cementation. It has the same carbonate content as the original sediment (point 14) and the same amount of compaction as the limestone sample (point 9). According to the compaction equation (Eq. 6-1), a porosity of 44.1% is obtained.

Another instructive example for the simulation of the diagenetic history is the carbonate content versus degree of compaction data for the interbedded marl – limestone sections already presented in Fig. 6-4. The cemented and dissolution-affected carbonate content versus degree of compaction data are plotted on two three-dimensional representations of the carbonate compaction equation, indicating carbonate-poor and carbonate-rich sections (Fig. 6-6). By determining the neutral carbonate content (C_n) between the dissolution-affected and cemented data sets, the curve of mechanical compaction is defined and, thus, the *mean* original composition of the former sediment is determined. The onset of lithification (composition II in Fig. 6-6) is obtained at the intersection of the mechanical compaction curve and cementation curve for the limestone layers; the latter curve is defined by the intersection at III. Chemical compaction in the interbedded marl beds is thought to simultaneously provide the carbonate for cementation in the limestone layers, as shown below. Thus, the curve of chemical compaction is defined by the line between points II and IV, denoting the compositions for the onset of lithification and maximum pressure dissolution, respectively.

Diagenetic influence on bedding rhythm as expressed by the carbonate compaction equation

Small-scale variations in compaction affect the bedding pattern of stratified calcareous rocks by the interaction of two mechanisms: differential compaction, which is associated with variations in carbonate content as described by the carbonate compaction equation (Eq. 6-1), and differential compaction, which is associated with variation in rock volume or thickness. Cemented layers have a higher carbonate content and are little to moderately compacted, whereas beds affected by carbonate dissolution have lower carbonate content and a smaller thickness. In alternations with primary differences in carbonate content, zones of

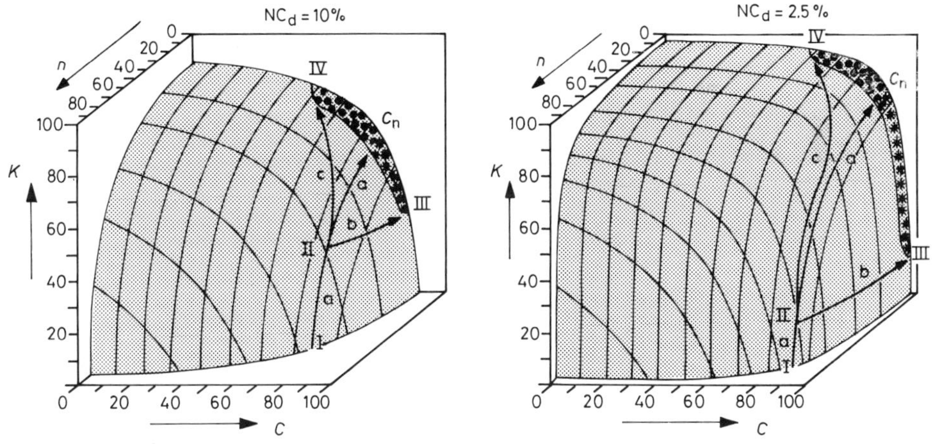

Fig. 6-6. Diagenetic overprint in carbonate-poor (left) and carbonate-rich (right) interbedded marl – limestone sections, using three-dimensional representation of the carbonate compaction equation. The three axes are: compaction (K, ‰), carbonate content (C, ‰), and porosity (n, ‰). Diagenetic overprint of mean original composition (*I*) involves a phase of mechanical compaction denoted by composition (*II*); thereafter, limestone layers undergo cementation, whereas marl beds are subjected to chemical compaction, resulting in compositions (*III*) and (*IV*), respectively. Note that the measured carbonate content and degree of compaction values (cemented = stars; dissolution-affected = dots) cannot be explained simply by mechanical compaction, because unrealistic variations in the initial porosity have to be assumed. A "neutral" carbonate content (C_n) is indicated to show the boundary between cemented and dissolution-affected values for carbonate content versus degree of compaction. (After Ricken, 1987.)

cementation and carbonate dissolution alternate, and the original bedding rhythm is considerably enhanced (Hallam, 1986; Ricken 1986, 1987; Bathurst, 1987).

In the interbedded limestone layers investigated, mechanical compaction before the onset of later cementation was measured to vary between a few and 45%, depending on the primary carbonate mineralogy and the amount of clay. Beds rich in carbonate and with some aragonite and Mg-calcite experienced the lowest degree of mechanical compaction, indicating an early onset of cementation, whereas higher degrees of compaction were observed in carbonate-poor beds. Compaction is lowest in the middle of the limestone layers, steadily increasing (up to 85%) towards the neighboring marl beds; consequently, cementation began in the middle of the present limestone layers. With increasing compaction, however, carbonate content decreased non-linearly towards the marl beds as shown by Eq. 6-1 and in Fig. 6-2. As a result, the originally sinusoidal carbonate content curves of the later limestone layers become highly angular with a steep drop in carbonate content towards the marl beds (Fig. 6-7). Additionally, the original variations in carbonate content are considerably enhanced. Related to cementation, the carbonate content in the middle of the limestone layers is relatively constant, because minor variations in the degree

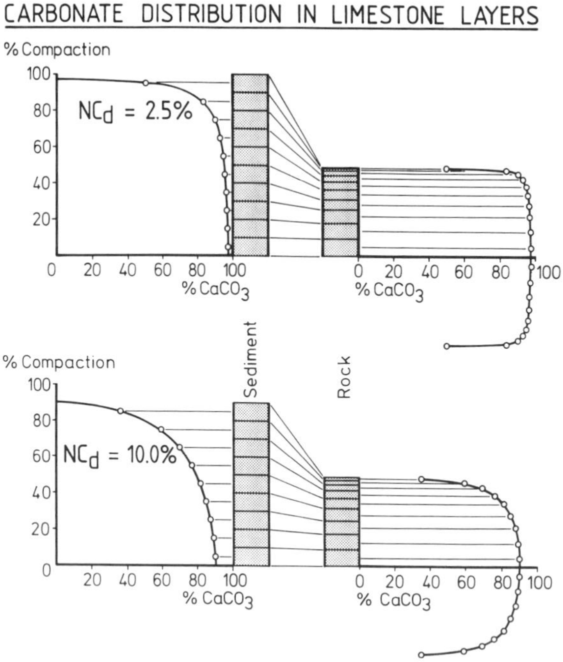

Fig. 6-7. Model for the carbonate content curves of diagenetically modified limestone layers. Degree of compaction increases steadily from the middle of the limestone layers towards the marl beds. Carbonate curves follow the theoretical relationship between carbonate content and degree of compaction according to Eq. 6-3 (see Fig. 6-2). Note that carbonate-rich limestone beds with low NC_d show angular carbonate curves, whereas limestone layers poorer in carbonate content (with lower NC_d value) show convex carbonate curves.

of compaction before the onset of cementation (i.e., mechanical compaction) have only a very small effect on the carbonate content, as depicted by the compaction equation (see Fig. 6-2). Thus, diagenetically-modified, carbonate-rich sequences become highly rhythmic: they consist of brick-like limestone layers, with a relatively constant carbonate content in the middle, interbedded with thin marl beds.

At low primary carbonate contents, mechanical compaction is usually so high that differential compaction and, thus, the enhancement of the bedding rhythmicity is small. On the other hand, there can be a significant enhancement of primary carbonate content variations and rhythmicity of beds. For a detailed treatment of the diagenetic influence on bedding rhythm, the reader is referred to Ricken (1986) and to Ricken and Eder (1991).

Determination of cement content

One of the most important questions associated with carbonate diagenesis is the quantity of additional cement introduced into a given carbonate rock. When a clear distinction between calcareous grains and the matrix cement is possible (e.g., in some bioclastic carbonates), the cement content can be estimated by employing peels or thin sections (Meyers and Hill, 1983). This is, however, difficult to accomplish in many fine-grained and micritic carbonates, where the amount of cement, sediment matrix, and their neomorphic modifications cannot be distinguished. A numerical method, therefore, is introduced, which is closely associated with the evaluation of compaction and thus the reduction of the original pore space, because the degree of compaction determines the amount of additional cement that can be precipitated in this reduced pore space. In a basic equation, the cement content is expressed as a percentage of the original bulk volume (i.e., the absolute cement content, Z_d, vol%). This equals the original porosity (n_o) minus the degree of compaction (K) and the absolute value of the present rock porosity (n_d; Fig. 6-8):

$$Z_d = n_o - K - n_d, \text{ with } n_d = n(1 - 0.01K) \tag{6-5}$$

In Eq. 6-5 the absolute porosity (n_d) is expressed as a percentage of the original sediment bulk volume, n being porosity of the sediment or rock volume at a particular stage of compaction. Upon substitution of the latter relation for n_d into Eq. 6-5, the absolute cement content is equal to:

$$Z_d = n_o - n + K(0.01n - 1) \tag{6-6}$$

Consequently, the absolute cement content depends on the original and present porosity and the degree of compaction. The latter can be replaced by solving Eq. 6-1 for K, thus introducing the postdiagenetic carbonate content (C) and the NC_d value:

$$Z_d = n_o - 100 + \left(\frac{NC_d}{1 - 0.01C}\right) \tag{6-7}$$

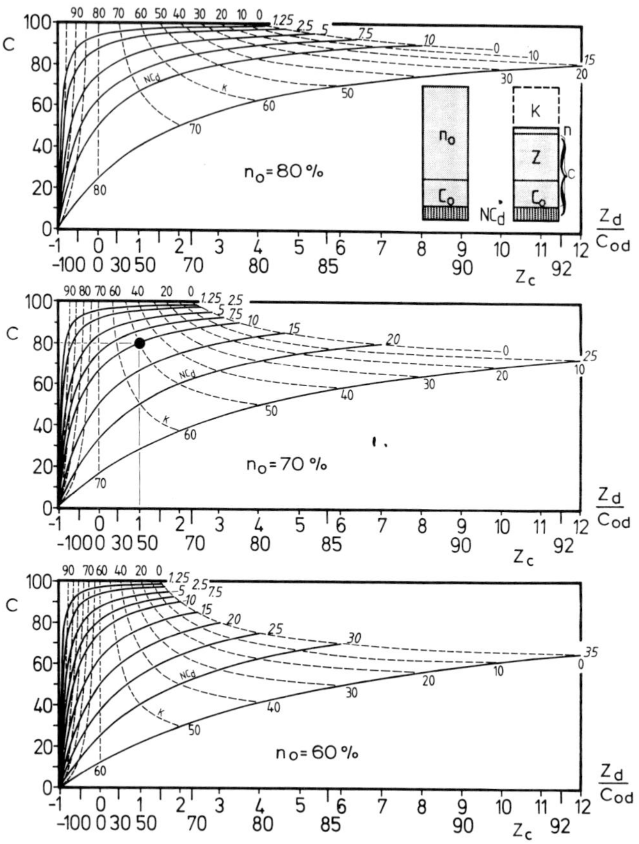

Fig. 6-8. Diagrams for the estimation of cement content in fine-grained, nonporous carbonate rocks, using the following parameters: original porosity (n_0, %), carbonate content (C, %), degree of compaction (K, %), and NC_d value (%). Example (see solid circle in middle diagram): A limestone sample having an original porosity of 70%, a carbonate content of 80%, and a compaction of 50% (curved dashed line) or a NC_d value of 10% (curved solid line), has a cement content (Z_c) of 50% of the total carbonate fraction. For such a cement content, the ratio of cement to primary carbonate (Z_d/C_{od}) equals 1. Inset shows volume changes in the course of cementation. Left: uncompacted sediment with original porosity (n_0) and original carbonate content (C_0). Right: compacted rock volume (K), the porosity (n) is largely reduced by precipitation of carbonate cement (Z). Note that the rock carbonate content (C) is composed of the original and the cemented carbonate fractions. (After Ricken, 1986.)

Equations 6-6 and 6-7 are the basic formulas for the determination of absolute cement values. Many geologists and geochemists, however, are not interested in this absolute value. Instead, they prefer to express the cement content (Z_c) as a percentage of the total carbonate fraction, which is equal to ($100 - K - n_d - NC_d$) of the original sediment bulk volume. The relative cement content (Z_c), therefore, can be expressed as follows:

$$Z_c = \frac{100\ Z_d}{(100 - K - n_d - NC_d)} \qquad (6\text{-}8)$$

On substituting Eq. 6-5 (n_d) and Eq. 6-6 (Z_d) into Eq. 6-8, the following expression, which defines the cement content as a percentage of the carbonate fraction, is obtained:

$$Z_c = \frac{[100n_o - 100n + K(n - 100)]}{[100 - NC_d - n + K(0.01n - 1)]} \qquad (6\text{-}9)$$

Because the degree of compaction is often difficult to determine, K in Eq. 6-1 is substituted into Eq. 6-9, resulting in the expression:

$$Z_c = \frac{[NC_d + n_o - 100 + C(1 - 0.01n_o)] \times 10^4}{NC_d C} \qquad (6\text{-}10)$$

The last two equations allow a relatively accurate estimation of the cement content, either as a function of n_o, n, K, and NC_d, or of n_o, C, and NC_d. As already mentioned, the carbonate content in volume % in Eq. 6-10 is practically equivalent to weight % carbonate.

A more practical estimation of the cement content in fine-grained carbonates (i.e., calcareous oozes, muds, and wackestones) is performed by using the diagrammatic expression of Eqs. 6-9 and 6-10 (Fig. 6-8). For various original porosities (n_o), the cement content can be estimated from the following variables: carbonate content, degree of compaction, or standardized content of noncarbonate fraction (NC_d). Negative cement content values indicate carbonate dissolution. Typical values for the NC_d (see above) or degree of compaction and the original porosity must be employed in order to obtain a reasonably good estimation of the cement content. Original porosities in fine-grained and bioclastic carbonate sediments are reported in the papers by Hamilton (1976), Keller et al. (1979), Enos and Sawatzky (1981), and Moore (1989). Original porosities in fine-grained calcareous sediments are found to be relatively high as a result of the large amount of intraparticle pore space. In carbonate muds and pelagic oozes, porosities around 70 – 75% are reported, whereas wackestones show porosities of 60 – 70% (Enos and Sawatzky, 1981).

For a given sample with an assumed original porosity and NC_d value, the primary carbonate content (C_o) can be estimated using the following formula:

$$C_o = \frac{(100 - NC_d - n_o)}{(1 - 0.01n_o)} \qquad (6\text{-}11)$$

Cement content of concretions

In concretions, the cement content (Z_{con}) can be more easily estimated and comto the rather complicated general procedure of cement estimation explained above. This is because in host rocks containing only a few concretions, it can be assumed that the original carbonate content is not significantly altered by the concretion growth. Consequently, the cement content (Z_{con}, vol%), which is equivalent to the pore space of the host rock during cement precipitation (n_h) (Lippmann, 1955; Seibold, 1962; Raiswell, 1971), can be calculated using the carbonate content of the host rock and that of the concretion as follows:

$$Z_{con} = n_h = C_{con} - \left[\frac{(100 - C_{con})\, C_h}{100 - C_h} \right] \tag{6-12}$$

where C_{con} is the carbonate content in the middle of the concretion and C_h is the carbonate content of the host rock. In Eq. 6-12 it is assumed that the present rock porosity is small enough to be neglected. Evaluation of more than 60 carbonate pairs, e.g., samples from centers of concretions and from surrounding rock, as reported in the literature, shows that porosities at the onset of cementation range from 30–90%, with a maximum of 80–90% (Ricken and Eder, 1991). As already shown by Raiswell (1971), Hudson (1978), Coleman and Raiswell (1981), Gautier and Claypool (1984), most concretions underwent near-surface cementation, whereas others experienced a relatively late onset of cementation during shallow burial. The early and late concretion types can be distinguished by different amounts of compactional draping, which can be determined by using the carbonate compaction equation (Eq. 6-1).

Decompaction and carbonate mass balance calculations

Mass balance calculations are designed to solve one of the most difficult problems in carbonate diagenesis, i.e., the origin and distribution of carbonate cement. Mass balances show whether or not a diagenetic carbonate system was closed. Cementation and dissolution processes within a closed system indicate small-scale carbonate redistribution, whereas in an open system carbonate is transported via diffusion or advection into or out of the diagenetic system (Ricken, 1986). Origin of the carbonate cement can be considered as one of the major unsolved problems in carbonate diagenesis (Bathurst, 1976).

The mass balance calculation is performed for a diagenetic system in a lithologic section a couple of meters thick. For this section, numerical decompaction is performed by transforming the section into an artificial sediment. The mean decompaction porosity of the entire section can be similar, lower or higher than that found in similar types of sediments from Recent environments. When these porosities are similar, no significant net transport of carbonate into or out of the section occurred, and the diagenetic system is assumed to be largely closed. On the other hand, when the porosity after decompaction is significantly lower or higher, carbonate was either transported into the section or removed, indicating an open system.

The first step in numerical decompaction is to establish a complete record of compaction for small thickness intervals (e.g., 1–5 cm) throughout the section, either by compaction measurements or by calculating the degree of compaction using carbonate content and NC_d values (Eqs. 6-1 or 6-3). Then, every interval is decompacted according to its specific compaction, and the mean decompaction porosity for the entire section (n_o^*, vol%) is calculated using the following equation:

$$n_o^* = \frac{[\Sigma\,(1 - 0.01K)\,n + K]}{N} \tag{6-13}$$

where K is the degree of compaction in vol%, n is the present porosity, and N is the number of decompacted intervals.

Assuming that the mean decompaction porosity would indicate a closed-system carbonate redistribution, the original mean composition of the cemented and dissolution-affected beds can be estimated through the actual carbonate mass-balance calculation (Fig. 6-9). Such a balance is based on the fact that dissolution-affected beds have a *higher* decompaction porosity than the mean of the whole section, whereas cemented beds have a *lower* decompaction porosity. The exchange of carbonate mass between the dissolution-affected and cemented beds is balanced when this difference in compaction porosity is adjusted. This is performed by assuming that original beds had an identical or nearly identical porosity, as schematically illustrated in the upper diagram of Fig. 6-9. For a more detailed description of the method the reader is referred to Ricken (1986).

An example of a carbonate mass balance is shown for a 3-m-thick subsection in an Eocene pelagic marl – limestone sequence of the Gubbio section, Italy (Fig. 6-9). The various rock volumes involved in the mass balance calculation are graphically represented in a histogram, where the frequencies of rock volumes are represented according to their various carbonate contents. Such rock volumes with low to high carbonate contents were decompacted as described above. Decompacted sediment volumes were then separated into cemented and dissolution-affected portions by assuming that the mean decompacted porosity was originally equal or similar in the dissolution-affected and cemented beds. The cement content of this example is, on average, of the order of one third of the postdiagenetic limestone carbonate content (Fig. 6-9). The original mean decompacted porosity amounts to 66%. Primary variations between original carbonate-rich and carbonate-poor sediment were, on average, only small; such variations were later significantly enhanced by differential diagenesis.

The investigated epicontinental and pelagic interbedded marl – limestone sections from Europe and North America have mean decompacted porosities ranging from 59 to 77%. These porosities are thus close to the above-mentioned porosities of Recent, fine-grained carbonate sediments (Hamilton, 1976; Keller et al., 1979; Enos and Sawatzky, 1981), indicating that differential cementation and dissolution between beds occurred predominantly as a closed-system redistribution. This differential diagenesis is thought to be related to stress differences in the grain structure between the interbedded marl and limestone layers (Ricken, 1986). Thus, small-scale carbonate redistribution by diffusion rather than transport by pore-water advection seems to be the dominant lithification process. Such arguments are supported by relatively high velocities of diffusional transport observed in interstitial waters

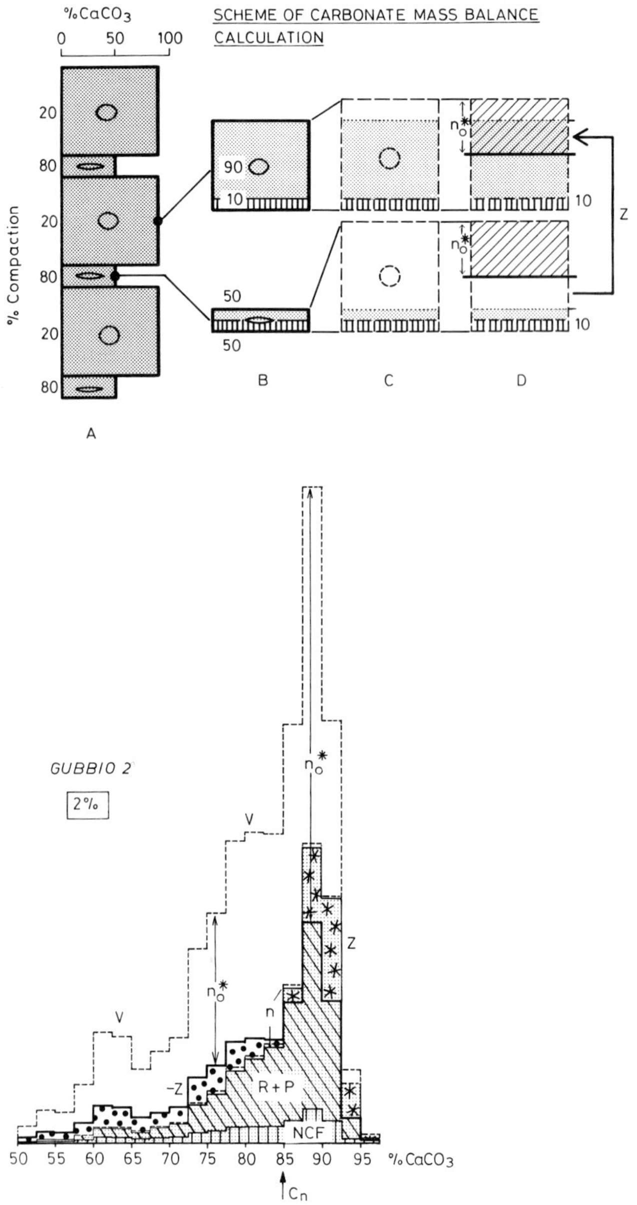

Fig. 6-9. How to perform a carbonate mass balance calculation.
Upper diagram: Scheme of a carbonate mass balance. (*A*) Interbedded limestone – marl sequence with different degrees of compaction as indicated by burrow flattening. (*B*) Composition of succeeding beds rich and poor in carbonate content. The carbonate and noncarbonate fractions are depicted by shaded and vertically hatched portions, respectively. (*C*) Numerical decompaction results in "original" volumes of sediments with restored circular burrows. Note differences in the carbonate volume and decompaction porosity for the two bed types. (*D*) The quantities of dissolved and reprecipitated carbonate (Z) can be

(Einsele, 1977; Berner, 1980; Hesse, 1986). On the other hand, Bathurst (1987) found that platform limestones show petrographic evidence for early cementation in the middle of limestone beds, followed by a late-phase pressure dissolution. Here, the diagenetic system was mainly open, as the pore space was already largely filled with introduced cement before the onset of significant pressure dissolution.

Compactional enrichment of sediment constituents, organic carbon, and minor elements

Compactional enrichment with various sediment constituents, such as larger grains and fossils or chemical components, is commonly observed. This enrichment is related to the reduction in the original volume causing the constituents to be concentrated within increasingly smaller sediment volumes (Eder, 1982). This compactional enrichment is only observed under the conditions of: (1) differential compaction with carbonate cementation and dissolution; (2) chemical compaction and pressure dissolution, with the consequent increase in the content of less-soluble components; or (3) combination of mechanical compaction followed by cementation. The number of sediment constituents contained in a given reference volume (P) increases with increasing compaction, which is equal to:

$$P = \frac{P_0}{(1 - 0.01K)} \qquad (6\text{-}14)$$

where P_0 is the primary content of particles, and K is the percentage of compaction.

In this context, it is especially important to mention that organic matter becomes enriched in the highly-compacted clay-rich beds of the investigated interbedded marl – limestone sequences. As four examples from the Western Interior of the U.S.A. show, organic carbon contents are enriched by a larger factor ranging from 2 to 10 (Fig. 6-10). All examples, where compaction and organic carbon were individually measured, show a fairly good fit between the measured data and the curves depicting compactional enrichment according to Eq. 6-14. The writer agrees with Shinn et al. (1984) that the compactional enrichment of organic matter probably promotes early hydrocarbon migration in carbonate rocks.

The processes causing compactional enrichment of minor elements, such as Mg and Fe, in the carbonate fraction are not well understood. In the studied interbed-

---

determined when an equal amount of decompaction porosity (n_0*, diagonally hatched) is assumed for both beds.
Lower diagram: Example of a carbonate mass balance. The frequency of the various volumes contained in the studied section is plotted versus the present carbonate content. Outer frame of histogram shows the decompacted sediment volume (V). Subtraction of mean decompaction porosity (n_0*) results in the amounts of dissolved ($-Z$) and cemented carbonate (Z). R and P denote the relic and primary carbonate fraction, respectively, whereas NCF stands for the noncarbonate fraction. n indicates the present rock porosity. C_n is the neutral carbonate content between cemented and dissolution-affected rock portions. Eocene marl – limestone sequence of the Gubbio section, Italy. (After Ricken, 1986.)

ded marls, either preexisting dolomite becomes passively enriched, or enrichment in Mg content in the pore fluid causes dolomitization at the strained calcite surfaces (e.g., Wanless, 1979; Mattes and Mountjoy, 1980; Jorgensen, 1983)* Enrichment factors relative to the cemented limestone layers range from 2 to 6 for the investigated interbedded marl – limestone sections (Fig. 6-10). Numerical description of trace-element enrichment is more difficult than expressed by Eq. 6-14, because trace elements can be only enriched as a portion of the dissolution-affected carbonate fraction. Numerical simulation of increasing Mg and Fe contents by increasing compaction (Fig. 6-10) and minor-element mass balances indicate that the marl beds are only incompletely closed, and that a certain portion of minor elements is reprecipitated in the carbonate cement in the limestone layers (Ricken, 1986). The degrees of closure for the marl beds in three investigated European interbedded marl – limestone sequences range, on average, between 24 and 4% for various elements with the following order from high to low degrees of enrichment: Mg, Fe, and Mn and Sr. As already observed by Wanless (1979), this indicates that dolomitization is an important process during chemical compaction and pressure dissolution in carbonates.

CONCLUSIONS

In this chapter, a new approach to carbonate diagenesis is made. Diagenesis is viewed in terms of changing volumes in carbonate content, porosity, and degree of compaction, using numerical interrelationships among these parameters. The results of the volume approach taken herein are the quantification of diagenetic processes and histories, of which the most important are evolution of compaction and porosity, carbonate content, cement content, and compactional enrichment of organic carbon and minor elements.

(1) A predictable relationship among the degree of compaction, porosity, volume of carbonate, and the standardized noncarbonate fraction content was established by the writer. Compaction measurements by evaluating the deformation of originally circular, horizontal bioturbation tubes showed the validity of the compaction equation derived. When rock porosity is below 15%, a simplified equation can be employed in order to determine the degree of compaction using the carbonate content and values for the standardized content of the noncarbonate fraction.

(2) Diagenetic processes and histories can be simulated by using a three-dimensional representation of the carbonate compaction equation. Three basic sediment – rock transformations and their interrelationships can be modeled in terms of changing values of degree of compaction, porosity, and carbonate content. Sediment – rock transformations include mechanical compaction, cementation by additional carbonate, and pressure dissolution. In addition, the formation of secondary porosity can be simulated.

* One of the editors also believes that interstitial pore waters become enriched in Mg upon compaction (Chilingarian and Rieke, 1976).

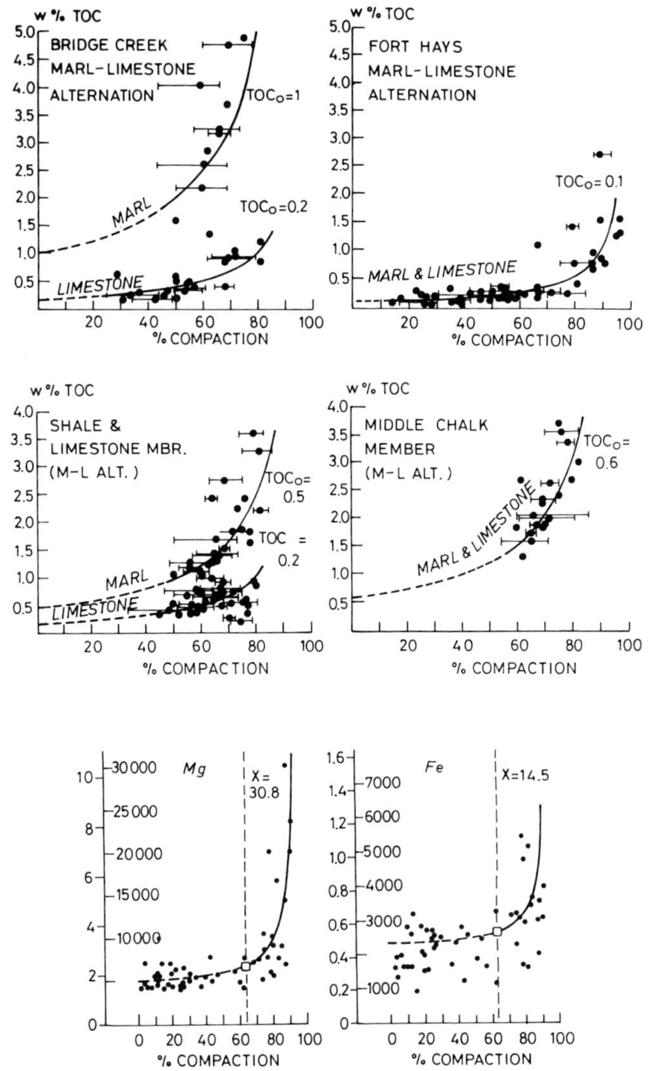

Fig. 6-10. Compactional enrichment of organic carbon and trace elements in various interbedded marl – limestone sections. Upper diagrams show relationship between measured organic carbon content (*TOC*, wt%) and compaction values (*K*), together with theoretical curves describing the compactional enrichment assuming various initial organic carbon concentrations (TOC_0). Sections exhibiting two enrichment curves have well-defined organic carbon differences between marl and limestone layers in the original sediment (Upper Cretaceous, Western Interior of the U.S.A.). Lower diagrams show Mg and Fe enrichment within limestone and subsequent marl beds of the Upper Jurassic interbedded marl – limestone sections of southern Germany. Mg and Fe concentrations on the vertical axes are given in percentage of $MgCO_3$ and $FeCO_3$ of the total carbonate fraction (numbers on the left), and in ppm of the total carbonate fraction (numbers on the right). Vertical, dashed lines divide graphs into cemented limestone layers (left-hand side) and dissolution-affected marl beds (right-hand side). The degree of closure of minor element concentration in the marl beds is indicated by *X* (%).

(3) Cement content and the original composition can be ascertained by using the degree of compaction and the original porosity of the sediment. Numerical decompaction and carbonate mass-balance calculations can be used to determine the mean original composition and the degree of closure of the diagenetic system. Degree of compaction was predictively related to the enrichment of less soluble substances, such as organic matter. In the Upper Cretaceous interbedded marl – limestone sections (Western Interior of U.S.A.), the organic carbon content was found to be enriched with increasing differential compaction by factors ranging from 2 to 10. Trace elements, e.g., Mg, are also enriched with increasing degree of differential compaction as a result of complicated dissolution – cementation processes.

(4) The volume approach developed by the writer was applied to interbedded marl – limestone sections from various locations in Europe and North America. After a phase of mechanical compaction, there is a considerable diagenetic redistribution of carbonates, resulting in a complete cementation of the limestone layers, whereas marl beds are affected by pressure dissolution. In carbonate-rich sediments, this process of differential cementation and dissolution enhances the bedding rhythms by reducing the thickness of the marl beds and by creating brick-like limestone layers with relatively constant carbonate contents in the middle.

ACKNOWLEDGEMENTS

The author would like to acknowledge the fruitful and stimulating discussions with R.G.C. Bathurst, G.V. Chilingarian, G. Einsele and other colleagues. Also, I like to thank G.V. Chilingarian and L. Hobert for critically reviewing and improving this manuscript.

REFERENCES

Arthur, M.A., Dean, W.E., Bottjer, D. and Scholle, P.A., 1984. Rhythmic bedding in Mesozoic – Cenozoic pelagic carbonate sequences: The primary and diagenetic origin of Milankovitch-like cycles. In: A.L. Berger, J. Imbrie, J. Hays, G. Kukla and B. Saltzman (Editors), *Milankovitch and Climate,* Part 1. Reidel, Dordrecht, pp. 191 – 222.

Baldwin, B., 1971. Ways of deciphering compacted sediments. *J. Sediment. Petrol.,* 41: 293 – 301.

Barron, E.J., Arthur, M.A. and Kauffman, E.G., 1985. Cretaceous rythmic bedded sequences: a plausible link between orbital variations and climate. *Earth Planet. Sci. Lett.,* 72: 327 – 340.

Bathurst, R.G.C., 1976. *Carbonate Sediments and Their Diagenesis.* Developments in Sedimentology, 12. Elsevier, Amsterdam, 658 pp.

Bathurst, R.G.C., 1980. Deep crustal diagenesis in limestones. *Rev. Inst. Invest. Geol.,* 34: 89 – 100.

Bathurst, R.G.C., 1983. Neomorphic spar versus cement in some Jurassic grainstones: significance for evaluation of porosity evolution and compaction. *J. Geol. Soc. London,* 140: 229 – 237.

Bathurst, R.G.C., 1987. Diagenetically enhanced bedding in argillaceous platform limestones: stratified cementation and selective compaction. *Sedimentology,* 34: 749 – 778.

Bathurst, R.G.C., 1991. Pressure dissolution and limestone bedding: the influence of stratified cementation. In: G. Einsele, W. Ricken, and A. Seilacher (Editors), *Cycles and Events in Stratigraphy.* Springer, Berlin (in press).

Beaudoin, B., Fries, G. and Pinoteau, B., 1984. Calcul des coefficients de décompaction et estimation des paléorecouvrements. *Doc. B.R.G.M., Progr. "Géologie Profonde de la France",* 11: 77 – 89.

Berger, A., Imbrie, J., Hays, J., Kukla, G. and Saltzman, B. (Editors), 1984. *Milankovitch and Climate.* NATO Ser., C 126, Reidel, Dordrecht, 895 pp.
Berner, R.A., 1980. *Early Diagenesis – a Theoretical Approach.* Princeton Series in Geochemistry. Princeton, N.J., 241 pp.
Brown, P.R., 1969. Compaction of fine-grained terrigenous and carbonate sediment – a review. *Bull. Can. Pet. Geol.,* 17: 486–495.
Carozzi, A., 1961. Distorted oolites and pseudoolites. *J. Sediment. Petrol.,* 31: 262–274.
Chanda, S.K., Bhattacharyya, A. and Sarkar, S., 1977. Deformation of ooids by compaction in the Precambrian Bhander Limestone, India: implications for lithification. *Bull. Geol. Soc. Am.,* 88: 1577–1585.
Chilingar, G.V., Bissell, H.J. and Wolf, K.H., 1979. Diagenesis of carbonate sediments and epigenesis (or catagenesis) of limestones. In: G. Larsen and G.V. Chilingar (Editors), *Diagenesis in Sediments and Sedimentary Rocks.* Developments in Sedimentology, 25A. Elsevier, Amsterdam, pp. 247–422.
Chilingarian, G.V. and Rieke, H.H., 1976. Compaction of argillaceous sediments. In: W.H. Fertl (Editor), *Abnormal Formation Pressures.* Developments in Petroleum Science, 2. Elsevier, Amsterdam, pp. 49–100.
Chilingarian, G.V. and Wolf, K.H., 1975. *Compaction of Coarse-Grained Sediments,* I. Elsevier, Amsterdam, 552 pp.
Choquette, P.W. and Pray, L.C., 1970. Geologic nomenclature and classification of porosity in sedimentary carbonates. *Bull. Am. Assoc. Pet. Geol.,* 54: 207–250.
Coleman, M.L. and Raiswell, R., 1981. Carbon, oxygen and sulphur isotope variations in concretions from the Upper Lias of N.E. England. *Geochim. Cosmochim. Acta,* 45: 329–340.
Eder, F.W., 1982. Diagenetic redistribution of carbonate, a process in forming limestone–marl alternations. In: G. Einsele and A. Seilacher (Editors), *Cyclic and Event Stratification.* Springer, Berlin, pp. 98–112.
Einsele, G., 1977. Range, velocity and material flux of compaction flow in growing sedimentary sequences. *Sedimentology,* 24: 639–655.
Einsele, G., 1982. Limestone–marl cycles: diagnosis, significance, causes – a review. In: G. Einsele and A. Seilacher (Editors), *Cyclic and Event Stratification.* Springer, Berlin, pp. 8–53.
Einsele, G. and Mosebach, R., 1955. Zur Petrographie, Fossilerhaltung und Entstehung der Gesteine des Posidonienschiefers im Schwäbischen Jura. *Neues Jahrb. Geol. Paläontol. Abh.,* 101: 319–430.
Einsele, G. and Ricken, W., 1991. Limestone–marl alternations – an overview. In: G. Einsele, W. Ricken, and A. Seilacher (Editors), *Cycles and Events in Stratigraphy.* Springer, Berlin, (in press).
Ekdale, A.A. and Bromley, R.G., 1984. Comparative ichnology of shelf-sea and deep-sea chalk. *J. Sediment. Petrol.,* 58: 322–332.
Ekdale, A.A., Muller, L.N. and Novak, M.T., 1984. Quantitative ichnology of modern pelagic deposits in the abyssal Atlantic. *Palaeogeogr. Palaeoclimatol. Palaeoecol.,* 45: 189–223.
Enos, P. and Sawatzky, L.H., 1981. Pore networks in Holocene carbonate sediments. *J. Sediment. Petrol.,* 51: 961–985.
Fischer, A.G., 1986. Climatic rhythms recorded in strata. *Annu. Rev. Earth Planet. Sci.,* 14: 351–376.
Fischer, A.G., De Boer, P.L. and Premoli Silva, I., 1990. Cyclostratigraphy. In: R.N. Ginsburg and B. Beaudoin (Editors), *Cretaceous Resources, Events and Rhythms.* NATO ASI, Ser. C, 304: 139–172.
Flügel, E., 1982. *Microfacies analysis of limestones.* Springer, Berlin, 633 pp.
Füchtbauer, H. and Richter, D.K., 1989. Karbonatgesteinte. In: H. Füchtbauer (Editor), *Sedimente und Sedimentgesteinte.* Schweizerbart, Stuttgart, pp. 233–434.
Gaillard, C. and Jautee, E., 1985. Compaction et déformation des structures de bioturbation. Abstract, *A.S.F.,* Paris.
Garrison, R.E., 1981. Diagenesis of oceanic sediments: a review of the DSDP perspective. *Spec. Publ. Soc. Econ. Paleontol. Mineral.,* 32: 181–207.
Garrison, R.E. and Kennedy, W.J., 1977. Origin of solution seams and flaser structures in the Upper Cretaceous chalks of southern England. *Sediment. Geol.,* 19: 107–137.
Gautier, D.L. and Claypool, G.E., 1984. Interpretation of methanic diagenesis in ancient sediments by analogy with processes in modern diagenetic environments. In: D.A. McDonald and R.C. Surdam (Editors), *Clastic Diagenesis. Am. Assoc. Pet. Geol., Mem.,* 37: 111–123.
Hallam, A., 1986. Origin of minor limestone–shale cycles: climatically induced or diagenetic? *Geology,* 14: 609–612.

Hamilton, E.L., 1976. Variations of density and porosity with depth in deep-sea sediments. *J. Sediment. Petrol.,* 46: 280 – 300.

Häntzschel, W., 1975. Trace fossils and problematica. In: R. Moore and C. Teichert (Editors), *Treatise on Invertebrate Paleontology,* Part W. Univ. Kansas Print. Serv., 269 pp.

Herbert, T.D. and Fischer, A.G., 1986. Milankovitch climatic origin of mid-Cretaceous black shale rhythms in central Italy. *Nature,* 321: 739 – 743.

Hesse, R., 1986. Early diagenetic sediment – water interactions: modern offshore basins. *Geosci. Can.,* 13 (3).

Hudson, J.D., 1978. Concretions, isotopes, and the diagenetic history of the Oxford Clay (Jurassic) of central England. *Sedimentology,* 25: 339 – 370.

Jorgensen, N.O., 1983. Dolomitization in chalk from the North Sea Central Graben. *J. Sediment. Petrol.,* 53: 557 – 564.

Kahle, C.F., 1966. Some observations on compaction and consolidation in ancient oolites. *Compass,* 44: 19 – 29.

Keller, G.H. and Bennett, R.H., 1970. Variations in the mass physical properties of selected submarine sediments. *Mar. Geol.,* 9: 215 – 223.

Keller, G.H., Lambert, D.N. and Bennett, R.H., 1979. Geotechnical properties of continental slope deposits – Cape Hatteras to Hydrographer Canyon. *Soc. Econ. Paleontol. Mineral., Spec. Publ.,* 27: 131 – 151.

Kendall, A.C., 1975. Post-compactional calcitization of molluscan aragonite in a Jurassic limestone from Saskatchewan, Canada. *J. Sediment. Petrol.,* 45: 399 – 404.

Kennedy, W.J., 1975. Trace fossils in carbonate rocks. In: R.W. Frey (Editor), *The Study of Trace Fossils.* Springer, Berlin, pp. 377 – 397.

Lippmann, F., 1955. Ton, Geoden und Minerale des Baremme von Hoheneggelsen. *Geol. Rundsch.,* 43: 475 – 502.

Lippmann, F., 1973. *Sedimentary Carbonate Minerals.* Springer, Berlin, 228 pp.

Logan, B.W. and Semeniuk, V., 1976. Dynamic metamorphism, processes and products in Devonian carbonate rocks, Canning Basin, Western Australia. *Spec. Publ. Geol. Soc. Aust.,* 6: 138 pp.

Marshall, J.D., 1987. *Diagenesis of Sedimentary Sequences. Geol. Soc., Spec. Publ.,* 36, London, 360 pp.

Matter, A., 1974. Burial diagenesis of pelitic and carbonate deep-sea sediments from the Arabian Sea. *Initial Report DSDP,* 23: 421 – 470.

Mattes, B.W. and Mountjoy, E.W., 1980. Burial dolomitization of the upper Devonian Miette buildup, Jasper National Park, Alberta. *Spec. Publ. Soc. Econ. Paleontol. Mineral.,* 28: 259 – 297.

Meyers, W.J. and Hill, B.E., 1983. Quantitative studies of compaction in Mississippian skeletal limestones, New Mexico. *J. Sediment. Petrol.,* 53: 231 – 242.

Moore, C.H., 1989. *Carbonate Diagenesis and Porosity.* Developments in Sedimentology, 46. Elsevier, Amsterdam, 338 pp.

Plessmann, W., 1966. Diagenetische und kompressive Verformung in der Oberkreide des Harz-Nordrandes sowie im Flysch von San Remo. *Neues Jahrb. Geol. Paläontol.,* Mh. 8: 480 – 493.

Pray, L.C., 1960. Compaction in calcilutites (abstract). *Bull. Geol. Soc. Am.,* 71: 1946.

Raiswell, R., 1971. The growth of Cambrian and Liassic concretions. *Sedimentology,* 17: 147 – 171.

Reeder, R.J. (Editor), 1981. *Carbonates: Mineralogy and Chemistry. Mineral. Soc., Am. Rev. Mineral.,* 11: 394 pp.

Ricken, W., 1985. Epicontinental marl – limestone alternations: Event deposition and diagenetic bedding (Upper Jurassic, southwest Germany). In: U. Bayer and A. Seilacher (Editors), *Sedimentary and Evolutionary Cycles.* Lecture Notes Earth Sciences, 1: 127 – 162.

Ricken, W., 1986. *Diagenetic Bedding: a Model for Marl – Limestone Alternations.* Lecture Notes Earth Sciences, 6: 210 pp.

Ricken, W., 1987. The carbonate compaction law: a new tool. *Sedimentology,* 34: 571 – 584.

Ricken, W. and Eder, F.W., 1991. Diagenetic overprint in calcareous rocks: modification of stratification and bedding rhythm – overview. In: G. Einsele, W. Ricken and A. Seilacher (Editors), *Cycles and Events in Stratigraphy.* Springer, Berlin (in press).

Rieke, H.H. and Chilingarian, G.V., 1974. *Compaction of Argillaceous Sediments.* Developments in Sedimentology, 16. Elsevier, Amsterdam, 424 pp.

Rittenhouse, G., 1971. Pore space reduction by solution and cementation. *Bull. Am. Assoc. Pet. Geol.,*

55: 80 – 91.
Schlanger, S.O. and Douglas, R.G., 1974. Pelagic ooze – chalk – limestone transition and its implications for marine stratigraphy. In: K.J. Hsü and C. Jenkyns (Editors), *Pelagic Sediments. Int. Assoc. Sedimentol., Spec. Publ.,* 1: 117 – 148.
Schneiderman, N. and P.M. Harris (Editors), 1985. *Carbonate Cements. Soc. Econ. Paleontol. Mineral., Spec. Publ.,* 36: 1 – 379.
Schwarzacher, W., 1987. The analysis and interpretation of stratification cycles. *Paleoceanography,* 2: 79 – 95.
Scoffin, T.P., 1987. *An Introduction to Carbonate Sediments and Rocks.* Blackie, London, 274 pp.
Seibold, E., 1962. Kalk-Konkretionen und karbonatisch gebundenes Magnesium. *Geochim. Cosmochim. Acta,* 26: 899 – 909.
Shinn, E.A., Halley, R.B., Hudson, J.H. and Lindz, B.H., 1977. Limestone compaction – an enigma. *Geology,* 5: 21 – 24.
Shinn, E.A., Robbin, D.M. and Claypool, G.E., 1984. Compaction of modern carbonate sediments: implications for generation and expulsion of hydrocarbons. In: J.G. Palacas (Editor), *Petroleum Geochemistry and Source Rock Potential of Carbonate Rocks. Am. Assoc. Pet. Geol., Studies in Geology,* 18: 197 – 203.
Simpson, J., 1985. Stylolite-controlled layering in an homogeneous limestone: pseudo-bedding produced by burial diagenesis. *Sedimentology,* 32: 495 – 505.
Steinen, R.P., 1978. On the diagenesis of lime mud: scanning electron microscopic observations on subsurface material from Barbados. W.I. *J. Sediment. Petrol.,* 48: 1139 – 1147.
Tucker, M.E. and Wright, V.P., 1990. *Carbonate Sedimentology.* Blackwell Scientific Publications, Oxford, London, 482 pp.
Wanless, H.R., 1979. Limestone response to stress: pressure solution and dolomitization. *J. Sediment. Petrol.,* 49: 437 – 462.
Wedepohl, K.H., 1970. Geochemische Daten von sedimentären Karbonaten und Karbonatgesteinen in ihrem faziellen und petrographischen Aussagewert. *Verh. Geol. Bundesanst. Wien,* 4: 692 – 705.
Weedon, G.P., 1989. The detection and illustration of regular sedimentary cycles using Walsh power spectra and filtering with examples of the Jurassic of Switzerland. *J. Geol. Soc. London,* 146: 133 – 144.
Wolf, K.H. and Chilingarian G.V., 1976. Compactional diagenesis of carbonate sediments and rocks. In: G.V. Chilingarian and K.H. Wolf (Editors), *Compaction of Coarse-Grained Sediments.* Developments in Sedimentology, 18B, Elsevier, Amsterdam, pp. 719 – 768.
Wolfe, M.J., 1968. Lithification of a carbonate mud: Senonian chalk in Northern Ireland. *Sediment. Geol.,* 2: 263 – 290.
Zankl, H., 1969. Structural textural evidence of early lithification in fine-grained carbonate rocks. *Sedimentology,* 12: 241 – 256.

Chapter 7

DIAGENETIC HISTORY OF THE AYMESTRY LIMESTONE BEDS (HIGH GORSTIAN STAGE), LUDLOW SERIES, WELSH BORDERLAND, U.K.

A. HAMID MOHAMAD and E. V. TUCKER

INTRODUCTION

The Aymestry Limestone Beds (Murchison, 1834; Alexander, 1936) represent the only significant carbonate facies in the Ludlow rocks of the Welsh Borderland. They occupy the upper part of the Bringewood Formation within the Gorstian Stage of the Ludlow Series (Holland et al., 1980; see Table 7-1). The facies is developed throughout much of the shelf area of the Welsh Borderland, but passes westwards into calcareous siltstones in the Ludlovian basin facies of Wales (Lawson, 1973). In the shelf areas, the sediments exhibit a mixed carbonate to fine clastic facies characteristic of a shallow-water epeiric sea.

The upper boundary of the Aymestry Limestone can be shown faunally to cross into the younger Ludfordian Stage, especially to the west of the town of Ludlow, but the base is no longer considered to be diachronous (Lawson, 1973).

The carbonate facies reaches its maximum development along the Main Outcrop between Aymestry and View Edge (Fig. 7-1). It persists as a true limestone into several of the Silurian inliers in the eastern area of the Welsh Borderland, but passes southward into more clastic sediments that are best regarded as highly calcareous siltstones (Usk and Tites Point). Sedimentation was interrupted in the late Gorstian in these southern areas producing, in places, a pronounced stratigraphical hiatus and absence of the Aymestry Limestone Beds.

A twofold sequence consisting of a lower silty facies and an upper carbonate facies, designated Megafacies A and Megafacies B, make up the Aymestry Limestone. Calcareous siltstones and bioturbated mudstones of subtidal origin and characterized by *Thalassinoides* sp., form the sites of secondary concretions, and are widespread at first. Stratigraphically higher levels are more calcareous with shoal banks consisting largely of the thick-shelled brachiopod *Kirkidium knightii,* coral – bryozoa biostromes, coral – cryptalgal laminites and ostracod grainstones along open shelf areas. Nodular limestones with concretions centered on *Thalassinoides* burrows and, towards the top of the Aymestry Limestone, based on *Ophiomorpha* and *Lingulichnus* sites, are common in a protected lagoon setting. Thin conglomeratic limestones terminate sedimentation over the region of the lagoon.

The Welsh Borderland area witnessed a change from subtidal to intertidal environments in an open embayment backed by protected lagoons. Interruption of sedimentation in the later stages is marked by hardgrounds with little evidence of extended exposure in the supratidal zone. The fabric and mineralogy of skeletal

TABLE 7-1

Classification of Ludlow Series in the Welsh Borderland, U.K. (based on Holland et al., 1980)

Ludlow Series	**Ludfordian stage**	
	Whitcliffe Formation	Upper Whitcliffe Beds
		Lower Whitcliffe beds
	Leintwardine Formation	Upper Leintwardine Beds
		Lower Leintwardine Beds
	Gorstian Stage	
	Bringewood Formation	Upper Bringewood Beds
		Lower Bringewood Beds
	Elton Formation	Upper Elton Beds
		Middle Elton Beds
		Lower Elton Beds

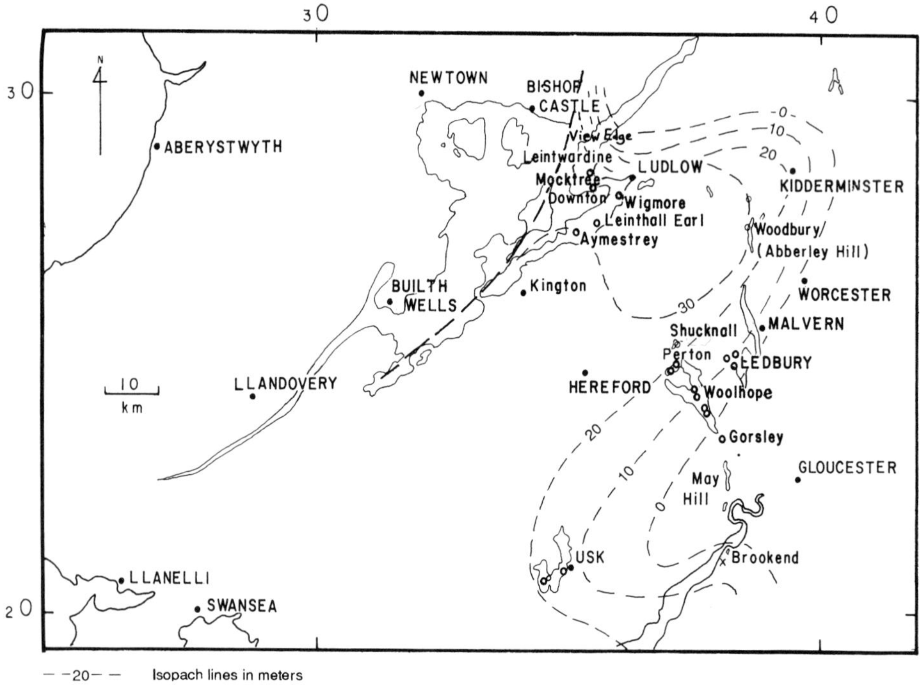

Fig. 7-1. Distribution of Ludlow Series in the Welsh Borderland area. (Important exposures of Aymestry Limestone are indicated by "o" and Brookend borehole by "×". Isopach lines are shown with the thickness given in meters.)

grains and the paragenesis of cements reflect salinity of the precipitating environments and support a shoaling and shallowing model for the late Gorstian.

STRATIGRAPHIC FRAMEWORK

A rapid transition from irregularly and thickly bedded siltstone to massively bedded, calcareous and nodular siltstones defines the base of the Aymestry Limestone. A marked numerical decrease in the strophomenid brachiopods (typifying the lower Bringewood Formation) and a concomitant increase in corals and bryozoa accompany the lithological change. *Atrypa reticularis, Strophonella euglypha* and *Sphaerirhynchia wilsoni* form a distinctive assemblage in the lower one-third of the upper Bringewood Formation (Aymestry Limestone). The remaining succession is dominated by a coral – bryozoa association with the addition of *Kirkidium knightii* along the Main Outcrop. The top of the upper Bringewood Formation (and hence Gorstian Stage) is defined by a rich assemblage of *Dayia navicula* and *Isorthis orbicularis*, together with several other distinctive brachiopods and the trepostome bryozoa *Orbignyella fibrosa*. The faunal change is usually accompanied by a lithological change to argillaceous siltstones except in the Abberley – Ludlow Area at Shelderton and on Wigmore Road (Ludlow), where a nodular limestone facies persists into the lowest Ludfordian Stage and the top of the Aymestry Limestone is diachronous.

LITHOLOGICAL CHARACTERISTICS

Mixed carbonate – fine clastic sediments dominate the Aymestry Limestone with a notable increase in carbonate content through the succession. A pure crystalline limestone facies is generally absent except within some bioclastic – biostromal units in the Main Outcrop. Elsewhere, impure carbonates occur as nodules, lenticles and thin flaggy beds of limited extent, and are characteristic of the formation.

Six lithofacies can be defined in the broad area spanning the eastern inliers (but excluding Usk and Tites Point) and these are typified in the continuous section at Shucknall Quarry (Fig. 7-2 and Table 7-2).

Contrasting with this succession there is a more uniformly silty facies with a low carbonate ratio, seen at Usk and Tites Point. Only Lithofacies 1 and 2 of Shucknall Quarry are well represented in these areas where they form the lower two-thirds of the total thickness. The upper part consists of thin beds of limey wackestone, and finely laminated siltstone. Small-scale scour channels are filled either with fine sandstone or shell coquinas.

Along the Main Outcrop the Aymestry Limestone is richly calcareous and dominated by either shell bank limestone up to 4 m in thickness or thinner coral – cryptalgal units.

The succession displayed at Downton Gorge (National Grid Ref: SO/43067313) and Mocktree (SO/41507540) (Fig. 7-3) consists of:

Unit 5 (**top**)	–	Thin, irregularly bedded grainstone (Lithofacies 6 of Shucknall)
Unit 4	–	Coral – cryptalgal biostromes
Unit 3	–	*Kirkidium* shell bank
Unit 2	–	Limey wackestone (Lithofacies 2 of Shucknall)
Unit 1	–	Nodular siltstone

Units 3 and 4 are typical of and largely confined to the outer shelf areas.

Summary of facies relationships

In the eastern inliers, Lithofacies 1 – 3 (Table 7-2) comprise Megafacies A, Lithofacies 4 – 6 comprise Megafacies B. Towards the western edge of the shelf Megafacies B is notable for the occurrence of storm-generated shell banks.

The two megafacies occur widely in the Welsh Borderland although the lower silt-dominated facies is lithologically more uniform than the upper carbonate-rich facies. Massively-bedded blue grey calcareous siltstones, structureless through bioturbation, pass upwards into concretionary siltstones with wackestone nodules based on *Thalassinoides* sites. The nodules assume lensoid and lenticular forms, some coalescing into irregular limestone beds. Bedding can be traced from the surrounding sediment into this type of nodule. The nodular horizons are spaced at regular intervals of about 0.25 m, alternating with thicker siltstone units imparting cyclicity to the sequence. Fine laminae are frequently disrupted by small *Chondrites* burrows. Epifaunal bryozoa and stromatoporoids are attached to the surface of some nodules indicating early diagenesis and exposure of nodules on the sea bed.

Three types of concretion exist:

(1) Ellipsoidal to sub-spherical forms less than 150 mm in length and no more than 50 mm thick.
(2) Lenticular forms with obvious channel geometry less than 250 mm wide, 70 mm deep, and with a variable length.
(3) Coalescing forms attaining a length of 3 m and commonly displaying internally dichotomous branching with bulbous chamber structures.

Mottled grey, structureless mudstones conclude the succession in Megafacies A. Small horizontal burrows up to 5 mm in diameter, with a maximum length of 30 mm, infilled with paler silt, are common. Benthic organisms are rare except for *Lingula* sp. and several gastropods with a grazing habit. Disseminated pyrite is present throughout the two structureless units, aggregating into concretions 40 mm in diameter in the top unit.

Smectite-rich clay beds of volcanic origin occur at several levels throughout the Aymestry Limestone. They are usually thin, but thicker beds range from 100 to 250 mm in thickness. These more prominent seams commonly separate distinct lithological units, and show current reworking of sediment in their upper part together with reworking by infaunal organisms.

Lithofacies 4 represents the acme of calcareous sedimentation coinciding with flourishing colonies of solitary corals and dominated by traces of *Lingulichnus* sp.

This represents the major lithology in Megafacies B. Although it appears massive when fresh weathering reveals a mass of closely-spaced limey packstone nodules in a subordinate matrix of calcareous siltstone. The nodules are usually ellipsoidal to sub-spherical and more rarely irregular, and up to 50 mm long. They constitute at least 80% of the sediment volume. In the lower part of this unit, the long axis of the nodules adopts a random orientation, succeeded by beds in which the majority of the nodules lie discordant to the bedding (*Lingula* sp. has been seen at the core

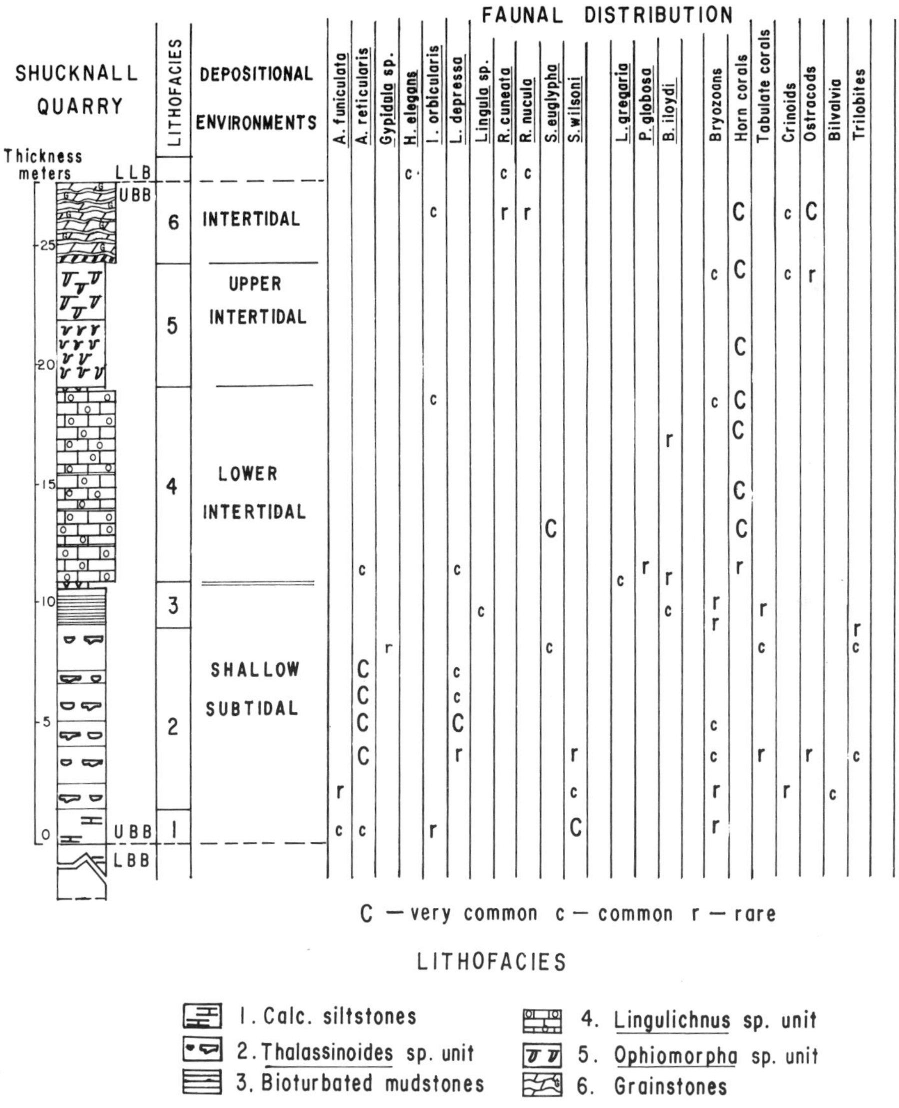

Fig. 7-2. A typical Aymestry Limestone sequence for the inshore shelf area, showing the distribution of lithofacies and fauna.

of many of these nodules) and finally by beds where nodules are orientated parallel to the bedding. Different organisms or environments must be responsible for these contrasts. At each point of change, concretions coalesce into a thin continuous bed suggesting an extended phase of diagenesis perhaps accompanying a hiatus.

TABLE 7-2

Generalized succession for the Aymestry Limestone of the protected inshore belt represented by the inliers of the eastern Welsh Borderland (thickness values given are for the Shucknall inlier)

Lithological character	Environment
Lithofacies 6 (3 m thick) TOP	
Crystalline limey grainstone interbedded with flaser-bedded siltstones carrying veneers of cryptalgal laminites	A sharp base and irregular top to the lenticular bedded grainstones, accompanied by microkarst solution features indicating erosion and intermittent exposure in a shallow inshore zone; algal drapes flourished at times on the silt substrate
Lithofacies 5 (6 m thick)	
Massive calcareous siltstones containing concretions orientated subvertically and obliquely to the bedding; this lithofacies is restricted to the Shucknall – Woolhope inliers	Concretions follow the club-shaped dwelling burrows of *Ophiomorpha,* belonging to a moderately high-energy regime
Lithofacies 4 (8 m thick)	
Nodular limestones consisting of relatively small concretions (average size 80 mm × 30 mm) in close contact; most nodules are vertical or subvertical; solitary corals reach their acme in this unit and the brachiopods *Lingula* sp. and *Strophonella euglypha* are common.	Nodules can be related to the dwelling burrows of the inarticulate brachiopod *Lingula;* solitary corals indicate a clear, shallow sea
Lithofacies 3 (3 m thick)	
Bioturbated mudstones devoid of bedding; fossils are rare.	A uniform argillaceous facies of a quiet-water setting with a rich infauna; organism activity has destroyed almost all traces of primary lamination
Lithofacies 2 (4 m thick)	
Blue grey, calcareous siltstone with discrete nodules of limey wackestone; nodules merge into lenses up to 350 mm in length; brachiopods are abundant	The loci of nodules are typically burrow traces of *Thalassinoides* and the lensoid concretions are based on the sites of shallow channels a few centimeters deep; the siltstones are homogeneous and represent a quiet-water, subtidal environment, washed by gentle currents
Lithofacies 1 (1.6 m thick) BASE	
Olive grey, massive siltstone with occasional nodules	This unit forms a transition from beds of the underlying lower Bringewood formation; it represents subtidal siltstones with a low carbonate content

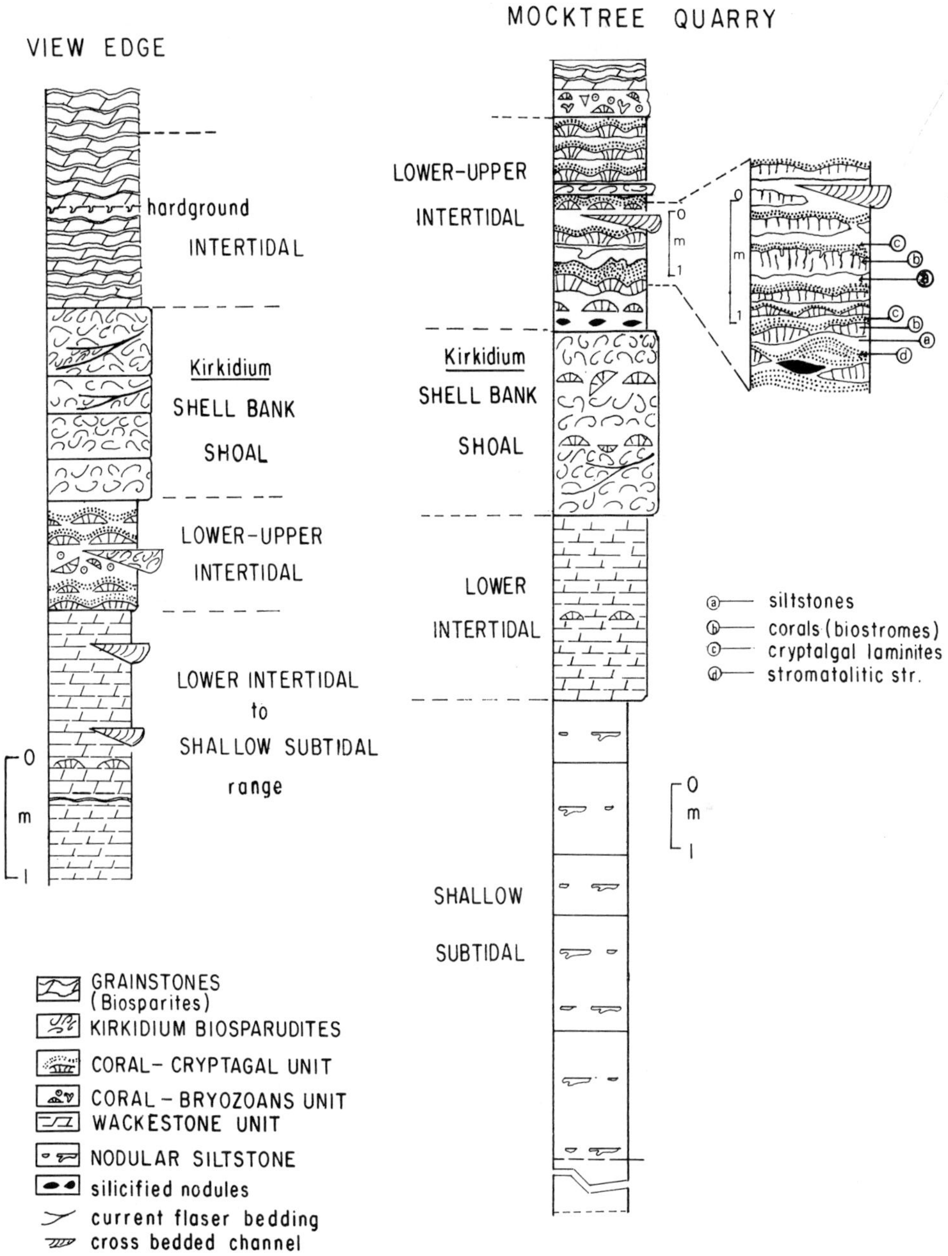

Fig. 7-3. Upward shoaling sequences from shallow subtidal to tidal flat conditions on the outer part of the shelf. *Kirkidium* shell banks and a coral – cryptalgal facies characterize the shoal environment.

The sediment between the nodules is commonly intensely bioturbated and preserves chondritic burrows about 3 mm in diameter. Where nodules are randomly orientated, these small burrows are predominantly horizontal. At the *Lingulichnus*-dominated level, where the nodules are mainly vertical, the small burrows have also a vertical orientation but avoid the nodules, suggesting early lithification, pre-dating the burrowing activity of the *chondrites* organism.

Lithofacies 5 is restricted geographically to the Shucknall inlier and the northern end of the Woolhope inlier; it is dominated by *Ophiomorpha* traces. This facies is not in a consistent stratigraphic position even in this small area, forming the lowest lithotope of Megafacies B at some localities (e.g., Tower Hill) and the topmost one elsewhere (Dean's Place).

Light grey nodular limestone of limey packstone – grainstone type set in a subordinate matrix of darker bioturbated calcareous siltstone typify the facies. Nodules have distinctive morphologies and their relationship with burrow sites is more exact than in Lithofacies 4. Burrow styles include club-shaped burrows with an inclined to horizontal tunnel up to 10 mm in diameter enlarged at its lower end, through relatively narrow vertical shafts with a terminal curvature, to vertical spiral shafts with the form of *Gyrolites* sp. extending downward for up to 300 mm. Some burrows have truncated upper ends and these can be filled with shells. The different styles of burrows are again stratigraphically arranged and at Shucknall Quarry the lower half of Lithofacies 5 is characterized by club-shaped concretions succeeded by a sub-unit of nodules based on the truncated burrows, and associated with periods of stronger current activity. The majority of burrow styles is attributed to *Ophiomorpha* sp.

Thinly-bedded grainstones with numerous ostracods form the youngest sedimentary unit in Megafacies B and these sediments stretch from the eastern inliers to the Main Outcrop. Only at Usk and Tites points the facies is not represented, being replaced by siltstones. Lenticular to flaggy limestones with partings of thinly laminated siltstone characterize this stratigraphic level. The limestones have frequently a smooth base but an irregular top surface, the product of either erosion or solution. Domal shapes of stromatolite form are produced within the laminated siltstone and some veneers are friable, consisting of dark brown organic laminite, probably of cryptalgal origin.

Thin intraformational conglomeratic limestones, up to 60 mm thick, interbedded with richly fossiliferous siltstones cap the grainstones. They form a significant stratigraphic horizon at the close of the Aymestry Limestone and continue into the earliest Ludfordian. This lithofacies forms a thin blanket overstepping Lithofacies 5 and 6 in the eastern inlier. In the Main Outcrop it is recorded only at Bengry Track (Lawson, 1973). It represents the terminal phase of shallowing prior to the onset of a transgression in the Ludfordian, and is preserved as transgressive lag deposits. These two processes cause the omission of the Aymestry Limestone and other older Ludlow deposits from the southern part of the Woolhope inlier to May Hill.

The conglomeratic limestones contain smooth ellipsoidal limestone clasts supported in a framework of comminuted shell debris. Pebbles are usually less than 50 mm in length and are from 5 to 30 mm thick. Small surface borings less than 5 mm in diameter have been attributed to *Trypanites* sp. (Cherns, 1980). At the southern end of the Woolhope inlier and at May Hill, pebbles in the basal Ludfordian are more variable in composition including lime mudstone or wackestone pebbles and phosphatic nodules, probably with a longer history of reworking (probably during storms) on the sea floor. Some show imbricate stacking.

Several minor lithofacies occur sporadically at many stratigraphic levels. The more significant lithofacies are cross-bedded calcarenites found in small-scale scour channels and cut-and-fill structures, typically 300 mm wide and 10 mm deep, lenticular coquina limestone, and sediments that show early stages in the development of omission surfaces and hardground. Cross-bedding in channel sediments is highlighted by the conspicuous alignment of shell debris. Graded bedding is developed in the synchronously deposited sediment of each cut-and-fill structure, with coarser bioclastic constituents at the base grading into finely laminated siltstone. These structures are seen most commonly on the surface of volcanic ash (bentonite) seams and produce discrete lenses of calcareous sediment. The limestone coquinas occur as lenticular bodies, also with channel geometry. They are less than 40 mm thick, but extend laterally up to 5 m. These thin beds of concentrated shells attain their maximum development in Lithofacies 6; they occur only rarely in Megafacies A. Disarticulated and comminuted brachipod valves predominate, probably concentrated in depressions by storm swells. Omission surfaces and hardgrounds provide evidence of interrupted sedimentation and their more common occurrence in the higher beds of the Aymestry Limestone is an indicator of shallower water conditions. Firm and even stony substrates were produced providing sites for epifaunal attachment and encrustation. The hardgrounds display borings subsequently filled with clay. Incipient dolomitization is known in some of these hardgrounds, which characterize the terminal phase of shallowing.

At Usk (Fig. 7-4) and Tites points a bioturbated siltstone facies contrasts with the carbonate sediments seen elsewhere. Calcareous sediments are rare, restricted to organic concentrates and thin limey wackestones, occurring also as nodules within the otherwise fine-grained clastic sediments. The fauna leaves little doubt that the Upper Gorstian Stage is represented in these two areas. Finely comminuted shell debris abounds, attributable to infaunal organism activity. Small-scale coral–bryozoa biostromes (not exceeding 150 mm × 5 m), based on *Favosites* sp. and encrusting bryozoa, are associated with limestone pebble seams (e.g., Coed-y-Ffern, Llandegfedd Reservoir, Brookend Borehole). Small channels not deeper than 100 mm and up to 3 m wide are filled with fine-grained current bedded arenites, shell fragments, and disarticulated valves. Imbricate packing of isorthid valves in the upper layers suggests shallow water, perhaps in an intertidal setting.

Conglomeratic limestones are a feature of the Tites Point area (Brookend Borehole), concentrated towards the top of the sequence. *Trypanites* sp. borings are abundant in the pebbles.

Along the strike in the Main Outcrop, the Aymestry Limestone varies substantially in thickness. The Leinthall Earls and Downton Gorge sections are exceptional for their thickness of about 40 m; elsewhere the thickness rarely exceeds 18 m.

Megafacies A occupies about half the total thickness consisting of bioturbated calcareous siltstone (Lithofacies 1) and nodular (wackestone) siltstone (Lithofacies 2 but lacking *Thalassinoides*) closely resembling the facies in the eastern inliers. The

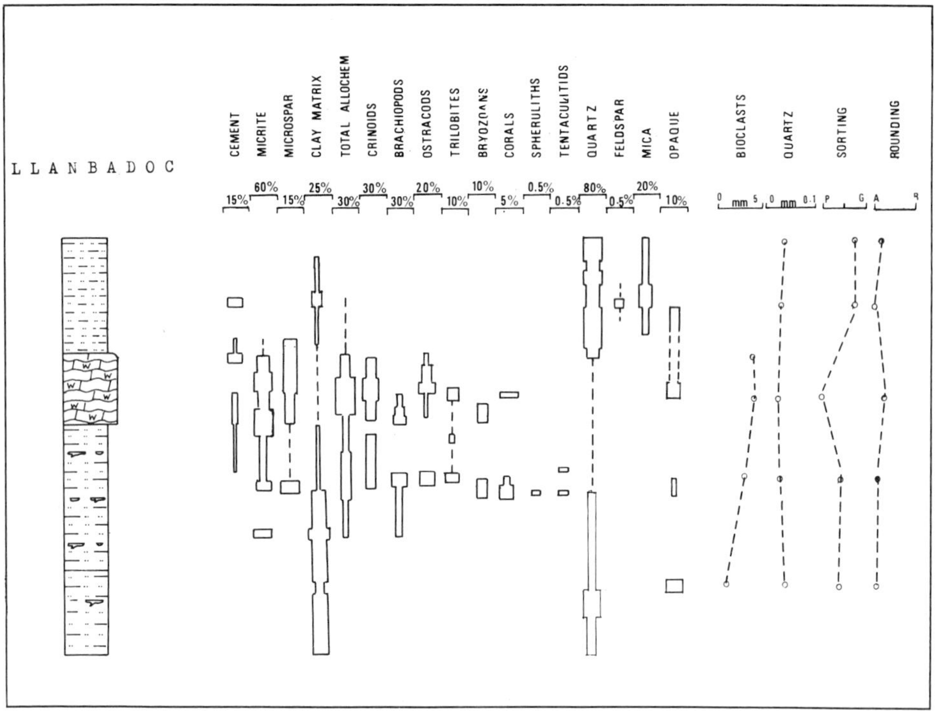

Fig. 7-4. A lithofacies and faunal profile of the Aymestry Limestone in the Usk inlier (*w* = wackestone).

carbonate-rich Megafacies B contrasts with these eastern inliers with thick *Kirkidium* shell bank units and coral – cryptalgal laminite units dominant. The flaggy-bedded ostracod grainstones that terminate the succession indicate a greater uniformity of sedimentary environment across the whole shelf. Up to three shell banks occur with, at Aymestry, seven conspicuous seams of *Kirkidium knightii*. The coquinas are incorporated into a framework of limey grainstones consisting of comminuted shell debris, algal fragments and spheruliths.

The shell banks represent the most turbulent conditions in the Aymestry Limestone developing as shoals in the tidal zone. Coral – cryptalgal units range from a single layer of coral to units 3 m thick, consisting of interbedded coral seams and finely laminated siltstones. They include laterally persistent colonies of *Favosites* sp., *Heliolites* sp., and stromatoporoids usually found in growth position; but they are sometimes overturned. Holes near the center of heliolitid corals indicate boring by lithophages and Newall (1966) demonstrated a symbiotic relationship with *Lingula* sp. Thin veneers of brown friable laminite (5 mm) drape the corals, overlain in turn by flaser-bedded siltstones (not more than 40 mm). Corals re-established themselves on this surface and the pattern is repeated. Laminites develop into domed stromatolites above some corals. Shallow channels 100 mm deep and 1.5 m across are common, filled with cross-bedded and well-sorted bioclastic grainstone (or biosparites).

The upward change from Megafacies A to Megafacies B illustrates replacement of relatively quiet mud-rich waters by turbulent clearer water and protected coralliferous environments. At Leinthall and Aymestry the two facies are repeated in four shoaling cycles.

At Ludlow, on an easterly arm of the Main Outcrop, shoaling sequences cannot be identified. The succession resembles more closely the eastern areas and the facies suggests protected environments in the lee of shell banks lying to the west.

PETROGRAPHY

The carbonate rocks of the Aymestry Limestones are distinguished compositionally from the siliciclastic sediments by containing at least 50% carbonate minerals. The rock is designated calcareous if the concentration of carbonate minerals is between 10 and 50% of the bulk composition. Because carbonate rocks contain both physically transported particles, and chemically precipitated in-situ carbonates or cements (ranging from micrite mud to the products of recrystallization and replacement), a dual scheme for the description of particle size is warranted, to distinguish between primary and secondary constituents. For example, calcirudite may be cemented with coarse crystalline calcite, or calcarenite may be supported by very fine crystalline calcite.

The descriptive term "extremely fine crystalline" (for crystal sizes between 0.005 and 0.004 mm), measured on the Scanning Electron Microscope, is preferred to "aphanocrystalline" (Folk, 1965).

Carbonate rocks are inherently complex (Chilingar et al., 1967a; Bricker, O.P., 1971; Wilson, J.L., 1975). The grains, although largely monomineralic, are texturally diverse and polygenetic. The basic textural components in carbonate rocks are grains, matrix, cement, and voids (Leighton and Pendexter, 1962), although voids have been totally occluded in the Aymestry Limestone through cementation by sparry calcite, and by microcrystalline calcite (VFxn) matrix and/or cement within intergranular spaces. A modified definition of micrite is adopted here to refer to a coherent crystal fabric with particle size less than 0.012 mm, irrespective of origin.

Allochems in the Aymestry Limestone are discrete particles, 0.016 mm or larger, usually originating within the depositional basin. The grain framework includes intraclasts derived from the breakup of penecontemporaneous carbonate sediments which are disrupted by: burrowing activity or desiccation, skeletal grains (bioclasts), peloids and pelletoids composed of microcrystalline calcite showing either an organic structure of fossil fragments (peloids) or composite quartz grains in a mud matrix (pellets), ooids and spheruliths, the latter attributable to phosphatic calculi originating within zooecial chambers of bryozoa (Oakley, 1936). Orthochemical components comprise sparry calcite, microspar and micrite (for the siltstone microfacies, the mud matrix constitutes detrital grains less than 0.020 mm in diameter). Other miscellaneous constituents of significance are secondary replacement components resulting from dolomitization, silicification, and pyritization. Figure 7-5 attempts to integrate the schemes of Folk (1959, 1962), Dunham (1962) and Plumley et al. (1962) in a form modified to be highly descriptive of rock texture,

DUNHAM (1962)				FOLKS (1959, 1962)					ENERGY INDEX Plumley *et al.* (1962)
TEXTURE				GRAIN TYPES					
DEPOSITIONAL TEXTURE RECOGNIZABLE	Original component not bound together during deposition	MUD SUPPORTED	Limy-MUDSTONE (10% grains)	ORTHOCHEMICAL ROCKS: Micrite, dismicrite, lime mud, calcilutites, chalk.					QUIET I (Deposition in quiet H_2O)
				ALLOCHEMICAL ROCKS	BIOCLAST	INTRACLAST	PELLETS	OOIDS	
			Limy-WACKESTONE (10% grains)		BIOMICRITE	INTRAMICRITE	PELMICRITE		INTERMITTENTLY AGITATED II (Deposition in alternately agitated and in quiet H_2O)
		GRAIN SUPPORTED	Limy-PACKSTONE (5% mud)		BIOMICSPARITE	INTRA-MICSPARITE	PELMICSPARITE	OOMICSPARITE	SLIGHTLY AGITATED III (Deposition in slightly agitated H_2O)
			Limy-GRAINSTONE (5% mud)		BIOSPARITE	INTRASPARITE	PELSPARITE	OOSPARITE	MODERATELY AGITATED IV (Deposition in moderately agitated H_2O)
	ORGANIC frame-builders		Limy-BOUNDSTONE	AUTOCHTHONOUS REEF ROCKS Biolithite					STRONGLY AGITATED V (Deposition and growth in strong agitated water)
DEPOSITIONAL TEXTURE OBLITERATED	Recrystallized Texture		Crystalline	MICROSPARITE (13 - 30μ) PSEUDOSPARITE (> 30μ)					

Fig. 7-5. Descriptive classification for limestones.

but with a salient genetic implication. The following examples illustrate the method used in assigning the rocks:

(1) grain-supported, shelly biosparite (grainstone);

(2) grain-supported, ostracod biomicsparite (packstone); and

(3) mud-supported, pellet, intramicrite (wackestone).

The examples offer a compromise between the classifications of Folk and Dunham to yield more flexible descriptions of the framework, compositional variation, and nature of the matrix. The energy index, based on Plumley et al. (1962), appraises the depositional environment, although here the grading spectrum of energy and water disturbance is not necessarily related to bathymetry or palaeogradient, because quiet-water and low-energy conditions can exist in both very-shallow protected water and basinal depocenters; great restraint should be exercised, however, on the usage of Plumley's hydrodynamic index. The common pitfall is the often misleading interpretation of micrite as a typical indicator of quiescent environment, because micrite can also be derived from degrading neomorphic recrystallization of precursor sparry calcites (cements).

The Aymestry Limestone commonly shows incomplete recrystallization with recognizable allochem components, except that precursor micrite matrix has recrystallized to 13 – 30 μm microspar. The suffix "microsparite" is added to the existing term, if the matrix is dominantly composed of microspar. The equivalent Dunham term is limey wackestone, because microspar evidently constitutes the mud-supported texture.

The rocks are analyzed using ultra-thin sections and acetate peels of etched polished surfaces. The cementation paragenesis and chemical composition are derived from the analyses using Alizarin Red S – Potassium Ferricyanide staining (Dickson, 1966), SEM, EDS, and XRD techniques. Dickson's method was modified for fine-grained carbonate sediment by reducing the recommended concentration of the etching solution to 0.5% HCl and the duration of staining. Stage 3 of Dickson's procedure was omitted, because it deepened the stain and obscured finer fabric. Staining time was critical and etching was restricted to 15 seconds with 0.5% HCl and 45 seconds for staining with Alizarin Red S – Potassium Ferricyanide.

MICROFACIES

Clastic rocks

Fine-grained clastic rocks are the dominant lithotopes of the lower half of the Aymestry Limestone and throughout the whole succession in the Usk inlier.

These fine-grained quartz wackes form the bulk of the more massively bedded deposits of Lithofacies 1 and 2 of both inner and outer shelf areas. Loosely packed equigranular, angular quartz grains are supported in a mud matrix. Mica, authigenic pyrite and comminuted shell are minor constituents. The relative abundances of constituents are: 40.9% quartz, 6% mica, 7% skeletal grains, 5% authigenic pyrite, 39% matrix, and 2% calcite cement. The sediments are bioturbated and the fill of *Chondrites* sp. and *Planolites* sp. burrows are particularly well-sorted. They display corona-like backfill structures as evidence of organism manipulation. Finely comminuted skeletal grains are attributable to the foraging activities of infaunal organisms.

The size of quartz grains is mainly that of fine silt, sometimes reaching that of very fine sand. In contrast to the angular silt grains, the coarser fraction includes idiomorphic crystal shapes along with rounded and ameboid shapes some with rare inclusions of microapatite. These grains are probably of volcanic origin as well as minute laths of biotite, a common constituent of volcanic bentonite clays.

Calcareous siltstone consists of fine-grained, poorly-sorted angular clasts with finely comminuted skeletal debris consituting up to 25% of the grains. These sediments are the major type in Megafacies B, forming the host sediments of the *Lingulichnus* and *Ophiomorpha* units. Silt-grade quartz is the dominant constituent, but grains seldom display grain-to-grain contact. The loosely packed grains are supported by an argillaceous matrix of clay and other sheet silicate minerals with calcite cements and authigenic pyrite. Other clasts include plagioclase feldspar, glauconite, muscovite and tourmaline crystals.

Fine-grained quartz arenite is restricted to the Usk inlier within the topmost beds of the Aymestry Limestone. Individual beds are usually lenticular or channel-shaped set within a muddy siltstone facies. The sands display micro-scale cross-bedding. Grains are closely packed and devoid of matrix. Particle size ranges from coarse silt (0.020 – 0.063 mm) to fine sand (0.063 – 0.125 mm). Quartz grains are usually clean and well-sorted with a sub-angular shape. In the presence of calcite cements, a greater angularity of grains exists attributable to the corrosive action of calcitic solutions. Transported skeletal fragments are concentrated at the base of channels, and the grains are mostly replaced by hematite. The proportions of constituent grains are as follows: quartz, 73%; muscovite, 0.1%; biotite, 10%; pyrite, 1.5%; tourmaline, 0.3%; lithic fragments (chert), 3%; feldspar, 0.5%; skeletal grains, 0.1%;

and calcite cement, 5.5%.

The high degree of sorting and lack of mud matrix indicates winnowing of sediments in nearshore, possibly tidal environments. Sedimentary features such as small-scale cross-bedding in shallow channels suggest a reduced energy setting. The lithic fragments originated on nearby landmasses that are likely to have existed from Wenlockian times (Cope and Bassett, 1987).

Carbonate rocks

Depositional textures range from a mud-supported framework to totally grain-supported frameworks consisting mainly of skeletal grains. Mud-supported wackestones characterize Megafacies A, whereas limey packstone and grainstone types typify Megafacies B and indicate increased energy levels.

Limey wackestone, mud-supported wackestone and wackestone – packstone are the major microfacies of the nodular limestones especially in Lithofacies 2 (*Thalassinoides* unit). Mud-supported bioclastic biomicsparite (wackestones) consists principally of micrite matrix and microspar derived from recrystallization of micrite. Bioclastic and detrital grains are subordinate allochemical elements. Sporadic patches of sparry calcite (pseudospar) are attributable to late-stage diagenetic recrystallization of micrite. Organic grains are derived from crinoids, trilobites, ostracods, brachiopods and rarely tentaculitids, with an average size of 0.2 mm. They have been pulverized by the foraging activity of *Chondrites* sp. and *Planolitis* sp. organisms. Quartz silt, fine-grained acicular muscovite, biotite, and glauconite are present also.

Wackestone – packstone (biomicrite – biomicsparite) occur as channel fills and show progressive textural gradation from one form to another in a single lenticle or scour channel. The rocks are supported by micrite and subordinate clay. Allochems consist of: (1) silt-grade angular quartz interclasts, occurring sparingly; (2) bioclastic grains constituting 50 – 60% of bulk volume; (3) colorless to pale yellow phosphatic spheruliths; (4) muscovite; (5) glauconite; and (6) authigenic pyrite. Micritization is common, especially involving crinoid bioclasts; bryozoan fragments are the most resistant. Borings in brachiopod fragments are also common and some of these micro-cavities are excavated beneath encrusting *Monotrypa* sp. bryozoa colonies. Algal biomicrite is found only at View Edge on the Main Outcrop occurring within a wackestone unit. It contains cryptocrystalline resinous algal fragments. Some micrite grains, however, display rosette and acicular forms (Fig. 7-6).

The wackestone – packstone associations are poorly sorted with a mud-supported framework, indicative of relatively calm conditions. Much of the micrite matrix has recrystallized to microspar and sometimes to pseudospar. Husseini and Mathew (1972) ascribed recrystallization and textural obliteration in muddy limestones to sedimentary environments characterized by restricted circulation conducive to high salinities and temperature. The interbedded algal biomicrite with abundant cryptocrystalline calcite in the matrix supports a quiet-water environment and shallow bathymetry. A substantial proportion of fossil remains have undergone post-mortem transport, however, suggesting intermittent disturbance of the water. These associated microfacies suggest alternating agitated and quiet water in a protected, possibly lagoon, situation. The poorly-sorted carbonate sediments are collected in pockets in front of migratory ripples occasioned by low-amplitude waves.

Limey packstone microfacies characterize both Lithofacies 4 (*Lingulichnus* unit) and Lithofacies 5 (*Ophiomorpha* unit) of the eastern inliers. Grain-supported skeletal packstone or biomicsparites form the infill of *Lingulichnus* and *Ophiomorpha* burrows. The allochems are principally highly-comminuted bioclastic grains, although the concentration of silt-grade quartz is significant in the vicinity of intense

Fig. 7-6. (A) Thin section photomicrograph of algal biomicrite showing algal fragments with a distinctive cellular structure.
(B) SEM photograph of the algal fragments. The original structure is preserved.
(C and D) SEM photographs of the matrix consisting of cryptocrystalline calcite with rosette and acicular forms.

bioturbation. Relative proportions of allochems and orthochems are as follows: bioclasts, 42%; quartz grains, 8%; sparry calcite, 28%; clay-grade matrix, 21%; and authigenic pyrite and accessory minerals, up to 1%. The bioclasts consist mainly of crinoid and bryozoan remains with an increase in the proportion of crinoids upwards, reflecting external ecological controls. *Fistulipora* sp., *Rhomboporella* sp., *Monotrypa* sp., and *Ptilodictya* sp. are common bryozoa. Other bioclasts include calcisponge spicules and brachiopods. Crinoid fragments have been preferentially micritized and microborings on brachiopod fragments are infilled with micrite cements. Phosphatic calculi are accessory constituents, and their abundance increases parallel with bryozoa concentrations. Quartz grains are sporadically distributed, and are notably more abundant at the site of chondritic burrows. They show two modal sizes, silt and fine sand (0.07 mm); the coarser grains tend to be better rounded and display corrosion reaction rims. The sparry calcite cement occurs both in the interskeletal and intraskeletal voids. In the former case, flat fragments of shell bridge the voids ("bridging and umbrella effect") providing a protective hood above the pore in which cement is precipitated. The intraskeletal voids produce geopetal structures with sediment filling the lower part. The matrix forms a significant proportion of the sediment bulk consisting of micrite (sometimes microspar) and clay minerals. The latter belong to the smectite group and are especially common in bioturbated areas.

Ostracod – bioclastic packstone (grain-supported ostracod – biomicrosparite) is associated with small-scale cut-and-fill channels with skeletal fragments aligned along current bedding. Microspar predominates over micrite in the matrix. This microfacies is seen chiefly in Lithofacies 4 and 6 in the Abberley Hills (Woodbury Quarry), with the channels frequently cutting down into volcanic smectite clay beds, and also at a similar stratigraphic level in the southern Woolhope inlier (Sleaves Oak, Dean's Place and Gwynne's Hill). The allochems consist largely of ostracod carapaces (average size 0.1 mm), most of which are altered to microcrystalline calcite; some are completely replaced by pyrite. Other bioclasts include bryozoa, gastropods, tentaculitids, trilobites, brachiopods and, rarely, sponge spicules. Relative abundances of constituents are: bioclasts, 29%; quartz grains, 12%; mica, 2%; micrite, 12%; sparry calcite, 30%; and authigenic pyrite, 15%. Micrite rind is common on shell surfaces and some grains are degraded to peloids. The ratio of micrite matrix (12%) to allochems (80%) indicates deposition in agitated water of an intertidal to subtidal setting.

Limey grainstones have a grain-supported fabric in which sparry calcite is common. Clay-grade material is absent except when introduced into borings in rock, notably in the topmost Aymestry Limestone.

A grain-supported shelly biosparite is seen as dense accumulation of shells within small channels (5 m wide and 80 mm deep): isorthid brachiopods are vertically stacked at Darren Farm (Usk inlier). The channels can be associated with several facies, for example, the *Thalassinoides* unit and the flaggy bedded grainstone of the Abberley Hills, whereas, at View Edge and Mocktree they form trough cross-bedded deposits within *Kirkidium* units. The bulk composition is as follows: bioclasts, 34%; detrital quartz, 12%; sparry calcite, 52.6%; matrix, up to 1%; and authigenic minerals (e.g., pyrite), 0.3%. The microfacies forms under moderate energy condi-

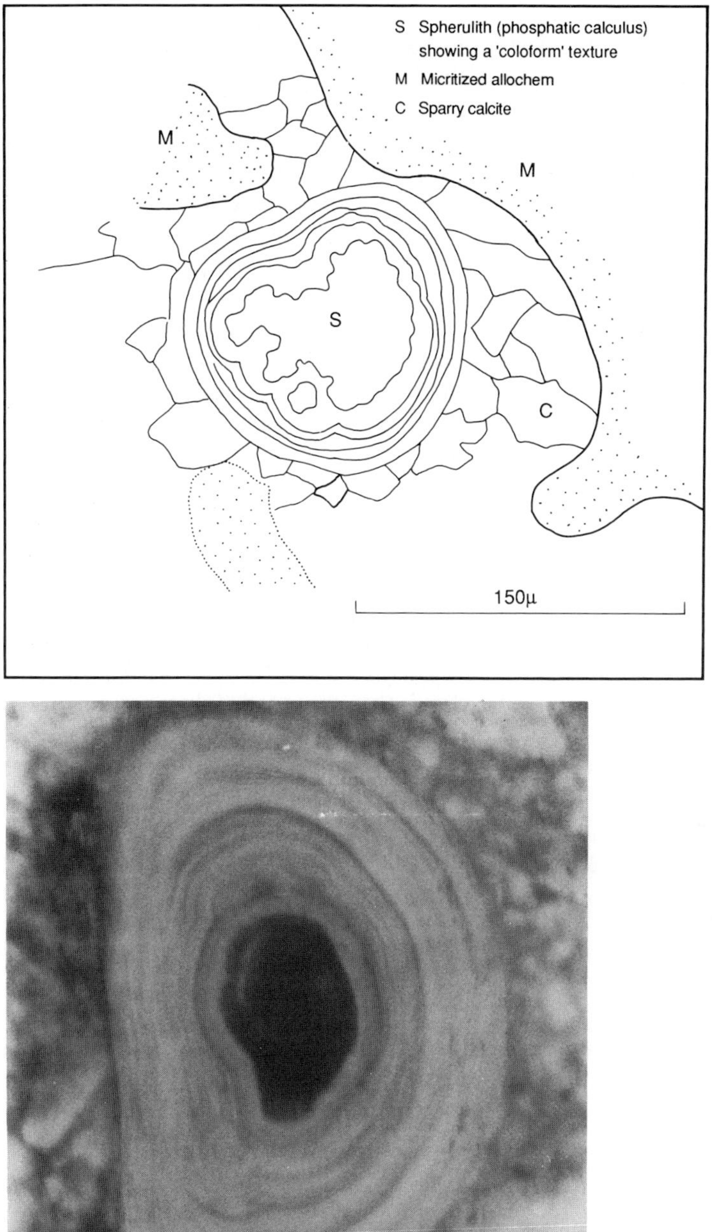

Fig. 7-7. Top: spherulith (phosphatic calculus: *S*) showing its colloform texture. (*M*: micritized allochem; *C*: sparry calcite.) Bottom: the reniform shape of a spherulith is illustrated in the photomicrograph.

tions possibly in the intertidal zone.

Grain-supported ostracod, bioclastic biosparites form major rock types accross the whole region and make up the bulk of the flaggy bedded grainstones (Lithofacies 6), the highest stratigraphical unit. Ostracod bioclasts constitute 50% of the allochems. The carapaces are disarticulated and arranged convex upward. Crinoid ossicles represent less than 10% and most have been developed into pseudospar though the mimetic microstructures are still visible. Phosphatic calculi (Fig. 7-7) are locally abundant and form a major constituent of the rock in the southern Woolhope inlier, and again in some outer shelf areas. Most ostracod valves display dark micrite rinds and pervasive micritization has continued to degrade some fragments into dark opaque grains. Micritization is fabric selective: the fibrous to foliated microstructures of punctate and pseudopunctate brachiopods are least affected. Sparry calcite cement is the major orthochemical constituent, occupying both intergranular and intragranular space. Intergranular calcite can be recrystallized to pseudospar especially in rocks that have undergone epigenetic changes (Wolf, 1965), illustrated by bored biosparites associated with hardgrounds. In such examples, where pseudospar is the dominant constituent, the term biopseudosparite is more appropriate.

Shelly biosparudite or coquina is confined to the outer shelf area where *Kirkidium* shell banks are found. The microfacies displays clean, well-sorted and rounded, closely-packed bioclastic frameworks, cemented with sparry calcite. The ratio of grain framework to cement is about 4:1. Micrite, rarely present as depositional matrix, is nevertheless common as part of the degraded fabric in bioclasts. Allochems consist of: bioclasts, 67%; spheruliths, 15%; peloid and pellet aggregates, up to 3%; ortho sparry calcite, 20%; and authigenic minerals, such as pyrite and opaque minerals; chert and dolomite together with idiomorphic quartz are also present.

Skeletal grains include a high proportion of worn and rounded undifferentiated forms (63%); the remainder consists mainly of brachiopods (12%), molluscs (4.5%), crinoids (9%), bryozoa (3.5%), and ostracods (8%). Bryozoan remains are predominantly of *Rhombopora* sp. with few other genera present. The remains are less abundant than in the eastern inliers but phosphatic calculi derived from bryozoa are relatively common. Crinoids are characterized by centripetal dark micrite rinds along with overgrowths of epitaxial sparry calcite. Peloids of microcrystalline calcite are relatively common, originating by total micritization of bioclasts. Pellets of fine quartz grains in a micrite or mud matrix probably have a faecal origin. Several generations of sparry calcite cement, showing ferroan and non-ferroan types, infill intergranular and intragranular voids. Micrite is present in some skeletal fragments as geopetal bottom-fill with overlying drusy calcite. In some cases these micrite patches have recrystallized to microspar.

Lithoclastic (conglomeratic) – skeletal biosparudite is characterized by lithoclasts of penecontemporaneous limestones. They are largely confined to the eastern areas, but occur also at Bengry Track (Lawson, 1973) near Aymestry. The lithotope forms a thin cover to the Aymestry Limestone and is the product of erosion of sediment recently lithified and transported only a short distance within the depositional basin. The clasts include elongate and flaky pieces, reminiscent of mud flakes in terrigenous sequences but of marine calcareous type. The host sediment consists of poorly-sorted, closely-packed allochemical grains supported by several generations of sparry calcite cement. Bioclasts and lithoclasts show a wide range of size and shape. The last generation sparry calcite occurs as fracture infills transecting the earlier-formed cement and carbonate grains. The skeletal fragments, substantially ostracods, are fragmentary although some bryozoa are well preserved. The lithoclasts were derived largely from rocks in the mud-supported limestone microfacies and range in size from granules to pebble and cobble classes (Wentworth Scale). Most of the clasts are well-rounded with oval shapes, others are flatter and more elongate. Some surfaces of the clasts are microbored suggesting pre-existence as hardgrounds. The composition of these lithoclasts varies from biomicrites (wackestones) to biomicsparites (packstones). Microspar is a common constituent and owes its origin to recrystallization of micrite. The microspar forms the core of the lithoclast, whereas outer zones are characterized by darker, cryptocrystalline calcite with appreciable amounts of pyrite. Much of the microspar has a ferroan calcite composition; the iron enrichment probably originates from the pyrite formed in microborings. Lithoclasts derived from older packstones contain finely comminuted bioclasts and display good sorting. In contrast, the bioclastic elements of the allochemical framework of the host rock is poorly sorted although most grains are well-rounded. They represent fragmentary remains (0.1 – 1.5 mm in size) of crinoids, sponge spicules, trilobites, bryozoa, corals, ostracods and brachiopods. More completely preserved bryozoa include *Leioclema* sp., *Rhombopora* sp., and *Fistulipora* sp., up to 2.5 mm in size.

The limey grainstones are well-sorted and generally lack micrite. Allochems are commonly closely packed and show imbricate stacking in channel-fill sediments. Ortho sparry calcite forms the cement in many examples. These features point to turbulent water conditions leading to the accumulation of onshore storm deposits in the form of *Kirkidium* shell banks. The poorly-sorted pebbles of the lithoclastic – skeletal biosparudite microfacies represent erosion of penecontemporaneously formed sediment transported away from the shore by storm swells or backwash.

Alteration of skeletal fragments through biologically-induced decay involves micritization of calcitic grains, together with the development of a surface crust. Algal and bacterial organisms are presumed to be responsible for the degradation, the process acting synchronously with deposition of the sediment. The process largely accompanies submarine diagenesis.

In the Aymestry Limestone a natural break in particle size distribution of cryptocrystalline calcite grains occurs between 0.012 and 0.014 mm (Fig. 7-8) and this conveniently divides micrite and microspar. The term micrite is restricted here to a coherent crystal fabric less than 0.012 mm in diameter and microspar refers to the size class between 0.013 and 0.03 mm, with pseudospar greater than 0.30 mm in size (Folk, 1959). Micritization involves alteration of pre-existing fabric by processes that may or may not be specific (Bathurst, 1966); no genetic connotations are implied (Milliman et al., 1985). The original fabric is transformed by destruction of its ordered arrangement (Alexandersson, 1972) and the original texture can be replaced completely. This has been expressed as "degrading recrystallization or

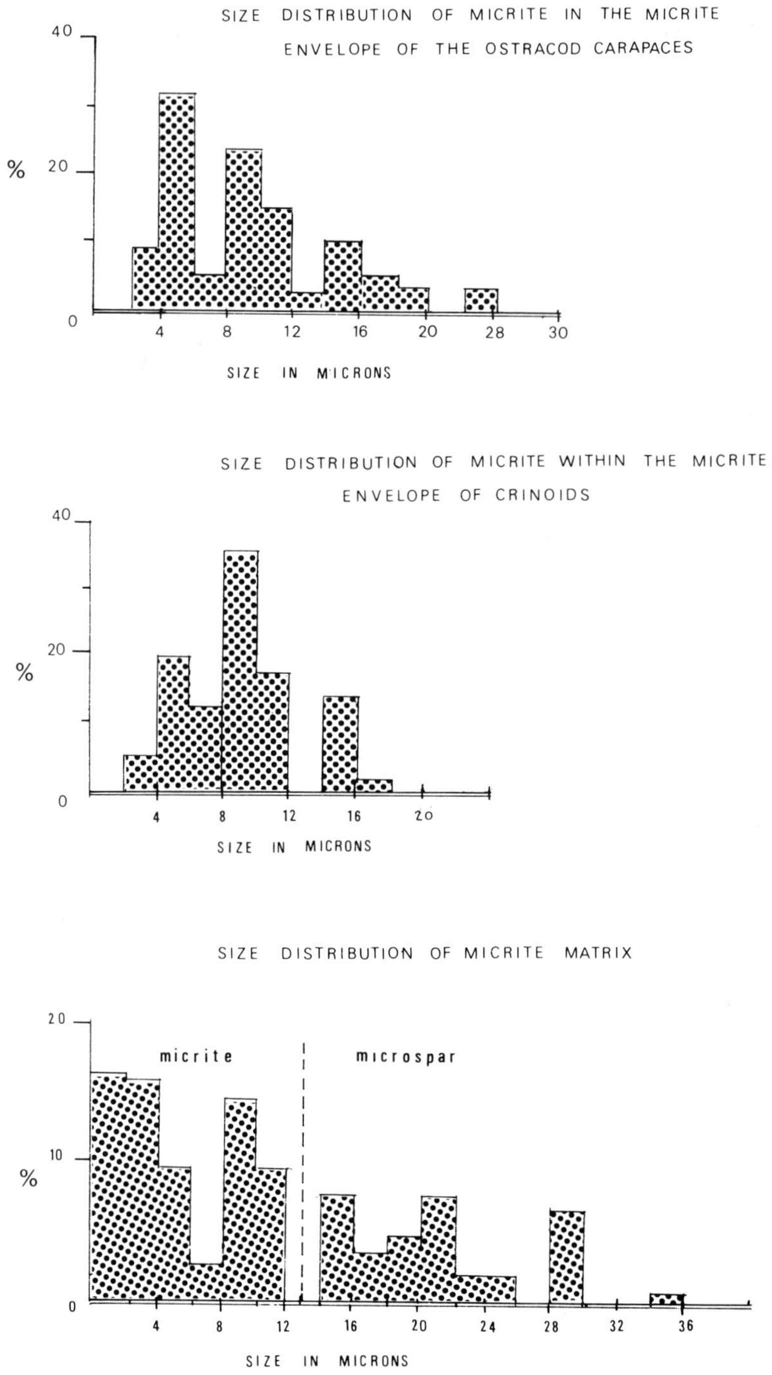

Fig. 7-8. Size distribution of micrite and microspar grains.

neomorphism" by Folk (1965) and "recrystallization to cryptocrystalline carbonate" by Purdy (1968). Micritization is selective, affecting some shells more than others. Skeletal fabric and to a lesser extent mineralogy influence susceptibility to micritization; shells with homogeneous prismatic microstructures more frequently show micrite rinds than shells with fibrous or foliated microstructures. Pervasive micritization of crinoid ossicles may relate to the inherent instability of high-Mg calcite in the exoskeleton. Stabilization from high-Mg calcite to low-Mg calcite in recent carbonates of the Persian Gulf has in many cases been accompanied by

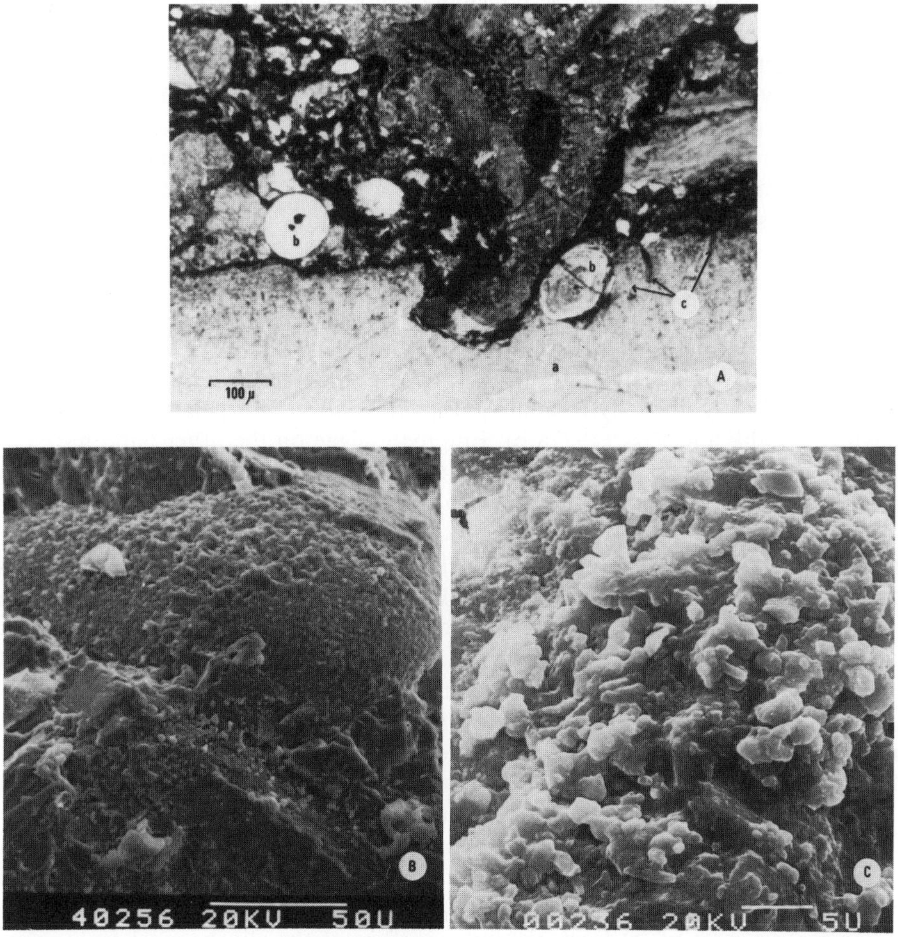

Fig. 7-9. (A) Photomicrograph showing allochems of *Atrypa (a)*, spherulith (*b*), quartz silt, and other undifferentiated bioclasts. The junction between the *Atrypa* shell and the other allochems is marked by darker material, probably algal-mat encrustation on the shell. Some of the allochems rest in a solution cavity on the shell surface. Microborings in the shell (*c*) are infilled with micrite.
(B) SEM photograph of the surface of a shell associated with algal "mucus", displaying a spongy and porous character.
(C) A detail of photo (B) showing the microcrystalline calcite of the porous zone.

micritization of skeletal grains (Chafetz et al., 1988). Other fossils present typically have low-Mg calcite in their shells. The rank order of decreasing susceptibility to micritization for the important fossil groups in the Aymestry Limestone is: crinoids, trilobites, ostracods, impunctuate brachiopods, punctate and pseudopunctate brachiopods, bryozoa, corals, and tentaculites.

The surface of skeletal grains with micrite rind envelopes shows: (1) shallow pits less than 0.04 mm in diameter; (2) shallow furrows formed from coalescing pits; or (3) a microporous surface, which leaves a ghost of the original particle. In the early stage of biological destruction, causal traces are recognizable, but for severely micritized grains the organisms responsible cannot be identified.

The boundary between the micritic envelope and the skeletal core is seldom regular. The envelope normally develops around the periphery of the grains and may progressively replace the core by centripetal replacement (Bathurst, 1966) until the grains are reduced to structureless opaque pseudomorphs. The envelope has a spongy or microporous texture and consists of a coherent crystal aggregate of cryptocrystalline calcite or micrite (Figs. 7-9 – 7-13). In ostracods, crinoids, trilobites and spheruliths the color may vary from golden resinous to grey and opaque with chocolate brown tints. For incomplete replacement, the boundary between the micrite and skeletal cores often preserves microtube structures penetrating into the skeletal substrate.

The relationship between algae and micritization is well known. Endolithic algae that bore into carbonate substrates (Carpenter, 1854; Lukas, 1973) include filamentous cyanophytes, chlorophytes and rhodophytes, all with a penetrative mode of behavior (Kobluk and Risk, 1977a,b). Epiphytic algae on the other hand have encrusting modes and may circumcrust skeletal grains as algal mucus. The chasmolithic algae are another form which thrive especially in cavities not of their own creation.

Girvanella sp. is a filamentous alga which belongs to the cyanophyta and is common in the Lower Palaeozoic (Johnson, 1961). In a thin section examined from Shucknall Quarry, the algal remains reveal a network of tubules 0.010 mm in diameter, calcified with a microsparry calcite or microspar infilling; the calcified filaments show no obvious external structures or ornaments. This *Girvanella* sp. encrusts and penetrates a crinoid ossicle which has degenerated into a structureless mass but still retains its primary form, preserved ghost-like.

Borings of *Girvanella* sp. are not restricted to skeletal grains but are also found on spheruliths. The surfaces of these superficially coated grains are frequently infested with circular pits and shallow furrows or grooves, occasionally producing deeper depressions and incisions that expose and weaken further the internal layers. Extensive borings of these endolithic algae partially or completely destroy the ordered arrangement of, at least, the outer surface of the allochemical grains converting it into micrite which, unlike lime mud or ooze, has a coherent fabric.

Micritization can also proceed beneath a layer of algal mucus and is not related to the activity of endolithic algae. The algal material shows up in thin section as brown superficial layers (on bioclastic grains) composed of calcium phosphate (from EDS). Under these layers bioclastic grains are adversely micritized (dark color) through etching, dissolution or boring. The particular example (Figs. 7-9B and

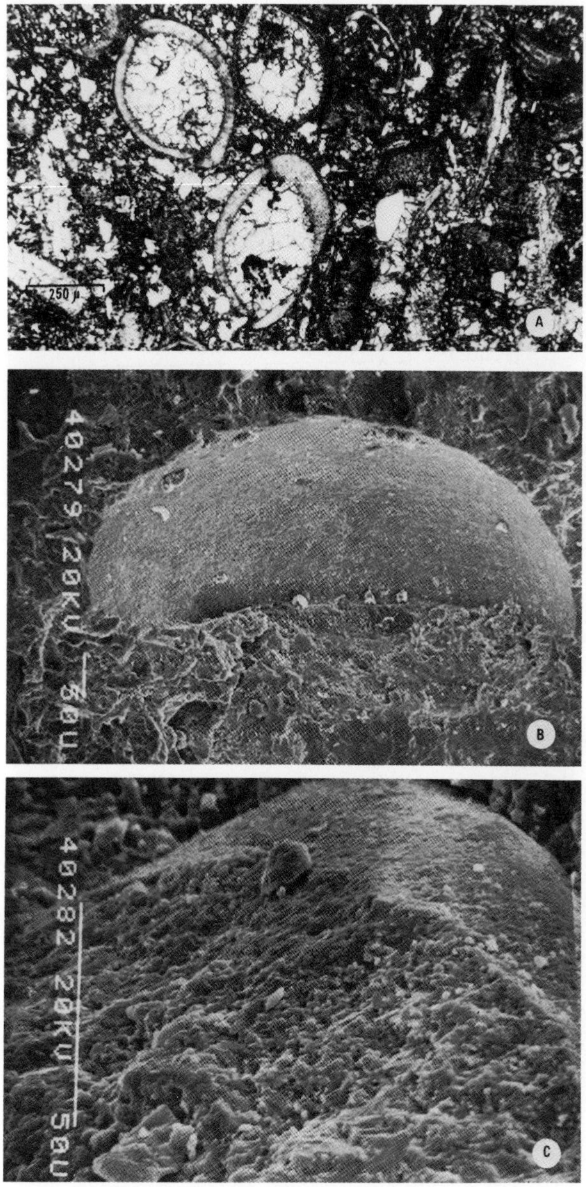

Fig. 7-10. (A) Photomicrograph showing ostracods in the fill of a chondritic burrow. There is a development of dark rinds on the outer surface of the ostracod shells.
(B) SEM photograph of an ostracod shell showing shallow pits representing either an etched surface or, more probably, sites of incipient microborings.
(C) A detail of photo (B) showing the spongy character of the micritized grains.

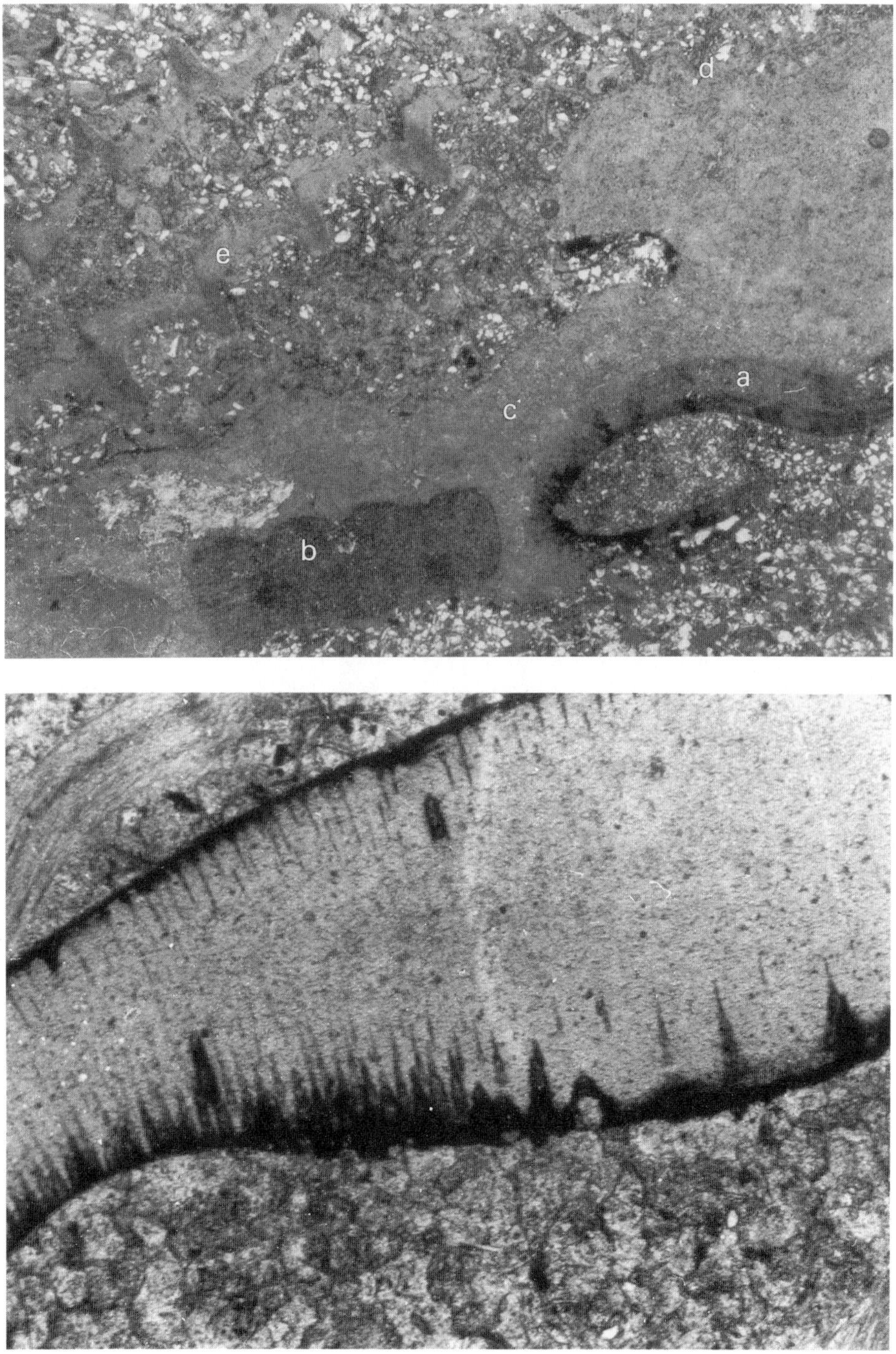

Fig. 7-11. Photomicrographs illustrating preferential micritization of a trilobite carapace (*a*) and crinoid ossicles (*b*). The trilobite shows evidence of centripetal replacement. Molluscan fragments (*c*), *Monotrypa* sp. (*d*), and tentaculitids (*e*) are resistant to micritization. The higher magnification image of the trilobite (bottom photomicrograph) shows microborings in the envelope.

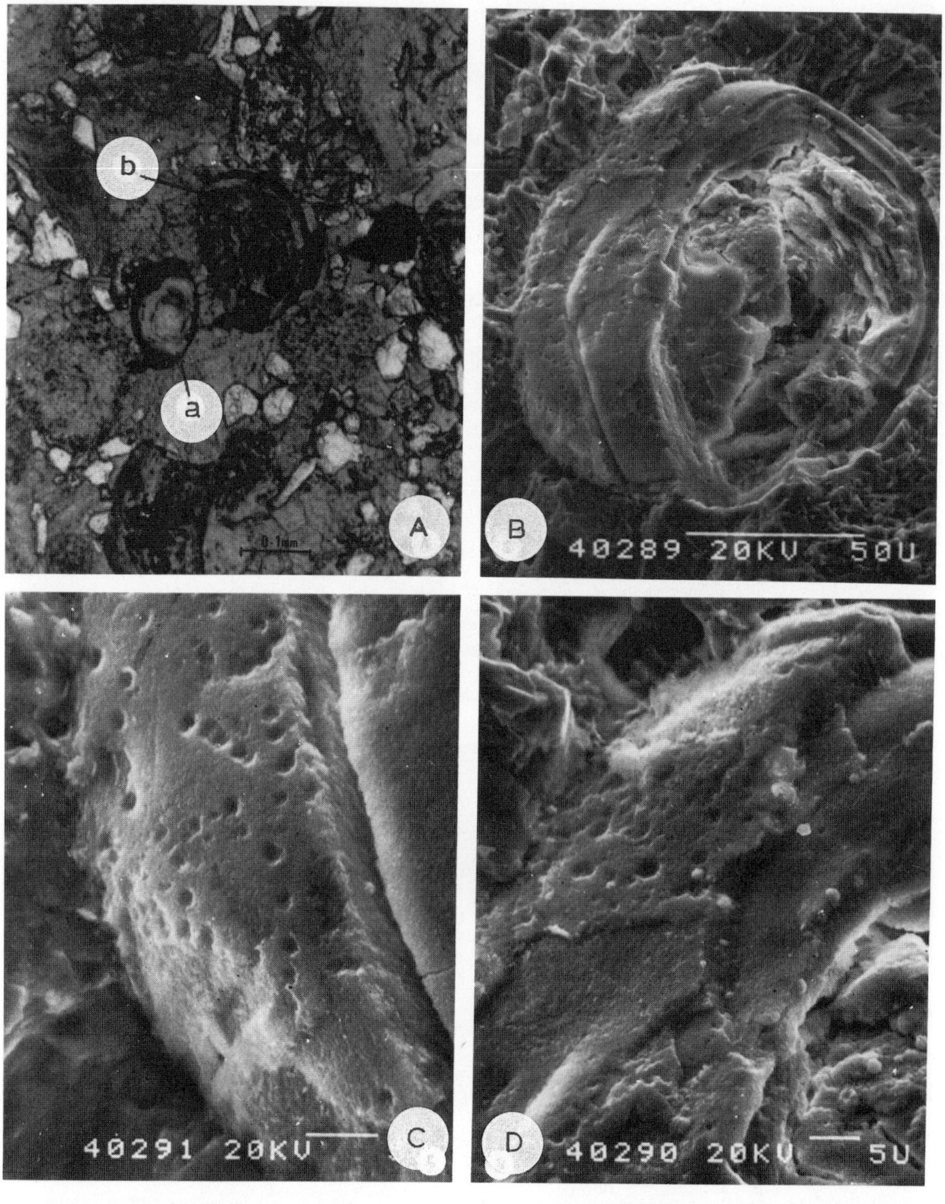

Fig. 7-12. (A) Photomicrograph of spherulithic biosparite (grainstone) showing two forms of spherulith: oval shapes lacking internal structure (*a*) and concentric laminae enclosing a group of spherules (*b*). (B) SEM photograph of a spherulith (phosphatic calculi) showing the concentric internal structure. (C and D) Surface morphology of a spherulith displaying microborings possibly of algal origin. Microborings sometimes coalesce to form a groove creating a greater surface area by exposing more of the internal layering.

C) of the pentamerid brachiopod *Kirkidium knightii* displays the relationship between the algal mat and the micritized marginal zone of the affected substrate immediately below. The zone has a ferroan calcite composition but the unaffected core of the shell interior consists exclusively of non-ferroan calcite. Ferroan calcite is also preferentially high in the microtubes or microcavities that extend into the substrate.

The micritized outer surface of the shells is a microporous aggregate of microcrystalline calcite probably produced by selective leaching, eventually converting the substrate into a micrite residue. This development parallels the formation

Fig. 7-13. (A) Photomicrograph of the endolithic alga *Girvanella* sp., penetrating the margin of a crinoid fragment. The crinoid is thoroughly micritized.
(B) SEM photograph of algal filaments resembling tubules of *Girvanella* sp.
(C) SEM photograph of micrite, present in a micritized crinoid substrate beneath *Girvanella* sp.

of Alexandersson's (1972) shell residue micrite.

Bathurst (1964, 1966, 1971, p. 388), Alexandersson (1972), Kobluk and Risk (1977b) and Schneider (1977) attributed the evolution of micrite envelopes to the process of boring – infilling. The mechanism involves repeated infillings of vacated algal borings on carbonate grains. The endolithic algae which are principally responsible for the perforation of grain surfaces are not directly responsible for the subsequent infillings of micrite cement. The micrite is presumed to precipitate in the boring posthumously after the algae decay (Bathurst, 1964; Kobluk and Risk, 1977a, 1977b). Random micrite growth might initially take place on the walls of algal borings and proceed until the cavities are practically infilled. In an intermediate stage of micritization the microborings are devoid of micrite infills and the grain's microstructure remains intact.

In a series of repeated boring – infillings the physiological activity of the existing algae, involving assimilation of carbon dioxide and bicarbonate can increase the level of $CaCO_3$ saturation and subsequently trigger the precipitation of $CaCO_3$ inorganically in these microenvironments. Goldman et al. (1972) have described microenvironments between algal filaments of marine and lacustrine milieux which have a high pH (> 10) attributable to algal assimilation of CO_2 and HCO_3^-. So it is feasible to assume that the CO_3^{2-} concentration can be raised to a level that $CaCO_3$ is precipitated. Precipitation of micrite cement in seawater may, however, be slightly reduced by inhibitors and crystal poisoning by Mg^{2+} and certain organic substances (Pytkowitz, 1969; Chave and Suess, 1970).

The process of repeated boring – infilling usually leads to the development of destructive micrite envelopes which are characterized by centripetal replacement of the substrate. Another form of envelope can be generated by the accretionary addition of $CaCO_3$ on the substrate under algal mucus or mats or by the "constructive" micrite envelope formed by calcification of exposed filaments of endolithic algae (Kobluk and Risk, 1977a).

Syngenetic micritization takes place at or near the sediment – water interface. Recent algae are known to subsist to a depth of up to 160 cm below the water – sediment interface (May and Perkins, 1979), where the metabolic activities of burrowing organisms increase the carbon dioxide budget in the sediment supporting algal life. Endolithic algae (*Girvanella* sp.) are believed to be the main agents for micritization of most allochemical grains in the Aymestry Limestone.

Generally the high Mg-calcite skeletons are most susceptible to micritization and such degradation of calcite has been ascribed to destructive micritization (Kobluk and Risk, 1977b).

The mineralogy of boring infills and/or micrite cement *per se* is influenced by the mineralogical composition of the substrate or host. This is illustrated by the intimate association of micrite cement in the stomapore of crinoid ossicles: both displaying a high ferroan content. The presence of ferroan calcites within the microstructure of crinoid exoskeleton as well as in the micrite cement suggests that both substrate and infill are the byproduct of replacement of original high-Mg calcite precursors (Richter and Füchtbauer, 1978). In modern analogues it has been documented that micrite mud essentially comprises high-Mg calcites (Alexandersson, 1972). The mutual association of high-Mg calcite precursors in both substrate and boring infill

TABLE 7-3

Comparison of micrite from Aymestry Limestone with other analogues

Aymestry Limestones	Other case studies
Fabrics	
Coherent crystals with sizes < 12 μm (microspar 12 – 20 μm, pseudospar > 30 μm)	Coherent fabric with size < 4 μm (Folk, 1959); < 30 μm (Leighton and Pendexter, 1962); and 4 – 40 μm (Macintyre, 1977)
Genesis	
(1) Repeated boring – infilling by endolithic algae (*Girvanella* sp.); precipitation of micrite cement in the vacated boring after algal decay, giving rise to micrite envelopes	Mainly algal boring – infilling (Bathurst, 1966; Friedman et al., 1971; Alexandersson, 1972; Kobluk and Risk, 1977a,b)
(2) Subordinate dissolution – reprecipitation under algal mat	Inorganic origin such as recrystallization (Purdy, 1968); dissolution – reprecipitation (Kendall and Skipworth, 1969); shell residue micrite (Alexandersson, 1972); partial diagenetic dissolution (Neugebauer, 1978)
Mineralogy	
Boring infill (micrite cement) has high-Mg calcite precursor; the mineralogy/chemistry of substrates influence the type of calcite nucleation in the vacated boring	Modern examples from the Mediterranean and Bahamas show aragonite and high-Mg calcite composition for micrite (Alexandersson, 1972); substrate influences the mineralogy of boring infill

(cement) in the Aymestry Limestone is related to the substrate's ability to nucleate precipitate of a similar chemical composition within its microenvironment (Table 7-3).

DIAGENESIS

The processes of diagenesis in the carbonate environments of the Aymestry Limestone are relatively complex. There is total occlusion of porosity through precipitation of several generations of cement, accompanying the changing chemistry of pore fluids produced during the burial history of the sediment. The shallow water environment in which sediment accumulated allowed periodic erosion and possible exposure, at times producing minor diastems. The latter manifest themselves as hardgrounds with organism borings and encrustration, degradation of cement, and incipient dolomitization.

Diagenesis in carbonate sediments

The main diagenetic process affecting carbonate sediments involves progressive cementation of intergranular and intragranular voids of the mud or grain-supported fabric (Chilingar et al., 1967b). By common convention, a cement is any crystalline encrustation precipitated from solution. Other changes such as recrystallization and replacement (especially dolomitization) are equally common, brought about by reaction between one mineral and another, or between one or several minerals and the supernatant pore fluids. Synsedimentary textures, involving micritization of allochemical grains, may accompany early diagenetic change.

Common carbonate diagenetic environments

Cementation in carbonate sediments can take place in various diagenetic environments ranging from subaerial conditions to fresh-water and marine conditions. Pore water exists in aquifers recharged from vadose and phreatic sources, or when present as saline pore fills from marine and littoral sources (Bathurst, 1958; Friedman, 1964). Longman (1980) reviewed these diagenetic environments and attempted a spectral division based on the fluid chemistry, the distribution of fluids in the pores and the fabric characteristics of the precipitated cement. Four major environments considered to be typical of the near-surface and shallow-subsurface are identified (Table 7-4, based on Longman, 1980). They are synthesized as follows with minimal specific citation.

The marine phreatic zone is characterized by the occurrence of normal marine water in intergranular spaces within the sediments. This zone occurs within the top 100 m of the sedimentary pile beneath shallow seas but well above the carbonate compensation depth, characterized by saturated water. Diagenesis of carbonates begins at a very early stage of the sedimentary history, shortly after deposition. Longman (1980) reviewed the fabric of cements formed in modern submarine environments and equivalent fabrics in more ancient rocks. He attributed their development to the rate of movement and chemistry of interstitial water. In a stagnant zone, characterized by slow water circulation, cement is generally lacking although micritization is common. In contrast, the active zone with active water circulation is characterized by the development of random acicular aragonite, syntaxial fibrous or botryoidal aragonite and magnesian-calcite cements.

At the opposing end of the water chemistry spectrum, is the vadose zone, comprising the water table and the land surface. Pore fluids are generally meteoric in origin and undersaturated in $CaCO_3$. Characteristically, the sediment is not water-saturated and pores invariably contain air and water phases, generating interstitial forces under surface tension that govern the movement of pore fluids. Meniscus cement fabric is a product of these forces (Dunham, 1971).

The freshwater phreatic zone, which is sandwiched between the vadose zone and the mixed marine – phreatic freshwater zone (or mixing zone), has pore fluids derived from meteoric water containing variable amounts of leached carbonate from the vadose zone. The water table marking the surface of water saturation, defines the upper limit of the phreatic zone, whereas the lower limit is transitional and grades into "marine" water especially at sites adjacent to the sea. The main characteristic of the phreatic zone is the increasing $CaCO_3$ saturation of water with depth. In this zone and above, aragonite solutions exist because the pore water is undersaturated with respect to dissolved $CaCO_3$. Below this zone dissolution of aragonite and

TABLE 7-4

A summary of subsurface diagenetic environments (based on Longman, 1980)

Marine phreatic zone (affecting sediments to no more than 100 m)	Mixing zone	Freshwater phreatic zone	Freshwater vadose zone
Stagnant zone	**Zone of precipitation**	**Zone of solution**	**Zone of solution**
Processes	*Processes*	*Processes*	*Processes*
(1) Little or no water circulation through sediment	(1) Mixing of marine and freshwater phreatic – brackish environment	(1) Solution by undersaturated meteoric water	(1) Solution by undersaturated meteoric water.
(2) Possibly bacterial control on cementation	(2) Active circulation due to tides		(2) Production of CO_2 in soil zone aiding solution
(3) Water saturated with $CaCO_3$	(3) Salinity variation due to seasonal rainfall		
Products	*Products*	*Products*	*Products*
(1) Little cementation except in skeletal micropores	(1) Zone adjacent to fresh-water lense (a) Micrite (b) Bladed calcite	(1) Development of moldic and/or vuggy porosity	(1) Extensive solution
(2) No leaching	(2) Zone adjacent to marine lense (a) Isopachous (syntaxial) cement (b) Mg-calcite	(2) Possible neomorphism of un-stable grains	(2) Preferential removal of aragonite if present
(3) No alteration of grains	(3) Dolomitization		(3) Formation of vugs in limestone
(4) Micritization			
Active zone		**Stagnant zone**	**Zone of precipitation**
Processes		*Processes*	*Processes*
(1) Random aragonite needles		(1) Little or no water movement	(1) Meniscus or pendant distribution of water
(2) Isopachous fibrous aragonite		(2) Water saturated with $CaCO_3$ products	(2) CO_2 loss or evaporation
(3) Botryoidal aragonite (4) Micritic Mg-calcite (5) Isopachous fibrous Mg-calcite			

(6) Mg calcite pseudo-pellets (7) Polygonal boundaries between isopachous cements (8) Interbedded cements and sediments (9) Borings in cements (10) Most cementation in reefs or surf zones		*Products* (1) Little cementation (2) Stabilization of Mg-calcite and aragonite (3) Little or no leaching (4) Preservation of porosity (5) Neomorphism of aragonite grains with some preservation of textures **Active zone** *Processes* (1) Active water circulation (2) Some leaching of aragonite; leaching may be accompanied by calcite replacement (3) Rapid cementation *Products* (1) Abundant equant calcite cement (2) Isopachous bladed calcite cement (3) Interlocking crystals (4) Crystals coarsen toward center of pores (5) Complete replacement of aragonite by equant calcite (6) Syntaxial overgrowths (7) Relatively low porosity	*Products* (1) Minor cementation (2) Meniscus cements (3) Pendant cements (4) Equant calcite (5) Preservation of most porosity

precipitation of calcite takes place before reaching the lowermost stagnant, saturated zones.

Pore water is chemically saturated with $CaCO_3$ in the stagnant zone and little or no cementation occurs, but neomorphic processes operate. At these deeper burial levels, inversion of aragonite or metastable high-Mg calcite grains to calcite or low-Mg calcite takes place; the neomorphic recrystallization involves dissolution and reprecipitation of magnesian-enriched fluids migrating as a thin front through the grains (Bathurst, 1971). The diagnostic fabrics for the phreatic freshwater cements are abundant equant calcite, syntaxial bladed calcite, interlocking crystals and neomorphic alteration of aragonite to calcite (Longman, 1980).

The mixing zone environment is characterized by brackish water resulting from the mixing of fresh-phreatic and marine-phreatic waters (Longman, 1980). Folk (1974) termed this the peritidal diagenetic environment which includes supratidal and intertidal settings. Salinities vary greatly at the freshwater end, where the spectrum of diagenetic processes include: formation of micrite and fibrous cements; leaching of aragonite; and neomorphism of both aragonite and magnesian-rich calcite to calcite. Towards the marine end of the mixing zone, a submarine cement type becomes increasingly common.

PETROGRAPHY OF THE CEMENTS IN AYMESTRY LIMESTONE

The important cement fabrics are found in grain-supported packstones and grainstone microfacies; and various types of intergranular and intragranular sparry calcite and micrite cements are defined. Six texturally different fabrics are described, which are characterized by morphology, substrate selectivity, and inclusion content.

These are: (a) syntaxial micrite cements; (b) crinoid-syntaxial (isopachous); (c) inclusion-rich syntaxial fibrous; (d) granular mosaic; (e) clear drusy (rhomb) mosaic; and (f) very coarsely-crystalline poikilotopic calcite. The identification of these fabrics is enhanced by the compositional variation as observed from staining with Alizarin Red S – Potassium Ferricyanide and microprobe analysis.

Syntaxial micrite

Micrite or microcrystalline calcite is attributed to early submarine diagenetic changes at the sediment – water interface. Micrite cements are precipitated in algal borings of allochemical grains, as a byproduct of the process of micritization developed along the margins of skeletal grains. A second form of micrite cement unrelated to biological activity, however, occurs within the fine stoma-pores of crinoid ossicles (Fig. 7-14), probably as the product of crystallization from a Mg-rich lime mud. These cements display a turquoise blue color with Alizarin Red S – Potassium Ferricyanide stains (Dickson, 1966; Lindholm and Finkelman, 1972) indicating ferroan calcitic composition. Under cross-polarized light, this micritic cement displays either unit extinction or remains in optical continuity with the crinoidal calcite. Microprobe analysis shows that the stoma-filling cement contains

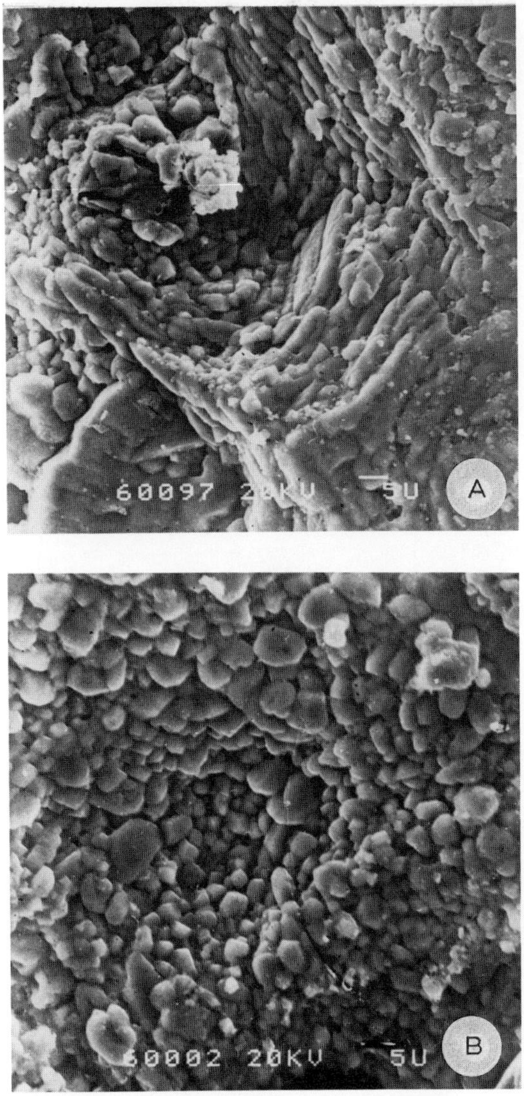

Fig. 7-14. (A) SEM photograph of a stoma-pore of a crinoid ossicle.
(B) SEM photograph of the micrite cement within a stoma-pore.

an average of 2.23 mol% $MgCO_3$. Strontium, however, is not detected, suggesting that these syntaxial micrite cements might once have been high-Mg lime muds which are not uncommon in modern marine precipitates (Friedman, 1986). Macqueen et al. (1974) observed that the fine stoma of Pleistocene calcite echinoid plates are partly filled with both syntaxial low-Mg calcite and high-Mg calcite cements. Although the high-Mg calcite is metastable, it survived because it is enveloped by the low-Mg calcite cement. The residual high-Mg calcite helps to support the view that low-Mg

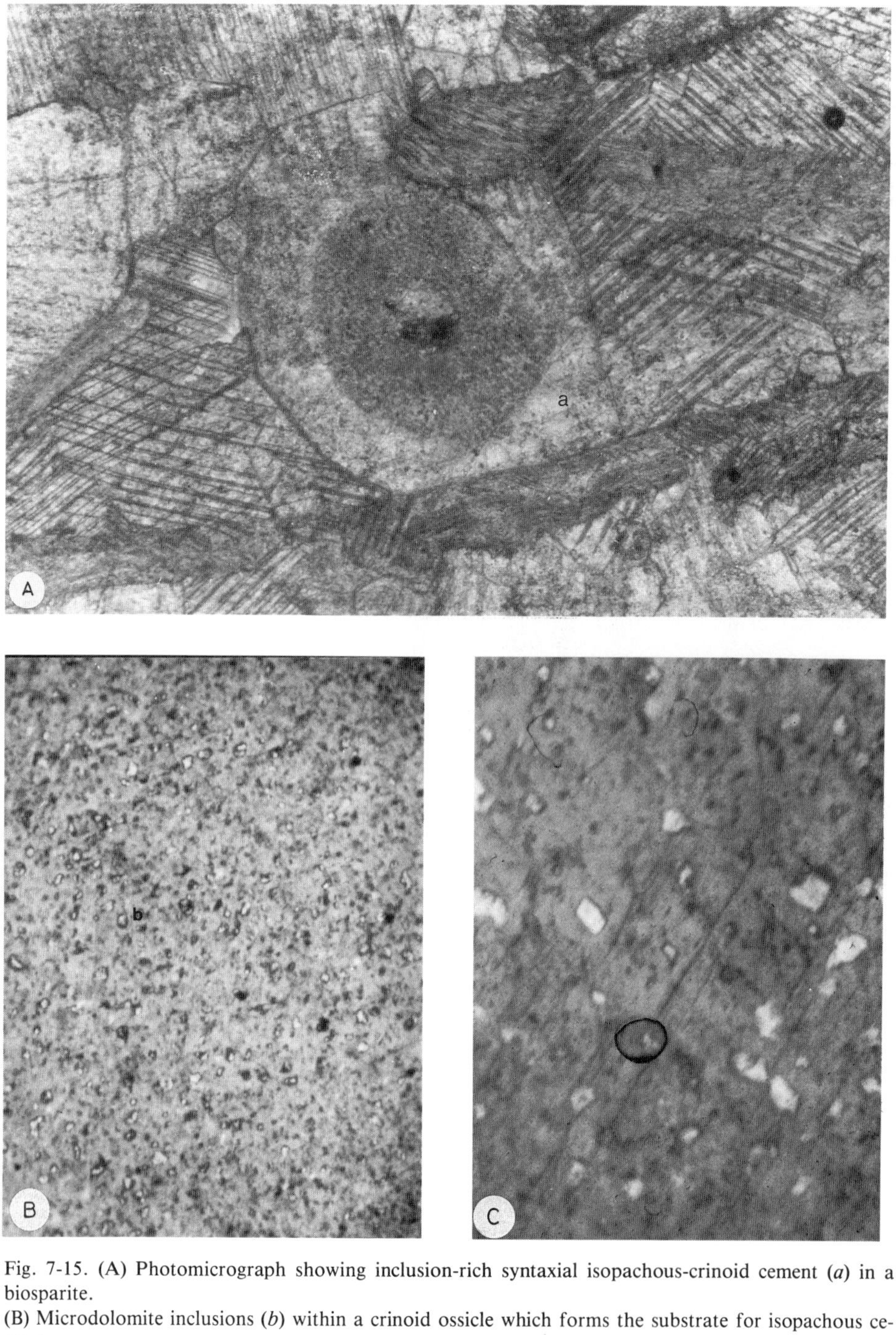

Fig. 7-15. (A) Photomicrograph showing inclusion-rich syntaxial isopachous-crinoid cement (*a*) in a biosparite.

(B) Microdolomite inclusions (*b*) within a crinoid ossicle which forms the substrate for isopachous cement.

(C) A detail of photo (B) showing microdolomite inclusions along cleavage traces of calcite in a crinoid.

calcite is derived from metastable high-Mg calcite.

Some crinoid ossicles contain no micrite cement but have discrete rhombs of microdolomite along the cleavage (Fig. 7-15). In contrast to the ossicles with micritic cements, crinoids that host microdolomite appear to lack the characteristic reticulate microstructures, possibly because of neomorphic recrystallization. Further, the microdolomite crystals are confined within the crinoid ossicle and are rarely found elsewhere, suggesting that the inclusions are derived from the high-Mg calcite.

Crinoid-isopachous cement

The crinoid-isopachous cements are especially common in the grain-supported crinoid – ostracod biosparite. The cement constitutes the isopachous fringing crust, smooth or irregular, and displays unit extinction with the crinoidal calcite substrate. Dusty inclusions produce a cloudy appearance. Alizarin Red S – Potassium Ferricyanide staining shows that most of the isopachous cements have a weakly ferroan-calcite composition.

Microprobe analysis of these ferroan-calcite cements gives a low-Mg calcite composition (average: 2.20 mol% $MgCO_3$).

Syntaxial fibrous calcite

This is another example of an early diagenetic fabric, especially in the biosparite microfacies. The fabric develops as a crystalline calcite crust on the immediate free surface of allochems or on the cavity walls of the intraskeletal voids. The voids represent depositional or early diagenetic cavities. Unlike the crinoid-isopachous cements, these calcite crusts have sharp boundaries with the substrate and their crystal elongation lies normal to the substrate.

The syntaxial fibrous calcite can be strictly grouped into two optic groups, though their crystal morphologies may show no striking dissimilarity.

Radial fibrous calcite

The basic radial fibrous fabric consists of fibrous to bladed calcite crystals with well-defined crystal faces. They coalesce to form a continuous lining or crust. The length of individual crystals varies from tens of microns to 0.4 mm. The boundary between the crystal foundations and the substrate is sharp and often characterized by a cloudy fringe or dust line. Crystal densities and the regularity of the orientation of crystals generally increase away from the substrate. The surface morphology of the substrate exerts a strong influence on the nucleation of the crystal, and the shape of the substrate imparts a measure of control on the size of the voids into which the crystals will grow: on curved surfaces crystal densities increase away from the convex substrate and crystals tend to fan out as bundles, whereas on the concave side crystals mostly converge away from the foundation with a corresponding decrease in crystal densities. Crystal boundaries are usually non-planar and the optic axes are commonly oriented parallel to the length of the crystal, or may vary slightly within 5° and diverge away from the substrate.

Fascicular calcite

Fascicular calcite (Figs. 7-16 and 7-17) occurs primarily within cavities and forms a sharp boundary with the cavity wall. The fascicular calcite exhibits a botryoidal form (Schroeder, 1972); Ginsburg and James (1976) use the term spherulithic cement instead of botryoidal cement. Kendall and Tucker (1973) and Kendall (1976) interpreted a similar fabric in ancient limestones as replacement of early diagenetic acicular submarine cement. Fascicular calcites may also occupy the upper space within a geopetal structure. The effect of bioerosion of these cements and substrate

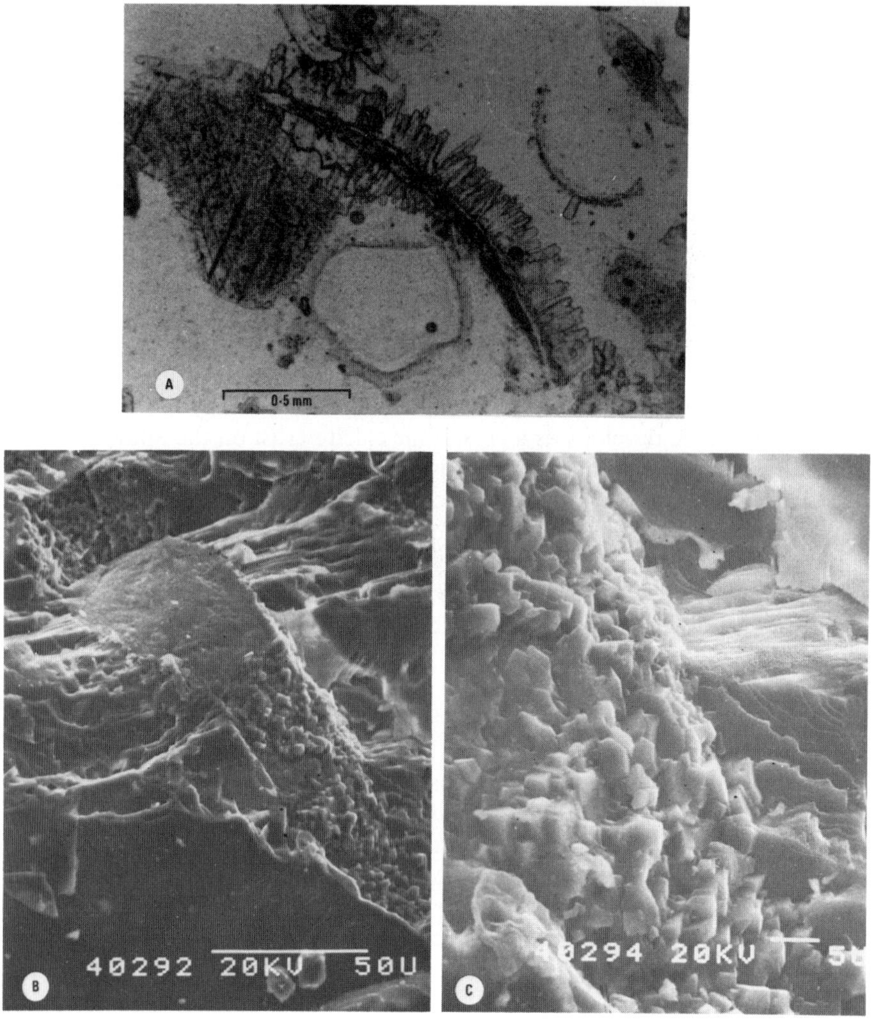

Fig. 7-16. (A) Photomicrograph of syntaxial fibrous (acicular) calcite. Part of the acicular calcite occurs as coalescent fibrous crust.
(B) SEM photograph of the syntaxial fibrous calcite with microsubcrystals in the marginal zone.
(C) A detail of photo (B) showing the microsubcrystals which represent first-generation cement.

by *Chondrites* sp. provides circumstantial evidence for the fabrics being formed during early diagenesis in a submarine environment, predating the burrowing activities.

The gross fabric of fascicular calcite cements does not differ significantly from the normal habit of radial fibrous calcite. Nonetheless these calcite crystals have distinctive optical properties which merit their separate treatment. The fabrics observed have the following characteristics:

(1) Crystals decrease in size toward the cavity walls or substrate and the marginal zones are characterized by the development of subcrystals. In the absence of subcrystals or crystallites, the larger crystals may abut directly against the substrate or seed epitaxially.

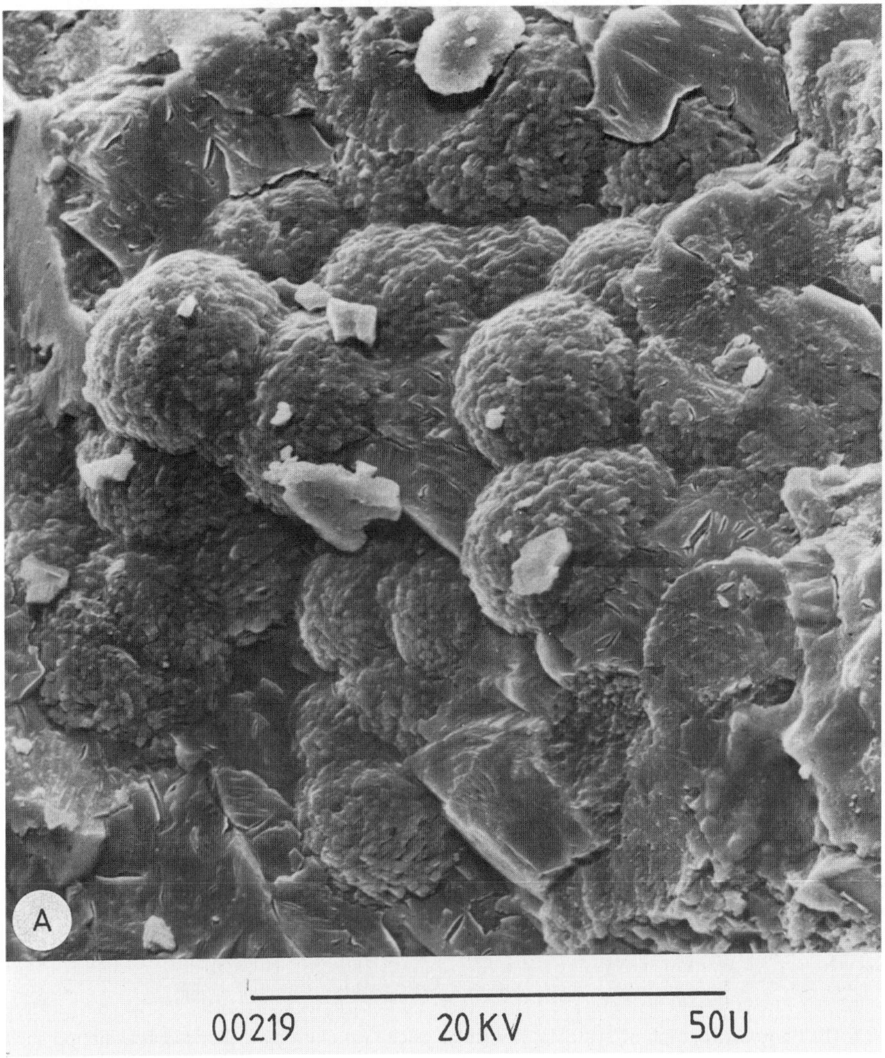

Fig. 7-17. (A) SEM photograph of botryoidal (syntaxial fascicular) calcite which grew originally as bundles of acicular calcite crystals.

(2) Elongation of the crystals is normal to the substrate.
(3) Intercrystalline boundaries are commonly nonplanar.
(4) The length-fast vibration usually diverges away from the crystal foundation and the arc of the curved twin lamellae has its convex margin facing away from the substrate.

The internal patterns of the fast vibration and curved twin lamellae expressed in (4) above are diagnostic optical properties of fascicular optic calcite (Kendall, 1976). In contrast, its counterpart the radiaxial fibrous calcite (Bathurst, 1971) has a similar fabric relationship; however, the fast vibration directions invariably converge away from the wall of the substrate, whereas the curved twins are arranged with concave surfaces facing away from the substrate. In addition, the single, large crystal, which forms the element of the fascicular fabric, often contains divergent subcrystals at the foundation. These subcrystals rarely orientate in the same direction as the younger crystals. The fascicular calcites may grade into clear blocky paraxial drusy calcite which occupies the remaining voids; they normally constitute late cements. The boundary between syntaxial fascicular and paraxial blocky calcite is represented by the domal growth front of the former (Figs. 7-18, 7-22, 7-24, 7-29). Within this zone, both calcites are optically continuous indicating no crystallographic dislocation in the subsequent phase of crystal growth and precipitation, i.e., later generation cement proceeds epitaxially or along the same axis into the remaining voids.

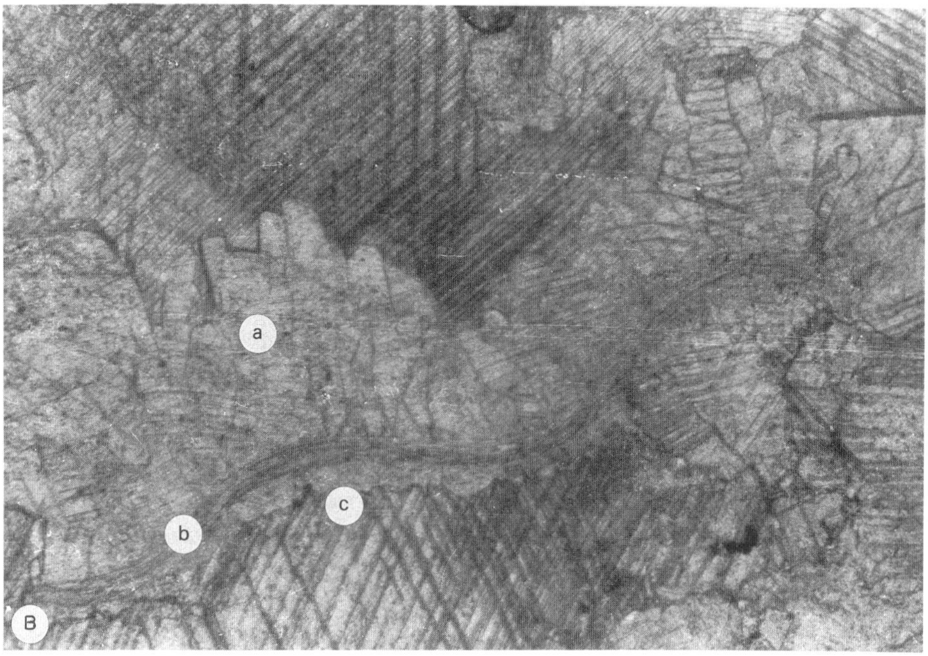

Fig. 7-17. (B) Photomicrograph of syntaxial fascicular calcite (*a*) abutting a chonetid brachiopod shell (*b*). The irregular surface on the other side of the shell is attributed to organic modification or early solution. (Width of photograph is 0.8 mm.)

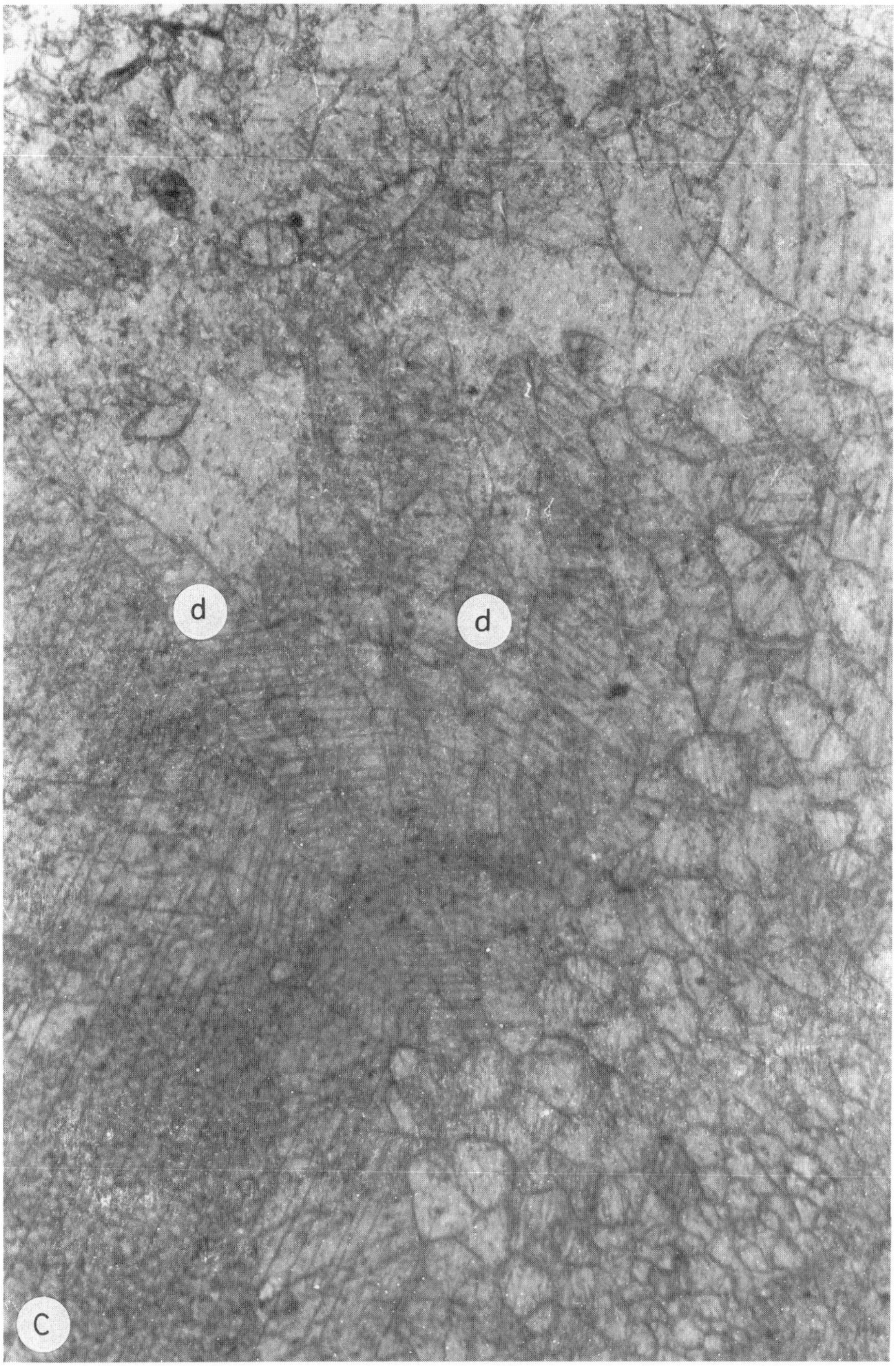

Fig. 7-17. (C) A detail of photo (B) illustrating fascicular calcite with arched twin lamellae (*d*). (Width of photograph is 0.175 mm.)

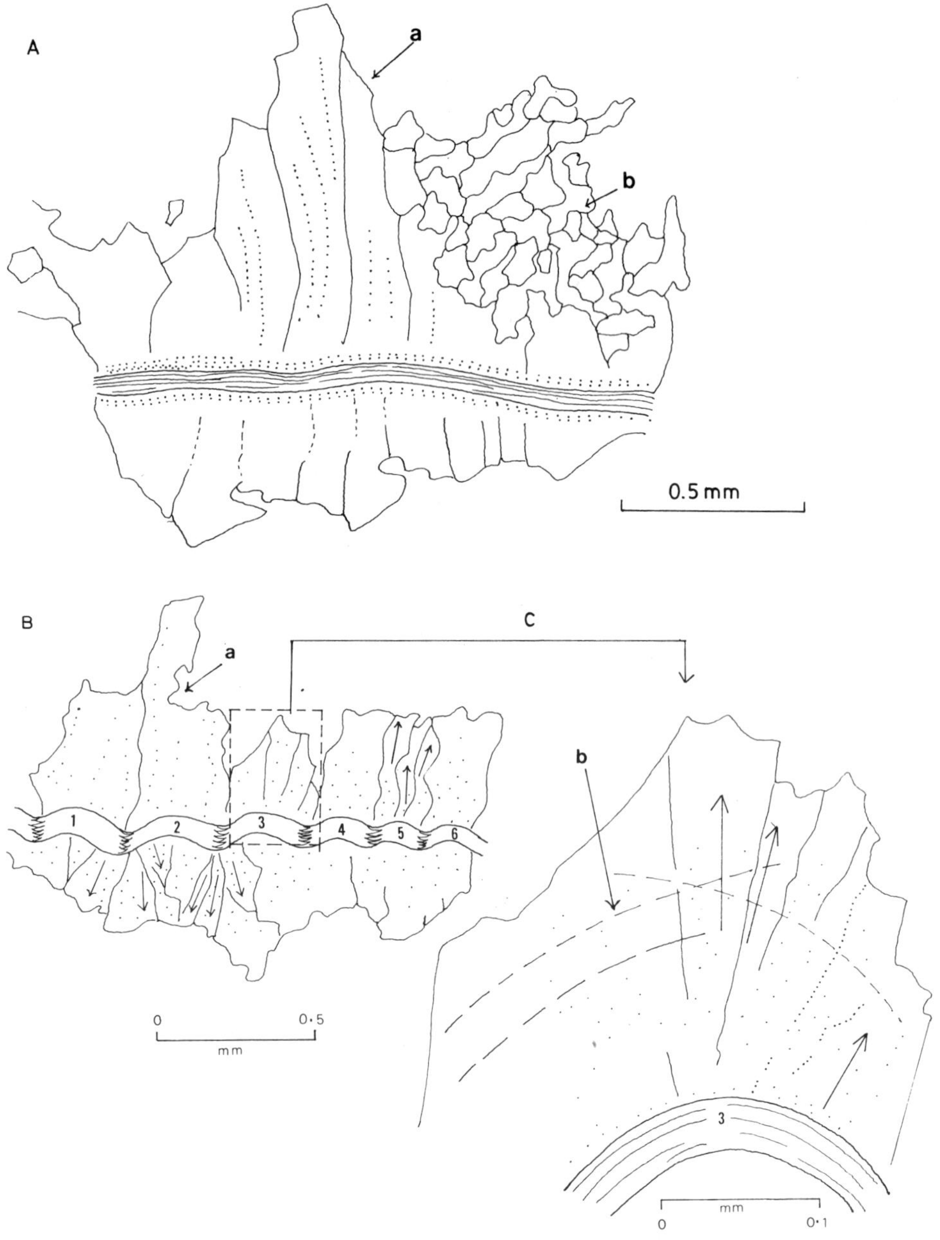

Fig. 7-18. (A) Radial fibrous (bladed) calcite cement (*a*). The irregular termination of the crystals is probably due to boring or solution. The ameboid texture (*b*) represents neomorphosed cement in the vicinity of a burrow.

(B) Inclusion-rich syntaxial fibrous cement seeded epitaxially onto a chonetid shell. Crystal terminations again show the effect of boring (*a*). Substrate morphology has influenced the growth direction of calcite crystals: arrows indicate the direction of length-fast vibrations.

(C) Part of the syntaxial fibrous calcite cement showing divergent optic axes and arched twin lamellae (*b*).

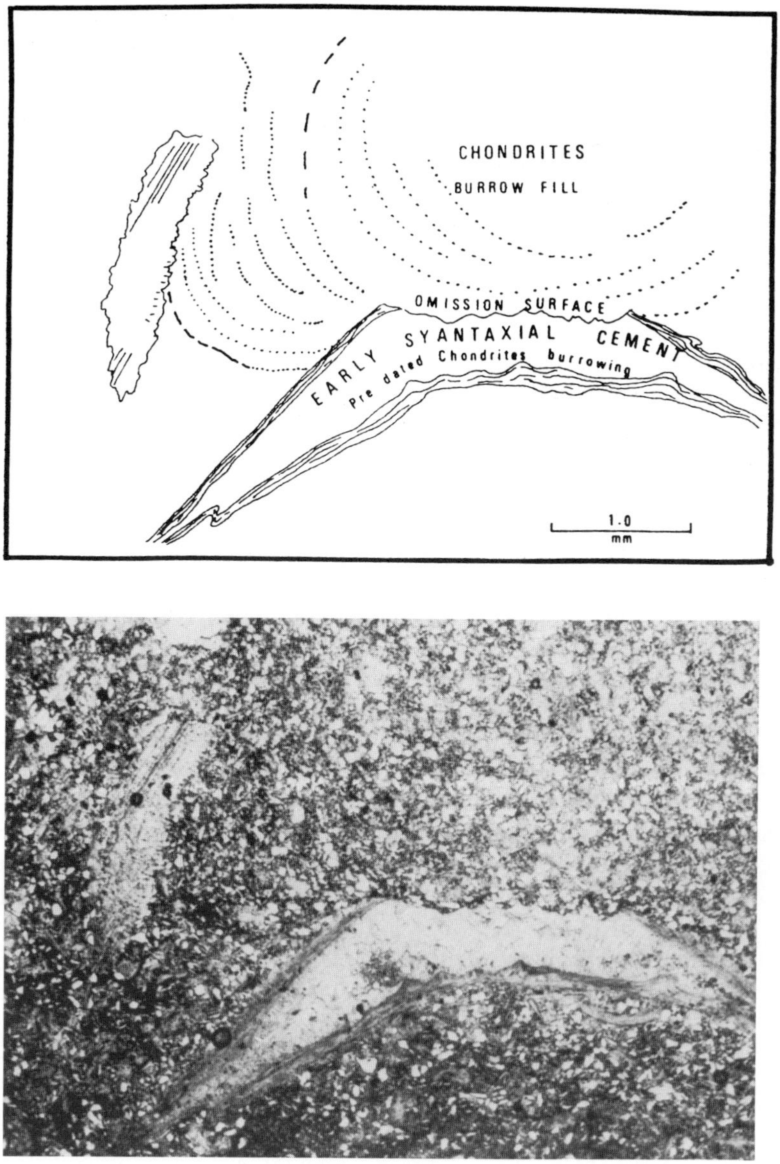

Fig. 7-19. Disruption of early syntaxial cement by *Chondrites* which penetrated the cemented fabric and created a micro-omission surface, illustrating early lithification of the sediment.

The deformed crystal lattice as shown by curved twin lamellae argues against the precipitational origin of these calcites and strongly supports neomorphic recrystallization. Kendall and Tucker (1973) and Kendall (1976), however, described the curved twin as attributable to replacement processes. They interpreted the fascicular-optic calcite as being produced by replacement of closely-packed acicular precursor cement. If this is indeed the case, then the present residual fabric represents a neomorphosed cement.

Features common to both syntaxial radial-fibrous and syntaxial fascicular calcite (Figs. 7-18 and 7-21)

Both syntaxial radial-fibrous and syntaxial fascicular calcites appear cloudy or turbid in thin section due to the widespread occurrence of inclusions. These inclusions are responsible for crystals displaying pseudo-pleochroic brownish colors. Some inclusions show a definite relationship to the present crystal fabric by following the crystal boundaries. Dust lines, which are confined to the marginal zones between the cement and substrate, probably represent mimetic or inherited fabrics of the precursor subcrystals which have now been neomorphosed (Kendall and Tucker, 1973). In some cases, the dust lines may be arranged along the length of the coalescive crystals normal to the substrate and coincide with individual crystals as observed under polarized light. This suggests that dust lines represent former boundaries of either acicular or bladed calcites. Garrison (1972) proposed that the development of randomly disseminated dust inclusions represents former argillaceous impurities incorporated into the cement, probably as a result of rapid crystallization in a submarine diagenetic environment and preserved during subsequent replacement processes.

Both the simple radial-fibrous and fascicular calcite crusts may be overlain by thin films of clay veneer, separating these cements from later cement. The crystals (see Fig. 7-18) show strong irregularities of the termination, possibly caused by organism modification (algal boring) or dissolution. The occurrence of microhardground as manifested from bioerosion of the cement by *Chondrites* sp. (Fig. 7-19) provides circumstantial evidence for early submarine cementation. Wolf (1965) observed similar features in Devonian algal reefs of New South Wales, Australia, and interpreted the development of fibrous crust or syntaxial fibrous calcite as examples of littoral cementation.

In recent submarine environments, for example, the coral – algal reefs of Funafuti (Cullis, 1904), Bermuda (Shinn, 1971; Ginsburg and Schroeder, 1973; Schroeder, 1972), Jamaica (Goreau and Land, 1974), Belize (James et al., 1976), and the Caribbean Fringing Reef of Panama (Macintyre, 1977), cement morphologies develop as syntaxial fibrous or isopachous generation parallelling those described here. The morphological similarities lend support to the idea that these calcite cements were originally fibrous submarine cement.

Inclusion-rich neomorphosed granular cement

Bathurst (1958) introduced the term "granular cement" to refer to sparry calcite cements precipitated between allochems. Although he discontinued using this term (Bathurst, 1971), it is resurrected here for those calcite cements lacking drusy fabric and characterized by equigranular or almost equigranular anhedral and/or subhedral crystals (Fig. 7-20). The cement is coarsely crystalline (size 0.1 – 0.5 mm) and, like the syntaxial fibrous calcite crystals, was made turbid by inclusions and

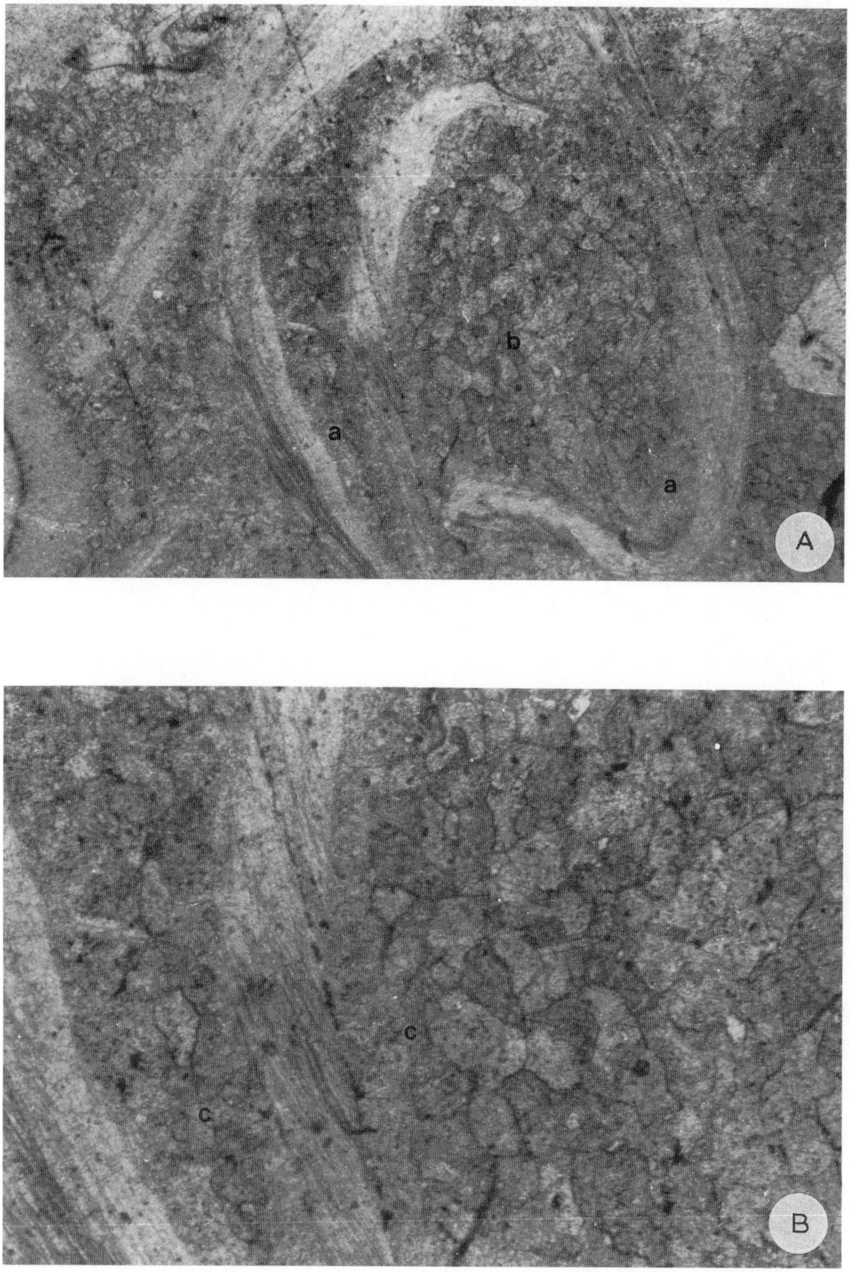

Fig. 7-20. (A) Relationship between inclusion-rich fibrous calcite (*a*) and inclusion-rich granular calcite (*b*). (Width of photograph is 0.8 mm.)
(B) A detail of (A) showing irregular crystal boundaries of the granular calcite and the sharp boundary between syntaxial fibrous and granular calcite demarcated by thin veneers of clay (*c*). (Width of photograph is 0.8 mm.)

displays brownish pseudopleochroism. The fabric exhibits a mosaic form with irregular crystal boundaries when viewed under plane polarized light. When seen under crossed polars, however, the irregular crystal boundaries often display planar interfaces. This feature indicates the existence of earlier or relict crystal faces prior to neomorphic recrystallization or overgrowth development. The fast vibration

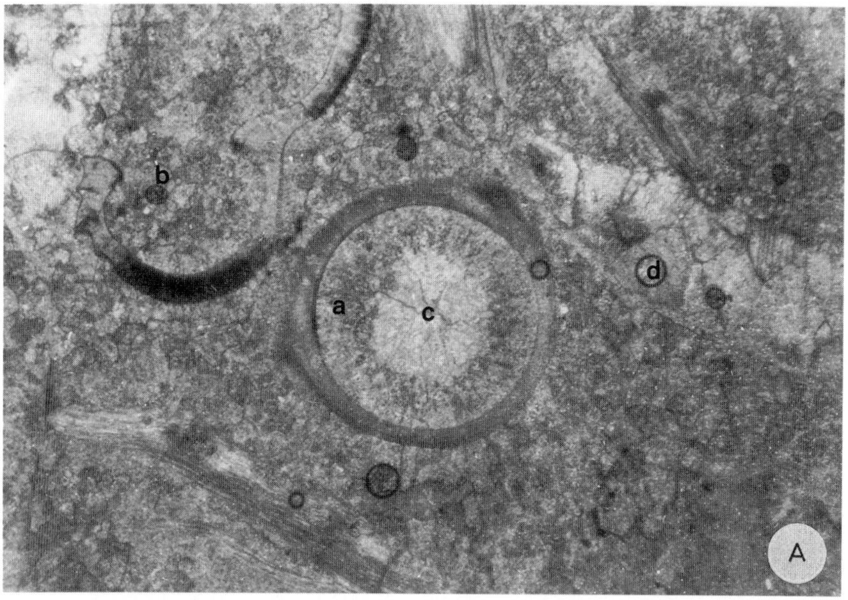

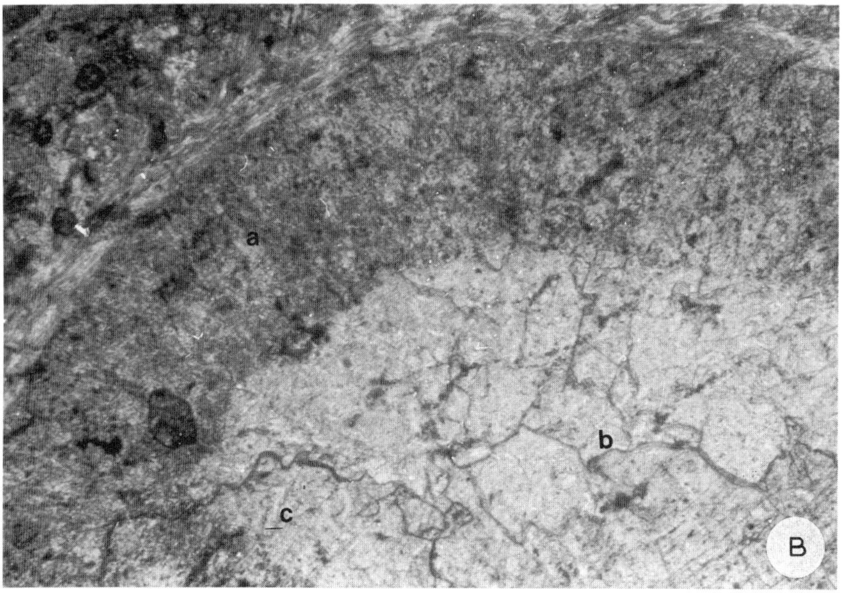

Fig. 7-21. (A) Relationship between various types of calcite: inclusion-rich syntaxial fibrous calcite (*a*), inclusion-rich granular calcite (*b*) and inclusion-free drusy calcite (*c*), which occupies the remaining void within the tentaculitid. The neomorphic calcite (*d*), which recrystallized from the bivalve shell, shows dusty patches representing mimetic shell microstructure. (Width of photograph is 1.5 mm.)
(B) Relationship between the inclusion-rich syntaxial fibrous calcite (*a*) and inclusion-free drusy calcite (*b*) associated with the umbrella texture. There is domal (spherical) growth front between (*a*) and (*b*); (*c*) represents the inclusion-free bladed calcite seeded syntaxially on (*a*) and displaying optical continuity under crossed nicols. (Width of photograph is 0.75 mm.)
(C) The development of spherical growth fronts of the syntaxial fibrous (fascicular) calcite cement. The marginal area of the front is characterized by high ferroan content. The inclusion-free drusy calcite occupying the remaining pore space represents the latest cement fabric. (Width of photograph is 0.75 mm.)

directions of the neomorphosed cements are randomly orientated and have undulose extinction. Granular cement may completely fill intergranular voids or is restricted to the central part only. In the latter case, the crystals are closely associated with syntaxial fibrous calcite and are separated by sharp boundaries which may be marked by thin clay films.

These neomorphosed granular cements stained pale red-purple, 5RP6/2 (using Lindholm and Finkelman's technique, 1972), indicating the presence of ferroan calcite (FeO-I: 0.5 – 1.5% of FeO). Probe analysis of the ferroan calcite crystals give an average composition of 1.6 mol% $MgCO_3$, a value which compares well with the determined level of residual magnesium in the low-Mg calcite precursor of ostracods and chonetid brachiopods. The granular fabric represents a neomorphosed cement probably resulting from the replacement of an original low-Mg calcite cement.

Drusy calcite mosaic

Drusy paraxial calcite mosaic consists of blocky calcites growing centripetally into the voids remaining after formation of first-generation cement or the intraskeletal cavities. The calcite is coarsely crystalline (size ≤ 0.5 mm) and seeded epitaxially on the depositional particles subsequent to the development of syntaxial fibrous cement. Crystal size normally increases away from the substrate indicating high original crystal nucleation. Some crystals remain small and equigranular especially in the residual pore spaces at the center of the intra- or intergranular voids. A low nucleation rate is not unexpected because in this late stage of cementation pore fluids are likely to be depleted of precipitating $CaCO_3$ and the availability of free surfaces for crystal nucleation will be limited. The overriding difference between these paraxial blocky calcite crystals and the earlier generation syntaxial fibrous calcite is the lack of inclusions. In plane light, the crystals are transparent and clear.

The fabric characteristics of this cement are in close accord with those described by Bathurst (1971). The important fabric characteristics are:
(1) Plane intercrystalline boundaries and enfacial junctions;
(2) the boundary between the drusy mosaic and the substrate is sharp;
(3) pendant geopetal structures occupy voids above internal sediments within shells;
(4) preferred orientation of the optic axes normal to the cavity walls;
(5) multi-generation, indicated by zoning; and
(6) the cement lacks relict structures.

Drusy calcite mosaic displays three distinct morphologies under SEM investigation:
(1) Rhombic calcite with a blocky shape in thin section. These rhombs (Fig. 7-22) have an irregular surface and crevasses, which are coincident with and controlled by cleavage traces, giving rise to "V" shapes or "flame" structures. The irregularities are attributable to surface etching and leaching which may preferentially attack weaker zones within the crystals to reveal the core. The preferential solution reflects the mineralogical variation of specific zones within the crystals, confirmed by the reaction to staining, which differentiates the various ferroan zones present.
(2) Elongate rhombs or scalenohedra which are resistant to etching (Fig. 7-23).
(3) Prismatic calcite, which constitutes the latest generation and shows etching and leaching along the C-axis.

In most crystals, the crystallographic axes and the optic axes coincide or vary only slightly. The optic axes of the calcite rhombs, as indicated by the C-axis (length-fast vibration direction), are orientated almost normal to the substrate. In a sequence of scalenohedral and rhombohedral calcites the fast vibration is almost parallel, which indicates that during crystal growth there is no optic or crystallographic dislocation.

Multi-phase zoning (Fig. 7-24) is commonly observed in stained thin sections of drusy calcite. The zoning represents precipitation under different compositional phases rather than growth zoning. In growth-rate zoning, the crystals would be expected to have crystalcentric or in this case rhombocentric equizonation; however,

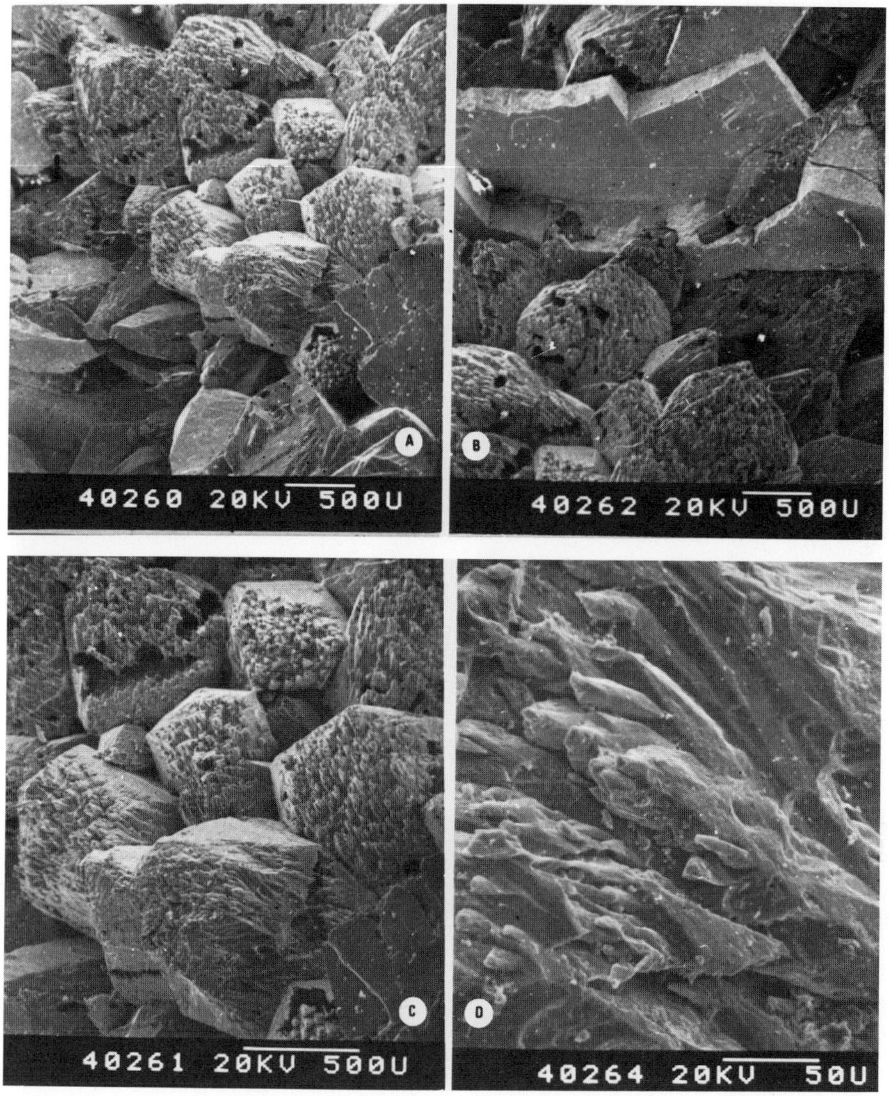

Fig. 7-22. (A) and (B). SEM photographs of drusy calcite showing rhombic and scalenohedral forms, more resistant to surface etching.
(C) Rhombic calcite with irregular surfaces attributable to surface etching along the cleavage.
(D) A detail of (C) showing V-shaped and irregular "flame" structures on the etched surface.

the zones are not restricted to single crystals and cut across crystal boundaries. There are a total of six compositional zones with alternation of various ferroan calcites. Using Lindholm and Finkelman's (1972) scheme these zones can be quantified into:

Zonation	Color stains		Calcite phase	Percent FeO
Zone 1	Purple	5P6/2	FeO II	1.5 – 2.5
Zone 2	Blue	5B5/6	FeO III	2.5 – 3.5
Zone 3	Purple blue	5P6/6	FeO II	1.5 – 2.5
Zone 4	Red purple	5RP6/2	FeO I	0.5 – 1.5
Zone 5	Purple blue	5P6/2	FeO II	1.5 – 2.5
Zone 6	Red purple	5R6/2	FeO I	0.5 – 1.5

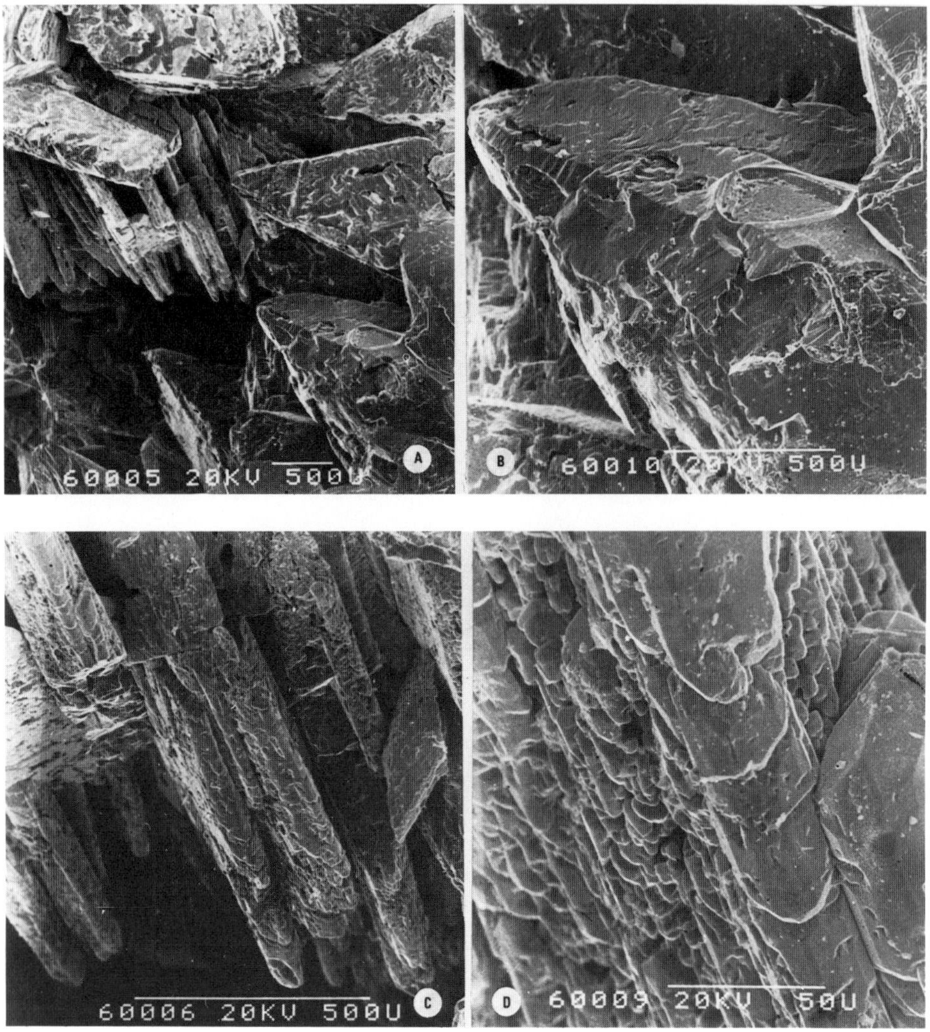

Fig. 7-23. (A) SEM photograph showing the relationship between scalenohedral calcites and a book of prismatic calcite crystals occupying the remaining void.
(B) Surface etching of scalenohedral calcite with crater-like structures. Exfoliation of the surface may result from the progressive growth of these structures.
(C and D) Fabrically-controlled leaching of the prismatic calcite, along cleavage traces.

Zones 1 and 2 occur within the syntaxial fibrous calcite which are characterized by heavy inclusions. Zone 2, however, represents the growth front of the syntaxial fibrous calcite before the commencement of the blocky paraxial calcite cementation. It is probable that the syntaxial fibrous cement had an initial nonferroan composition later modified by the mineralizing fluids responsible for calcite veining.

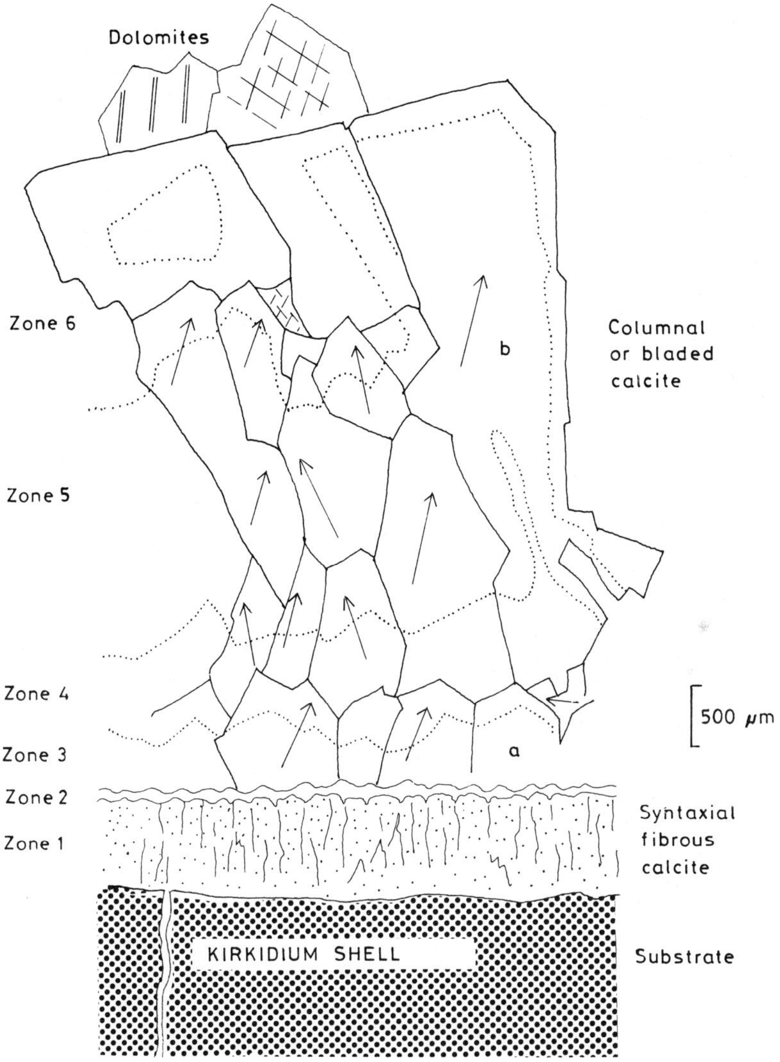

Fig. 7-24. The development of zoned calcite cement. Inclusion-rich syntaxial fibrous calcite displays compositional zoning with Zone 1 and Zone 2 characterized by nonferroan calcite and a ferroan growth front, respectively.
The inclusion-free columnal (drusy) calcite, with rhombic forms (*a*) and scalenohedral forms (*b*), falls into four zones, boundaries of which do not coincide with crystal faces. The different phases of crystal growth involve no disruption of the crystal axis.

Fluctuations in the ferroan iron content from 0.5 to 2.5% FeO indicate changes in Fe saturation of the pore fluids in freshwater phases and that chemically similar pore fluids would be present throughout the rock, assuming that the effective porosities are maintained.

Poikilotopic calcite

Poikilotopic calcite refers to large calcite crystals containing other crystals or minerals. The fabric produced is confined to hardground bioclastic grainstone and postdates all fabrics described in the preceding sections.

The cement fabric is extremely coarse crystalline with a crystal size of < 1.0 mm. The characteristic features are:

1) Inclusions of other pre-existing cements such as neomorphosed syntaxial fibrous calcite and crinoid-isopachous calcite within the poikilotopic calcite.
(2) Crystal boundaries are planar with enfacial junctions. The twin lamellae are straight and undeformed.
(3) The fast vibration of the crystal is parallel to the longer axis or direction of elongation.
(4) Occurrence of dolomite rhombs (size 0.02 – 0.06 mm) as inclusions within the crystal and in the intercrystalline boundaries.

These features show that the calcite crystals are undoubtedly cements and postdate neomorphism of the early-formed cements. Staining by Alizarin Red S – Potassium Ferricyanide reveals a weakly ferroan calcite composition: ferroan calcite I (5RP 6/2). Because the poikilotopic fabric is confined to hardground lithologies, it is likely to reflect partly subaerial diagenesis, possibly vadose to freshwater phreatic. The existence of earlier-formed cements, i.e., syntaxial fibrous calcites and crinoid-isopachous cements engulfed within the poikilotopic calcite, enables the sequence of events or paragenesis to be interpreted. The occurrence of syntaxial cements as inclusions within the poikilotopic fabric suggests a paragenetic sequence from submarine to freshwater – vadose diagenesis. This sequence of events is always present in the hardgrounds and points to the probable lowering of the sea level.

Minor miscellaneous cements

Leaching and solution processes are usual in diagenesis. Solution of allochems creates mouldic porosity or voids in which subsequent precipitation of lime-rich pore solution can occur. The leaching and solution appear to be selective and preferentially affect gastropods and bivalve shells, whereas other varieties such as brachiopods, ostracods, corals, tentaculitids and crinoids are rarely affected. The aragonitic shells of gastropods and bivalves go readily into solution in contrast to the calcitic shells of the other groups.

The mouldic calcite cements are inclusion-free and ferroan-rich. The fabric displays characteristics of cement and is similar to drusy calcite except that it occupies solution cavities retaining the general outline of the dissolved shell.

Vein calcite occurs as fracture-healing infills and postdates all other cement types.

It has a clear appearance and is augmented with a low ferroan calcite content. This late cement is related to a deeper-burial connate environment.

CEMENT ASSEMBLAGES

The various cement fabrics discussed in the preceding sections can be attributed mainly to three diagenetic environments:

(I) Mixing Zone Assemblage which includes the micritic cement, syntaxial fibrous calcite, and crinoid syntaxial cement.

(II) Freshwater Phreatic Assemblage which postdates the syntaxial fabric and comprises the inclusion-rich granular cement (neomorphosed cement), drusy calcite and mouldic calcite cements.

(III) Vadose Assemblage which develops as a result of extreme shallowing, probably resulting in intermittent subaerial exposure (hardground) and the creation of shrinkage fractures. As a consequence, the vadose assemblage displays a spectrum of fabrics and contains elements of the earlier assemblages (I) and (II) preserved as inclusions or poikilotopes.

(I) Mixing Zone Assemblage

The Mixing Zone Assemblage results from the interaction of freshwater phreatic and marine fluids and the fabric evolves during early stages of diagenesis. The fabric resembles closely those described for submarine cementation (Garrison, 1969; Shinn, 1971; Ginsburg and Schroeder, 1973; Schroeder, 1972; James et al., 1976; Davies, 1977; Lohmann and Meyers, 1977; Longman, 1980) and includes an association of micritic cement, crinoid-isopachous cements and syntaxial fibrous cement.

Staining by Alizarin Red S – Potassium Ferricyanide shows that most of these calcite cements have nonferroan compositions though some ferroan calcite may be present. The cements were analyzed using a microprobe analyser (EDS) to establish the concentration and distribution of Ca, Mg, Fe, Mn and Sr within the crystal lattice. For a particular case of inclusion-rich syntaxial fibrous calcite, traverses from the crystal foundation or substrate through the crystal terminations into the adjacent blocky calcites (Fig. 7-25) were made with 40-micrometer rester. A more restricted 1 – 2-micrometer rester or spot was used for the analysis of the microinclusions including microdolomite inclusions in crinoidal calcite.

The elemental composition of calcite constituting the substrate and their sparry calcite crust or cements which developed syntaxially are given in Table 7-5. $MgCO_3$ content is similar. In ostracod skeletal grains (substrate), the average $MgCO_3$ content is 2.09 mol%. In the ostracods and chonetid brachiopods, the shells contain an average of 1.55 mol% and 1.40 mol% $MgCO_3$, respectively. In contrast, crinoids have a relatively higher Mg content, averaging 4.5 mol% $MgCO_3$. The crinoid-isopachous cement has a much lower $MgCO_3$ with an average of 2.23 mol%. The iron content of the syntaxial cement is relatively low, however, ranging from 0.05 mol% to 0.42 mol% $FeCO_3$.

Results of this traverse from a chonetid substrate through the fabric of the syntaxial fibrous calcite in the direction of crystal elongation, show a significant increase in the Mg content away from the substrate with a maximum value of 2.0 mol% $MgCO_3$. A spot nearest to the substrate has a minimum value of 1.15 mol% $MgCO_3$. Strontium is undetected near the base, but away from the substrate its content increases from 18 ppm to a maximum of 86 ppm (Table 7-5). These concentrations of Sr are close to the detection limit of the EDS so that, although the trend

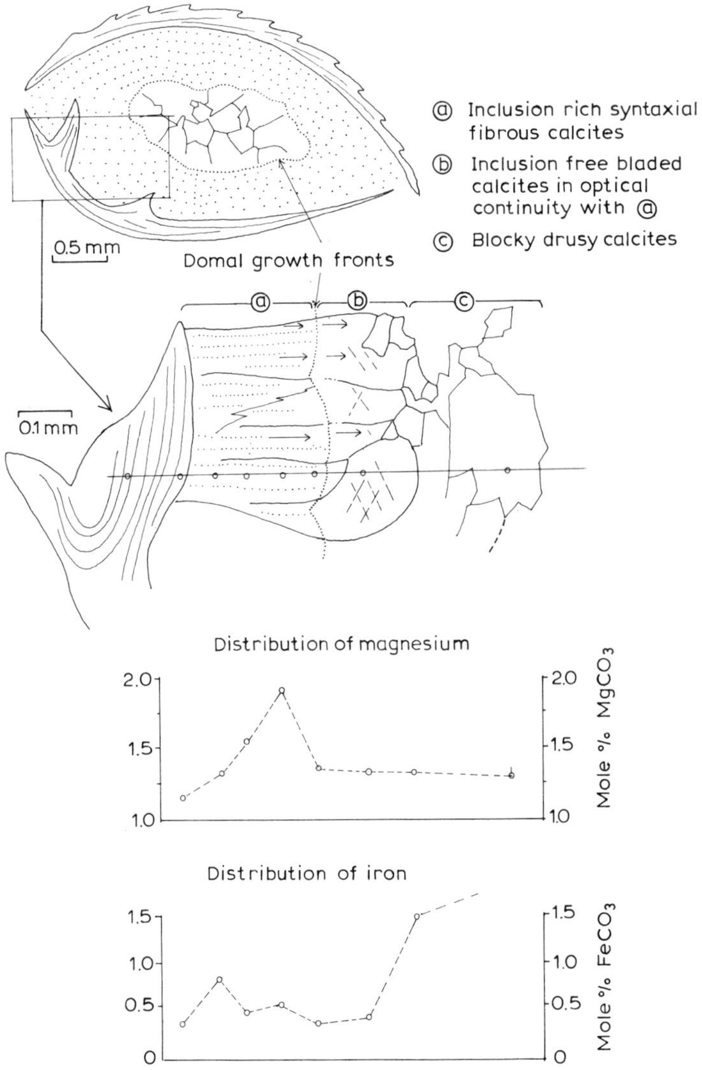

Fig. 7-25. EDS spot analyses on a traverse from the inclusion-rich syntaxial fibrous calcite cement to the inclusion-free blocky paraxial calcite druse (*c*). Contents of $MgCO_3$ are higher in the syntaxial calcite than in the drusy calcite. In contrast, the mol% of $FeCO_3$ is higher in the drusy calcite cement.

in concentration is meaningful, the absolute value may not be exact. In these calcite cements, the iron content never exceeds 1 mol% $FeCO_3$; the maximum recorded value is 0.8 mol% $FeCO_3$.

It is pertinent to note that living brachiopods and ostracods utilize low-Mg calcite for their skeleton, whereas crinoids normally have a high-Mg calcite content (Milliman, 1974). Fossil crinoids, however, have low-Mg calcite (Richter and Fuchtbauer, 1978). The comparatively low Mg content in most of the syntaxial cements examined (fibrous calcite, crinoid-isopachous calcite and micrite cements) is consistent with the Mg contents of their substrate, i.e., brachiopods and ostracods (low-Mg calcite) and crinoids (formerly high-Mg calcite), suggesting that the Mg content in the cement conforms to the relict level of Mg, especially in the crinoid calcite residue. The average content of Mg in these cements is also within the known maximum limit of the Mg content found in modern aragonite cement.

The remarkable correlation of the level of relict Mg in the fossil crinoid and all the forms of syntaxial cement associated with it, lend credence to the idea that the cements were also originally magnesium-enriched. From the fabric criteria described earlier, however, these cements are believed to have been neomorphosed and, if this is really so, it appears that neomorphism was accompanied by loss of magnesium. The depletion in Mg could only arise from a high-Mg calcite precursor rather than from an aragonite precursor because the relict Mg level is close enough to the maximum magnesium content of modern aragonite cement. A possible aragonite precursor for the low-Mg syntaxial calcite, therefore, seems unlikely. The conversion of high-Mg calcite to low-Mg calcite is inevitable because the low-Mg calcite is a more

TABLE 7-5

Elemental composition of calcite in the substrate and sparry calcite crust

Skeletal grains	Mol% of total carbonate*				Mol% of syntaxial calcite*			
	$CaCO_3$	$MgCO_3$	$FeCO_3$	$MnCO_3$	$CaCO_3$	$MgCO_3$	$FeCO_3$	$MnCO_3$
Ostracods	97.30	2.09	0.34	0.36	98.07	1.58	0.13	0.20
Crinoids	92.60	4.50	2.10	0.79	97.59	2.23	0.05	0.08
Chonetids	97.65	1.55	0.26	0.55	97.88	1.39	0.42	0.12

Spot (1 – 2 μm)	$CaCO_3$ (mol%)	$MgCO_3$ (mol%)	$FeCO_3$ (mol%)	$MnCO_3$ (mol%)	Sr^{2+} (ppm)
1	98.280	1.146	0.316	0.258	–
2	97.716	1.288	0.781	0.153	62.3
3	97.850	1.545	0.480	0.011	18.4
4	97.176	1.988	0.568	0.205	86.5
5	96.812	1.320	0.400	0.042	85.6
6	97.946	1.327	0.493	0.179	54.9

* Average value.

stable polymorph (Milliman, 1974, p.267). From consideration of thermodynamic criteria, Bickle and Powell (1977) suggested that only calcite with less than 1 mol% $MgCO_3$ and $FeCO_3$ is stable. But according to Chave (1954), calcite containing up to 4 mol% $MgCO_3$ is common in the diagenetic realm and the work of Lindholm and Finkelman (1972) showed that calcite with up to 4 mol% $FeCO_3$ is not uncommon. The concentrations of Mg and Fe in this assemblage of cements in the Aymestry Limestone appear to be consistent with the findings of Chave (1954) and Lindholm and Finkelman (1972).

Various mechanisms for the converson of high-Mg to low-Mg calcite have been discussed (Bathurst, 1971, p. 331 – 556; Lohmann and Meyers, 1977; Richter and Füchtbauer, 1978). The main mechanisms invoked in the past are dissolution – reprecipitation and incongruent dissolution (or solid diffusion). Dissolution – reprecipitation involves microscale dissolution through migrating magnesium-enriched thin solution films followed by reprecipitation of low-Mg calcite on the migrating front. In contrast, solid diffusion does not require the intermediate aqueous phase. Gomberg and Bonatti (1970) and Land and Epstein (1970) indicated from the stable isotope criteria that the process of replacement of high-Mg calcite by low-Mg calcite involves CO_3^{2-}, which suggests that at least some of the $MgCO_3$ exists in the aqueous phase. This line of evidence tends to discount incongruent dissolution or solid diffusion as an effective mechanism because the CO_3^{2-} has a large molecular size, which renders diffusion difficult. Consequently, dissolution – reprecipitation probably constitutes the main process involved in neomorphic change of the residual syntaxial calcite, especially the syntaxial fibrous form which conceivably is derived from the lateral coalescence of acicular cement. This neomorphic replacement involves wet processes (Bathurst, 1971, p. 475 – 477), as exemplified by a rare geological case in which aragonite (unstable polymorph) protected from moisture by hydrocarbons remained unchanged (Fyfe and Bischoff, 1965; Kinsman, 1969).

In the case of microdolomite inclusions, which are restricted to the high-Mg crinoid residue, the $MgCO_3$ may have a local or autochthonous source within the crinoid microstructure. The fact that microdolomite inclusions tend to be associated with crinoid fragments devoid of the characteristic reticulate meshwork, suggests that obliteration of the fabric occurred when the high-Mg crinoid-calcite precursor was neomorphosed to low-Mg calcite and microdolomite (Fig. 7-15). The destruction of the fabric may be attributable to dissolution – precipitation processes in contrast to epitaxial replacement (Richter and Füchtbauer, 1978), which involves no visible change in the skeletal microstructures.

(II) Freshwater Phreatic Assemblage

The inclusion-rich granular cement, inclusion-free drusy mosaic, and mouldic calcite are morphologically different from the early submarine (Mixing Zone) Cement Assemblage characterized by syntaxial fibrous forms and magnesium-enriched precursors. The majority of cements in the phreatic assemblage are equant in shape and mostly inclusion free. An exception is the inclusion-rich granular calcite which has undergone neomorphic recrystallization, as evidenced by the ghost crystalline planar boundaries observed under crossed polars. Nevertheless, the Phreatic Cement Assemblage is composed characteristically of ferroan-rich calcite.

In the phreatic diagenetic realm, intergranular pores or voids are saturated with meteoric water and are in hydraulic continuity with each other (Longman, 1980). The degree of cementation depends largely on the rate of fluid migration within the interconnected pore system. The increase in supply of water in a permeable rock has been known to promote rapid diagenetic changes (Land, 1970). Longman (1980), however, indicated that neomorphic recrystallization of aragonite to calcite is common in phreatic environments and this can be substantiated by the occurrence of many paramorphic calcites, which are the by-products of recrystallization of aragonitic shells to (ferroan) calcite. The availability of water may promote rapid diffusion of Ca^{2+} and CO_3^{2-} from one locus to another in the dissolution – reprecipitation process of converting aragonite to calcite.

The elemental composition within this cement assemblage shows an insignificant magnesium content (average: 0.58 mol% $MgCO_3$) but a relatively significant iron content with an average of 3.20 mol% $FeCO_3$. Lindholm and Finkelman (1972) found that calcite with up to 4 mol% $FeCO_3$ is more common during the late diagenetic phase. Richter and Füchtbauer (1978) discussed the implications of the presence of ferroan calcite in diagenetic environments and concluded that ferroan calcite is characteristic of a meteoric – phreatic environment rather than being formed as an early submarine cement. Work in the modern submarine setting of the Atlantic off America (Manheim and Bischoff, 1969) and off N.W. Africa (Hartman et al., 1976) shows that the FeO content in pore fluids of marine sediments is extremely low: the mole ratio of Fe^{2+} to Ca^{2+} is usually less than 0.001 (Hartman et al., 1976). These workers demonstrated that in the uppermost meter of sediment below the sediment – water interface, Fe^{2+} was continually depleted, due to Fe^{2+} uptake by sulphate-reducing bacteria in the low-Eh environment within the sediment (Berner, 1971, p. 199). Hence, the Fe^{2+} uptake by calcite in marine connate water may occur late in diagenesis or may not happen at all. In other words, ferroan calcite has little chance of being formed under the unfavorable marine conditions.

In addition to criteria which support the formation of ferroan calcite in the phreatic zone, the evidence is provided by the zoning fabric associated with the compositional variation. The development of well-defined and sharp multi-phase zonation of calcite cements indicate sudden changes in the composition of the pore fluid. Such changes are rarely possible in submarine environments because of the stability of the physicochemical conditions.

A possible source of iron are the volcanic bentonite clays which are rich in biotite. The iron may be derived by leaching from the biotite or from iron oxide or sulphide present within associated fine-grained sediments during diagenesis. Oldershaw and Scoffin (1967) suggested that in the Wenlock Limestone, which has a similar development of shale and/or clay interbedded with limestone, clay constitutes the principal donor sediment for iron during precipitation of late cement. Both clays and shales are rich in ferromagnesian minerals.

(III) Vadose Assemblage

Poikilotopic calcite is the only significant vadose cement resulting from subaerial exposure. Calcite veining can be considered as fracture healing in the same environment.

The occurrence of extremely coarse-crystalline poikilotopic cement engulfing earlier mixing zones and phreatic assemblages, in hardground biosparite samples, illustrates a rapid transition from submarine to vadose environments. The bioerosion phase, resulting in boring or excavation of the limestone during the subaerial stage (epidiagenetic hardground stage), produced a secondary solution porosity which allowed subsequent precipitation of meteoric vadose cement. In this setting, it is quite common to find a reversal of these processes involving degradation of cement, especially around the aureole of the *Trypanite* sp. borings, such as the occurrence of a grumous or clotted micritic texture (Fig. 7-26). Micrite clots or patches develop non-selectively in sparry calcite cement and in neomorphosed skeletal grains, in contrast to the early diagenetic or submarine micrite which develops as a micrite envelope mainly attributable to algal boring-infill. The process of micritization of sparry calcite in this vadose environment is considered as degrading recrystallization involving concomitant dissolution – reprecipitation of calcite to calcite. This is analogous to the recrystallization of high-Mg calcite to low-Mg calcite. Kahle (1977) described similar features in the subaerial Holocene calcareous crust and regarded the process of micritization of sparry calcite as sparmicritization. Accordingly, he suggested that micritization of the vadose caliche proceeded by concomitant dissolution – precipitation sparmicritization or CDP sparmicritization.

INCIPIENT DOLOMITIZATION

Dolomitization, unlike calcitization, is not a common feature in the Upper Bringewood Beds and the process is normally restricted to incipient hardgrounds. Generally, it represents secondary dolomitization though an early-formed primary dolomite also occurs: complete conversion of biosparite to dolostone has been observed.

Microdolomite inclusions are confined to crinoidal calcite. In hardground biosparite, however, relatively small calcitized crystals (0.02 – 0.06 mm) of former dolomite rhombs are found as pseudomorphs distributed within the extremely coarse-crystalline poikilotopic cements. Poikilotopic calcite also shows evidence of minor secondary dolomitization, manifested as unstained patches or zones when subjected to Alizarin Red S – Potassium Ferricyanide stains. This incipient dolomitization is confined to the fine rhombs of precursor dolomite. The latter is a product of dedolomitization of the primary dolomite rhombs which occur as random, clear and unstained patches. The occurrence of these precursor primary dolomites (or dedolomite) in poikilotopic calcite cement, both of which were later weakly dolomitized, suggests that a reverse diagenetic process was maintained.

The fine rhombs of precursor dolomite originate as part of the primary fabric in an early stage of diagenesis, and microfacies and ichnofacies criteria suggest that they developed in the peritidal zones (Mazzullo and Reid, 1988).

In contrast to primary dolomites, late diagenetic dolomitized calcite crystals are much larger in size ($\leq$ 0.8 mm) and usually lack well-defined shapes. The coarse crystallinity is attributable to a slow rate of formation. There are three main modes of occurrence:

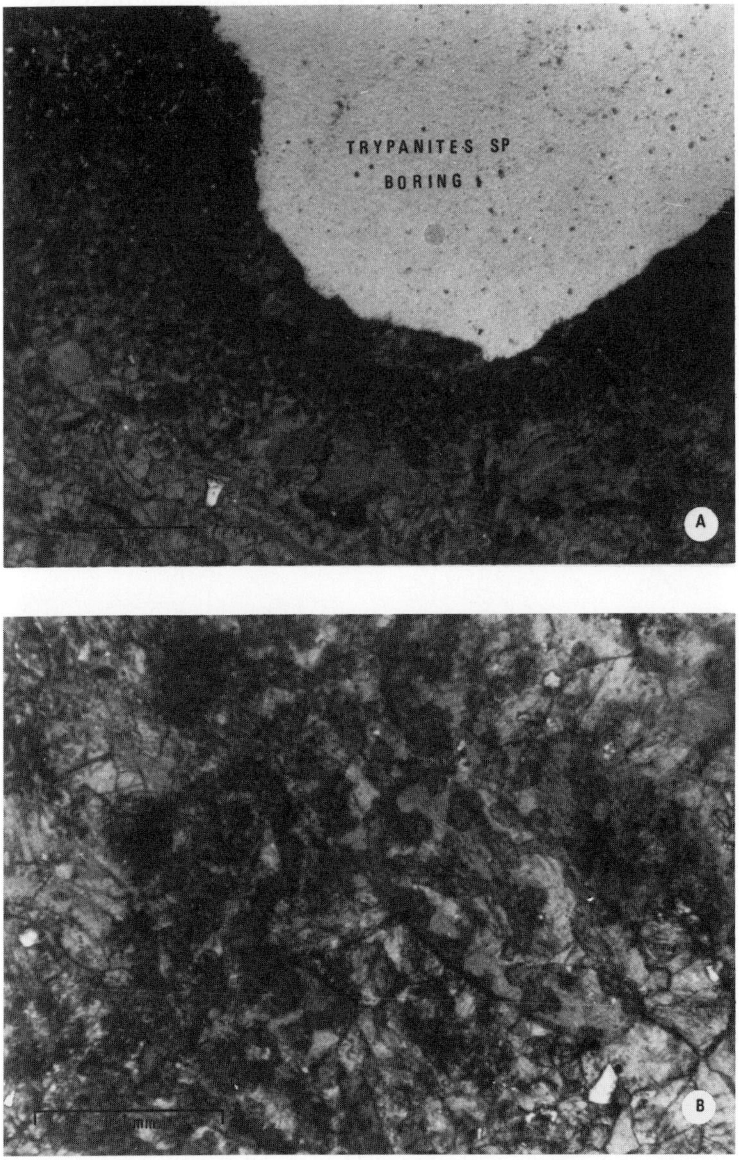

Fig. 7-26. (A) Photomicrograph showing grumous or clotted texture lining the wall of a boring. (B) A detail of (A) illustrating the clotted texture attributable to sparmicritization.

- dolomitization of calcite within *Trypanites* sp. borings;
- replacement of drusy calcite cement; and
- homoaxial replacement of crinoid-isopachous pseudospar. Typically, the dolomitized pseudospar displays twin lamellae parallel to the shorter diagonal of the rhombs.

In the case of dolomitized calcite within the *Trypanites* sp. burrows, the presence of clay may promote dolomitization because the replacement of adsorbed Mg^{2+} within clays by Na^{+} and Ca^{2+} during clay diagenesis is known to occur (Kahle, 1965). Such a mechanism has been suggested for Upper Jurassic limestones where clay minerals are thought to have triggered dolomitization (Schmidt, 1965). In the case of homoaxial replacement of crinoid-isopachous pseudospar, the magnesium

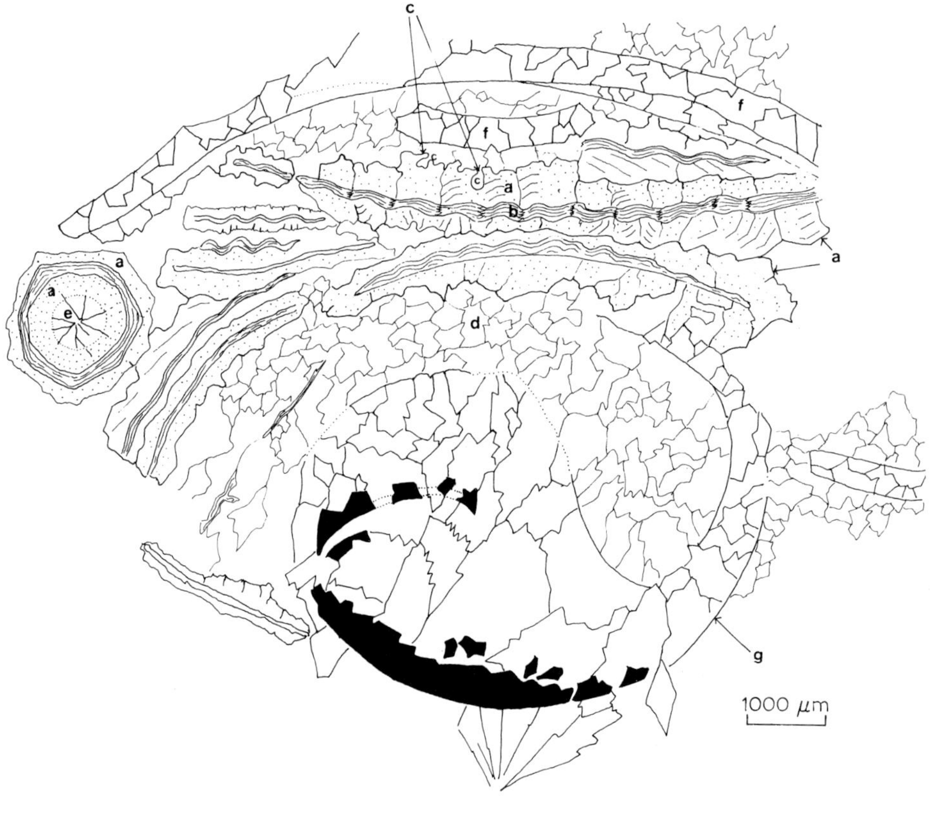

Fig. 7-27. The relationship between the various sparry calcites (cements and recrystallized calcites) in shelly (chonetid) biosparite. This example demonstrates the paragenetic sequence of cementation. The inclusion-rich syntaxial fibrous calcite crusts (*a*) seeded epitaxially onto the chonetid brachiopod shell (*b*) constitute the earliest-formed cement. This syntaxial cement shows microborings (*c*) and forms an irregular but sharp boundary with the subsequent neomorphosed granular cement (*d*). The inclusion-free, ferroan-rich drusy calcite cement (*e*) in a tentaculitid constitutes the third-generation cement. The neomorphic (paramorphic) calcite, replacing after a bivalve (*f*) and gastropod (*g*), is ferroan-rich and lacks the typical enfacial junction (crystal boundary), which characterizes most cements. The exact date of neomorphism of these paramorphic calcites is uncertain, but probably occurred much later than the granular cement or is synchronous with the drusy calcites.

is probably derived from within the precursor high-Mg calcite skeleton of crinoids and replacement may have proceeded by concomitant dissolution – reprecipitation.

CEMENTATION SEQUENCE: SUMMARY

The destruction of primary intergranular porosity by progressive cementation leaves various signatures on the rock during diagenesis. Though by no means simple to interpret, the fabric relationships of these cements record vividly the sequence of events or the paragenesis of the lithified sediments, as summarized in Figs. 7-21, 7-27, and 7-28, and in Table 7-6.

Stage 1

Stage 1 represents the initial stage of diagenesis below the sediment – water interface in which the sediments are constantly bathed in submarine water. Stage 1 is characterized by micritization of allochems and precipitation of high-Mg lime mud (or micritic cement) in fine pores of skeletal microstructures, e.g., crinoids.

Stage 2

Stage 2 reflects early diagenetic changes involving the reduction of primary intergranular porosity by the development of acicular Mg-enriched calcite. These cements are preserved as syntaxial fibrous calcite and isopachous rims on all skeletal grains. The fabric and mineralogy of these early-formed cements are greatly influenced by the substrate characteristics. The common occurrence of dense dusty inclusions indicates a rapid rate of crystallization. In addition, the formation of fine-grained dolomite rhombs takes place from the hypersaline fluids of the Mixing Zone.

Stage 3

Stage 3 marks increased burial and the influence of freshwater in the phreatic realm. Crystals in the cements are equant and coarsely crystalline, suggesting a slow rate of crystallization. The relict level of Mg is extremely low, but the proportion of iron in the connate water is relatively high giving rise to ferroan calcite cement.

Stage 4

Stage 4 reflects continued diagenesis from Stage 3, involving leaching and solution of aragonitic shells (mostly gastropods and bivalves) creating solution voids, which allow precipitation of an equant calcite mosaic pseudomorph after the shells. Their high-Mg calcitic counterparts are unaffected by solution but are susceptible to neomorphic recrystallization to low-Mg calcite. The microdolomite inclusions, which are confined to the crinoidal calcite domain, may recrystallize in this stage by concomitant dissolution – reprecipitation.

Stage 5

Stage 5 represents an epidiagenetic stage (hardground) which may involve solution and reprecipitation. In this subaerial vadose realm, leaching and bioerosion

Table7-6

A summary of diagenetic textures, environments and paragenesis in the Aymestry Limestone

CEMENT TYPES \ PROPERTIES	HABITS	SIZE	INCLUSIONS	COLOR	CRYSTAL BOUNDARY	COMPOSITION	MISCELLANEOUS
MICRITES	1. Envelopes (boring - infilling) 2. Syntaxial in crinoid stomapores	crypocrystalline (≤ 12μ)	absent	dark to opaque	not visible in thin section	ferroan with 2.23 mole% Mg CO_3	mineralogy of precipitates influenced by substrates
ISOPACHOUS RIMS CALCITES	Fringing crust on crinoid ossicles (epitaxial)	crust thickness: 50 –250μ)	dusty	turbid	coalescing, non-planar	weakly ferroan, 2.20 mole% Mg CO_3	optical continuity between cement and substrate
SYNTAXIAL FIBROUS CALCITES	Radial fibrous, fascicular and botryoidal	10–400μ (crystal length parallel to c-axis)	dusty	turbid	coalescing, non-planar	weakly ferroan, ≤ 2.23 mole % Mg CO_3	- boring on cement - domal growth fronts
INCLUSION-RICH GRANULAR CALCITES	Equant to inequant	100–500μ (diameter)	dusty	turbid and pseudo pleochroic to brownish tinge	non-planar	ferroan with 0.5–1.5% FeO; 1.6 mole % Mg CO_3	Neomorphosed cement probably influenced by prescence of clay introduced by boring activities
DRUSY MOSAIC CALCITES	Paraxial, blocky (Rhombic to scalenohedral)	≤500μ (diameter) crystal enlarges centripetally	absent	clear	planar with enfacial junction	ferroan with 0.5–2.5% FeO; 1.32 mole% Mg CO_3	rhombocentric zonation of ferroan and non-ferroan calcites
POIKILOTOPIC CALCITES	Poikilotopic	≤1000μ (length parallel to c-axis)	incorporation of pre-existing fabric (syntaxial calcites and dedolomites)	clear	planar with enfacial junction	very weakly ferroan 0.5–1.5% FeO	Recrystallization and dedolomization of inclusions pre-date the poikilotopic cement
LATE CEMENT eg. Fracture infilling	Discordant, cross-cutting older fabrics	≤500μ (length parallel to c-axis)	absent	clear	planar	ferroan	due to compaction or deep burial

CEMENT MORPHOLOGIES

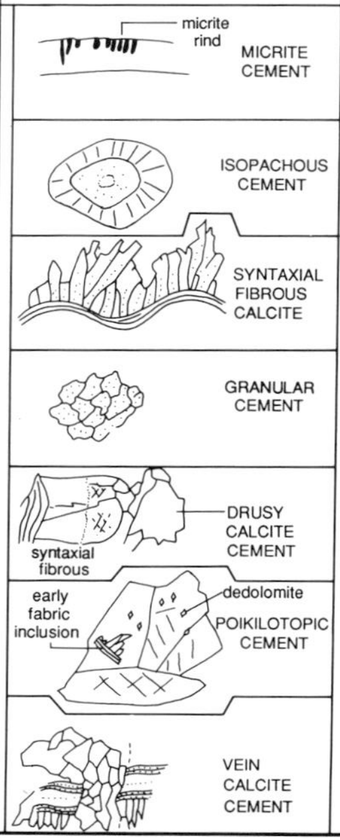

PARA-GENESIS | DIAGENETIC ENVIRONMENTS

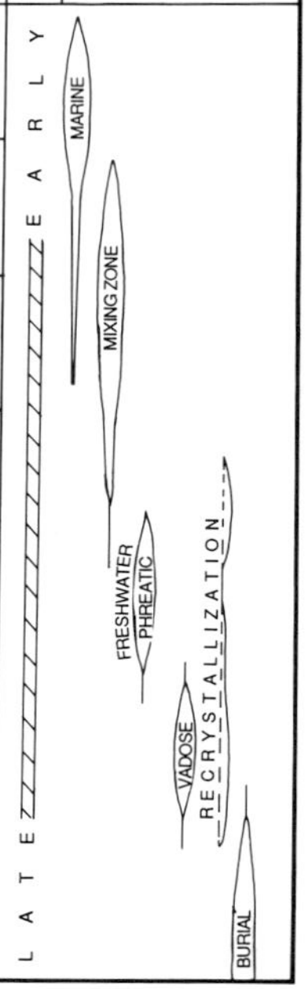

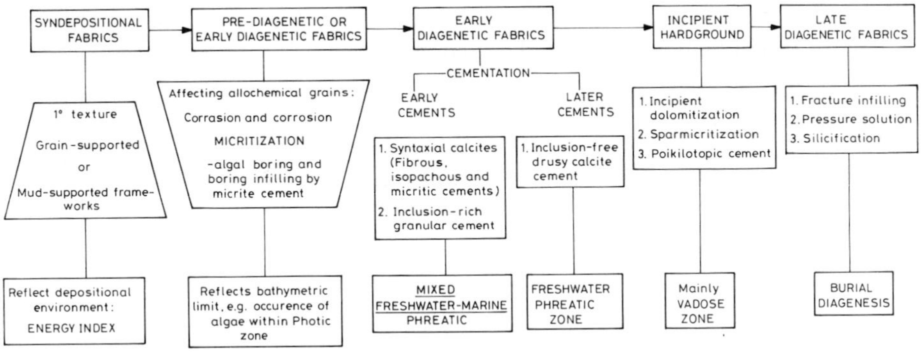

Fig. 7-28. Stages in lithification of limestone lithologies.

may obliterate the signature of early diagenetic cements. The epidiagenetic stage is known to induce incipient dolomitization.

The above stages involve reduction in porosity by progressive cementation, invariably completing lithification of the carbonates. It represents a simple model, however, which attempts to illustrate diagenetic changes within the carbonate microfacies; various other factors, such as the exact timing of neomorphism, remain unresolved.

POST-DIAGENETIC FABRIC IN VERY FINE-GRAINED CLASTICS

The carbonate microfacies, although monomineralic, provides diverse diagenetic fabrics. In contrast, fine-grained siliciclastic rocks such as the fine-grained quartz arenites and muddy siltstones have a limited variety of postdepositional fabrics. The most commonly observed textures can be described as follows.

Cement

Most of the fine-grained clastics have a dominant mud matrix, but calcite and silica cements are occasionally found in well-sorted, very fine-grained quartz arenites. The calcite cement is probably derived from the leaching of skeletal material in adjacent fossiliferous beds. With increasing depth of burial, the pH (pH < 7.8) and Eh of the connate water decreases (Tucker and Van Straaten, 1970) and may cause dissolution of comminuted shell fragments which are disseminated throughout the surrounding muddy sediment. With increasing overburden pressure, the dissolved carbonate can migrate upward into the high-Eh or oxidation zone where the higher pH causes the precipitation of calcium carbonate in more porous, cleanly-washed quartz arenite.

Corrosion textures

Corrosion and corrasion textures occur on the surface of detrital quartz grains when the grains are juxtaposed with calcite cement. This is probably due to etching of the surface of the grains by the high-pH carbonate-rich connate water, enhancing the depositional angular fabric of the quartz detritus.

Compaction textures

Compaction textures are extremely common in very fine-grained arenites. The features displayed include bending of plastic mica grains and fracturing of grains of brittle minerals such as tourmaline, apatite and idiomorphic quartz.

Pressure solution (horsetail) textures

A pressure solution fabric is common in calcareous siltstone and is characterized by whisps (horsetails) of irregular thin clay veneers running parallel with each other (Fig. 7-29). These clay-rich seams probably represent the residual insoluble components of pressure-solution fronts formed in the late diagenetic stage under overburden pressure, when some of the calcite cement recrystallizes and insoluble residues are concentrated along the solution fronts.

SILICIFICATION

The silicification of limestones (especially chertification of nodules) is rarely seen except in the outer shelf areas. It is most commonly developed in the upper part of the Aymestry Limestone.

Chertification is a very late diagenetic phenomenon involving the replacement of the allochems and orthochems. The silicified biosparite contains intergrowths of microcrystalline quartz and chalcedonic quartz.

Microcrystalline quartz (Midgley, 1951; Folk and Weaver, 1952) occurs as randomly-oriented equant crystals displaying characteristic undulose extinction, a pressure effect probably caused by superimposition of crystals. The chalcedonic quartz usually postdates microcrystalline quartz. Under crossed polars the chalcedonic polymorph is seen to occur in a radiating fibrous or fascicular optic form with a characteristic ''brush'' extinction. The fibers are an optical illusion because they are physically inseparable; nevertheless this variety of silica cement may parallel the development of syntaxial fibrous calcite cement. When these silica polymorphs occur together, they are normally separated by a well-defined curved boundary.

In addition to the two polycrystalline quartz polymorphs, chert nodules may contain crystals of drusy quartz (Fig. 7-29). These are more coarsely crystalline than the matrix and preferentially line intraskeletal cavities within chalcedony. The growth of coarse drusy quartz reflects the slow rate of precipitation probably due to a diminishing rate of supply of silica-rich solution during diagenesis.

A number of factors seem to determine the type of quartz polymorph formed. Chalcedonic quartz usually postdates the crystallization of microcrystalline quartz and commonly nucleates on the spherical growth front of the microcrystalline variety, growing in the remaining space within the cavity. Folk and Weaver (1952) suggested that the factor which governs crystallization of the two polymorphs is the fabric of the nucleating centers and that microcrystalline quartz require numerous, closely-spaced centers arranged in three-dimensional arrays for their nucleation. In contrast, chalcedony nucleates from a few widely-spaced centers lying along a surface. The affinities for the respective fabric explains the predominance of microcrystalline quartz in the earlier phase of precipitation, producing a spherical or botryoidal structure with crystals radiating in many directions. The chalcedonic quartz invariably develops only as continuous epitaxial forms and constitutes a natural successor to the microcrystalline polymorphs.

In the Lower Greensand Formations (Aptian) of the Weald in southern England, Middlemiss (1975, 1978) described two forms of chert: replacement cherts which originate by replacement of calcarenites and calcareous sandstone; and spicule-bed cherts which comprise siliceous sponge spicules. In both cherts, sponge spicules form a major constituent and Middlemiss (1978) considered the cherts to have a secondary origin, the spicules providing the authochthonous source of silica. In the Aymestry Limestone siliceous spicules are rare, though calcareous spicules are quite common. Watkins and Aithie (1979) claim that the majority of these calcareous spicules were originally siliceous stauracts of the protospongiid *Phormosella ovata* subsequently calcitized. Such spicules may have yielded the authochthonous silica, although another possible source of silica is bentonite which formed by devitrification of volcanic ash. Although bentonite occurs commonly throughout the Aymestry Limestone, it appears to be a less likely silica source, because in the inner shelf areas where bentonites are ubiquitous chertification is lacking.

Fig. 7-29. (A) Photomicrograph illustrating a postdiagenetic solution feature paralleling original bedding. Thin clay veneers form "horsetail" structures representing the residual product of pressure solution. (Width of photograph is 1.5 mm.)

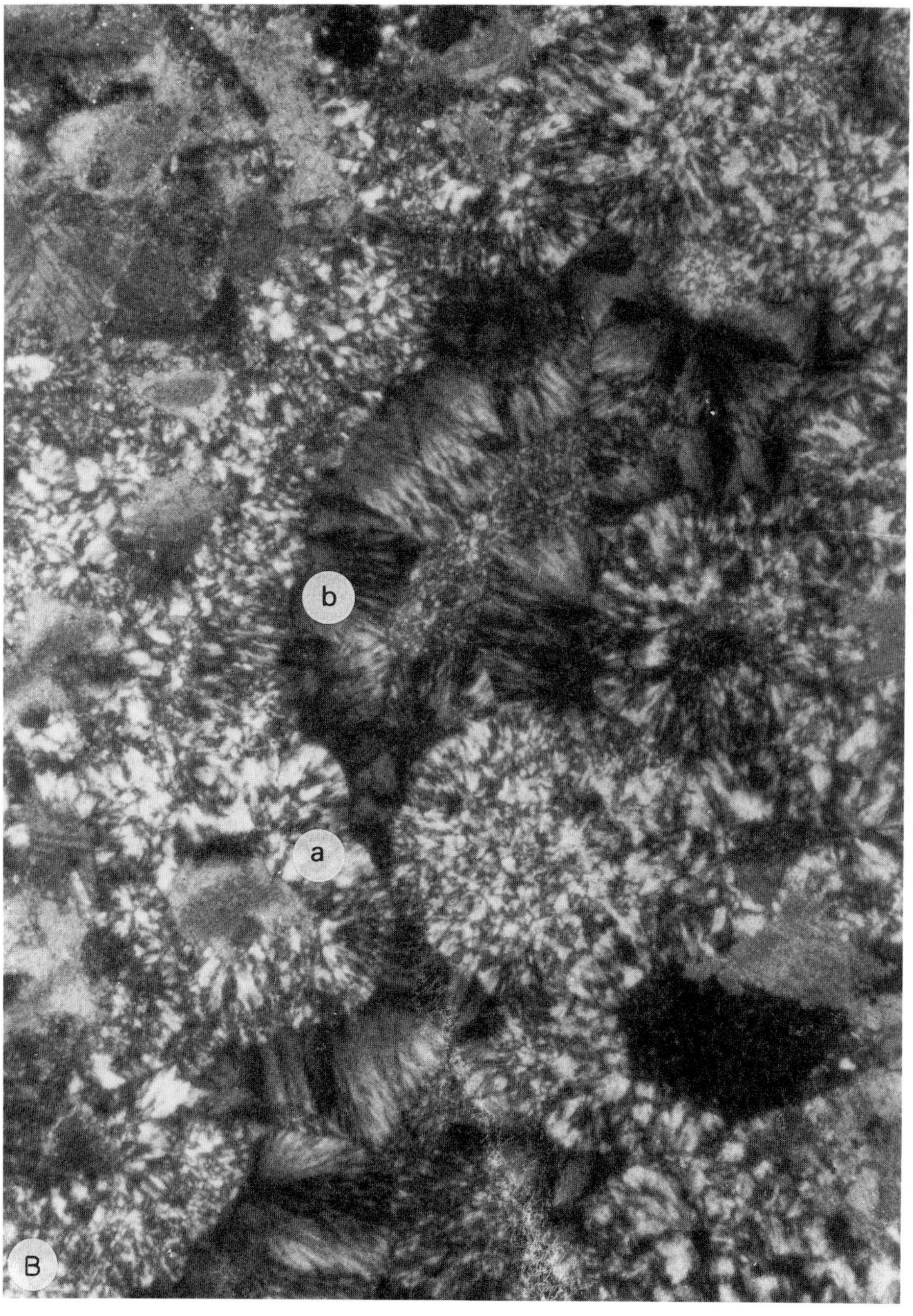

Fig. 7-29. (B) Photomicrograph of replacement chert consisting of botryoidal microcrystalline quartz (*a*) and fibrous chalcedonic quartz (*b*). (Width of photograph is 0.4 mm.)

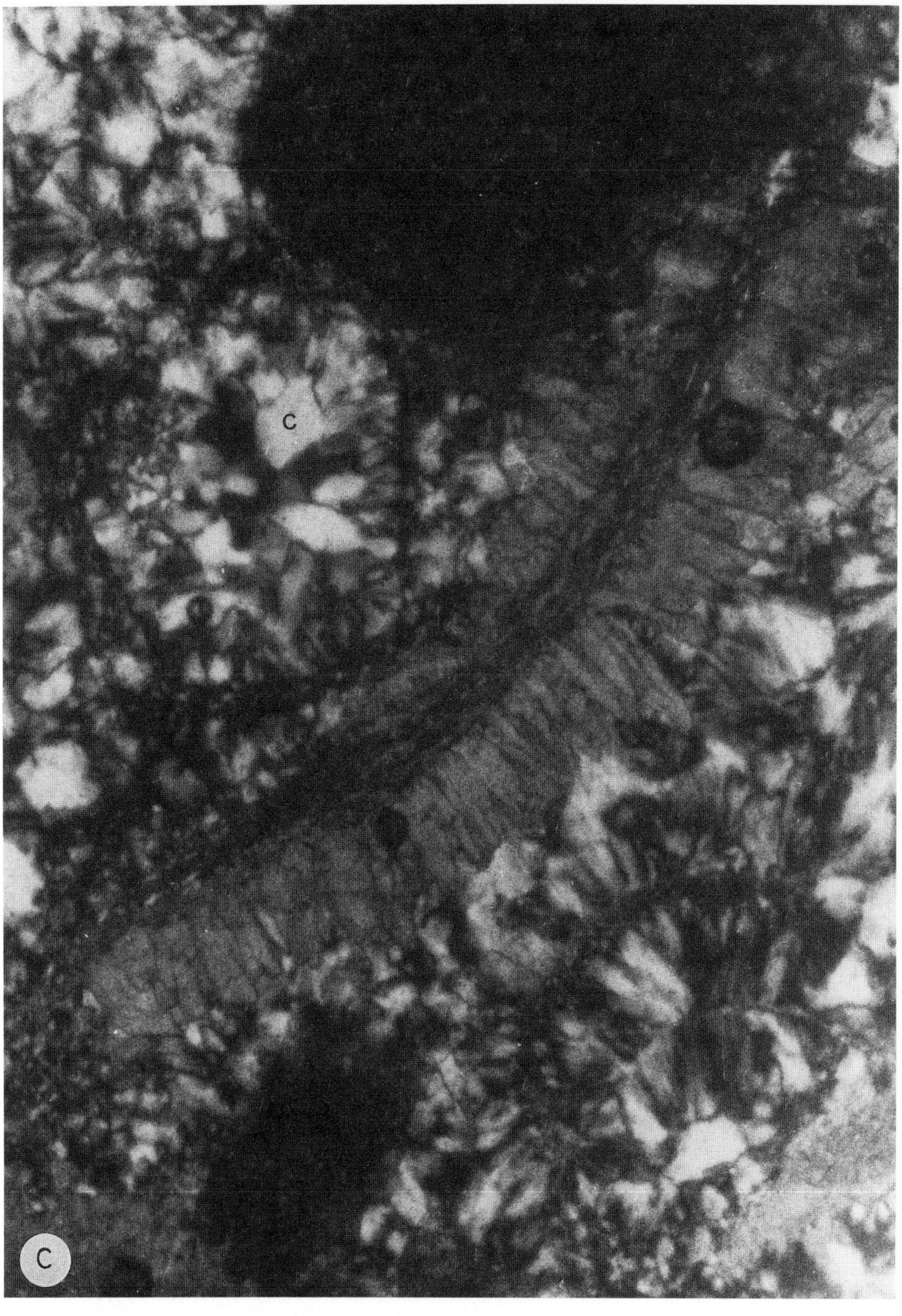

Fig. 7-29. (C) Microcrystalline quartz and equant drusy quartz (*c*). Unsilicified syntaxial fibrous calcite cement displays acicular form. (Width of photograph is 0.4 mm.)

ACKNOWLEDGEMENTS

This chapter is based on research carried out for A.H.M.'s doctorate at Queen Mary College, University of London. A.H.M. is indebted to the National University of Malaysia for financing the research. The authors are also grateful to Drs. George V. Chilingarian and K.H. Wolf for valuable suggestions.

REFERENCES

Alexander, F.E.S., 1936. The Aymestry Limestone of the Main Outcrop. *Q.J. Geol. Soc. London,* 92: 103 – 115.

Alexandersson, T., 1969. Recent littoral sublittoral high-Mg calcite lithification in the Mediterranean. *Sedimentology,* 12: 47 – 61.

Alexandersson, T., 1972. Micritization of carbonate particles: processes of precipitation and dissolution in modern shallow marine sediment. *Uppsala Univ. Geol. Inst. Bull.,* 3: 201 – 236.

Bathurst, R.G.C., 1958. Diagenetic fabrics in some British Dinantian limestones. *Liverpool Manchester Geol. J.,* 2 (1): 11 – 36.

Bathurst, R.G.C., 1964. The replacement of aragonite by calcite in the molluscan shell wall. In: J. Imbrie and N. Newell (Editors), *Approaches to Palaeoecology.* Wiley, New York, N.Y., pp. 357 – 376.

Bathurst, R.G.C., 1966. Boring algae, micrite envelopes and lithification of molluscan biosparite. *Liverpool Manchester Geol. J.,* 5: 15 – 32.

Bathurst, R.G.C., 1971. *Carbonate Sediments and Their Diagenesis.* Elsevier, Amsterdam, 658 pp.

Berner, R.A., 1971. *Principles of Chemical Sedimentology.* McGraw-Hill, New York, N.Y., 240 pp.

Bickle, M.J. and Powell, R., 1977. Calcite – dolomite geothermometry for iron-bearing carbonates. *Contrib. Mineral. Petrol.,* 59: 281 – 292.

Bricker, O.P. (Editor), 1971. *Carbonate Cement.* The Johns Hopkins Press, Baltimore, Md., 376 pp.

Carpenter, W., 1845. On the microscopic structure of shells. *Br. Assoc. Adv. Sci. Rep.,* 14: 1 – 24.

Chafetz, H.S., McIntosh, A.G. and Meyers, W.J., 1988. Freshwater phreatic diagenesis in the marine realm of recent Arabian Gulf carbonates. *J. Sediment. Petrol.,* 58: 433 – 440.

Chave, K.E., 1954. Aspects of the biogeochemistry of magnesium 1. Calcareous marine organisms. *J. Geol.,* 62: 266 – 283.

Chave, K.E. and Suess, E., 1970. Calcium carbonate saturation in sea water. *Limnol. Oceanogr.,* 15: 633 – 637.

Cherns, L., 1980. Hardgrounds in the Lower Leintwardine Beds (Silurian) of the Welsh Borderland. *Geol. Mag.,* 117 (4): 311 – 401.

Chilingar, G.V., Bissell, H.J. and Fairbridge, R.W. (Editors), 1967a. *Carbonate Rocks: Physical and Chemical Aspects.* Elsevier, Amsterdam, 413 pp.

Chilingar, G.V., Bissell, H.J. and Wolf, K.H., 1967b. Diagenesis of carbonate rocks. In: G. Larson and G.V. Chilingar (Editors), *Diagenesis in Sediments.* Elsevier, Amsterdam, pp. 179 – 322.

Cope, J.C.W. and Bassett, M.G., 1987. Sediment sources and Palaeozoic history of the Bristol Channel area. *Proc. Geol. Assoc.,* 98: 315 – 330.

Cullis, C.G., 1904. The mineralogical changes observed in the cores of Funafuti. The Atoll of Funafuti. *R. Soc. London, Section V,* pp. 61 – 124.

Davies, G.R., 1977. Former magnesian calcite and aragonite submarine cements in upper Palaeozoic reefs of the Canadian Arctic; A summary. *Geology,* 5: 11 – 15.

Dickson, J.A.D., 1966. Carbonate identification and genesis as revealed by staining. *J. Sediment. Petrol.,* 36: 491 – 505.

Dunham, R.J., 1962. Classification of carbonate rocks according to depositional texture. In: W.E. Ham (Editor), *Classification of Carbonate Rocks. Mem. Am. Assoc. Pet. Geol.,* 1: 108 – 121.

Dunham, R.J., 1971. Meniscus sediments. In: O.P. Bricker (Editor), *Carbonate Cement.* Studies in Geology, 19. Johns Hopkins, Baltimore, Md., pp. 297 – 300.

Folk, R.L., 1959. Practical petrographic classification of limestone. *Am. Assoc. Pet. Geol.,* 43: 1 – 38.

Folk, R.L., 1962. Spectral subdivision of limestone types. In: W.E. Ham (Editor), *Classification of Carbonate Rocks. Mem. Am. Assoc. Pet. Geol.,* 1: 62–84.

Folk, R.L., 1965. Some aspect of recrystallization in ancient limestones. In: L.C. Pray and R.C. Murray (Editors), *Dolomitization and Limestone Diagenesis. Soc. Econ. Paleontol. Mineral., Spec. Publ.,* 13: 14–48.

Folk, R.L., 1974. The natural history of crystalline calcium carbonate: effect of magnesium content and salinity. *J. Sediment. Petrol.,* 44: 40–53.

Folk, R.L. and Weaver, C.E., 1952. A study of the texture and composition of chert. *Am. J. Sci.,* 250 (7): 498–510.

Friedman, G.M., 1964. Early diagenesis and lithification in carbonate sediments. *J. Sediment. Petrol.,* 34: 777–813.

Friedman, G.M., 1965. Occurrence and stability relationships of aragonite, high-magnesian calcite and low-magnesian calcite under deep-sea conditions. *Bull. Geol. Soc. Am.,* 76: 1191–1196.

Fyfe, W. S. and Bischoff, J.L., 1965. The calcite aragonite problem. In: L.C. Pray and R.C. Murray (Editors), *Dolomitization and Limestone Diagenesis. Soc. Econ. Paleontol. Mineral., Spec. Publ.,* 13: 3–13.

Garrison, R.E., 1967. Pelagic limestones of the Oberalm beds (Upper Jurassic–Lower Cretaceous) Austrian Alps. *Bull. Can. Pet. Geol.,* 15: 21–49.

Garrison, R.E., 1972. Inter and intra pillow limestones of the Olympic Peninsula, Washington. *J. Geol.,* 80: 310–322.

Ginsburg, R.N. and James, R.P., 1976. Submarine botryoidal aragonite in Holocene reef limestones, Belize. *Geology,* 4: 431–436.

Ginsburg, R.N. and Schroeder, J.H., 1973. Growth and submarine fossilization of algal cup reefs, Bermuda. *Sedimentology,* 20: 575–614.

Goldman, J.C., Porcella, D.B., Middlebrookes, E.J. and Toerien, D.F., 1972. The effect of carbon dioxide on algal growth – its relationship to eutrophication. *Water Res.,* 6: 637–679.

Gomberg, D.N. and Bonatti, E., 1970. High-magnesian calcite: leaching of magnesium in the deep sea. *Science,* 168: 1451–1453.

Goreau, T.F. and Land, L.S., 1974. Fore reef morphology and depositional process, North Jamaica. In: L.F. Laporte (Editor), *Reefs in Time and Space. Soc. Econ. Paleontol. Mineral., Spec. Publ.,* 18: 77–90.

Hamid Mohamad, A., 1981. The petrology and depositional environment of the Upper Bringewood Beds of the south eastern part of the Welsh Borderland. Ph.D. thesis (unpubl.), Univ. London, 260 pp.

Hartmann, M., Muller, P.J., Suess, E. and Van der Weijden, C.H., 1976. Chemistry of late Quaternary sediments and their interstitial waters from the N.W. African continental margin. *Meteor – Supply Pap. Geol. Surv. C,* 24: 1–67.

Heckel, P.H., 1983. Diagenetic model for carbonate rocks in midcontinent Pennsylvanian eustatic cyclothems. *J. Sediment. Petrol.,* 53: 733–759.

Holland, C.H., Lawson, J.D. and Walmsley, V.G., 1968. The Silurian rocks of the Ludlow district, Shropshire. *Bull. Bur. Mus. Nat. Hist. (Geol.),* 8: 95–171.

Holland, C.H., 1980. Silurian series and stages: decisions concerning chronostratigraphy. *Lethaia,* 13: 238.

Holland, C.H., Lawson, J.D., Walmsley, V.G. and White, D.E., 1980. Silurian Stages. *Lethaia,* 13: 268.

Husseini, S.I. and Matthew, R.K., 1972. Distribution of Mg calcite in lime muds of the Great Bahama Bank: Diagenetic implications. *J. Sediment. Petrol.,* 42: 179–182.

James, N.P., Ginsburg, R.W., Marszalek, D.S. and Choquette, P.W., 1976. Facies and fabric specificity of early subsea cements in shallow Belize (British Honduras) reefs. *J. Sediment. Petrol.,* 46: 523–544.

Johnson, J.H., 1961. Limestone-building algae and algal limestones. *Q. Colo. Sch. Mines,* 297 pp.

Kahle, C.F., 1965. Possible roles of clay minerals in the formation of dolomite. *J. Sediment. Petrol.,* 35: 448–453.

Kahle, C.F., 1977. Origin of subaerial Holocene crusts: role of algae, fungi and sparmicritization. *Sedimentology,* 24: 413–436.

Kendall, A.C., 1976. Origin of fibrous calcite cements that apparently replace foraminiferal tests. *J. Sediment. Petrol.,* 46: 545–547.

Kendall, A.C. and Tucker, M.E., 1973. Radiaxial fibrous calcite: a replacement after acicular carbonate. *Sedimentology,* 20: 365–389.

Kinsman, D.J.J., 1969. Interpretation of Sr^{2+} concentrations in carbonate minerals and rocks. *J. Sediment. Petrol.,* 39: 486 – 508.

Kobluk, D.R. and Risk, M.J., 1977a. Calcification of exposed filaments of endolithic algae, micrite envelope formation and sediment production. *J. Sediment. Petrol.,* 47: 517 – 528.

Kobluk, D.R. and Risk, M.J., 1977b. Micritization and carbonate grain binding by endolithic algae. *Bull. Am. Assoc. Pet. Geol.,* 61: 1069 – 1082.

Land, L.S., 1970. Phreatic versus vadose meteroic diagenesis of limestones: evidence from fossil water table. *Sedimentology,* 14: 175 – 185.

Land, L.S. and Epstein, S., 1970. Late Pleistocene diagenesis and dolomitization, North Jamaica. *Sedimentology,* 14: 187 – 200.

Lawson, J.D., 1973. Facies and faunal changes in the Ludlovian rocks of Aymestrey, Herefordshire. *Geol. J.,* 8: 247 – 278.

Leighton, M.W. and Pendexter, C., 1962. Carbonate rock types. In: W. E. Ham (Editor), *Classification of Carbonate Rocks. Mem. Am. Assoc. Pet. Geol.,* 1: 33 – 62.

Lindholm, R.C. and Finkelman, R.B., 1972. Calcite staining: Semiquantitative determination of ferrous iron. *J. Sediment. Petrol.,* 42: 239 – 242.

Lohmann, K.C. and Meyers, W.J., 1977. Microdolomite inclusions in cloudy prismatic calcites: A proposed criterion for former high-magnesium calcites. *J. Sediment. Petrol.,* 47: 1078 – 1088.

Longman, M.W., 1980. Carbonate diagenetic textures from near surface diagenetic environments. *Bull. Am. Assoc. Pet. Geol.,* 64: 461 – 487.

Lukas, K.J., 1973. Taxonomy and ecology of the endolithic microflora of reef corals, with a review of the literature on endolithic microphytes. Ph.D. thesis, Univ. Rhode Island, 159 pp.

Macintyre, I.G., 1977. Distribution of submarine cements in a modern Carribean fringing reef, Galeta Point, Panama. *J. Sediment. Petrol.,* 47 (2): 503 – 517.

Macqueen, R.W., Ghent, E.D. and Davies, G.R., 1974. Magnesium distribution in living and fossil specimens of the echinoid *Peronella lesueure* Agassiz, Shark Bay, Western Australia. *J. Sediment. Petrol.,* 44: 60 – 69.

Manheim, F.T. and Bischoff, J.L., 1969. Geochemistry of pore waters from Shell Oil Company drill holes on the continental slope of the Northern Gulf of Mexico. *Chem. Geol.,* 4: 63 – 82.

May, J.A. and Perkins, R.D., 1979. Endolithic infestation of carbonate substrate below the sediment – water interface. *J. Sediment. Petrol.,* 49: 357 – 378.

Mazullo, S.J. and Reid, A.M., 1988. Sedimentary textures of recent Belizean peritidal dolomite. *J. Sediment. Petrol.,* 58: 479 – 488.

Middlemiss, F.A., 1975. Studies on the sedimentation of the Lower Greensand of the Weald. 1875 – 1975: a review and commentary. *Proc. Geol. Assoc.,* 86: 457 – 473.

Middlemiss, F.A., 1978. The cherts in the Hythe Beds (Lower Cretaceous) south-east England. *Proc. Geol. Assoc.,* 89: 283 – 298.

Midgley, H.G., 1951. Chalcedony and flint. *Geol. Mag.,* 88: 179 – 184.

Milliman, J.D., 1974. *Recent Sedimentary Carbonates, Part 1. Marine Carbonates.* Springer, Berlin, 375 pp.

Milliman, J.D., Hook, J.A. and Golubic, S., 1985. Meaning and usage of micrite cement and matrix – Reply to discussion. *J. Sediment. Petrol.,* 55: 777 – 784.

Murchison, R.I., 1834. On the structure and classification of the Transition Rocks of Shropshire, Herefordshire and part of Wales, and on the lines of disturbance which have affected that series of deposits, including the Valley of Elevation of Woolhope. *Proc. Geol. Soc. London,* 2: 13 – 18.

Oldershaw, A.E. and Scoffin, T.P., 1967. The source of ferroan and non-ferroan calcite cements in the Halkin and Wenlock limestones. *Geol. J.,* 5: 309 – 320.

Plumley, W.J., Risley, G.A., Graves, W.R., Jr. and Kaley, M.E., 1962. Energy index for limestone interpretation and classification. In: W.E. Ham (Editor), *Classification of Carbonate Rocks. Mem. Am. Assoc. Pet. Geol.,* 1: 85 – 108.

Purdy, E.G., 1968. Carbonate diagenesis: an environmental survey. *Geol. Romana,* III: 183 – 228.

Pytkowicz, R.M., 1969. Chemical solution of calcium carbonate in sea water. *Am. Zool.,* 9: 673 – 679.

Richter, D.K. and Füchtbauer, H., 1978. Ferroan calcite replacement indicates former magnesian calcite skeleton. *Sedimentology,* 25: 843 – 860.

Schmidt, V., 1965. Facies, diagenesis, and related reservoir properties in Gigas Beds (Upper Jurassic), north-western Germany. In: L.C. Pray and R.C. Murray (Editors), *Dolomitization and Limestone*

Diagenesis: A Symposium. Soc. Econ. Paleontol. Mineral, Spec. Publ., 13: 124 – 168.
Schneider, J., 1977. Carbonate construction and decomposition by epilithic and endolithic micro-organisms in salt and fresh water. In: E. Flugel (Editor), *Fossil Algae.* Springer, Berlin, pp. 248 – 260.
Schroeder, J.H., 1972. Submarine and vadose cements in Pleistocene Bermuda reef rock. *Sediment. Geol.,* 10: 179 – 204.
Shinn, E.A., 1971. Aspects of diagenesis of algal cup reefs in Bermuda. *Gulf Coast Assoc., Geol. Soc. Trans.,* 21: 387 – 394.
Shukla, V. and Friedman, G.M., 1983. Dolomitization and diagenesis in a shallowing-upward sequence: the Lockport Formation (Middle Silurian), New York State. *J. Sediment. Petrol.,* 53: 703 – 717.
Tucker, M.E. and Van Straaten, P., 1970. Conodonts and Facies on the Chudleigh Schwelle. *Extract Proc. Ussher Soc.,* 2 (3).
Watkins, R. and Aithie, C.J., 1980. Carbonate shelf environments and faunal communities in the Ludlow Beds of the British Silurian. *Palaeogeogr. Palaeoclimatol. Palaeoecol.,* 29: 341 – 368.
Wilson, J.L., 1975. *Carbonate Facies in Geologic History.* Springer, New York, N.Y., 471 pp.
Wolf, K.H., 1965. Littoral environment indicated by open-space structures in algal limestones. *Palaeogeogr. Palaeoclimatol. Palaeoecol.,* 1: 183 – 223.

Reprinted from: *Diagenesis, III. Developments in Sedimentology, 47.*
Edited by K.H. Wolf and G.V. Chilingarian.

Chapter 8

GEOCHEMICAL AND ISOTOPIC CONSTRAINTS ON SILICA AND CARBONATE DIAGENESIS IN THE MIOCENE MONTEREY FORMATION, SANTA MARIA AND VENTURA BASINS, CALIFORNIA

RICHARD W. HURST

INTRODUCTION

Modern ideas regarding the origin of the Monterey Formation emerged in the 1960s with the recognition that the Miocene diatomites are direct analogues of laminated and massive diatomaceous muds currently accumulating off California, Mexico, and Peru under the influence of biologically productive surface waters and low-oxygen bottom waters (Byrne and Emery, 1960; Emery, 1960; Bandy, 1961; Ingle, 1967, 1980; Soutar, 1971; Soutar and Crill, 1977; Isaacs, 1981; Pisciotto and Garrison, 1981). More recent studies have further developed the association between oceanographic setting, Miocene climatic events, and deposition of the Monterey Formation, including important contributions by Graham (1976), Pisciotto (1978), Isaacs (1982), Govean (1980), and Lagoe (1982), with general models for deposition of the Monterey diatomites (Fig. 8-1) calling for coincident interaction among Neogene tectonic, climatic, and oceanographic events (Ingle, 1980, 1981a, 1981b; Isaacs, 1981; Pisciotto and Garrison, 1981). Garrison and Douglas (1981) provided comprehensive reviews of Monterey Formation lithologies, biostratigraphy, diagenetic history, and depositional history. Many of the papers in this cited volume emphasize the analogous relationship between modern sites of diatomaceous sedimentation in the Gulf of California and southern California continental borderland and lithofacies patterns displayed within the Miocene Monterey Formation.

This chapter first reviews the geology and geochemistry of the Monterey Formation as it relates to silica and carbonate diagenesis. New data utilizing Sr isotopic and trace/major element analyses of both siliceous and carbonate phases are then presented in order to propose a model of cogenetic silica – carbonate diagenesis and its bearing on petroleum production and migration in the Monterey Formation.

PALEOCEANOGRAPHY

It is now clear that water mass characteristics and dynamics play a large role in creating appropriate settings for deposition and preservation of organic-rich diatomaceous muds in the modern ocean, and must have done so in the past at least as far back as the Cretaceous when diatoms first appeared in the geologic record.

Bramlette (1946) suggested that the laminated character of much of the Monterey Formation might be due to the exclusion of large burrowing organisms in oxygen-

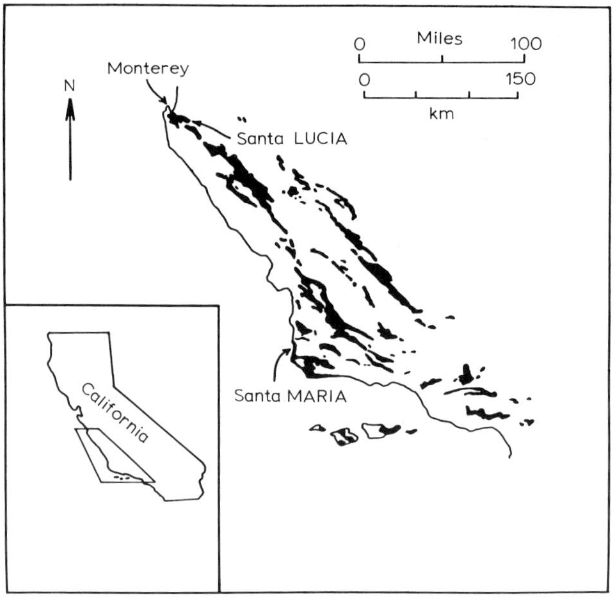

Fig. 8-1. Outcrop distribution of Neogene siliceous rocks (Monterey Formation and equivalents) in California. (After Pisciotto, 1978).

poor bottom water. Modern research has repeatedly confirmed Bramlette's foresight and firmly established the association among the oxygen minimum layer, exclusion of bottom infauna, and consequent preservation of organic-rich laminated sediments both beneath Recent upwelling systems and by analogy in rocks of Cenozoic, Mesozoic, and Paleozoic age (Fischer and Arthur, 1977; Parrish, 1982).

Origin and role of the oxygen minimum layer

The water filling the ocean basins of the world is density-stratified with each water mass characterized by a unique set of properties imparted by physical and biological conditions at its latitude of formation. Density stratification largely is a function of variations in temperature and salinity reflected in a basic three-layer system involving: (1) a well-mixed surface layer (0 – 100 m); (2) a permanent thermocline layer encompassing intermediate water (~ 100 – 1000 m,); and (3) cold, relatively saline deep water (~ 1000 m +) having an origin in high-latitude regions of the ocean. Other less conservative properties including oxygen, phosphate, and carbonate contents are affected by both physical and biological processes leading to variations of these properties with latitude, depth, and time. Modifications of original water mass character occur as a function of circulation, rate of mixing, and residence time at a given location. Impingement of discrete water masses against continental margins ultimately affects the properties of underlying sediments and associated biotas leaving an imprint of water mass character in the stratigraphic record: a record which until recently has been largely overlooked.

An important feature of the water column impinging against the continental margins of the eastern Pacific Ocean is the oxygen minimum layer. This feature is defined by water containing less than 1.0 ml l^{-1} of dissolved oxygen and commonly occurring at intermediate depths between 200 and 1500 m (Ingle, 1981b). The depletion of oxygen within the intermediate water masses of this region is a function of both biochemical processes and slow circulation (Wyrtki, 1962). The intensity and areal extent of the oxygen minimum layer reflect the rates of upwelling and production of organic materials in the overlying surface water with sluggish circulation and consumption of dissolved oxygen through oxidation of organic debris at depths below the effective photic and surface layers (150 m) in which wind-driven mixing and photosynthesis constantly recharge oxygen.

Vigorous upwelling induced by seasonal winds and the Coriolis effect lead to high primary productivity of phytoplankton in surface waters of the eastern Pacific as nutrient-rich water is brought to the surface from the upper portions of the oxygen minimum layer. The resulting rain of organic debris consumes the available dissolved oxygen during its decomposition and oxidation at intermediate depths below the photic zone. The intensity of oxygen depletion is primarily a function of biochemical action and the rate of supply of organic material which are, in turn, governed by rates of upwelling tuned to climatically regulated variations in zonal wind systems.

In addition, the oxygen minimum layer coincides with phosphate and nitrate maxima in the water column. Consequently, key nutrients regulating productivity of phytoplankton, including diatoms and coccolithophores, accumulate in the oxygen minimum layer and are periodically transported into the photic zone via upwelling, which completes the recycling process and triggers an acceleration of primary productivity.

The combined effect of productivity and circulation in creating the oxygen minimum layer is well illustrated by the latitudinal differences in the geometry of this feature along the eastern Pacific margin. The oxygen minimum layer is 1250 m thick off Mexico where upwelling is vigorous but motion of North Pacific Intermediate Water is sluggish, whereas this layer is only 350 m off Peru due to equatorial penetration of oxygen-rich Antarctic Intermediate Water which prevents expansion of the oxygen minimum layer despite high rates of productivity in the overlying water (Ingle, 1981b).

Deposition of laminated diatomites

Studies by Calvert (1964, 1966), Rhodes and Morse (1971), and others indicate that laminated diatomaceous muds most commonly form beneath bottom waters associated with well-developed oxygen minima with less than 0.20 ml l^{-1} of dissolved oxygen. These suboxic to near-anoxic waters exclude large invertebrates which would normally destroy bedding through bioturbation of sediments. The core of the well-developed oxygen minimum layer in the eastern Pacific Ocean contains less than 0.25 ml l^{-1} of dissolved oxygen. Hence, seasonal laminae are commonly preserved as varve-like couplets where these waters impinge against the shelf edge, slopes, bank tops, or basin plains, as exemplified by the deposition of laminated diatomites in the Guaymas Basin (Ingle, 1981b). It is these sorts of deposits which are thought to represent modern analogues of laminated diatomites of the Miocene Monterey Formation.

Examples of modern oxygen minima from the Santa Barbara Basin, Guaymas Basin, and the Indian Ocean illustrate the relationships between dissolved oxygen levels and preservation of laminated diatomites. Impingement of a well-developed oxygen minimum layer against an open continental slope results in a simple depositional pattern in which: (1) laminated diatomaceous muds are present beneath the near-anoxic core of this feature, (2) bioturbated to partially laminated organic-rich diatomaceous muds occur beneath suboxic waters of the oxygen minimum layer, and (3) homogeneous or massive bioturbated sediments containing relatively low amounts of organic matter characterize oxic facies at depths above and below the oxygen minimum (Ingle, 1981b).

Where the oxygen minimum layer intersects an irregular margin topography, more complex interactions ensue with dissolved oxygen values and lithofacies patterns regulated by depth of basin sills (Emery, 1960; Douglas et al., 1981). Basin sills represent the shallowest closed bathymetric contour defining a basin's geometry and control the flow and character of water to the subsill portion of a given basin regardless of the depth of the basin floor. The sensitivity of sill depth control of dissolved oxygen and related lithofacies patterns is clearly demonstrated in the Guaymas and Santa Barbara basins (Ingle, 1981a). The effective sill depth in the Guaymas Basin lies below the core of the oxygen minimum layer, which restricts deposition of laminated diatomites to basin slopes. These modern water mass – lithofacies patterns have obvious potential for interpretation of laminated through massive diatomites of the Monterey Formation as discussed by Pisciotto and Garrison (1981).

Source bed potential of oxygen minima lithofacies

The impingement of low-oxygen water against the sea floor leads to reducing conditions and enhanced preservation of the organic matter that is the end-product of major upwelling systems (see for example Parrish, 1982). This association is typified by the relatively high organic content of diatomaceous muds accumulating beneath oxygen minima in the Santa Barbara and Guaymas basins (Emery, 1960; Van Andel, 1964). The lipid-rich character of marine organic matter in these sediments marks them as ideal potential source beds for generation of petroleum (Didyk et al., 1978; Dow, 1978; Tissot and Welte, 1978; Demaison and Moore, 1980). Studies of diatomaceous facies in the Miocene Monterey Formation indicate that these rocks contain up to 24% organic matter by weight, which is consistent with their presumed origin (Isaacs, 1983). Differences in the abundance of organic matter, however, may be related to factors such as the presence of bacterial mats (Williams and Reimers, 1983) or complex interactions involving preferential adsorption of organic matter on detrital clays (Isaacs, 1983).

In any event, the prolific production of petroleum from the Monterey Formation

(Taylor, 1976) presents a dramatic manifestation of the high rates of biologic productivity, development of intense oxygen minima, and the ensuing capacity of Miocene borderland basins to collect and preserve exceptional amounts of organic matter over a 10-million-year period.

DEPOSITIONAL HISTORY

The Monterey Formation is the most lithologically distinct unit within the Neogene stratigraphic column of California, constituting a widespread biogenous deposit in basinal sequences otherwise dominated by terrigenous clastics (see Fig. 8-1). Whereas individual facies relationships within this formation reflect local basin configuration and oceanography, the depositional history of the unit as a whole involves the interplay between major Neogene climatic, oceanographic, and tectonic events (Ingle, 1973, 1981b; Pisciotto and Garrison, 1981; Isaacs et al., 1983). Furthermore, the Monterey Formation forms the best-studied example of a remarkable belt of diatomaceous rocks deposited around the margin of the North Pacific Ocean during Mid and Upper Miocene time. Despite local differences in thickness, age range, and silica accumulation and diagenesis among these deposits, they display clear similarities in terms of depositional and paleobathymetric history, and commonly occur within similar stratigraphic successions, implying synchronous control by major tectonic and paleoceanographic events. Thus analysis of the Monterey Formation in any given basin along the California margin must take into account that these sediments represent a local expression of a ubiquitous Miocene lithofacies extending from the Tres Marias Islands off central Mexico to the Korean Peninsula (Ingle, 1981b).

Basinal stratigraphies

The typical stratigraphic sequence filling many Neogene basins in California and elsewhere around the North Pacific margin invariably records three major depositional phases, which reflect regional synchroneity of both tectonic and oceanographic processes. These control the individual basin histories as follows: (1) initial margin subsidence and deposition of Oligo – Miocene volcanic, continental, and/or littoral through bathyal marine units of terrigenous composition; followed by (2) development of silled marginal basins and deposition of Mid to Upper Miocene diatomaceous sediments in anoxic and suboxic shelf, slope, and basin environments relatively starved of terrigenous clastics; and terminating with (3) the introduction of rapidly-deposited terrigenous clastics which diluted diatomaceous sediments and capped underlying diatomites with prograding fan, slope, and shelf units as rates of basin subsidence were overwhelmed by increasing rates of sediment accumulation during Plio – Pleistocene time (Ingle, 1981b).

This pattern is well documented within the western Ventura Basin of southern California (Fig. 8-2) and many other basins along the California margin formed during Miocene time in conjunction with the collision of the American and Pacific plates and the creation of a transform margin (Ingle, 1973, 1980; Blake et al., 1978;

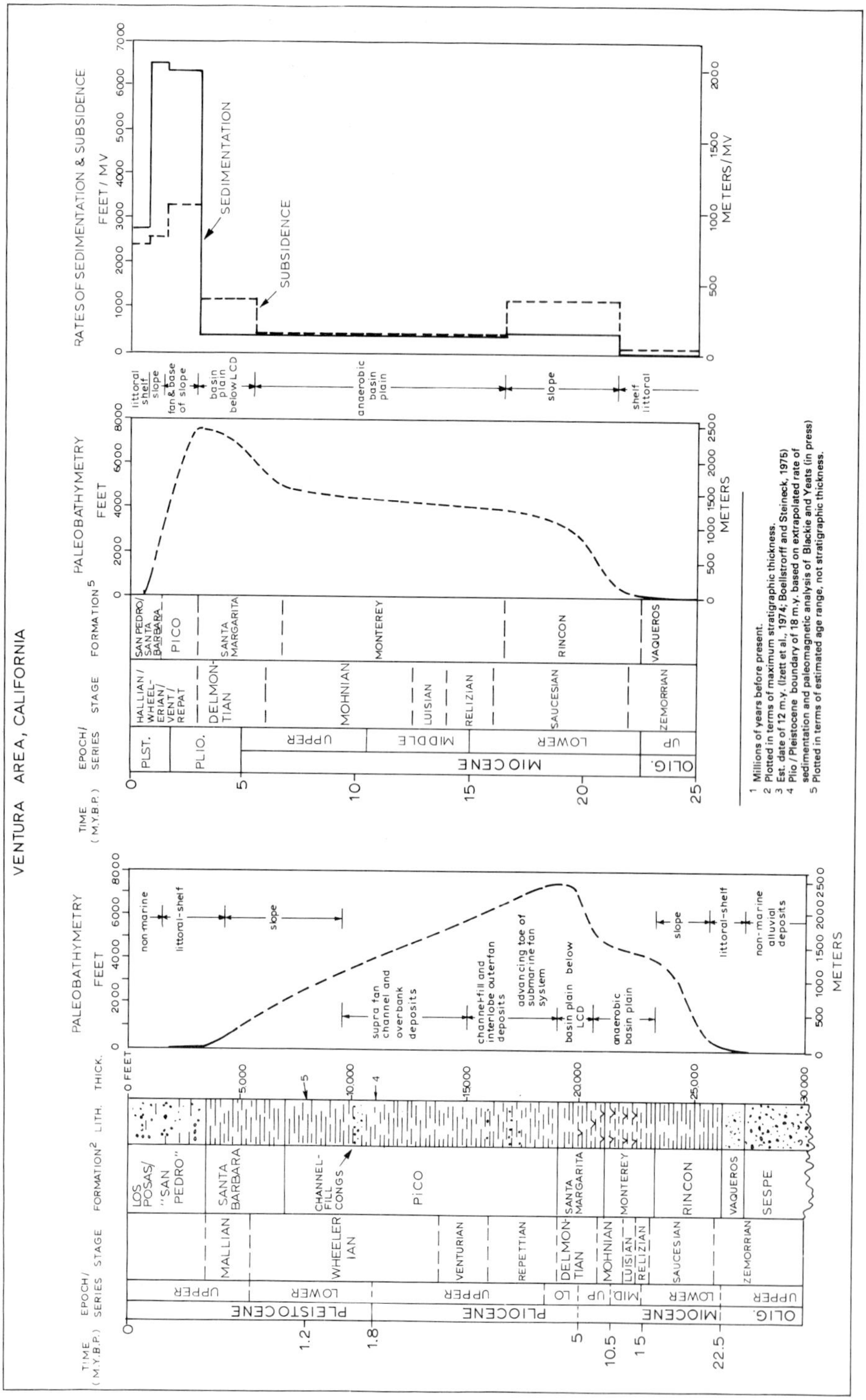

Fig. 8-2. Paleobathymetric and depositional history of the western Ventura Basin, California. (After Ingle, 1981a,b.)

Howell et al., 1980). Individual basin histories, particularly in southern California, are subject to varied interpretations, especially where paleomagnetic data suggest tectonic rotation and/or translation (e.g., Kamerling and Luyendyk, 1979; Hornafius et al., 1982). The typical three-phase stratigraphic history, however, can be recognized even in the more complex basinal settings, such as the Santa Maria Basin where the Point Sal, Monterey, Sisquoc, Foxen and Careaga formations record basin subsidence, Mid to Upper Miocene diatomaceous sedimentation, and Plio – Pleistocene basin filling. The diatomaceous deposits of the Monterey Formation call for controls on basin sedimentation that superceded local differences in margin character.

Regional relationships and controls

The distinctive two-fold sedimentary package of Miocene diatomites and capping Plio – Pleistocene terrigenous units, common to California basinal sequences, stands as the stratigraphic norm around much of the North Pacific margin regardless of provincial differences in depositional and tectonic settings, as illustrated by the essentially identical depositional histories of the Ventura Basin of California and the Akita Basin of Japan (Ingle, 1981b).

As presented by Ingle (1981a), three facts stand out with respect to the occurrence of Miocene diatomaceous units in California and the entire Pacific region:

(1) The significant thickness of Miocene diatomaceous sediment accumulated and preserved within North Pacific marginal basins requires prolific rates of diatom productivity and, in turn, vigorous upwelling circulation and nutrient supply necessarily linked to intensified atmospheric circulation.

(2) Marginal basins containing the Miocene diatomaceous lithofacies display relatively synchronous initial periods of subsidence and later development during Neogene time, despite the fact that they were formed at various convergent (e.g., Sea of Japan), divergent (e.g., Gulf of California), and translational (e.g., California borderland) plate junctures.

(3) The dominantly biogenous composition of the Miocene diatomaceous units demands synchronous reductions in the delivery of terrigenous clastics to the various continental and insular margins despite the adjacency of the Miocene strandline.

Both climatic and tectonic controls have been suggested separately as explanations for the widespread distribution of the Pacific Miocene diatomites (Lipps, 1969; Orr, 1972). A combination of these two factors, however, appears to best explain their occurrence wherein: (1) a mid-Cenozoic tectonic episode controls the timing and formation of basins; and (2) climatically induced acceleration of diatom productivity in mid-Miocene time provides the required volumes of diatomaceous sediment, as well as the intensification of oxygen minima favoring the preservation of organic matter of the sea floor, with both tectonic and climatic events responsible for a reduction in deposition of terrigenous clastics (Ingle, 1973, 1981b).

MIOCENE CLIMATIC AND TECTONIC EVENTS

Climate

Isotopic and faunal evidence derived from the study of Deep Sea Drilling Project materials has provided an unprecedented record of the evolution of global climate and ocean circulation over the past 100 million years (Savin, 1977; Kennett, 1977, 1982; Berger et al., 1981). These studies demonstrate that Cenozoic climate has cooled through the combined effects of changes in ocean/continent configurations and ocean/atmosphere dynamics (Berger, 1982; Kennett, 1982). The Miocene portion of this history is most clearly recorded by variations in the stable isotopic record of ^{18}O analyzed at Deep Sea Drilling Site 289 in the Central Pacific Ocean. This record and associated analyses of faunal changes among temperature sensitive groups, including planktonic foraminifera (Keller, 1981) and diatoms (Barron, 1981), suggest that the global climate shifted from a nonglacial to glacial mode in mid-Miocene time as a function of rapid buildup of the Antarctic ice cap between 16 and 13 million years ago (Ma) (Woodruff et al., 1981). This major climatic shift produced an increasingly steep pole-to-equator thermal gradient and consequent acceleration of atmospheric and oceanic circulation. The massive increase in diatom productivity, signaled by the synchronous appearance of mid-Miocene diatomites in California and elsewhere in the Pacific, appears to be a direct expression of increasingly vigorous upwelling in the California Current Province and other boundary currents in the North Pacific induced by mid-Miocene polar refrigeration.

In addition, global changes in deep-sea circulation, induced by both climatic events and tectonic changes in ocean gateways, apparently resulted in a transfer of silica from the Atlantic to the Pacific and Indian Oceans about 16 – 15 my. This led to the expansion of siliceous ooze deposition around Antarctica, increased diatomaceous sedimentation in the east equatorial Pacific, and the massive deposition of diatomite around the margin of the North Pacific, accompanied by decreasing siliceous productivity in the North Atlantic (Keller and Barron, 1983).

Eustatic events

The origin of episodic changes in sea level, originally documented by Vail et al. (1977), remain enigmatic. Nevertheless, it is apparent that eustatic changes in sea level have impacted global patterns of marine sedimentation. Increased flux of terrigenous debris occurred during low stands of sea level, whereas ponding of terrigenous sediments in estuaries and lagoons during high stands drastically reduced the flux of these materials to adjacent continental margins (Vail et al., 1977). The mid-Miocene high stand or transgression recorded by Vail et al. (1977) and Vail and Hardenbol (1979) would most certainly have aided starvation of marginal basins in the North Pacific. The magnitude of this high stand, however, is unknown. Thus, it seems more likely that rapid tectonic drowning of the Miocene continental margins during initial stages of basin subsidence were responsible for ponding of terrigenous sediments, allowing relatively undiluted deposition of diatomites during Mid and Upper Miocene time (Ingle, 1981b). At least two severe eustatic falls of sea level, associated with Late Miocene polar glaciation, appear to have been responsible for increasing rates of delivery of terrigenous sediments to nearshore basins (Ingle, 1978).

Tectonic events

Paleobathymetric analysis of Neogene basinal sequences around the North Pacific rim indicates that initial basin subsidence occurred almost simultaneously throughout this region during latest Oligocene through early Miocene time (Dott, 1969; Ingle, 1973). Apparent synchroneity of marginal basin formation and occurrence of Miocene diatomites prevailed despite fundamental differences in the tectonic setting of the Sea of Japan (convergent back-arc basins), the Gulf of California (divergent rift basins), and the California Miocene borderland (transform margin basins). These widespread tectonic events call for a mechanism affecting all margins of the Pacific Plate simultaneously. The simplest and most readily available mechanism of appropriate scale involves the apparent increase in rate of spreading on the East Pacific Rise beginning in Late Oligocene – Early Miocene time, which presumably resulted in increased rates of subduction and back-arc spreading in the western Pacific, acceleration of motion along the evolving transform margin in California, and direct rifting in the Gulf of California in Late Miocene time (Ingle, 1981a). Other possible mechanisms of similar scale include changes in the relative motions between oceanic and continental plates in the Pacific Basin as a whole (Engerbretson, 1982; Barron, 1986; Hurst, 1986a).

MONTEREY FORMATION LITHOFACIES

General subdivisions

Monterey sediments record the interplay of Miocene tectonic, volcanic, and oceanographic events, all of which varied from basin to basin. For this reason, the vertical succession of Monterey lithofacies is not identical in every basin (see Epstein and Nary, 1982; Conrad and Ehlig, 1983), but some widespread trends are evident. For example, in Early to Middle Miocene time, Monterey hemipelagic sediments of the present Coast Range area were dominated by a mixture of siliceous diatom frustules and calcareous components (i.e., coccoliths and foraminifera). Thus, the lower section of the Monterey Formation is partly calcareous in many areas of the Coast Ranges; in some areas, these basinal calcareous sediments sometimes are assigned to separate formations. Starting in Middle Miocene time, about 15 my, global climatic cooling led to intensified coastal upwelling systems, higher nutrient levels in surface waters, and proliferation of diatoms, which respond much more productively to elevated nutrients than do calcareous plankton. Consequently, the upper part of the Monterey Formation is highly siliceous in most areas (see discussions in Pisciotto and Garrison, 1981; Garrison, 1981; Isaacs et al., 1983).

These vertical variations of Monterey lithofacies have led to different schemes for subdividing the unit (Fig. 8-3). The most generalized approach is that of Pisciotto

TABLE 8-1

Summary of characteristics of the Miocene sequence along the Santa Barbara coast (from Isaacs, 1981; dolomite may locally replace calcite in all calcareous rock types)

Formation	Common rock types	Mean mineral abundance (range in common rock types)						
		Silica (wt%)	Detrital minerals (wt%)	Carbonate minerals (wt%)	Apatite (wt%)	Organic matter (wt%)	Average thickness (m)	Age (my)
Sisquoc Formation	Clayey siliceous rock (siliceous mudstone and shale; diatomaceous mudstone and shale)	30	62	6	0	2	> 500	5.5 – 3.5
Monterey Formation; Clayey – siliceous member	Siliceous rock (porcelanite, siliceous mudstone chert; diatomaceous shale and mudstone, diatomite)	55 (15 – 90)	39 (10 – 80)	0	0 (0 – 1)	6 (2 – 12)	150	8 – 5.5
Upper calcareous-siliceous member	Calcareous-siliceous rock (calcareous porcelanite, shale chert: calcareous diatomaceous shale and diatomite)	47 (10 – 90)	25 (5 – 65)	22 (2 – 75)	0 (0 – 3)	6 (2 – 10)	30	
Transitional marl-siliceous member	Calcareous-siliceous rock and carbonaceous marl	40 (5 – 90)	26 (5 – 65)	23 (5 – 75)	1 (0 – 20)	10 (2 – 20)	40	11 – 8
Carbonaceous marl member	Carbonaceous marl (commonly phosphatic) and calcareous-siliceous rock	16 (3 – 90)	23 (5 – 65)	42 (5 – 75)	6 (0 – 20)	13 (2 – 24)	75	15 – 11
Lower calcareous-siliceous member	Calcareous-siliceous rock	41 (5 – 90)	20 (5 – 50)	32 (5 – 75)	0 (0 – 10)	7 (2 – 18)	120	18 – 15
Rincon Shale	Clay-rich						500	25 – 18

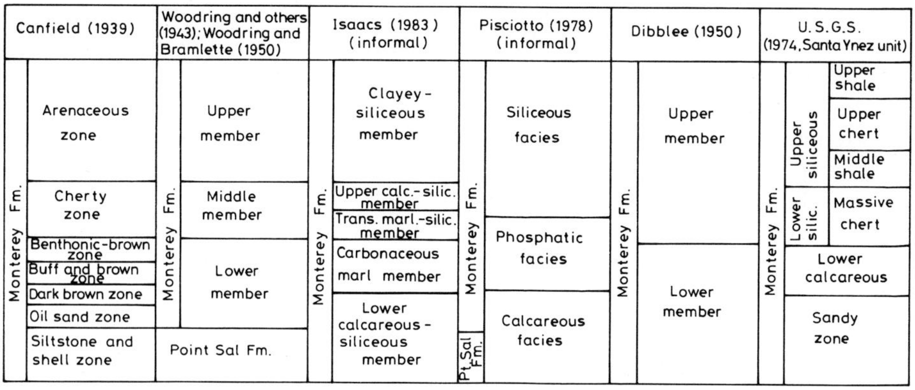

Fig. 8-3. Stratigraphic subdivisions of the Monterey Formation by various authors.

(1978; see also Pisciotto and Garrison, 1981), who noted the occurrence of widespread phosphatic rocks in the middle part of the formation and divided the unit into a lower calcareous facies, a middle phosphatic facies, and an upper siliceous facies (Fig. 8-3). Although these categories provide a basis for the discussion which follows, it must be emphasized that all three are not present everywhere. In particular, the middle phosphatic facies is absent in many areas where conditions for intense phosphatization probably were not present. Also, some workers have found it advantageous to make more detailed subdivisions (see Canfield, 1939; Isaacs, 1980; and Fig. 8-3 and Table 8-1). In addition, some basins contain volcanic rocks or siliciclastic units that are interbedded or otherwise stratigraphically associated with the Monterey Formation.

Calcareous facies

The lower part of the Miocene basinal sequence, in some places assigned to the Monterey Formation, contains platy and fissile, organic-rich limestones, calcareous siltstones, and mudrocks. In addition to abundant coccoliths and foraminiferal tests, most of these rocks also have diatom frustules and diagenetic opal-CT or quartz, indicating that the original sediment was a calcareous siliceous mud. The mean composition of such rocks in the Santa Barbara area is 41% silica, 32% carbonates, 20% detrital minerals, and 7% organic matter (Table 8-1; Isaacs, 1980, 1983). Localized concretions and lenticular beds of dolomite are abundant.

Phosphatic facies

At the top of the calcareous facies and within the transitional interval to the upper siliceous facies, organic-rich phosphatic shales and mudstones are present in many parts of the Coast Ranges (Dickert, 1966, 1971; Pisciotto, 1978).

The phosphate, mainly in the form of scattered, small authigenic nodules and sand-size peloids of cryptocrystalline carbonate fluorapatite, usually occurs in laminated, organic-rich (up to 20% TOC) calcareous shales and limestones. Bending of laminae around nodules as a result of compaction and scattered conglomeratic beds of reworked nodules indicate that phosphatization occurred just below the sea floor, during very early diagenesis. The most intensely phosphatized rocks that could be dated accurately were found to be of Relizian to Luisian age – a time span of approximately 12 – 15 m.y. (Dickert, 1971; Pisciotto, 1978; Mertz and Garrison, 1983). Table 8-1 shows compositional data on this facies (see "Carbonaceous marl member") in the Santa Barbara area.

Similar modern phosphatized sediments occur in several parts of the present oceans, most notably off the west coast of South America and off Southwest Africa (Namibia) (Burnett, 1977, 1980, 1982; Birch et al., 1983). These are regions of intense year-round upwelling, very high productivity, and well-developed oxygen minimum zones. The phosphatization occurs where the oxygen minimum zone intersects the sea floor on the shelf and outer slope. The low levels of dissolved oxygen there allow substantial amounts of organic matter to accumulate in bottom sediments (Demaison and Moore, 1980). Oxidation of this organic matter during sulfate reduction apparently is a major source of dissolved phosphate in pore waters (Burnett, 1977, 1980), but the specific mechanisms leading to the formation of authigenic apatite are poorly understood.

In addition to the scattered nodules and peloids of phosphate in organic-rich rocks, distinctive thin beds of phosphatic sandstones of a very different character occur in the Monterey Formation at scattered localities (Dickert, 1966, 1971; Pisciotto, 1978). The phosphate grains typically are peloids or ooids, or a combination of the two. These phosphatic sandstones may also contain glauconite or larger nodules of phosphate which incorporate glauconite and earlier-formed phosphatic grains, thereby indicating multiple episodes of phosphatization. These sandstones, which range in age from Relizian to Mohnian, occur in two kinds of setting: (1) as thin (a few meters) intervals forming part of a condensed banktop or starved outer shelf sequence (Dickert, 1971; Graham, 1976; Pisciotto, 1978; Garrison et al., 1979), and (2) as thinner (a few centimeters) graded layers that are interbedded with hemipelagic deep-water deposits (in both the calcareous and siliceous facies). The latter are probably turbiditic redeposits from nearby banktops (Graham, 1976; Pisiciotto, 1978; Younse, 1979).

Siliceous facies

The siliceous facies is the thickest and most widespread lithofacies in the Monterey Formation (Bramlette, 1946). These diatomaceous rocks (and their diagenetically equivalent cherts, porcelanites, and siliceous mudrocks) record intense coastal upwelling and attendant primary productivity during Middle Miocene to Pliocene time. The main compositional variation in these rocks resides in the amount of detrital components (Table 8-1). This probably is a reflection of the existence of both detrital-rich proximal basins and detrital-poor distal basins. Compilations of compositions for rocks of this facies were made by Isaacs (1980, 1983) for the central Santa Barbara Basin and by Williams (1982) for the central San Joaquin Basin. The most significant sedimentological features are sedimentary structures and cycles the origins of which are currently under investigation.

Volcanic rocks

Chiefly of Early or Middle Miocene age, volcanic rocks are present only locally and occur either below the Monterey Formation or as interbedded tuffs, most commonly in the lower sections. These rocks are present as shallow intrusions, air-fall tuffs, subaerial agglomerates, and submarine ash flows which vary in composition from basalt to rhyolite. From a sedimentological standpoint, the best-studied of the volcanic units is the Obispo Formation, thought by Hall (1981) to be the product of an eruption along a leaky transform fault. Studies by Fisher (1977) and Surdam and Stanley (1981) show that Obispo rocks were deposited largely by submarine ash flows and volcaniclastic turbidity currents on the flank of a volcanic ridge. The localized nature of these Miocene volcanics and their usefulness in the reconstruction of basin tectonics is emphasized by Epstein and Nary (1982) and Hurst (1982).

Similar Miocene volcanic rocks form important petroleum reservoirs in northern Japan (Aoyagi and Iijima, 1983). Although not yet widely exploited, these rocks may have this potential also in California, especially in some offshore areas.

Associated siliciclastic rocks

A few scattered thin turbiditic sandstone beds of siliciclastic composition occur in most Monterey sequences. In proximal basins and distal basins fed longitudinally by sand dispersal systems, Miocene basinal successions may be dominated by turbidite fans interbedded with hemipelagic successions. Well-known examples include the Modelo Formation of the Los Angeles Basin and the various fan systems in the San Joaquin Basin (Webb, 1981; Graham et al., 1982).

SILICEOUS SEDIMENT DIAGENESIS IN THE MONTEREY FORMATION

Diagenesis of silica

The general course of inorganic silica phase transformations in deep-sea sediments and, by implication, in many of the siliceous sediments, porcelanites, and cherts in continental margin settings (e.g., Monterey Formation) now is fairly well understood (e.g., Bramlette, 1946, Calvert, 1966, 1977; Murata and Nakata, 1974; Murata and Larson, 1975; Keene, 1976; Kastner et al., 1977; Hein et al., 1979a,b; Pisciotto, 1978, 1981a,b; Isaacs, 1981, 1982; Kastner, 1981; Kastner and Gieskes, 1983). Amorphous silica, opal-A, is the most abundant phase in many of these sediments (primarily as diatom tests) and is the most reactive inorganic phase. In response to a variety of physical and chemical factors, it undergoes changes, such as congruent dissolution and/or phase transformations to varieties of opal-CT and to quartz.

The order of decreasing solubility of the most common silica phases is opal-A, opal-CT (cristobalite/tridymite; Jones and Segnit, 1971), chalcedony, and quartz. The solubility of each of these phases increases with temperature and pressure (e.g., Kennedy, 1950; Krauskopf, 1956; Fournier and Rowe, 1962; Jones and Pytkowicz,

1973; Duedall et al., 1976; Walther and Helgeson, 1977; Fournier and Potter, 1982). The dissolution rates and solubilities of the silica phases also are influenced by the chemistry of the fluid phase and its ionic strength (e.g., Van Lier et al., 1960; Jorgensen, 1968; Wirth and Gieskes, 1979; Marshall, 1980a,b; Fournier et al., 1982).

Hence, variations in the nature of the original silica phase(s), in the sediments and pore fluids, which are controlled by the geologic, tectonic, and geochemical histories of the sediments, are responsible for the observed local and regional differences in the diagnetic histories of siliceous Monterey rocks. Silica phases differ vertically and laterally in sedimentary rocks of the same geologic age (Isaacs, 1980).

Bramlette (1946) observed the overall silica phase transformation sequence of diatomite to porcelanite to chert in the Monterey Formation. A summary of the most important physical and chemical parameters which control these silica reactions is presented in Table 8-2 (Kastner et al., 1977).

For example, as described by Kamatami (1971), Hurd and Theyer (1975), Lawson et al. (1978), and Hurd and Birdwhistell (1983), the physical and chemical properties of mature opal-A are distinctly different from those of immature opal-A. Unfortunately, no data are available on either opal-A maturation in the Monterey Formation or on the diagenetic processes related to opal-A maturation, and to the resultant shift in the *d*(101) reflection of opal-CT. Similarly, research on the effect of organic matter on the mechanisms and rates of these phase transformations is in its infancy and no definitive data are available. Surdam and Stanley (1981), Kablanow and Surdam (1983), and Hurst (1986b) suggest a physical relationship between silica diagenesis and organic matter migration; the release of water during silica phase transformations may provide the fluid drive for hydrocarbon migration. This is an important observation and will be addressed later in this chapter when interrelated silica – carbonate diagenesis is discussed.

TABLE 8-2

The most important physical and chemical parameters which control silica diagenesis

	Opal-A to Opal-CT	Opal-CT to varieties of quartz
Temperature	I	I
Pressure	i	i
pH	I	I
Ionic strength	i	i
Concentration of dissolved silica in fluid phase	I	I
Availability of Mg^{2+} (or Fe^{3+}, Al^{3+}) and OH^-	i	–
Absence of competing diagenetic reactions which require Mg^{2+} and OH^-	i	–
Cl^-/SO_4^{2-} ratio in fluid phase	–	i

i: important.
I: very important.

Physical properties

Much of the recovered hydrocarbons is produced from fractured rocks the occurence of which varies regionally. In the Santa Maria Basin and San Joaquin Valley, for example, 70 to 80% of the hydrocarbons produced is from fractured rocks, primarily porcelanites and cherts, then carbonates, and to a lesser extent from porcelaneous mudstones or shales (Regan and Hughes, 1949; Regan, 1953; Crawford, 1971). According to Taylor (1976), in the Los Angeles and Ventura basins, however, only about 5% of the oil recovered through 1975 has been found in fractured siliceous shales. This suggests that carbonates may be volumetrically important fractured reservoirs in these and, perhaps, the Santa Maria Basin (Redwine, 1981; Hurst, this chapter).

The fracturability of siliceous sedimentary rocks depends on their mineralogical and chemical composition and diagenetic history, especially degree of cementation/lithification. Diatomaceous sediments do not fracture easily, and the origin of fractures in the diagenetic porcelaneous and cherty rocks is unclear, although they most probably are the result of mechanical forces and/or diagenetic reactions. Silty and sandy siliceous rocks tend to be more resistant to fracturing (Surdam and Stanley, 1981).

As shown in Table 8-3, cherty and porcelaneous rocks have significantly lower porosities (and higher densities) than diatomaceous sediments (Pisciotto, 1978; Isaacs, 1980, 1981).

Two depth zones of abrupt reductions in porosity (within 50 m depth) are observed by Isaacs (1980). These zones correspond to the opal-A to opal-CT and opal-CT to quartz transformations. Similar observations are reported by Hein et al. (1979a) in Bering Sea siliceous sediments. The magnitude of the porosity changes across these mineralogical boundaries is controlled strongly by the amount of admixed clay minerals. The overall trend is that of porosity decreases with increasing clay content.

The drastic change in porosity occurs when opal-A transforms to opal-CT, as shown on Table 8-4. Accordingly, the most important lithologic boundary in the Monterey Formation and similar siliceous rocks is the diagenetic transformation from opal-A to opal-CT, as observed by Isaacs (1981) along the Santa Barbara coast, by Hein et al. (1979a) in the Bering Sea, and by Grechin et al. (1981) in

TABLE 8-3

Physical properties of siliceous rocks

	Quartz porcelanite, chert and porcelaneous mudrock	Opal-CT porcelanite, chert and porcelaneous mudrock	Diatomite, muddy diatomite
Porosity (%)	5 – 20	25 – 40	55 – 70
Average dry bulk density (g cm^{-3})	1.8 – 2.1	1.4 – 1.7	0.7 – 1.1

TABLE 8-4

Relationship between porosity and percent silica in siliceous rocks, Monterey Formation (From Isaacs, 1980)

Silica	~85% silica	~70% silica	~40% silica
Porosity (%) of diatomaceous sediments	~70	~65	~60
Porosity (%) of opal-CT rocks	~25	~35	~30
Porosity (%) of quartz rocks	~ 5	~10	15–20

Neogene siliceous rocks in the outer California continental borderland and off Baja California. In areas where significant local differences in the composition of the siliceous rocks exist, the diagenetic front of opal-A to opal-CT is smeared out, resulting in the porosity loss over a broader depth interval.

According to Isaacs (1980), the mechanism of porosity reduction is compaction in conjunction with solution and reprecipitation of silica. This causes the collapse of the framework of diatom frustules, which gives rise to the high porosity of diatomaceous sediments. Pisciotto (1978) observed that diatom frustules break mechanically prior to extensive dissolution. Compaction may be partially responsible for the observed breakage followed by increased dissolution due to the increase in surface area due to breakage.

Opal-A to opal-CT transformation

Silica diagenesis, which is thoroughly documented in the Monterey Formation (Murata and Nakata, 1974; Murata and Randall, 1975; Murata et al., 1977; Pisciotto, 1978, 1981a; Isaacs, 1980, 1982; Grechin et al., 1981; Kastner, 1981; Surdam and Stanley, 1981; and references therein), proceeds through two distinct mineralogical steps: (1) opal-A to opal-CT, and (2) opal-CT to quartz.
It is appropriate, therefore, to discuss the two steps separately.

The most extensive published research on silica diagenesis in the Monterey Formation was carried out in the following regions: the Temblor Range (Murata and Larson, 1975; Murata and Randall, 1975; Murata et al., 1977, 1979); the Santa Maria region (Pisciotto, 1978, 1981a); the California coast, west of Santa Barbara (Isaacs, 1980, 1981, 1982); the Pismo and Huasna basins (Surdam et al., 1980; Surdam and Stanley, 1981b; Kablanow and Surdam, 1983); the Lost Hills oil field (Kruger, 1983), and the outer California Borderland and off Baja California (Grechin et al., 1981). Generally, these studies, as well as the classical work by Bramlette (1946), indicate that, although temperature is the most important factor controlling diagenetic phase transformations in the silica system, it is not the only one (Table 8-2).

According to Murata and Larson (1975) and Murata et al. (1977), the transformation temperature range of opal-A to opal-CT in the Monterey Formation is 28° – 56°C; according to Pisciotto (1981a,b), the minimum temperature range is 24° – 38°C and the maximum is 37° – 52°C. The depth of transformation is shallower in areas of high present-day thermal gradients. In the Pismo Basin, the opal-A to opal-CT transformation preceded hydrocarbon generating reactions (Surdam and Stanley, 1981).

These studies demonstrate that the transformations from pure diatomite to porcelanite and chert occur earlier and at lower temperatures than that of diatomaceous mudrocks to siliceous mudrocks. Additionally, in the coastal area west of Santa Barbara, rocks with progressively more detritus have progressively smaller initial opal-CT *d*(101) spacings. This can be explained on the basis of experimental studies (Kastner et al., 1977; Kastner and Gieskes, 1983), which have demonstrated the inhibiting effect of admixed clay minerals (e.g., smectite) on this silica transformation reaction. The clay minerals compete with the silica system for the Mg^{2+} and OH^-, inhibiting the formation of a magnesium hydroxide (silicate) compound which serves as a nucleus for opal-CT crystallization.

Despite the extent of the aforementioned studies, very little is known about the actual mechanism of the opal-A to opal-CT transformation. It appears that the mechanical breakage of the diatom frustules either precedes or at least catalyzes their dissolution (Pisciotto, 1978) and, on the basis of oxygen isotope analyses (Murata et al., 1977), the reaction proceeds through a solution and precipitation mechanism. Strontium isotopic analyses of Monterey siliceous rocks further suggest that isotopic equilibrium is achieved between the reprecipitated silica phase and the interstitial water (Hurst, 1986b; this chapter).

Opal-CT to quartz transformation

Opal-CT has a characteristic X-ray diffraction pattern (Jones and Segnit, 1971) of three broad peaks near 4.3, 4.1 and 2.5 Å. Jones and Segnit (1971) have also shown that stacking disorder in opal-CT causes shifts in the *d*(101) cristobalite peak at about 4.1 Å. The progressive peak sharpening and shift toward smaller *d*(101) spacing of opal-CT with depth observed in the Monterey Formation (e.g., Murata and Nakata, 1974; Murata and Randall, 1975; Pisciotto, 1978, 1981b; Isaacs, 1982; Kruger, 1983) was interpreted subsequently by Murata and Nakata (1974), Murata and Larson (1975), and Murata et al. (1977) to be the result of solid-state ordering of a disordered opal-CT structure. The near constant $\delta^{18}O$ value of opal-CT from various depths (thus temperatures), which lead to their mechanistic explanation, does not exclude another interpretation. On the basis of low-temperature diffusion coefficients in solids, it is unlikely that the ordering of opal-CT observed in the relatively young Monterey sediments occurred through solid-state reactions. The opal-CT to quartz transformation is a solution – precipitation reaction (Hurst, 1986b; this chapter). The oxygen isotopic temperature range of opal-CT to quartz transformation is 57 – 80°C according to Murata and Larson (1975) and Murata et al. (1977); and 31 – 80°C according to Pisciotto (1981b).

Opal-CT occurs in the form of lepispheres in cavities and void spaces; otherwise massive opal-CT predominates. Diagenetic quartz observed in the Monterey Formation occurs as crypto- and microcrystalline quartz, chalcedony, and quartzine. Unfortunately, no systematic study of the distribution, in time and space, of quartz varieties in the Monterey Formation is available. Such information as well as the characterization of other quartz varieties (e.g., length-slow chalcedony) could be useful in identifying various geochemical environments of quartz crystallization in the Monterey Formation.

Generally, opal-CT with a high *d*(101) spacing does not transform directly to quartz. It first orders into an opal-CT of lower *d*(101) spacing. For example, in the Temblor Range rocks the *d*(101) spacing of opal-CT varies with depth from 4.12 to 4.04 Å over 50 to 80°C (Murata and Nakata, 1974; Murata and Larson, 1975); in the Santa Maria Basin, variations of 4.10 to 4.04 Å are reported by Pisciotto (1981a); and in the Lost Hills oil field it changes from 4.09 to 4.04 Å (Kruger, 1983). In the Santa Barbara area, however, quartz was observed to form from opal-CT with a *d*(101) spacing of 4.07 Å (Isaacs, 1982). Similarly, in the outer California continental borderland, variations in the *d*(101) spacing of opal-CT of only 4.12 to 4.10 and 4.07 Å were observed by Grechin et al. (1983). The above differences in the shift of the *d*(101) value of the opal-CT which transform to quartz are significant, and could be useful in estimating thermal and diagenetic histories.

When dissolved silica reaches supersaturation, quartz should precipitate. The dissolved silica content of pore fluids is controlled primarily by the lithology, temperature, types and rates of diagenetic silicate reactions (mainly dissolution, precipitation, and adsorption) and pH. In Monterey Formation sections in which quartz precipitated from a relatively disordered opal-CT, the dissolution rate of opal-A must have been the rate-controlling step, whereas in the other sections (e.g., Temblor Range) the formation of opal-CT was the rate-controlling step.

It is interesting that in the rocks from the Santa Barbara area, Isaacs (1980, 1982) observed that the transformation temperature of opal-CT to quartz is correlated positively to the ratio of silica to detrital material. In rocks with progressively more detrital content, quartz formed progressively earlier and at lower temperatures. The opposite relationship between the amount of detrital material and temperature of quartz formation was observed by Pisciotto (1978) in the Santa Maria region. On the basis of oxygen isotope determinations of opal-CT and quartz, he determined that in porcelanites (with higher detrital content) the temperatures of formation of both opal-CT and quartz are greater than in cherts.

Isaacs (1982), on the basis of her observation in the Santa Barbara area, concluded that detrital material (e.g., clay minerals) enhanced the opal-CT to quartz transformation reaction. This conclusion is not supported by either Pisciotto's (1978) observations or by experimental work (Kastner, 1983). Instead, the initial *d*(101) value of opal-CT is a more important factor controlling the rate of opal-CT to quartz transformation. In the Santa Barbara coastal area, rocks with low silica to detrital material ratios have smaller opal-CT initial *d*(101) spacings than rocks with higher silica to detrital material ratios. Table 8-5 gives a summary of the relationships between lithology and *d*(101) spacing of opal-CT in a marine environment.

In the Temblor Range siliceous rocks, Murata and Nakata (1974) and Murata et al. (1979) showed that the *d*(101) spacing of opal-CT in porcelanites is consistently 0.01 Å lower than in associated cherts, which formed earlier. In the Santa Maria region, however, the *d*(101) spacing of opal-CT is the same in both porcelanites and associated cherts (Pisciotto, 1978). The reason(s) for the different behavior of opal-CT in Monterey Formation porcelanites and cherts of different areas in California is not yet understood.

In the Santa Barbara coastal region, in rocks with silica to detrital material ratios greater than eight, the presence of carbonate affected the rate of quartz formation significantly (Isaacs, 1980, 1982). Diagenetic dolomite beds are occasionally replaced by silica, but always in the form of quartz and not opal-CT, even in an opal-CT diagenetic zone (Murata et al., 1977). Murata et al. (1977) and Pisciotto (1978) have shown that the $\delta^{18}O$ of opal-CT is always greater than that of coexisting quartz in the deep-sea environment, indicating that a higher temperature is required for the formation of quartz.

SECONDARY CARBONATES IN THE MONTEREY FORMATION

Bramlette (1946) first documented the widespread occurrence of secondary carbonates in Monterey rocks, and recent reviews are found in Pisciotto (1981c, d) and Isaacs et al. (1983). Similar Miocene rocks occur in contemporaneous offshore sediments of the California borderland (Pisciotto and Mahoney, 1981).

Once considered a minor lithology in the Monterey Formation, such rocks, particularly dolomites, now are recognized as important fractured reservoirs in the Santa Maria Basin (Redwine, 1981; Roehl, 1981). Similar rocks may hold this potential

TABLE 8-5

Effect of sediment composition on opal-CT formation

	Marine sediments		Predicted "type" of opal-CT to form
(I)	Pure opal-A (> 50%) + carbonate		Medium *d*(101) spacing
(II)	Pure opal-A (< 50%) + carbonate		Low *d*(101) spacing
(III)	Pure opal-A		Medium to high *d*(101) spacing
(IV)	Opal-A (> 50%) + clay minerals	Smectite and smectite/illite	High to medium *d*(101) spacing
		Illite, chlorite, kaolinite	Medium to high *d*(101) spacing
(V)	Opal-A (< 50%) + clay minerals	Smectite and smectite/illite	Medium to low *d*(101) spacing
		Illite, chlorite, kaolinite	Medium *d*(101) spacing

The rate of opal-CT to quartz transformation strongly depends on the "type" *d*(101) spacing of opal-CT.

The type of opal-CT strongly depends on dissolved silica concentrations.

in other basins. Additional interest has centered on secondary carbonates for two other reasons. First, as initially recognized by Bramlette (1946), secondary carbonate concretions may contain well-preserved diatoms, which are not preserved in the surrounding host rocks, thereby enhancing biostratigraphic age dating of the unit, particularly its cherty and porcelaneous parts (Grivetti, 1982). Second, some dolomite concretions and beds yield strong paleomagnetic signals, including records of both normal and reversed polarities, whereas surrounding host rocks do not (Hornafius et al., 1981).

Both secondary calcite and dolomite are present in the Monterey Formation, with dolomite being much more abundant, based on present knowledge. Because of this and because of its importance as a reservoir, most of the recent research has focused on dolomite, which is also the focus of the discussion which follows. Diagenetic siderite, ankerite and, possibly, rhodochrosite, must be present in the Monterey Formation as well, but have not received attention, as yet.

Mineralogy, petrology and geologic occurrence of Monterey dolomites

Calcium-rich dolomite (49 – 56 mol% $CaCO_3$), with iron content as high as 16 mol% $FeCO_3$, is the predominant secondary carbonate mineral. The dolomite is present in various forms: concretions and lenticular beds, graded turbiditic sandstones, brecciated zones, cements, and disseminated in other lithologies, such as calcareous mudrocks or porcelanites. The concretions and lenticular beds represent the most abundant form of dolomite. They are restricted to specific stratigraphic horizons and commonly show a rhythmicity, with dolomitic layers occurring at intervals of 5 to 15 m. Dolomite occurs throughout the Monterey Formation, but is particularly abundant in organic-rich and coccolith-rich rocks of the calcareous and phosphatic facies in the lower Monterey Formation.

Field and petrologic evidence indicate that the dolomites formed by cementation and replacement of a variety of host sediments, most commonly calcareous sediments in the lower Monterey Formation, but also siliceous deposits of the upper Monterey Formation (Bramlette, 1946; Murata et al., 1969a, 1972; Pisciotto, 1978, 1981d; Friedman and Murata, 1979). Laminations from adjacent host sediments can be traced into dolomite concretions where they are perfectly preserved as layers three to four times thicker than their counterparts outside the concretions (Bramlette, 1946). Other primary sedimentary structures preserved within dolomite layers include slump structures, micro-unconformities, and burrows. Few primary features can be observed in thin-sections. The dolomites consist of a mosaic of small anhedral to subhedral crystals containing scattered siliciclastic grains and an occasional foraminiferal test, which may be dolomitized. Coccoliths, abundant in many of the host sediments, are never preserved in the dolomites and, presumably, are completely replaced. Some individual dolomite crystals contain inclusions of organic matter and show zoning and syntaxial overgrowths, indicating several stages of growth.

Several lines of evidence reveal that some of the dolomitic bodies began to form very early during diagenesis, within a few meters below the sea floor. The noted thickness differences of laminae inside and outside of concretions suggest precompactional growth. Murata et al. (1969b) calculated that some Monterey carbonate concretions may have had porosities as high as 80% when carbonate cementation began. At several localities in the Salinas and Santa Barbara basins, reworked dolomite concretions occur as clasts in sandstone gravity flow deposits and in intraformational slump breccias, indicating erosion from just below the sea floor and downslope resedimentation (Garrison and Graham, 1984). One consequence of this early dolomitization is that many Monterey dolomites preserve compositions, textures, and structures of the host sediments, which were destroyed during the burial diagenesis of these sediments.

Stable isotope geochemistry

Carbon and oxygen isotope analyses of dolomites in Monterey and correlative rocks along the Pacific margin have been reported and interpreted in Friedman and Murata (1979), Hein et al. (1979a,b), Murata et al. (1969, 1972), Pisciotto (1978, 1981d), Pisciotto and Mahoney (1981), Roehl (1981), and Spotts and Silverman (1966). Strontium isotopes are discussed in a later section of this chapter.

Carbon isotopes. Carbon isotope values in Monterey dolomites have an exceptionally broad range: from −26 to +21 permil (‰) relative to the Peedee Belemnite Standard (PDB). This has been interpreted to be the result of dolomitization in different zones of organic matter decay (Table 8-6; Irwin et al., 1977; Pisciotto and Mahoney, 1981). Light-carbon dolomites could be derived from light-carbon carbon dioxide in shallow zones of aerobic microbial oxidation and anaerobic sulfate reduction (Zone II, Table 8-6). Later and deeper during burial diagenesis, where methane generation by carbon dioxide reduction due to microbial metabolic processes produces heavy-carbon biocarbonates in pore waters, dolomites with more positive $\delta^{13}C$ values could form (Zone III). Alternatively, heavy-carbon dioxide values might be inherited from carbon dioxide, which formed along with methane during acetate fermentation. Thermal decarboxylation at greater depths would reverse the trend toward light carbonates (Zone IV).

Complexities in the overall isotopic patterns of individual dolomite bodies may reflect the fact that, within any sedimentary section, dolomite can form simultaneously at several different depths or form at different times within the same interval (Pisciotto and Mahoney, 1981). Murata et al. (1972) recognized the following three groups of Monterey dolomites based on chemical and isotopic properties, and these were interpreted by Pisciotto (1981d) in terms of diagenetic zones (Table 8-6):

Light-carbon – low-iron dolomites ($\delta^{13}C$ values of −26 to −4‰ PDB, < 2 mol% $FeCO_3$) are formed, according to Pisciotto (1981d), either at shallow depths in the anaerobic zone of sulfate reduction (Zone II, Table 8-6), where capture of iron in pyrite within a finite iron reservoir would account for their low-iron content, or as a result of aerobic respiration.

Heavy-carbon – low-iron dolomites ($\delta^{13}C$ values of +5 to +21‰ PDB, < 2 mol% $FeCO_3$) are formed, as suggested by Pisciotto (1981d) in the anaerobic zone of active methane production (Zone III), with the low-iron content reflecting its earlier depletion in the zone of sulfate reduction.

Heavy-carbon – high-iron dolomites ($\delta^{13}C$ values of −2 to +10‰ PDB, > 2

TABLE 8-6

Diagenetic zones and oxidation – reduction reactions for subsurface microbial metabolic processes (from Pisciotto and Mahoney, 1981)

	Diagenetic zones and oxidation – reduction reactions	Approximate depth range of zone below sea floor (m)	Observed or estimated $\delta^{13}C$ range (permil PDB)		
			CO_2*	CH_4	carbonate
(I)	Microbial oxidation (aerobic respiration): $CH_2O + O_2 \rightarrow CO_2 + H_2O$	0 – 0.01	– 18 to – 28	– **	No ppt (Curtis, 1978)
(II)	Microbial sulfate reduction (anaerobic): $2CH_2O + SO_4^{2-} \rightarrow S^{2-} + 2\,CO_2 + 2H_2O$	0.01 – 10	– 15 to 30	– **	0 to – 25
(III)	Methanogenesis: methane production and carbonate precipitation (anaerobic): $Me^{2+} + 2HCO_3^- + 8H^+ \rightarrow CH_4 + MeCO_3 + 3H_2O$ where Me = Ca, Mg, Fe, etc.	10 – 1000	– 20 to + 10	– 47 to – 90	– 10 to + 15
(IV)	Thermocatalytic decarboxylation, generally: fatty acids → *n*-alkanes + fatty acids + CO_2	50 – 2500	– 10 to – 20	– 60 to – 80	0 to – 10 (– 25?)

Note: Reactions, depth intervals, and isotopic data compiled from Cooper and Bray (1963), Eisma and Jung (1969), Claypool et al. (1973), Claypool (1974), Claypool and Kaplan (1974), Irwin et al. (1977), Curtis (1978), Hein et al. (1979a).

* Total dissolved CO_2, mainly as bicarbonate.
** Methane not present.

mol% $FeCO_3$) are interpreted by Pisciotto (1981d) to be products of dolomitization in deeper zones of methanogenesis and thermocatalytic decarboxylation (Zone IV), with the higher iron content reflecting gradual depletion of Mg^{2+} in pore waters and progressive substitution of Fe^{2+} for Mg^{2+} in later-formed dolomites.

Oxygen isotopes. Reported oxygen isotope values in Monterey dolomites have a relatively narrow range of 23 – 38‰ relative to Standard Mean Ocean Water (SMOW) (Pisciotto, 1981d), but their interpretation remains in doubt.

One approach is to use $\delta^{18}O$ values to estimate the temperatures of mineral formation, applying the experimental fractionation expressions for either dolomite – water or protodolomite – water, and assuming the $\delta^{18}O$ of the water to be 0.0‰ SMOW. Using this method, Pisciotto (1978) calculated temperatures of formation for Monterey dolomites of 17 – 78°C for offshore Miocene dolomites recovered from the California borderland. Pisciotto and Mahoney (1981) obtained temperatures of 10 – 50°C, which suggest dolomite formation at burial depths of 86 – 658 meters.

Calculation of temperatures of formation is difficult, however, because of uncertainties in applying the dolomite – water expression, because the $\delta^{18}O$ of the equilibrating fluid is unknown and inasmuch as dolomitic bodies may continue to grow during progressive burial and with age (Pisciotto, 1978, 1981d; Pisciotto and Mahoney, 1981; Hurst, this work). This latter possibility has been evaluated by serial sampling and isotopic analyses of single dolomite concretions and layers (Kushnir and Kastner, 1981; Hurst, this chapter).

Taking a different viewpoint, Friedman and Murata (1979) measured $\delta^{18}O$ for dolomites in the Monterey Formation on the western side of the San Joaquin Basin. Their values range from 31.5 – 34.0‰ (SMOW) and show a general increase from Lower to Upper Miocene dolomites. This trend parallels that of Miocene benthic foraminifera from the North Pacific Ocean (Douglas and Savin, 1973), which likewise show progressive increases in $\delta^{18}O$, reflecting decreasing bottom-water temperatures during the Miocene. Friedman and Murata suggested that the similarity between the two trends means that the dolomites formed in isotopic equilibrium with pore waters which were isotopically similar to the bottom waters in which the benthic foraminifera lived. This implies that the dolomite formed at relatively shallow burial depths (< 200 m) in the host sediment and maintained its original isotopic composition during subsequent burial diagenesis. These conflicting viewpoints should be resolved through systematic analyses of dolomitic sections in different areas.

EXPERIMENTAL AND OCEANOGRAPHIC OBSERVATIONS

Within the past few years, a combination of experimental work and sampling of Quaternary dolomites from modern depositional settings has provided important new information on the origin of Monterey dolomites. Kelts and McKenzie (1982) reported dolomite in samples from several DSDP sites in the Gulf of California. Dolomites with positive $\delta^{18}C$ values occur in Quaternary sediments on the Guyamas slope (Site 479). These lithified dolomites occur at depths as shallow as

~ 80 m below the sea floor within unlithified diatomaceous ooze, which is identical to the upper Monterey diatomaceous sediments. Kelts and McKenzie attribute the heavy-carbon dolomite to a biogenic source, probably bacterial fermentation, which fractionated carbon into isotopically light methane and heavy carbon dioxide. Individual dolomite beds display decreasing $\delta^{18}O$ values outward from the center of the bed, a trend attributed by Kelts and McKenzie to dolomitization during the increase in temperature resulting from progressive burial.

Dolomites and dolomitic limestones associated with similar organic-rich sediments have been dredged from the continental slope off Peru (Kulm et al., 1981; Suess et al., 1982). Like the Monterey dolomites, they have variable $\delta^{13}C$ values, ranging from −28 to +14‰, a variability attributed by Suess et al. (1982) to different stages of sedimentary organic matter decomposition.

Parelleling these discoveries is the experimental work of Baker and Kastner (1981) which showed that dolomitization takes place readily when the concentrations of dissolved sulfate, an inhibitor, are low. Bicarbonate ions can be supplied by sulfate reduction, methanogenesis, or fermentation; magnesium can come from seawater, ion-exchange sites on silicates, and exchange with NH_4^+ formed during sulfate reduction. Calcium, therefore, might be the limiting factor in this dolomitization environment. Sulfate reduction in organic-rich sediments thus favors dolomitization for three reasons: (1) inhibiting sulfate ions are removed; (2) alkalinity is increased; and (3) NH_4^+, which later released adsorbed Mg^{2+}, is produced. Further, their experiments indicate that the conversion of opal-A to opal-CT may retard dolomitization because of the formation of Mg^{2+} and OH^--bearing nuclei which promote opal-CT genesis. Mg^{2+} although taken up, would be released during the conversion of opal-CT to quartz, possibly generating a later phase of dolomitization. The Mg content of opal-CT-bearing lithologies, however, may not be sufficient for dolomitization. In addition, dolomite occurrence increases in the opal-CT zones of the Mussel Rock section; this observation does not support the conclusion of Baker and Kastner (1981).

Dolomites and hydrocarbons

Important fractured reservoirs in the Santa Maria Basin (Regan and Hughes, 1949) have been shown by Redwine (1981) and Roehl (1981) to consist in part of intensely fractured dolomitic rocks, which are, in their view, the result of repeated episodes of tectonic rock dilation followed by natural hydraulic fracturing. Dolomitization acts to embrittle calcitic rocks, because the mineral dolomite, unlike calcite, does not twin readily in response to stress and, therefore, behaves in a nonductile manner. Thus, fractured dolomitic rocks are particularly significant in the lower calcareous part of the Monterey Formation.

In addition to their fracture porosity, Monterey dolomites may contain significant intercrystalline porosity, but no systematic studies of this aspect have been published. Pisciotto (1981d) has suggested that early-formed dolomites may act to impede vertical migration of fluids, including hydrocarbons. But the early diagenetic dolomites might, in contrast, become intensely fractured because they are subjected to repeated episodes of tectonic and hydraulic stresses. Furthermore, cogenetic silica – carbonate events may enhance the physical properties of Monterey sediments required for hydrocarbon migration (Hurst, this chapter).

GEOCHEMISTRY OF MONTEREY FORMATION CARBONATES AND SILICEOUS SEDIMENTS

The Sr isotopic evolution of seawater and marine carbonates

The Sr isotopic composition of seawater at any given time is established by exchange reactions between seawater and marine volcanic rocks, marine carbonates, and continental detritus transported to the ocean. The long residence time of Sr in the oceans (4 m.y.) coupled with the rapid mixing of the oceans (1000 y) results in Sr isotopic homogenization throughout the oceans, i.e., at any given time the $^{87}Sr/^{86}Sr$ ratio of seawater is the same everywhere in the oceans. Calcium carbonate precipitating from seawater, or extracted by organisms from seawater, incorporates Sr in the crystal structure. Because the Sr isotopic composition does not fractionate as a result of these processes, the carbonate precipitates and tests assume the isotopic composition of seawater at the time of deposition. This property of the marine geochemistry of Sr has been used to determine the Sr isotopic evolution of seawater throughout geologic time (Peterman et al., 1970; Dasch and Biscaye, 1971; Veizer and Compston, 1974; Tremba et al., 1975; Brass, 1976; Clauer, 1976; Faure et al., 1978; Kovach, 1980; Burke et al., 1982; DePaolo, 1986).

Carbonate Sr isotopic results

The strontium isotopic ratios of Monterey Formation carbonates are presented in Tables 8-7 (Goleta Slough) and 8-8 (Mussel Rock). These dolomite samples were chosen in order to compare the Sr isotopic compositions of an early diagenetic dolomite layer (Goleta Slough; located in diatomaceous mudrocks with a short

TABLE 8-7

Sr isotopic compositions and concentrations of an early diagenetic dolomite, Goleta Slough, Santa Barbara, California

Sample no.	0.1 *N* HCl	Dolomite	Sr(ppm) in dolomite
13 (top)	0.70880	0.70880	639.9
12	0.70886	0.70883	601.2
11	0.70883	n.d.	
9	n.d.	0.70880	1269
7	n.d.	0.70883	346.5
5	0.70885	0.70883	595.1
4	0.70881	0.70882	
3 (bottom)	0.70883	0.70882	

n.d.: not determined.

diagenetic history) with those of a dolomite concretion the diagenetic history of which spans the silica phase transitions (Mussel Rock; located in opal-CT/quartz cherts).

The results indicate that the early diagenetic dolomite layer (Table 8-8) incorporated Sr from contemporaneous seawater, because its Sr isotopic chronostratigraphic age, 8.3 ± 0.7 m.y. (Hurst, 1987), is in excellent agreement with biochronologic results (7.6 – 8.4 m.y.; J.A. Barron, pers. commun.). With the exception of two samples (0.1 *N* HCl leachates 5 and 12), the Sr isotopic composition ranges from 0.70880 to 0.70883 – the entire range is within the analytical error at the 95% confidence level (± 0.00002) relative to the average Sr isotopic composition. The dolomite layer is within diatomaceous sediments (opal-A) which did not experience deep burial. Hence, chemical modification via ion exchange, diffusion, and other diagenetic reactions after dolomitization was minimal. Strontium concentrations are also typical of early diagenetic dolomites which form from seawater that has not been modified as a result of water – rock reactions (Burns and Baker, 1985). In spite of the author's concern to remove any calcite present using 0.1 *N* HCl, XRD data indicated each sample was pure dolomite with no evidence of other mineralogic phases present – the 0.1 *N* HCl leach did not remove more than 2% of the dolomite, which further indicates that the Sr isotopic composition of the individual layers varied little during their growth.

The results from this early diagenetic dolomite contrast with those of the dolomite concretion in interlayered opal-CT/quartz cherts at Mussel Rock (Table 8-8). Isotopic variations in calcite range from 0.70860 to 0.70879, whereas those in dolomite range from 0.70860 to 0.70912. Samples located in the interior of the concretion (MR 4BCS – 3 and 4; Fig. 8-4) yield a Sr isotopic chronostratigraphic age of 11.5 ± 0.8 m.y., which is consistent with the biochronologic ages (Garrison et

TABLE 8-8

Sr isotopic compositions of a Mussel Rock dolomite concretion (MR 4BCS series)

Sample no.	Calcite	Dolomite	T_{fd} (°C)
2a,b	0.70863	0.70864	28
3	0.70874	0.70870	
4	0.70874	0.70869	22
5	0.70879	0.70860	
6	0.70876	0.70894	
7	0.70863	n.d.	24
8	0.70871	0.70912	
9a	0.70857	0.70910	27
10	0.70875	0.70875	30
11	0.70866	0.70860	
12	0.70860	0.70873	
13	n.d.	0.70874	
14	0.70875	0.70873	42
15	0.70861	0.70863	

n.d.: not determined.

al., 1984). This age is based upon Sr isotopic compositions ranging from 0.70869 to 0.70874 (average 0.70872 ± 2) and a composite seawater Sr isotopic evolution curve (NESSIE; Hurst, 1987). Isotopic compositions lying outside of this range indicate that other Sr reservoirs were available to the dolomite as it grew. Temperatures of formation based upon oxygen isotopes (Garrison et al., 1984) vary from 22° to 42°C and substantiate a longer, more complex diagenetic history for this concretion than that of the dolomite layer at Goleta Slough.

The isotopic compositions of the dolomite from this concretion do not vary systematically relative to the coexisting calcite. The difference in the strontium isotopic composition of coexisting calcites and dolomites is at a maximum for the samples midway between the top center of the concretion and where the concretion "pinches out". These variations are well outside analytical errors and suggest that the dolomite concretions may have grown episodically, responding to variations in their local geochemical environment (Hurst and Davis, 1981; Hurst, 1986a,b).

Carbonate and silica diagenetic relationships

Mussel Rock

The processes involved in dolomitization and their relation to silica diagenesis are of importance because both lithologies form fractured reservoirs. The data for the Mussel Rock dolomites are summarized in Fig. 8-5. The fields depict the relationships among the carbonates (Cc = calcite; DOLO = dolomite), Miocene seawater (MSW), the silica polymorphs (QC = quartz chert; CTC + PORC = CT − chert + porcelanite), and the initial strontium isotopic ratios (IR) from the

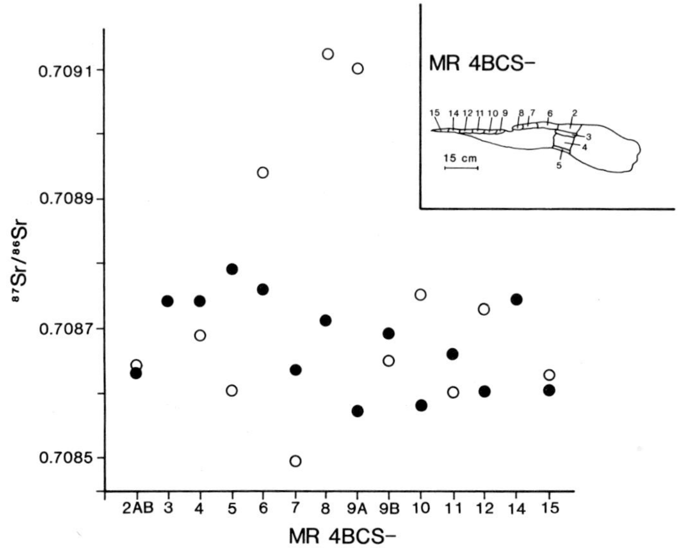

Fig. 8-4. $^{87}Sr/^{86}$ Sr ratios of a dolomite concretion (see inset for sample locations) from Mussel Rock, Santa Maria, California. Solid circles are calcite analyses and open circles are co-existing dolomite analyses.

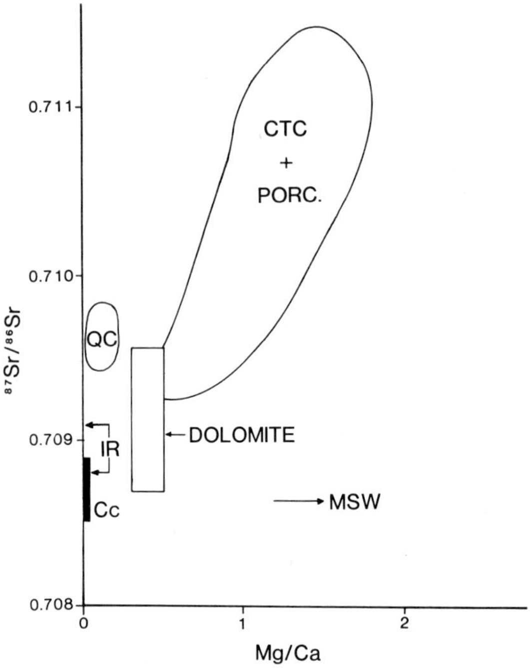

Fig. 8-5. $^{87}Sr/^{86}Sr$ versus Mg/Ca ratios for dolomites and siliceous sediments in the Santa Maria area. Fields are as follows: *CTC* + *Porc* = CT – chert + porcelanite; *QC* = quartz chert; *IR* = initial $^{87}Sr/^{86}Sr$ ratios from silica polymorphic linear arrays (Brueckner and Snyder, 1985); calcite (*Cc*) and *DOLOMITE* data are from Hurst (1986a); *MSW* = Miocene seawater. Dolomites have $^{87}Sr/^{86}Sr$ ratios which are indicative of being in equilibrium with Miocene seawater or are derived via exchange reactions with the surrounding siliceous sediment.

rubidium – strontium studies by Brueckner and Snyder (1985) on silica polymorphs in the Santa Maria area. The Mg/Ca ratios of dolomites from the Mussel Rock section have been measured by Pisciotto (1978) and are utilized in Fig. 8-5. The strontium isotopic ratios of the carbonates from Mussel Rock are reported in this chapter. Two-component mixtures on this diagram would plot on a line connecting the two end-member components, whereas ternary mixtures would lie within a three-component triangle.

The Mg – Ca – Sr isotopic geochemistry of Monterey Formation dolomites is a ternary mixture of authigenic marine calcite, Miocene seawater (Mg/Ca = 5), and CT-chert plus porcelanite. The data indicate that the formation of the Monterey Formation dolomites involves the precipitation of calcite from seawater followed by the incorporation of Mg (from seawater ± sediment) and, in some instances, more radiogenic strontium from siliceous sediments. Although Mg, Ca, and Sr can be derived from seawater to explain the formation of early diagenetic dolomites, a more radiogenic source of Sr must be present to produce the epigenetic dolomites. This requires a source of strontium with a higher Sr isotopic composition than that of seawater during Miocene time.

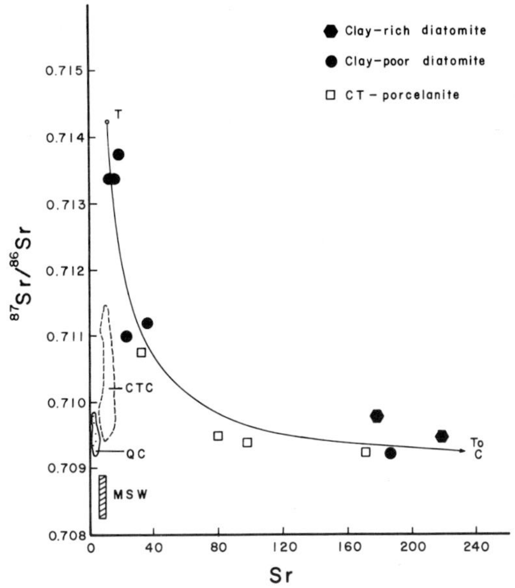

Fig. 8-6(A). $^{87}Sr/^{86}Sr$ versus Sr concentration mixing relationships of diatomite and CT-porcelanite, Monterey Formation, Santa Maria, California. (Data from Brueckner and Snyder, 1985.) Labelled fields are identical to those in Fig. 8-5. End-members include terrestrial (*T*) and carbonate (*C*) components as explained in the text.

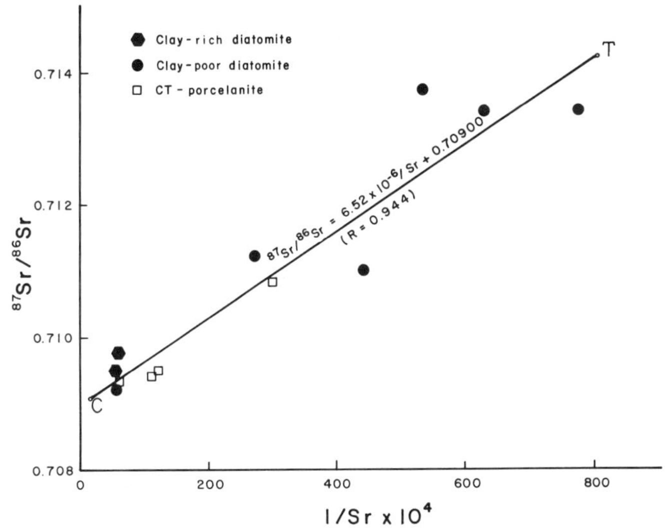

Fig. 8-6(B). $^{87}Sr/^{86}Sr$ versus 1/Sr mixing relationships for diatomite and CT-porcelanite, Monterey Formation, Santa Maria, California. (Data from Brueckner and Snyder, 1985.) The linear relationship between end-member components *T* and *C* substantiates a mixing model to explain the Rb/Sr systematics of these Monterey Formation lithologies.

The data in Fig. 8-5 also suggest that epigenetic dolomites and silica diagenesis are related geochemically. This conclusion arises from the similarities between the initial strontium isotopic ratios (IR) determined from the siliceous phases in the Mussel Rock section (Brueckner and Snyder, 1985) and the strontium isotopic compositions of the more radiogenic dolomites. It appears that these two mineralogic groups, the silica polymorphs and dolomites, were attempting to equilibrate with the same interstitial fluid in the Monterey Formation sedimentary pile (see next section).

Rb/Sr systematics of diatomites and CT-porcelanites

Brueckner and Snyder (1985) interpret the Rb/Sr linear arrays derived from Monterey Formation diatomites and CT-porcelanites as mixing lines. The CT-chert and quartz chert linear arrays are believed to have resulted from diagenetic reactions (H.K. Brueckner, pers. commun.). To evaluate the possibility of mixing, their data have been plotted on $^{87}Sr/^{86}Sr$ versus Sr and $^{87}Sr/^{86}Sr$ versus 1/Sr diagrams. If mixing is indeed the case, the data can be fit to a hyperbolic curve on the former diagram and a linear array on the latter diagram; both plots must be satisfied to substantiate mixing. Diatomite and CT-porcelanite data appear on Figs. 8-6A and 8-6B, whereas CT-chert and quartz chert data appear on Figs. 8-7A and 8-7B. Of

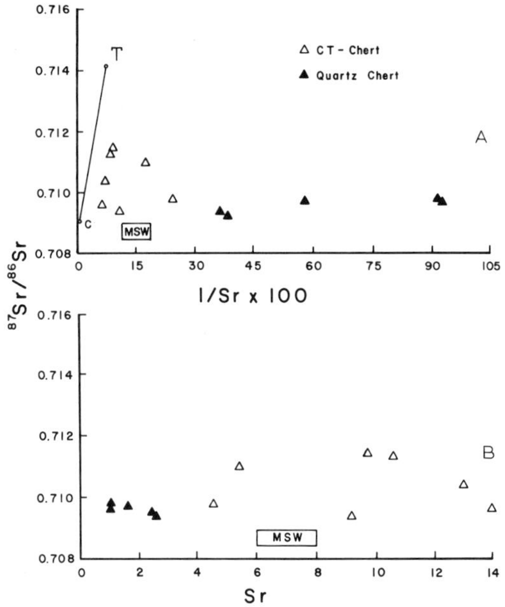

Fig. 8-7(A en B) Mixing relationships in CT- and quartz cherts from the Monterey Formation, Santa Maria, California. (Data from Brueckner and Snyder, 1985.) Mixing between the diatomites and CT-porcelanites (Fig. 8-6B) is shown for reference in Fig. 8-7B. Note that the dual criteria for mixing are not satisfied; the Rb/Sr system of the CT- and quartz cherts is recording diagenetic reactions associated with the silica polymorphic transitions; specifically, dehydration reactions.

the 26 whole rock samples plotted, only two samples (No. 108 CT-porcelanite from Sweeny Road; No. 103B quartz chert from Mussel Point) are outliers when compared to the other data and are excluded from the discussion. The presence of detrital plagioclase is suspected in one sample (No. 103B) and may explain the discrepancy (H.K. Brueckner, pers. commun.).

The clay-poor and clay-rich diatomites as well as the CT-porcelanites substantiate the mixing model very nicely. A hyperbolic curve can be fit to the data, when the Sr isotopic composition is plotted against Sr concentration (Fig. 8-6A). A well-correlated ($R = 0.944$) linear array results when the Sr isotopic composition is plotted against the reciprocal Sr concentration (Fig. 8-6B). Geologically reasonable end-member components can be chosen to fit the mixing curves. These include a clay-rich terrigenous source (T in Figs. 8-6A and 8-6B), with an average Sr isotopic composition of 0.7142 and Sr concentration of 13 ppm and a dolomitic carbonate (C in Figs. 8-6A and 8-6B), with an average Sr isotopic composition of 0.70913 and Sr concentration of 500 ppm. Both end-member compositions are consistent with the known geology and Rb/Sr geochemistry of the Monterey Formation (Brueckner and Snyder, 1985; Hurst, unpubl.).

The figures also clarify the role of carbonate as a buffering medium in the system because the clay-rich diatomites and some CT-porcelanites have less radiogenic Sr isotopic compositions despite the presence of more radiogenic clays. Inasmuch as $^{87}Sr/^{86}Sr$ ratios are inversely related to concentration, it is clear that a marine carbonate phase must buffer the system. Furthermore, as diagenetic grade increases through the silica phase transitions (opal-A, opal-CT, quartz), the Sr isotopic compositions of the CT- and quartz cherts (CTC and QC fields in Figs. 8-6A and 8-6B) approach that of the carbonate end-member component. None of the siliceous samples, however, have Sr isotopic compositions identical to that of contemporaneous Miocene seawater (Burke et al., 1982; MSW in Figs. 8-6A and 8-6B).

Rb/Sr systematics of CT-cherts and quartz cherts

The corresponding plots for both the CT- and quartz cherts are presented in Figs. 8-7A and 8-7B. Whereas mixing of terrigenous and carbonate end-members satisfies the diatomite and CT-porcelanite data, the CT- and quartz chert data do not yield hyperbolic curves (Fig. 8-7A) and linear arrays (Fig. 8-7B). Thus, the dual criteria to test for true mixing are not satisfied. As noted in the diatomite and CT-porcelanite data, Miocene seawater (MSW) does not appear to be a viable component in the Monterey Formation Rb/Sr system. This indicates that Monterey sediments were chemically isolated from seawater rapidly.

RESPONSE OF THE Rb/Sr SYSTEM TO DIAGENESIS

Both experimental (Mizutani, 1977) and O isotope studies (Murata and Larson, 1975) support a solution – reprecipitation model to explain silica phase transitions. M. Kastner (pers. commun.) has also determined that O isotopes in the siliceous phases reequilibrate with interstitial waters during the silica phase transitions.

Isotopic exchange and equilibration with interstitial waters further suggest that Monterey sediments became chemically isolated from the overlying oceanic water column. This closure is in part related to the drastic decrease in sediment porosity as silica diagenesis proceeds (Isaacs, 1981). Thus, sediments – interstitial water equilibration during solution – reprecipitation reactions is to be expected. Two conclusions can be deduced from these data: (1) Sr isotopes also reequilibrate with interstitial waters during the silica phase transitions, and (2) the Rb/Sr system should respond in some systematic manner to the phase changes.

The systematic response of the Rb/Sr system in Monterey sediments to diagenesis can be observed in the consistent decrease in the apparent age (t_a) of the Rb/Sr linear arrays and the increase in the initial Sr isotopic ratios (I_0) as silica diagenetic grade increases (Fig. 8-8; data from Brueckner and Snyder, 1985). Furthermore these two parameters are well correlated:

$$I_0 = -8.38 \times 10^{-6} (t_a) + 0.7097 \qquad (R = 0.987)$$

If the Monterey Formation had behaved as an open system *at all times*, a linear relationship (including those of the Rb/Sr evolution diagrams) with such a high degree of correlation would not be expected. Thus, the Rb/Sr linear arrays of the

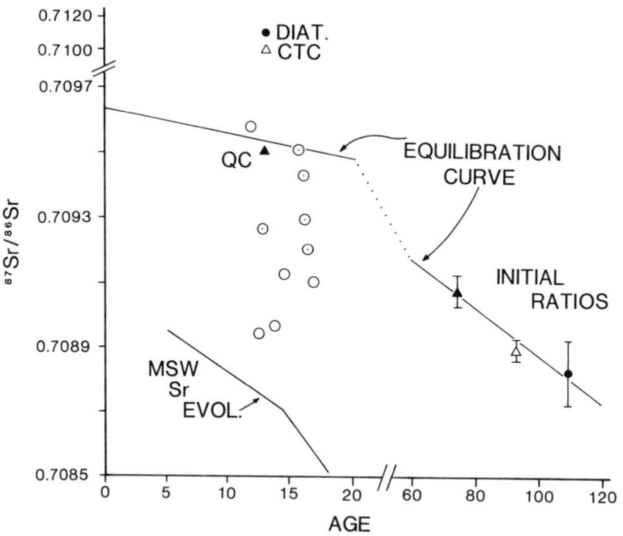

Fig. 8-8. Sr isotopic evolution diagram of interstitial waters in the Monterey Formation, Santa Maria, California. Symbols and abbreviations are as follows: solid circles = diatomites (*DIAT*); open triangle = CT-chert (*CTC*); solid triangle = quartz chert (*QC*); open circle = radiogenic Monterey Formation dolomites; dotted circles = radiogenic Pt. Sal Formation dolomites; *MSW Sr Evol.* = Miocene seawater Sr evolution curve; *INITIAL RATIOS* are those of the silica polymorphic linear arrays (from Brueckner and Snyder, 1985). The *EQUILIBRATION CURVE* shows the $^{87}Sr/^{86}Sr$ evolution of interstitial waters in the Monterey Formation. Dolomites precipitated from solutions expelled during silica polymorphic dehydration reactions will be in isotopic equilibrium with the Initial Ratio of that transition – this is the case for the four least radiogenic Monterey Formation dolomites.

diatomites, CT-cherts and quartz cherts are demonstrating the principle of "isochron" rotation in response to the silica phase transitions just as true isochrons rotate and reequilibrate in response to later metamorphism.

The relationship between I_0 and t_a is interpreted by Hurst (1986b) to represent the Sr isotopic evolution of interstitial waters in the Monterey Formation within portions of the Santa Maria Basin. One possible scenario which explains the geochemical evolution of the interstitial waters is as follows: As Monterey Formation sedimentation proceeded and silica phase transitions began, the porosity decreased (Isaacs, 1981) and alkalis, such as Rb, were expelled (Brueckner and Snyder, 1985). Although the interstitial waters which remained are expected to become increasingly more radiogenic due to the decay of ^{87}Rb in clays, the buffering action of the carbonates and the resistance of clays to rapid equilibration with interstitial waters (Dasch and Biscaye, 1971) prohibited the rapid increase expected in the Sr initial ratios during closed system evolution. Furthermore, the preferential expulsion of alkalis and, possibly, radiogenic Sr during the opal-A to opal-CT transition (Figs. 8-9 to 8-12) resulted in a substantial decrease in the rate of evolution of the ^{87}Sr/^{86}Sr ratio in sediments and interstitial waters. This was reflected in the initial ratio evolution.

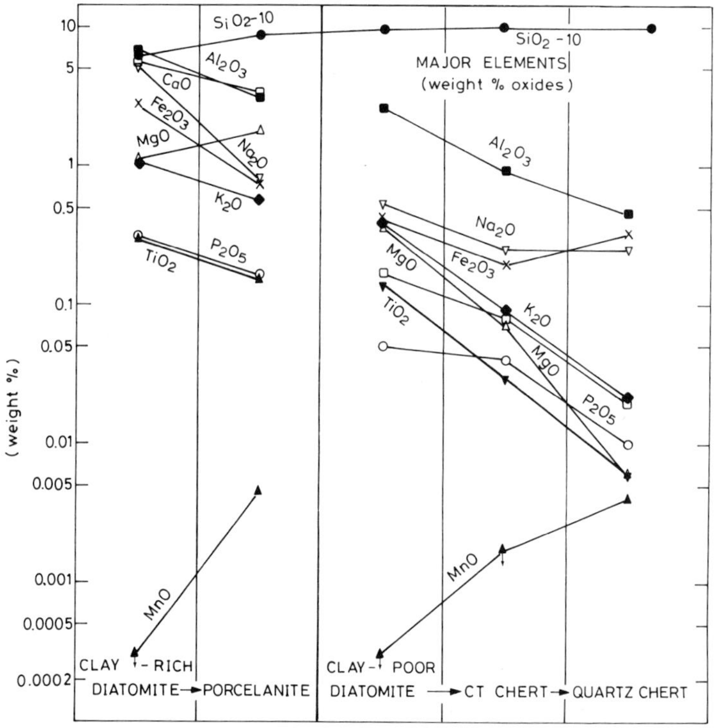

Fig. 8-9. Expulsion of major oxides during silica polymorphic transitions. More refractory elements show relative enrichments.

Phase changes occurring by solution – reprecipitation allowed the silica polymorphs (which formed) to equilibrate with the interstitial waters. Although the sediments responded to the diagenetic reactions accompanying the silica phase transitions, the Rb/Sr system maintained a record of the older, provenancial age (Brueckner and Snyder, 1985). This may be related to the greater mobility of Rb (and perhaps radiogenic Sr) in an environment which is carbonate buffered, i.e., the terrigenous component is being preferentially expelled from the system. This

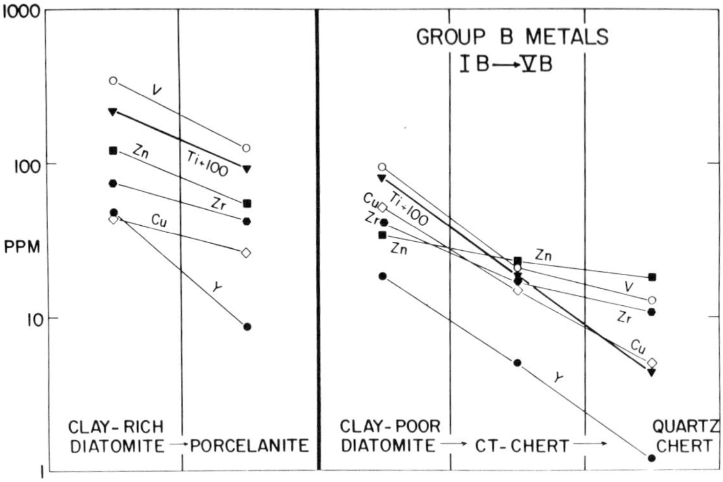

Fig. 8-10. Expulsion of Group B metals (IB – VB) during silica polymorphic transitions.

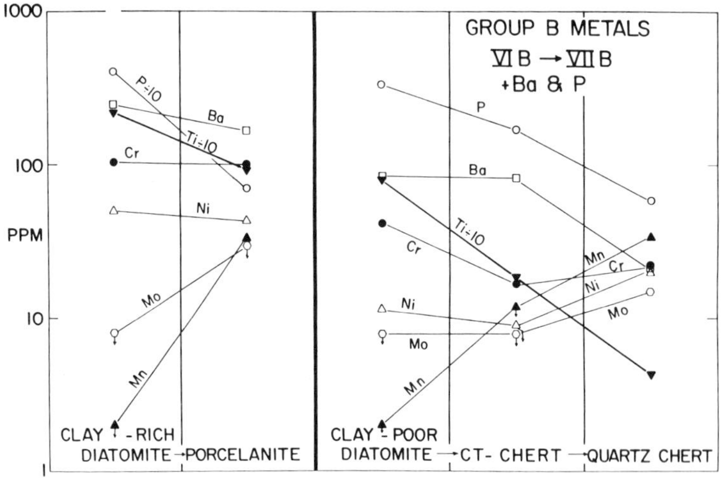

Fig. 8-11. Expulsion of Group B metals (VIB – VIIB), Ba, and P during silica polymorphic transitions. Note that Ni, Mo, Mn, and Cr are more refractory and are enriched in the solid phase during the transitions.

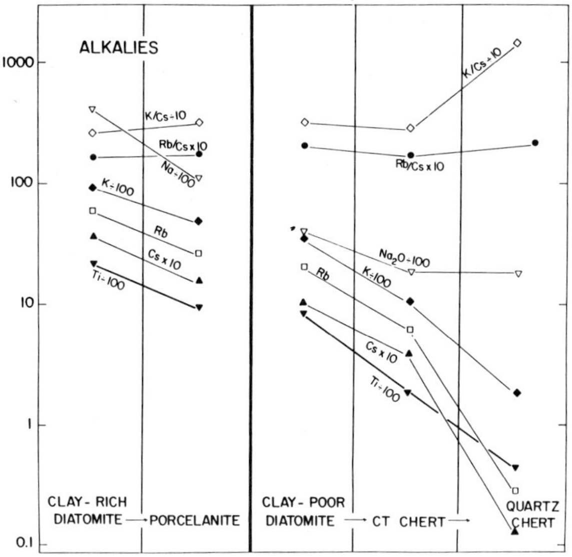

Fig. 8-12. Expulsion of alkalis during the silica polymorphic transitions. Alkalies with larger ionic radii are more readily expelled as diagenetic grade increases.

scenario views the Monterey Formation as a partially open system during the silica phase transitions, due to the brittle nature of the polymorphs being produced, and as a closed system the remainder of the time. This concept is extremely pertinent to cogenetic silica – carbonate diagenesis in the Monterey Formation discussed later in this chapter.

Timing of the silica phase transitions

The Rb/Sr data may also be used to calculate the approximate time of the silica polymorphic transitions in this area of the Santa Maria Basin. Two assumptions are necessary for this model: (1) The Sr initial ratios of the diatomite, CT-chert and quartz chert linear arrays are the Sr isotopic compositions of the interstitial waters from which the silica phases reprecipitated during phase transitions (as discussed earlier), and (2) the average $^{87}Rb/^{86}Sr$ ratios of the siliceous whole rocks (Table 8-9; from Brueckner and Snyder, 1985) are relatively representative of the Monterey Formation. The average $^{87}Rb/^{86}Sr$ ratio (Table 8-9) weights each datum point equally and, therefore, the clay-poor diatomites (six samples) are weighted by a factor that is three times that of the clay-rich diatomites (two samples). Inasmuch as exact volumetric proportions are not precisely known for each lithology, the results include calculated age ranges based upon the statistical average and standard deviation of the $^{87}Rb/^{86}Sr$ ratios from Brueckner and Snyder (1985).

In order to calculate the postdepositional timing of the silica phase transitions, the time required to produce the observed increase in the intial Sr isotopic composition, $\Delta\ I_0$, has been calculated using the relationship:

$$\Delta I_0 = {}^{87}Rb/{}^{86}Sr \times \lambda \times \Delta t$$

where λ is the decay constant for ^{87}Rb ($1.42 \times 10^{-11} y^{-1}$), $^{87}Rb/^{86}Sr$ is the average ratio of this parameter in the sediments prior to the transition, and Δt is the elapsed time between the silica phase transitions. The results and parameters are shown in Table 8-9.

The calculated postdepositional timing of both the opal-A to opal-CT (3.5 m.y.) and opal-CT to quartz (11 m.y. after the opal-A to opal-CT transition) transitions, based upon the Rb/Sr data, are consistent with the results of the thermal gradient – O isotope results of Pisciotto (1981) for this area of the Santa Maria Basin. Taking 17 m.y. as the inception of Monterey sedimentation, a 14.5 m.y. model age for the opal-A to opal-CT transition and a 2.5 m.y. model age for the opal-CT to quartz transition result.

It is significant that these two different isotopic systems, which vary in their response to the physicochemical environment due to the presence (O) or absence (Sr) of fractionation, should yield such consistent results. The Rb/Sr isotopic system appears to be recording the solution – reprecipitation event associated with the phase changes. The O-isotope systematics identify the peak periods of opal-CT and quartz production as 3 m.y. and 12 m.y. after deposition, respectively (Pisciotto, 1981), compared to the 3.5 m.y. and 14.5 m.y. after deposition based on this work using the Rb/Sr systematics. The discrepancy may lie in the continual response of the O-isotopic system to variations in temperature, whereas the Sr isotopic system is interpreted to be responding to a dehydration event accompanying a particular silica phase transition. In geologic terms, these events, from the standpoint of the Sr isotopes, are nearly instantaneous.

Dolomitization and silica phase transformations

The isotopic data reported in this and earlier studies continue to support an early diagenetic origin for dolomites with continued growth during diagenesis. This is particularly true for dolomites subjected to a changing geochemical environment as silica phase transitions proceed. Dolomite sample MR4BCS from the Mussel Rock Section (Table 8-8) is an excellent example: it accreted over a temperature interval of ~20°C and exhibits Sr isotopic compositions which are not necessarily those ex-

TABLE 8-9

Rb/Sr data summary (from Brueckner and Snyder, 1985)

Sample	t_a*	I_o*	$^{87}Rb/^{86}Sr$ (avg.)**
Diatomites	110	0.7088	2.01 ± 1.10
CT-chert	93	0.7089	1.30 ± 0.59
Quartz chert	74	0.7091	0.46 ± 0.22

* t_a = apparent age in m.y.; I_o = initial Sr-isotopic ratio.
** Average $^{87}Rb/^{86}Sr$ ratio ± 1 standard deviation; the error limits are used to calculate age ranges for the Monterey Formation fracturing event.

pected in carbonates formed in ~11.5-m.y. old seawater (0.70872 ± 2). The question arises as to the origin of the Sr isotopic anomalies observed in this concretion.

Strontium isotopic anomalies for this concretion are defined as those which lie outside the range expected for 11.5-m.y. seawater (stratigraphic age of MR4BCS) at the 95% confidence level, i.e., 0.70872 ± 4 (Hurst, 1987). Eleven calcite and dolomite analyses lie below this range (0.70857 – 0.70866) and three dolomite analyses lie above this range (0.70894 – 0.70912).

The anomalies which are observed to be less radiogenic (average 0.70862) may have derived some Sr from the recrystallization of foraminifera in the lower member of the Monterey Formation. Recrystallization of older calcareous microfossils provides a potential source of less radiogenic Sr; furthermore, the Sr isotopic range, 0.70857 – 0.70866, is that observed for the lower Monterey Formation in the Santa Maria area (Hurst, 1986a). Available petrographic data suggest the recrystallization of foraminifera and subsequent upward migration of carbonate-rich fluids from the lower member (Garrison et al., 1984).

The Sr isotopic analyses of three dolomites which are more radiogenic (0.70894 – 0.70912), are identical within error limits to the initial ratios of the linear arrays derived by Brueckner and Snyder (1985) for the opal-CT cherts (0.70889) and quartz cherts (0.70909). They are also indistinguishable from measured $^{87}Sr/^{86}Sr$ ratios of dolomite veins in dilation breccias located in the Colson – Tepusquet Canyon area of the Santa Maria Basin (Table 8-10). Given the geographic spread between the Mussel Rock and Colson – Tepusquet Canyon areas (~50 km), the similarity in Sr isotopic compositions is quite remarkable and has implications perti-

TABLE 8-10

Sr isotopic compositions (ratios) of secondary carbonates, Santa Maria Basin

Sample description	Sample no.	Dolomite	Average
Dilation breccia			
Late stage	CC1	0.70909	
	CC2	0.70905	
	CC3	0.70908	0.70908 ± 1
	CC4	0.70907	
	CC5	0.70911	
Early stage	CC 6964	0.70885	
	TC 6965	0.70884	
	TC2	0.70891	0.70888 ± 2
	TC3	0.70890	
Santa Maria Valley	BL6, 184	0.70909	
dolomite with evidence	L10, 192	0.70907	0.70907 ± 1
of secondary alteration	L10, 194	0.70906	
Initial Sr isotopic ratios*:			
Diatomites		0.70880	
CT-cherts		0.70889	
Quartz cherts		0.70909	

* Data from Brueckner and Snyder (1985).

nent to the basin's geochemical history. Is it possible that one diagenetic process is responsible for this similarity?

This question was addressed by Hurst (1986b) at which time the similarity in Sr isotopic compositions was attributed to the basinwide circulation of a fluid expelled during dehydration reactions associated with silica phase transformations. More recent data, however, are not consistent with this interpretation. Although fluid expulsion during silica phase transformations is the likely source of more radiogenic Sr, basinwide circulation is not required. The control of the diagenetic and geochemical history of the basin could be accomplished by a similarity in the sediment mixture entering the Santa Maria Basin during early diagenesis. This allows the basic starting material to remain relatively homogeneous throughout the basin thereby producing virtually identical geochemical environments on a local scale (approximately meters to tens of meters), throughout the basin.

Silica phase transitions and other dehydration reactions could occur at very different times and depths in the basin. These local diagenetic reactions would produce similar geochemical products throughout the basin, as the basic starting materials were the same. Garrison et al. (1984) reached this conclusion based upon data from four wells which penetrated the lower Monterey Formation and were distributed throughout the Santa Maria Basin. Even though the entire sequence of silica phase transitions were encountered, both the inorganic (relative proportions of major elements) and organic data indicated that the starting material was remarkably similar.

Hence, the identical Sr isotopic ratios of dolomite concretions, initial ratios of linear arrays defined by the Si polymorphs, and fracture-fill dolomite (which are separated by ~ 50 km) are interpreted as recording the migration of fluids expelled during dehydration reactions associated with silica phase transitions. Because the starting sedimentary composition was similar throughout the Santa Maria basin, the expelled fluids were geochemically similar and their Sr isotopic compositions were identical. Although the exact timing of the silica phase transitions cannot be determined throughout the basin, the Sr isotopic compositions of epigenetic or fracture-fill dolomites can be correlated to either the opal-A to opal-CT transition, if its Sr isotopic ratio is ~ 0.7089, or the opal-CT to quartz transition, if its Sr isotopic composition is ~ 0.7091. These isotopic ratios are the initial ratios of the opal-CT and quartz linear arrays of Brueckner and Snyder (1985). The data further suggest that dolomite growth continues for millions of years after diagenesis begins although, volumetrically, major growth occurs within the first 0.5 – 1 m.y.

Relation to petroleum migration

The fracture-fill dolomites in the Colson Canyon – Tepusquet region are associated with hydrocarbons (Redwine, 1981; Roehl, 1981). Dolomite layers in the Goleta Slough area near Santa Barbara also contain hydrocarbons in fractures even though the silica diagenetic grade is low (opal-A). One question which has been posed is related to the importance of dolomites as fractured reservoirs in the Monterey Formation. Inasmuch as most elements are expelled during the silica phase transitions (Figs. 8-9 – 8-12) and given that oil production/migration occurs within the temperature range of silica polymorphic transitions, one could expect that a significant portion of the petroleum is also expelled. Rather than being trapped in either opal-CT or quartz cherts, it is conceivable that a large portion of the petroleum eventually resides in fractured dolomite layers (e.g., Colson Canyon – Tepusquet area). The implications for petroleum production and exploration in the Monterey Formation include: (1) dolomite reservoirs may be correlated with either the transition to opal-CT or quartz based upon their Sr isotopic composition; depending on time – temperature conditions, dolomites associated with one or the other silica phase may have a higher probability of being a producer, (2) carbonate-filled fractures may act as major hydrocarbon traps, and (3) dolomite reservoirs may be far more important than previously envisioned in the Monterey Formation.

ACKNOWLEDGEMENTS

The author wishes to thank the following colleagues for informative discussions concerning various aspects of the Monterey Formation: R. Garrison, J. Ingle, C. Isaacs, I.R. Kaplan, and M. Kastner. J. Barron kindly assessed published biochronologic data. R.E. Dennison, T. McCulloch, and W. Burke provided unpublished information on Sr in seawater and ran samples after one of the author's mass spectrometers was inadvertently realigned by the Whittier Narrows Earthquake. Earlier phases of this work were supported by the following petroleum industry companies via a research grant to Global Geochemistry Corporation where the author served as a senior scientist and project manager: Amoco, Gulf Oil, Mobil Oil, Union Oil, Sohio, Pennzoil, Texaco, Sun Oil, Marathon Oil, Quintana, Phillips Petroleum, Husky Oil, Arco, Argo, BP Alaska, Amerada Hess, Champlin, Coastal Oil and Gas, Conoco, Elf Aquitaine, Getty Oil, Koch Exploration, Reading and Bates, Schlumberger, Shell, Superior Oil, and Union Texas. The author wishes to thank Chempet Research Corporation for allowing him to publish certain proprietary information. The manuscript was typed efficiently despite poor handwriting and numerous disappearing discs by Mrs. Geri Higley. Drafting was orchestrated by Mr. Scott Thompson. G.V. Chilingarian and K.H. Wolf graciously extended an invitation to submit this paper and had the patience to await its completion. Finally, the author would like to thank his wife, Maria, and sons, William and Christopher for their love, support, and understanding throughout this project.

REFERENCES

Aoyagi, K. and Iijima, A., 1983. Reservoir characteristics and petroleum migration in the Miocene Onnagaawa Formation of Akita, Japan. In: C.M. Isaacs and R.E. Garrison (Editors), *Petroleum Generation and Occurrence in the Miocene Monterey, California*. Pacific Section, SEPM, Spec. Publ., pp. 75 – 84.

Baker, P.A. and Kastner, M., 1981. Constraints on the formation of sedimentary dolomite. *Science*, 213: 214 – 216.

Bandy, O.L., 1961. Distribution of foraminifera, radiolaria, and diatoms in sediments of the Gulf of California. *Micropaleontology*, 17: 1 – 26.

Barron, J.A., 1981. Late Cenozoic diatom biostratigraphy of the middle-latitude eastern North Pacific.

Initial Rep. Deep Sea Drilling Project, 63: 507 – 538.
Barron, J.A., 1986. Paleoceanographic and tectonic controls on deposition of the Monterey Formation and related siliceous rocks in California. *Palaeogeogr., Palaeoclimatol., Palaeoecol.*, 53: 27 – 45.
Berger, W.H., Vincent, E. and Thierstein, H.R., 1981. The deep-sea record: major steps in Cenozoic evolution. *SEPM, Spec. Publ.*, 32: 489 – 504.
Berger, W.H., 1982. Climate steps in ocean history – lessons from the Pleistocene. In: W.H. Berger and J. Crowell (Editors), *Climate in Earth History*. Natl. Acad. Sciences Press, Washington, D.C., pp. 43 – 53.
Birch, G.F., Thompson, J., McArthur, D.M. and Burnett, W.C., 1983. Pleistocene phosphorites off the west coast of South Africa. *Nature*, 302: 601 – 603.
Blake, Jr. M.C., Campbell, R.H., Dibble, Jr. T.W., Howell, D.G., Nilsen, T.H., Normark, W.R., Vedder, J.G. and Silver E.A., 1978. Neogene basin formation in relation to plate-tectonic evolution of the San Andreas Fault system, California. *Am. Assoc. Pet. Geol., Bull.*, 62: 344 – 372.
Boellstorff, J.D. and Steineck, P.L., 1975. The stratigraphic significance of fission-track ages on volcanic ashes in the marine late Cenozoic of southern California. *Earth Planet. Sci. Lett.*, 27: 143 – 154.
Bramlette, M.N., 1946. The Monterey Formation of California and the origin of its siliceous rocks. *U.S. Geol. Surv. Prof. Pap.*, 212: 57 pp.
Brass, G.W., 1976. The variation of the marine $^{87}Sr/^{86}Sr$ ratio during Phanerozoic time: Interpretation using a flux model. *Geochim. Cosmochim. Acta*, 40: 721 – 730.
Brueckner, H.K. and Snyder W.S., 1985. Chemical and Sr isotopic variations during diagenesis of Miocene siliceous sediments of the Monterey Formation, California. *J. Sediment. Petrol.*, 55: 553 – 568.
Burke, W.H., Denison, R.E., Hetherington, E.A., Koepnik, R.B., Nelson, H.F. and Otto, J.B., 1982. Variation of seawater $^{87}Sr/^{86}Sr$ throughout Phanerozoic time. *Geology*, 10: 516 – 519.
Burnett, W.C., 1977. Geochemistry and origin of phosphorite deposits from off Peru and Chile. *Geol. Soc. Am. Bull.*, 88: 813 – 823.
Burnett, W.C., 1980. Apatite-glauconite associations off Peru and Chili: paleoceanographic implications. *J. Geol. Soc. London*, 137: 757 – 764.
Burnett, W.C., 1982. Growth rate of phosphate nodules from the continental margin off Peru. *Science*, 215: 1616 – 1010.
Burns, S.J. and Baker, P.A., 1987. A geochemical study of dolomite in the Monterey Formation, California. *J. Sediment. Petrol.*, 57 (1): 128 – 139.
Byrne, J.V. and Emery, K.O., 1960. Sediments in the Gulf of California. *Geol. Soc. Am. Bull.*, 71: 983 – 1010.
Calvert, S.E., 1964. Factors affecting the distribution of laminated diatomaceous sediments in the Gulf of California. In: Tj.H. van Andel and G.G. Shor (Editors), *Marine Geology of the Gulf of California. Am. Assoc. Petrol. Geol., Mem.*, 3: 311 – 330.
Calvert, S.E., 1966. The accumulation of diatomaceous silica in the sediments of the Gulf of California. *Geol. Soc. Am. Bull.*, 77: 569 – 596.
Calvert, S.E., 1977. Mineralogy of silica phases in deep-sea cherts and porcelanites. *Phil. Trans. R. Soc. London*, A. 286: 239 – 252.
Canfield, C.R., 1939. Subsurface stratigraphy of Santa Maria Valley oil field and adjacent parts of Santa Maria Valley, California. *Am. Assoc. Pet. Geol., Bull.*, 23: 45 – 81.
Clauer, N., 1976. $^{87}Sr/^{86}Sr$ composition of evaporitic carbonates and sulphates from Miocene sediment cored in the Mediterranean Sea (DSDP, Leg 13). *Sedimentology*, 23: 133 – 140.
Claypool, G.E., 1974. Anoxic diagenesis and bacterial methane production in deep sea sediments. Ph.D. Thesis, Univ. California, Los Angeles, Calif., 276 pp.
Claypool, G.E. and Kaplan, I.R., 1974. The origin and distribution of methane in marine sediments. In: I.R. Kaplan (Editor), *Natural Gases in Marine Sediments*. Plenum Press, New York, N.Y., pp. 99 – 139.
Claypool, G.E., Presley, B.J. and Kaplan, I.R., 1973. Gas analyses of sediment samples from Legs 10, 11, 13, 14, 15, 18, and 19. In: J.S. Creager and D.W. Scholl (Editors), *Initial Reports of the Deep Sea Drilling Project*, 19: 879 – 884.
Conrad, C.L. and Ehlig, P.L., 1983. The Monterey Formation of the Palos Verdes Peninsula, California – an example of sedimentation in a tectonically active basin within the California Continental

Borderland. In: D.K. Larue and R.J. Stell (Editors), *Cenozoic Marine Sedimentation, Pacific Margin, U.S.A.*, Pacific Section SEPM, Spec. Publ., pp. 1103 – 1116.

Cooper, J.E. and Bray, E.E., 1963. A postulated role of fatty acids in petroleum formation. *Geochim. Cosmochim. Acta,* 27: 1113 – 1127.

Crawford, F.D., 1971. Petroleum potential of Santa Maria province, California. *Am. Assoc. Pet. Geol., Mem.*, 15: 316 – 328.

Curtis, C.D., 1978. Possible links between sandstone diagenesis and depth-related geochemical reactions occurring in enclosing mudstones. *J. Geol. Soc. London,* 135: 107 – 117.

Dasch, E.J. and Biscaye, P.E., 1971. Isotopic composition of strontium in Cretaceous to Recent pelagic foraminifera. *Earth Planet. Sci. Lett.*, 11: 201 – 204.

Demaison, G.J. and Moore, G.T., 1980. Anoxic environments and oil source bed genesis. *Am. Assoc. Pet. Geol., Bull.*, 64: 1179 – 1209.

DePaolo, D.J., 1986. Detailed record of the Neogene Sr isotopic evolution of seawater from DSDP site 590B. *Geology*, 14: 103 – 106.

Dibblee, T.W., Jr., 1950. Geology of the southwestern Santa Barbara County, California. *Calif. Div. Mines Geol. Bull.,* 150: 95 pp.

Dickert, P.F., 1966. Tertiary phosphatic facies of the Coast Ranges. *Calif. Div. Mines Geol., Bull.*, 190: 289 – 304.

Dickert, P.F., 1971. Neogene phosphatic facies in California. Ph.D. thesis, Stanford Univ., Stanford, Calif., 305 pp.

Didyk, B.M., Simoneit, B.R.T., Brassell, S.C. and Eglington G., 1978. Organic geochemical indicators of paleoenvironmental conditions of sedimentation. *Nature*, 272: 216 – 222.

Dott, R.H., 1969. Circum-Pacific Late Cenozoic structural rejuvenations; implications for sea floor spreading. *Science*, 166: 874 – 876.

Douglas, R.G. and Savin, S., 1973. Oxygen and carbon isotopes of Cretaceous and Cenozoic foraminifera from the central North Pacific Ocean. *Initial Rep. Deep Sea Drilling Project*, 17: 591 – 605.

Douglas, R.G., Colburn, I.P. and Gorsline, D.S. (Editors), 1981. *Depositional Systems of Active Margin Basins.* Short Course Notes, Pacific Section SEPM, 165 pp.

Dow, W.G., 1978. Petroleum source beds on continental slopes and rises. *Am. Assoc. Pet. Geol., Bull.*, 62: 1584 – 1606.

Duedall, I.W., Dayal, R. and Wiley, J.D., 1976. The partial molal volume of silicic acid in 0.725 *M* NaCl. *Geochim. Cosmochim. Acta*, 40: 1185 – 1189.

Eisma, E. and Jung, J.W., 1989. Fundamental aspects of the generation of petroleum. In: G. Eglinton and M.T.J. Murphy (Editors), *Organic Geochemistry*. Springer, New York, N.Y., pp. 676 – 698.

Emery, K.O., 1960. *The Sea off Southern California*. Wiley, New York, N.Y., 366 pp.

Engerbretson, D.C., 1982. Relative motions between oceanic and continental plates. Ph.D. thesis, Stanford Univ., Stanford, Calif., 211 pp.

Epstein, S.A. and Nary, Q.L., 1982. A look at the geology of the hot Santa Maria basin area off California. *Oil Gas J.*, 80 (346): 162 – 168.

Faure, G., Assereto, R. and Tremba, E.L., 1978. Strontium isotope composition of marine carbonates of Middle Triassic to Early Jurassic age, Lombardic Alps, Italy. *Sedimentology*, 25: 523 – 543.

Fischer, A. and Arthur, M.A., 1977. Secular variations in the pelagic realm. In: H. Cook and P. Enos (Editors), *Deep-Water Carbonate Environments*. SEPM, Spec. Publ., 25: 19 – 50.

Fisher, R.V., 1977. *Geologic Guide to Subaqueous Volcanic Rocks in the Nipomo, Posmo Beach and Avila Beach areas.* Field trip guidebook for the Geol. Soc. Am. Penrose Conf. on Subaqueous Volcanic Rocks, Santa Barbara, Calif., 79 pp.

Fournier, R.O. and Rowe, J.J., 1962. The solubility of cristobalite along the three-phase curve, gas plus liquid plus cristobalite. *Am. Mineral.*, 47: 897 – 902.

Fournier, R.O. and Potter, R.W., 1982. An equation correlating the solubility of quartz in water from 25° to 900°C at pressures up to 10,000 bars. *Geochim. Cosmochim. Acta*, 46: 1969 – 1974.

Fournier, R.O., Rosenbauer, R.J. and Bischoff, J.L., 1982. The solubility of quartz in aqueous sodium chloride solution at 350°C and 180 to 500 bars. *Geochim. Cosmochim. Acta*, 46: 1975 – 1978.

Friedman, I. and Murata, K.J., 1979. Origin of dolomite in Miocene Monterey Shale and related formations in the Temblor Range, California. *Geochim. Cosmochim. Acta*, 43: 1357 – 1365.

Fritz, P. and Smith, D.G.W., 1970. The isotopic composition of secondary dolomites. *Geochim.*

Cosmochim. Acta, 34: 1161 – 1173.

Garrison, R.E., 1981. Pelagic and hemipelagic sedimentation in active margin basins. In: R.G. Douglas and R.E. Garrison (Editors), *Depositional Systems of Active Continental Margin Basins*. Short Course Notes, Pacific Section SEPM, pp. 15 – 38.

Garrison, R.E. and Douglas, R.G. (Editors), 1981. *The Monterey Formation and Related Siliceous Rocks of California*. Pacific Section SEPM, Spec. Publ., 327 pp.

Garrison, R.E. and Graham, S.A., 1984. Early diagenetic dolomites and the origin of dolomite-bearing breccias, lower Monterey Formation, Arroyo Seco, Monterey County, California. In: R.E. Garrison, M. Kastner and D.H. Zenger (Editors), *Dolomites of the Monterey Formation and Other Organic-Rich Units*. Pacific Section, SEPM, 41: 87 – 102.

Garrison, R.E., Stanley, R.G. and Horan, L.J., 1979. Middle Miocene sedimentation and the southwestern edge of the Lockwood High, Monterey County, California. In: S.A. Graham (Editor), *Tertiary and Quaternary Geology of the Salinas Valley and Santa Lucia Range, Monterey County, California*. Pacific Section, SEPM Field Guide No. 4: 51 – 65.

Garrison, R.E., Hurst, R.W., Ingle, J.C., Kastner, M. and Kaplan, I.R., 1984. *Geochemical and Paleoenvironmental History of the Monterey Formation: Sediments and Hydrocarbons*. Final Report Petroleum Industry, 1058 pp. (unpublished).

Govean, F.M., 1980. Some paleoecologic aspects of the Monterey Formation, California. Ph.D. thesis, Univ. California, Santa Cruz, Calif., 278 pp.

Graham, S.A., 1976. Tertiary sedimentary tectonics of the central Salinian Block of California. Ph.D. thesis, Stanford Univ., Stanford, Calif., 510 pp.

Graham, S.A., Williams, L.A., Bate, M. and Weber, L.S., 1982. Stratigraphic depositional framework of the Monterey Formation and associated coarse clastic rocks, central San Joaquin basin, California. In: L.A. Williams and S.A. Graham (Editors), *Monterey Formation and Associated Coarse Clastic Rocks, Central San Joaquin Basin, California*. Pacific Section, SEPM, Annu. Fieldtrip Guidebook, pp. 3 – 16.

Grechin, V.I., Pisciotto, K.A., Mahoney, J.J. and Gordeevan, S.N., 1981. Neogene siliceous sediments and rocks off southern California and Baja California, Deep Sea Drilling Project Leg 63. In: R.S. Yeats et al. (Editors), *Initial Reports of the Deep Sea Drilling Project*, 63 pp. 579 – 593.

Grivetti, M.C., 1982. *Aspects of Stratigraphy, Diagenesis, and Deformation in the Monterey Formation near Santa Maria – Lompoc, California*. M.S. thesis, Univ. California, Santa Barbara, Calif., 154 pp.

Hall, C.A., Jr, 1981. San Luis Obispo transform fault and Middle Miocene rotation of the western Transverse Ranges, California. *J. Geophys. Res.*, 86: 1015 – 1031.

Hein, J.R., O'Neil, J.R. and Jones, M.G., 1979a. Origin of authigenic carbonates in sediments from the deep Bering Sea. *Sedimentology*, 26: 681 – 705.

Hein, J.R., Vanek, E. and Allen, M.A., 1979b. X-ray mineralogy and diagenesis. In: H.E. Cook (Editor), *Geologic Studies of the Point Conception Deep Stratigraphic Test Well OCS-CAL 78 – 164 No. 1, Outer Continental Shelf, Southern California, United States*. U.S. Geol. Surv. Open File Rep., 79 – 1218: 79 – 96.

Hornafius, J.S., Luyendyk, B.P., Weaver, D.W., Wornardt, W.W. and Fuller, M., 1981. Paleomagnetic results from the Monterey Formation in the western Transverse Ranges, California. *EOS, Am. Geophys. Union, Trans.*, 62: 855 (abstract).

Hornafius, J.S., Kamerling, M., Terres, R. and Luyendyk, B., 1982. Differential tectonic rotation within the western Transverse Ranges. *Geol. Soc. Am., Cordilleran Section, Abstracts with Programs*, 14: 173 – 174.

Howell, D.G., Crouch, J.K., Greene, H.G., McCulloch, D. and Vedder, J.G., 1980. Basin development along the Late Mesozoic and Cenozoic California margin: a plate tectonic margin of subduction and transform tectonics. In: P.F. Ballance and H.G. Reading (Editors), *Int. Assoc. Sedimentol., Spec. Publ.*, 4.

Hurd, D.C. and Birdwhistell, S., 1983. On producing a more general model for biogenic silica dissolution. *Am. J. Sci.*, 283: 29 – 47.

Hurd, D.C. and Theyer, F., 1975. Changes in the physical and chemical properties of biogenic silica from the central equatorial Pacific. I. Solubility, specific surface area, and solution rate constants of acid-cleaned samples. In: T. Gibb (Editor), *Analytical Methods in Chemical Oceanography*. Advances in Chemistry, Ser. 147, Am. Chem. Soc., pp. 211 – 230.

Hurst, R.W., 1982. Petrogenesis of the Conejo Volcanic suite, southern California: Evidence for mid-ocean ridge – continental margin interaction. *Geology*, 10: 267 – 272.

Hurst, R.W., 1986a. Sr isotopic chronostratigraphy of the Miocene Monterey Formation in the Ventura and Santa Maria Basins, CA. *Geology*, 14: 459 – 462.

Hurst, R.W., 1986b. Chemical and Sr isotopic variations during diagenesis of Miocene siliceous sediments of the Monterey Formation: A discussion. *J. Sediment. Petrol.*, 56: 567 – 573.

Hurst, R.W., 1987. The Anatomy of NESSIE (Neogene seawater Sr isotopic evolution): Paleoceanographic and paleoclimatic implications. *Fourth Int. Congr. Pacific Neogene, Stratigraphy.*

Hurst, R.W. and Davis, T.E., 1981. *Pb and Sr Isotopes in Cores and Groundwater.* San Juan Basin Research Site, New Mexico. DOE Open File Rep., 42 pp.

Ingle, J.C., Jr, 1967. Foraminiferal biofacies variation and the Miocene – Pliocene boundary in southern California. *Bull. Am. Paleontol.*, 52 (236): 210 – 394.

Ingle, J.C., Jr, 1973. Summary comments on Neogene biostratigraphy, physical stratigraphy, and paleoceanography in the marginal northeastern Pacific Ocean. In: *Initial Reports of the Deep Sea Drilling Project*, 18: 949 – 959.

Ingle, J.C., Jr, 1978. Evidence of latest Miocene refrigeration and associated eustatic events in California. *Geol. Soc. Am., Abstracts with Programs*, 10: 427.

Ingle, J.C., Jr, 1980. Cenozoic paleobathymetry and depositional history of selected sequence within the southern California Continental Borderland. *Cushman Found. Foraminiferal Res., Spec. Publ.*, 19: pp. 163 – 195.

Ingle, J.C., Jr, 1981a. Cenozoic depositional history of the northern continental borderland of southern California and the origin of associated Miocene diatomites. In: C.M. Isaacs (Editor), *Guide to the Monterey Formation in the California Coastal Area, Ventura to San Luis Obispo.* Pacific Section AAPG, Spec. Publ., 52: 1 – 8.

Ingle, J.C., Jr, 1981b. Origin of Neogene diatomites around the North Pacific rim. In: R.E. Garrison et al. (Editors), *The Monterey Formation and Related Siliceous Rocks of California.* Pacific Section SEPM, pp. 159 – 179.

Irwin, H., Curtis, C.D. and Coleman, M., 1977. Isotopic evidence for source of diagenetic carbonates formed during burial of organic-rich sediments. *Nature*, 269: 209 – 213.

Isaacs, C.M., 1980. Diagenesis in the Monterey Formation examined laterally along the coast near Santa Barbara, California. Ph.D. thesis, Stanford, Univ., Stanford, Calif., 329 pp.

Isaacs, C.M., 1981. Porosity reduction during diagenesis of the Monterey Formation, Santa Barbara coastal area, California. In: R.E. Garrison et al. (Editors), *The Monterey Formation and Related Siliceous Rocks of California.* Spec. Publ. Pacific Section SEPM, pp. 257 – 271.

Isaacs, C.M., 1982. Influence of rock composition on kinetics of silica phase changes in the Monterey Formation, Santa Barbara area, California. *Geology*, 10: 304 – 308.

Isaacs, C.M., 1983. Compositional variation and sequence in the Miocene Monterey Formation, Santa Barbara coastal area, California. In: D.K. Larne and R.J. Steel (Editors), *Cenozoic Marine Sedimentation, Pacific Margin, U.S.A.* Pacific Section SEPM.

Isaacs, C.M., Pisciotto, K.A. and Garrison, R.E., 1983. Facies and diagenesis of the Miocene Monterey Formation, California: a summary. In: A. Iijima, J.R. Hein and R. Siever (Editors), *Siliceous Deposits in the Pacific Region.* Developments in Sedimentology, 36. Elsevier, Amsterdam, pp. 247 – 282.

Izett, G.A., Naeser, C.W. and Obradovich, J.D., 1974. Fission track age of a zircon from an ash bed in the Pico Formation near Ventura, California. *Geol. Soc. Am., Abstracts with Programs,* 6: 197.

Jones, J.B. and Segnit, E.R., 1971. The nature of opal. I. Nomenclature and constituent phases. *J. Geol. Soc. Aust.*, 18: 56 – 68.

Jones, M.M. and Pytkowicz, R.M., 1973. The solubility of silica in seawater at high pressures. *Soc. R. Sci. Liege, Bull.*, 42: 118 – 120.

Jorgensen, S.S., 1968. Solubility and dissolution kinetics of precipitated silica in 1 *M* NaCl, at 25°C. *Acta Chem. Scand.*, 22: 335 – 341.

Kablanow, R.I. and Surdam, R.C., 1983. Diagenesis and hydrocarbon generation in the Monterey Formation, Huasna Basin, California. In: C.M. Isaacs, R.E. Garrison, S.A. Graham and W.A. Jensky (Editors), *Petroleum Generation and Occurrence in the Miocene Monterey Formation, California.* SEPM Pacific Section, pp. 53 – 68.

Kamatami, A., 1971. Physical and chemical characteristics of biogenous silica. *Mar. Biol.*, 8: 89 – 95.

Kamerling, M.G. and Luyendyk, B.P., 1979. Tectonic rotations of the Santa Monica Mountains region,

western Transverse Ranges, California, suggested by paleomagnetic vectors. *Geol. Soc. Am. Bull.*, 90: 331 – 337.
Kastner, M., 1981. Authigenic silicates in deep-sea sediments: formation and diagenesis. In: C. Emiliani (Editor), *The Sea,* Vol. 7. Wiley, New York, N.Y., pp. 915 – 980.
Kastner, M. and Gieskes, J.M., 1983. Opal-A to opal-CT transformation: a kinetic study. In: A. Iijima, J.R. Hein and R.Siever (Editors), *Siliceous Deposits in the Pacific Region*, Developments in Sedimentology, 36. Elsevier, Amsterdam, pp. 211 – 227.
Kastner, M., Keene, J.B. and Gieskes, J.M., 1977. Diagenesis of siliceous oozes, I. Chemical controls on the rate of opal-A to opal-CT transformation – an experimental study. *Geochim. Cosmochim. Acta*, 41: 1041 – 1059.
Keene, J.B., 1976. The distribution, mineralogy, and petrography of biogenic and authigenic silica from the Pacific Basin. Ph.D. thesis, Scripps Inst. Oceanography, 264 pp.
Keller, G., 1981. Miocene biochronology and paleoceanography of the North Pacific. *Mar. Micropaleontol.*, 6: 535 – 551.
Keller, G. and Barron, J.A., 1983. Paleoceanographic implications of Miocene deep-sea hiatuses. *Geol. Soc. Am. Bull.*, 94: 590 – 613.
Kelts, K. and McKenzie, J.A., 1982. Diagenetic dolomite formation in Quaternary anoxic diatomaceous muds of Deep Sea Drilling Project Leg 64, Gulf of California. In: J.R. Curry et al. (Editors), *Initial Reports of the Deep Sea Drilling Project,* 64: 553 – 569.
Kennedy, G.C., 1950. A portion of the system silica – water. *Econ. Geol.*, 45: 629 – 653.
Kennett, J.P., 1977. Cenozoic evolution of Antarctic glaciation, the circum-Antarctic Ocean, and their impact of global paleoceanography. *J. Geophys. Res.*, 82: 3843 – 3860.
Kennett, J.P., 1982. *Marine Geology*. Prentice-Hall, Englewood Cliffs, N.J., 813 pp.
Kovach, J., 1980. Variations in the strontium isotopic composition of seawater during Paleozoic time determined by the analysis of conodonts. *Geol. Soc. Am., Abstracts with Program,* 12: 465.
Krauskopf, K.B., 1956. Dissolution and precipitation of silica at low temperatures. *Geochim. Cosmochim. Acta*, 10: 1 – 26.
Kruger, M.A., 1983. Diagenesis of Miocene biogenic sediments in Lost Hills oil field, San Joaquin Basin, California. In: C.M. Isaacs et al. (Editors), *Petroleum Generation and Occurrence in the Miocene Monterey Formation, California*. Pacific Section SEPM, pp. 39 – 51.
Kulm, L.D., Schrader, H., Resig, J.M., Thornburg, T.M., Masias, A. and Johnson, L., 1981. Late Cenozoic carbonates on the Peru continental margin: lithostratigraphy, biostratigraphy, and tectonic history. In: L.D. Kulm et al. (Editors), *Nazca Plate: Crustal Formation and Andean Convergence. Geol. Soc. Am., Mem.,* 154: 469 – 507.
Kushnir, J. and Kastner, M., 1984. Two forms of dolomite occurrences in the Monterey Formation, California: Concretions and layers – a comparative mineralogical, geochemical, and isotopic study. In: R.E. Garrison, M. Kastner and D.H. Zenger (Editors), *Dolomites of the Monterey Formation and Other Organic-rich Units.* Pacific Section SEPM, Spec. Publ., 41: 171 – 184.
Lagoe, M., 1982. Stratigraphy and paleoenvironments of the Monterey Formation and associated rocks, Cuyama Basin, California. Ph.D thesis, Stanford Univ., Calif., 193 pp.
Lipps, J.H., 1969. Ecology and evolution of silicoflagellates. *Proc. North American Paleontology Convention*, Part G: pp. 965 – 993.
Lawson, D., Hurd, D.C. and Pankratz, H.S., 1978. Silica dissolution rates of decomposing phytoplankton assemblages at various temperatures. *Am. J. Sci.*, 278: 1373 – 1393.
Marshall, W.L., 1980a. Amorphous silica solubilities – I. Behavior in aqueous sodium nitrate solutions: 23 – 300°C, 0 – 6 molal. *Geochim. Cosmochim. Acta*, 44: 907 – 913.
Marshall, W.L., 1980b. Amorphous silica solubilities: III. Activity coefficient relations and predictions of solubility behavior in salt solutions, 0 – 350°C. *Geochim. Cosmochim. Acta*, 44: 915 – 924.
Matsuhisa, Y., Goldsmith, J.R. and Clayton, R.N., 1976. Oxygen isotope fractionation in the system quartz – albite – anorthite – water (abstract). *Geol. Soc. Am., Abstracts with Program,* 8: 999.
Matsuhisa, Y., Goldsmith, J.R. and Clayton, R.N., 1978. Mechanisms of hydrothermal crystallization of quartz at 250°C and 15 Kb. *Geochim. Cosmochim. Acta*, 62: 173 – 182.
Mertz, K.A., Jr. and Garrison, R.E., 1983. Notes on the Sandholt Member of the Monterey Formation in the Arroyo Seco Area, Monterey County, California. In: J.C. Ingle and R.E. Garrison (Editors), *Field Guide to the Monterey Formation, Northern Santa Lucia Mountains and Monterey Area, Monterey County, California*. Pacific Section, SEPM, pp. 89 – 95.

Mizutani, S., 1977. Progressive ordering of cristobalitic silica in the early stage of diagenesis. *Contrib. Mineral. Petrol.*, 61: 129 – 140.

Murata, K.J. and Larson, R.R., 1975. Diagenesis of Miocene siliceous shales, Temblor Range, California. *J. Res. U.S. Geol. Surv.*, 3: 553 – 566.

Murata, K.J. and Nakata, J.K., 1974. Cristobalitic stage in the diagenesis of diatomaceous shale. *Science*, 184: 567 – 568.

Murata, K.J. and Randall, R.G., 1975. Silica mineralogy and structure of the Monterey Shale, Temblor Range, California. *J. Res. U.S. Geol. Surv.*, 3: 567 – 572.

Murata, K.J., Friedman, I. and Madsen, B.M., 1969a. Isotopic composition of diagenetic carbonates in Miocene formations of California and Oregon. *Science*, 156: 1484 – 1486.

Murata, K.J., Friedman, I. and Madsen, B.M., 1969b. Isotopic composition of diagenetic carbonates in marine Miocene formations of California and Oregon. *U.S. Geol. Surv., Prof. Pap.*, 614-B: 24 pp.

Murata, K.J., Friedman, I. and Cremer M., 1972. Geochemistry of diagenetic dolomites in Miocene marine formations of California and Oregon. *U.S. Geol. Surv., Prof. Pap.*, 724-C: 12 pp.

Murata, K.J., Friedman, I. and Gleason, J.D., 1977. Oxygen isotope relations between diagenetic silica minerals in Monterey Shale, Temblor Range, California. *Am. J. Sci.*, pp. 277 – 272.

Murata, K.J., Dibblee, T.W., Jr. and Drinkwater, J.L., 1979. Thermal effects of large bodies of intrusive serpentinite on overlying Monterey Shale, southern Diablo Range, Cholame area, California. *U.S. Geol. Surv. Prof. Pap.*, 1089: 1 – 8.

Orr, W.N., 1972. Pacific northwest siliceous phytoplankton. *Palaeogeogr., Palaeoclimatol., Palaeoecol.*, 12: 95 – 114.

Parrish, J.T., 1982. Upwelling and petroleum source beds with reference to the Paleozoic. *Am. Assoc. Pet. Geol., Bull.*, 66: 744 – 750.

Peterman, Z.E., Hedge, C.E. and Tourtelot, H.A., 1970. Isotopic composition of seawater throughout Phanerozoic time. *Geochim. Cosmochim. Acta*, 34: 105 – 120.

Pisciotto, K.A., 1978. Basinal sedimentary facies and diagenetic aspects of the Monterey Shale, California. Ph.D. thesis, Univ. of Calif., Santa Cruz, Calif., 450 pp.

Pisciotto, K.A., 1981a. Diagenetic trends in siliceous facies of the Monterey Shale in the Santa Maria region, California. *Sedimentology*, 28: 547 – 571.

Pisciotto, K.A., 1981b. Distribution, thermal histories, isotopic composition, and reflection characteristics of siliceous rocks recovered by the Deep Sea Drilling Project. In: J.E. Warme, R.G. Douglas and E.L. Winterer (Editors), *The Deep Sea Drilling Project: a Decade of Progress. Soc. Econ. Paleontol. Mineral., Spec. Publ.*, 32: 129 – 148.

Pisciotto, K.A., 1981c. Notes on Monterey rocks near Santa Maria, California, In: C.M. Isaacs (Editor), *Guide to the Monterey Formation in the California Coastal Area, Ventura to San Luis Obispo. Pacific Section. Am. Assoc. Pet. Geol.*, 52: 73 – 81.

Pisciotto, K.A., 1981d. Review of secondary carbonates in the Monterey Formation, California. In: R.E. Garrison et al. (Editors), *The Monterey Formation and Related Siliceous Rocks of California. Pacific Section SEPM, Spec. Publ.*, pp. 273 – 283.

Pisciotto, K.A. and Garrison, R.E., 1981. Lithofacies and depositional environments of the Monterey Formation, California. In: R.E. Garrison et al. (Editors), *The Monterey Formation and Related Siliceous Rocks of California. Pacific Section SEPM, Spec. Publ.*, pp. 97 – 122.

Pisciotto, K.A. and Mahoney, J.J., 1981. Isotopic survey of diagenetic carbonates, DSDP Leg 63. In: R.S. Yeats, B.U. Haq et al. (Editors), *Initial Reports of the Deep Sea Drilling Project*, 63: 595 – 609.

Redwine, L.E., 1981. Hypothesis combining dilation, natural hydraulic fracturing, and dolomitization to explain petroleum reservoirs in Monterey Shale, Santa Maria area California. In: R.E. Garrison et al. (Editors), *The Monterey Formation and Related Siliceous Rocks of California, Pacific Section SEPM, Spec. Publ.*, pp. 221 – 248.

Regan, L.J., Jr., 1953. Fractured shale reservoirs of California. *Am. Assoc. Pet. Geol., Bull.*, 37: 201 – 216.

Regan, L.J., Jr. and Hughes, A.W., 1949. Fractured reservoirs of Santa Maria district, California. *Am. Assoc. Pet. Geol., Bull.*, 33: 32 – 51.

Rhodes, D.C. and Morse, J.W., 1971. Evolutionary and ecologic significance of oxygen deficient marine basins. *Lethaia*, 4: 413 – 428.

Roehl, P.O., 1981. Dilation brecciation – a proposed mechanism of fracturing, petroleum expulsion and dolomitization in the Monterey Formation, California. In: R.E. Garrison et al. (Editors), *The*

Monterey Formation and Related Siliceous Rocks of California. Pacific Section SEPM, Spec. Publ., pp. 285 – 315.

Savin, S., 1977. The history of the earth's surface temperature during the past 100 million years. *Earth Planet. Sci., Annu. Rev.*, 5: 319 – 355.

Soutar, A., 1971. Micropaleontology of anaerobic sediments and the California Current. In: B. Funnel and W. Reidel (Editors), *The Micropaleontology of Oceans*. Cambridge Univ. Press, London, pp. 223 – 230.

Soutar, A. and Crill, P.A., 1977. Sedimentation and climatic patterns in the Santa Barbara Basin during the 19th and 20th centuries. *Geol. Soc. Am. Bull.*, 88: 1161 – 1172.

Spotts, J.H. and Silverman, S.R., 1966. Organic dolomite from Point Fermin, California. *Am. Mineral.*, 51: 1144 – 1155.

Suess, E., Kulm, L.D. and Killingley, J.S., 1982. Mechanism of dolomitization of Peru convergent margin sediments: isotope and mineral record. *EOS, Am. Geophys. Union, Trans.*, 63: 999.

Surdam, R.C. and Stanley, K.O., 1981. Stratigraphic and sedimentologic framework of the Monterey Formation, Pismo Syncline, California. In: C.M. Isaacs (Editor), *Guide to the Monterey Formation in the California Coastal Area, Ventura to San Luis Obispo. Pacific Section, Am. Assoc., Pet. Geol.*, 52: 83 – 91.

Surdam, R.C., Hall, C.A. and Murphy, J., 1980. Diagenetic reactions in the Monterey Formation, Pismo Syncline, California. *Am. Assoc. Pet. Geol. Mtng*, Denver, Colo.

Taylor, J.C., 1976. Geologic appraisal of the potential of offshore Southern California; the borderland compared to onshore basins. U.S. *Geol. Surv., Circ.*, 730: 43 pp.

Tissot, B.P. and Welte, D.H., 1978. *Petroleum Formation and Occurrence: a New Approach to Oil and Gas Exploration.* Springer-Verlag, New York, N.Y., 538 pp.

Tremba, E.L., Faure, G., Katsikatsos, G.C. and Summerson, C.H., 1975. Strontium isotope composition in the Tethys Sea, Euboea, Greece. *Chem. Geol.*, 16: 109 – 120.

United States Geological Survey, 1974. Proposed plan of development, Santa Ynez unit, Santa Barbara Channel, off California – A final environmental statement. *U.S. Geol. Surv., F.E.S.*, 74-20, 1: II-118 – II-174.

Vail, P.R., Mitcum, R.M. and Thompson, S., III, 1977. Seismic stratigraphy and global changes of sea level. Part 4: global cycles of relative changes of level. In: C. Payton (Editor), *Seismic Stratigraphy – Applications to Hydrocarbon Exploration. Am. Assoc. Pet. Geol., Mem.*, 26: 83 – 97.

Vail, P.R. and Hardenbol, J., 1979. Sea level changes during the Tertiary. *Oceanus*, 22: 71 – 79.

Van Andel, Tj.H., 1964. Recent sediments of the Gulf of California. In: Tj.H. van Andel and G.G. Shor (Editors), *Marine Geology of the Gulf of California. Am. Assoc. Pet. Geol., Mem.*, 3: 216 – 310.

Van Lier, J.A., De Bruyn, P.L. and Overbeck, J.Th.G., 1960. The solubility of quartz. *J. Phys. Chem.*, 64: 1675 – 1682.

Veizer, J. and Compston, W., 1974. $^{87}Sr/^{86}Sr$ composition of seawater during the Phanerozoic. *Geochim. Cosmochim. Acta*, 38: 1461 – 1484.

Walther, J.V. and Helgeson, H.D., 1977. Calculation of the thermodynamic properties of aqueous silica and the solubility of quartz and its polymorphs at high pressures and temperatures. *Am. J. Sci.*, 277: 1315 – 1351.

Webb, G.W., 1981. Stevens and earlier Miocene turbidite sandstones, southern San Joaquin Valley, California. *Am. Assoc. Pet. Geol., Bull.*, 65: 438 – 465.

Williams, L.A., 1982. Lithology of the Monterey Formation (Miocene) in the San Joaquin Valley of California. In: L.A. Williams and S.A. Graham (Editors), *Monterey Formation and Associated Coarse Clastic Rocks, Central San Joaquin Basin, California.* Pacific Section SEPM, Volume and Guidebook, pp. 17 – 36.

Williams, L.A. and Reimers, C., 1983. Role of bacterial mats in oxygen-deficient marine basins and coastal upwelling regimes: preliminary report. *Geology*, II: 267 – 269.

Wirth, G.S. and Gieskes, J.M., 1979. The initial kinetics of the dissolution of vitreous silica in aqueous media. *J. Colloid Interface Sci.*, 68: 492 – 500.

Woodring, W.P. and Bramlette, M.N., 1950. Geology and paleontology of the Santa Maria District, California. *U.S. Geol. Surv. Prof. Pap.* 222: 185 pp.

Woodring, W.P., Bramlette, M.N. and Lohman, K.E., 1943. Stratigraphy and paleontology of the Santa Maria Distrinct, California. *Am. Assoc. Pet. Geol. Bull.*, 27: 1335 – 1360.

Woodruff, F., Savin, S. and Douglas, R.G., 1981. Miocene stable isotope record: a detailed Pacific Ocean study and its paleoclimatic implications. *Science*, 212: 665 – 668.
Wyrtki, K., 1962. The oxygen minimum in relation to ocean circulation. *Deep-sea Res.*, 9: 11 – 23.
Younse, G.A., 1979. The stratigraphy and phosphoritic rocks of the Robinson Canyon-Laureles Grade Areas, Monterey County, California. M.Sc. thesis, San Jose State Univ., San Jose, Calif., 127 pp.

Reprinted from: *Diagenesis, III. Developments in Sedimentology, 47.*
Edited by K.H. Wolf and G.V. Chilingarian.

Chapter 9

MATURATION OF ORGANIC MATTER AS REVEALED BY MICROSCOPIC METHODS: APPLICATIONS AND LIMITATIONS OF VITRINITE REFLECTANCE, AND CONTINUOUS SPECTRAL AND PULSED LASER FLUORESCENCE SPECTROSCOPY

P.K. MUKHOPADHYAY

INTRODUCTION

During burial and heating, the organic matter within the network of dispersed organic-rich sediments or coal progressively loses oxygen, nitrogen and hydrogen from its hetero-atomic structure in order to form clusters of aromatic layers and chemical ordering, and is eventually transformed into a polycondensed lamella (Fig. 9-1). This type of chemical ordering is known to coal petrographers as "coalification" or increase in rank, to palynologists as "carbonization", and to organic geochemists as "maturation". Maturation parameters are extremely important in delineating important aspects of geological sciences: such as oil and gas windows and their limitations; checking parameters for mathematical modeling for hydrocarbon generation and destruction; and determinations of intrusive (deep-seated) bodies, reverse and normal faulting, unconformities, and coking characteristics of coal.

Organic matter is highly susceptible to small changes in temperature, which result in the irreversible reaction of maturation. Maturation parameters, therefore, are dependent on maximum temperature (related to heat flow in the basin) and effective heating time. Because of variability in organic facies, however, each maturation parameter is related to reaction kinetics of individual or bulk organic matter. Depending on temperatures and chemical ordering owing to reaction kinetics, there are four processes or zones of maturation: diagenesis, catagenesis, metagenesis, and metamorphism (Larsen and Chilingar, 1979; Teichmüller and Teichmüller, 1979; Tissot and Welte, 1984).

At the present state of art, there are more than twenty important maturation parameters, including chemical parameters (carbon content, aromaticity, etc.); geochemical parameters (T_{max}, sterane isomerization, methylphenanthrene index, etc.); and microscopic parameters (vitrinite reflectance, thermal alteration index, spectral fluorescence, etc.; Hunt, 1979; Tissot and Welte, 1984). Table 9-1 shows the most important microscopic parameters.

Of the various maturation parameters, vitrinite reflectance, spectral fluorescence, and laser fluorescence utilize incident-light microscopy. The present chapter focuses on these three selected incident-light microscopic parameters (especially vitrinite reflectance) in order to illustrate in detail the methods, problems, limitations and applications. A brief description of all other microscopic parameters are given in order to help correlation of various microscopic maturation parameters. Vitrinite

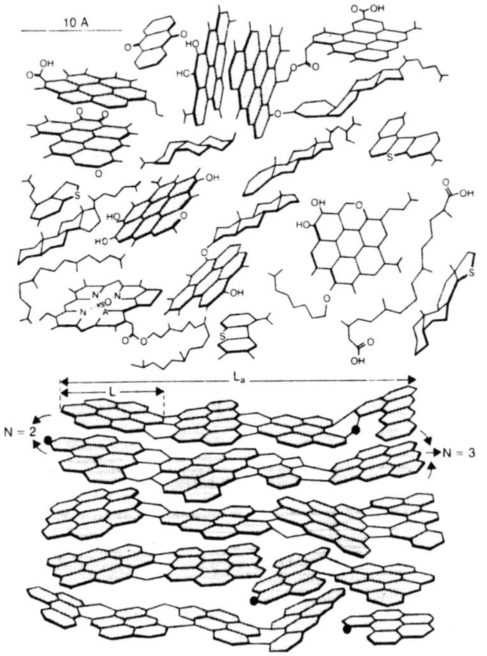

Fig. 9-1. Molecular structure of coals in two rank stages (A – lignite – subbituminous, B – anthracite) showing the changes affecting layered aromatic structural units. N = number of layers per stack, L = diameter of aromatic units. (After Teichmüller, 1987a.)

TABLE 9-1

Important microscopic parameters with reflected-light and transmitted-light microscopy

Methods	Measurement object
Reflected-light microscopy	
(1) Vitrinite reflectance (% R_0, mean of random or maximum reflectance)	Vitrinite
(2) Graptolite reflectance	Graptolite
(3) Chitinozoa/Scelecodont reflectance	Chitinozoa and Scelecodont
(4) Solid bitumen ("migrabitumen") reflectance	Various solid bitumens
(5) Spectral fluorescence	Sporinite, cutinite, alginite, resinite, suberinite, groundmass, bitumen, crude oil, condensate and solid bitumen
(6) Absolute fluorescence intensity	Sporinite, alginite and solid bitumen
(7) Pulsed laser fluorescence	Sporinite, resinite, alginite, crude oil and condensate
Transmitted-light microscopy	
(1) Thermal alteration index (*TAI*)	Sporinite (spores and pollen)
(2) Conodont alteration index (*CAI*)	Conodont
(3) Translucency of palynomorphs	Sporinite (spores and pollen)

reflectance is determined for the specific maceral vitrinite (kerogen part) in dispersed organic matter and coal. On the other hand, spectral fluorescence and laser fluorescence can be used for the determination of maturity of both kerogen (explained later) (mainly on individual maceral group like liptinite) and bitumen fractions of coal and dispersed organic matter as well as maturity of crude oil or condensate.

The objectives of this chapter are to:

(a) illustrate the complexity, usefulness and limitations of vitrinite reflectance measurements using both coal and dispersed organic matter;

(b) demonstrate the advantages and limitations of spectral fluorescence parameters for vitrinite-lean kerogen, bitumen, crude oil and condensate;

(c) illustrate the relationship between vitrinite reflectance, spectral fluorescence and other maturation parameters; and

(d) introduce pulsed laser fluorescence for the evaluation of maturation of crude oils and condensates.

VITRINITE REFLECTANCE

In spite of its limitations and problems, vitrinite reflectance is still the most widely used maturation parameter as evidenced by the published geological literature, because it shows a more or less uniform change during maturation compared to liptinite and inertinite macerals. Vitrinite is present in most of the sedimentary rocks, and the reflectance of vitrinite was very well standardized by coal petrographers during the past 40 years (especially for the last 20 years). Vitrinite reflectance is measured from the physicochemical property of a specific maceral group vitrinite, found in both dispersed organic matter and coal. In recent years, the vitrinite reflectance measurement was oversimplified, which resulted in confusion for proper evaluation of the various geological phenomena in a sedimentary sequence. In order to properly evaluate vitrinite reflectance in a dispersed organic matter, it is necessary to know the origin of "vitrinite".

Origin and diversity of vitrinite

Both dispersed organic matter and coal are not homogeneous. Just as various minerals constitute a rock, organic-rich rocks contain several types of organic constituents which are called macerals (ICCP, 1971; Stach et al., 1982; Teichmüller, 1987b). In an organic rock, various macerals are intermixed with different minerals. Often the terms "maceral" and "kerogen" (in a dispersed organic matter) are used interchangeably. Kerogen is a chemical term indicating that the organic fraction is insoluble in organic solvent (Tissot and Welte, 1984), whereas macerals contain both kerogen and bitumen fractions (Mukhopadhyay et al., 1985a). During isolation of organic matter (or so-called kerogen) from a shale or carbonate by acid digestion and heavy liquid separation, however, the bitumen fraction is not totally extracted. Moreover, except for a few, most macerals do not contain more than 10% bitumen. For isolated organic matter, therefore, the terms kerogen and maceral can be used interchangeably.

According to petrographic and chemical properties, macerals are divided into three major groups: vitrinite, liptinite and inertinite. *Huminite* (low-rank counterpart of vitrinite which releases humic acid when treated with alkali) or *vitrinite* is derived mainly from the ligno-cellulose components of plants (mainly from woody tissue, bark, stem, etc.). Ligno-cellulose components can be derived from both arboreal plants (example: Nyssa, Taxodium, etc.) or non-arboreal plants (example: reed, grasses, floating marsh, weeds, sedges, etc.), from non-marine and marine environments and vegetations, or even from the phytoplankton source (non-marine, brackish and marine environments). *Liptinite* is mainly derived from plant and animal lipid, exine, cutine and suberin of non-marine, brackish, and marine environments. *Inertinite* is mainly derived from both ligno-cellulose and lipid components of plants and animals which are oxidized by water circulation, bacteria and forest fire. Huminite or vitrinite is formed via formation of fulvic and humic acid, and humin from the depolymerization of plant biopolymers. Lignin is mostly preserved in any environment; whereas cellulose is mostly destroyed during early diagenesis under normal conditions, except in some partially anoxic environments.

Huminite/vitrinite in coal

According to coal petrographic classification, based on morphology, the huminite/vitrinite maceral group has three subgroups: telinite (preserved plant cell structure), detrinite (detrital part of both arboreal and non-arboreal plants formed by mechanical or bacterial disintegration), and collinite (homogenized part of both

TABLE 9-2

Classification of huminite/vitrinite macerals in low- and high-rank coals (modified after Stach et al., 1982)

Brown coal – lignite				Bituminous coal		
Maceral group	Maceral subgroup	Maceral	Maceral type	Maceral type	Maceral	Maceral group
huminite	humotelinite	textinite		telinite 1	telinite	vitrinite
			texto-ulminite	telinite 2		
		ulminite				
			eu-ulminite			
				telocollinite		
			levigelinite			
		gelinite		gelocollinite	collinite	
			porigelinite			
	humocollinite					
		corpohuminite	phlobaphinite			
			pseudo-phlobaphinite	corpocollinite		
	humodetrinite	attrinite densinite		desmocollinite	vitrodetrinite	
	detromixinite	mixinite or humosapropelinite		saprocollinite		

arboreal and herbaceous plants formed by biochemical gelification). Table 9-2 shows variability of different vitrinite maceral and maceral types (submaceral) as seen in coal in low and high maturation (modified from Stach et al., 1982). For the definition of most of the macerals or submacerals, see appendix A (this chapter) or refer to Stach et al. (1982).

A new subgroup "detromixinite", which is intermediate between the vitrinite and liptinite groups has been introduced recently (Mukhopadhyay, 1989b). The reason is that fine-grained liptinitic matrix (lipid part of the plant and animal) is often associated with vitrinitic matrix forming these macerals. A new maceral (humosapropelinite/mixinite or saprovitrinite) are included in the subgroup detromixinite. Morphologically, these macerals are somewhat similar to huminite/vitrinite. Mixinite or humosapropelinite are distinctly different from attrinite/densinite or vitrodetrinite – they are more similar to desmocollinite. Plate 9-1 illustrates the morphology of some of the vitrinite macerals discussed earlier, such as texto-ulminite (Plate 9-1A), telinite (Plate 9-1C), telocollinite (Plate 9-1D), gelinite (Plate 9-1B), gelocollinite (Plate 9-1E), desmocollinite (Plate 9-1E), corpocollinite (Plate 9-1F), attrinite/densinite (Plate 9-1G), mixinite (Plate 9-1H), saprocollinite (Plate 9-1I), and bitumen-impregnated telocollinite (Plate 9-1J). For definition of these macerals and submacerals, see appendix A.

In spite of the variability of huminite and vitrinite macerals in coal, it is not difficult to select correct measurable huminite/vitrinite grains, because, in most cases, vitrinite constitutes more than 60% of all the macerals, and coal rarely contains a second-cycle vitrinite or vitrinite-like solid bitumen.

Huminite/vitrinite in dispersed organic matter

In dispersed organic matter, as in oil shales and petroleum source rocks, plant substances forming huminite/vitrinite macerals are complex. Huminite is formed from various complicated processes in shallow-marine, deep-marine, aquatic, and deltaic environments. For example, in a shallow marine environment, the plant substances forming huminite can be derived from bark, woods and roots of arboreal vegetation transported from fluvial swamp, or from herbaceous plants transported from the deltaic marsh, or from the grasses and sea weeds of shallow-marine environment. These huminite-forming first-cycle plants can be mixed with existing second-cycle (allochthonous) huminite or vitrinite derived from the erosion of pre-existing peat, lignite, or bituminous coal, or organic-rich shales. Apart from these two types of huminite or vitrinite, some secondary macerals (designated as solid bitumen or migrabitumen; Jacob, 1989), which have some morphological features similar to those of vitrinite, begin to be formed from the liptinitic macerals during catagenesis. These macerals occur either within or outside the main kerogen network. The macerals which occur outside the kerogen network are called solid bitumen; these macerals are generally hydrogen-rich. Typical examples of these macerals are albertite and ozocerite, which are formed in dispersed organic matter or exsudatinite in coal. The macerals, which are formed within the kerogen network become more hydrogen deficient than the associated primary macerals (mainly amorphous liptinite) and are semi-inert to inert in nature. Typical examples of these macerals are granular vitrinite, micrinite, and rank inertinite, etc. (Teichmüller and

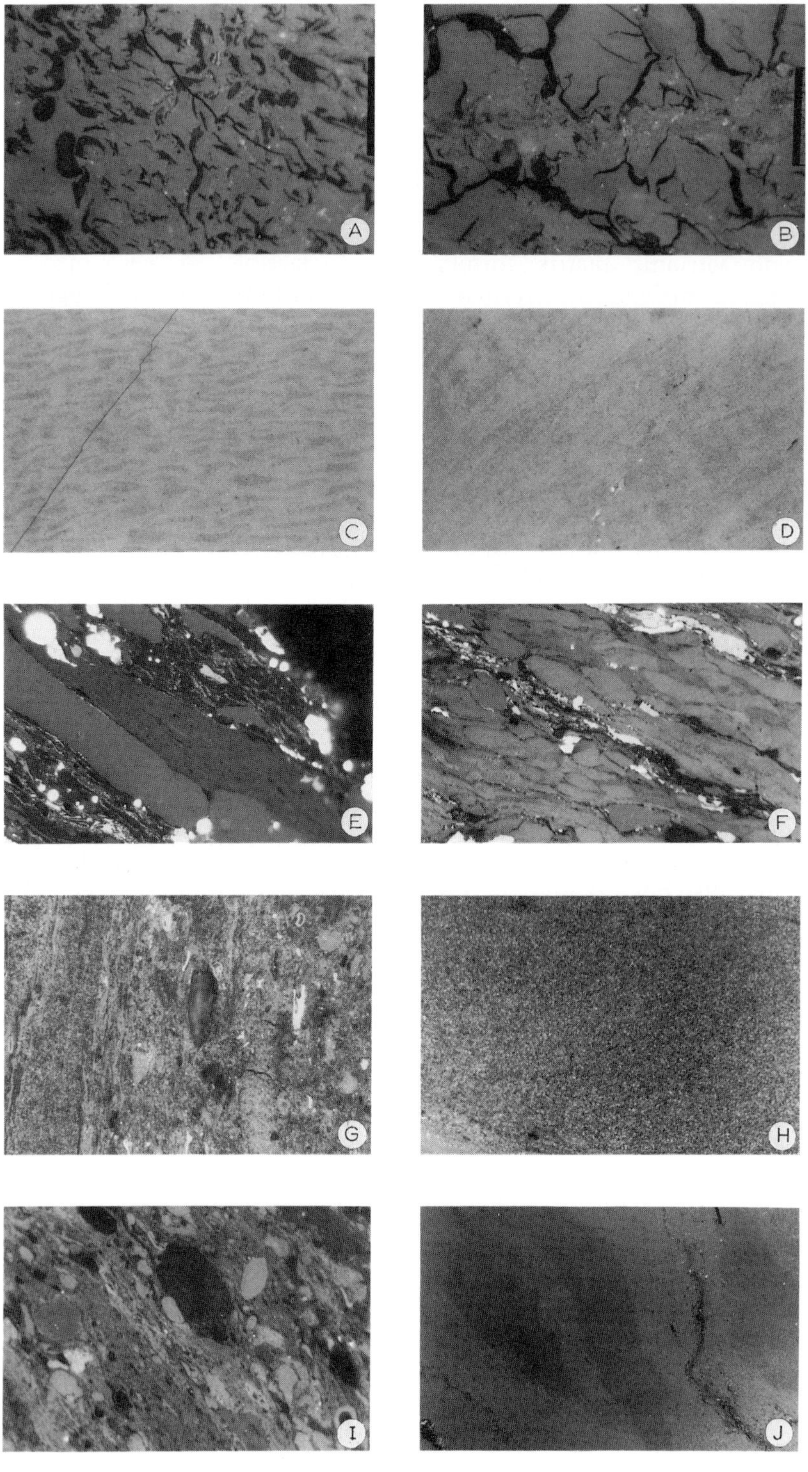
A
B
C
D
E
F
G
H
I
J

Ottenjann, 1977; Mukhopadhyay et al., 1985a).

Table 9-3 lists the different huminite/vitrinite macerals and vitrinite-like macerals observed in a dispersed organic matter. The list demonstrates the complexity of huminite/vitrinite macerals. Some researchers, therefore, emphasize the difficulty in identifying various vitrinite macerals in dispersed organic matter, because of lack of structure, morphology and fine-grained texture (Durand et al., 1987). For vitrinite reflectance measurements, only primary autochthonous (first-cycle) vitrinite grains must be used.

Vitrinite macerals are best identified in reflected white-light microscopy. In transmitted light, all vitrinite types are dark and do not show any variability (Plate 9-2C).

Plate 9-2 illustrates photomicrographs of various huminite/vitrinite and vitrinite-like macerals, such as primary ulminite in isolated kerogen (Plate 9-2A, B) and in whole rock (Plate 9-2E); first-cycle telecollinite in kerogen concentrate (Plate 9-2B); vitrinite in transmitted light (Plate 9-2C); gelinite or gelocollinite in isolated kerogen (Plate 9-2B, D); and gelocollinite in whole rock (Plate 9-2F); granular vitrinite (Plate 9-2B); second-cycle telocollinite and gelocollinite (Plate 9-2A, B, and F); second-cycle corpocollinite (Plate 9-2F); albertite (Plate 9-2G); and asphalt (Plate 9-2H).

Sample preparation

Figure 9-2 represents a schematic presentation of samples prepared for the vitrinite reflectance measurement. For the measurement of vitrinite reflectance, coal samples are crushed to −20 mesh, impregnated in epoxy resin, ground, and polished according to the method described by Stach et al. (1982).

For vitrinite reflectance measurement in dispersed organic matter three types of preparation can be made:
(1) polished whole-rock pellet,
(2) polished pellet of isolated kerogen, and
(3) polished smear slide (without cover glass) of isolated kerogen (Amoco type).

PLATE 9-1
Photomicrographs of various huminite/vitrinite macerals in coal in normal reflected light (scale indicates 50 μm).
(A) Texto-ulminite, Eocene lignite, Wilcox Group (Texas); R_o = 0.33%.
(B) Gelinite, Eocene lignite, Wilcox Group (Texas); R_o = 0.35%.
(C) Telinite, Carboniferous high-volatile coal, Cumberland Coalfield (Nova Scotia); R_o = 0.75%.
(D) Telocollinite, Carboniferous high-volatile bituminous coal, Cumberland Coalfield (Nova Scotia); R_o = 0.9%.
(E) Gelocollinite and desmocollinite, Carboniferous high-volatile bituminous coal, Cumberland Coalfield (Nova Scotia); R_o = 0.75%.
(F) Corpocollinite and gelocollinite, Carboniferous high-volatile bituminous coal, Cumberland Group (Nova Scotia); R_o = 0.9%.
(G) Attrinite and densinite, Eocene lignite, Wilcox Group (Texas); R_o = 0.35%.
(H) Mixinite, Eocene lignite, Jackson Group (Texas); R_o = 0.30%.
(I) Saprocollinite and corpocollinite, Eocene sub-bituminous coal, Claiborne Group (Texas); R_o = 0.49%.
(J) Suppression of vitrinite reflectance in gelocollinite due to association of alginite and cutinite, Carboniferous bituminous coal, Cumberland Basin (Nova Scotia); R_o = 0.9%.

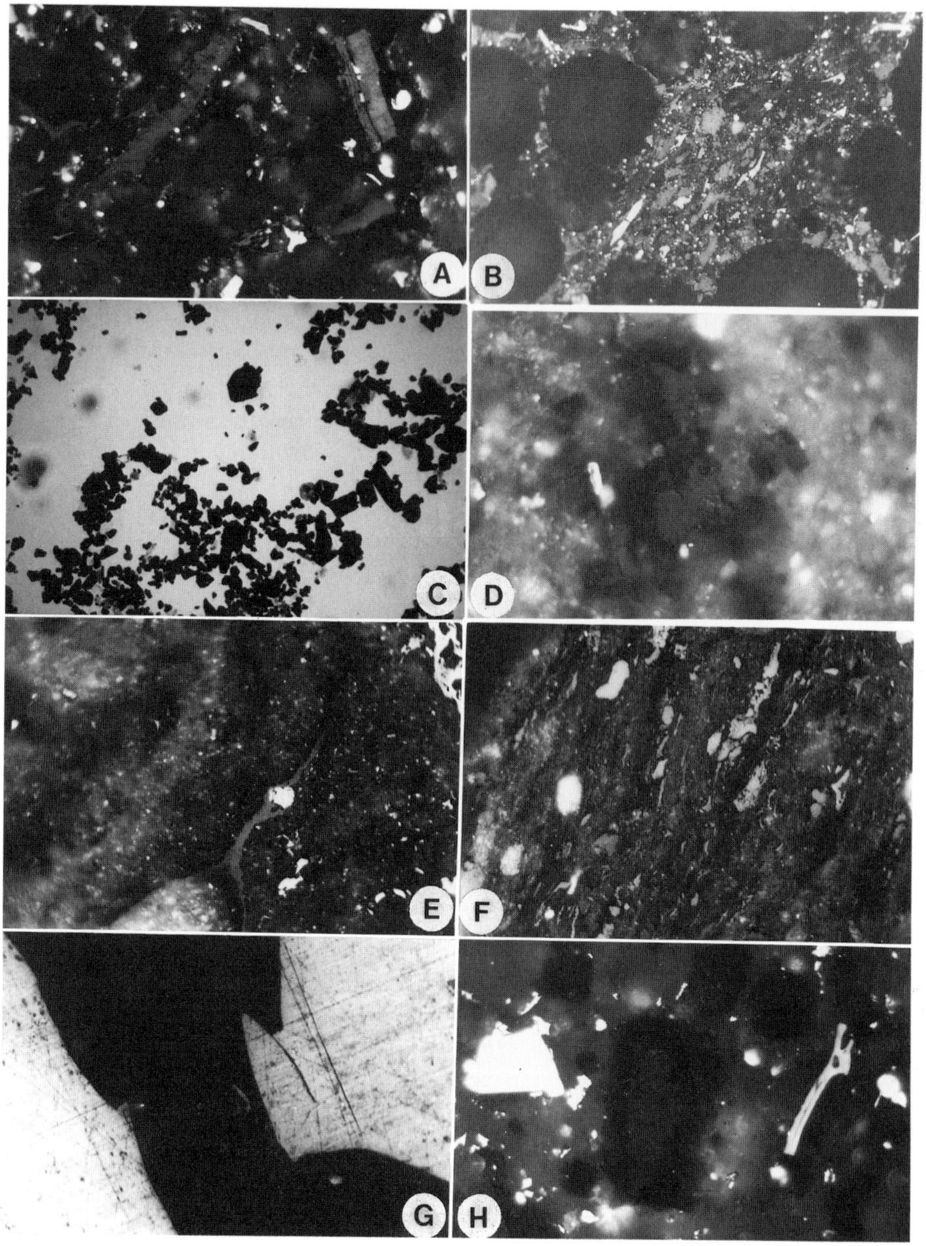

PLATE 9-2
Photomicrographs of huminite/vitrinite macerals in dispersed organic matter; all samples (except 3C) are in white reflected light.

(A) First-cycle ulminite and second-cycle telocollinite. (Kerogen concentrate; Jurassic, Scotian Shelf; × 500; R_o = 0.35%.)

(B) First-cycle telocollinite and gelocollinite, second-cycle telocollinite and gelocollinite, and granular vitrinite (rank-inertinite). (Kerogen concentrate, Jurassic, Scotian Shelf; × 500; R_o = 0.80%.)

For the first type of preparation, whole-rocks are crushed to −20 to +40 mesh or −20 mesh, impregnated in epoxy resin and ground and polished similar to the method adopted for coal (Stach et al., 1982). For the second and third types of preparation, organic matter is isolated using, successively, HCl, HF, centrifugation, and heavy liquid separation (using $ZnBr_2$ at a specific gravity of 1.9 g cm^{-3}) Durand, 1980; Mukhopadhyay et al., 1985a). During kerogen isolation, it is important to take every precaution not to cause any oxidation of the organic matter. In both techniques 2 and 3, isolated organic matter is impregnated in cold-set epoxy resin either in a pellet form, concentrating the samples in a small area (technique 2) or in a glass slide (technique 3). After impregnation, grinding and polishing are done using a slide holder for technique 3, and automatic polishing machine for both

TABLE 9-3

Classification of huminite/vitrinite macerals (both primary and secondary) and vitrinite-like macerals observed in dispersed organic matter

	Diagenesis $R_o < 0.55\%$	Catagenesis $R_o > 0.55\%$
Important huminite/vitrinite for reflectance measurement		
(1)	Textinite**	Telinite**
(2)	Texto-ulminite/ulminite*	Telocollinite*
(3)	Gelinite*	Gelocollinite*
(4)	Corpohuminite**	Corpocollinite**
(5)	Attrinite/densinite**	Desmocollinite**
		Vitrodetrinite+
Unimportant huminite/vitrinite for reflectance measurement		
(1)	2nd-cycle huminite+	2nd-cycle vitrinite+
(2)	Mixinite or humosapropelinite+	Saprovitrinite+
Problematic macerals for huminite/vitrinite reflectance		
(1)	Solid bitumen+	Solid bitumen+
(2)		Rank-inertinite+ or granular vitrinite+
(3)	Bitumen-impregnated huminite+	Bitumen-impregnated vitrinite+
(4)	Huminite derived from cavings+	Vitrinite derived from cavings+

* Huminite/vitrinite used for the calculation of mean reflectance.
** Only used when neither telocollinite nor gelocollinite are present.
\+ Macerals should be discarded in the calculation of mean huminite/vitrinite reflectance.

PLATE 9-2 (continued).
(C) Vitrinite in transmitted light. (Kerogen concentrate, Jurassic, Scotian Shelf; × 200; R_o = 1.2%.)
(D) First-cycle gelinite in whole rock. (Cretaceous, Scotian Shelf; × 500; R_o = 0.35%.)
(E) First-cycle ulminite in association with amorphous liptinite; whole rock. (Woodford Shale, Oklahoma; × 500; R_o = 0.45%.)
(F) Second-cycle gelocollinite and corpocollinite in association with amorphous liptinite; whole rock. (Messel Shale, West Germany; × 500; R_o = 0.3%.)
(G) Albertite; whole solid bitumen impregnated in epoxy resin. (Carboniferous, New Brunswick, Canada; × 500; R_o = 0.7%.)
(H) Asphalt in isolated kerogen. (Jurassic, Scotian Shelf; × 500; R_o = 0.75%.)

techniques 2 and 3, adopting a method similar to that of Stach et al. (1982) and Bustin et al. (1985). Some laboratories make plugs of isolated kerogens from three or four samples (Fig. 9-2) in one polished pellet for cost effectiveness. Polishing of kerogen depends on the variation in kerogen type and maturation. Some researchers use kerogen smear slide (covered with glass slide) without much polishing (Schwab, 1985; Lo, 1988). This method is not acceptable for vitrinite reflectance, because the quality of the reflectance measurement is dependent on proper polishing.

Vitrinite reflectance measured on whole-rock is often known as the "European Method", because it was developed by coal petrographers in Europe and on polished isolated kerogens as the "North American Method", as it was developed by the American/Canadian organic petrographers (especially in oil companies) (Bostick,

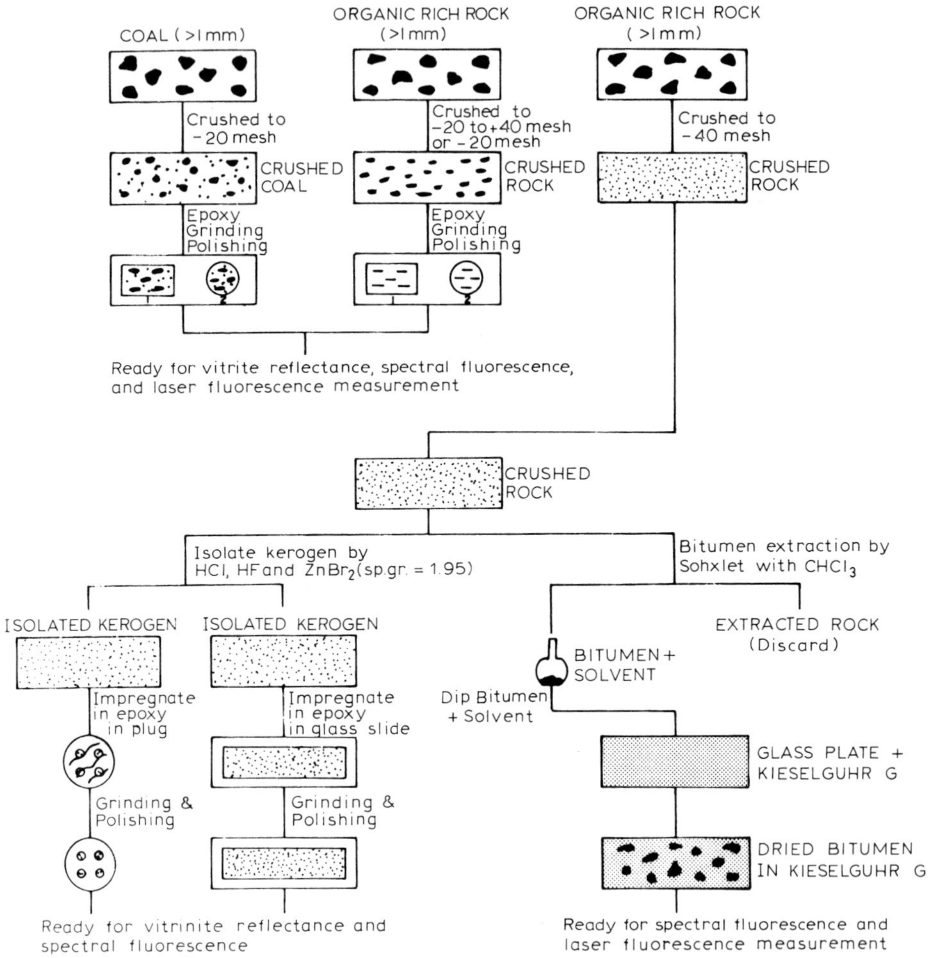

Fig. 9-2. Schematic diagram of preparation procedure of samples for vitrinite reflectance and fluorescence measurement.

1973; Castano and Sparks, 1974; etc.). Both methods have advantages and disadvantages. In an organic-rich sediment containing Types IIA – IIB (source rock between classical kerogen Type II and III), IIB, and III kerogen (Mukhopadhyay, 1989a), whole-rock preparation is most useful inasmuch as one can clearly differentiate autochthonous and allochthonous populations and solid bitumen, and bitumen-impregnated vitrinites. On the other hand, in the case of an organic-lean sediment and a typical Type-IIA kerogen (anoxic marine), isolated kerogen plug preparations may be more useful; because they result in more measurable grains. In the case of typical Types-I and -IIA (classical Type II; Tissot and Welte, 1984) kerogen, however, both types of preparation should be used if possible.

Principles, instrumentation and important requirements

Reflectance of vitrinite is measured as the percentage of the incident light intensity which is reflected from the polished surface of vitrinite relative to a standard substance (such as glass, sapphire, etc.). This measurement uses immersion oil and is related to the refractive index and absorptive index of immersion oil and standard which follows the Fresnal – Beer equation:

$$R_o = \frac{(\mu - \mu_o)^2 + \mu k^2}{(\mu + \mu_o)^2 + \mu^2 k_o^2} \quad (1)$$

where:

μ, μ_o = refractive index of vitrinite and immersion oil, respectively;
k, k_o = absorption index of vitrinite and immersion oil, respectively.

$$\% R_o \text{ (standard)} = \frac{(\mu_{standard} - \mu_{oil})^2}{(\mu_{standard} + \mu_{oil})^2} \times 100 \quad (2)$$

where μ = refractive index.

The earliest use of vitrinite reflectance as a rank parameter was made by Hoffmann and Jenkner (1932). During this early period, the Berek Photometer (a two-beam system) was used for vitrinite reflectance measurements. A detailed description of the Berek Photometer instrumentation was given by Davis (1978) and Bustin et al. (1985). McCartney (1952) introduced vitrinite reflectance measurement by the photomultiplier system (using a single-light beam), which after several modifications became the standard technique used at the present time. Details of the microscope system are discussed in Davis (1978), Stach et al. (1982), Bustin et al. (1985), and Robert (1985). During 1950 – 1970, some coal petrographers measusured vitrinite reflectance in air instead of oil and later calibrated to oil immersion (Stach et al., 1982; Bustin et al., 1985). The modern method of vitrinite reflectance determination, however, involves measurement under immersion oil only.

Vitrinite reflectance is measured by the ratio of reflected light to incident light (using an oil immersion objective), with an incident-light microscope fitted with a halogen lamp with stabilized power supply, a photomultiplier, a digital display unit with or without a computer, and a 546 nm (green light) filter in the measuring path. It also uses a six-in-one glass standard (reflectance between 0.3 and 1.68% R_0) or sapphire standard and an immersion oil having a refractive index of 1.516. Accordingly, the reflectance of vitrinite is measured as a percentage of reflectance in oil immersion compared to a known standard. The following five points should be taken into consideration before measurement of vitrinite reflectance, because each one can create errors in reflectance measurement: (a) polishing, (b) instrument standardization, (c) cleaning of the standard, (d) choice of the objective and aperture diaphragm, and (e) choice of the vitrinite grains.

(a) The polished pellets of coal or dispersed organic matter should be scratch-free and without relief. Vitrinite reflectance measured on micro-scratches can lower the mean vitrinite reflectance by 0.1 – 0.2% R_0. Before measurement, the polished sample should be kept in a desiccator for at least four hours, because moisture can lower the reflectance value.

(b) The measurement should be made in a uniformly dim-lighted or partially dark room. The linearity of the instrument should be checked at least once every day by using two or three different reflectance standards. Before reflectance measurements are made, all parasite reflectance and zero correction should be checked. After every 50 measurements, the instrument should be calibrated once again with a known standard.

(c) The reflectance standard should be kept in a dust-free place. Once every week, the standard (especially a glass standard) should be cleaned by Kodak photoflo solution with distilled water, cotton balls and compressed air, after wiping the immersion oil from the top of the standard with a soft tissue paper.

(d) Generally a 25× oil immersion objective is used for coal and 40 – 50× objective for dispersed organic matter. The measuring diaphragm should be 10 – 20 micrometers for coal and 5 – 15 micrometers for dispersed organic matter depending on the nature of the particle. For dispersed organic matter, it is very important that the measuring diaphragm is not larger than the measured vitrinite grain. If the measuring diaphragm is changed because of grain size, the instrument should be calibrated with the known standard.

(e) In dispersed organic matter, varieties of huminite or vitrinite macerals, maceral types (submacerals), and vitrinite-like macerals are generally observed. Table 9-3 indicates that nomenclature used for huminite/vitrinite macerals below 0.55% R_0 (lignite to subbituminous in rank – possible range of diagenesis) are different than those above 0.55% R_0 (bituminous to anthracite in rank – possible range of catagenesis). Some of the macerals (example: solid bitumen) are not within the vitrinite group; however, their morphology sometimes shows close similarity with vitrinite submacerals. For description of these macerals, see appendix A.

In coal, second-cycle vitrinites are not present (except for some Gondwana coals). On the other hand, pseudovitrinites are common in coal.

The following trends of reflectance of important huminite/vitrinite macerals are common: reflectance of pseudovitrinite or second-cycle vitrinites > corpocollinite > gelocollinite > telocollinite > telinite > desmocollinite > saprovitrinite. Figure 9-3 illustrates the range of huminite reflectance between textinite and corpohuminite (Mukhopadhyay, 1989b). At higher maturation, this varia-

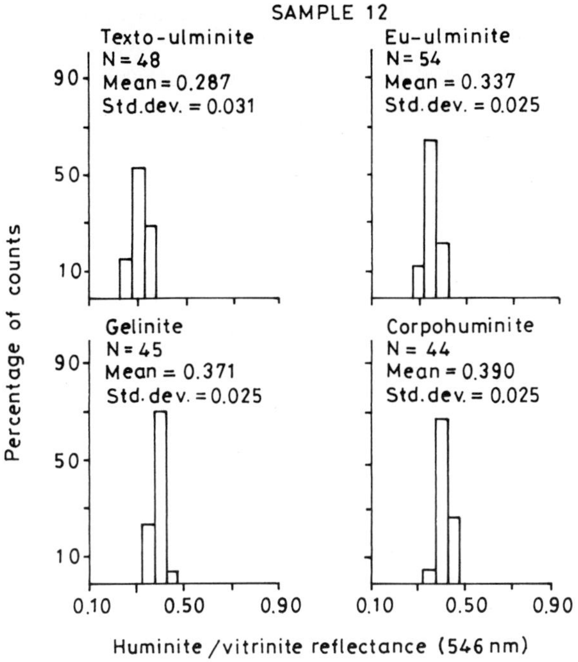

Fig. 9-3. Reflectance histograms for different huminite macerals from a typical lignite (R_o = 0.35%), Wilcox Group, Texas. (After Mukhopadhyay, 1989b.)

tion of vitrinite reflectance within various macerals of the vitrinite group increases two to three times. It is important, therefore, to measure only telocollinite or gelocollinite (if available) grains for determining the maturity of a source rock or coal. Second-cycle vitrinites will always have higher reflectance than first-cycle corpocollinite. On the other hand, saprovitrinite and bitumen-impregnated vitrinite will always have lower reflectance than desmocollinite below a maturity of 1.3% R_o. The reflectivity of solid bitumen is generally lower than desmocollinite and saprovitrinite below a maturity of 1.0% R_o (Jacob, 1989).

Two types of vitrinite reflectance are measured:

(1) random reflectance (% R_{random} or % R_o or % R_m), which uses no polarizer in the light path; and

(2) maximum reflectance (R_{max}) and minimum reflectance (R_{min}), which use a polarizer oriented at 45° to the plane of symmetry and rotating the specimen-holding stage through 360°.

The mean and standard deviation of both random and maximum reflectances are calculated measuring 50 – 100 points or grains (if available) in a sample. For the calculation of the mean and standard deviation, see Stach et al. (1982) and Robert (1985).

For the maturation study of dispersed organic matter, the mean of the random reflectance (% R_o) is preferred, because it is less time consuming, measures very small particles without problem, and is useful for measuring biaxial vitrinite (Hevia

Rodriguez, 1977; Teichmüller, 1987b). However, for coal, where larger vitrinite grains are usually measured, maximum and minimum reflectances are generally preferred beyond 0.7% R_o. Empirically, the following relation between random and maximum/minimum reflectances exists according to different authors:

(a) $R_{random} = \frac{2R_{max} + R_{min}}{3}$ (Ragot, 1977)

(b) $R_{max} = 1.066\ R_{random}$ (Ting, 1978)

(c) $R_{random} = \frac{1}{2}(R_{max} + R_{min})$ (Bustin et al., 1985)

Huminite/vitrinite are generally uniaxial, but progressively become anisotropic and, sometimes, biaxial due to load pressure and, thus, develop bireflectance. Bireflectance is calculated as the difference between R_{max} and R_{min} and is dependent on pressure rather than temperature.

As stated earlier, there are four stages of maturation of dispersed organic matter: diagenesis, catagenesis, metagenesis, and metamorphism. The diagenetic stage lies between 0.1 and 0.45/0.55% R_o; catagenesis, between 0.45/0.55 and 2.0% R_o; metagenesis, between 2.0 and 3.5/4.0% R_o; and metamorphism starts beyond 4.0 – 5.0% R_o. These four stages of maturation in dispersed organic matter are correlated to subbituminous, bituminous, anthracite, and meta-anthracite/semi-graphite stages in coal. In early diagenesis, various vitrinite macerals are distinctly different from each other in morphology. During diagenesis and catagenesis, the various huminite/vitrinite macerals are progressively homogenized by biochemical and geochemical gelification (Stach et al., 1982) and almost completely homogenized in the metagenesis stage. The measured reflectance values of autochthonous vitrinite grains assume a wider range, because of the progressive development of anisotropy and homogenization. Beyond 1.3% R_o, it is sometimes difficult to differentiate between telinite, telocollinite, gelocollinite, desmocollinite and saprocollinite.

Figure 9-4A shows the reflectance of coal in polarized light such as R_{max} and R_{min} compared to reflectance in non-polarized light, called R_{random} or R_{av} (Sharkey and McCartney, 1981) at various rank. Figure 9-4B illustrates the relation between R_{max} and R_m of coal at various rank stages (Teichmüller and Teichmüller, 1981). Both of these figures demonstrate the spreading out of vitrinite reflectance at a maturity beyond 1.5% R_o, because of anisotropy and measurement of different vitrinite macerals which lost identification criteria due to homogenization by gelification.

Vitrinite reflectogram and identification

The mean of the vitrinite reflectance for both dispersed organic matter and coal is generally calculated by plotting the frequency of huminite/vitrinite population against every 0.05% reflectance steps (called V-steps) in a histogram and selecting the autochthonous vitrinite population. For coal, all the measured vitrinite populations are considered for the calculation of the mean vitrinite reflectance and their

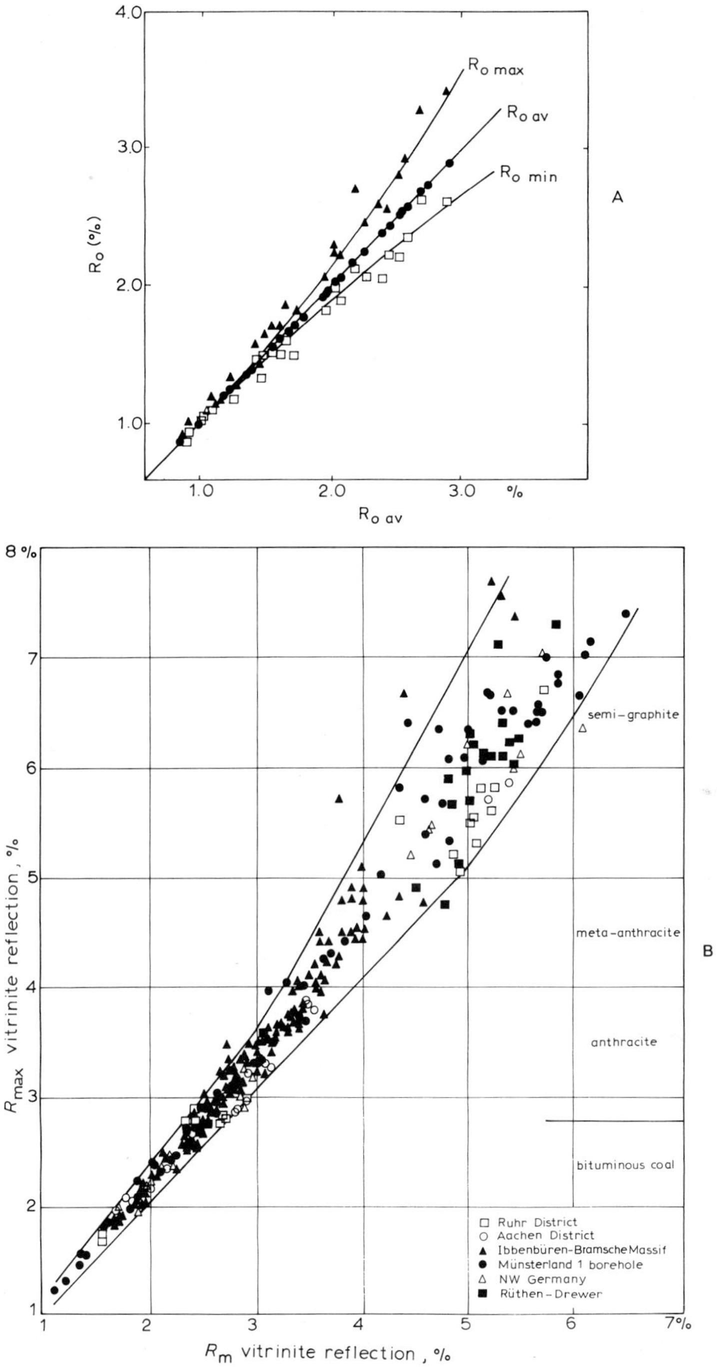

Fig. 9-4. Reflectance of coal in polarized light ($R_{o\ max}$ and $R_{o\ min}$) as a function of reflectance in non-polarized light ($R_{o\ av}$) or $R_{o\ random}$. R_m = mean random reflectance.
(A) After Sharkey and McCartney, 1981.
(B) After Teichmüller and Teichmüller, 1981.

standard deviation. Precaution should be taken to avoid measuring desmocollinite, saprocollinite and pseudovitrinite in the reflectance profile. Figure 9-5 illustrates three histograms and photomicrographs of huminite/vitrinite populations from three different rank coals: (A) lignite, (B) medium-volatile bituminous coal, and (C) semi-anthracite. The standard deviation and the nature of the histogram shows the increase of dispersion of vitrinite reflectance between 0.3 – 1.09% and 1.09 – 2.16% R_o. The photomicrographs show the progressive homogenization.

Morphologically, homogenization of vitrinite reflectance is difficult to detect in the dispersed organic matter because of the complexity and similarity of vitrinite and vitrinite-like macerals and their small size. In vitrinites studied in isolated

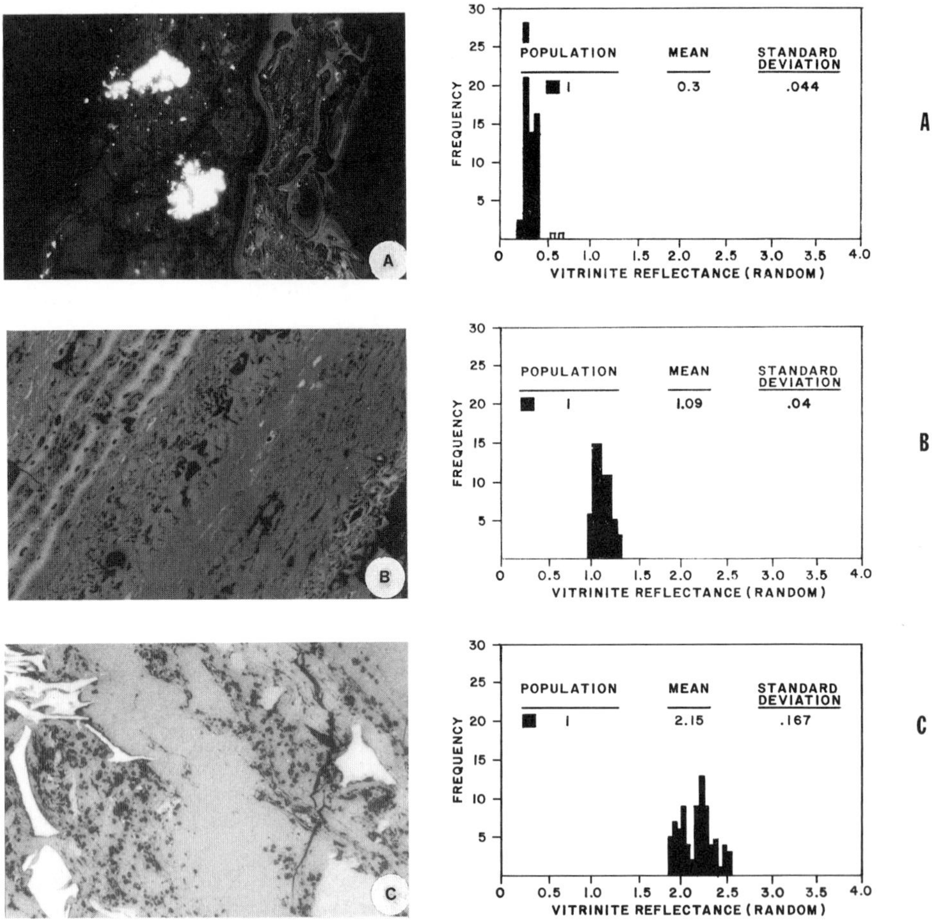

Fig. 9-5. Combination of photomicrographs and reflectance histograms (random reflectance) of vitrinite (mainly from telocollinite and gelocollinite) from coal in three maturation stages.
(A) Lignite from Eocene of Jackson Group, Texas; R_o = 0.3%.
(B) Medium-volatile bituminous coal of Barakar Formation, Permian, India; R_o = 1.09%.
(C) Semi-anthracite coal from Allan Hill, Permian, Antarctica; R_o = 2.15%.

kerogens having sufficient measurable grains, however, one can distinguish various macerals of vitrinite group (Plate 9-2). The reflectance histograms of vitrinite and vitrinite-like macerals in a dispersed organic matter can be divided into three populations (Figs. 9-6A, B). Cavings from the younger strata or mud additives in drill cuttings are difficult to recognize before selecting these three vitrinite populations; cavings and mud additives have to be excluded from the vitrinite reflectance profile before choosing these three populations. Whole-rock polished plugs are most suitable to identify cavings and drilling mud additives. Within the three populations in a reflectance histogram, the lowest reflectance is observed in the solid bitumen and bitumen-impregnated vitrinites and highest in the second-cycle vitrinites; first-cycle or autochthonous vitrinite reflectance lies in between. There may or may not be any clear boundary between these three populations. Reflectance of the autochthonous populations should be judged by viewing the morphology and reflectance of autochthonous inertinites and can be identified by their angularity, smoothness in color, and their lower reflectance than primary inertinites. Solid bitumen can be differentiated from autochthonous vitrinite by polishing difficulty, fluorescence, granular nature, and association of granular pyrite. Bitumen-impregnated vitrinite can be identified by fluorescence, presence of Newton's ring, smearing and sharp variation in reflectivity within one vitrinite grain. Allochthonous vitrinite is identified by its association with bimacerite grain (recycled coaly fragment), oxidation features, rugged or smooth edges, and variation in reflectivity in one grain, and its higher reflectance than autochthonous inertinite (especially sclerotinite; Mukhopadhyay et al., 1985b).

Some laboratories follow a different procedure in finding out the mean huminite/vitrinite reflectance. All vitrinite and vitrinite-like grains are measured and plotted on the histogram. The mean of the huminite/vitrinite reflectance is chosen according to the shape of the lowest measured vitrinites of discrete populations. This type of calculation may only be useful in a Type-IIB kerogen (after excluding the cavings and drilling mud additives), which contains enough first-cycle vitrinite and not much solid bitumen. In a typical oil-source rock of Type-IIA or -IIA – IIB kerogen (Type-II kerogen; Tissot and Welte, 1984), however, evaluation of the mean R_0 based on histogram and lowest reflectance is often misleading and shows either lower or higher reflectance than the autochthonous population. It has been observed in some deltaic and shallow marine sediments (shales and carbonates containing Type-IIB kerogen) that from one hundred measured vitrinite and vitrinite-like grains, only one or two grains of first-cycle vitrinite can be traced. In those cases, proper identification of first-cycle telocollinite (or gelocollinite) is essential and it needs more precise measurements than most observers are capable of.

Figures 9-6A and B demonstrate a possible variability of histograms for source rocks having typical Type-IIA (mostly marine organic matter) and Type-IIB (mostly terrestrial organic matter) kerogen. In Type-IIA kerogen (Type II of Tissot and Welte, 1984), not many first-cycle populations are observed, which contain more solid bitumen. On the other hand, with a typical Type-IIB (Type II – III of Tissot and Welte, 1984) kerogen, the source rock contains more vitrinite grains of different populations. It is, therefore, necessary to be careful when selecting a first-cycle vitrinite population in any one of the kerogen types.

Chemistry of vitrinite reflectance

As discussed earlier, physicochemical changes of vitrinite due to increasing organic maturity (such as changes in vitrinite reflectance and refractive and absorp-

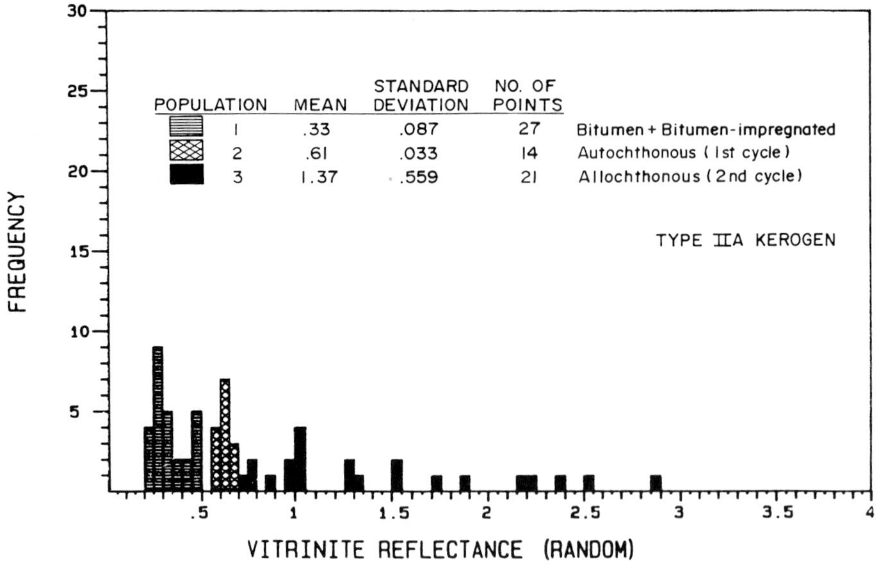

A

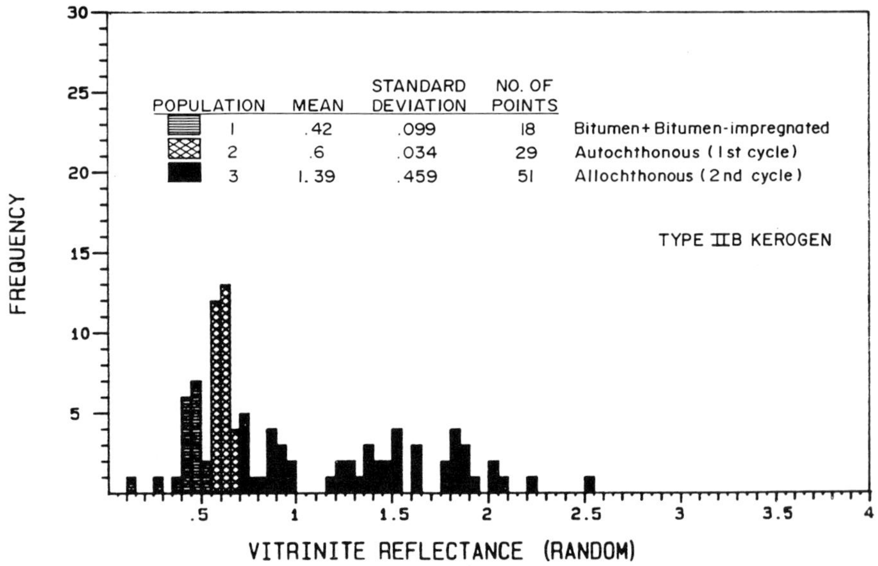

B

Fig. 9.-6. Typical vitrinite reflectance histogram including three populations of vitrinite or vitrinite-like macerals in dispersed organic matter.
(A) Type-IIA kerogen;
(B) Type-IIB kerogen.

tion indices, etc.), are related to an increasing aromatization and organization of vitrinite structure and an increase in aromatic-cell surface areas (Van Krevelen, 1961; Sharkey and McCartney, 1981). Sharkey and McCartney (1981) showed the increase of aromaticity, f_a, from 0.41 to 0.83; and of hydrogen related to aromatic structures, H_{ar}, from 0.07 to 0.50 in a series of pyridine-extracted coal having maturity of 61 – 90% carbon content (% maf).

The increase in vitrinite reflectance is related to three basic factors (Durand et al., 1987):

(a) the ratios of the atoms comprising vitrinite (mainly C, H, and O);
(b) the way the atoms are linked (chemical structure); and
(c) the way the chemical structures are spatially disposed (microtexture related to vitrinite anisotropy).

In the peat stage (R_o < 0.18%), huminite evolves by selective degradation of cellulose compounds and preservation of lignin-like compounds. As maturation advances to the lignite stage (R_o ranging from 0.18 to 0.36%), the lignin structural units (aromatic lamellae less than 10 Å in size) undergo a series of defunctionalization reactions (Hatcher et al., 1988). The first of these involves loss of methoxyl groups and replacement by hydroxyl groups. Maturation from lignite to the sub-

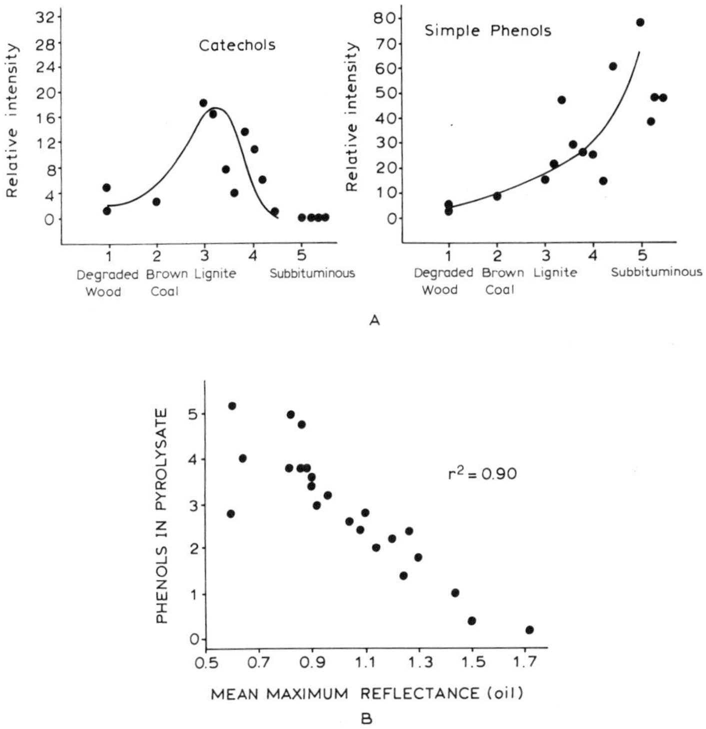

Fig. 9-7. (A) Sum of the relative intensities of catechols and simple phenols in pyrolysis-g.c. products of xylem tissue. (After Hatcher et al., 1988.) (B) Percentage of phenols in pyrolysate from vitrinite concentrate in relation to vitrinite reflectance (random); r = correlation coefficient. (After Larter, 1989.)

bituminous stage (R_0 = 0.36 – 0.55%) involves the replacement of phenolic hydroxyls by simple phenol and methylated phenols (Hatcher et al., 1988) Fig. 9-7A). At this stage, increase of vitrinite reflectance is related mainly to the loss of carbon dioxide and water from the phenolic structure. If the huminite particle is hydrogen-rich, there is no increase in reflectance.

During catagenesis, the increase of vitrinite reflectance is related to the loss of hydrogen by the removal of hydrogen in the functional groups and oxygenated aromatic species or from the cleavage of alkyl – oxygen bonds in aryl – alkyl ether (Larter, 1989). Senftle et al. (1986) and Larter (1989) showed the relation between the increase of vitrinite reflectance and loss of phenol in pyrolysate obtained from a pyrolysis experiment (Fig. 9-7B). This undergoes condensation – polymerization from dimerization of micro-biologically-degraded aromatic phenols. If vitrinite is hydrogen-rich (example: saprovitrinite), carbon – carbon bonds of the long-chain aliphatics are broken. At this stage, the basic structural units may be preferentially oriented because of load pressure. Anisotropy increases with a marked increase in vitrinite reflectance.

In metagenesis, the vitrinite becomes hydrogen- and oxygen-poor as a consequence of the release of all the functional groups and formation of polycondensed aromatic layers (up to 500 Å). Increase in lithostatic pressure results in orientation of aromatic clusters (Oberlin et al., 1980) and in an increase in vitrinite reflectance anisotropy.

In metamorphism, fusion of aromatic clusters, development of crystal growth, rapid loss of hydrogen, and the reorganization of vitrinite structure are the main features. These changes result in an extreme increase of vitrinite anisotropy and reflectance.

A schematic presentation of aromatic ordering in coal with R_0 ranging from 0.55% to >5% was demonstrated by Teichmüller (1987a).

Applications

Maturation boundaries

The most common use of vitrinite reflectance is, by plotting it against depth, to define the boundaries of maturation of a source rock or the rank of a coal. The empirical boundaries of maturation are often related to oil, wet gas, and dry gas generation. Figure 9-8 shows the vitrinite reflectance profile of Wilcox sediments according to depth from Texas Gulf Coast and its relation to present-day subsurface borehole temperatures. Figure 9-8 also shows another maturation parameter, T_{max} (°C) from Rock-Eval pyrolysis, correlated to vitrinite reflectance data. As discussed, the boundaries of various hydrocarbon generation zones (oil, wet gas, and dry gas) are shown. The sediments above the oil zone and below the dry zone are considered immature and metamorphosed. According to the present concept, major hydrocarbon occurrences do not exist in those two zones.

A typical mean vitrinite reflectance versus depth can be plotted either on a log scale (Fig. 9-8) or on an arithmetic scale (Fig. 9-9; Cardott and Lambert, 1985). Figures 9-8 and 9-9 (A and B) demonstrate the difference in the nature of the curve. Dow (1977) and Dow and O'Conner (1982) showed that R_0 data (plotted against depth) on a log scale usually fall close to a straight line, because the reaction rate for the increase in vitrinite reflectance (mainly aromatization and condensation reactions) follows an exponential increase. This situation is only true in the case of: (a) constant deposition rate, (b) constant paleoheat-flow, (c) constant thermal conductivity, and (d) uniform kerogen type (Falvey and Deighton, 1982), as in the case of a type area of Wilcox Group of Texas Gulf Coast (Fig. 9-8). On a R_0 versus depth plot, the reflectance can be plotted either with only the mean R_0 values (Figs. 9-8 or 9-9A) or with the mean R_0 values and standard deviations (Fig. 9-9B). For the determination of the reflectance gradient (with the best-fit curve), it is advisable to plot the mean R_0 and the standard deviation. Figure 9-9A shows the reflectance versus depth profiles (and their best-fit curves) in two boreholes of the Scotian Shelf. The heat flows and paleo/present-day temperatures of these two boreholes are the same. The variation in reflectance between the depths of 3000 m and 5500 m is possibly controlled by thermal conductivities determined by different lithologies

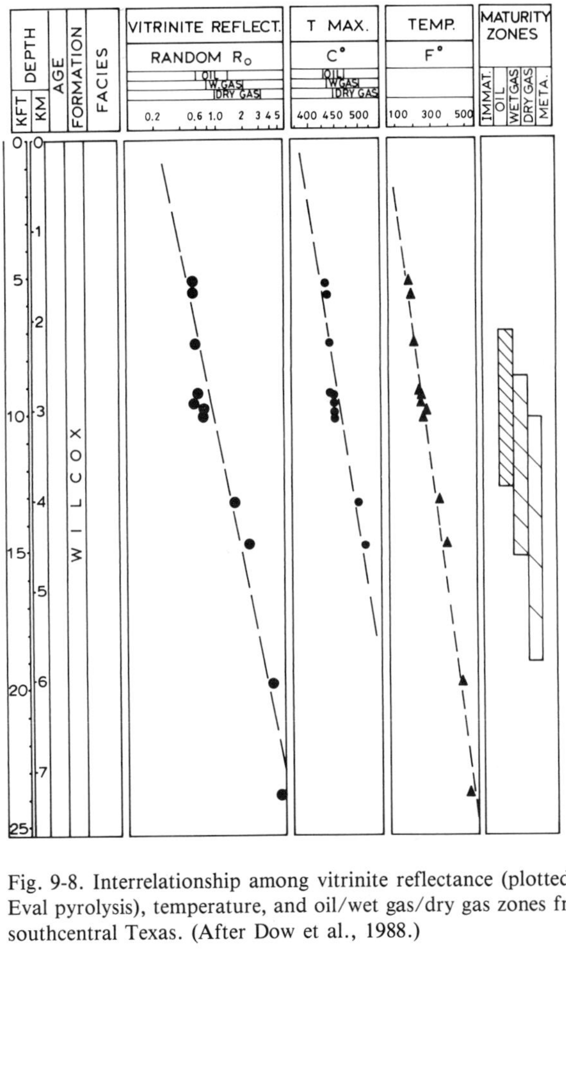

Fig. 9-8. Interrelationship among vitrinite reflectance (plotted on semilog scale), depth, T_{max} (Rock-Eval pyrolysis), temperature, and oil/wet gas/dry gas zones from the deep Wilcox (Eocene) sediments, southcentral Texas. (After Dow et al., 1988.)

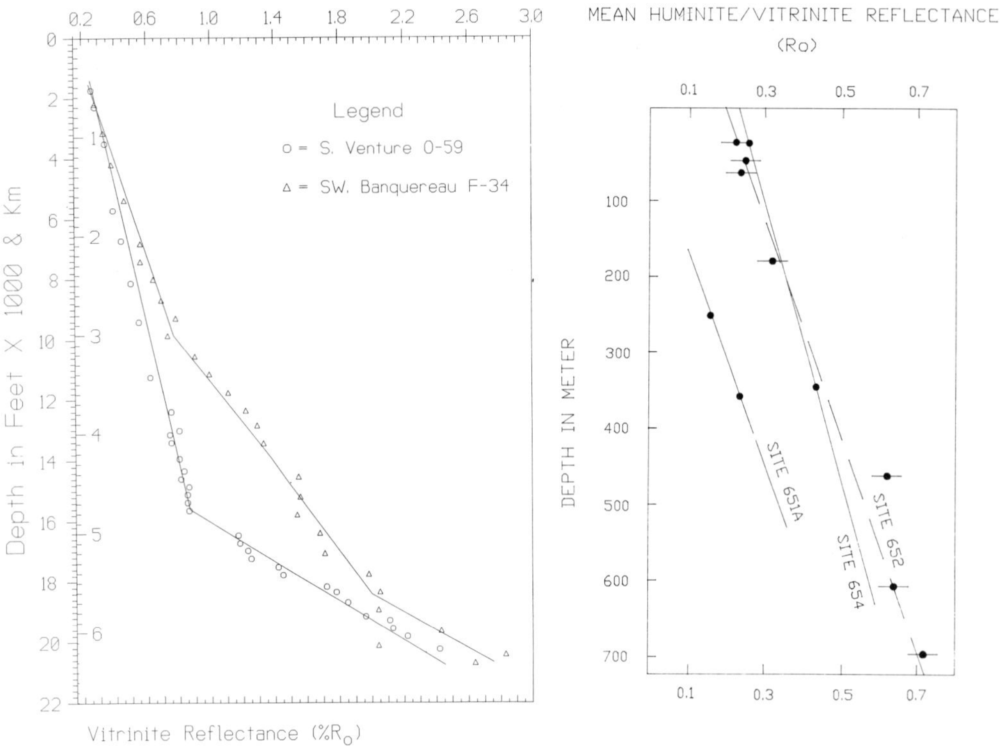

Fig. 9-9. Relationship between huminite/vitrinite reflectance and depth plotted on an arithmetic scale. (A) Mean reflectance plotted without standard deviation from two boreholes of the Scotian Shelf. (After M. Avery, pers. comm.)
(B) Mean reflectance plotted with standard deviation from the three boreholes of ODP Leg 107. (After Mukhopadhyay, 1990a.)

in two boreholes. A similar observation was made by Falvey and Deighton (1982; Fig. 9-10A). Figure 9-9B demonstrates the reflectance gradient in three boreholes from ODP Sites. Sites 651A and 652 have higher reflectance gradients than Site 654 because of a higher heating rate. In view of the complexities of vitrinite reflectance, it is preferable to plot vitrinite reflectance data on an arithmetic scale instead of a semilog scale. This will enhance opportunities to understand the problems of vitrinite reflectance.

The increase of vitrinite reflectance is dependent mainly on temperature and geological time. Temperature is dependent on the thinning and stretching of the continental crust, heat flux, convection, and thermal conductivity. The influence of time increases with temperature increase. For example, at temperatures below 50°C, the time factor can be neglected (Teichmüller, 1987a). A high heat flux does not yield high temperatures unless the system has an ability to retard heat flow (Yukler and Kokesh, 1984). Figure 9-10A shows vitrinite reflectance versus depth plots for two sets of samples (Falvey and Deighton, 1982). More resistive shales show higher paleoheat-flow and higher vitrinite reflectance. In a uniformly subsiding basin, present-day temperatures can be related to paleoheat-flow. Quigley et al. (1987) demonstrated a near linear relation between the present-day borehole temperature and measured vitrinite reflectance from various basins of the world (Fig. 9-11A). Teichmüller (1979) showed the relation between temperature and vitrinite reflectance in the upper Rhine Graben; the steeper reflectance gradient is related to the lower thermal gradient. Dow (1977) illustrated the influence of geological time on a series of Cretaceous to Pleistocene sediments from the Louisiana Gulf Coast (Fig. 9-11B). In this case, the reflectance gradient is steeper when the time effect is lower. A similar effect of geological time on vitrinite reflectance was observed in the Jurassic to Cretaceous sediments of the Scotian Shelf, where the temperature history is similar (Mukhopadhyay and Wade, unpubl. data). Yukler and Thompsen (1989) demonstrated that hydrocarbon generation from a source rock has a higher temperature dependency, whereas vitrinite reflectance is more time dependent. Using a chemical kinetic model, Burnham and Sweeney (1989) indicated that the relationship between the extent of oil generation and vitrinite reflectance is nearly independent of heating rate.

Vitrinite reflectance can measure temperatures over a range of 25 to 400°C and has an added advantage over other paleogeothermometers, such as clay mineralogy, fluid inclusion, fission-track annealing, and stable isotopes, because vitrinite reflectance reflects the maximum temperature to which the entire system has been exposed (Price and Barker, 1985).

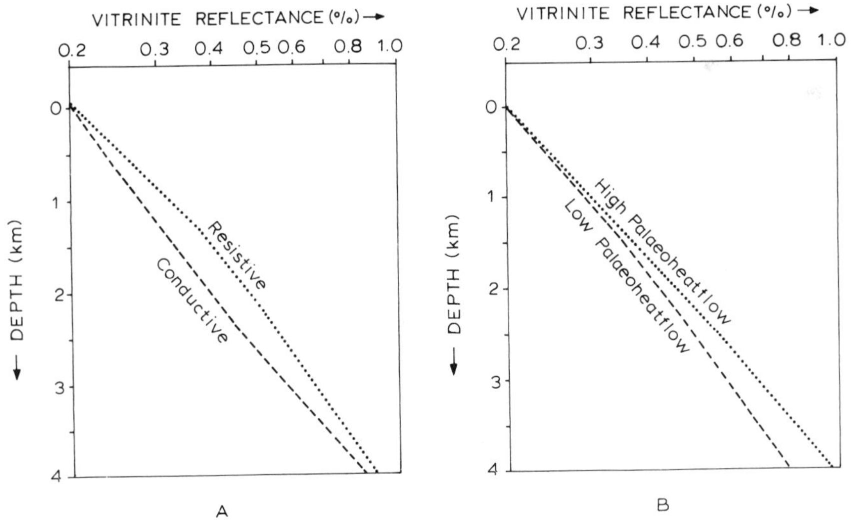

Fig. 9-10. Relation between vitrinite reflectance and depth (after Falvey and Deighton, 1982), showing the effect on vitrinite reflectance due to:
(A) resistive and conductive strata, and
(B) variation of paleoheat-flow.

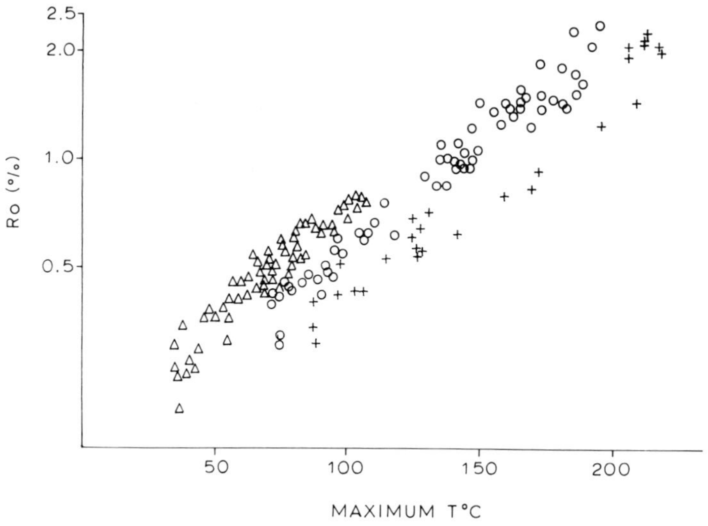

A

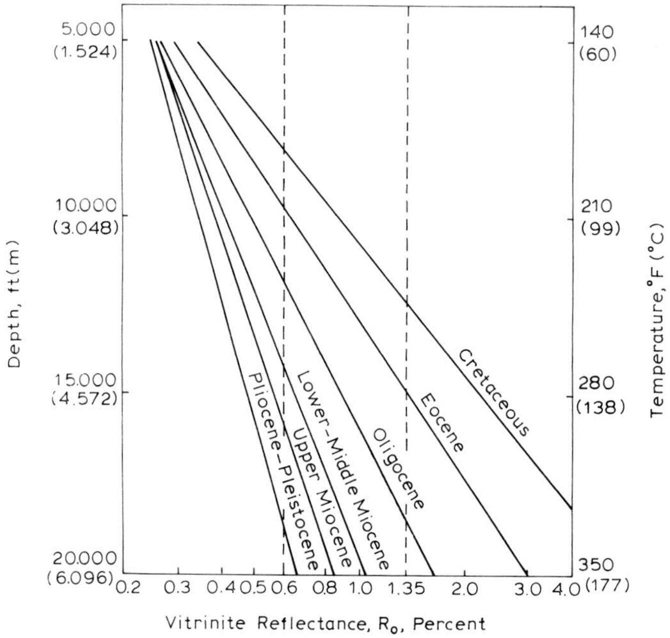

B

Fig. 9-11. Relationship between vitrinite reflectance and (A) present-day borehole equilibrium temperature (after Quigley et al., 1987); and (B) different geological times (after Dow, 1977).

Hydrocarbon generation

As discussed earlier, one of the important uses of vitrinite reflectance is to define the boundary of hydrocarbon generation of various kerogen types, although the kinetics of hydrocarbon generation (mainly from the liptinite macerals) and vitrinite reflectance are possibly different. Oil and gas are formed at elevated temperatures from the remains of dead organisms (biopolymer) via a very high molecular weight solid component (geopolymer) called kerogen. During advanced maturation, labile and refractory kerogen parts break down into oil and gas leaving behind the inert part of the kerogen (Quigley et al., 1987; Tissot and Welte, 1984). Figure 9-12 shows the hydrocarbon generation curves for different kerogen types in a plot of hydrogen index (Rock-Eval pyrolysis) versus vitrinite reflectance (%R_0) between 0.3 and 1.5% R_0. Figure 9-12 also shows the diagenesis and catagenesis boundaries. Hydrocarbon potential of various kerogen types below a maturation level of 0.5% R_0 shows a broad range (Fig. 9-12). Accordingly, as maturation advances to 1.3% R_0 the labile part of the kerogen loses hydrocarbons from the kerogen network. The hydrocarbon generation curves of various kerogen types thereby converge. The convergence of hydrocarbon generation curves is steeper for kerogen Types IIA and IIA – IIB than for kerogen Type III. This is due to the greater release of hydrocarbons from the oil-prone kerogens network (Types I, IIA, IIA – IIB of Mukhopadhyay, 1989a) compared to other less oil-prone or non-source rocks. Kerogen Type IIA of this diagram is comparable to kerogen Type II; and kerogen Types IIA – IIB and IIB are comparable to kerogen Type II – III of Tissot and Welte (1984). Kerogen of Types I and IIA behaves similar to kerogen Types IIB and III beyond R_0 greater than 1.3%. An "oil window" is generally considered to exist within a vitrinite reflectance of 0.5 to 1.35% R_0 (Pusey, 1973; Hunt, 1979; Tissot and Welte, 1984). Recent data on mathematical basin modelling (Yukler and

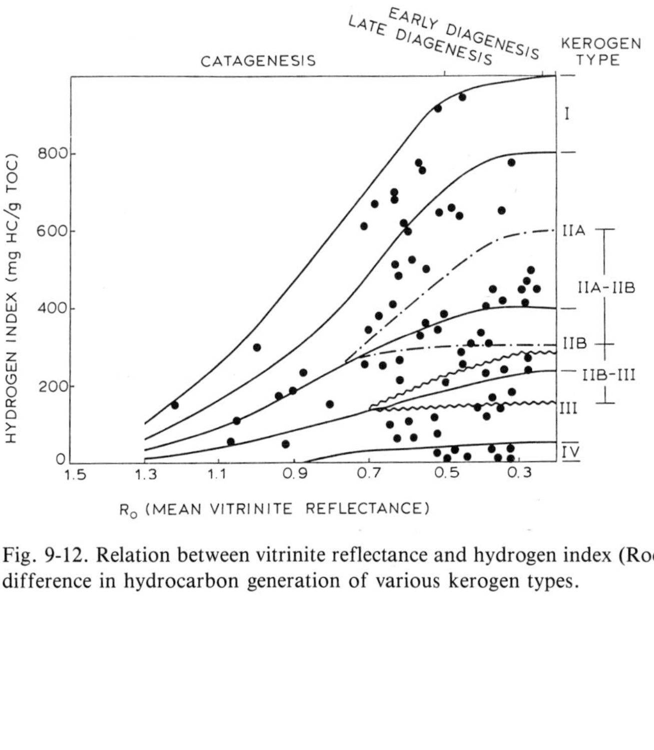

Fig. 9-12. Relation between vitrinite reflectance and hydrogen index (Rock-Eval pyrolysis), showing the difference in hydrocarbon generation of various kerogen types.

Kokesh, 1984; Tissot and Welte, 1984; Welte and Yalcin, 1988), however, indicate that the "oil window" concept is valid only in some restricted sense and can be changed according to heating rate, kerogen types, etc.

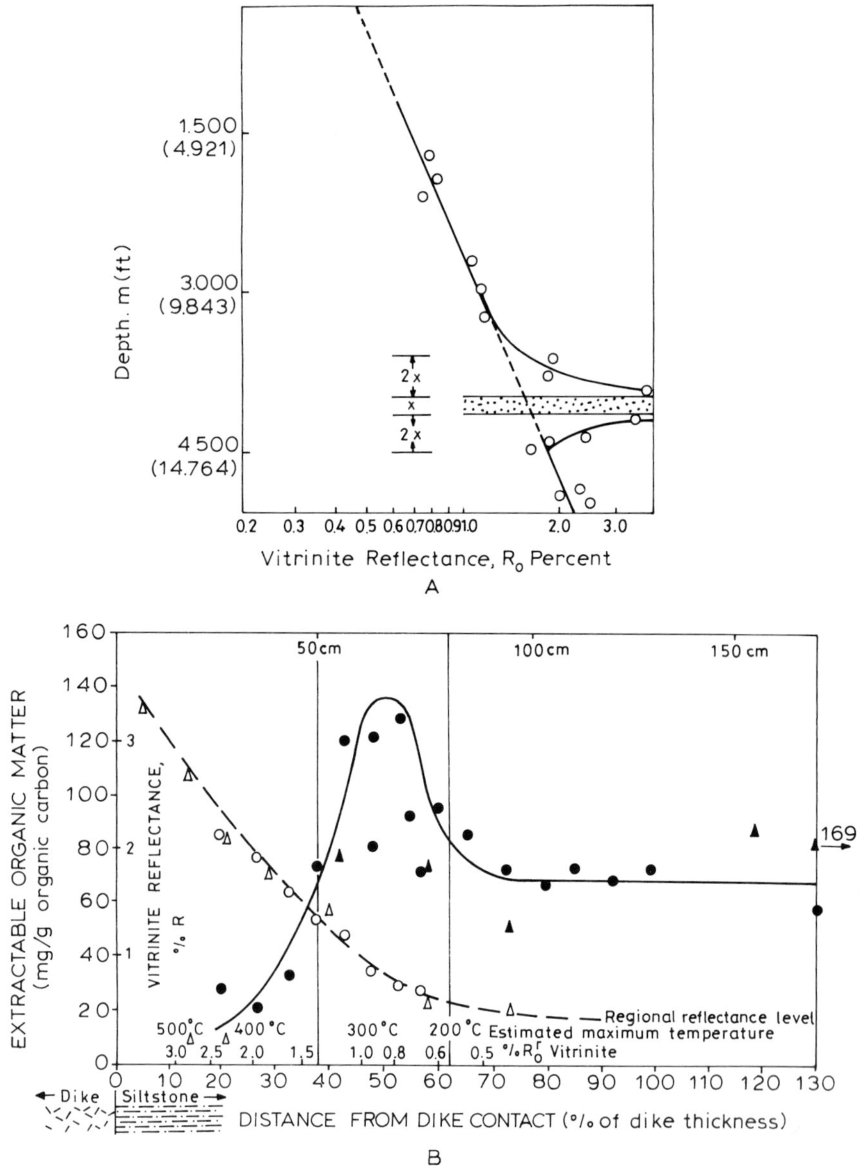

Fig. 9-13. (A) Relation between vitrinite reflectance and depth showing the effect of igneous intrusion. (After Dow, 1977.) (B) Interrelationship among extractable hydrocarbons, distances from the intrusive dyke, and vitrinite reflectance, showing the influence of the igneous intrusion on maturation parameters. (After Clayton and Bostick, 1986.)

Identification of geological phenomena

Intrusive bodies: Intrusive bodies (especially dykes) in contact with organic-rich sediments show pronounced effects resulting in a rapid increase in vitrinite reflectance (Dow, 1977; Clayton and Bostick, 1986; Figs. 9-13A and B). From the vitrinite reflectance data, the variation of temperature can be calculated (Clayton and Bostick, 1986; Fig. 9-13B). The change in vitrinite reflectance depends on the thickness of the intrusive bodies. As an example, near the contact zone with thicker intrusive bodies, an organic-rich sediment with 1.4% R_0 can be changed to 12% R_0 (Teichmüller, 1987a) or from 0.8% R_0 to 3.0% R_0 (Clayton and Bostick, 1986). In a deeper borehole, vitrinite reflectance could be helpful in identifying very thin intrusive bodies (even 30 to 40 cm thick) by the sudden increase in reflectance.

Unconformity: Vitrinite reflectance plotted against depth on a semilog scale may exhibit offset, which is often interpreted as caused by an unconformity (Dow, 1977). It was suggested that the offset can be used to estimate the thickness of the eroded section (Dow, 1977; Price and Barker, 1985). Katz et al. (1988), from time – temperature modeling, however, demonstrated no statistically significant difference in vitrinite reflectance across the erosional unconformity. They suggested further that offsets at great depths may be caused by igneous and metamorphic events and complications associated with editing and interpreting vitrinite reflectance.

Structural history: Faults can be recognized by abnormality of vitrinite reflectance if the throw is greater than 100 m. An empirical calculation of the throw of the fault can be calculated from the reflectance gradient (Dow, 1977; Teichmüller, 1987b). The use of vitrinite reflectance in some of the structural features from the Sydney, Bowen-Surat, and Gippsland basins in eastern Australia have been evaluated by Middleton (1982) using mathematical modeling.

Sediment near a shear zone or thrust plane can show an abnormal increase of vitrinite reflectance (3% from 0.8% R_0) due to the locally-generated frictional heat (Bustin, 1983).

A vitrinite reflectance against depth profile in a recently-studied borehole in the Swiss Alps shows an interesting phenomenon. Down to a depth of 2500 m, the reflectance progressively increases from 0.4 to 0.9% R_0. In the next 500 m interval, the vitrinite reflectance shows a sudden reversal (decrease) to 0.5 – 0.4% R_0. At depths greater than 3000 m, however, the sediments show a normal reflectance gradient from 0.9 to 1.3% R_0. The anomalous reflectance between 2500 and 3000 m is explained as an effect of a younger allochthon or thrust sheet. Vitrinite reflectance, in this case, detected the presence of a thrust sheet of younger origin.

A regional maturation pattern as evidenced by vitrinite reflectance of dispersed organic matter and coal in structurally complicated regions, such as in Alpine foredeep or at the Rocky Mountain front ranges and foothills, may provide evidence of crustal thicknesses and thrusting (Teichmüller and Teichmüller, 1979; Kalkreuth and Langenberg, 1986; Kalkreuth and McMehan, 1988). Figures 9-14A and B demonstrate how maturation is related to pre-, syn-, and postorogenic deformation. These data show the maturity difference between imbricated molasse of the Alpine foreland (very low reflectance gradient: 0.03 – 0.09% R_0 per km; Teichmüller, 1987a) and imbricated Hercynian molasse (0.5 – 1.0% R_0 per km). These variations in maturation gradient and present-day geothermal gradient are related to crustal

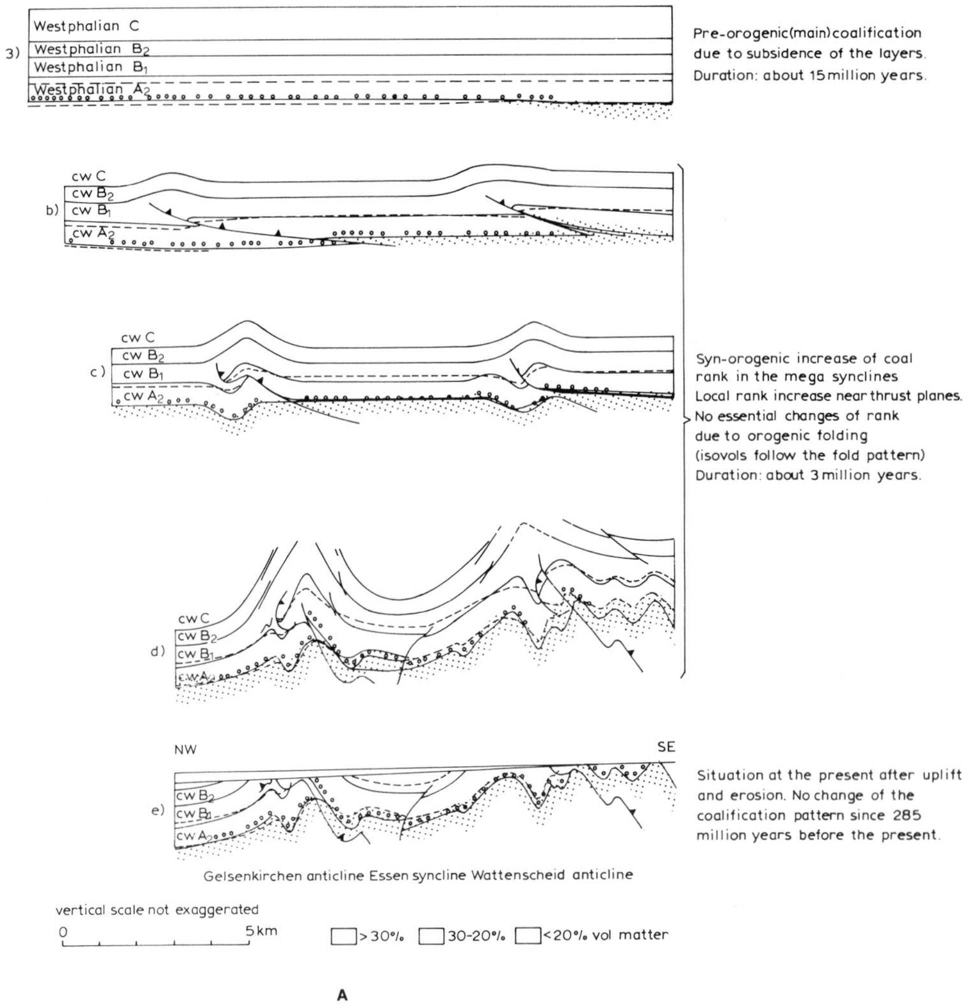

Fig. 9-14A. For caption see next page.

thickness. Higher crustal thickness lowers the maturation gradient because of low heat flux due to the convection and fluid movement. This type of regional maturation study related to structural deformation is significant for future oil and gas exploration, because one can expect some low-maturity source rocks in an overmature sequence as in the Rocky Mountain foothills of Alberta and British Columbia (Kalkreuth and McMechan, 1988) and imbricated nappe under the Po Plain (Teichmüller, 1987a). From the regional maturation study, it is evident that: (a) lower maturation in the foredeep is related to subsidence, compression, and thickening of the crust due to the low thermal stress; and (b) higher maturation in the backdeep is related to the uplift, tension, and plutonics, which, in turn, are related to higher thermal stresses.

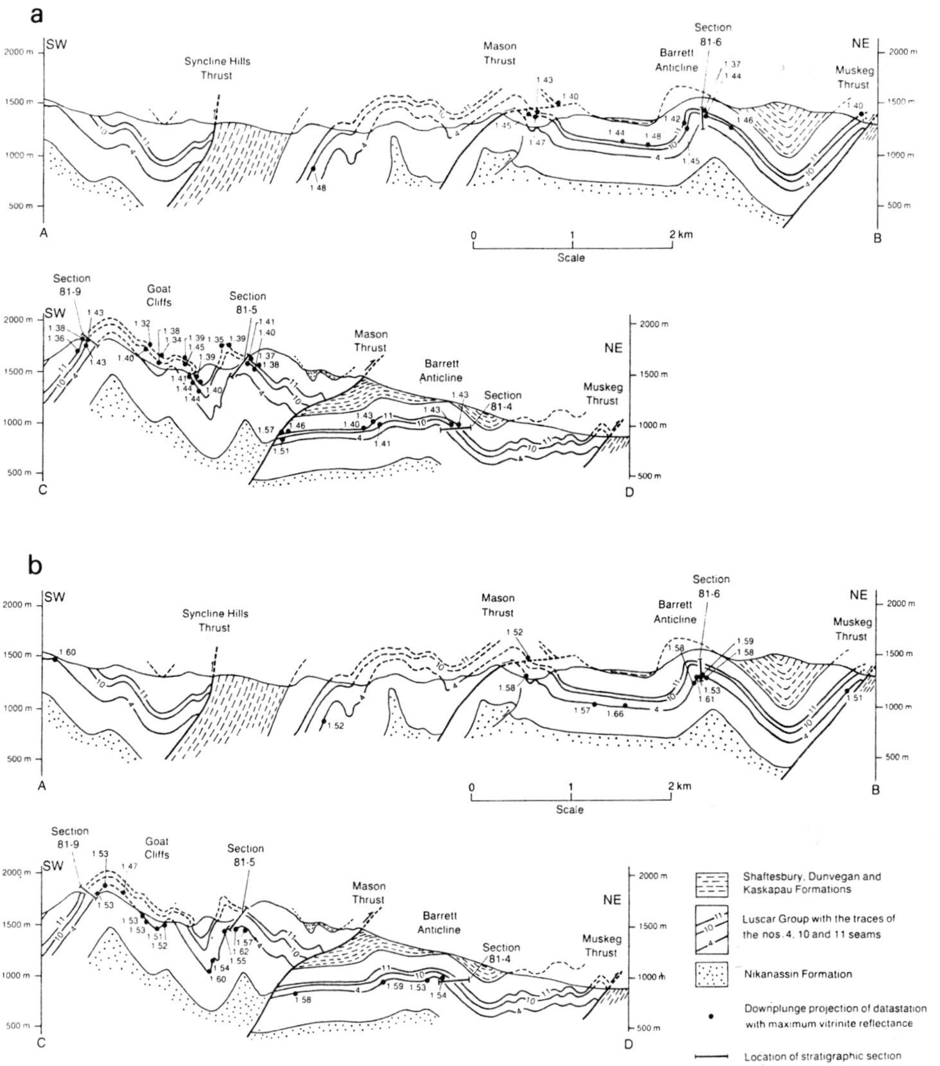

B

Fig. 9-14. Regional maturation profiles in structurally disturbed areas showing the relation between vitrinite reflectance and structural features in:
(A) Alpine forelands (after Teichmüller, 1987a); and
(B) Rocky Mountain foothills (after Kalkreuth and Langenberg, 1986; Kalkreuth and McMehan, 1988).

Paleoheat measurement

Using the thickness of overburden, thermal conductivity, and vitrinite reflectance gradient, an empirical formula can be deduced to define paleoheat-flow density (Bunterbarth and Teichmüller, 1982) or paleoheat-flux (Armagnac et al., 1988) through a first-order time – temperature integral.

Defining a pressure effect

In general, a confined static pressure lowers the vitrinite reflectance, whereas atmospheric pressure increases the vitrinite reflectance (Horvath, 1983; Goodarzi, 1985). Tectonic stress induces strong bireflectance. Overpressure (abnormal) pressure is believed to have several causes, such as rapid sedimentation and disequilibrium compaction resulting in entrapment of pore water, hydrocarbon generation, chemical osmosis, tectonic stresses, etc. In most cases, overpressure sharply increases vitrinite reflectance (Hunt, 1979). In the overpressure zone of Texas Gulf Coast (probably caused by disequilibrium compaction), the geothermal gradient increases from 2.7°C/100 m to 4.7°C/100 m with a sudden increase in vitrinite reflectance (pressure gradient increases from 0.46 psi ft^{-1} to 0.85 psi ft^{-1}; Dow et al., 1988).

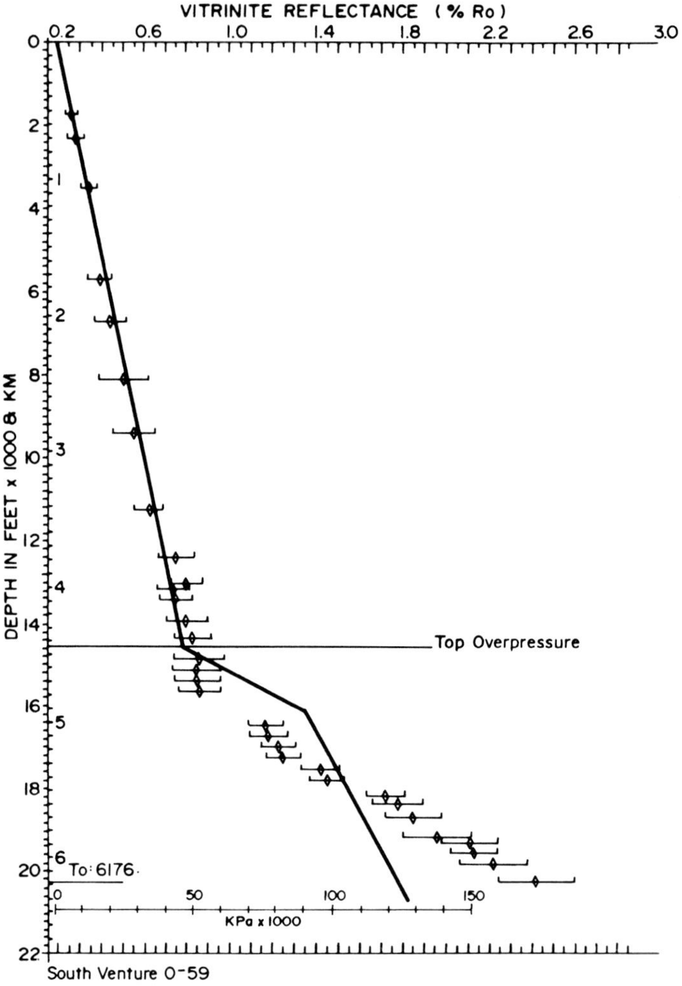

Fig. 9-15. Relationship between vitrinite reflectance (and pressure data; KPa × 1000) and depth for borehole South Venture 0-59 in the Scotian Shelf. (After J. Wade, pers. commun.)

Similarly, the overpressure zone in the Scotian Shelf (possibly caused by hydrocarbon generation) shows an abnormal increase in vitrinite reflectance trend in some wells. Figure 9-15 shows the relation between vitrinite reflectance (%R_0) and the measured pressure (KPa × 1000) from borehole South Venture 0-59. The abnormal pressure started at 4500 m in that well. The vitrinite reflectance gradient increased in South Venture 0-59 from 0.26% R_0 per km (up to 4500 m – lie within normal hydrostatic pressure) to 0.9% R_0 per km (between 4500 and 6000 m – lie within abnormal pressure). The pressure increased from 50 (KPa × 1000) at 4500 m to 128 (KPa × 1000) at 6000 m in this well. Similar observations are being made by Law et al. (1989) from the Rocky Mountain Forland basins which are associated with low-permeability gas-bearing sequence. Law et al. (1989) related the "kinky" vitrinite reflectance profile with conductive heat transfer processes rather than convective processes in normal pressured sequence. Proper evaluation of vitrinite reflectance may, therefore, reveal the prior existence of paleo-overpressures in a geological sequence.

Oxidation in sediments

The abnormality of vitrinite reflectance caused by oxidation halos in the cracks of vitrinite, pseudovitrinite, and oxidized bimacerite grains can reveal whether a sediment was oxidized during deposition or early diagenesis, or during sample storage.

Basin modeling

Two types of numerical models (empirical and kinetics) have been proposed to simulate the time and temperature (rate) in maturation (coalification) and hydrocarbon generation (Karweil, 1956; Lopatin, 1971; Bostick, 1973; Tissot and Espitalié, 1975; Kanstler et al., 1978; Waples, 1980; Welte and Yukler, 1981; Price, 1983; Ritter, 1984; Yukler and Kokesh, 1984; Burnham et al., 1988; Tissot et al., 1988; Burnham and Sweeney, 1989; Larter, 1989). Both types of models are based on the Arrhenius equation of first-order reactions and utilize maturation profiles from vitrinite reflectance measurements. Yukler and Kokesh (1984), Tissot et al. (1988), Welte and Yalcin (1988), and Larter (1989) reviewed the fundamental principles of both model types. The predicted thermal histories and hydrocarbon generation models were based on computer simulation of natural systems using parameters like lithology, porosity, kerogen type, heat flow, thermal conductivity, time of rifting, etc. These models determine the degree of maturation and hydrocarbon generation at different time periods. Figure 9-16 shows the burial curves and the maturity lines (calibrated with vitrinite reflectance) of the Wilcox Group (depositional environment: continental slope) from the Victoria County of Texas Gulf Coast using the time – temperature index of Lopatin (1971, modified by Waples, 1980) (Dow et al., 1988). According to the *TTI* model of Lopatin – Waples, the oil window in Fig. 9-16 was set at *TTI* between 15 and 160 (equivalent vitrinite reflectance range of 0.65 to 1.30% R_0). The dry gas window begins at a *TTI* of about 1500 (R_0 = 2.2%). The predicted vitrinite reflectance (Fig. 9-16) correlates well with the measured vitrinite reflectance (Fig. 9-18). Other mathematical models using chemical kinetics generally calibrate the predicted maturity (in most cases vitrinite reflectance) with a real

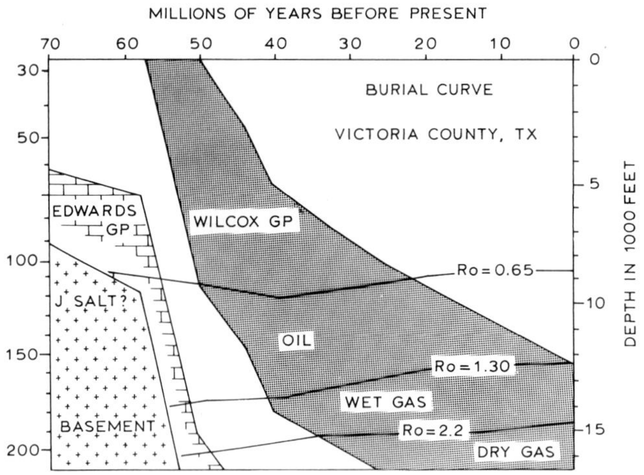

Fig. 9-16. Burial curve and predicted vitrinite reflectance (from *TTI*) in oil, wet gas and dry gas zones in sediments of the Wilcox and Edwards Groups, Victoria County, southeast Texas. (After Dow et al., 1988.)

system of measured data to check the validity of the model parameters as shown in Figs. 9-17 and 9-18 (Welte and Yukler, 1981; Yukler and Kokesh, 1984; Welte and Yalcin, 1988; Larter, 1989; Sweeney and Burnham, 1990). This demonstrates the importance of vitrinite reflectance in mathematical basin modeling.

Problems

In spite of accurate standardization of the vitrinite reflectance measurements on coal, reflectance measured on dispersed organic matter is often influenced by the following factors:
(a) sample contamination and drilling effect,
(b) sample preparation,
(c) vitrinite identification,
(d) bitumen impregnation,
(e) lithology and kerogen type,
(f) organic facies, and
(g) activation energy related to kerogen type.

Some of the resulting problems have been discussed previously (Dow, 1977; Price and Barker, 1985). These problems, however, have to be re-evaluated to obtain a better understanding of the vitrinite reflectance. A combined effect of all these problems resulted in an anomalous vitrinite reflectance scattering in a round robin analysis of eight unknown samples by Dembicki (1984; Fig. 9-19).

Sample contamination and drilling effect

In general, well cuttings are subjected to serious contamination problems due to caving, mud additives, oxidation due to turbodrilling, etc.

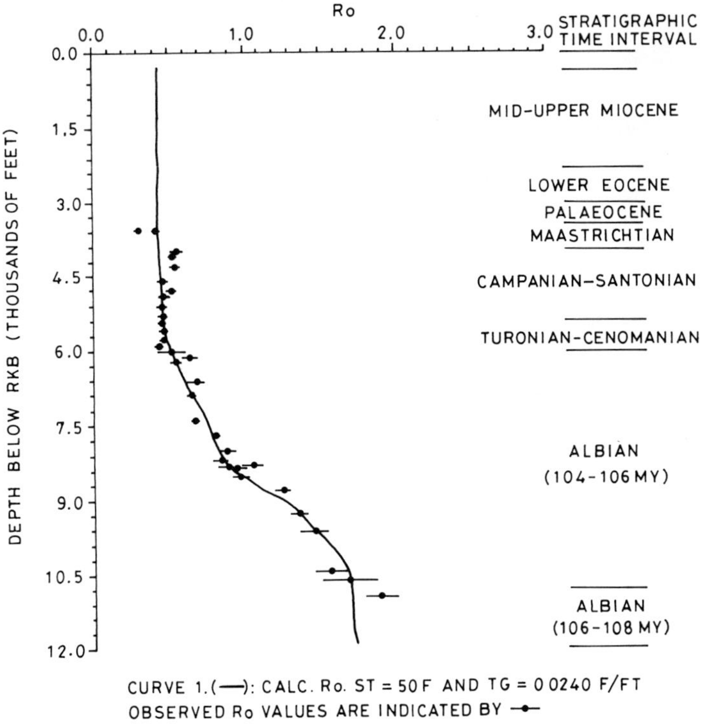

Fig. 9-17. Variation of predicted (from mathematical modeling) and measured vitrinite reflectance with depth and age of rock sequence. (After Yukler and Kokesh, 1984.)

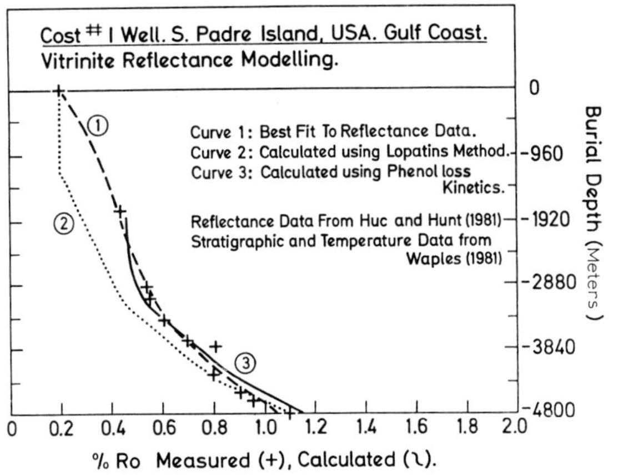

Fig. 9-18. Variation of predicted (from kinetics model of vitrinite pyrolysis) and measured vitrinite reflectance with depth. (After Larter, 1989.)

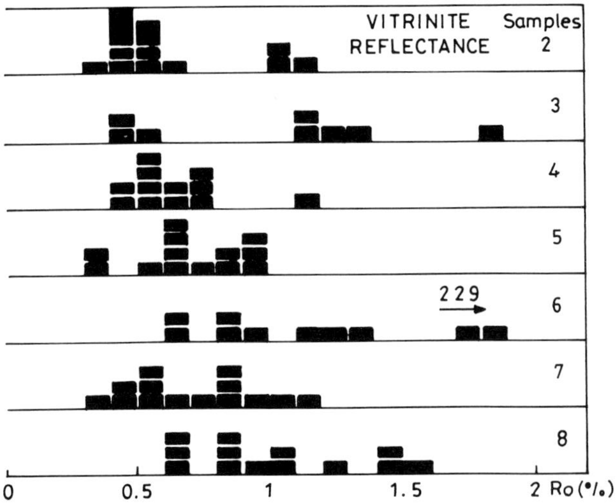

Fig. 9-19. Vitrinite reflectance histograms of eight unknown samples measured during an inter-laboratory exchange study, showing the problem in vitrinite reflectance measurements. (After Dembicki, 1984.)

Persistent caving from the uphole section (especially if it consists of coal) can present a serious threat to the validity of R_0 data, because primary vitrinite in cavings will show the same R_0 values throughout the section. Consequently, R_0 values will always be lower than expected.

Some mud additives like "resinex" contain grains of primary ulminite (vitrinite) and other vitrinite macerals. The influence of "resinex" on the vitrinite reflectance was observed in several boreholes in the North Sea, Scotian Shelf, and Beaufort Sea. As a result of this contamination, R_0 values will range from 0.3 to 0.4%, irrespective of depth. A detailed description of various mud additives similar to different organic matters has been discussed by Jordon (1983) (for details see Vorabutr and Chilingarian, 1983). The only way to eliminate the problem caused by drilling mud additives is to hand-pick small-sized cuttings under binocular microscope and compare them to petrophysical logs (e.g., gamma ray and sonic).

In various offshore wells in the North Sea, Scotian Shelf, Hibernia area, etc. sediments have been drilled into with oil-base mud systems in order to avoid formation damage, to obtain a better core sample quality, to lower drilling cost because of faster drilling, etc. Drilling with an oil-base mud system, however, creates tremendous problems in evaluating organic geochemical parameters (Brown, 1988; Mukhopadhyay and Birk, 1989; Mukhopadhyay, 1990b). In order to evaluate the influence of oil-base drilling mud on vitrinite reflectance, a bomb experiment was performed using Type-IIA kerogen and simulating drilling (Mukhopadhyay and Birk, 1989) with an oil-base mud. Figure 9-20 shows the reflectance histogram of this sample before the experiment (Fig. 9-20A), after the bomb experiment with oil-base mud (Fig. 9-20B), and after cleaning the oil-base mud-soaked sample with detergent (Fig. 9-20C). Vitrinite reflectance sharply decreased after the use of oil-

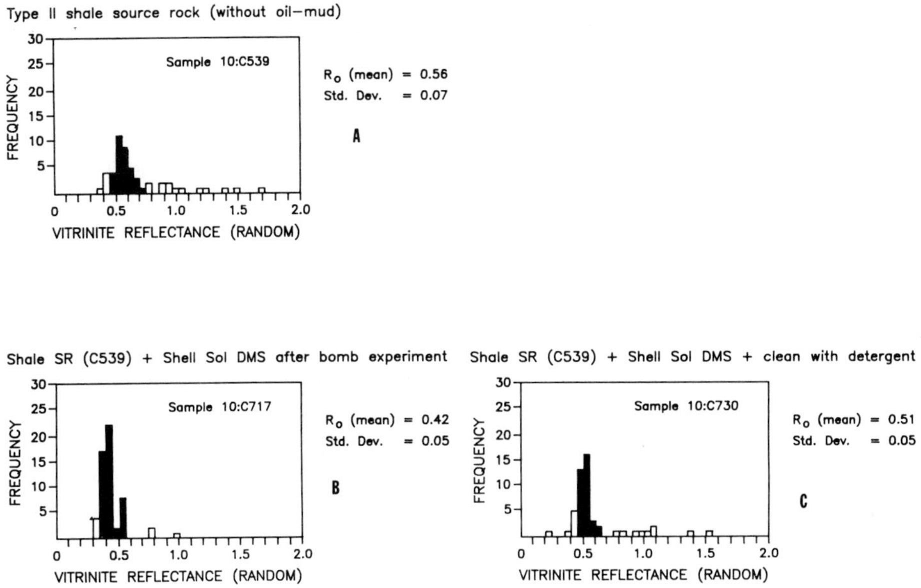

Fig. 9-20. Histograms of vitrinite reflectance (autochthonous vitrinite is shaded) measured from a Type-II kerogen source rock with or without the influence of oil-base mud. (After Mukhopadhyay and Birk, 1989.) Shell Sol DMS = commonly used base oil for drilling mud.

base mud (R_o = 0.56 – 0.42%), which could only be partially improved after cleaning. Similar suppression of vitrinite reflectance was observed, when drill cuttings (samples) of two nearby boreholes (Terra Nova K-08: drilled with oil-base mud; Terra Nova C-09: drilled with water-base mud) from Jean d'Arc Basin of offshore Newfoundland were analyzed (Mukhopadhyay, 1990b).

Turbodrilling can generate localized high temperature and oxidation, which may increase vitrinite reflectance in immature sediments.

Sample preparation

As discussed earlier, proper kerogen isolation and polishing (in a kerogen plug) are extremely important in vitrinite reflectance measurements. It has been observed that some laboratories do not take necessary precaution in polishing, which can be documented even in some published literature (Buiskool Toxopeus, 1983; Massoud and Kinghorn, 1986). The effect would be lowering of vitrinite reflectance. It is unfortunate that a kerogen isolation procedure has not yet been standardized. For both kerogen isolation and pellet preparation, temperature should not exceed 60°C. It is advisable to use cold set resin for plugging of whole rock or isolated kerogen for vitrinite reflectance measurements.

Vitrinite identification

As already discussed, the accuracy of vitrinite reflectance in dispersed organic matter depends on the identification of primary vitrinite. It is necessary to identify accurately primary vitrinite from allochthonous vitrinite, solid bitumen, and vitrinite-like primary and secondary macerals.

Bitumen impregnation

Within the principal phase of oil generation, vitrinites, which are associated with liptinite macerals in coal and dispersed organic matter, are impregnated with bitumen (Plate 9-2J). These bitumens were generated from liptinite macerals and redistributed while in a liquid form within the other organic matter networks during primary migration. Vitrinite grains measured in the principal phase of oil generation are often soaked with bitumen, because vitrinite has a molecular sieve-like structure. This effect is more pronounced when the vitrinite grains are floating within highly oil-prone amorphous liptinite (sapropelinite I and IIA of Mukhopadhyay, 1989a). Bitumen can soak vitrinite grains if it had migrated from a different source. The effect will be similar to the effect of oil-base mud, resulting in lowering of the vitrinite reflectance by 0.3 to 0.4 % (R_0) compared to the normal reflectance of vitrinite at that rank. The best way to separate bitumen is by solvent extraction before kerogen isolation. Solvent extraction, however, is generally done on a pulverized sample (– 60 mesh), which may crush some of the primary vitrinite and possibly create identification problems.

Lithology effect

The effect of lithology on vitrinite reflectance was demonstrated by Jones et al. (1972), Bostick and Foster (1975), and Fermont (1988) in coals surrounded by various lithologies and from dispersed organic matter in shales, sandstones and limestones. It was observed that vitrinite reflectance is dependent on thermal conductivity. Vitrinite reflectance, measured on isolated kerogens and coals from various lithologies and kerogen types from a borehole in the North Sea, showed an anomalous variability of vitrinite reflectance (Fig. 9-21). The highest reflectance was observed in coal and Type-III kerogen (terrestrial organic matter), whereas the lowest reflectance of vitrinite was found in sandstones. Shales with Type-IIA kerogen (mainly marine organic matter) and Type-IIB kerogen (mixed organic matter) have intermediate reflectances. Type-IIA kerogen in shale had a lower reflectance than Type-IIB kerogen in both shale and limestone. This variation in vitrinite reflectance is possibly due to both heat capacity and thermal conductivity of the sediments and variation of biodegradation during early diagenesis (Fermont, 1988). This variability in vitrinite reflectance can be evaluated more easily if the reflectance is plotted against depth using an arithmetic R_0 scale. Figure 9-9A shows the variation of vitrinite reflectance gradient from two Scotian Shelf boreholes. The variation is caused by the difference in lithologic assemblages.

Organic facies

It is widely documented that vitrinite grains associated with an anoxic environment and higher liptinite content exhibit suppressed reflectance values (Hutton and Cook, 1980; Kalkreuth, 1982; Newman and Newman, 1982; Price and Barker, 1985; Kalkreuth and Macaley, 1987). Figure 9-22 demonstrates the suppression or abnormality of vitrinite reflectance due to various causes: (a) related to hydrogen-rich (desmo

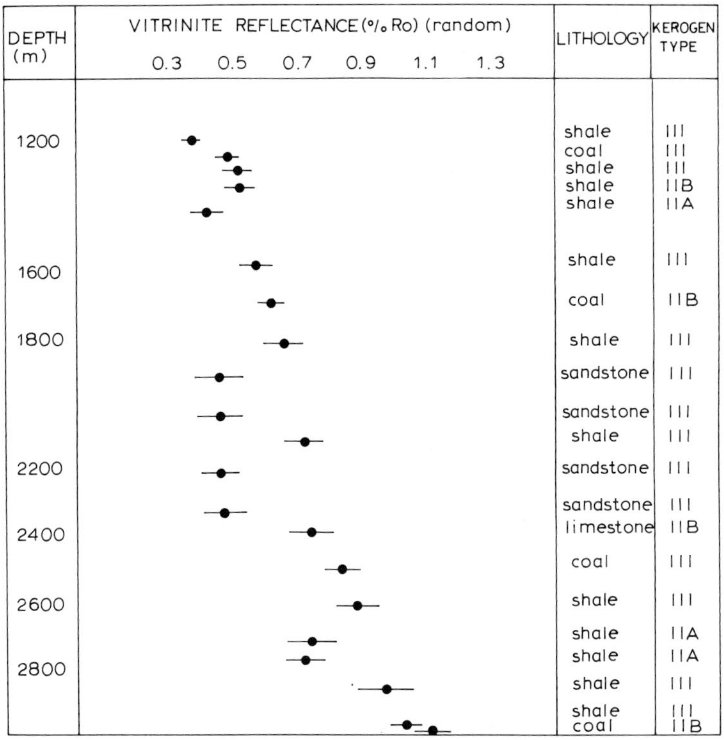

Fig. 9-21. Influence of kerogen type and lithology on vitrinite reflectance as seen from the measured vitrinite reflectance of various sedimentary rocks from a borehole in North Sea.

collinite) and hydrogen-poor vitrinite macerals (Fig. 9-22A; Buiskool Toxopeus, 1983); (b) related to organic facies and kerogen types (Fig. 9-22B; Price and Barker, 1985); and (c) related to an increase of liptinite (alginite in this case) macerals (Fig. 9-22C; Hutton and Cook, 1980; Kalkreuth, 1982; Kalkreuth and Macaley, 1987).

Vitrinite in a marine environment contains a mixture of organic constituents, such as plant debris transported from swamps and marshes of alluvial and delta plains, in-situ marine grasses and weeds, and peat, lignite and bituminous coal deposits eroded from the surrounding land masses (Mukhopadhyay et al., 1985b, 1990a). In a marine environment, huminite/vitrinite also contain some part of algal remains. In a marine anoxic environment, most of the humic substances from all these plant substances are either pulverized or destroyed; only the resistant humic substances mixed with lipid parts of the plants, which have protective coverings, are preserved. This results in more hydrogen-rich huminite/vitrinite and mixed macerals, such as desmocollinite or saprocollinite, and possibly corpocollinite. The condensation reaction for the increase in vitrinite reflectance for these hydrogen-rich macerals should be different than for oxygen-rich telocollinite and gelocollinite (well-preserved or gelified wood, bark and other woody tissues). These hydrogen-rich marine vitrinites contain more straight-chain aliphatics in their hydroaromatic structure, whereas oxygen-rich vitrinites of alluvial- or delta-plain environments are more

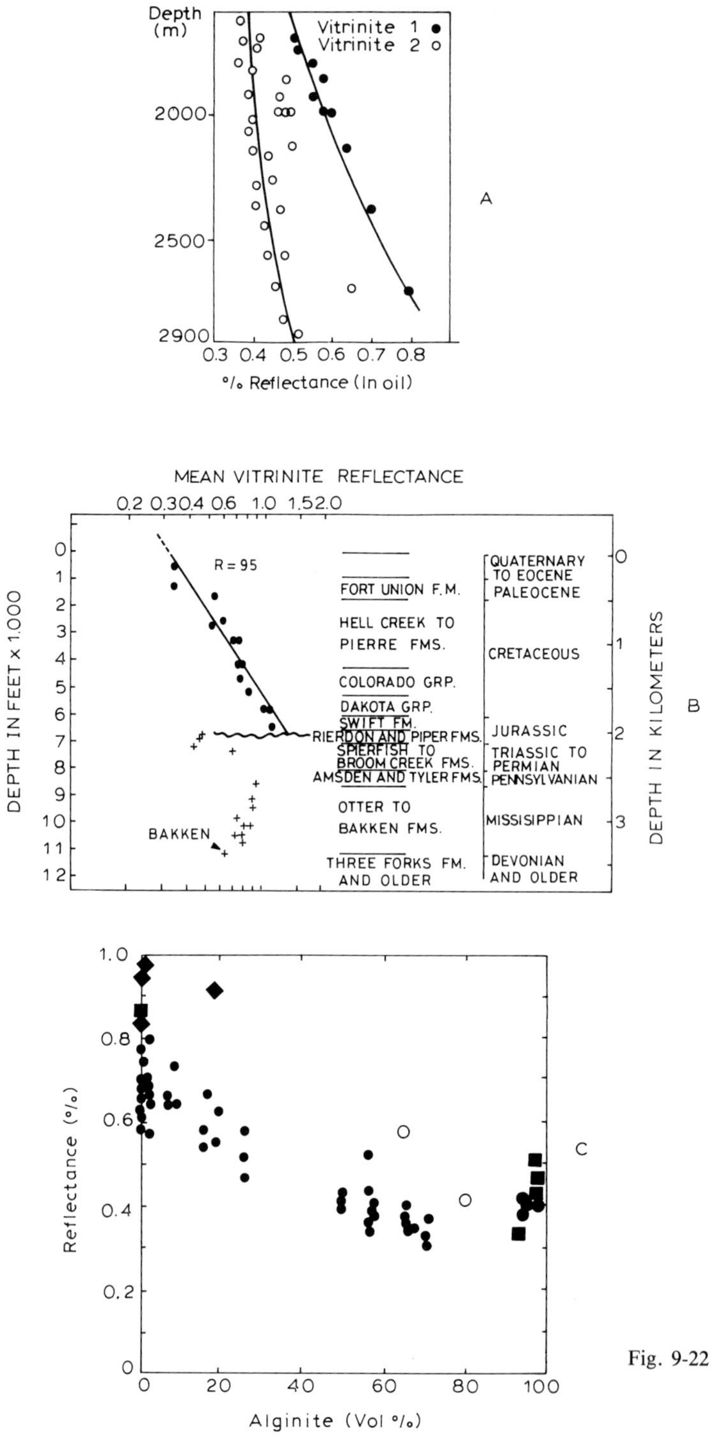
Depth (m)
Vitrinite 1
Vitrinite 2
2000
2500
2900
0.3 0.4 0.5 0.6 0.7 0.8
% Reflectance (In oil)
A
MEAN VITRINITE REFLECTANCE
0.2 0.3 0.4 0.6 1.0 1.5 2.0
DEPTH IN FEET x 1,000
R = 95
BAKKEN
FORT UNION F.M.
HELL CREEK TO PIERRE FMS.
COLORADO GRP.
DAKOTA GRP.
SWIFT FM.
RIERDON AND PIPER FMS.
SPIERFISH TO BROOM CREEK FMS.
AMSDEN AND TYLER FMS.
OTTER TO BAKKEN FMS.
THREE FORKS FM. AND OLDER
QUATERNARY TO EOCENE
PALEOCENE
CRETACEOUS
JURASSIC
TRIASSIC TO PERMIAN
PENNSYLVANIAN
MISSISIPPIAN
DEVONIAN AND OLDER
DEPTH IN KILOMETERS
B
Reflectance (%)
Alginite (Vol %)
C

Fig. 9-22.

phenolic. To increase the maturation for the hydrogen-rich vitrinites, higher temperatures are needed to crack carbon – carbon bonds. As discussed earlier, however, bitumen generated in liptinite-rich sediments can suppress the vitrinite reflectance. The best way to evaluate the relation between hydrocarbon generation and vitrinite maturation is to measure telocollinite grains free from bitumen, if available.

Activation energy and heating rate

As previously discussed, the activation energy required to break chemical bonds for hydrocarbon generation is related to kerogen type (Tissot et al., 1988). Accordingly, three coal types of different vegetation types were pyrolyzed (anhydrous pyrolysis) at different heating rates and the pseudo-activation energies were calculated (Mukhopadhyay, 1989b; Mukhopadhyay et al., 1989). Humic coal contains mainly telocollinite and gelocollinite (Plate 9-1D, E); mixed coal, mainly mixinite, vitrodetrinite, and desmocollinite (Plate 9-1E, B); and sapropelic coal, mainly saprocollinite, desmocollinite and corpocollinite (Plate 9-1F, I). The activation energy of alkane generation from vitrinite, as calculated by Burnham and Sweeney (1989), was 50 kcal $mole^{-1}$. Mukhopadhyay (1989) and Mukhopadhyay et al. (1989), however, showed that the pseudo-activation energy of hydrocarbon generation of high-molecular-weight hydrocarbons from three coal types are different: humic coal – 58.6 kcal $mole^{-1}$; mixed coal – 42.6 kcal $mole^{-1}$; and sapropelic coal – 46.8 kcal $mole^{-1}$. The variation of activation energy in the three coal types may explain why the vitrinite reflectance of mixed coal (R_0 = 0.89%) is lower than that of both humic (R_0 = 1.13%) and sapropelic coals (R_0 = 0.97%), after hydrous pyrolysis using similar temperature and pressure conditions (Mukhopadhyay, 1989). The vitrinite reflectance before hydrous pyrolysis for the three coal types was similar: humic coal – 0.34% R_0; mixed coal – 0.36% R_0; and sapropelic coal – 0.49% R_0.

Based on the study of 30 different sedimentary basins with variable heating rates, Mukhopadhyay (unpubl. data) established a relation between vitrinite reflectance and hydrocarbon generation for a Type-IIA – IIB kerogen. On comparing measured vitrinite reflectance and organic geochemical data (extract yield) in three basins with different geothermal gradients (heating rate), it becomes obvious that the so-called "oil window" (0.5 – 1.3% R_0) (Pusey, 1973; Dow, 1977) is not valid for all basins with different heating rates (Fig. 9-23) (Mukhopadhyay, unpubl. data). In a basin with high heat flow (e.g., Santa Maria Basin, California), the vitrinite reflectance of Type-IIA – IIB kerogen lags behind hydrocarbon maturity. The morphology of liptinites and geochemical data suggest that the peak generation of hydrocarbons oc-

---

Fig. 9-22. Changes in vitrinite reflectance due to:
(A) variation of maceral types in coal with depth (after Buiskool Toxopeus, 1983);
(B) variation of kerogen type from type III to type II with depth (after Price and Barker, 1985);
(C) variation of the amount of alginite content in coal and oil shale (after Kalkreuth and Macaley, 1987). ♦● = coals and oil shales from Stellarton Basin, Nova Scotia; ■ = oil shales from New Brunswick; ○ = other oil shales, Nova Scotia; ● = coals and oil shales from Australia. (From Hutton and Cook, 1980.)

curs at a much lower vitrinite reflectance than that indicated by the oil window concept (0.8 – 1.0% R_0). On the other hand, in a basin with a slow heating rate (such as Georges Bank, offshore Bengal), the vitrinite reflectance attains maturity faster than hydrocarbon generation. This suggests two possibilities: either (a) maturity of vitrinite is more dependent on time, whereas hydrocarbon generation from liptinite macerals is more dependent on temperature (Yukler and Thompsen, 1988), or (b) a higher activation energy is needed for the maturation of saprocollinite (major vitrinite in marine environment) than for the amorphous liptinite (the main hydrocarbon-generating maceral) in a marine environment.

The early, peak, and final phase of hydrocarbon generation of a source rock (especially with kerogen Type IIA and kerogen Type IIA – IIB) can be defined by using fluorescence microscopy and the mass-balance between primary and secondary macerals (Mukhopadhyay et al., 1985a; Dow et al., 1988; Mukhopadhyay and Wade, 1990). Correlation of these data and vitrinite reflectance will enhance the concept of maturation of vitrinite macerals and hydrocarbon generation from liptinite macerals. Figure 9-24 illustrates the hydrocarbon generation limits for dif-

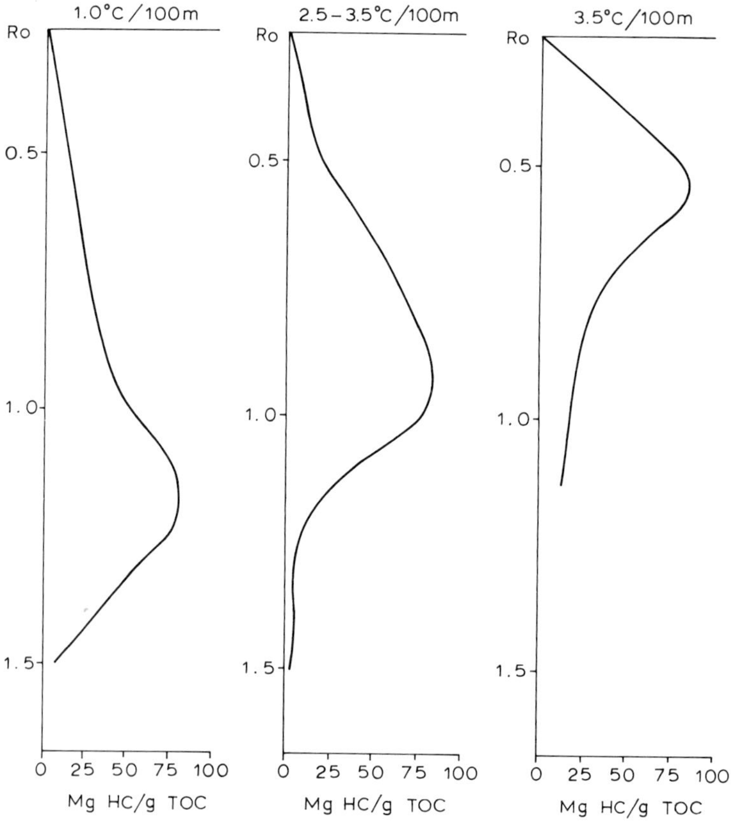

Fig. 9-23. Influence of geothermal gradients or heating rates on hydrocarbon generation curves for Type-IIA – IIB kerogen; vitrinite reflectance was plotted versus HC (mg)/*TOC* (g) ratio. *TOC* = total organic carbon.

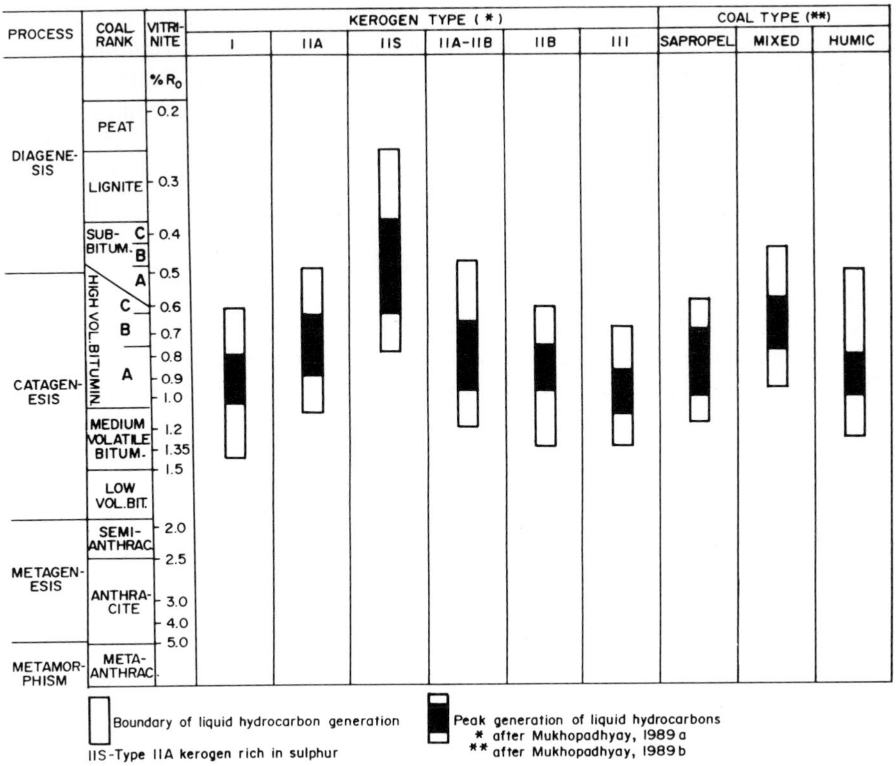

Fig. 9-24. Oil and gas windows as related to vitrinite reflectance and kerogen types.

ferent kerogen types related to vitrinite reflectance in a similar thermal regime or geothermal gradient (20 – 30°C per km).

OTHER MATURATION PARAMETERS

Reflectance of phytoclasts and zooclasts

Higher land plants did not evolve until Late Silurian. Most pre-Devonian rocks are, therefore, devoid of vitrinite. Reflectance of blue-green algae, graptolite, chitinozoa and scolecodont are used extensively for maturity determination of Ordovician and Silurian sediments (Teichmüller, 1978, 1987; Goodarzi, 1984; Goodarzi and Norford, 1985; Bertrand and Heroux, 1987a). Goodarzi and Norford (1985) demonstrated that reflectances of both graptolite and chitinozoa increases with increasing depth, although the graptolites have higher reflectances than those of chitinozoans. Bertrand and Heroux (1987), on the other hand, found that at low maturity, chitinozoa had higher reflectances than graptolites. The reflectances of graptolites and vitrinite can be compared directly in some Late Devonian and younger sediments. The vitrinite reflectance is generally lower than the graptolite reflectance by (within) 1.5% R_0. Similar to vitrinite reflectance, graptolite reflectance is influenced by temperature, rock matrix, organic facies, weathering, and thermal anomalies; graptolites exhibit anisotropy (bireflectance) beyond 2.0% R_0 (Goodarzi et al., submitted). Vitrinite-like algal filaments are observed even in the Proterozoic sediments, which can be used for maturity determination (Teichmüller, 1987a). For details of graptolite, chitinozoa, and scelecodont reflectance, see Goodarzi et al. (submitted). In pre-Cambrian sediments of Australia (1200 m.y.), reflectances of the "vitrinite-like" maceral (VLM) (possibly derived from algae) show good correlation with lambdamax and the red-green quotient of the alginite (Mukhopadhyay, unpubl. data). Buchardt and Lewan (1990) documented the trend of "vitrinite-like macerals (VLM)" in pre-Silurian sediments from southern Scandinavia. They concluded that "VLM" was derived from cellulose and related polysaccarides.

Reflectance of solid bitumen

In a petroleum source rock, solid bitumens are generated mostly from the liptinite macerals and redistributed by primary migration within the rock matrix during late diagenetic and early catagenetic stages. These solid bitumens are sometimes expelled from the source rock and form a separate organic rock lying parallel or perpendicular to other synsedimentary rocks. Jacob (1967, 1989) extensively studied various properties of solid bitumens including their reflectance. He observed three series of solid bitumens. The procedure for the measurement of solid bitumen reflectance is similar to that of vitrinite reflectance. The reflectance of solid bitumen is useful in vitrinite-lean sediments (Bertrand and Heroux, 1987). Figure 9-25 illustrates the relation between reflectances of vitrinite and solid bitumen. Solid bitumens appear to start forming when vitrinite reflectance reaches 0.45% R_0. The reflectance of solid bitumen is lower than the corresponding vitrinite reflectance below a maturity of 1.0% R_0 (Jacob, 1989; Fig. 9-25). For details on the geochemistry and other petrographic nomenclatures of solid bitumen, see Abraham (1945) and Curiale (1986).

Thermal Alteration Index

The Thermal Alteration Index *(TAI)* of sporinite, which was first introduced by Gutjahr (1966) as spore carbonization measurements, was later correlated to vitrinite reflectance and hydrocarbon generation by Staplin (1969). The Thermal Alteration Index *(TAI)* of Staplin (1969) is based on a numerical scale of 1 to 5 using subjective spore/pollen coloration. A similar scale called translucency of palynomorphs was developed later by Grayson (1975). The Thermal Alteration Index *(TAI)* of Staplin (1969) or its modified form is presently being used in most laboratories. Jones and Edison (1978) presented the most illustrated and modified *TAI* scale, which was correlated with vitrinite reflectance. Similar to vitrinite reflectance, mean coloration was determined from several autochthonous spores or pollens for establishing the maturity of sediments. During advanced maturation,

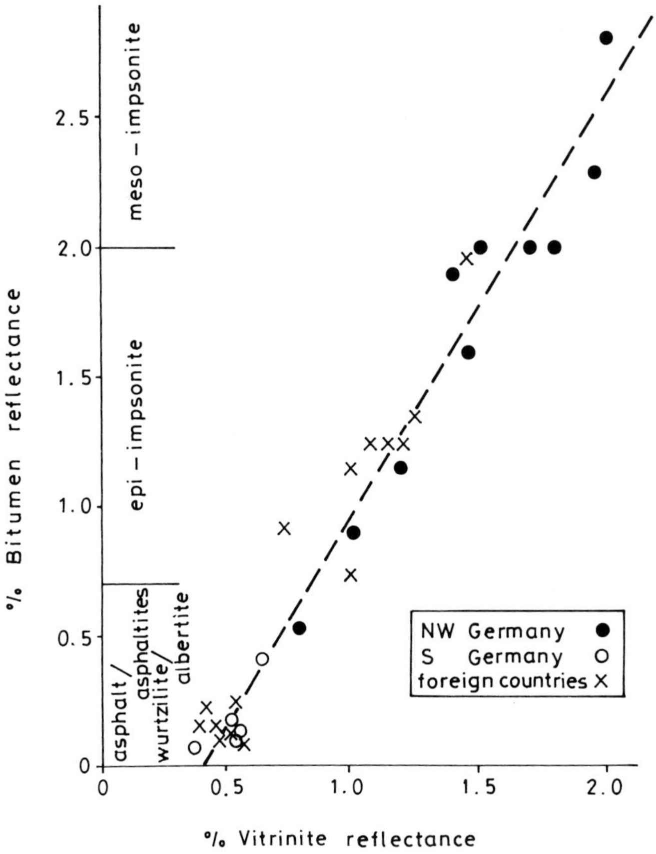

Fig. 9-25. Relationship between reflectance of solid bitumen and vitrinite reflectance in dispersed organic matter. (After Jacob, 1989.)

transparent green or yellow spores become translucent or opaque brown to black. Table 9-8 illustrates the relation between vitrinite reflectance and Thermal Alteration Index *(TAI)*. Unlike vitrinite reflectance, the main problem of *TAI* is that various laboratories use different scales for maturity determination.

Conodont Alteration Index

As conodonts are most abundant in marine carbonates, where vitrinites are rare and palynomorphs are poorly preserved, the Conodont Alteration Index *(CAI)* was introduced by Epstein et al. (1977). It offers a good alternative (to vitrinite reflectance) maturation parameter, especially in the metagenetic and metamorphic stages. Similar to *TAI*, *CAI* uses a numerical scale: *TAI* ranges from 1 to 5, whereas *CAI* ranges from 1 to 8. On the other hand, *CAI* covers diagenetic to metamorphic

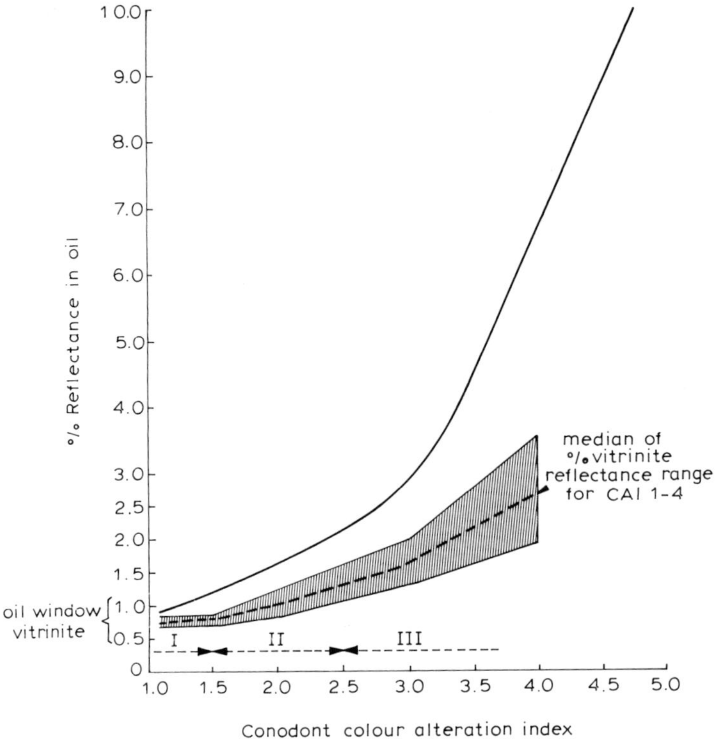

Fig. 9-26. Relationship between vitrinite reflectance and conodont color alteration index (*CAI*). (After Goodarzi and Norford, 1989.) I, II and III are the oil, wet gas, and dry gas zones, respectively.

(greenschist facies) stages, like vitrinite reflectance. The *CAI* stages 1 – 5 are generally used, because they cover the maturation between diagenesis and early greenschist facies. The *CAI* stages 6 – 8 cover different metamorphic stages (Teichmüller, 1987a). Figure 9-26 illustrates the relation between *CAI* (stages 1 – 5) and vitrinite reflectance (R_0: 0.2 – 5%).

FLUORESCENCE MICROSCOPY

Fluorescence microscopy became popular because of problems involved in measuring the vitrinite reflectance and in order to develop some maturation parameters which can be utilized for the determination of the maturity of kerogen (in coal and dispersed organic matter), bitumen (both liquid and solid), and crude oil in one single microscopic system. Moreover, in order to correlate hydrocarbon generation and maturation of organic matter, fluorescence microscopy has an advantage over vitrinite reflectance measurements, because it utilizes various macerals of the liptinite group. The reaction kinetics of hydrocarbon generation and maturation of liptinites should be somewhat similar, because hydrocarbons are mainly generated from liptinite macerals and not from vitrinites.

CONTINUOUS SPECTRAL FLUORESCENCE

The incident-light fluorescence method for the study of bituminous coal was first applied by Schohardt (1936). It was further utilized for the study of peat and lignite by Jacob (1952), and for fresh pollen and spores by Van Gijzel (1967). The more precise instrumentation for the determination of the maturity of liptinite macerals (in coal and dispersed organic matter), bitumen and crude oil, however, was introduced much later (Jacob, 1972; Ottenjann et al., 1974; Teichmüller, 1974; Van Gijzel, 1975; Teichmüller and Ottenjann, 1977; Hagemann and Hollerbach, 1981; Ottenjann, 1981; Stach et al., 1982; Bustin et al., 1985; Bensley and Davis, 1988).

At present, a wealth of data is available on different aspects of fluorescence microscopy of coal, dispersed organic matter, bitumen and crude oil (e.g., see Crelling and Dutcher, 1979; Leythaeuser et al., 1980; Ottenjann, 1981; Van Gijzel, 1982; Crelling, 1983; Spiro and Mukhopadhyay, 1983; Mukhopadhyay and Gormly, 1984; Bertrand et al., 1986; Diessel, 1986; Mukhopadhyay and Rullkötter, 1986; Teichmüller, 1986; Von der Dick and Kalkreuth, 1984; Thompson-Rizer and Woods, 1987; Ottenjann, 1988; Lin and Davis, 1988; Jacob, 1989; Davis et al., 1990).

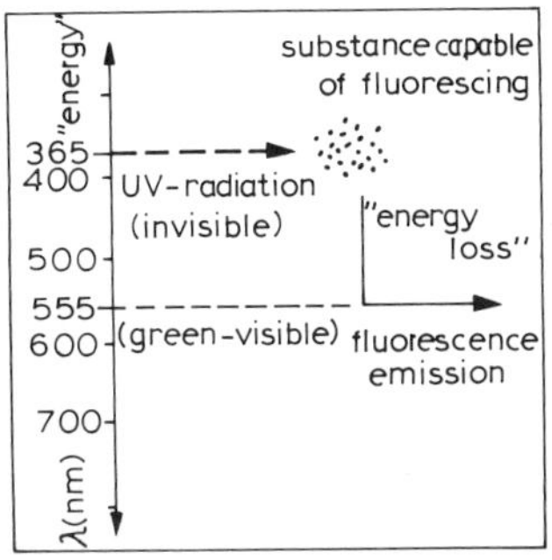

A

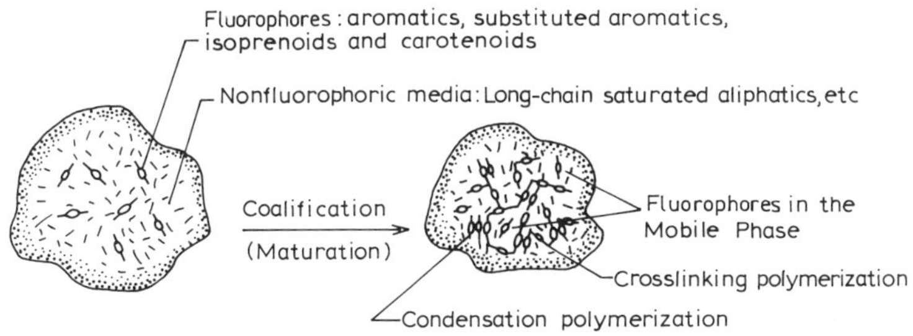

B

Fig. 9-27. Conceptual models of the origin of fluorescence: (A) after Zeiss microscope brochure; and (B) after Lin and Davis (1988).

Chemistry of fluorescence

Fluorescence is the luminescence of a substance excited by radiation. In other words, electrons from the ground state (S_0) are excited by absorption irradiation to higher-energy orbits (S_1, S_2, etc.). Fluorescence signals are emitted when the excited electrons return to the ground state (Lin and Davis, 1988). Accordingly, the wave length of the emitted fluorescence of substances called fluorophores is longer than that of the excitation (Fig. 9-27A). Some of the macerals in coal and dispersed organic matter, some of the solid and liquid bitumens, crude oil and condensates are autofluorescent as a result of ultraviolet and blue light excitation.

Figure 9-27B shows a schematic structural model of organic matter fluorescence. Fluorescence is emitted by the presence of aromatics, substitute aromatics (heterocomponents), isoprenoids, and carotenoids present within a maceral (both kerogen and bitumen fraction) or in crude oil. The fluorescence emission of these fluorophores in any rank is enhanced by the increased supply of nonfluorophoric media (dilution effect caused by substances like long-chain *n*-alkanes), whereas fluorescence is reduced by an increased supply of polycondensed fluorophores which cause intramolecular quenching (Bertrand et al., 1986; Gangopadhyay et al., 1988; Lin and Davis, 1988). The absence of fluorescence of some of the macerals (like pyrofusinite) at a very immature stage indicates that pi-electrons are too delocalized to show any visible fluorescence.

Instrumentation, fluorescent colors and parameters

Instrumentation

A schematic representation of fluorescence excitation and emission using a chromatic beam splitter in incident light, is shown in Fig. 9-28A. Figure 9-28B illustrates the salient microscopic components needed for a continuous fluorescence microscopy. The main components are: an incident-light microscope (like a Zeiss Photomicroscope or UMSP or Leitz MPV 3), a light source (Xenon or mercury), exciter and barrier filters, monochromator, photomultiplier and a computer. For 3-D fluorescence two monochromators are needed. Other requirements are:

(a) measurement should be done in a dark room;
(b) dry or water objective must be provided with a high numerical aperture (example: neofluor 40/0.9);
(c) condensor aperture diaphragm should be completely opened until it is used for measurement;
(d) ground glass should not be present in the illuminating beam path; and
(e) before sample measurement, the instrument has to be standardized by using a standard lamp and correcting stray fluorescence. For details, see Ottenjann (1981).

The filter set combinations used, are shown in Table 9-4.

The details of the microscopic system were given by Ottenjann (1981), Stach et al. (1982), Bustin et al. (1985), and Thompson-Rizer and Woods (1987).

Fluorescent color of various organic matter

All primary and secondary liptinite macerals (like alginite, albertite, etc.; for definition see Stach et al., 1982; and Jacob, 1989), which contain a lipid fraction of plants and zooplankton, show fluorescence colors of various shades. These colors change at various ranks (between 0.1 and 1.5% R_0). The fluorescence color of most

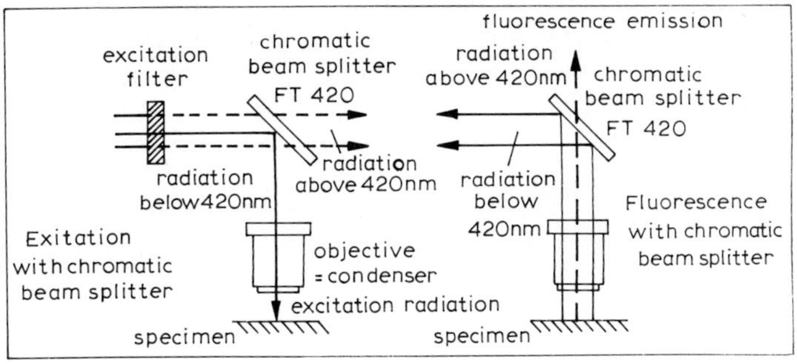

A

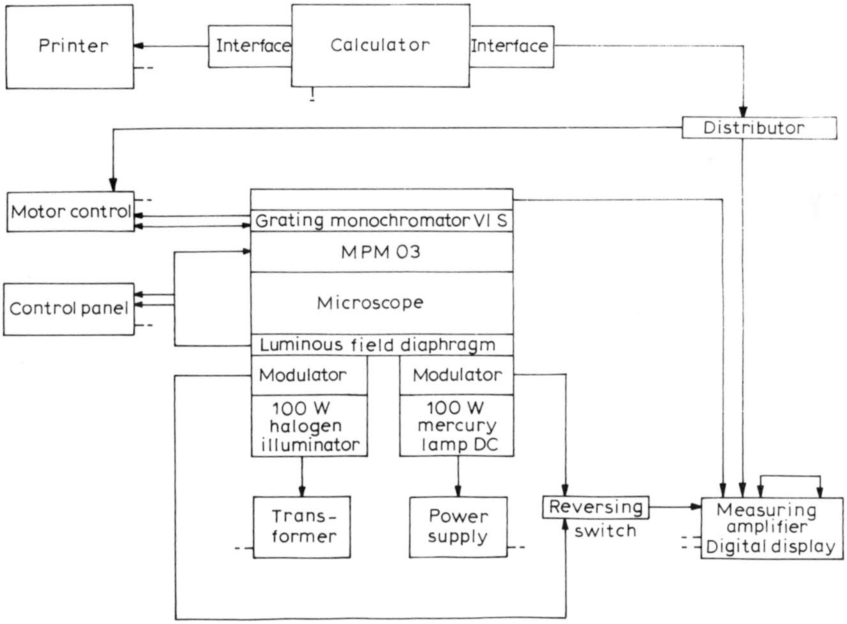

B

Fig. 9-28. Schematic diagrams showing:
(A) the importance of a chromatic beam splitter in fluorescence excitation and emission (after Zeiss microscope brochure); and
(B) various components used for continuous fluorescence spectroscopy. (After Ottenjann, 1981.)

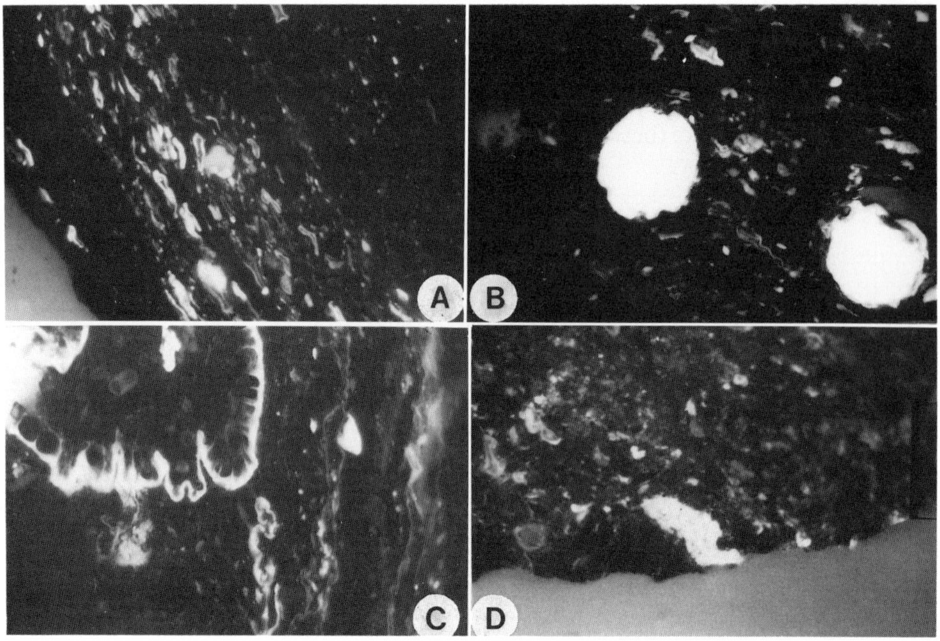

Fig. 9-29. Photomicrographs of liptinite macerals in blue light excitation using black and white film. Liptinites exhibit white morphologies in a dark background of vitrinite and inertinite macerals.
(A) Sporinite, resinite and liptodetrinite (Eocene, Wilcox Group, Texas; R_0 = 0.35%).
(B) Resinite (large round shape), sporinite and liptodetrinite (Eocene, Wilcox Group, Texas; R_0 = 0.34%).
(C) Cutinite and liptodetrinite (Eocene, Jackson Group, Texas; R_0 = 0.30%).
(D) Alginite (telalginite), liptodetrinite and sporinite (Eocene, Jackson Group, Texas; R_0 = 0.28%).

TABLE 9-4

Filter set combinations used in the microscopic system

Excitation range	Exciter filter	Chromatic beam splitter	Barrier filter
UV	BP365/10	FT390	LP395
UV	G365	FT420	LP418
Violet	G405	FT460	LP475
Violet	BP390-440	FT460	LP475
Blue-violet	G436	FT510	LP515
Blue	G436	FT510	LP520

G = solid glass filter; BP = band pass filter; FT = chromatic beam splitter; LP = longwave pass filter.

of the liptinite macerals are more intense than that of vitrinite and inertinite macerals.

Huminite/vitrinite macerals are fluorescent either at a very low maturity (below 0.3% R_0 – primary fluorescence) or between 0.8 and 1.5% R_0 (secondary fluorescence; Ottenjann et al., 1982; Lin and Davis, 1988). Inertinite macerals are in general nonfluorescent by UV to blue-light excitation and visible-light emission. Diessel (1986), however, observed fluorescence in reactive semi-fusinite and macrinite (inertinite macerals) of high-volatile to medium-volatile bituminous coals, when blue-light excitation was used. This phenomenon is possibly related to secondary fluorescence (Lin and Davis, 1988). Secondary inertinite (rank-inertinite) macerals are all nonfluorescent. Figure 9-29 shows an example of four fluorescent (shown white due to the use of black and white film) primary liptinite macerals in low-rank coals from Texas (R_0 is between 0.25 and 0.40%). The fluorescent color identifies the morphology of sporinite (Figs. 9-29A, D), liptodetrinite (Figs. 9-29A – D), resinite (Figs. 9-29A, B), cutinite (Fig. 9-29C), and alginite (Fig. 9-29D) compared to nonfluorescent vitrinite groundmass. Appendix B will identify fluorescence colors of primary and secondary macerals of various rank under blue-light excitation from normal nonoxidized sediments.

Fluorescent parameters

Parameters generally used for fluorescence study are as follows:

Lambdamax	=	peak of the fluorescence emission curve, between 400 and 700 nm (Fig. 9-30A).
Red/green quotient (Q)	=	ratio of the relative excitation at 650 nm and 500 nm (Fig. 9-30A).
Alteration	=	increase (positive) or decrease (negative) or first increase/decrease then decrease/increase (ambivalent) (Ottenjann, 1988) in fluorescence intensity relative to a standard (Fig. 9-30B).
Alteration A	=	$\dfrac{\sum_{700\ \mathrm{nm}}^{400\ \mathrm{nm}} (I_{30} - I_0)}{\sum_{700\ \mathrm{nm}}^{400\ \mathrm{nm}} I_0} \times 100$, and
Alteration B	=	$\dfrac{546\ \mathrm{nm}\ (I_{30} - I_0)}{546\ \mathrm{nm}\ (I_0)} \times 100$
where I_0, I_{30}	=	fluorescence intensity measured with reference to a standard after initial intensity and intensity after 30 minutes.

Fluorescence alteration (Ottenjann et al., 1974) or fading (Van Gijzel, 1975) is a photochemical reaction caused by prolonged ultraviolet excitation. Alteration or fading is called positive when the intensity of fluorescence increases. On the other hand, it is called negative when the intensity decreases. (For the possible cause of the origin of fluorescence alteration, see p. 489.) Ambivalent alteration is caused by the combination of negative and positive alteration. A three-dimensional measure-

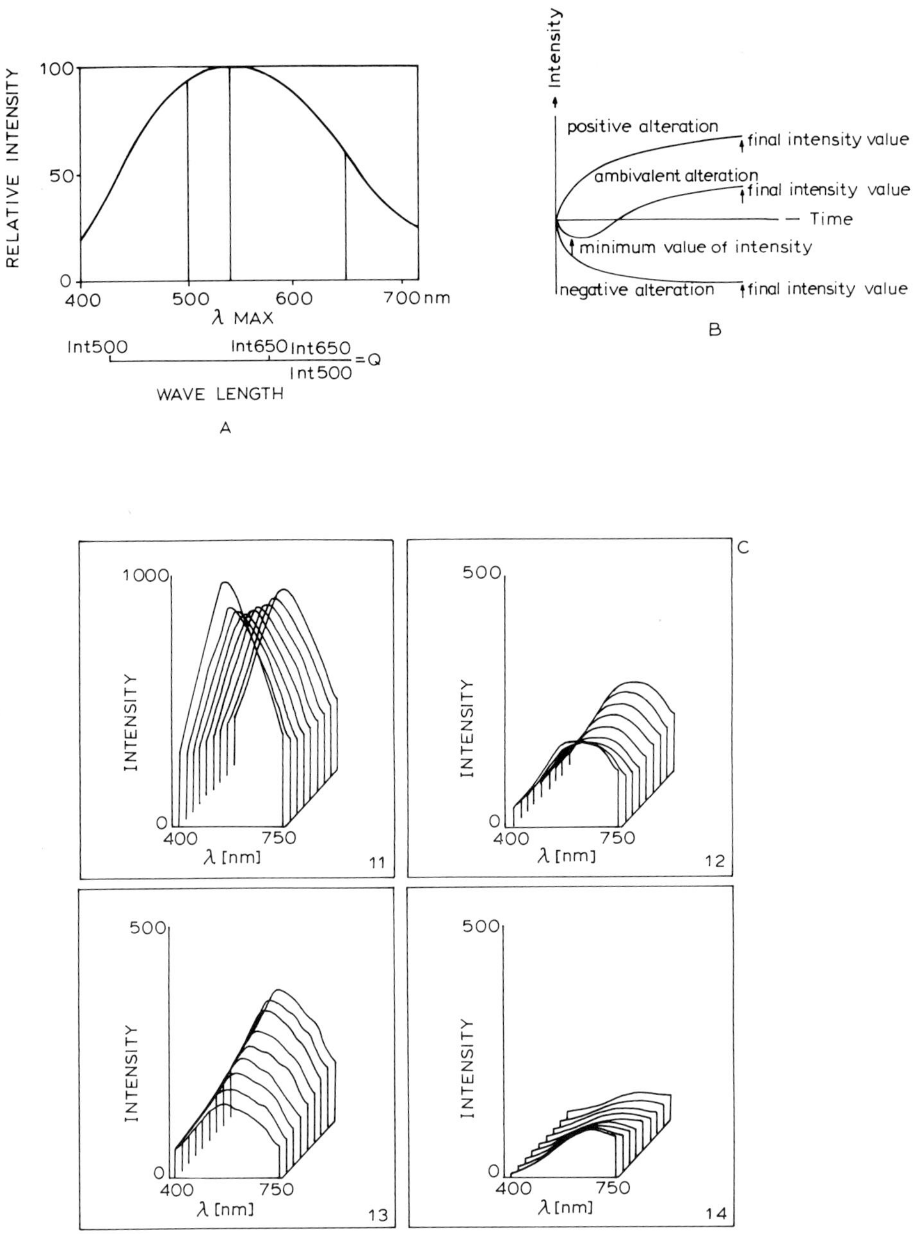

Fig. 9-30. Illustration of fluorescence parameters.
(A) Fluorescence emission spectra of sporinite, showing the relation between wavelength and relative intensity.
(B) Schematic diagram showing three types of fluorescence alteration (after Ottenjann, 1988).
(C) 3-D fluorescence alteration (after Ottenjann, 1988), showing: 11 = strong negative alteration; 12 = ambivalent alteration; 13 = positive alteration; and 14 = weak negative alteration.

ment of fluorescence alteration of alginite, sporinite, and bitumens was described by Ottenjann (1988; Fig. 9-30C).

Figure 9-30A shows typical fluorescence emission spectra of an organic matter and the method of calculation for lambdamax and the red/green quotient. Figure 9-30B illustrates a schematic presentation of positive, negative, and ambivalent alteration. Alteration (relative intensity) can be measured relative to a standard (e.g., uranyl glass) or without any standard (Ottenjann, 1988). Jacob (1972) measured the fluorescence intensity taking masked uranyl glass (standard) as 100% (1.00 in his scale). The absolute fluorescence intensity of various coal macerals was first measured by Bensley and Davis (1988) using calibrated light-emitting diodes (Fig. 9-31C).

Apart from the above parameters, Hagemann and Hollerbach (1981) and Thompson-Rizer and Woods (1987) used a colorimetric system (using x, y, z axes) for the maturity determination of source rocks and coal. Some other parameters, like relative intensity at 490 nm (Thompson-Rizer and Woods, 1987) and violet/blue ratio (Von der Dick and Kalkreuth, 1984) for determining the maturity of kerogen and bitumen, suggest that a relative decrease in blue color is a sensitive maturation parameter. Lo (1988) showed a relation between vitrinite reflectance and lambdamax or Q-value from a series of coal macerals by using violet excitation. From the combined multivariate analysis of fluorescence spectra and nonlinear numerical modeling of liptinite macerals and extractable hydrocarbons, Michelson and Khorasani (1990) developed a new fluorescence maturity parameter called "Omega Factor". This parameter has not yet been properly established.

Similar to the vitrinite reflectance, at least 10 – 25 grains/points of fluorescent objects are measured. Mean and standard deviation should be used for maturity determination.

Sample preparation

Coal and dispersed organic matter (kerogen fraction)

The sample preparation for the fluorescence measurement is the same as that for the vitrinite reflectance measurement (Fig. 9-2). The samples, however, do not need good polishing. For the study of maturation and migration phenomena of coal and organic sediments, it is advisable to use polished core samples without epoxy resin (if possible). Cuttings samples should be plugged in epoxy. If possible, for quantitative fluorescence measurement, whole rock or whole coal should be used instead of isolated kerogens in order to avoid the influence of chemicals needed for kerogen isolation. Within reasonable limits, in determining maturity by fluorescence one may use smear slide of kerogen concentrate (without cover glass) or polished kerogen plug.

Bitumen, crude oil and condensate

Kieselgur G (as a thick paste with water) should be spread on a rough glass slide (3 inches by 1 inch) and air dried. This dried glass slide with Kieselgur should be dipped in a container filled with solvent-extracted bitumen or crude oil mixed with toluene and methanol (3 : 1 ratio) and dried (Hagemann and Hollerbach, 1981).

After preparation, the slide containing the crude oil or bitumen should be kept in a dark place and studied within half an hour. In another method, the sample is simply spread on a glass slide and air-dried (Bertrand et al., 1986; Ottenjann, 1988).

Applications

Fluorescence of organic matter is observed only within a narrow range of maturation: 0.15 – 1.5% R_0. Fluorescence parameters can be used in identifying the diagenetic and part of the catagenetic stages of organic matter maturation. Because of the lack of proper standardization up to date, however, spectral fluorescence of organic matter has not yet been utilized for various geological applications.

Fluorescence as maturation and migration parameter

Earlier researchers demonstrated that, with increasing rank of both kerogen and bitumen, fractions of coal and petroleum source rocks, as well as crude oil and condensate, show progressive changes in lambdamax, *Q*-value, and alteration (Ottenjann et al., 1974; Teichmüller, 1974; Van Gijzel, 1975; Teichmüller and Wolf, 1977; Teichmüller and Ottenjann, 1977; Alpern and Cheymol, 1978; Ottenjann, 1981; Hagemann and Hollerbach, 1981; Van Gijzel, 1982; Crelling, 1983; Von der Dick and Kalkreuth, 1984; Bertrand et al., 1986; Thompson-Rizer and Wood, 1987; Ottenjann, 1988).

In the case of kerogen fraction, lambdamax shifts towards red in higher maturation, which was explained to be due to the elimination of nonfluorophoric media and the increase in intramolecular quenching (Lin and Davis, 1988). Lambdamax and the *Q*-value obtained from ultraviolet excitation of both kerogen and bitumen fractions of various organic matter, are often correlated with increasing vitrinite reflectance, which is the most standardized maturation parameter (Ottenjann, 1980; Teichmüller, 1982; Thompson-Rizer and Wood, 1987) (Fig. 9-33A). Goodarzi et al. (1989) showed a similar relation between vitrinite reflectance and lambdamax of *Tasmanite* algae and dinoflagellate from a series of source rocks from the Canadian Arctic Archipelago (Fig. 9-31B). Few data exist on the maturation of kerogen and bitumen fractions from a single organic matter sample (Mukhopadhyay and Rullkötter, 1986; Ottenjann, 1988). Absolute fluorescence intensity of various macerals from the same coal sample often shows different fluorescence spectra (Fig. 9-31C). This suggests that for the accurate determination of a maturity profile of a series of samples of different rank, the same maceral type should be measured (Fig. 9-31A).

Fig. 9-31. (A) Relation between vitrinite reflectance and lambdamax (left side) and *Q*-value (right side) of liptinite macerals from the various coal samples (after Ottenjann, 1981).
(B) Relation between vitrinite reflectance and lambdamax of *Tasmanales* (telalginite) and dinoflagellate (lamalginite) (after Goodarzi et al., 1989). ○: samples of *Tasmanales* algae, Posidonia Shale (West Germany); ●: *Tasmanales* algae, Eden Bay Formation (Arctic Canada); and □: dinoflagellates, Eden Bay Formation (Arctic Canada).
(C) Absolute fluorescence intensities of various macerals measured in different coals (after Bensley and Davis, 1988).

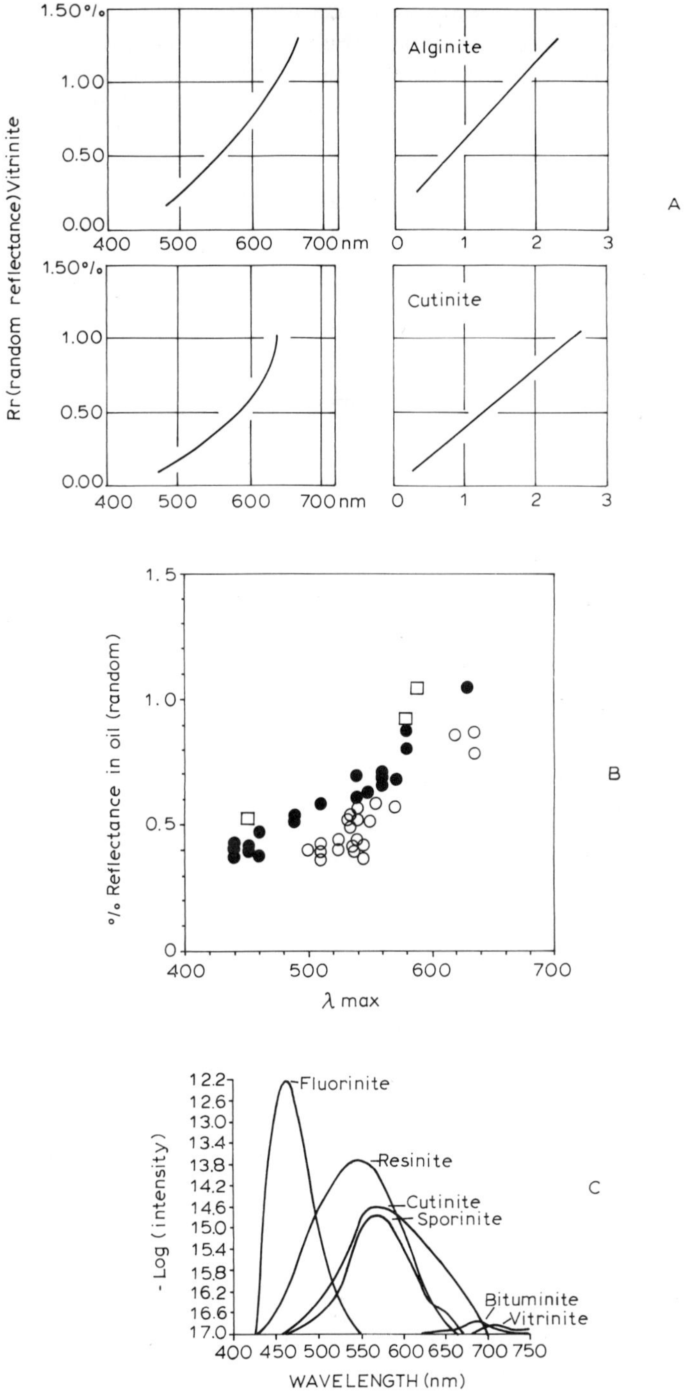
Rr (random reflectance) Vitrinite
1.50%
1.00
0.50
0.00
400
500
600
700 nm
0
1
2
3
Alginite
Cutinite
A
% Reflectance in oil (random)
1.5
1.0
0.5
0
400
500
600
700
λ max
B
- Log (intensity)
12.2
12.6
13.0
13.4
13.8
14.2
14.6
15.0
15.4
15.8
16.2
16.6
17.0
400 450 500 550 600 650 700 750
WAVELENGTH (nm)
Fluorinite
Resinite
Cutinite
Sporinite
Bituminite
Vitrinite
C

Fig. 9-31.

Considering the lack of data on the maturation of kerogen and bitumen fractions of a single sample, two complete suites of samples (coals from Cretaceous of Alberta, Canada, and shales from Lias Epsilon, Jurassic of West Germany) of variable maturity (R_0 ranging from 0.5 to 1.45%) were analyzed using ultraviolet excitation (Mukhopadhyay and Rullkötter, 1986). The analyzed coals are of kerogen Type IIB

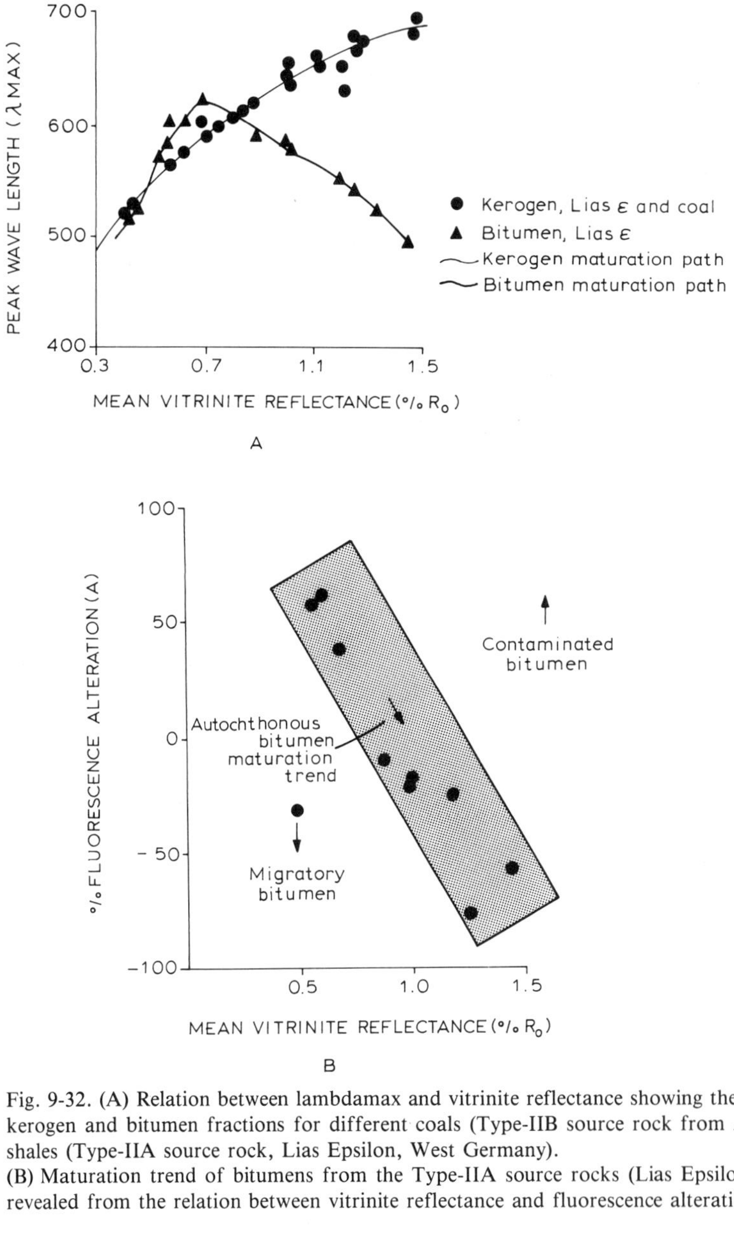

Fig. 9-32. (A) Relation between lambdamax and vitrinite reflectance showing the maturation trends of kerogen and bitumen fractions for different coals (Type-IIB source rock from Alberta, Canada) and shales (Type-IIA source rock, Lias Epsilon, West Germany).
(B) Maturation trend of bitumens from the Type-IIA source rocks (Lias Epsilon, West Germany) as revealed from the relation between vitrinite reflectance and fluorescence alteration.

(Type II – III of Tissot and Welte, 1984) and contain some *Botryococcus* (telalginite; Hutton, 1987) algae. The Jurassic shales are of kerogen Type IIA (Type II of Tissot and Welte, 1984) and contain abundant *Tasmanales* (telalginite; Hutton, 1987) algae. In the kerogen fraction, the lambdamax of both types of telalginite showed a progressive shift towards the red spectrum when correlated with vitrinite reflectance (R_0 ranging from 0.5 to 1.45%; (Fig. 9-32A). The *Q*-value progressively increased from 0.75 to 2.75 compared to a vitrinite reflectance of 0.5 to 1.45% R_0. Up to 0.7% R_0, the fluorescence alteration of the telalginites (kerogen fraction) is positive. Between 0.7 and 1.0% R_0, telalginites show both positive and negative alterations. Beyond 1.0% R_0, the alteration is negative. The lambdamax and *Q*-value of the bitumen fractions of shale samples show a different trend when compared with vitrinite reflectance (Fig. 9-32A). Bitumen shows a red shift up to 0.7% R_0. Beyond 0.7% R_0, bitumens show a reverse trend and shift towards blue color again. This reverse trend of maturation of the bitumen fraction compared to that of the kerogen fraction of the same sample may be explained by the formation of redistributed bitumen (secondary bitumen of Ottenjann, 1988) due to the primary migration within the organic matter network. This phenomenon can be explained by the formation of more nonfluorophoric substances through the cracking of asphaltene-rich primary bitumen and thereby hindering intramolecular quenching. In the kerogen network of alginite, the structure becomes more polycondensed during the red shift after releasing more nonaromatic hydrocarbons and thereby enhancing intramolecular quenching. Figure 9-32B shows the relation between fluorescence alteration of the bitumen extracts and vitrinite reflectance of Lias Epsilon samples. This relation illustrates the maturation trend of the autochthonous bitumen and helps to detect contaminated or allochthonous bitumen.

A study of twelve crude oil and condensate samples from North Slope, Alaska (Mukhopadhyay and Rullkötter, 1986), showed that fluorescence microscopic parameters (lambdamax, *Q*-value and alteration), used for coal and dispersed organic matter, can reveal the maturity difference between normal crude oil and condensates and biodegraded crude oils (biodegradation is identified by gas chromatographic fingerprinting). Figure 9-33A shows how fluorescence spectra of a crude oil changes after 30 minutes of ultraviolet excitation (alteration). Figures 9-33B and C illustrate the relation between fluorescence alteration and lambdamax or lambdamax at the initial (0 – 1 minute) and final (30 minutes) phases of excitation. These two figures demonstrate the progressive change in lambdamax and alteration from biodegraded or immature crude oil to normal crude oil to condensate, which illustrates the difference in maturity. The highest alteration was recorded for the condensate samples, and the least for the immature oil. The positive alteration of immature crude oil may be related to the creation of additional absorption centers (double bonds) and increased molecular rigidity as a result of photochemical oxidation and enhancement of crosslinking (R. Lin, personal commun.; Davis et al., 1990), possibly by cracking of asphaltene. Negative alteration of mature crude oil and condensate may be related to polymerization or condensation, or caused by oxygen quenching and subsequent oxygen incorporation (Davis et al., 1990).

Van Gijzel (1982) distinguished solid bitumens and crude oils according to various fluorescence parameters and showed the general characteristics of paraffinic, in-

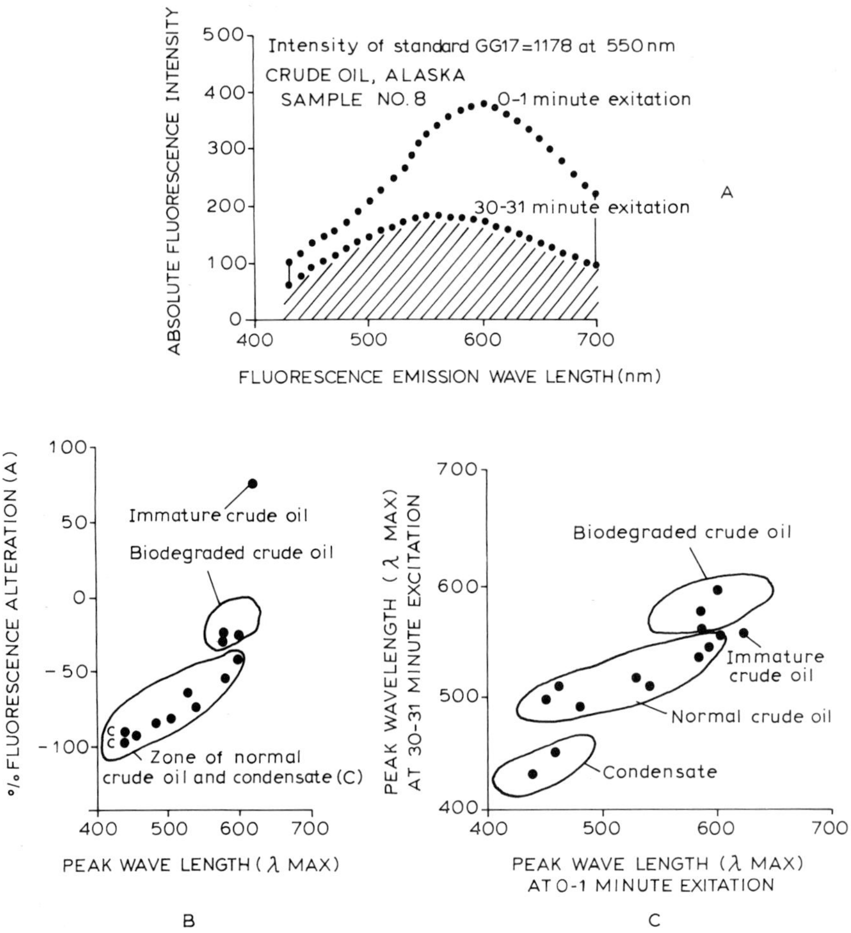

Fig. 9-33. (A) Fluorescence alteration (negative in this case) of a typical crude oil from Alaska.
(B) Maturation trend of various oils and condensates as revealed by the relation between fluorescence alteration and lambdamax.
(C) Maturation trend of various crude oils and condensates as seen from the relation between lambdamax at initial measurements and measurements after 30 minutes of excitation.

termediate and naphthenic oils. Hagemann and Hollerbach (1986), using a series of crude oil and condensate samples, compared the fluorescence spectra of whole oil/condensate samples and aromatic and heterocomponent fractions of the oils and condensates. Similar to the observation of Mukhopadhyay and Rullkötter (1986), Hagemann and Hollerbach (1986) demonstrated the maturation of crude oils from a continuous shift of lambdamax from 550 nm to 440 nm. Heavy biodegraded crude oils have a lambdamax around 590 – 610 nm. Bertrand et al. (1986) demonstrated that the lack of fluorescence in vitrinite macerals is related to a high concentration of aromatic hydrocarbons. They also pointed out that loss of fluorescence of liptinite macerals with advanced maturation is related to an increasing aromatization

TABLE 9-5

Relationship between fluorescence maturation parameter (lambdamax), maturity boundaries and nature of hydrocarbon generation

Maceral	Lambdamax (nm)	Maturity	Hydrocarbon generation
Alginite	490 – 510	Immature	Minor HC
	540 – 560	Mature	Beginning of oil window
	600 – 640	Mature	Principal phase of HC generation
	670 – 690	Overmature	End of oil window
	690 – 700	Overmature	Wet gas zone
Amorphous liptinite IIA	520 – 560	Immature	Minor HC
	560 – 600	Mature	Beginning of oil window
	600 – 640	Mature	Major HC generation
	650 – 660	Overmature	End of oil window
	660 – 700	Overmature	Wet gas zone
Cutinite	500 – 520	Immature	Minor HC
	530 – 560	Mature	Beginning of oil window
	560 – 610	Mature	Major HC generation
	640 – 660	Overmature	End of oil window
	660 – 700	Overmature	Wet gas zone

of the matrix. Bertrand et al. (1986) also compared whole oil and deasphaltized crude oil samples and showed that asphaltene quenches fluorescence.

Defining hydrocarbon generation

Unlike vitrinite reflectance, the degree of maturity measured by fluorescence can be related to hydrocarbon generation from the source rock, because most crude oils are generated from liptinite macerals (Tissot and Welte, 1984; Mukhopadhyay et al., 1985a). Liptinite maturation and hydrocarbon generation, therefore, follow a similar trend in spite of various heating rates or activation energies provided only macerals like alginite (telalginite and lamalginite), cutinite and amorphous liptinite IIA (Mukhopadhyay, 1989a) are measured. Table 9-5 illustrates the relation between lambdamax of liptinite macerals (alginite, cutinite and amorphous liptinite IIA) and hydrocarbon generation.

Identifying maceral types and solid bitumen

Fluorescence parameters (lambdamax and *Q*-value) are utilized to differentiate various maceral types within a single liptinite maceral (example: resinite) (Crelling, 1983; Mukhopadhyay and Gormly, 1984; Pasley and Crelling, 1988). Absolute fluorescence intensity identifies characteristic features of various macerals (Bensley and Davis, 1988; Fig. 9-31C).

From the relationship of the solubility in carbon disulphide and absolute fluorescence intensity, Jacob (1967, 1989) distinguished three series of solid bitumens having different chemical compositions: (1) ozocerite; (2) asphalt, gilsonite, etc.; and (3) wurzelite, albertite, etc.

Identification of oxidation and geological phenomena

Oxidized sporinite, alginite and oxidized bitumen or crude oil (biodegraded) generally show a red shift (lambdamax) compared to the autochthonous population.

In areas of igneous intrusion, nonfluorescent liptinite macerals exist in surrounding fluorescent liptinites. Spectral fluorescence, however, has not yet been utilized to identify the effects of an unconformity or faulting, as does vitrinite reflectance.

On examining a series of overmature source rocks (R_o = 1.4 – 2.0%) associated with overpressuring, Mukhopadhyay and Wade (1990) observed that alginite retains red fluorescence at that stage of maturity. They suggested that the retardation of red shift in an overmature sequence was due to overpressuring.

Problems related to measuring spectral fluorescence

The problems related to spectral fluorescence measurement have not been fully established yet because of lack of a data base similar to that for vitrinite reflectance.

The main problem for spectral fluorescence or absolute intensity measurement is the lack of a uniform standard and instrument calibration. Thompson-Rizer et al. (1988) showed the variation of spectral fluorescence measurements by various laboratories using three plexiglass standards. Jacob (1967) utilized masked uranyl glass as a standard for intensity measurements, whereas Bensley and Davis (1988) used light-emitting diodes for absolute fluorescence intensity measurements.

Another problem lies in the measurement of amorphous liptinite. Spiro and Mukhopadhyay (1983) showed that biodegraded algae (bituminite or sapropelinite or amorphous liptinite) showed an earlier red shift than the corresponding non-

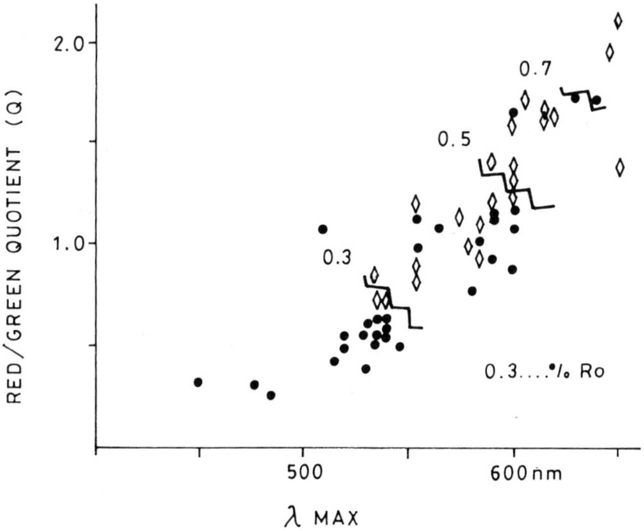

Fig. 9-34. Relationship between lambdamax and Q (red/green quotient) showing the variation of maturation in algae (solid circles) and degraded algae (diamond signs) showing vitrinite reflectance boundaries. (After Spiro and Mukhopadhyay, 1983.)

biodegraded algae (Fig. 9-34). The lambdamax and *Q*-value of degraded algae at maturities below 0.7% R_0 is similar to nondegraded structured algae of higher maturation. For the determination of maturity by spectral fluorescence, only non-biodegraded liptinite macerals should be used.

PULSED LASER FLUORESCENCE

The use of pulsed laser fluorescence microscopy is an extension of continuous fluorescence microscopy applied for the characterization of coal, kerogen, bitumen, crude oil and condensate. This method uses a nitrogen dye laser for excitation and for the measurement of the resultant fluorescence decay times in the nanosecond or subnanosecond range and spectra. Landis et al. (1987) used this technique to study liptinite macerals in coal, whereas Mukhopadhyay et al. (1987) and Gangopadhyay et al. (1988) utilized this method for the first time to characterize a series of crude oils and condensates with variable maturity and type.

Principle, instrumentation and sample preparation

Laser fluorescence utilizes nitrogen laser excitation at a fixed wavelength, which gives the resultant decay of each individual fluorophore. Each fluorophore has a

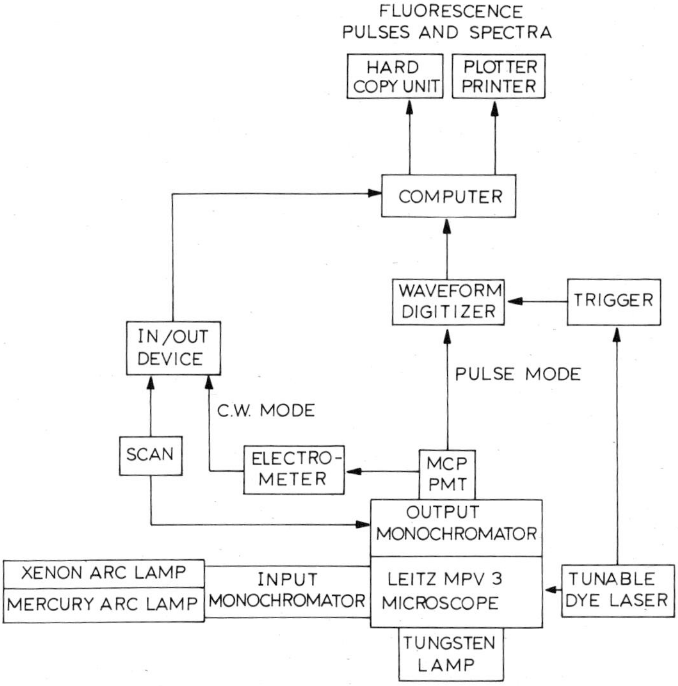

Fig. 9-35. Schematic diagram of the various components of time-resolved laser fluorescence spectroscopy. (After Gangopadhyay et al., 1988.)

characteristic decay time which is measured in nanoseconds. This technique allows nondestructive in-situ observation of fluorescing solids and liquids. For the liquid, it is possible to identify the individual components by comparing their time-resolved spectra with those of reference solutions.

A schematic diagram of the time-resolved fluorescence microscopic unit is shown in Fig. 9-35. The details of the basic data for the instruments are shown in Table 9-6 (Landis et al., 1987). A nitrogen-pumped tunable dye laser using an emission peak at 373 nm gives 13 mJ pulses of 700 ps full-width at half-maximum at a rate of 10Hz (Gangopadhyay et al., 1988). The light pulses are directed to the sample via a liquid light guide using Leitz MPV3 microscope. The emitted fluorescence pulses from the sample are directed by a two-stage proximity-focussed microchannel plate (MCP) photomultiplier tube. The individual output pulses of the MCP are acquired and signal-averaged by a Tektronix digitizer. The measured signal is the convolution of the actual fluorescence decay associated with the instrument response.

Time-resolved fluorescence spectra are recorded in 10-nm intervals by scanning the emission monochromator of the microscope from 400 to 800 nm acquiring the fluorescence decay at each wavelength. The overall emission spectrum is separated

TABLE 9-6

Basic data of the laser instrument for time-resolved fluorescence spectroscopy

Diameter of analyzed area	5 μm
Tuning range of emission monochromator	220 – 800 nm
Range used	380 – 800 nm
Monochromator bandwidth	1 – 7 nm
Continuous fluorescence excitation	
Tuning range of excitation monochromator	220 – 800 nm
Typical excitation wavelength	365 nm
Monochromator bandwidth	1 – 7 nm
Fluorescence excitation by pulsed laser	
Dye laser pulse duration (FWHM)	0.7 ns
Pulse energy (BDBP dye, 373 nm, 10Hz)	10 μJ
Peak power	10 kW
Laser bandwidth (FWHM)	0.04 nm
Dye laser tuning range	360 – 800 nm
Excitation wavelength used	373 nm
Pulse repetition rate	1 – 100 Hz
Number of photons emitted per pulse (BDBP dye)	2×10^{13}
Photons onto sample	1×10^{13}
Photons onto measured region of sample (– 5 μm)	10^{11} – 10^{12}
Typical fluorescence yield	0.001
Photons reaching MCP photomultiplier after passing emission monochromator	10^4
Typical number of photoelectrons per pulse	$10^2 – 10^3$
Instrument function risetime (laser pulse + photomultiplier + digitizer preamplifier)	0.7 ns
FWHM of instrument function	1.1 ns
Single pulse digitization rate	10 – 100 GHz
Number of pulses signal averaged, typical	64

into its component spectra by plotting the product $A_i t_i$ for each individual fluorescence decay at 10-nm intervals. A_i is the pre-exponential coefficient and t_i is the fluorescence decay time of the ith component. The system spectral response corrects the A-coefficients. The At spectra are the fluorescence spectra of the component fluorophore. The mathematical deduction of the A-coefficients is given by Gangopadhyay et al. (1988).

The sample preparation for the whole rock and kerogen concentrate are similar to the method adopted for continuous fluorescence. For crude oil and condensate samples, the method adopted was similar to that of Hagemann and Hollerbach (1981), which was described in the section on continuous fluorescence spectroscopy. A part of this study has been published earlier by Gangopadhyay et al. (1988).

The fluorescence emission in oils and condensate is mainly controlled by aromatics like benzene, naphthalene, phenanthrene, anthracene, pyrene and benzonaphthalene. The asphaltenes present in crude oil act as quenchers to the fluorescence emitted by aromatics. The fluorescence decay time measurement of crude oils is controlled by the interaction rate between a fluorophore and a quencher; individual oils have different interaction rates.

TABLE 9-7

List of samples analyzed for time-resolved fluorescence spectroscopy showing location, age, formation, API gravity and type of oil and condensate samples

Sample	Location	Age/formation	°API	Type
A	Smith County, Texas	Cretaceous, Travis Peak Formation, Chapel Hill	58	Condensate
B	Offshore South Africa	Cretaceous	57	Condensate
C	Smith County, Texas	Cretaceous, Travis Peak Formation, Chapel Hill	54	Condensate
D	Houston County, Texas	Oligocene, Frio Formation, A-Sand, Delee no. 2	49.5	Condensate
E	Smith County, Texas	Cretaceous, Travis Peak Formation, Chapel Hill	45	Crude oil
F	Santa Maria Basin, California	Miocene, Monterey Formation	41	Crude oil
G	Chamousca Field, Duval County, Texas	Eocene, Wilcox Group	35.4	Crude oil
H	Bee County, Texas, FRL 3010 – Mobil Oil	Eocene, Wilcox Group	31.6	Crude oil
I	Lovetts Creek, 3440–008, Alabama	Jurassic, Smackover Formation	30.3	Crude oil
J	Duval County, Texas	Eocene, Wilcox Group	20.7	Crude oil
K	Officina Reservoir, Venezuela	Lower Miocene	15	Heavy crude oil
L	Santa Maria Basin, California	Miocene, Monterey Formation	15	Heavy crude oil
M	Soner Ranch, San Miguel tar belt, South Texas	Cretaceous	5	Asphalt

Results

The location, formation, age and API gravity of the analyzed thirteen samples of crude oils and condensates are given in Table 9-7. The emitted fluorescence of all samples was heavily quenched due to strong interaction between quenchers and emitters. The fluorescence of these samples is due to the presence of several fluorophores possessing characteristic spectra. The combined effect of these spectra yield one broad composite continuous fluorescence spectrum when working in spec-

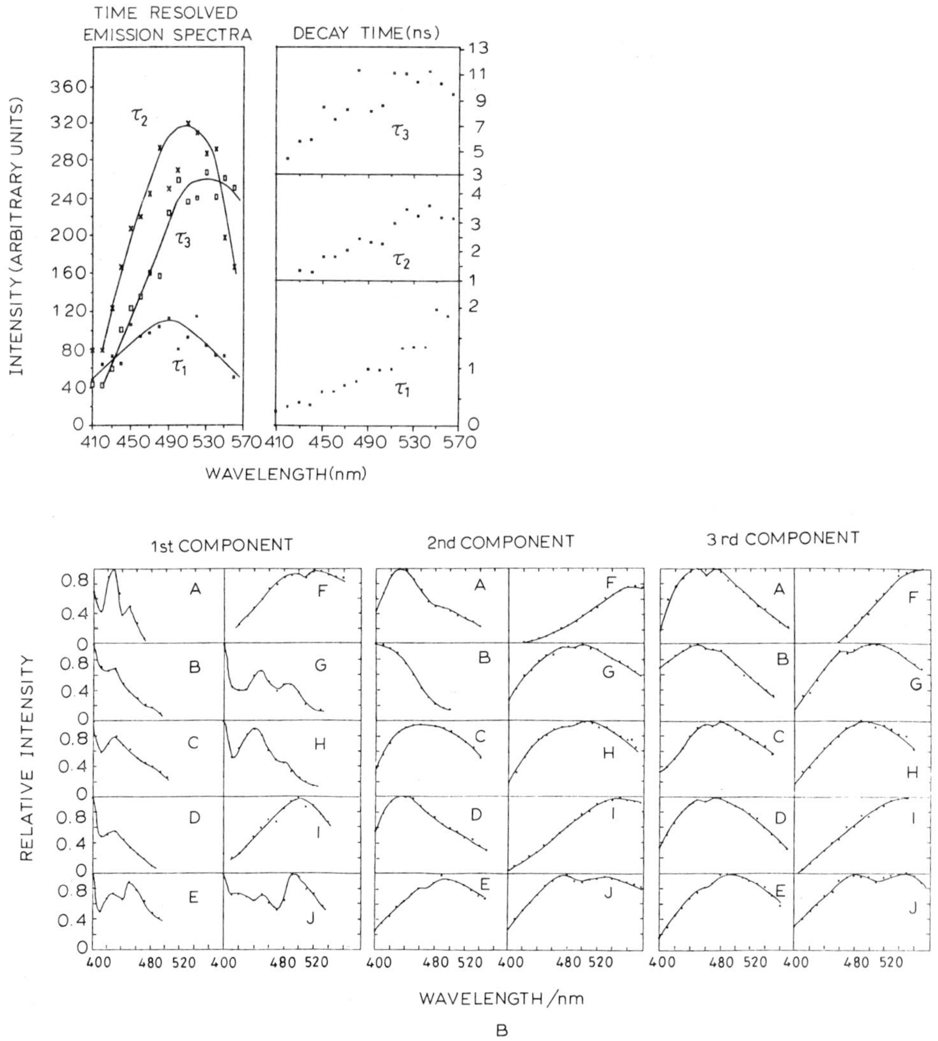

Fig. 9-36. (A) Time-resolved fluorescence emission spectra and decay times of heavy crude oil K. (After Gangopadhyay et al., 1988.)
(B) Pulsed fluorescence data of the decay times of the three components of all analyzed crude oils and condensate samples; t_1, t_2, t_3 = decay times of three different fluorophores.

tral fluorescence. The time-resolved study of these oil samples can be decomposed into three dominant fluorophores (Fig. 9-36A).

Figure 9-36A shows the time-resolved spectra ($A_i t_i$) versus wavelength and the corresponding fluorescence times of the three components of heavy crude oil sample K. In contrast to the fluorescence decay of condensate sample C, the lifetimes of the three components increased continuously with increasing wavelength. This change indicates that a three-component fit is not appropriate, and that the sample most probably consists of more fluorophores than can be resolved. Similar results were obtained from heavy crude oil sample M.

The component spectra of the three time-resolved fluorophores are distinct for various condensates and crude oils (Fig. 9-36B). South African condensate (sample B) and Texas condensates (samples A, C, and D) can be distinguished based upon their different time-resolved spectra. On the other hand, similar compositions are suggested for crude oils E and G, as seen from their six time-resolved spectra. The time-resolved spectra of California crude oil (sample F) matches that of Alabama crude oil (sample I), but their spectra differ much more than those of crude oils E and G. Sample J, a biodegraded crude oil defined by gas chromatography, has a spectrum which indicates some abnormality.

Figure 9-37 shows the plot of decay times of three components versus API gravity of all samples. This represents a significant correlation between API and fluorescence lifetimes from these diverse samples.

The fluorescent compounds present in these samples resembled the time-resolved

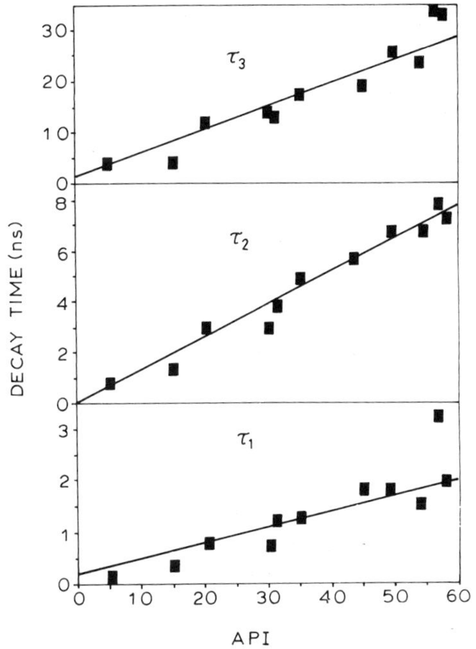

Fig. 9-37. Relation between three fluorescence decay times and API gravity of various crude oils and condensate samples.

spectra and decay times of two organic compounds, fluorescent properties of which are well documented. The spectrum and decay time of the short-lived component are similar to those of anthracene, whereas the long-lived (third) component spectrum and decay time resemble those of pyrene. The second component spectrum was different for different samples. No suitable reference compound, therefore, was found to match this component (Gangopadhyay et al., 1988).

Applications

Maturation of crude oil

According to organic geochemical data (GC pattern, stable carbon isotopes of saturated and aromatics fractions, etc.), samples A, B, C, D, E, G, H, I, and L are normal crude oils and condensates, whereas samples J, F, and M are biodegraded oils. A plot of two decay times (t_1 and t_2) shows some distinct trend (Fig. 9-38). If the trend of the normal crude oils and condensates are considered, this diagram may represent a maturation trend. In that case, it indicates that sample M, which is a biodegraded heavy crude oil (5° API) was originally derived from a matured crude oil (between 31° API and 41° API). Biodegraded crude oil F similarly was originally a normal crude oil (32° API).

Terrestrial versus marine origin of oil

Crude oil sample E and condensate samples A and C are derived from the similar reservoir rock of the Travis Peak Formation (East Texas). From the organic geochemical parameters (GC pattern, stable carbon isotopes of the saturate and

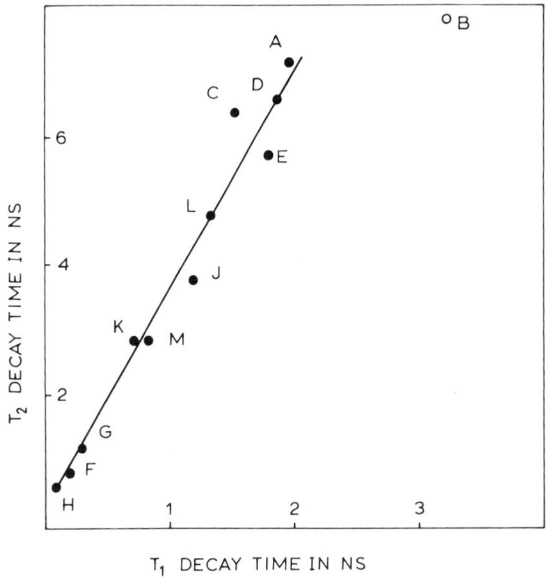

Fig. 9-38. Linear relation between time-resolved fluorescence decay times (first and second, which have shortest and intermediate decay times), showing the maturation trend of the samples. Letters *A, B, C,* etc. designate numbers of various crude oil and condensate samples.

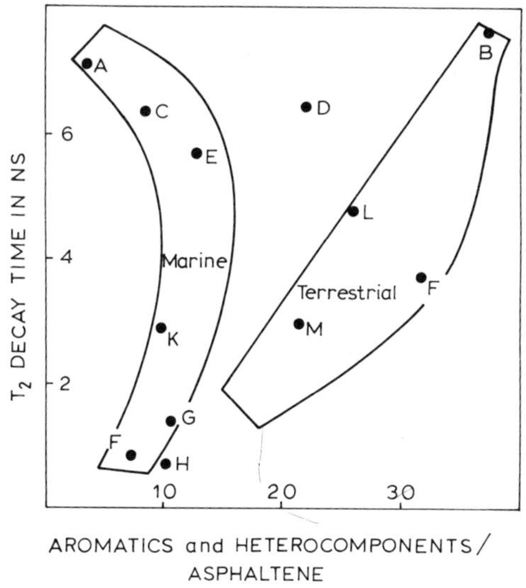

Fig. 9-39. Relationship between intermediate decay time (t_2) and the (aromatics + heterocomponents)/(asphaltenes) ratio. The former two are fluorescence emitters, whereas the latter is the fluorescence quencher. The graph shows trends for oils and condensates of terrestrial and marine origin.

aromatics fractions, etc.), it is known that condensate samples A and C and crude oil sample E were derived from the same marine clastic source rock and are very similar. The three spectra and corresponding decay times (especially t_2) of these three samples are similar. The slight difference in decay time and spectra of t_3 is due to the difference in the degree of maturity. These are totally different from the condensate sample B and crude oil sample G, indicating a different origin. These data indicate that the pulsed laser fluorescence method can correlate oils of the same origin in a basin in spite of differences in the degree of maturity.

Figure 9-39 shows the plot of the t_2 decay time of the oils with the ratio of aromatics + heterocomponents (fluorescence emitter) to asphaltenes (fluorescence quencher) derived from liquid chromatography. Accordingly, oil and condensate samples A, C, E, K, G, and F fall totally on a different path than oil and condensate samples B, F, L, and M. From the geochemical analysis, it is known that oil samples A, C, E, and K are derived from marine clastic source rock, whereas B, L, and F are derived from terrestrial source rocks. These preliminary data indicate that the pulsed laser fluorescence method may help in oil – oil correlation (especially condensate – oil correlation).

CORRELATION OF MATURATION PARAMETERS

As discussed earlier from the kinetic standpoint, measured maturation parameters of different organic matters should be correlated with extreme care. Each type of organic matter reacts differently to heat and has a different conductivity to resist heat. The reaction to different heating rates and effective heating times would be

TABLE 9-8

Various microscopic maturation parameters as related to transformation of macerals and hydrocarbon generation at different stages of maturation

PROCESS	COAL RANK	VITRINITE % Ro	THERMAL ALTERATION INDEX TAI *	TAI **	CONODONT ALTERATION INDEX (CAI)	FLUORESCIENCE PARAMETERS: COLOUR OF ALGINITE ***	λ MAX (NM)***	ALTERATION (%)***	TASMAN. ALG. (Q) ****	MINERAL MATRIX**** Q	% I	TRANSFORMATION OF MACERALS: AMORPHOUS LIPTINITE (I,IIA)	VITRINITE	ZONES OF HC GENERATION
DIAGENESIS	PEAT	0.2	1 YELLOW	1.5	1 YELLOW	GREENISH YELLOW	(+ 10nm)					Formation of solid bitumen	Free cellulose; recognizable plants	Biogenic methane & early condensate
	LIGNITE	0.3					500	+20 to +50				Formation of liquid bitumen in pore spaces	No free cellulose; recognizable plants	
	SUB-BITUM C	0.4		2.3					0.5	0.8	75		Geochemical gelication, formation of vitrinite and exsudatinite	
CATAGENESIS	SUB-BITUM B, A	0.5	2 ORANGE	2.5		GOLDEN YELLOW	540	+10 to +20	0.7	1.0		Formation of liquid bitumen, semi-inert and inert macerals like micrinite, etc.		Oil, wet gas
	HIGH VOL. BITUMIN C	0.6							1.0	1.2	25–75			
	HIGH VOL. BITUMIN B	0.7							1.3				major exsudatinite, formation of micrinite	Oil window
		0.8		2.8	2 LIGHT BROWN	DULL YELLOW	600	0 to +10	1.5	1.4				
	HIGH VOL. BITUMIN A	0.9							1.8	1.6				Start of major dry gas
		1.0	3 BROWN	3.0		ORANGE	640	−10		1.8	5	Formation of pryobitumens and clusters of inert macerals	Major anistropy and micrinite development	
	MEDIUM VOLATILE BITUMIN	1.2								2.0				
		1.35			3 BROWN	RED	680	−10 to −70		2.2				
	LOW VOL. BIT.	1.5		3.5		NON FLUORESCENT					2	Mainly pyrobitmens and inert macerals like clustered micrinite		Wet gas & dry gas
METAGENESIS	SEMI-ANTHRAC.	2.0	4 BLACK	3.7					NO VISIBLE ORGANIC FLUORESCENCE			Clusters of inertodetrinite and clustered micrinite	Anistropy increased, development of fine pores	Dry gas
	ANTRACITE	2.5			4 DARK BROWN									
		3.0	5 BLACK											
		4.0		4.0	5 BLACK									
METAMORP-HISM	META-ANTHRAC.	5.0												

* TAI after Staplin (1969)

*** Fluorescent parameters after Mukhopadhyay & Rullkotter (1986) different. The common practice of the organic petrographer and geochemist is, however, to present a table correlating various maturity parameters in order to define the boundaries of hydrocarbon generation and changes that occur within the organic matter network (Gutjahr, 1966; Staplin, 1969; Hunt, 1979; Bostick, 1979; Heroux et al., 1979; Tissot and Welte, 1984; Teichmüller, 1988).

Table 9-8 shows the correlation of some of the important microscopic organic maturation parameters with (1) vitrinite reflectance; (2) coal rank; (3) fluorescence parameters; (4) Thermal Alteration Index; (5) Conodont Alteration Index; (6) morphological changes within the huminite/vitrinite and amorphous liptinite (I, IIA); and (7) the probable hydrocarbon generation at different stages of diagenesis, catagenesis, metagenesis and early phase of metamorphism. This correlation applies to different basins where the heating rate is more or less the same (20 – 25°C per 100 m) and which have a mixed Type-IIA – IIB organic matter. This correlation, however, is not for a universal correlation of organic maturation parameters.

SUMMARY AND CONCLUSIONS

The extensive review in this chapter explains the genesis, chemistry, and maturation of organic matter. The applications and limitations of two major microscopic maturation parameters, i.e., vitrinite reflectance and fluorescence spectrometry are presented in detail. Pulsed laser fluorescence microscopy is also introduced as a future maturation tool for the characterization of crude oils and condensates as well as kerogens in coal and petroleum source rocks. The main conclusions can be summarized as follows:

(1) Vitrinite reflectance should be considered as a useful maturation indicator provided:

(a) standardization of the instruments and sample preparation techniques are properly followed;
(b) chemistry of vitrinite reflectance is clearly understood;
(c) cavings and drilling muds with organic additives are taken into consideration;
(d) the morphologic characters of various vitrinite macerals are properly understood; proper knowledge of solid bitumen, saprocollinite, and bitumen-impregnated vitrinite, which must be excluded in the calculations of mean reflectance, is essential;
(e) proper selection of autochthonous vitrinite is made under the microscope and not from the histogram pattern of all measured grains;
(f) vitrinite reflectance is plotted versus depth on an arithmetic scale and not on a semilog scale, so that the cause of anomalous reflectance can be properly understood; and
(g) it is understood that reflectance of vitrinite is dependent on organic facies, heating rate, kerogen type, chemical kinetics of various vitrinite macerals, and bitumen impregnation.

(2) Maturation of vitrinite (increase in reflectance) and hydrocarbon generation are different processes and should not be universally correlatable phenomena. Vitrinite reflectance is time and temperature dependent, whereas hydrocarbon

generation is more temperature dependent. Vitrinite reflectance is mainly a condensation reaction, whereas hydrocarbon generation is generally a dehydrogenation and cracking reaction in the late diagenetic and catagenetic stages.

(3) Vitrinite reflectance is useful in (a) delineating a regional maturation profile, (b) as a checking parameter for geochemical basin modeling, (c) in distinguishing important geological phenomena, (d) in determining paleoheat-flow in a basin, and (e) in establishing the degree of oxidation in sediments.

(4) Fluorescence parameters (lambdamax, *Q*-value, and alteration) are more useful in determining maturity than vitrinite reflectance for diagenetic and catagenetic stages, especially for carbonates and shales of Type-I and -IIA kerogen and for pre-Devonian sediments. This is due to the fact that changes in liptinite fluorescence and hydrocarbon generation are related processes.

(5) Fluorescence parameters can provide information about maturation of kerogen in coal and source rock as well as bitumen from a source rock, and crude oil or condensate in a single microscopic system. Fluorescence alteration can identify bitumens of various types, which has direct bearing on migration studies or the identification of allochthonous and contaminated bitumens in a source rock.

(6) Pulsed laser fluorescence method utilizes decay time and component spectra of at least three individual fluorophores, which determine maturity of normal and biodegraded crude oil. From the ratio of the decay times and the ratio of asphaltenes to aromatics, various sources of crude oil and condensate can be identified. In the future, it would become an important method for oil – oil (especially condensate – oil) correlation and for determining maturity of biodegraded crude oil. This method will be useful in determining various maceral types (especially liptinite) within a single maceral.

ACKNOWLEDGMENTS

The author is extremely grateful to John H. Calder, W.D. Smith, and the drafting section of the Nova Scotia Department of Mines and Energy, Halifax, Nova Scotia, for preparing most of the drafted figures and other assistance. The manuscript was reviewed by G.V. Chilingarian, University of Southern California, Los Angeles, California; Karl H. Wolf, Woden, Canberra, Australia; K.D. McAlpine and John A. Wade, Atlantic Geoscience Centre, Bedford Institute of Oceanography, Dartmouth, Nova Scotia; and D. Birk, Geofuel Research Inc., Sydney, Nova Scotia. The author is grateful to Wallace G. Dow, D.G.S.I., The Woodlands, Texas, and John A. Wade, Atlantic Geoscience Centre, Geological Survey of Canada, Dartmouth, Nova Scotia, for providing some pressure data in boreholes, and to W.L. Borst and S. Gangopadhyay of Texas Tech University, Lubbock, Texas, for laser fluorescence data. The author also acknowledges the help of M.P. Avery, Atlantic Geoscience Centre, Geological Survey of Canada, Dartmouth, Nova Scotia, for providing microphotography equipment and drafting (by computer) of one figure. The author acknowledges Dr. L.R. Snowdon, ISPG, GSC, Calgary, Canada, for his permission to use data on oil-base drilling fluid.

APPENDIX A: GLOSSARY

First-cycle huminite/vitrinite: These are macerals and maceral types within the huminite and vitrinite group which are formed from plant substances during deposition or early diagenesis.

In coals of bituminous rank, there are three macerals within the vitrinite group: telinite, collinite, and vitrodetrinite. The telinite maceral has two submacerals (telinite 1 and telinite 2), whereas the collinite maceral has four submacerals (telocollinite, gelocollinite, corpocollinite, and desmocollinite). In reflected light, all these macerals and submacerals are gray. As shown earlier, the reflectance of desmocollinite is lower than that of telinite, the reflectance of telinite is lower than that of telocollinite, and the reflectance of telocollinite is lower than that of gelocollinite. Corpocollinite has the highest reflectance. The following are the characteristics of the five vitrinite macerals and submacerals:

Telinite: this maceral shows nongelified plant tissues with well-preserved cells, often filled with corpocollinite or resinite or colloresinite (Plate 9-1C). The cell walls are called telinite. In high-volatile bituminous coal, this maceral is nonfluorescent under ultraviolet excitation. *Textinite* or *Texto-ulminite* (Plate 9-1A) is the low-rank counterpart of telinite ($R_0 < 0.55\%$).

Telocollinite: this submaceral (maceral type in Table 9-2) shows partially homogenized and gelified cells, when cell walls and cell infillings have similar reflectance, or cell lumens are not filled (Plate 9-1D). This maceral is produced from biochemical and geochemical gelification and often shows relict cell structures (Plate 9-1D). Similar to telinite, it is nonfluorescent in ultraviolet excitation. *Eu-ulminite* or *ulminite B* is the low-rank counterpart of telocollinite (for photomicrographs see Mukhopadhyay, 1989b).

Gelocollinite: this is a completely homogenized and gelified submaceral showing a smooth gray color and contains no relict cell structures (Plate 9-1E). In high-volatile bituminous rank, it is nonfluorescent to ultraviolet excitation. In medium-volatile bituminous rank, it shows dark brown secondary fluorescence in blue-light excitation (Ottenjann et al., 1982; Lin and Davis, 1988). *Porigelinite* and *levigelinite* (Plate 9-1B) are the low-rank counterparts of gelocollinite.

Corpocollinite: similar to gelocollinite, corpocollinite is also a completely homogenized submaceral. It, however, always has a round or oval shape and occurs either as isolated bodies or as cell fillings (Plate 9-1F) and, generally, originates from the humic gel. Fluorescent properties of corpocollinite are similar to those of gelocollinite. *Corpohuminite* is the low-rank counterpart of this maceral (for photomicrographs see Mukhopadhyay, 1989b).

Desmocollinite: this maceral shows fine-grained detrital fragments, which are completely homogenized and gelified. This submaceral is possibly derived from humic detritus mixed with a minor amount of liptinitic groundmass (Plate 9-1E). In high- to medium-volatile bituminous rank, this maceral shows brown fluorescence. In reflected white light, it is often associated with bands or patches of granular micrinite in high- to low-volatile bituminous coal rank. *Attrinite* and *densinite* (Plate 9-1G) are the low-rank ($R_0 < 0.55\%$) counterpart of this maceral.

Vitrodetrinite: this maceral shows detrital fragments of telocollinite, gelocollinite, and other vitrinite types. Unlike corpocollinite, this maceral is fragmentary and of angular shape. This maceral is nonfluorescent in ultraviolet excitation and generally shows higher reflectance than corresponding telocollinite or gelocollinite. This maceral is formed by mechanical degradation of plant parts during deposition.

Pseudovitrinite: this maceral shows oriented or nonoriented small slits within vitrinite submacerals (telocollinite and gelocollinite) and has higher reflectance than all other vitrinite macerals or submacerals. This maceral is possibly formed by oxidation in a late diagenetic stage or during sample storage. Pseudovitrinite is nonfluorescent to ultraviolet excitation.

Detro-mixinite: a subgroup of macerals that are fragmented to smaller particles and their identity destroyed by mechanical or bacterial degradation. According to fluorescence, these organic macerals are mainly derived from ligno-cellulose components of plants with a varying mixture of plant lipids.

Mixinite/humosapropelinite: these macerals are granular and gray in color in normal reflected light, granular dark brown in transmitted light, and dark-brown to nonfluorescent in blue-light excitation (grain size less than 5 μm). This maceral is possibly formed by mechanical degradation or bacterial degradation of mainly huminite and minor terrestrial exinite or alginite. *Saprocollinite (saprovitrinite)* is the high-rank ($R_0 > 0.55\%$) counterpart of mixinite or humosapropelinite.

Second-cycle huminite/vitrinite: these macerals are incorporated in sediment during deposition by erosion of existing sedimentary rocks, peat, lignite, high-rank coals and oil shale.

Secondary macerals: these macerals or products are formed at the expense of liptinite macerals (especially amorphous liptinites) during advanced maturation. One of these macerals (called "granular vitrinite" by Mukhopadhyay et al., 1985a) closely resembles desmocollinite. It is gray and granular in reflected white light and shows some brown fluorescence. (For detailed definition of this maceral, see Mukhopadhyay et al., 1985a.) Other vitrinite-like secondary macerals are some varieties of *solid bitumen* ("migra-bitumen" of Jacob, 1989). *Granular vitrinite* is an example of hydrogen-deficient secondary maceral, whereas *albertite* or *ozocerite* (solid bitumen) is a hydrogen-enriched secondary maceral compared to the original primary liptinites.

Solid bitumen: these macerals (example: albertite, asphalt, gilsonite, etc.) originated as secondary products or migration products of maturation of liptinite macerals. For detailed characteristics of these macerals see Jacob (1967, 1989).

APPENDIX B: FLUORESCENCE COLORS OF VARIOUS MACERALS AT TWO MATURATION STAGES

Maceral/submaceral	Color ($\approx$ 0.5% R_0)	Color ($\approx$ 1.1% R_0)
Liptinite (primary)		
Sporinite	Yellow	Orange, red
Cutinite	Yellow	Red
Suberinite	Yellow	Dark brown – nonfluorescent
Alginite (telalginite)	Green-golden, yellow	Diffuse orange, red
Particulate liptinite A (lamalginite)	Yellow	Orange, red
Resinite	Yellow	Nonfluorescent – dark brown, red
Liptodetrinite	Yellow	Red
Amorphous liptinite (bituminite)	Orange – nonfluorescent	Nonfluorescent – dark brown, red
Liptinite vitrinite (primary)		
Mixinite	Dark brown	
Saprocollinite		Red brown
Vitrinite		
Textinite/telinite	Nonfluorescent – yellow	Nonfluorescent
Ulminite/telocollinite	Nonfluorescent	Nonfluorescent
Gelinite/gelocollinite	Nonfluorescent	Dark brown
Attrinite/densinite	Nonfluorescent	
Desmocollinite		Orange – dark brown
Vitrodetrinite	Nonfluorescent	Nonfluorescent
Pseudo-vitrinite	Nonfluorescent	Nonfluorescent
Inertinite		
Fusinite	Nonfluorescent	Nonfluorescent
Semifusinite	Nonfluorescent	Nonfluorescent – dark brown
Sclerotinite	Nonfluorescent	Nonfluorescent
Macrinite	Nonfluorescent	Nonfluorescent – dark brown
Inerto-detrinite	Nonfluorescent	Nonfluorescent
Micrinite	Nonfluorescent	Nonfluorescent – dark brown
Migrabitumen		
Ozocerite	Yellow-green	Nonfluorescent
Asphalt	Yellow-brown	Nonfluorescent
Grahamite	Brown	Nonfluorescent
Albertite	Dark brown	Nonfluorescent
Impsonite	–	Nonfluorescent
Exsudatinite	Yellow-brown	Red – nonfluorescent
Crude oil	Yellow	Blue

REFERENCES

Abraham, H., 1945. *Asphalts and Allied Substances.* Van Nostrand-Rheinhold, Holland, 2142 pp.

Alpern, B. and Cheymol, D., 1978. Réflectance et fluorescence des organoclasts du Toarcien du bassin de Paris en fonction de la profondeur et de la température. *Rev. Inst. Fr. Pet.,* 33 (4): 515 – 535.

Armagnac, C., Kendell, C.St.G.C., Kuo, C., Lerche, I. and Pantano, J., 1988. Determination of paleoheat flux from vitrinite reflectance data, and from sterane and hopane isomer data. *J. Geochem. Explor.,* 30: 1 – 28.

Bensley, D.F. and Davis, A., 1988. The use of light-emitting diodes as fluorescence standards and fluorescence intensity of macerals. *Org. Geochem.,* 12: 345 – 349.

Bertrand, P., Pittion, J.L. and Bernaud, C., 1986. Fluorescence of sedimentary organic matter in relation to its chemical composition. *Org. Geochem.,* 10 (1 – 3): 641 – 647.

Bertrand, R. and Heroux, Y., 1987. Chitinozoan, graptolite, scolecodont reflectance as an alternative to vitrinite reflectance and pyrobitumen reflectance in Ordovician and Silurian strata, Anticosti Island, Quebec, Canada. *Bull. Am. Assoc. Pet. Geol.,* 71(8): 951 – 957.

Bostick, N.H., 1973. Time as a factor in thermal metamorphism of phytoclasts (coal particles). *C.R. 7me Congr. Int. Stratigr. Géol. Carbonif.,* 2: 183 – 193.

Bostick, N.H., 1979. Microscopic measurement of the level of catagenesis of solid organic matter in sedimentary rocks to aid exploration for petroleum and to determine former burial temperature – a review. *Soc. Econ. Paleontol. Mineral., Spec. Publ.,* 26: 17 – 43.

Bostick, N.H. and Foster, J.N., 1975. Comparison of vitrinite reflectance in coal seams and in kerogens of sandstones, shales, limestones in some part of a sedimentary section. In: B. Alpern (Editor), *Pétrographie de la Matière Organique.* C.N.R.S., Paris, pp. 13 – 25.

Brown, J.C., 1988. Oil-base drilling in the Sable Basin from the standpoint of the explorations. *Abstract CSPG-GAC Annu. Meet.,* St. John's, Nfld., A14.

Buchardt, B. and Lewan, M.D., 1990. Reflectance of vitrinite-like macerals as a thermal maturity index for Cambrian-Ordovician Alum shale, southern Scandinavia. *Bull. Am. Assoc. Pet. Geol.,* 74 (4): 394 – 406.

Buiskool Toxopeus, J.M.A., 1983. Selection criteria for the use of vitrinite reflectance as a maturity tool. In: J. Brooks (Editor), *Petroleum Geochemical Exploration in Europe.* Blackwell, Oxford, pp. 295 – 307.

Bunterbarth, G. and Teichmüller, M., 1982. Ancient heat flow density estimated from the coalification of organic matter in the borehole Urach 3 (SW-Germany). In: *Urach Geothermal Project,* pp. 89 – 95.

Burham, A.K. and Sweeney, J.J., 1989. A chemical kinetic model of vitrinite maturation and reflectance. *Geochim. Cosmochim. Acta,* 53: 2649 – 2657.

Burnham, A.T., Braun, R.L. and Samoun, A.M., 1988. Further comparison of methods for measuring kerogen pyrolysis rates and their fitting kinetic parameters. *Org. Geochem.,* 13 (4 – 6): 839 – 845.

Bustin, R.M., 1983. Heating during thrust faulting in the Rocky Mountains: Friction or fiction? *Tectonophysics,* 95: 308 – 328.

Bustin, R.M., Cameron, A.R., Grieve, D.A. and Kalkreuth, W.D., 1985. Coal petrology, its principles, methods, and applications. *Geol. Assoc. Can., Short Course Note,* 3: 273 pp.

Cardott, B.J. and Lambert, L.W., 1985. Thermal maturation of vitrinite reflectance of Woodford shale, Anadarko Basin, Oklahoma. *Bull. Am. Assoc. Pet. Geol.,* 69 (11): 1982 – 1998.

Castano, J.R. and Sparks, D.M., 1974. Interpretation of vitrinite reflectance measurements in sedimentary rocks and determination of burial history using vitrinite reflectance and authigenic minerals. *Geol. Soc. Am., Spec. Pap.,* 153: 31 – 52.

Clayton, J.L. and Bostick, N.H., 1986. Temperature effects on kerogen and on molecular and isotopic composition of organic matter in Pierre Shale near an igneous dyke. *Org. Geochem.,* 10 (1 – 3): 135 – 144.

Crelling, J.C., 1983. Current uses of fluorescence microscopy in coal petrology. *J. Microsc.,* 132 (3): 251 – 266.

Crelling, J.C. and Dutcher, R.R., 1980. Fluorescent macerals in Colorado coaking coals. In: L.M. Carter (Editor), *Proc. Fourth Symp. Rocky Mountain Geology,* pp. 58 – 61.

Curiale, J.A., 1986. Origin of solid bitumens with emphasis on biological marker results. *Org. Geochem.,* 10 (1 – 3): 559 – 580.

Davis, A., 1978. The reflectance of coal. In: C. Karr (Editor), *Analytical Methods of Coal and Coal Pro-*

duction, 1. Academic Press, New York, N.Y., pp. 27 – 81.
Davis, A., Rathbone, R.F., Lin, R. and Quick, J.C., 1990. Observation concerning the nature of maceral fluorescence alteration with time. *Org. Geochem.,* 16 (1 – 3): 897 – 906.
Dembecki, H., Jr., 1984. An interlaboratory comparison of source-rock data. *Geochim. Cosmochim. Acta,* 48: 2641 – 2649.
Diessel, C.F.K., 1986. Fluorometric analysis of inertinite. *Fuel,* 64: 1542 – 1546.
Dow, W.G., 1977. Kerogen studies and geological interpretations. *J. Geochem. Explor.,* 7: 79 – 99.
Dow, W.G. and O'Conner, D.I., 1982. Kerogen maturity and type by reflected light microscopy applied to petroleum exploration. *Soc. Econ. Paleontol. Mineral., Short Course Note,* 7: 133 – 157.
Dow, W.G., Mukhopadhyay, P.K. and Jackson, T., 1988. Source-rock potential and maturation of deep Wilcox from southcentral Texas. *Bull. Am. Assoc. Pet. Geol.,* 72 (2): 179.
Durand, B., 1980. *Kerogen. Insoluble Organic Matter from Sedimentary Rocks.* Technip, Paris, 519 pp.
Durand, B., Alpern B., Pittion, J.L. and Pradler, B., 1987. Reflectance of vitrinite as a control of thermal history of sediments. In: B. Durand (Editor), *Thermal History of Sedimentary Basins.* Technip, Paris, pp. 441 – 474.
Epstein, E., Epstein, J.B. and Harris, L.D., 1977. Conodont color alteration – an index to organic metamorphism. *U. S. Geol. Surv., Prof. Pap.,* 995.
Falvey, D.A. and Deighton, I., 1982. Recent advances in burial and thermal geohistory analysis. *J. Australas. Pet. Explor. Assoc.,* 22: 65 – 81.
Fermont, W.J.J., 1988. Possible causes of abnormal vitrinite reflectance in paralic deposits of the Carboniferous, in the Achterhoek area, The Netherlands. *Org. Geochem.,* 12 (4): 401 – 412.
Gangopadhyay, S., Landis, C.R., Pleil, M.W., Borst, W.L. and Mukhopadhyay, P.K., 1988. Time-resolved fluorescence spectroscopy of crude oils and condensates. *Fuel,* 67: 1674 – 1679.
Goodarzi, F., 1984. Organic petrography of graptolite fragments from Turkey. *Mar. Pet. Geol.,* 1: 202 – 210.
Goodarzi, F., 1985. Optical properties of vitrinite carbonized at different pressures. *Fuel,* 84: 156 – 162.
Goodarzi, F. and Norford, B.S., 1985. Graptolites as indicators of the temperature histories of rocks. *J. Geol. Soc. London,* 142: 1089 – 1099.
Goodarzi, F. and Norford, B.S., 1989. Variation of graptolite reflectance with depth of burial. *Int. J. Coal Geol.,* 11: 127 – 141.
Goodarzi, F., Brooks, P.W. and Embry, A.F., 1989. Regional maturity as determined by organic petrography and geochemistry of the Schei Point Group (Triassic) in the Sverdrup Basin, Canadian Arctic Archipelago. *Mar. Pet. Geol.,* 6 (4): 290 – 302.
Grayson, J.F., 1975. Relationship of palynomorph translucency to carbon and hydrocarbons in clastic rocks. In: B. Alpern (Editor), *Pétrographie Organique et Potentiel Pétrolier.* C.N.R.S., Paris, pp. 261 – 273.
Gutjahr, Ch.C.M., 1966. Carbonization measurements of pollen grains and spores and their application. *Leidse Geol. Meded.,* 38: 1 – 29.
Hagemann, H.W. and Hollerbach, A., 1981. Spectral fluorometric analysis of extracts, a new method for the determination of the degree of maturity of organic matter in sedimentary rocks. *Bull. Cent. Rech. Explor. Prod. Elf-Aquitaine,* 5: 635 – 650.
Hagemann, H.W. and Hollerbach, A., 1986. The fluorescence behavior of crude oils with respect to their thermal maturation and degradation. *Org. Geochem.,* 10 (1 – 3): 473 – 480.
Hatcher, P.G., Lerch, H.E., III., Kotra, R.K. and Verheyen, T.V., 1988. Pyrolysis-g.c.-m.s. of a series of degraded woods and coalified logs that increase in rank from peat to bituminous coals. *Fuel,* 67: 1067 – 1075.
Héroux, Y., Chagnon, A. and Bertrand, R., 1979. Compilation and correlation of major thermal maturation indicators. *Bull. Am. Assoc. Pet. Geol.,* 63: 2128 – 2144.
Hevia Rodriguez, V., 1977. Le concept de la réflectivité moyenne statistique des substances anisotropes. Relations avec la rang des anthracites et de la matière organique dispersé dans les sediments. In: R. Camposs and E.J. Goni (Editors), *Advances in Organic Geochemistry.* Endisma, Madrid, pp. 655 – 673.
Hoffmann, E. and Jenkner, A., 1932. Die Inkohlung und ihre Erkennung im Mikrobild. *Glückauf,* 68: 81 – 88.
Horvath, Z.A., 1983. Study on maturation process of huminitic organic matter by means of high-pressure experiments. *Acta Geol. Hung.,* 26: 137 – 148.

Hunt, J.M., 1979. *Petroleum Geochemistry and Geology*. Freeman, San Francisco, Calif., 617 pp.

Hutton, A.C., 1987. Petrographic classification of oil shales. *Int. J. Coal Geol.,* 8: 203 – 231.

Hutton, A.C. and Cook, A., 1980. Influence of alginite in the reflectance of vitrinite from Joadja, N.S.W., and some other coals and oil shales containing alginite. *Fuel,* 59: 711 – 716.

International Commission of Coal Petrology, 1971. *International Handbook of Coal Petrography,* 2nd ed. C. N. R. S., Paris.

Jacob, H., 1952. Fortschritte auf dem Gebiet der Braunkohlen-Lumineszenz - Mikroskopie. *Bergakademie,* 4: 337 – 347.

Jacob, H., 1967. Petrologie von Asphaltiten und asphaltischen Pyrobitumina. *Erdöl Kohle,* 20: 393 – 400.

Jacob, H., 1972. Mikroskop-Photometrie der organischen Stoffe von Boden I. Organopetrographische Nomenklatur und Mikroskop – Photometrische Methodik. *Die Bodenkultur,* 23: 217 – 226.

Jacob, H., 1989. Classification, structure, genesis, and practical importance of natural solid oil bitumen ("migrabitumen"). *Int. J. Coal Geol.,* 11: 65 – 79.

Jones, R.W. and Edison, T.A., 1978. Microscopic observations of kerogens related to geochemical parameters with emphasis on thermal maturation. In: D.F. Oltz (Editor), *Low Temperature Metamorphism of Kerogen and Clay Minerals. Soc. Econ. Paleontol. Mineral, Pacific Sect.,* Los Angeles, Calif., pp. 1 – 12.

Jones, J.M., Murchison, D.G. and Saleh, S.A., 1972. Variation of vitrinite reflectivity in relation to lithology. In: H.R. Gaertner and H. Werner (Editors), *Advances in Organic Geochemistry.* Pergamon, Oxford, pp. 601 – 612.

Jordon, N., 1983. Efectos contaminantes de aditivos de perforacion en estudios de petrografia organica. *Rev. Tec. Intevep.,* 3 (2): 161 – 166.

Kalkreuth, W., 1982. Rank and petrographic composition of selected Jurassic – Lower Cretaceous coals of British Columbia, Canada. *Bull. Can. Pet. Geol.,* 30: 112 – 139.

Kalkreuth, W. and Macaley, G., 1987. Organic petrology and geochemical (Rock-Eval) studies on oil shales and coals from the Pictou and Antigonish areas, Nova Scotia, Canada. *Bull. Can. Pet. Geol.,* 35 (3): 263 – 295.

Kalkreuth, W. and McMechan, M., 1988. Burial history and thermal maturity, Rocky Mountain Front Ranges, Foothills, and Foreland, east-central British Columbia and adjacent Alberta, Canada. *Bull. Am. Assoc. Pet. Geol.,* 72 (11): 1395 – 1410.

Kalkreuth, W. and Langenberg, C.W., 1986. The timing of coalification in relation to structural events in Cache area, Alberta, Canada. *Can. J. Earth Sci.,* 23: 1103 – 1116.

Kanstler, A.J., Smith, G.C. and Cook, A.C., 1978. Lateral and vertical rank variation: implications for hydrocarbon exploration. *J. Australas. Pet. Explor. Assoc.,* 18: 143 – 156.

Karweil, J., 1956. Die Metamorphose der Kohlen vom Standpunkt der physikalischen Chemie. *Z. Dtsch. Geol. Ges.,* 107: 132 – 139.

Katz, B.J., Pheifer, R.N. and Schunk, D.J., 1988. Interpretation of discontinuous vitrinite reflectance profiles. *Bull. Am. Assoc. Pet. Geol.,* 72 (8): 926 – 931.

Landis, C.R., Sullivan, G.W., Pleil, M.W., Borst, W.L. and Crelling, J.C., 1987. Pulsed laser fluorescence microscopy of coal macerals and dispersed organic material. *Fuel,* 66: 984 – 991.

Larsen, G. and Chilingar, G.V., 1979. *Diagenesis in Sediments and Sedimentary Rocks.* Developments in Sedimentology, 25A. Elsevier, Amsterdam, 579 pp.

Larter, S.R., 1989. Chemical models of vitrinite reflectance evolution. *Geol. Rundsch.,* 78 (1): 349 – 359.

Law, B.E., Nuccio, V.F. and Barker, C.E., 1989. Kinky vitrinite reflectance well profiles: evidence of paleopore pressure in low-permeability, gas-bearing sequences in Rocky Mountain Foreland Basins. *Bull. Am. Assoc. Pet. Geol.,* 73 (8): 999 – 1010.

Leythaeuser, D., Hagemann, H.W., Hollerbach, A. and Schaefer, R.G., 1980. Hydrocarbon generation in source rock beds as a function of type and maturation of their organic matter: a mass balance approach. In: *Proc. World Petroleum Congress,* 2: 31 – 41.

Lin, R. and Davis, A., 1988. A fluorogeochemical model of coal macerals. *Org. Geochem.,* 12 (4): 363 – 374.

Lo, H.B., 1988. Photometric methods for measuring the thermal maturity on strewn-mounted kerogen slides. *Org. Geochem.,* 12 (4): 303 – 307.

Lopatin, N.V., 1971. Temperature and geologic time as factors in coalification. *Akad. Nauk S.S.S.R., Ser. Geol.,* 3: 95 – 106.

Massoud, M.S. and Kinghorn, R.R.F., 1985. A new classification for the organic components of kerogen. *J. Pet. Geol.,* 8 (1): 85 – 100.

McCartney, J.T., 1952. A study of Seyler theory of coal reflectance. *Econ. Geol.,* 47: 202 – 210.

Michelson, J.K. and Khorasani, G.K., 1990. Monitoring chemical alteration of individual oil-prone macerals by means of microscopical fluorescence spectrometry combined with multivariate data analysis. *Org. Geochem.,* 15 (2): 179 – 192.

Middleton, M.F., 1982. Tectonic history from vitrinite reflectance. *Geophys. J.R. Astron. Soc.,* 68: 121 – 132.

Mukhopadhyay, P.K., 1989a. Characterization of amorphous and other organic matter types by microscopy and pyrolysis-gas chromatography. *Org. Geochem.,* 14 (3): 269 – 284.

Mukhopadhyay, P.K., 1989b. Organic petrography and organic geochemistry of Tertiary coals from Texas in relation to depositional environment and hydrocarbon generation. *Rep. Invest., Texas Bur. Econ. Geol.,* 188: 118 pp.

Mukhopadhyay, P.K., 1990a. Organic facies and maturation of sediments from the Tyrrhenian Sea, Ocean Drilling Program, Leg 107. In: K.A. Kastens, J. Mascle et al. (Editors), *Proc. Ocean Drilling Program, Scientific Results,* 107: 579 – 590.

Mukhopadhyay, P.K., 1990b. Oil-base Drilling Mud Research – Phase II. D.S.S. Canada, Rep. XSG89-00208-(605), 18 pp.

Mukhopadhyay, P.K. and Birk, D., 1989. Influence of Oil-base Mud on Organic Geochemical Parameters. D.S.S. Rep. 03AF.12193-8-1067, 18 pp.

Mukhopadhyay, P.K. and Gormly, J.R., 1984. Hydrocarbon potential of two types of resinite. *Org. Geochem.,* 6: 439 – 454.

Mukhopadhyay, P.K. and Rullkötter, J., 1986. Quantitative microscopic spectral fluorescence measurement of crude oil, bitumen, kerogen, and coal. *Bull. Am. Assoc. Pet. Geol.,* 70 (5): 624.

Mukhopadhyay, P.K. and Wade, J.A., 1990. Organic facies and maturation of sediments from three Scotian Shelf wells. *Bull. Can. Soc. Pet. Geol.,* 38 (4): 407 – 425.

Mukhopadhyay, P.K., Hagemann, H.W. and Gormly, J.R., 1985a. Characterization of kerogens as seen under the aspect of maturation and hydrocarbon generation. *Erdöl Kohle,* 38 (1): 7 – 18.

Mukhopadhyay, P.K., Rullkötter, J., Schaefer, R.G. and Welte, D. H., 1985b. Facies and diagenesis of organic matter in sediments from the Nankai Trough, DSDP, Leg 87A. In: H. Kagami et al. (Editors), *Initial Report DSDP, 87.* U.S. Gov. Printing Office, Washington, D.C., pp. 877 – 889.

Mukhopadhyay, P.K., Gangopadhyay, S., Landis, C., Pleil, M. and Borst, W.L., 1987. Characterization of crude oil and condensate by time-resolved laser fluorescence microscopy – preliminary study. *Bull. Am. Assoc. Pet. Geol.,* 71 (5): 597.

Mukhopadhyay, P.K., Gormly, J.R. and Zumberge, J.E., 1989. Generation of hydrocarbons from Tertiary coals of Texas-coal as potential source rock for liquid hydrocarbons in a deltaic basin? *Org. Geochem.,* 14 (3): 351 – 352.

Newman, J. and Newman, N.A., 1982. Reflectance anomalies in Pike River coals: evidences of variability in vitrinite type, with implications for maturation studies and "Suggate rank". *N. Z. J. Geol. Geophys.,* 25: 233 – 243.

Oberlin, A., Boulmier, J.L. and Villey, M., 1980. Electron microscopic study of kerogen microtexture. Selected criteria for determining the evolution path and evolution stage of kerogen. In: B. Durand (Editor), *Kerogen.* Editions Technip, Paris, pp. 191 – 242.

Ottenjann, K., 1981. Improved microphotometric fluorescence measurements of coal macerals. *Zeiss Inform.,* 26 (93E): 40 – 46.

Ottenjann, K., 1988. Fluorescence alteration and its value for studies of maturation and bituminization. *Org. Geochem.,* 12 (4): 309 – 322.

Ottenjann, K., Teichmüller, M. and Wolf, M., 1974. Spektrale Fluoreszenzmessungen an Sporiniten mit Auflicht-Anregung, eine mikroskopische Methode zur Bestimmung des Inkohlungsgrades gering inkohlter Kohlen. *Fortschr. Geol. Rheinl. Westfalen,* 24: 1 – 36.

Ottenjann, K., Wolf, M. and Wolff-Fischer, E., 1982. Das Fluoreszenzverhalten der Vitriniten zur Kennzeichnung der Kohlungseigenschaften von Steinkohlen. *Glückauf,* 43 (4): 2 – 8.

Pasley, M.A. and Crelling, J.C., 1988. Fluorescent spectral types of selected Colorado bituminous coals. *Org. Geochem.,* 12 (4): 333 – 344.

Price, L.C., 1983. Geologic time as a parameter in organic metamorphism and vitrinite reflectance as an absolute paleogeothermometer. *J. Pet. Geol.,* 6: 5 – 38.

Price, L.C. and Barker, C.E., 1985. Suppression of vitrinite reflectance in amorphous rich kerogen – a major unrecognized problem. *J. Pet. Geol.,* 8 (1): 59 – 84.

Pusey, W.C., III., 1973. How to evaluate potential gas and oil source rock. *World Oil,* 176: 71 – 75.

Quigley, T.M., Mackenzie, A.S. and Gray, J.R., 1987. Kinetic theory of petroleum generation. In: B. Doligez (Editor), *Migration of Hydrocarbons in Sedimentary Basins.* Edition Technip, Paris, pp. 649 – 665.

Ragot, J.P., 1977. Contribution à l'étude de l'évolution des substances carbonées dans les formations géologiques. Thèse, l'Université Paul Sabatier de Toulouse, 150 pp.

Ritter, U., 1984. The influence of time and temperature on vitrinite reflectance. *Org. Geochem.,* 6: 473 – 480.

Robert, P. 1985. The study of organic matter in reflected light and fluorescence. In: P. Robert (Editor), *Organic Metamorphism and Geothermal History.* Elf-Aquitaine and Reidel, Dordrecht, pp. 61 – 115.

Schohardt, M., 1936. Zur Praxis der Lumineszens-Mikroskopie mit dem Epi-Kondensor. *Zeiss Nachr.,* 2: 16.

Schwab, K.W., 1985. Vitrinite reflectance values derived from Staplin's (1969) visual kerogen standards. *Abstr. Soc. Org. Petrol.,* Annu. Meet., Houston, Tex., pp. 21 – 22.

Senftle, J., Larter, S.R., Bromley, B.W. and Brown, J.W., 1986. Quantitative chemical characterizations of vitrinite concentrates using pyrolysis-gas chromatography. Rank variation in pyrolysis products. *Org. Geochem.,* 9 (6): 345 – 350.

Sharkey, A.G. and McCartney, J.T., 1981. Physical properties of coal and its products. In: M.A. Elliot (Editor), *Chemistry of Coal Utilization.* Wiley, New York, N.Y., pp. 159 – 284.

Spiro, B. and Mukhopadhyay, P.K., 1983. Effects of degradation on spectral fluorescence of alginite. *Erdöl Kohle,* 36 (7): 297 – 299.

Stach, E., Mackowsky, M. Th., Teichmüller, M., Taylor, G.H., Chandra, D. and Teichmüller, R., 1982. *Textbook of Coal Petrology.* Gebruder Borntraeger, Stuttgart, 536 pp.

Staplin, F.L., 1969. Sedimentary organic matter, organic metamorphism, and oil and gas occurrence. *Bull. Can. J. Pet. Geol.,* 17: 47 – 66.

Sweeney, J.J. and Burnham, A.K., 1990. Evaluation of a simple model of vitrinite reflectance based on chemical kinetics. *Bull. Am. Assoc. Pet. Geol.,* 74 (10): 1559 – 1570.

Teichmüller, M., 1974. Enstehung und Veränderung bituminöser Substanzen in Kohlen in Beziehung zur Entstehung und Umwandlung des Erdöls. *Forschr. Geol. Rheinl. Westfalen,* 24: 65 – 112.

Teichmüller, M., 1978. Nachweis von Graptoliten – Periderm in geschiefernten Gesteinen mit Hilfe kohlenpetrologischen Methoden. *Neues. Jahrb. Geol. Paleontol. Monatsh.,* 7: 430 – 447.

Teichmüller, M., 1979. Die Diagenese der Kohligen Substanzen in den Gesteinen des Tertiärs und Mesozoikums des mittleren Oberrhein-Grabens. *Fortschr. Geol. Rheinl. Westfalen,* 27: 19 – 49.

Teichmüller, M., 1982. Fluoreszenzmikroskopische Änderungen von Liptiniten und Vitriniten mit zunehmendem Inkohlungsgrad und Ihre Beziehungen zu Bitumenbildung und Verkokungsverhalten. *Geologisches Landesamt Nordrhein – Westfalen, Spec. Publ.,* Krefeld, Federal Republic of Germany.

Teichmüller, M., 1986. Organic petrology of source rocks, history and state of art. *Org. Geochem.,* 10 (1 – 3): 581 – 600.

Teichmüller, M., 1987a. Recent advances in coalification studies and their application to geology. In: A.C. Scott (Editor), *Coal and Coal-Bearing Strata: Recent Advances. Geol. Soc. London, Spec. Publ.,* 32: 127 – 169.

Teichmüller, M., 1987b. Organic material and very low grade metamorphism. In: M. Frey (Editor), *Low Temperature Metamorphism.* Chapman-Hall, Glasgow, pp. 115 – 160.

Teichmüller, M. and Ottenjann, K., 1977. Art und Diagenese von Liptiniten und lipoid Stoffen in einem Erdölmuttergestein auf Grund fluoreszenzmikroskopischer Untersuchungen. *Erdöl Kohle,* 30: 387 – 398.

Teichmüller, M. and Teichmüller, R., 1979. Diagenesis of coal (coalification). In: G. Larsen and G.V. Chilingar (Editors), *Diagenesis in Sediments and Sedimentary Rocks.* Developments in Sedimentology, 25A. Elsevier, Amsterdam, pp. 207 – 246.

Teichmüller, M. and Teichmüller, R., 1981. The significance of coalification studies to geology – a review. *Bull. Cent. Rech. Explor. Prod. Elf-Aquitaine,* 5: 491 – 534.

Teichmüller, M. and Wolf, M., 1977. Application of fluorescence microscopy in coal petrology and oil exploration. *J. Micros.,* 109 (1): 49 – 73.

Thompson-Rizer, C.L. and Woods, R.A., 1987. Spectrofluorescence measurements of coals and petroleum source rocks. *Int. J. Coal. Geol.,* 7: 85 – 104.

Thompson-Rizer, C.L., Woods, R.A. and Ottenjann, K., 1988. Quantitative fluorescence results from sample exchange studies. *Org. Geochem,* 12 (4): 323–337.
Ting, F.C., 1978. New techniques for measuring maximum reflectance of vitrinite and dispersed vitrinite in sediments. *Fuel,* 57: 717–721.
Tissot, B. and Espitalié, J., 1975. L'évolution thermique de la matière organique des sédiments: applications d'une simulation mathématique. *Rev. Inst. Fr. Pét.,* 30: 743–777.
Tissot, B. and Welte, D.H., 1984. *Petroleum Formation and Occurrence.* Springer, Berlin, 699 pp.
Tissot, B., Pelet, R. and Ungerer, Ph., 1988. Thermal history of sedimentary basins, maturation indices, and kinetics of oil and gas generation. *Bull. Am. Assoc. Pet. Geol.,* 71 (12): 1445–1466.
Van Gijzel, P., 1967. Autofluorescence of fossil pollen and spores with special reference to age determination and coalification. *Leidse Geol. Meded.,* 40: 263–317.
Van Gijzel, P., 1975. Polychromatic UV-fluorescence microphotometry of fresh and fossil plant substances, with reference to location and identification of dispersed organic matter in rocks. In: B. Alpern (Editor), *Pétrographie de la Matière Organique des Sédiments, Relations avec la Paléotemperature et le Potential Pétrollier.* C.N.R.S., Paris, pp. 67–91.
Van Gijzel, P., 1982. Characterization and identification of kerogen and bitumen and determination of thermal maturation by means of qualitative and quantitative microscopical techniques. *Soc. Econ. Paleontol. Mineral., Course Note,* 7: 159–216.
Van Krevelen, D.W., 1961. *Coal.* Elsevier, Amsterdam, 514 pp.
Von der Dick, and Kalkreuth, W., 1984. Three-dimensional and synchronous fluorescence spectroscopy – a new approach to coalification studies. *Fuel,* 63: 1636–1640.
Vorabutr, P. and Chilingarian, G.V., 1983. Oil-based drilling fluids. In: G.V. Chilingarian and P. Vorabutr (Editors), *Drilling and Drilling Fluids.* Elsevier, Amsterdam, pp. 365–397.
Waples, D.W., 1980. Time and temperature in petroleum formation: Application of Lopatin's method to petroleum exploration. *Bull. Am. Assoc. Pet. Geol.,* 64 (6): 916–926.
Welte, D.H., 1987. Migration of hydrocarbons – Facts and theory. In: B. Boligez (Editor), *Migration of Hydrocarbons in Sedimentary Basins.* Technip, Paris, pp. 393–414.
Welte, D.H. and Yalcin, M.N., 1988. Basin modeling – a new comprehensive method in petroleum geology. *Org. Geochem.,* 13: (1–3): 141–152.
Welte, D.H. and Yukler, M.A., 1981. Petroleum origin and accumulation in basin evolution – a quantitative model. *Bull. Am. Assoc. Pet. Geol.,* 65: 589–617.
Yukler, M.A. and Thompsen, E., 1988. Effect of heating rates of hydrocarbon generation and occurrence. *Bull. Am. Assoc. Pet. Geol.,* 73 (3): 428–429.
Yukler, M.A. and Kokesh, F., 1984. A review of models used in petroleum resource estimation and organic geochemistry. In: J. Brooks and D.H. Welte (Editors), *Advances in Petroleum Geochemistry 1.* Academic Press, London, pp. 69–113.
Vorabutr, P. and Chilingarian, G.V., 1983. Oil-based drilling fluids. In: G.V. Chilingarian and P.

Chapter 10

DIAGENESIS AND ITS RELATION TO MINERALIZATION AND HYDROCARBON RESERVOIR DEVELOPMENT: GULF COAST AND NORTH SEA BASINS

MALCOLM P. R. LIGHT and HARRY H. POSEY

INTRODUCTION

Hydrologic systems in major sedimentary basins such as the North Sea and U.S. Gulf Coast Basins are dynamic and long-lived, and fluid transport within them is extensive and complex. Mineral deposits are considered to be a normal aspect of sedimentary basin evolution (Jackson and Beales, 1967). They, along with the formation and concentration of secondary minerals, the creation of secondary porosity, and the localization of mineral-defined facies formed by diagenesis and metamorphism, are controlled as much by the avenues of fluid migration and the composition of diagenetic fluids as by long-lived thermal conditions (Garven and Freeze, 1984a,b; Bethke, 1985).

Salt dome cap rocks and their commonly-associated components, hydrocarbons and base metals, represent a complex diagenetic assemblage of evaporite-bearing basins that record many of the fluid – rock reactions and hydrocarbon-maturation characteristics of the sedimentary column both above and below the evaporites. An integrated hydrothermal and structural model explaining such an assemblage was devised for the U.S. Gulf Coast Basin and is applied here to the North Sea Basin.

The North Sea Basin consists of a faulted basement overlain, respectively, by a deformed sediment shell that has been distorted by reactivated basement faulting, and a weakly to undeformed sedimentary carapace which developed above during thermal subsidence (Gibbs, 1987). In contrast, the U.S. Gulf Coast Basin consists mainly of a sedimentary mantle, deformed mostly by growth faulting, that probably lies above a highly faulted, folded and metamorphosed Paleozoic basement (see, for instance, Gregory et al., 1979). In the Gulf Coast, deep thermal waters and gases, particularly methane, are presently escaping along structurally-controlled channels (faults peripheral to grabens and salt diapirs) forming overpressured brine and hydrocarbon accumulations, producing diagenetic changes and depositing ore minerals and cap rocks (Muehlberger et al., 1987).

The interrelationships among diagenetic, structural, and mineralizing events has been investigated in the Gulf Coast Basin using seismic, geochemical, and isotopic studies of outcrops and cores cut in layered evaporites, salt pillows, salt domes and salt dome cap rocks by Seni and Jackson (1983), Jackson and Seni (1983), Kyle and Price (1985, 1986), Posey (1986), Posey et al. (1987a and b), and Land et al. (1988). The extent to which variations in the timing of salt diapirism in the North Sea Basin are controlled by the nature, orientation and movement history of the underlying crust has been examined by Gibbs (1987). Times of major uplift of salt domes are determined by interpreting unconformity-bound sequences on relevant seismic lines (see, for instance, Seni and Jackson, 1983; Jackson and Seni, 1983). An integrated hydrothermal model developed for the Gulf Coast Basin (Figs. 10-1 – 10-3) (Light, 1985; Kyle and Price, 1986; Light et al., 1987) appears to be applicable to the North Sea Basin, where similar studies have been undertaken (Schmidt et al., 1977; Carstens and Finstad, 1981; Dypvik, 1983; Watts, 1983; Larese et al., 1984; Lee et al., 1989).

Residual anhydrite in salt dome cap rocks was formed by salt dissolution (Goldman, 1933; Taylor, 1938). Anhydrite dissolves and sulfate is reduced by hydrocarbon-oxidizing bacteria (Feeley and Kulp, 1957; Milner et al., 1977) to calcite, water, and hydrogen sulphide with or without carbon dioxide (Ruckmick et al., 1979). Layered early-dark, late-light calcite pairs have formed consecutively in a time-reversed stratigraphy (Posey, 1986). Carbon isotopes of calcite cap rock become more depleted with increasing depth (Posey, 1986), a result of increased input of depleted organic carbon from zones of higher organic maturity (Larese et al., 1984), or higher methane content (Posey, 1986; Posey et al., 1987b). Precipitation of calcite, sulphur, and iron and zinc sulphides indicates a complex interaction between hydrocarbon- and metal-bearing deep-basin fluids and shallow oxidizing groundwaters (Price et al., 1983; Seni, 1987; Kyle and Agee, 1988). Salt dome cap rocks record within them a sequence of depositional, diagenetic, and tectonic events that show the various stages of salt diapirism, fluid (water and hydrocarbon) generation, overpressure development, and fluid migration and mixing phenomena (Kyle et al., 1987; Posey et al., 1987a,b; Seni, 1987; Light et al., 1987; Posey and Kyle, 1988).

Anhydrite crystals and halite that occur in major saline deposits record the trace element and isotopic compositions of the fluids from which they formed (see

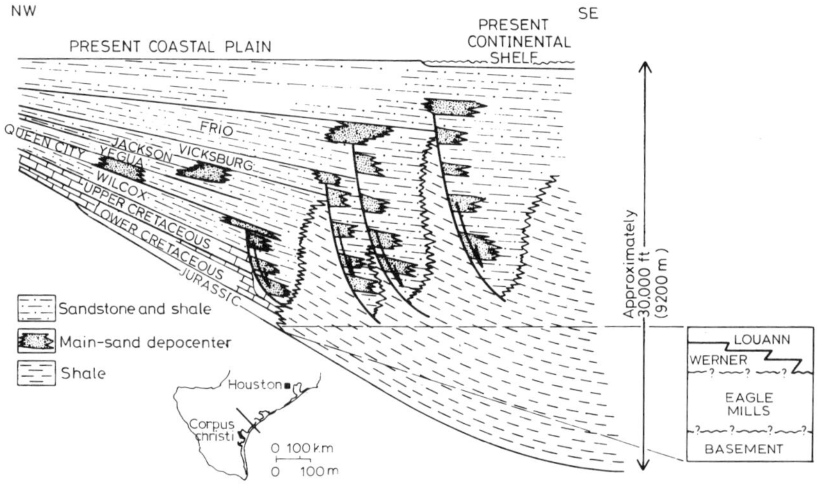

Fig. 10-1. Depositional and structural style of the Tertiary along the Texas Gulf Coast. (From Gregory et al., 1979.)

Holser, 1979). Although these crystals which are now found in salt domes are commonly assumed to be of marine origin, formed in epicontinental evaporite deposits, many anhydrites within the Gulf Coast have been found to contain radiogenic strontium and most salt dome halite contains depleted levels of bromide. These studies lead to the remarkable deduction that marine waters mixed with nonmarine fluids from continental clastic rocks prior to or during diapirism (Posey, 1986; Posey et al., 1987a; Land et al., 1988; Posey and Kyle, 1988).

The timing of evaporite diapirism in the Gulf Coast coincides with burial of the underlying clastics to a zone where they should have experienced high overpressures and hydrofracturing (Palciauskas and Domenico, 1980). Subsequent fluid expulsion from the clastics, upward into the evaporites (Burst, 1969; Powers, 1967) is probably related to increasing temperatures and pressures with burial (Palciauskas and Domenico, 1980; Light et al., 1987). This feature is important because the incorporation of fluids in the salt would have a major effect on its shear strength and buoyancy (Jenyon, 1985; Talbot and Jackson, 1987), inducing diapiric activity (Garfunkel and Almagor, 1987; Posey et al., 1987a,b), reactivating fault systems and allowing upward escape of geopressured fluids with consequent diagenesis of invaded sediments (Hanor, 1984; Light et al., 1987).

Anhydrite recrystallization induces isotopic mixing between pre-existing sulfates and new fluids (Posey, 1986), so the times of fluid migration can be related to regional tectonic and diapiric events (Light et al., 1987). Later fluids have formed oil – gas and metallic mineral deposits in salt dome cap rocks along the Gulf Coast (Price et al., 1983; Ulrich et al., 1984; Kyle and Price, 1985, 1986). Direct evidence of the times of fluid migration have been shown through paleomagnetic studies of anhydrite cap rock, which contains traces of pyrrhotite (Ulrich et al., 1984; Gose et al., 1985; Kyle et al., 1987).

A detailed knowledge of the rates of uplift of salt domes is important because of the marked effects such movements can produce on the surface elevations of the sea floor, particularly in the vicinity of offshore production platforms (Prof. A.J. Smith, pers. commun., 1988). The safety of large hydrocarbon and hazardous and toxic waste storage facilities located in salt domes (Seni et al., 1984, 1985) may also be jeopardized by slight diapiric movements, which could shear seals in storage systems, produce leaks into surrounding aquifers, or cause fires on the surface.

DEEP FLUID SOURCES IN BASINAL SETTINGS

Crustal fluids and their hydrodynamics strongly influence the deformation of minerals and rocks at all scales in the crust, embracing plate tectonics, convective energy, mass transport, the return of subducted fluids back to the upper crust, and degassing of the lower crust and mantle (Behr and Frentzel-Beyme, 1987). Most of the natural gas forms thermogenically from biological material at depths greater than liquid petroleum (Hunt, 1979; Tissot and Welte, 1984). Kerogen in some deep high-temperature wells can still have ample petroleum production potential to at least 350°C (Price, 1982) and deep high-temperature aquifers may contain methane derived from this source. Water and carbon dioxide released by prograde metamor-

Fig. 10-2. Model depicting the development of diagenetic geopressured – geothermal and hydrocarbon generation zones, salt domes, salt-dome cap rock formation and salt-dome mineralization for the coastal diapir province of the U.S. Gulf Coast Basin (from Light et al., 1987). Diapiric model from Woidt (1978); cap rock model from Price et al. (1983); diagenetic temperatures from Loucks et al. (1981); hydrocarbon generation temperatures from Hunt (1979) and Kharaka et al. (1985a,b).

Temperature – depth ranges for carbon dioxide generation, smectite-to-illite transformation, hydrocarbon generation, thermogenic gas generation, and albitization of plagioclase are taken from Gulf Coast Basin case studies. Mineral reactions that depend on fluid composition (smectite-to-illite transformation, for instance) or salinity will vary from basin to basin. Feldspar destruction (not shown), which takes place from the surface to any depth and is probably an important contributor of metal ions, is particularly susceptible to solution chemistry (Posey et al., 1987a).

(A) Deposition of the Mid-Jurassic Louann Evaporite (mostly halite) Formation above Triassic (conjectured) red beds. Nature of basement beneath the red beds is not known, although both metamorphosed Paleozoic Ouachita sedimentary facies and transitional crust have been proposed (Salvador and Buffler, 1982).

(B) By the beginning of Middle Cretaceous (or perhaps earlier), the evaporites were deforming beneath Late Jurassic to Early Cretaceous marine carbonates and fluvio-marine clastics, such that diapiric formations began to develop. By this time, red beds containing plagioclase had reached a thermal level wherein albite began to replace plagioclase.

(C) Red bed fluids beneath the evaporites became overpressured, exceeding lithostatic pressures, and hydrofractured surrounding rocks. Fluid escape routes formed, preferentially, in or near the cores of salt anticlines, walls, and diapirs, promoting the first major phase of diapiric halokinesis.

(D) By 25 Ma, organic-rich source rocks (as demonstrated locally in drill-hole example, PB No.2; see Light et al., 1987) were buried to temperature levels sufficient for generating high amounts of carbon dioxide. Fluids from overpressured zones escaped upward, preferentially, along the brecciated, fractured margins of salt domes and along growth faults. Carbonate rocks, which lay in the path of escaping CO_2, would have dissolved. Meteoric or other shallow, low-temperature, low-salinity fluids, shown by thick black arrows, mixed with upward-migrating formation fluids in convection cell fashion (see Hanor and Workman, 1986). Halite at the tops of diapirs dissolved leaving a residue of anhydrite, which formed anhydrite cap rock by an underplating mechanism.

(E) By 20 Ma, hydrocarbons (as demonstrated locally in drill-hole example PB No. 2) migrated upward along the salt stock margins and, in a cool ($< 70°C$) bacteria-charged environment, formed calcite cap rocks, also by an underplating mechanism. Excess CO_2 present in this environment occasionally dissolved some of the previously-formed calcite cap rock. The breakdown of kaolinite (*K*) in a few specific examples may have added metals to fluids as well as radiogenic strontium.

(F and G) As more sediments deposited in the basin, the Louann mother salt and units above continued to subside while the Louann continued to feed the core of salt stocks. The tops of diapirs remained at about the same elevation, relative to sea level, as sediments slipped like a sleeve alongside the diapir. In the latest stages of diapirism, hydrocarbon maturation and fluid migration appear as today, and hydrocarbons have been trapped alongside salt domes and in other traps (generalized here as a growth fault trap). Some hydrocarbon traps have become water washed (as depicted by the local example drill well Prets No. 1), but the migration of fluids continues. Thermogenically-produced hydrogen sulfide, from temperature – depth levels exceeding 250°C at 3 km adds relatively heavy reduced sulfur that occurs in some of the anhydrite-hosted sulfide minerals (see Kyle and Agee, 1988). Uranium deposits may form on the margins or above salt-dome cap rocks along redox boundaries. Metal sulfide and barite form in a zonal arrangement, with barite forming generally higher in the section than the metal sulfides.

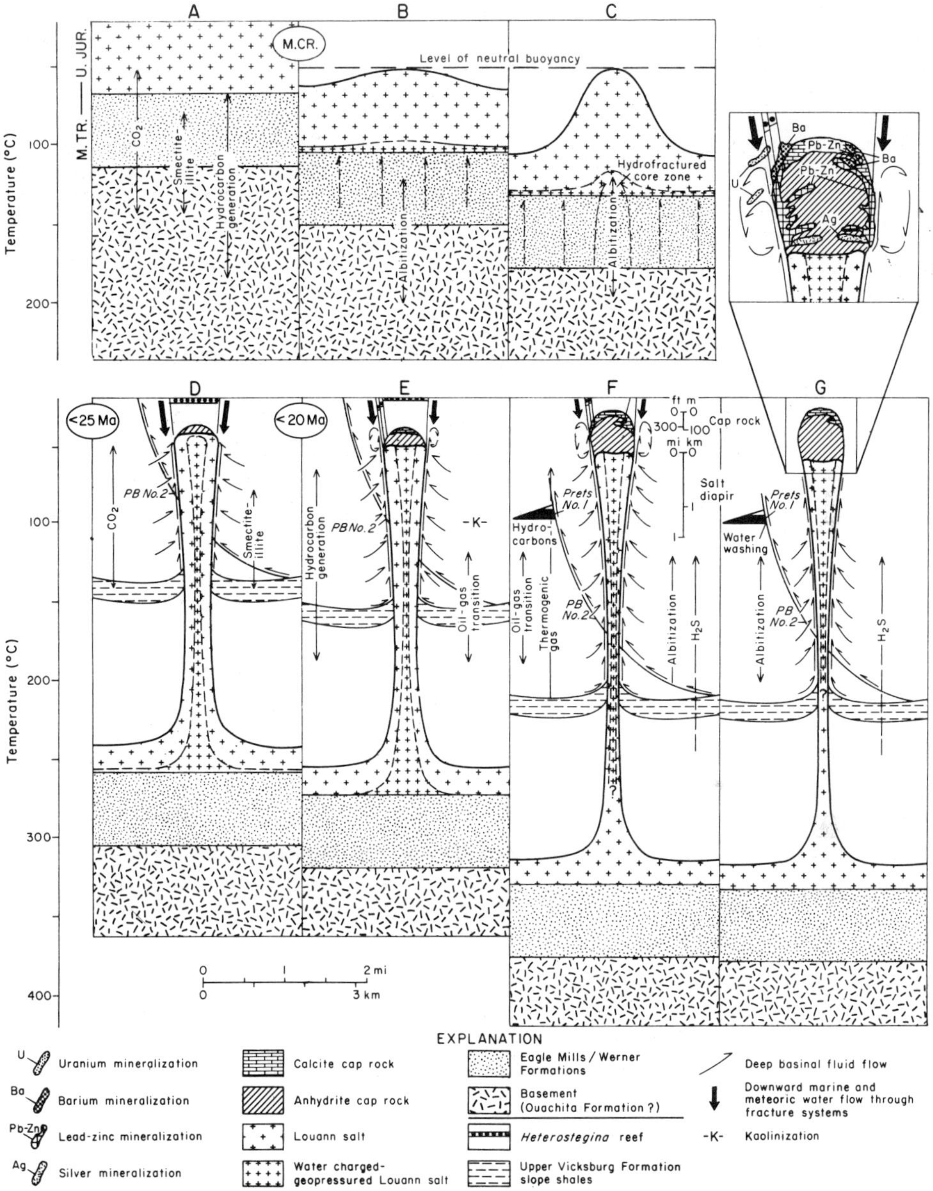
A
B
C
M.TR. —— U. JUR.
M.CR.
Level of neutral buoyancy
Temperature (°C)
100
200
CO_2
Smectite-illite
Hydrocarbon generation
Albitization
Hydrofractured core zone
Ba
Pb-Zn
U
Ag
D
E
F
G
<25 Ma
<20 Ma
PB No.2
Smectite-illite
Hydrocarbon generation
-K-
Oil-gas transition
ft m
0 0
300 100
mi km
0 0
Cap rock
Salt diapir
Prets No.1
Hydrocarbons
Water washing
Thermogenic gas
Albitization
H_2S
300
400
0 1 2 mi
0 3 km
EXPLANATION
Uranium mineralization
Barium mineralization
Lead-zinc mineralization
Silver mineralization
Calcite cap rock
Anhydrite cap rock
Louann salt
Water charged-geopressured Louann salt
Eagle Mills / Werner Formations
Basement (Ouachita Formation ?)
Heterostegina reef
Upper Vicksburg Formation slope shales
Deep basinal fluid flow
Downward marine and meteoric water flow through fracture systems
Kaolinization

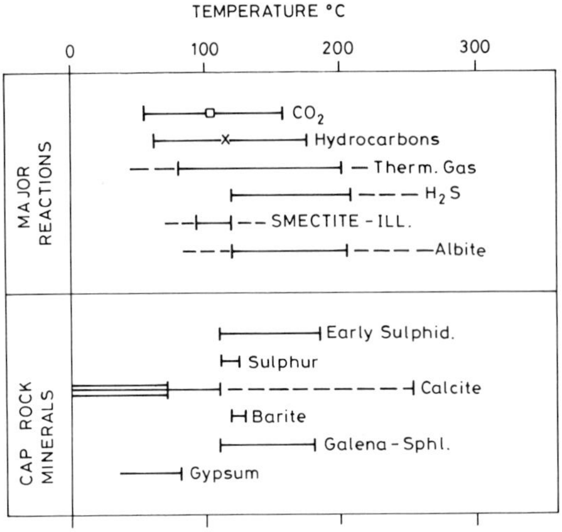

Fig. 10-3. Temperature range of major hydrocarbon and mineral reactions and measured temperatures of diagenetic minerals in salt-dome cap rocks. (Fluid inclusion and mineral stability data from Price et al., 1983; Kreitler and Dutton, 1983; Kaiser and Richmann, 1981; Loucks et al., 1981; Posey, 1986; Weaver, 1980; Knauth et al., 1980; hydrocarbon generation temperatures from Kharaka et al., 1985b; and Hunt, 1979.)

phic reactions at depth (Winkler, 1976) and mantle-derived carbon-dioxide-rich fluids that coexist with highly saline brines in the lower crust (Touret, 1986) are being continually added to meteoric and marine formation waters that evolve to connate waters (see Hanor, 1987).

Helium of mantle origin having high $^3He/^4He$ ratios is associated with methane along active rifts in the crust (Craig et al., 1975; Tolstikin, 1978). Helium and hydrogen degassing occurs along faults (Wakita et al., 1980) and these gases migrate upward in the earth's crust with carbon dioxide and water (Gluckhauf and Paneth, 1946) and occasionally accumulate in natural gas fields (Tade, 1967).

The lower crust tends to be wet in some crustal settings and dry in others (Dewey, 1986). The mantle, however, contains carbon-dixode-rich fluids; hence, the water-rich fluids of the lower crust probably originated in earlier subduction, crustal accretion, arc-magmatic, or deep penetrating meteoric environments (Dewey, 1986).

Carbon dioxide and saline brines coexist in the lower crust where they form immiscible fluids under amphibolite-grade conditions (Touret, 1986; Trommsdorff and Skippen, 1986). Textures of some metamorphic rocks imply that during amphibolite metamorphism salt-saturated fluids separated from a less-dense carbon-dioxide-rich fluid phase, which boiled off and may have escaped up fractures that contained fluids at less than lithostatic pressure (Trommsdorff and Skippen, 1986).

Most of the earth's crust down to depths of 10 to 20 km is pervaded by regions of vertical, parallel, liquid-filled (extensive dilatancy anisotropy) cracks aligned by tectonic stress (Lovell, 1988). At depths of 7 to 12 km, porous zones have developed along microcracks and channelways to form pervasive permeability (Behr and Frentzel-Beyme, 1987). Major listric faults that bound extensional basins appear to

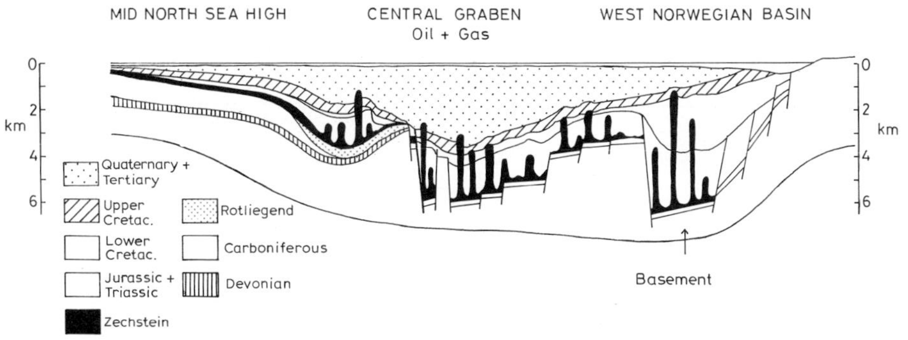

Fig. 10-4. Southwest-northeast cross-section through the North Sea showing the relationship of salt diapirs to sediment loading, faulting and oil/gas deposits. (After Faber and Stahl, 1984.)

flatten and extend into the top of the lower crust, whereas mantle thrusts detach from the Moho (Stesky and Brace, 1973; Warner and McGeary, 1987). Studies of oxygen isotopes suggest that surface-derived waters have migrated down listric faults to depths of some 12 km in the Hercynian orogenic belt (Wickham and Taylor, 1987). Pathways must, therefore, exist for carbon dioxide and hot water, generated at depth, to rise into shallow reservoirs (Gluckhauf and Paneth, 1946) driven up extensional fault systems by high overpressures (Figs. 10-4 and 10-5).

In the subsurface, fluid migration can occur through microfractures, fault planes or permeable rocks (Neglia, 1979). In normally pressured zones at shallow depths, faults may act as barriers to lateral fluid movement, whereas at greater depths they tend to form conduits for vertical passage of abnormally pressured fluids (Price, 1976; Carstens and Finstad, 1981).

Carbon dioxide remains trapped in the lower crust, except during periods of crustal stretching when it boils off and escapes up fracture systems, leaving a residual concentration of brine (Trommsdorff and Skippen, 1986). Loss of lower-crust carbon dioxide up fracture systems may explain some fault-related secondary porosity at depth and carbonates at shallower levels (Selley, 1979). Carbon dioxide release may also induce salt wall/diapir activity above faults which underlie layered evaporites (Light et al., 1987). Confined gas rising up vertical fractures will microfracture rocks at shallower levels when the gas pressure exceeds the fracture pressure of surrounding formations and can result in fracture propagation (Palciauskas and Domenico, 1980). Rising carbon dioxide will acidify shallower formation waters and may result in widespread dissolution phenomena (Huang and Keller, 1970; Schmidt and McDonald, 1979). Carbon dioxide released during a crustal extensional phase may thus induce salt wall/diapir activity above faults which underlie layered evaporites and diagenetic activity in surrounding basinal sediments (Figs. 10-4 and 10-5) (Light et al., 1987).

The northern passive continental margin of the northward subducting Hercynian orogenic belt (Windley, 1986) is characterized by the presence of salt structures in the North Sea and northern Europe which are more abundant in basins closer to the thrusts (Fig. 10-6). Hercynian age Mississippi-valley-type lead – zinc mineralization has developed in carbonates along this margin (Gabelman, 1976; Russel, 1976).

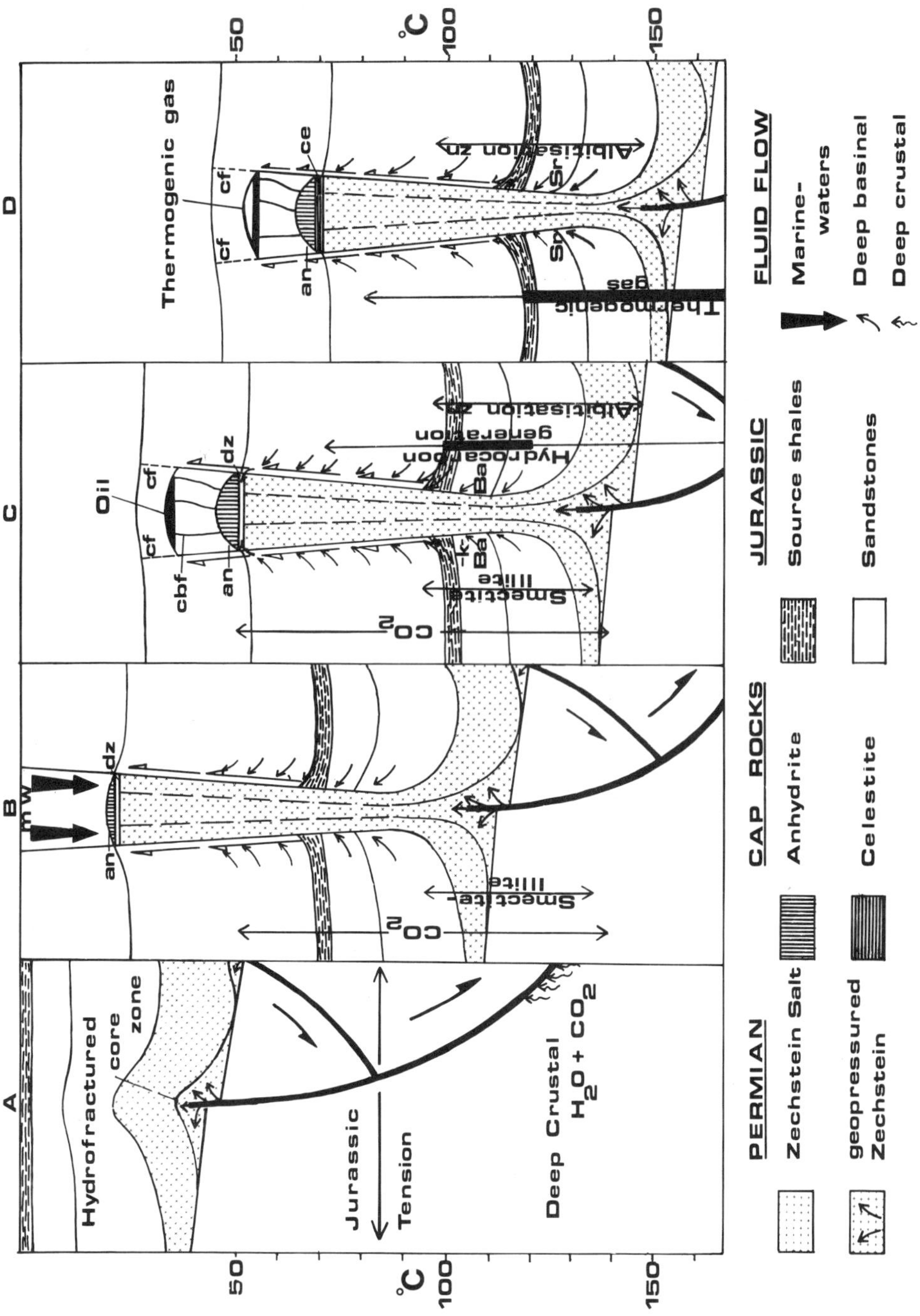
A
B
C
D
Hydrofractured
core
zone
Jurassic
Tension
Deep Crustal
$H_2O + CO_2$
°C
50
100
150
mw
an
dz
CO_2
Smectite-
Illite
cbf
cf
Oil
K
Ba
Hydrocarbon
generation
Albitisation zn
Thermogenic gas
ce
Sr
Thermogenic
gas
PERMIAN
Zechstein Salt
geopressured
Zechstein
CAP ROCKS
Anhydrite
Celestite
JURASSIC
Source shales
Sandstones
FLUID FLOW
Marine-
waters
Deep basinal
Deep crustal

Diapiric activity in the North Sea Basin and northern Europe may be related to the release, during the Jurassic period of extension of the North Sea and Atlantic Basins (Glennie, 1984a,b), of carbon dioxide and water trapped during earlier (Hercynian?) orogenic events.

LONG-DURATION DIAPIRIC UPLIFTS

The Zechstein evaporite deposits in the North Sea Basin comprise five major sedimentary cycles, sometimes beginning with widespread bituminous shales (Fig. 10-5; Taylor, 1981). The extent to which variations in the timing of salt diapirism in the North Sea Basin are controlled by the nature, orientation, and movement history of the underlying crust has been examined by Gibbs (1987) (Fig. 10-4).

Small diapirs, pillows, and swells occur with decaying amplitudes towards the edges of the North Sea Basin, whereas large walls and diapirs are confined to the central area, where salt and overburden are thickest (Fig. 10-4; Gussow, 1968). Although this relationship suggests that overburden thickness is a major factor in diapirism (Trusheim, 1960), the maturity of hydrocarbon source intervals is also highest in the Central Graben and decays toward the margins (Fig. 10-4) (Faber and Stahl, 1984; Baird, 1986). This suggests that hydrocarbon-generated overpressure could also be an important factor in initiating diapirism.

Knowledge of fracture intensity and orientation in low-matrix permeability reservoirs in the North Sea is important in primary production, depletion strategy, ultimate recovery, and stimulation design (Van der Vlis et al., 1979; Watts, 1983). Two fracture systems occur in Cretaceous chalks in the Ekofisk region: early conjugate shears healed by calcite, kaolinite and barite, and younger, open-tension fractures sporadically sealed by calcite and clay (Watts, 1983).

Tension fractures formed during the Miocene from a combination of halokinetic

---

Fig. 10-5. Hypothetical model depicting the development of diagenetic, geopressured – geothermal and hydrocarbon generation zones, salt domes and salt-dome cap rocks in the Central Graben of the North Sea. (Diagenetic temperatures from Loucks et al., 1981; and Lee et al., 1985, 1989; periods of diapiric activity from Dunn, 1975, and Watts, 1983; salt pillow model from Gibbs, 1984.) Compare with Fig. 10-2.

(A) Initial formation of salt swells and pillows during the Late Jurassic Cimmerian Orogeny.

(B) Diapiric uplift of salt and formation of fractures extending to seafloor. Leakage of marine waters (*mw*) down fractures to form a salt dissolution zone (*dz*) and residual anhydrite cap rock (*an*). Carbon dioxide and organic-acid-rich shale waters rise up faults peripheral to the salt dome forming secondary dissolution porosity at depth and precipitating calcite at shallower levels, sealing faults.

(C) Kaolinite precipitates in sandstones when temperatures exceed 100°C. Buildup of overpressure from hydrocarbon generation and shale dewatering of source rocks, subsequent rock dilation and migration results in further diapiric uplift of salt dome. Oil becomes trapped in reservoirs above the dome. When fluid temperatures exceed 110°C, kaolinites begin to alter to illite. Biogenic calcite cap rocks have not formed either because surface faults are sealed by calcite (*cf*) and barite (*cbf*) or prevailing cap rock temperatures are too high. Barite (*cbf*) precipitates in early fractures in reservoirs. Barium has been released to formation waters by albitization of feldspars.

(D) Formation of celestite (*ce*) in salt dome cap rock from mixing of strontium-rich fluids sourced in the albitization zone with sulphate-saturated cap rock waters.

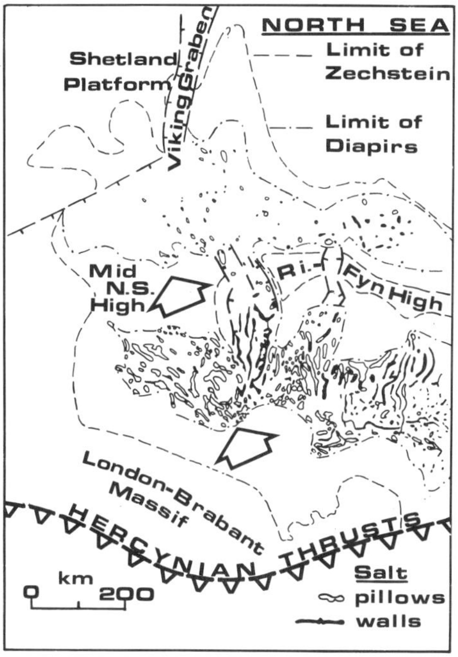

Fig. 10-6. Areas of salt dome diapirism (after Glennie, 1984) showing that salt domes increase in abundance toward basin centers and the Hercynian thrust front (after Bard et al., 1980). This region lies in a zone of Hercynian Mississippi-valley lead–zinc mineralization (Gabelmann, 1976). (The Hercynian age compression direction in the North Sea, from Gabelmann, 1976.)

doming and source-rock maturation overpressuring (Watts, 1983). A relationship thus exists between early diagnetic fracture-healing from initial carbon-dioxide-rich fluids (Hunt, 1979), formation of a pressure seal, overpressure buildup from subsequent hydrocarbon maturation and migration (Watts, 1983), and the initiation of halokinesis and tension fracture formation (Fig. 10-5).

Zechstein salt domes are oriented parallel to ancient structural trends (Christian, 1969; Brunstrom and Walmsley, 1969) and were initiated at a time when Palaeozoic fault lines in the North Sea were under tension during the opening of the central North Atlantic Ocean (Anderton et al., 1979) (Fig. 10-6). During periods of tension, fault lines are more likely to act as conduits for the vertical flux of fluids that may initiate salt diapirism (Talbot and Jackson, 1987) and, respectively, fill traps, form ore deposits, and recrystallize anhydrites in salt dome cap rocks (Posey et al., 1987). Aquathermal and hydrocarbon maturation (gas generation)-induced overpressuring (Jones, 1980) may have led to the recrystallization of salt-hosted anhydrites and initiation of salt diapirism in the Gulf Coast Basin (Light, 1985; Light et al., 1987; Posey et al., 1987a). Paleomagnetic evidence that pyrrhotitic anhydrite cap rocks formed shortly after the evaporites (Ulrich et al., 1984; Gose et al., 1985; Kyle et al., 1987) indicates that fluids came from beneath the evaporites, as there was no other place for metals to have come from at that time.

SHORT-DURATION DIAPIRIC UPLIFTS

Knowledge of the rate of short duration diapiric uplifts of salt stocks is important to both basic research and the safety of hydrocarbon and toxic chemical storage facilities in salt domes. Minor uplift of diapirs may shear cap-rock seals on storage systems, produce major leaks into surrounding aquifers, or cause fires on the surface.

The intrusion of salt diapirs results in the formation of several types of crestal, flank, tangential, and circumferential faults and shear zones (Fig. 10-2; Cloos, 1928; Jackson and Seni, 1984; Talbot and Jackson, 1987). Evidence for widespread diapiric uplift of Gulf Coast Basin diapirs at the end of the Oligocene suggests that concomitant removal of mother salt occurred. These reactivated Frio Formation growth faults dip seawards into the Houston Diapir Province, and caused regional subsidence (Light et al., 1987). Fault reactivation produced conduits through which deep saline brines migrated upward, converted smectite to illite, introduced hydrocarbons and mineralized salt dome cap rocks (Fig. 10-2) (Light et al., 1987).

The rhythmic interbeds of shale and reef at Damon Mound (Frost and Schafersman, 1979) and the Flower Garden Banks (Rezak, 1985) provide information for calculating uplift rates. The reefs have formed on the crests of rising diapirs to be subsequently drowned by fine clastics during periods of cessation in uplift or increased basin subsidence (Rezak, 1985). Coral growth rates suggest that periods of diapirism are of relatively short duration (8 – 20 years), whereas the periods of hiatus represented by the interlaminated muds may be up to 8000 years in duration (average rate of deposition of Oligocene muds = 100 – 120 m $m.y.^{-1}$; T.E. Ewing, pers. commun., 1985).

The average rate of Oligocene uplift of Damon Mound estimated from coral growth rates (0.08 m per 1000 years) is the same as the rate of uplift of Damon Mound since Oligocene time defined by stratigraphic parameters (Collins, 1985). Uplift rates at Damon Mound (0.6 m per 1000 years) since late Pleistocene times (Collins, 1985) are an order of magnitude greater than Oligocene rates, implying that Recent periods of diapirism are more frequent or have higher velocities.

By coupling the coral growth rate with the combined thicknesses of shale and reef measured in outcrop or core, the rate of basin subsidence can be determined provided the ages of the intervening shales or carbonates are known.

Detailed investigations of structural, petrographic, fluid inclusion, isotopic, and geomagnetic characteristics of salt domes/cap rocks can result in the development of predictors, which will indicate the likelihood of a resurgence of diapiric activity in a particular diapir (Fig. 10-7). The age of a reef can be calculated paleontologically, or by matching $^{87}Sr/^{86}Sr$ ratios with the strontium isotope seawater curve (Burke et al., 1982). Layered calcite cap rocks contain several percent of clay (Posey et al., 1987b), the ages of which may be determined by Rb/Sr and K/Ar dating (Lee et al., 1989). Layered anhydrite cap rocks may be dated palaeomagnetically because of their frequent content of disseminated pyrrhotite (Ulrich et al., 1984; Gose et al., 1985; Kyle et al., 1987).

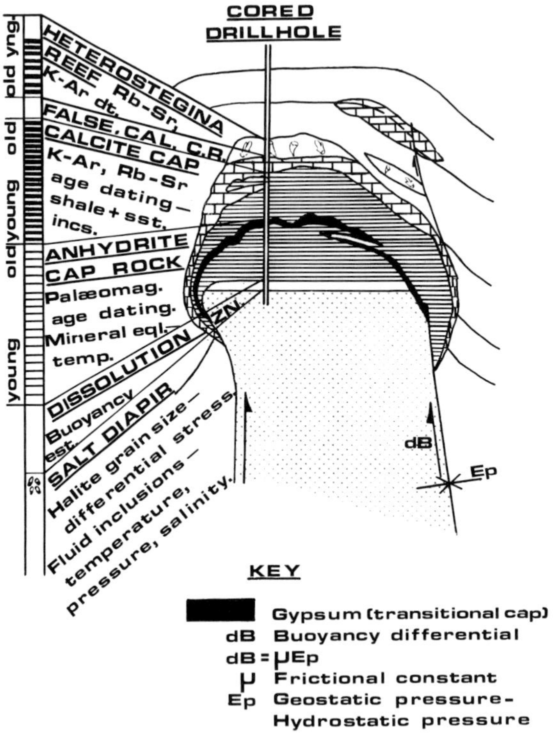

Fig. 10-7. Schematic cross-section of a salt dome and cap rock (after Posey, 1986) indicating possible structural, petrographic, fluid inclusion, isotopic and paleomagnetic investigations that can result in the development of predictors for short-duration diapiric uplifts.

Left-hand column shows the relative ages of various cap rocks. Marine calcite cap rocks, such as the Heterostegina reef found above Damon Mound, Texas, form in normal stratigraphic sequence – older at the base than at the top. False calcite cap rocks and true (banded) calcite cap rocks form sequentially from top to base by the successive replacement of anhydrite (Posey, 1986). Anhydrite cap rock, the first cap rock to develop, has a reverse stratigraphic sequence which forms through a succession of halite dissolution and anhydrite underplating events. Pyrrhotite, which may form during each underplating event, inherits the pole position present at the time of formation (Ulrich et al., 1984). Fossils, mostly spores and pollen, found in salt can be used to determine the age of the salt.

Various mechanisms for determining event ages have been used and others seem to be possible. Fossil ages have been used to determine ages of false calcite cap rocks. Isotopic means, such as Rb/Sr or K/Ar, may provide clay mineral ages, and whole-rock isochrons of metamorphosed clayey inclusions may indicate minimum times of movement. Paleomagnetic ages have been successfully utilized at Winnfield Dome, La. (see above) to determine the ages of anhydrite cap-rock formation.

The stability of salt domes can be imperfectly determined through study of material beneath the cap rocks. Occurrence of a zone of dissolution is strong indication that the diapir is active, and that cap rock is still forming. Buoyancy estimates can be made to determine whether the combined weight of the cap rock and overlying sediments are large enough to keep the cap rock section from experiencing further diapirism. Because halite undergoes recrystallization by the action of differential stresses near the salt/cap-rock boundary, halite grain-size studies can be used to determine the relative stability of a dome. Fluid inclusion studies indicate past conditions as they record the conditions of formation of primary and secondary minerals.

POROSITY PRESERVATION

The preservation of high porosities in the North Sea reservoirs has been variously explained by: (1) early hydrocarbon introduction (Van den Bark and Thomas, 1980); (2) early overpressuring of reservoirs (Blanche and Whitacre, 1978); (3) early cementation with subsequent secondary dissolution (Schmidt et al., 1977); (4) meteoric leaching (Kirk, 1980); and (5) the presence of magnesium-rich pore fluids (Neugebauer, 1974). Halokinesis during chalk deposition may also preserve porous nannofossils on highs (Van den Bark and Thomas, 1980). In the Gulf Coast Basin model (Light, 1985; Light et al., 1987) overpressuring caused by carbon dioxide and hydrocarbon generation and migration (Watts, 1983) initiates halokinesis and causes diagenetic reactions to proceed.

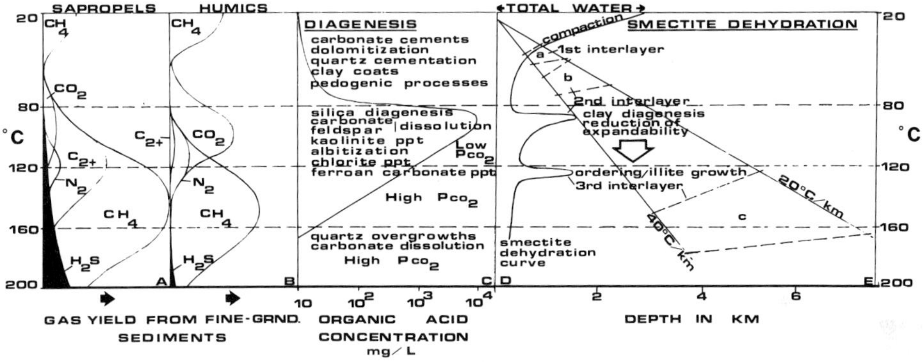

Fig. 10-8. Relative yields of major hydrocarbons from sapropelic (A) and humic (B) source materials (from Hunt, 1979). At shallow levels, only methane is generated, mostly by rotting of organics near the surface. With burial, the major concentrations of CO_2 are generated first, followed by the C_{2+} fraction (liquid hydrocarbons), nitrogen, and methane. Thermogenic H_2S, which forms by the reduction of sulfate, mostly from evaporites, develops in the deeper part of the methane-generation zone. Greater amounts of CH_4 and H_2S form from sapropelic sources than humic sources.

(C) Relative concentrations of organic acids (from Surdam et al., 1989) and major diagenetic reactions (from Crossey, 1985). The highest concentrations of organic acids (solid curve) occur where the concentrations of CO_2, which are high in the deeper section, change from high concentrations to low concentrations. The occurrence of high P_{CO_2} beneath the zone of CO_2 generation (seen in A and B) may indicate the influx of CO_2 from deeper (basement?) sources. Major diagenetic processes are shown for the zone within which they occur. Local solution chemistry will alter the relative depths of the zone boundaries and, of course, control the resulting mineralogy.

(D and E) Smectite dehydration curve (from Johns and Simoyama, 1972), and geothermal gradient lines. Water released from smectite dehydration through burial may be important as it comprises 10–15% of the compacted bulk volume of the shales (Burst, 1969; Mooney et al., 1952). Three major zones of structural water expulsion correlate with first (*a*), second (*b*), and third (*c*) interlayer destruction. Pressure–depth ranges in which loss of water from each water-layer complex might occur (from Colten-Bradley, 1987) vary with the geothermal gradient. Loss of fluid from smectite will accompany increases in fluid silica, calcium, magnesium, and iron (Boles and Franks, 1979). Ore metals will probably accompany iron.

Carbon dioxide is generated by: (1) hydrous pyrolysis reactions between organics and water (Lundegard, 1985); (2) decarboxylation of acetic and other organic acids between 80° and 200°C (Kharaka et al., 1985b; Fig. 10-8); and (3) thermal alteration of organic matter in the oil-generation window (Hunt, 1979). The net effect of carbon dioxide generation, irrespective of its mode of formation, is to enhance overpressuring and rock dilation (Light et al., 1987), acidify formation waters, and form secondary porosity by dissolution of carbonates and authigenic minerals (Selley, 1979; Posey et al., 1987b).

Formation fluids in the Gulf Coast Basin below a depth of 3600 m are acidic to neutral with pH values of 4 to 6.5 (Galloway, 1982). Rising carbon dioxide and organic-acid-rich fluids will leach carbonates at depth and precipitate calcite cements at shallower levels in lower temperature and pressure regimes (Selley, 1979). Some such secondary calcite may occur as salt dome cap rocks (Posey et al., 1987a) or as carbonate cements adjacent to salt domes (McManus and Hanor, 1988).

Secondary pore space in Jurassic sediments in the Norwegian North Sea forms 3 – 80% of the porosity (Larese et al., 1984). Secondary porosity has formed from dissolution of detrital and authigenic constituents (Larese et al., 1984). Inasmuch as leaching of feldspar, lithic fragments, and organic material represents less than 20% of secondary pore space (Larese et al., 1984), porosity enhancement by carbonate dissolution must be a major contributing factor. In the Norwegian North Sea, Jurassic sediments underwent an early period of porosity destruction by authigenic cementation which evolved, with continued burial, to secondary leaching due to increased input of acidic formation waters containing carbonic and/or organic acids (Huang and Keller, 1970; Schmidt and McDonald, 1979; Larese et al., 1984). Morton (1982) has shown that increased dissolution of epidote and amphibole in the more deeply buried Palaeocene sandstones in the Viking Graben is a consequence of leakage of isotopically light carbon-dioxide-rich pore fluids along bounding faults of the graben margin.

A pre-hydrocarbon carbon-dioxide-rich fluid flush has also been evoked to explain isotopically heavy carbon in calcite filling fractures at the Albuskjell field in the Central Graben of the North Sea Basin (Watts, 1983). The isotopically heavy carbon formed by bacterial action or precipitation of residual carbon dioxide generated by isotopically light hydrocarbon production at depth (Watts, 1983).

DIAGENETIC REACTIONS IN THE GULF COAST AND NORTH SEA BASINS

Loucks et al. (1986) have summarized the common diagenetic sequence found in Lower Tertiary sandstones along the Gulf Coast (Fig. 10-8). From the surface to depths of 1200 m clay coats form on framework grains, and feldspar is dissolved and replaced by calcite. Minor kaolinite, feldspar overgrowths and iron-poor calcite precipitate. Compaction and cementation reduce porosity by 30%.

At intermediate depths (1200 – 3400 m), early-formed carbonate cements undergo dissolution with subsequent formation of quartz overgrowths and carbonate cements. Later dissolution of lithic fragments and carbonate cements has restored porosity by 30%. Porosity is reduced by additional precipitation of kaolinite, iron-rich dolomite, and ankerite.

At depths greater than 3350 m late-stage iron-rich and iron-poor carbonate cements continue to precipitate; also within this depth range, plagioclase is altered to albite. Fibrous illite locally replaces kaolinite in geopressured aquifers in the Lower Frio Formation.

Initial porosity in eolian sandstones in the southern Permian Basin of the North Sea has been progressively reduced by compaction, pressure solution, and authigenic mineral growth, which included successively early carbonates, feldspar overgrowths, kaolinite, quartz overgrowths, chlorite, and fibrous illite (Glennie et al., 1978). In the northern North Sea, initial crystallization of pyrite was succeeded by quartz overgrowths, kaolinite, and subsequent dolomite and siderite cements (Carstens and Dypvik, 1981). Complete conversion of layered smectite – illite to 70% illite occurs between 4000 and 5000 m in the Norwegian North Sea (Dypvik, 1983).

Calcite and siderite cements in the Permian succession of the northern North Sea have depleted carbon and oxygen isotopic ratios (Larese et al., 1984). The oxygen isotopic depletion is a result of recrystallization at high burial temperatures (Fig. 10-9; Larese et al., 1984). The carbon isotopic depletion is a consequence of increased input of depleted carbon derived from organic maturation (Fig. 10-9) (Larese et al., 1984), a process that generates carbonic and organic acids (Fig. 10-8; Hunt, 1979). Late carbonate cements are ideal targets for subsequent dissolution by carbon dioxide and organic-acid-rich fluids that have flushed sediments in the northern North Sea (Huang and Keller, 1970; Schmidt and McDonald, 1979; Larese et al., 1984).

Kaolinite crystallization in the Frio Formation, Texas Gulf Coast, post-dates quartz overgrowths and calcite cementation and is one of the last authigenic minerals to form (Fig. 10-8; Loucks et al., 1981). Kaolinite cementation occurred after or simultaneously with major calcite leaching that resulted from flushing of geopressured formations by carbon dioxide/organic-acid-bearing waters (Loucks et al., 1981).

Kaolinite precipitated after feldspar overgrowths in Brent sandstones and coats pore walls and pore throats (Hallett, 1981), whereas rare late illite has formed from kaolinite and chlorite (Hallett, 1981). The preservation of kaolinite above the oil – water contact implies that it formed before or simultaneously with oil emplacement. Kaolinitization has been related to migration of meteoric waters through reservoirs during the Cimmerian uplift (Sommer, 1975, 1978). Kaolinite could equally well have formed during a period of flushing of formations by deeply-sourced carbon dioxide/organic-acid-bearing solutions (Hunt, 1979; Figs. 10-5 and 10-8).

Illites from the Groningen Gas Field and Broad Fourteens Basin (Netherlands North Sea) have been dated using the K/Ar method (Lee et al., 1985). Authigenic illite crystallized during Mid-Jurassic gas emplacement beneath gas – water contacts, and illite growth ceased as gas traps expanded (Lee et al., 1985; Eslinger and Pevear, 1988). Consequently, K/Ar ages of the finest-grained illites become younger with increasing depths in the reservoir (Eslinger and Pevear, 1988).

The temperature of intense oil generation in source rocks in many areas roughly coincides with the zone of abrupt illitization (Eslinger and Pevear, 1988), but post-dates kaolinitization of reservoirs that is a result, in the Gulf Coast, of flushing of formations by carbon dioxide and organic-acid-rich fluids (Loucks et al., 1981; Light et al., 1987). Clearly, the general sequence of burial metamorphism, carbon dioxide/organic acid generation, hydocarbon generation, and diagenetic mineral formation is similar in the North Sea and Gulf Coast (Figs. 10-2, 10-5 and 10-8).

INTEGRATED HYDROTHERMAL MODEL

The integrated hydrothermal model developed for the Gulf Coast Basin shows that the sequence of salt diapirism, cap-rock formation and cap-rock mineralization is related to the sequential generation and destruction of diagenetic minerals, hydrothermal fluids, hydrocarbons, and overpressures that form as a consequence of source rock and mother-salt burial (Light, 1985; Light et al., 1987; Posey et al., 1987b; Fig. 10-2). This model is applicable to other subsiding basins, such as the North Sea, which are undergoing hydrocarbon maturation and salt deformation.

In the North Sea Basin, the smectite – illite transformation and hydrocarbon generation (Dypvik, 1983) are shown to be related in temperature, and salt diapirism

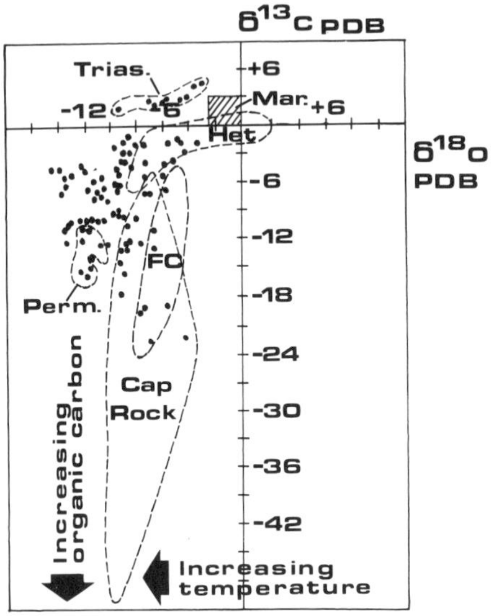

Fig. 10-9. Range of carbon-13 and oxygen-18 isotopic analyses of calcite from Gulf Coast salt dome cap rocks (Posey et al., 1987b) compared to isotopic analyses of carbonate cements from the Norwegian Sea and North Sea, Norway (Larese et al., 1984). Carbon isotopes for North Sea and cap-rock carbonates become lighter with increasing depth from increased input of more mature isotopically light organic carbon (Larese et al., 1984). "*Mar*" = marine limestone; "*Het*" = heterostegina reef; *FC* = false calcite cap rock; "*Cap Rock*" = true banded calcite cap rock. *Dots* = North Sea carbonate cements; Perm. = Permian calcite and siderite cements in North Sea.

and overpressuring are related in time (Watts, 1983). Furthermore, several clay mineral transformations in the North Sea Basin release ions necessary for late diagenetic cementation (Schmidt et al., 1977) and dissolution processes (Morton, 1982; Dypvik, 1983).

In the northern North Sea, hot fluids and hydrocarbons generated in deep source rocks were emplaced at temperatures greater than 110°C into shallow reservoirs (Liewig et al., 1987). This process caused illitization and simultaneous late silicification in Eocene time (Fig. 10-3) (Liewig et al., 1987). Diagenetic illite in the Rotliegendes Formation of the southern North Sea and northeast Netherlands formed between 100 and 175 m.y. ago (K/Ar age) and correlates with two phases of tectonic activity: the Jurassic Cimmerian orogenesis and Late Cretaceous – Early Tertiary inversion (Lee et al., 1989). Oxygen isotope data imply that pure illites formed from fluids at temperatures between 95° and 135°C prior to the emplacement of gas in the Jurassic (150 Ma ago; Van Wijhe et al., 1980; Lee et al., 1985, 1989). Formation fluids, however, could have become enriched in the heavy oxygen isotope from the leakage of old highly-evaporated brines from the overlying Zechstein evaporites (see Knauth and Beeunas, 1986).

Abundant illite has developed in close proximity to a major fault, tens of millions of years after illite growth in other wells (Lee et al., 1989), indicating a relationship between the migration of hot fluids up fault systems and illite formation. Tectonic processes during Jurassic time have controlled thermal and hydrologic parameters that, in turn, controlled diagenetic processes in the rocks (Lee et al., 1989).

Diagenetic fluids resulting in the formation of illite 105 to 120 Ma ago in the Rotliegendes of the southern North Sea and northeast Netherlands acquired lower oxygen isotope ratios, which suggests a meteoric water influx into formations uplifted by the Early Cretaceous inversion (Lee et al., 1989).

APPLICATION OF AN INTEGRATED MODEL TO A NORTH SEA CAP ROCK

Anhydrite and celestite-bearing salt dome cap rocks occur in the North Sea Basin (British Petroleum, pers. commun., 1987). Celestite, strontium sulphate, is not among the major cap-rock minerals accounted for in the Gulf Coast integrated hydrothermal model (Light et al., 1987), but its development may result from late, high-temperature reactions in source rocks (Fig. 10-5).

Excessive overpressures develop when source shales (Kimmeridge?) enter the carbon dioxide and light hydrocarbon generation zone (Figs. 10-2 and 10-5) (80 and 200°C; Hunt, 1979; Kharaka et al., 1985a). Overpressures are enhanced by waters released through the conversion of smectite to illite (Palciauskas and Domenico, 1980). Illite incorporates more strontium than smectite, so dewatering cannot release significant quantities of radiogenic strontium (Perry and Turekian, 1974).

Rocks dilate as a consequence of overpressure buildup (Palciauskas and Domenico, 1980). Overpressure buildup culminates in successive pre-hydrocarbon carbon dioxide/organic-acid-rich flushes of overlying formations (Loucks et al., 1981; Morton, 1982). Major secondary porosity forms in reservoirs as a consequence of leaching of carbonate cements and lithic fragments by acidic fluids

(Huang and Keller, 1970; Schmidt and MacDonald, 1979) with associated formation of authigenic kaolinite (Loucks et al., 1981). The carbonate cements have depleted oxygen isotope ratios, because of high temperatures and meteoric water, and depleted carbon isotope ratios, because of increased input of depleted organic carbon (methane) from organic maturation (Fig. 10-9) (Larese et al., 1984; Posey, 1986; Light et al., 1987; Posey et al., 1987b; Prikryl et al., 1988).

Continued burial of source shales into the zone of intense oil generation with associated release of shale waters will sustain or increase overpressure that can initiate diapirism (Watts, 1983; Garfunkel and Almager, 1987). Chalks were domed and fractured in the Central Graben of the North Sea by a strong (Miocene age) halokinetic pulse that was associated with maturation overpressuring (Fig. 10-5; Watts, 1983).

Diapiric movement of salt stocks will form fractures from great depths to the sea bed when the level of neutral buoyancy is reached (Jackson and Seni, 1984). Salt-undersaturated meteoric or marine waters can leak down fractures to form a salt dissolution cavity above the salt dome within which residual anhydrite accumulates (Fig. 10-5; Murray, 1966). Residual anhydrite is compacted onto the base of the forming cap rock by periodic diapiric uplift or settling of cap rock; consequently, cap-rock layering is inverted (Murray, 1966).

Maturation overpressuring of source rocks results in an upward migration of oil or gas-bearing brines through anhydrite cap rocks. This environment precipitates calcite and generates hydrogen sulphide from the action of bacteria on hydrocarbons (Fig. 10-2; Murray, 1966; Posey, 1986) that fill traps adjacent to diapirs. Layered calcite cap rocks may show progressive carbon and oxygen-isotope depletion with increasing depth (Fig. 10-9). This phenomenon is a consequence of calcite crystallizing at increasingly higher temperatures in fluids that contain depleted carbon isotopic ratios sourced from deeper, more mature, methane-rich levels (Loucks et al., 1981; Light et al., 1986; Posey, 1986; Posey et al., 1987b; Prikryl et al., 1988).

Hot hydrocarbon-bearing fluids convert kaolinite to illite at temperatures above 110°C with contemporaneous formation of silica (Lee et al., 1989). Fluids within the zone of salt dissolution will become saturated in sulphate because of anhydrite dissolution (Figs. 10-2 and 10-5) (Posey, 1986).

Feldspars undergo major albitization in the 100° – 150°C temperature range, although incipient albitization can occur at temperatures as low as 60°C (Loucks et al., 1981; Kaiser, 1983). Albitization of calcium plagioclase feldspar is an important source of calcium and non-radiogenic strontium, whereas albitization of K-feldspar will release rubidium and attendant radiogenic strontium (Figs. 10-2 and 10-5) (Posey, 1986; Light et al., 1987). Oil-bearing formation waters have high strontium contents (Collins, 1975), because the zone of oil generation (Hunt, 1979) and feldspar albitization (Kaiser, 1983) overlap (Fig. 10-5). Strontium is readily soluble only in fluids with low sulphate concentrations (Collins, 1975); so strontium-rich geothermal brines (Kharaka et al., 1980) will precipitate strontium sulphate (celestite) if brought into contact with sulphate-rich fluids in the anhydrite cap rock or salt dissolution zone (Posey, 1986). Mixing of hydrocarbon-bearing brines with sulphate-rich seawaters (Mason, 1966) can also precipitate strontium sulphate. Celestite crystals that have precipitated from fluid mixing in salt dissolution zones will accumulate and become compacted in layers by the periodic uplift of salt diapirs or subsidence of overlying cap rocks (Fig. 10-5).

In the Gulf Coast Basin, late-stage metallic sulphide and barite veins that cross-cut salt dome cap rocks were probably generated as a result of feldspar albitization and feldspar destruction (Posey, 1986). They have formed from the upward leakage of relatively hot, saline metalliferous brines along margins of rising salt diapirs (Fig. 10-2; Price et al., 1983; Light et al., 1987).

IDENTIFICATION OF HYDROCARBONS IN HALOKINETICALLY FORMED TRAPS

Uranium and thorium constitute trace elements in many common minerals, they form as minerals in their own right, are adsorbed onto clays or organic material, or are coprecipitated in sediments with iron-oxy-hydroxide (Durrance, 1987). Uranium is easily oxidized into the extremely soluble uranyl ion (UO_2^{2+}) by bacterial action and is, therefore, very mobile (Schlumberger, 1976). Thorium is fixed by adsorption on clays and its concentration may remain constant around 8 to 26 ppm in spite of thermal diagenesis (Hassan and Hossin, 1975).

A positive correlation generally exists between organic matter and the concentration of uranium and other heavy metals (Zr, Be, V, Cr, Mn, Ni, Co, Cu, Mo, Fe, Zn, Pb; Alekseyev et al., 1961; Kondratov et al., 1978; Taliyev et al., 1978; Gallager, 1980; and many others). Uranium accumulates in petroleum deposits because of the presence of a reducing environment or chelation by organic molecules (Philp and Crisp, 1982).

The uranium halo is often directly related to petroleum deposits that contain hydrogen-sulphide-rich fluids (Philp and Crisp, 1982). As petroleum becomes more asphaltic or oxidized, its capacity to extract uranium from groundwaters increases (Armstrong et al., 1972; Gallager, 1980; Philp and Crisp, 1982). Migrating uranium-bearing groundwater can be stripped of its uranium by viscous degraded petroleum that formed from the alteration of crude oil introduced into shallower formations from depth (Curiale et al., 1983).

Uranium deposits may form where convective fluid flow systems, which occur on the fractured and sheared flanks of salt diapirs (Fig. 10-2; Workman and Hanor, 1985; Bennet and Hanor, 1987; McManus and Hanor, 1988), bring highly reduced formation fluids into contact with oxygen-rich meteoric water. In this environment, hot, deeply sourced, hydrogen-sulphide-bearing brines are brought into contact with oxidizing, alkaline meteoric fluids, or marine fluids that contain higher uranium concentrations (Galloway, 1982b; Light et al., 1987).

Hydrogen sulphide generated at depth in the Gulf Coast Basin, is enriched in ^{34}S and formed early, isotopically heavy pyrite at the Fletcher uranium deposit in Texas (Ludwig et al., 1982). Uranium precipitated from the mixing of uranium-bearing meteoric waters with hot saline brines that had leaked up fault systems (Galloway, 1977).

Calcite-bearing salt-dome cap rocks that formed biogenically from methane (Posey, 1986; Posey et al., 1987b) should have lower uranium, vanadium, and nickel concentrations and higher Th/U ratios than cap rocks formed biogenically from oil (Fig. 10-10; Sassen, 1987). Hence, determination of the relative concentrations of these elements and the Th/U ratio in salt-dome calcite cap rocks should indicate whether the main migration event/events were gaseous or oily.

The concentration of radioactive and associated elements in Gulf Coast formation waters are shown in Figs. 10-11A and 10-11B (Kharaka et al., 1977, 1980 and 1985a,b; Kraemer, 1981, 1985; Kraemer and Reid, 1984). This diagram is designed to identify the nature of hydrocarbon accumulation.

Radium-226 is the daughter product of uranium-238, whereas radium-228 forms from the decay of thorium-232 (Kraemer, 1985). Variations in the high solubility of radium in a brine are controlled by formation-water salinity under geopressured –

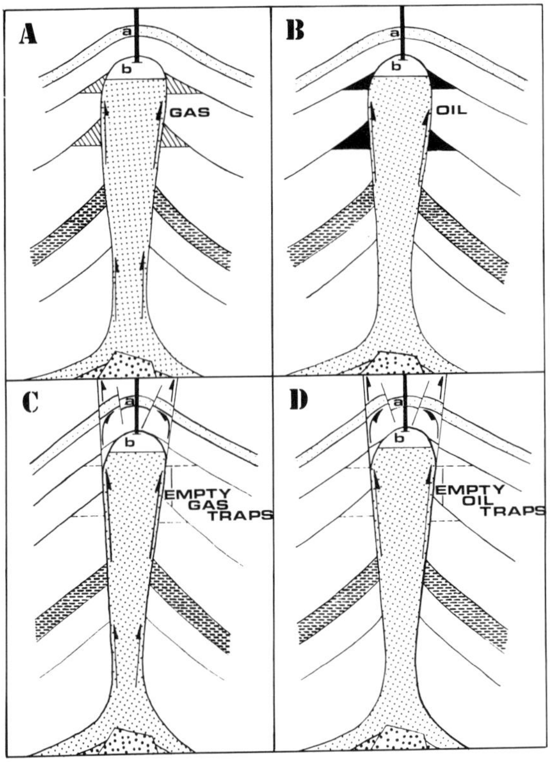

Fig. 10-10. Determination of the presence or absence of oil or gas in halokinetically-formed hydrocarbon traps.

(A) High Th/U ratio in calcite cap rock suggests biogenic formation from methane. Low $^{222}Rn/^{4}He$ and $^{222}Rn/^{226}Ra$ ratios in brines in shallower formations imply that salt-dome-bounding faults are sealing and gas may still be trapped.

(B) Low Th/U ratio in calcite cap rock suggests biogenic formation from oil. Low $^{222}Rn/^{4}He$ and $^{222}Rn/^{226}Ra$ ratios in brines in shallower formations imply that salt-dome-bounding faults may have sealed off and oil may still be trapped.

(C – D) High $^{222}Rn/^{4}He$ and $^{222}Rn/^{226}Ra$ ratios in brines saturating formations above methane or oil-generated calcite cap rocks indicate that fault systems peripheral to salt domes are leaking and traps are empty.

geothermal conditions (Fig. 10-12; Kraemer, 1985). Thorium is, however, extremely insoluble (Gutsalo, 1964; Kraemer, 1985).

High radium concentrations are a function of the geochemical environment of petroleum reservoirs or uranium occurrences associated with oil pools (Bloch and Key, 1981). Figure 10-11A is a ternary plot that shows the relative abundance of barium, radium-228 and radium-226 (Bloch and Key, 1981). To enable direct field use of this figure, the concentrations of radioactive species are given in radiometric activity: disintegrations per minute per liter (dpm l^{-1}). Radium-rich brines

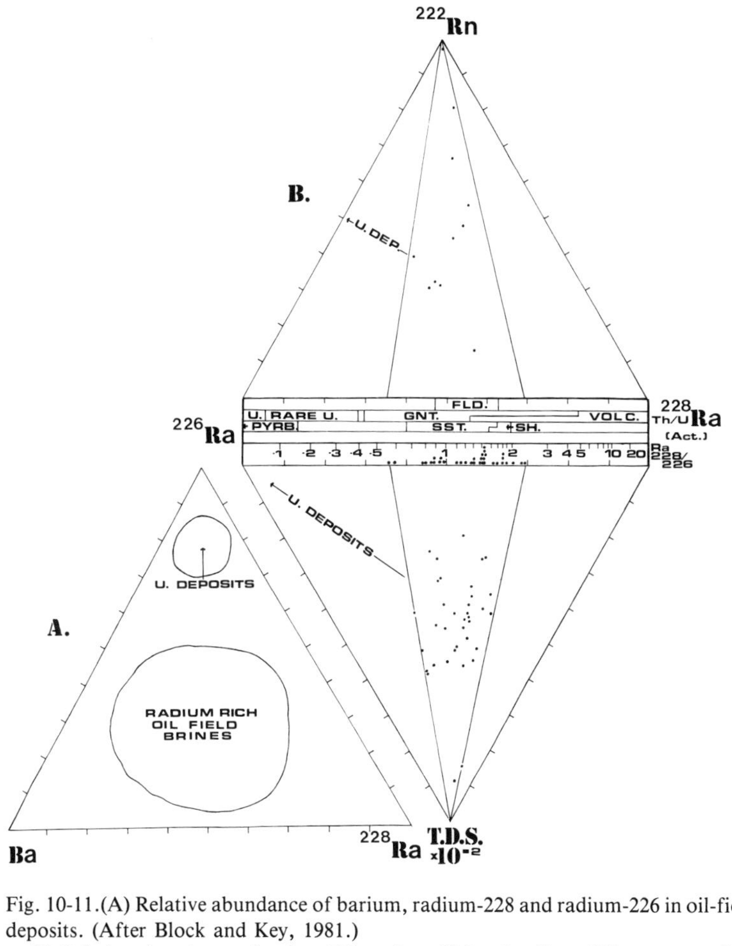

Fig. 10-11.(A) Relative abundance of barium, radium-228 and radium-226 in oil-field brines and uranium deposits. (After Block and Key, 1981.)

(B) Relative abundance of radon-222, radium-226 and radium-228 versus total dissolved solids (TDS) (from Kraemer, 1981, 1985; Kraemer and Reid, 1984; and Kharaka et al., 1977, 1980, 1985a,b). $^{228}Ra/^{226}Ra$ ratios in Gulf Coast Basin formation waters compared to Th/U ratios of common minerals, rocks and uranium deposits (data from Kraemer, 1985; and Schlumberger, 1981). *FLD.* = normal formation waters; *GNT.* = granitic rocks; *PYRB.* = uraniferous pyrobitumens; *RARE U.* = rare-uranium minerals; *SH.* = shales; *SST.* = sandstones; *U.* = uranium deposits; and *VOLC.* = volcanic rocks.

associated with normal mineral assemblages fall relatively close to the Ba/^{228}Ra side of the diagram, whereas brines associated with uranium plot toward the radium-226 apex (Bloch and Key, 1981).

Isotopic ratios of ^{228}Ra/^{226}Ra in formation water reflect the Th/U ratios of the aquifer (Kraemer, 1985). Figure 10-11B compares measured ^{228}Ra/^{226}Ra ratios in Gulf Coast Basin formation waters with Th/U ratios of common minerals, rocks and uranium deposits (Schlumberger, 1981). ^{228}Ra/^{226}Ra ratios of locally-sourced formation waters are similar to Th/U ratios of aquifers with which they are in equilibrium (Kraemer, 1985). Formation waters that show much lower ^{228}Ra/^{226}Ra ratios than aquifer Th/U ratios indicate the likelihood of uranium concentrations and/or oil deposits.

Helium-4 and radon-222 form from radioactive decay of uranium-238 and thorium deep within the earth and accumulate with petroleum in reservoirs (Dyck, 1976; Philp and Crisp, 1982; Durrance, 1987). Radon-222 anomalies are associated with oil and gas deposits (Dyck and Smith, 1969), and overpressured reservoirs may cause bulk movement of radon-bearing formation waters or diffusion of radon to the surface (Fig. 10-10; Gates and McEldowney, 1977; Philp and Crisp, 1982). Radon-222 has formed by decay of dissolved radium in formation waters and exhalation from decay of radium within the aquifer matrix (Kraemer, 1985). Radon-222 has a half life of 3.285 days (Philp and Crisp, 1982; Durrance, 1987); thus, it is not able to diffuse more than 40 – 60 m from its point of origin (Dyke and Smith, 1969).

Radon-222 is released by voids and fractures in rocks and is most efficiently car-

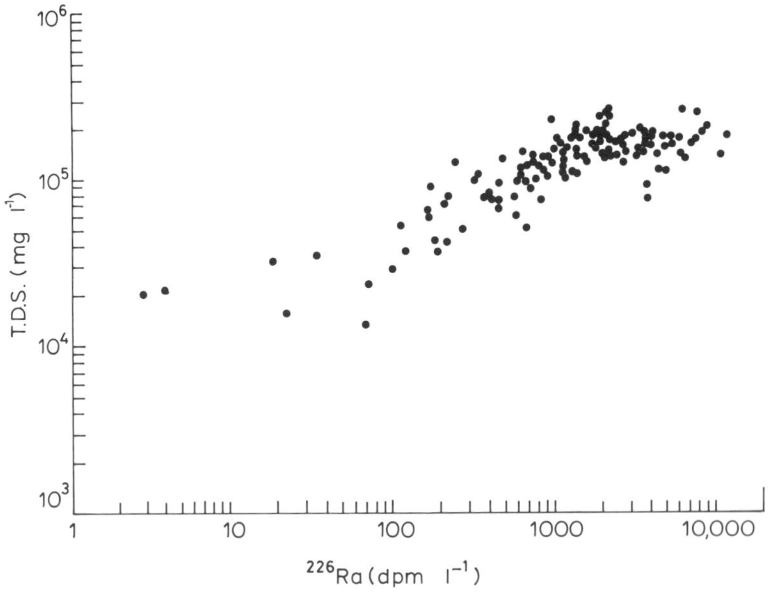

Fig. 10-12. Salinity – radium relation for data from the U.S. Gulf Coast region (from Kraemer and Reid, 1984) and U.S.S.R. (from Gutsalo, 1964).

ried in flowing formation waters or in exsolved gas, such as methane and carbon dioxide, through motionless fluids (Durrance, 1987). Stable helium-4 can migrate slowly up fractures from considerable depths (Durrance, 1987). Radon-222 and helium-4 fluxes are highest over major fractures where formation waters carry a rapidly moving gas phase, whereas low radon-222 and helium-4 fluxes occur in unfractured regions (Fig. 10-10; Durrance, 1987).

Formation waters with total dissolved solids in excess of 30,000 ppm, which have dissolved $^{222}Rn/^{226}Ra$ ratios greater than 7 – 10, contain radon-222 that has been introduced by fracture systems (Fig. 10-13). Abundant matrix-exsolved radon-222 is present in formation waters with less than 30,000 ppm total dissolved solids (Kraemer, 1985).

The $^{222}Rn/^{4}He$ ratio is larger for shallow gas origins or fast rates of leakage because radon from deeper sources would have decayed (Durrance, 1987). Fault and shear systems above or peripheral to salt domes are probably sealing where radon-222 concentrations are low and within the field of geopressured fluids (Figs. 10-10A and 10-10B). Formation fluids in such aquifers will show lower $^{222}Rn/^{226}Ra$ and $^{222}Rn/^{4}He$ ratios (Fig. 10-13).

Formation waters will show high radon-222 and helium-4 concentrations, high

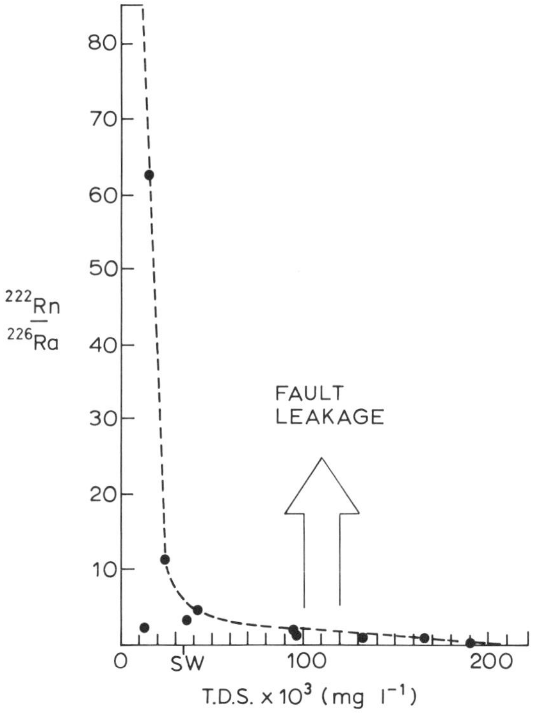

Fig. 10-13. Salinity – $^{222}Rn/^{226}Ra$ ratio relation. (Data for geopressured formation waters: from Kraemer, 1985). High $^{222}Rn/^{226}Ra$ ratios imply that faults are probably leaking. *SW* = seawater.

$^{222}Rn/^{226}Ra$ and $^{222}Ra/^{4}He$ ratios if hydrocarbon related fluids are rapidly leaking from the periphery or crest of diapirs along fractures or shear systems (Figs. 10-10C and 10-10D). This method of detecting leaking hydrocarbon accumulations should also be effective in non-salt environments.

CONCLUSIONS

An integrated hydrothermal model developed for the Gulf Coast Basin (Light, 1985; Kyle et al., 1986; Light et al., 1987) has been successfully applied to the North Sea Basin. This comprehensive model relates initiation of salt diapirism, cap-rock formation and cap-rock mineralization to sequential generation of diagenetic minerals, hydrothermal fluids, hydrocarbons, and overpressures that formed as a consequence of source rock and mother-salt burial (Light, 1985; Light et al., 1987; Posey et al., 1987a,b). In the North Sea Basin, additional account has been taken of deeply sourced carbon-dioxide-charged fluids that may have been released by the Jurassic stretching (Glennie, 1984a) of the North Sea and Atlantic Basins.

A modified integrated hydrothermal model can be employed in other actively filling or old/exposed basins. Comprehensive modelling may reveal novel techniques for locating salt-dome-related metal deposits, hydrocarbon fields, fractured reservoirs, or deep overpressured geothermal aquifers, which have significance for future geopressured-geothermal energy production.

Knowledge of rates of short-duration uplift of salt domes is important because of the effects such movements can produce on surface elevations of the seafloor, particularly in the vicinty of large offshore production platforms. Hydrocarbon and hazardous and toxic waste storage facilities located in salt domes may be jeopardized by slight diapiric uplifts, which could shear the seals in storage systems and cause leaks in the surrounding aquifers or fires at the surface. Detailed isotopic, diagenetic, and structural analyses of salt domes/cap rocks can lead to the development of predictors of short-duration diapiric uplifts.

The close association of petroleum accumulations with uranium is used to distinguish between salt-dome calcite cap rocks that have been preferentially formed from oil or from methane gas. Radon-222 and helium-4 concentrations in formation waters peripheral to and above salt-dome cap rocks will define whether fracture or fault systems adjacent to salt domes are leaking and whether reservoirs contain hydrocarbons. This method can be applied to petroleum reservoirs in evaporite-free environments.

ACKNOWLEDGEMENTS

Various drafts of this manuscript were reviewed by Prof. A.J. Smith, Royal Holloway and Bedford New College, University of London, Egham, whom the authors gratefully acknowledge. The first two illustrations were drafted under the direction of Richard L. Dillon, Bureau of Economic Geology, University of Texas at Austin. The authors have benefitted from discussions with colleagues throughout the U.S. Gulf Coast, especially Richard Kyle, Martin Jackson, Lynton Land, Jeff Hanor, Roger Sassen, and Peter Price. The review of the manuscript by Drs. George V. Chilingarian and K.H. Wolf is also appreciated.

REFERENCES

Alekseyev, F.A., Hottih, R.P. and Slindukova, G.D., 1961. Results of radio-chemical investigations of the Kyu-rov-Dag oil field. *Nucl. Geoph., Moscow,* pp. 160 – 175.

Anderton, R., Bridges, P.H., Leeder, M.R. and Sellwood, B.W., 1979. *A Dynamic Stratigraphy of the British Isles.* Allen and Unwin, London, 310 pp.

Armstrong, F.C. and Heemstra, R.H., 1972. Radiometrics proposed for exploration. *Oil Gas J.,* 70: 88 – 97.

Baird, R.A., 1986. Maturation and source rock evaluation of Kimmeridge clay, Norwegian North Sea. *Bull. Am. Assoc. Pet. Geol.,* 70: 1 – 11.

Bard, J.P., Burg, J.P., Matte, P.L. and Ribeiro, A., 1980. La chaîne hercynienne d'Europe occidentale en termes de tectonique que des plaques. In: J. Cogne and M. Slansky (Editors), *Geology of Europe.* BRGM, Orléans, France, pp. 233 – 246.

Behr, H.J. and Frentzel-Beyme, K., 1987. Permeability and palaeoporosity in crystalline bedrocks of the Central European basement – studies of cathodoluminescence. In: A. Boden and K.G. Eriksson (Editors), *Deep Drilling in Crystalline Bedrock.* Springer, Berlin, pp. 477 – 497.

Bennet, S.S. and Hanor, J.S., 1987. Dynamics of subsurface salt dissolution at Welsh dome, Louisiana Gulf Coast. In: I. Lerche and J.J. O'Brien (Editors), *Dynamical Geology of Salt Related Structures.* Academic Press, Orlando, Fla., pp. 653 – 677.

Blanche, J.B. and Whitacre, J.H.M., 1978. Diagenesis of part of the Brent Sand Formation (Middle Jurassic) of the northern North Sea basin. *J. Geol. Soc. London,* 135: 73 – 82.

Bethke, C.M., 1985. Compaction driven groundwater flow and heat transfer in intracratonic sedimentary basins and genesis of the Upper Mississippi Valley Mineral District. Doctoral thesis. Univ. Illinois, Urbana, Ill., 132 pp.

Bloch, S. and Key, R.M., 1981. Modes of formation of anomalously high radioactivity in oil-field brines. *Bull. Am. Assoc. Pet. Geol.,* 65(1): 154 – 159.

Boles, J.R. and Franks, S.G., 1979. Clay diagenesis in Wilcox sandstones of Southwest Texas. Implications of smectite diagenesis on sandstone cementation. *J. Sediment. Petrol.,* 49 (1): 55 – 70.

Brunstrom, R.G. and Walmsley, P.J., 1969. Permian evaporites in the North Sea Basin. *Bull. Am. Assoc. Pet. Geol.,* 53: 870 – 883.

Burke, W.H., Denison, R.E., Hetherington, E.A., Koepnick, R.B., Nelson, H.F. and Otto, J.B., 1982. Variation of seawater $^{87}Sr/^{86}Sr$ throughout Phanerozoic time. *Geology,* 10: 516 – 519.

Burst, J.K., 1969. Diagenesis of Gulf Coast clayey sediments and its possible relation to petroleum migration. *Bull. Am. Assoc. Pet. Geol.,* 53: 73 – 93.

Carstens, H. and Dypvik, H., 1981. Abnormal formation pressure and shale porosity. *Bull. Am. Assoc. Pet. Geol.,* 65 (2): 344 – 350.

Carstens, H. and Finstad, K.G., 1981. Geothermal gradients of the northern North Sea Basin, 50° – 62°N. In: L.V. Illing and G.D. Hobson (Editors), *Petroleum Geology of the Continental Shelf of North-West Europe.* Inst. Petroleum, London, pp. 152 – 161.

Christian, H.E., 1969. Some observations on the initiation of salt structures in southern British North Sea. In: P. Hepple (Editor), *The Exploration for Petroleum in Europe and North Africa.* Elsevier, Amsterdam, pp. 231 – 250.

Cloos, H., 1928. Uber antithetische Bewegungen. *Geol. Rundsch.,* 19(3): 249 – 251.

Collins, A.G., 1975. *The Geochemistry of Oil Field Waters.* Elsevier, New York, N.Y., 496 pp.

Collins, E.C., 1985. A review of the geology and Plio – Pleistocene deformations at Damon mound salt dome, Texas. In: *Phase III, Examination of Texas Salt Domes as Potential Sites for Permanent Storage of Toxic Chemical Waste.* Univ. Texas, Bur. Econ. Geol. Report prepared for the Texas Water Commission, contract IAC (84-85)-2203, pp. 275 – 305.

Colten-Bradley, V.A., 1987. Role of pressure in smectite dehydration – effects on geopressure and

smectite-to-illite transformation. *Bull. Am. Assoc. Pet. Geol.,* 71: 1414–1427.
Craig, H., Clarke, W.B. and Beg, M.A., 1975. Excess ^{3}He in deep water on the East Pacific rise. *Earth Planet. Sci. Lett.,* 26: 125–132.
Crossey, L.J., 1985. The origin and role of water-soluble organic compounds in clastic diagenetic systems. Ph.D. diss., Univ. Wyoming, Laramie, Wyo., 134 pp.
Curiale, J.A., Bloch, S., Rafalska-Bloch, J. and Harrison, W.E., 1983. Petroleum-related origin for uraniferous organic-rich nodules of Southwestern Oklahoma. *Bull. Am. Assoc. Pet. Geol.,* 67(4): 588–608.
Dewey, J.F., 1986. In: J.B. Dawson, D.A. Carswell, J. Hall and K.H. Wedepohl (Editors), *The Nature of the Lower Continental Crust. Geol. Soc., Spec. Publ.,* 24: 71–78.
Dunn, W.W., 1975. North Sea basinal area, Europe – an important oil and gas province. *Nor. Geol. Unders. Univ. Forl.,* 316: 69–97.
Durrance, E., 1987. Radon: an unobtrusive geological hazard. *Brit. Geol.,* 13: 72–76.
Dypvik, H., 1983. Clay mineral transformations in Tertiary and Mesozoic sediments from North Sea. *Bull. Am. Assoc. Pet. Geol.,* 67(1): 160–165.
Dyck, W., 1976. The use of helium in mineral exploration. *J. Geochem. Explor.,* 5: 3–20.
Dyke, W. and Smith, A.Y., 1969. The use of radon-222 in surface waters in geochemical prospecting for uranium. *Colorado School Mines, Mag.,* 64: 223–236.
Eslinger, E. and Pevear, D., 1988. Clay Minerals for Petroleum Geologists and Engineers. *Soc. Econ. Palaeontol. Mineral., Short Course Notes,* 22: 1-1-A-28.
Faber, E. and Stahl, W., 1984. Geochemical surface exploration for hydrocarbons in North Sea. *Bull. Am. Assoc. Pet. Geol.,* 68: 363–386.
Feeley, H.W. and Kulp, J.L., 1957. Origin of Gulf Coast salt dome sulphur deposits. *Bull. Am. Assoc. Pet. Geol.,* 41: 1802–1853.
Frost, S.M. and Schafersman, S.D., 1979. *Upper Oligocene Coral Reef on the Anahuac Formation in Damon Mound, Texas.* Houston Geol. Soc. Guidebook, pp. 26–44.
Gabelman, J.W., 1976. Orogenic and taphrogenic mineralization belts at continental margins. In: D.F. Strong (Editor), *Metallogeny and Plate Tectonics. Geol. Assoc. Can. Spec. Pap.,* 14: 273–300.
Gallagher, B.C., 1980. Genesis and significance of surficial geochemical anomalies in petroleum exploration: an overview. Colorado School Mines, Int. Rep., 39 pp.
Galloway, W.E., 1977. Catahoula Formation of the Texas Coastal Plain: depositional systems, composition, structural development, groundwater flow history, and uranium distribution. *The University of Texas at Austin, Bur. Econ. Geol., Rep. Invest.,* 87: 53 pp.
Galloway, W.E., 1982a. Epigenetic zonation and fluid flow history of the uranium-bearing fluvial aquifer systems. South Texas Uranium Province. *The University of Texas at Austin. Bur. Econ. Geol., Rep. Invest.,* 119: 31 pp.
Galloway, W.E., 1982b. Uranium mineralization in a coastal-plain fluvial aquifer system; Catahoula Formation, Texas. In: J.T. Nash (Editor), *Uranium Geology in Resource Evaluation and Exploration. Econ. Geol.,* 73(8): 1655–1676.
Garfunkel, Z. and Almager, G., 1987. Active salt dome development in the Levant basin, Southeast Mediterranean. In: I. Lerche and J.J. O'Brien (Editors), *Dynamical Geology of Salt and Related Structures.* Academic Press, Orlando, Fla., pp. 263–300.
Garven, G. and Freeze, R.A., 1984a. Theoretical analysis of the role of groundwater flow in the genesis of stratabound ore deposits; 1. Mathematical and numerical model. *Am. J. Sci.,* 284 (10): 1085–1124.
Garven, G. and Freeze, R.A., 1984b. Theoretical analysis of the role of groundwater flow in the genesis of stratabound ore deposits; 2. Quantitative results. *Am. J. Sci.* 284 (10): 1125–1174.
Gates, T.M. and McEldowny, R.C., 1977. Uranium exploration method may help find gas and oil. *World Oil,* 184: 55–57.
Gibbs, A.D., 1984. Clyde field growth fault secondary detachment above basement faults in North Sea. *Bull. Am. Assoc. Pet. Geol.,* 68: 1029–1039.
Gibbs, A., 1987. Development of extension and mixed mode sedimentary basins. In: M.P. Coward, J.F. Dewey and P.L. Hancock (Editors), *Continental Extensional Tectonics. Geol. Soc., Spec. Publ.,* 28: 19–33.
Glennie, K.W., 1984a. The structural framework and the pre-Permian history of the North Sea area. In: K.W. Glennie (Editor), *Introduction to the Petroleum Geology of the North Sea.* Blackwell Scientific Publ., Oxford, pp. 133–150.

Glennie, K.W., 1984b. Early Permian-Rotliegendes. In: K.W. Glennie (Editor), *Introduction to Petroleum Geology of the North Sea.* Blackwell Scientific Publ., Oxford, pp. 41 – 60; 133 – 150.

Glennie, K.W., Mudd, G.C. and Nagtegaal, P.J.C., 1978. Depositional environment and diagenesis of Permian Rotliegendes sandstone in Leman Bank and Sole Pit areas of the U.K. southern North Sea. *Geol. Soc. London J.,* 135: 25 – 34.

Gluckhauf, E. and Paneth, F.A., 1946. *Proc. R. Soc. London, Ser. A,* 185: 89 pp.

Goldman, M.I., 1933. Origin of the anhydrite cap rock of American salt domes. *U.S. Geol. Surv., Prof. Pap.,* 175: 83 – 114.

Gose, W.A., Kyle, J.R. and Ulrich, M.R., 1985. Preliminary paleomagnetic investigations of the Winnfield salt dome cap rock, Louisiana. *Gulf Coast Assoc. Geol. Soc. Trans.,* 35: 97 – 106.

Gregory, A.R., Dodge, M.A. and Posey, J.S. 1979. Progress report on evaluation of entrained methane in deep reservoirs, Texas Gulf Coast. In: M.H. Dorfman and W.L. Fisher (Editors), *Proc. 4th Congr. Geopressured Geothermal Energy, Austin 1.* Vol. 1: 414 – 510.

Gussow, W.C., 1968. Salt diapirism; importance of temperature, and energy source of emplacement. *Mem. Am. Assoc. Pet. Geol.,* 8: 16 – 52.

Gutsalo, L.K., 1964. On the distribution of radium in ground waters of the central part of the Dnieper-Donets basin. *Geokhimiya,* 12: 1305 – 1312.

Hallett D., 1981. Refinement of the geological model of the Thistle field. In: L.V. Illing and G.D. Hobson (Editors), *Petroleum Geology of the Continental Shelf of North-West Europe.* Inst. Pet., Heydon and Son, Plymouth, pp. 315 – 325.

Hanor, J.S., 1984. *Salinity and Geochemistry of Subsurface Brines in South Louisiana and Their Potential for Reaction with Injected Hydrothermal Waste Waters.* U.S. Dep. Energy, Washington, D.C., Rep. DOE/NV/10174-3, pp. 277 – 346.

Hanor, J.S., 1987. *Origin and Migration of Subsurface Sedimentary Brines. Soc. Econ. Paleontol. Mineral., Short Course,* 21, 247 pp.

Hanor, J.S. and Workman, A.L., 1986. Distribution of dissolved fatty acids in some Louisiana oil field brines. *Appl. Geochem.,* 1: 37 – 46.

Hassan, M. and Hossin, A., 1975. *Contribution à l'Etude des Comportements du Thorium et du Potassium dans les Roches Sédimentaires.* C.R. Acad. Sci., Paris, 280 pp.

Holser, W.T., 1979. Trace elements and isotopes in evaporites. In: R.G. Burns (Editor), *Marine Minerals. Mineral. Soc. Am., Short Course Notes,* 6: 295 – 336.

Huang, W.H. and Keller, W.D., 1970. Dissolution of rock-forming silicate minerals in organic acids: simulated first stage weathering of fresh mineral surfaces. *Am. Mineral.,* 55: 2076 – 2094.

Hunt, J.M., 1979. *Petroleum Geochemistry and Geology.* W.H. Freeman, San Fransico, Calif., 617 pp.

Jackson, M.P.A. and Seni, S.J., 1983. Geometry and evolution of salt structures in a marginal rift basin of the Gulf of Mexico, East Texas. *Geology,* 11: 131 – 135.

Jackson, M.P.A. and Seni, S.J., 1984. Atlas of Salt Domes in the East Texas Basin. *The University of Texas at Austin, Bur. Econ. Geol., Rep. Invest.,* 140: 102 pp.

Jackson, S.A. and Beales, F.W., 1967. An aspect of sedimentary basin evolution: the concentration of Mississippi Valley type ores during late stages of diagenesis. *Canad. Soc. Pet. Geol., Bull.,* 15: 383 – 433.

Jenyon, M.K., 1985. Basin edge diapirism and updip salt flow in Zechstein of southern North Sea. *Bull. Am. Assoc. Pet. Geol.,* 69 (1): 53 – 64.

Johns, W.D. and Simoyama, A., 1972. Clay minerals and petroleum-forming reaction during burial and diagenesis. *Bull. Am. Assoc. Pet. Geol.,* 56: 2160 – 2167.

Jones, P.H., 1980. Role of geopressure in hydrocarbon and water systems. In: W.H. Roberts (Editor), Problems of Petroleum Migration. *Stud. Geol. Am. Assoc. Pet. Geol.,* 10: 207 – 216.

Kaiser, W.R., 1983. Predicting reservoir quality and diagenetic history in the Frio Formation (Oligocene) of Texas. In: R.A. Morton, T.E. Ewing, W.R. Kaiser and R.J. Finley (Editors), *Consolidation of Geologic Studies of Geopressured Geothermal Resources in Texas.* The University of Texas at Austin, Bur. Econ. Geol., Rep. prepared for U.S. Dep. Energy, Contract DE-AC08-79ET27111: 137 – 170.

Kaiser, W.R. and Richmann, D.L., 1981. Predicting diagenetic history and reservoir quality in the Frio Formation of Brazoria County, Texas, and Pleasant Bayou test wells. *Proc. 5th Conf. Geopressured Geothermal Energy,* Austin, Tex., pp. 67 – 74.

Kharaka, Y.K., Callender, E. and Carothers, W.W., 1977. Geochemistry of geopressured waters of the northern Gulf of Mexico basin, Part 1, Brazoria and Galveston Counties, Texas. In: H. Paquet and Y. Tardy (Editors), *Proc. 2nd Int. Symp. Water-Rock Interaction,* Sect. II. Strasbourg, pp. 32 – 41.

Kharaka, Y.K., Lico, M.S., Wright, V.A. and Carothers, W.W., 1980. Geochemistry of formation waters from Pleasant Bayou No. 2 well and adjacent areas in coastal Texas. In: M.H. Dorfman and W.L. Fisher (Editors), *Proc. Fourth U.S. Gulf Coast Geopressured Geothermal Energy Conf. The University of Texas, Center for Energy Studies,* 1: 168 – 199.

Kharaka, Y.K., Carothers, W.W. and Law, L.M., 1985a. Origin of gaseous hydrocarbons in geopressured geothermal water (abstract). *Sixth U.S. Gulf Coast Geopressured Geothermal Energy Conf.* The University of Texas at Austin, 16 pp.

Kharaka, Y.K., Carothers, W.W. and Law, L.M., 1985b. Origin of gaseous hydrocarbons in geopressured geothermal waters. In: M.H. Dorfman and R.A. Morton (Editors), *Proc. 6th U.S. Gulf Coast Geopressured Geothermal Energy Conf.* Pergamon, New York, N.Y., 125 pp.

Kirk, R.H., 1980. Statfjord field, a North Sea giant. In: M.T. Halbouty (Editor), *Giant Oil and Gas Fields of the Decade, 1968 – 1978. Mem. Am. Assoc. Pet. Geol.,* 30: 95 – 116.

Knauth, L.P., Kumar, M.B. and Martinez, J.D., 1980. Isotope geochemistry of water in Gulf Coast salt domes. *J. Geophys. Res.,* 85: 4863 – 4871.

Knauth, L.P. and Beeunas, M.A., 1986. Isotope geochemistry of fluid inclusions in Permian halite with implications for the isotopic history of ocean water and the origin of saline formation waters. *Geochim. Cosmochim. Acta,* 50: 419 – 434.

Kondratov, L.S., Kosolapov, A.I., Ovsyanikov, V.M. and Sarkisyan, I.S., 1978. Study of trace elements in sedimentary rocks aimed at searching for oil and gas. *8th Int. Congr. Organic Chemistry,* 1: 49 – 50.

Kraemer, T.F., 1981. ^{234}U and ^{238}U concentration in brine from geopressured aquifers of the northern Gulf of Mexico basin. *Earth Planet. Sci. Lett.,* 56: 210 – 216.

Kraemer, T.F., 1985. Natural radioelement behavior in geopressured aquifers. In: M.B. Dorfman and R.A. Morton (Editors), *Geopressured Geothermal Energy.* Pergamon, New York, N.Y., pp. 127 – 136.

Kreitler, C.W. and Dutton, S.P., 1983. *Origin and Diagenesis of Caprock, Gyp Hill and Oakwood Salt Domes, Texas.* The University of Texas at Austin, Bur. Econ. Geol., Rep. Invest., 131, 58 pp.

Kraemer, T.F. and Reid, D.F., 1984. The occurrence and behavior of radium in saline formation water of the U.S. Gulf Coast region. *Isotope Geoscience,* 2: 153 – 174.

Kreitler, C.W. and Dutton, S.P., 1983. Origin and diagenesis of cap rock, Gyp Hill and Oakwood salt domes, Texas. *The University of Texas at Austin, Austin Bur. Econ. Geol., Rep. Invest.,* 131: 58 pp.

Kyle, J.R. and Agee, W.N., Jr., 1988. Evolution of metal ratios and $\delta^{34}S$ composition of sulfide mineralization during anhydrite cap rock formation, Hockley Dome, Texas, U.S.A. In: H.H. Posey (Editor), *Fluid – Rock Interactions in the Salt Dome Environment. Chem. Geol.,* 74 (1/2): 37 – 55.

Kyle, J.R. and Price, P.E., 1985. Mineralogical investigation of sulphide concentrations in the Hockley salt dome cap rock. Texas. In: W.C. Park, D.M. Hausen and R.D. Hagni (Editors), *Applied Mineralogy.* Metal. Soc. AIME, pp. 1065 – 1082.

Kyle, J.R. and Price, P.E., 1986. Metallic sulphide mineralization in salt-dome cap rocks, Gulf Coast, U.S.A. *Trans. Inst. Min. Metall.,* 95B: B6 – B16.

Kyle, J.R., Ulrich, M.R. and Gose, W.A., 1987. Textural and paleomagnetic evidence for the mechanism and timing of anhydrite cap rock formation, Winnfield salt dome, Louisiana. In: I. Lerche and J.J. O'Brien (Editors), *Dynamical Geology of Salt and Related Structures.* Academic Press, Orlando, Fla., pp. 497 – 542.

Land, L.S., Kupecz, J.A. and Mack, L.E., 1988. Louann salt geochemistry (Gulf of Mexico Sedimentary Basin, U.S.A.); a preliminary synthesis. In: H.H. Posey (Editor). *Fluid – Rock Interactions in the Salt Dome Environment. Chem. Geol.,* 74 (1/2): 25 – 35.

Larese, R.E., Haskell, N.L., Prebindowski, D.R. and Beju, D., 1984. Porosity development in selected Jurassic sandstones from Norwegian and North Seas, Norway – an overview. In: Graham and Trotman (Editors), *Petroleum Geology of the North European Margin. Norw. Pet. Soc.,* pp. 81 – 95.

Lee, M., Aronson, J.L. and Savin, S.M., 1985. K/Ar dating of time of gas emplacement in Rotliegendes sandstone, Netherlands. *Bull. Am. Assoc. Pet. Geol.,* 69: 1381 – 1385.

Lee, M., Aronson, J.L. and Savin, S.M., 1989. Timing and conditions of Permian Rotliegendes sandstone diagenesis, Southern North Sea: K/Ar and oxygen isotopic data. *Bull. Am. Assoc. Pet. Geol.,* 73(2): 195 – 215.

Liewig, N., Clauer, N. and Sommer, F., 1987. Rb – Sr and K – Ar dating of clay diagenesis in Jurassic sandstone oil reservoir, North Sea. *Bull. Am. Assoc. Pet. Geol.*, 71: 1467 – 1474.

Light, M.P.R., 1985. Integrated hydrothermal model for a proposed deep crustal borehole on the Texas Gulf Coast – origins of geopressured brines and lead – zinc, uranium, hydrocarbon and cap-rock deposits (abstract). *Thirteenth Meeting, Cornell Progr. Study of the Continents, Book of Abstracts.* (unnumbered).

Light, M.P.R., Posey, H.H., Kyle, J.R. and Price, P.E., 1987. Model for the origins of geopressured brines, hydrocarbons, caprocks and metallic mineral deposits, Gulf Coast, U.S.A. In: I. Lerche and J.J. O'Brien (Editors), *Dynamical Geology of Salt Related Structures.* Academic Press, Orlando, Fla., pp. 787 – 832.

Loucks, R.G., Richman, D.I. and Milliken, K.L., 1981. *Factors Controlling Reservoir Quality in Tertiary Sandstones and Their Significance to Geopressured Geothermal Production. The University of Texas at Austin, Bur. Econ. Geol., Rep. Invest.*, III: 41 pp.

Loucks, R.G., Dodge, M.M. and Galloway, W.E., 1986. *Controls on Porosity and Permeability of Hydrocarbon Reservoirs in Lower Tertiary Sandstones Along the Texas Gulf Coast. The University of Texas at Austin, Bur. Econ. Geol., Rep. Invest.*, 149: 78 pp.

Lovell, J.H., 1988. Symposium on a technique for earthquake prediction and monitoring in situ stress. *Br. Geol.*, 14/1: 16 pp.

Ludwig, K.R., Goldhaber, M.B., Reynolds, R.L. and Simmons, K.R., 1982. Uranium-lead isochron ages and preliminary sulfur isotope systematics of the Felder uranium deposit, south Texas. *Econ. Geol.*, 77: 557 – 563.

Lundegard, P.D., 1985. Carbon dioxide and organic acids: origin and role in diagenesis (Texas Gulf Coast Tertiary). Ph.D. diss., The University of Texas at Austin, Austin, Tex., 145 pp.

Mason, B., 1966. *Principles of Geochemistry.* Wiley, New York, N.Y., 329 pp.

McManus, K.M. and Hanor, J.S., 1988. Calcite and iron-sulfide cementation of Miocene sediments flanking the West Hackberry salt dome, southwest Louisiana, U.S.A. In: H.H. Posey and J.R. Kyle (Guest Editors), *Special Issue on Fluid-Rock Interactions in the Salt Dome Environment. Chem. Geol.*, 74: 99 – 112.

Milner, C.W.D., Rogers, M.A. and Evans, C.R., 1977. Petroleum transformations in reservoirs. *J. Geochem. Explor.*, 7(2): 101 – 153.

Miodrag, S., 1975. Should we consider geochemistry as an important exploratory technique. *Oil Gas J.*, 73 (31): 106 – 110.

Mooney, R.W., Keenan, A.G. and Wood, L.A., 1952. Adsorption of water vapor by montmorillonite. 1. Heat of desorption and application of BET theory. *J. Am. Chem. Soc.*, 74: 1367 – 1371.

Morton, A.C., 1982. Lower Tertiary sand development in Viking Graben, North Sea. *Bull. Am. Assoc. Pet. Geol.*, 66: 1542 – 1559.

Morton, J.P., 1983. Age of clay diagenesis in the Oligocene Frio Formation, Texas Gulf Coast. Ph.D. diss., The University of Texas at Austin, Austin, Tex., 33 pp.

Muehlberger, W.R., Galloway, W.E., Land, L.S., Sharp, J.M., Sawyer, D.S. and Light, M.P.R., 1987. Deep crustal drilling, Texas Gulf Coast, United States of America. In: A. Boden and K.G. Eriksson (Editors), *Deep Drilling in Crystalline Bedrock,* Vol. 2. Springer, Berling, pp. 155 – 171.

Murray, G.E., 1966. Salt structures of Gulf of Mexico basin – a review. *Bull. Am. Assoc. Pet. Geol.*, 50: 439 – 478.

Neglia, S., 1979. Migration of fluids in sedimentary basins. *Bull. Am. Assoc. Pet. Geol.*, 63: 573 – 597.

Neugebauer, J., 1974. Some aspects of cementation in chalk. In: K.J. Soo and H.C. Jenkins (Editors), *Pelagic Sediments on Land and under Sea. Int. Assoc. Sedimentol., Spec. Publ.*, pp. 149 – 176.

Palciauskas, V.V. and Domenico, P.A., 1980. Microfracture development in compacting sediments, relation to hydrocarbon maturation kinetics. *Bull. Am. Assoc. Pet. Geol.*, 64: 927 – 937.

Perry, C. and Turekian, K.K., 1974. The effect of diagenesis on the redistribution of strontium isotopes in shales. *Geochim. Cosmochim. Acta,* 38: 929 – 935.

Philp, R.P. and Crisp, P.T., 1982. Surface geochemical methods used for oil and gas prospecting – a review. *J. Geochem. Explor.*, 17: 1 – 34.

Posey, H.H., 1986. Regional characteristics of strontium, carbon and oxygen isotopes in salt dome cap rocks of the Western Gulf Coast. Ph.D. diss., Univ. North Carolina at Chapel Hill, N.C., 239 pp.

Posey, H.H., Kyle, J.R., Jackson, T.J., Hurst, S.D. and Price, P.E., 1987a. Multiple fluid components of salt diapirs and salt dome cap rocks, Gulf Coast U.S.A. *Appl. Geochem.*, 2: 523 – 534.

Posey, H.H., Price, P.E. and Kyle, R.J., 1987b. Mixed carbon sources for calcite cap rocks of Gulf

Coast salt domes. In: I. Lerche and J.J. O'Brien (Editors), *Dynamical Geology of Salt and Related Structures*. Academic Press, Orlando, Fla., pp. 593 – 630.

Posey, H.H. and Kyle, J.R., 1988. Fluid – Rock interactions in the salt dome environment: an introduction and review. In: H.H. Posey and J.R. Kyle (Guest Editors), *Fluid – Rock Interactions in the Salt Dome Environment. Chem. Geol.,* 74: 1 – 24.

Powers, M.C., 1967. Fluid release mechanisms in compacting marine mudrocks and their importance in oil exploration. *Am. Assoc. Pet. Geol. Bull.,* 51: 1240 – 1254.

Price, L.C., 1976. Aqueous solubility of petroleum as applied to its origin and primary migration. *Bull. Am. Assoc. Pet. Geol.,* 60: 213 – 244.

Price, L.C., 1982. Organic geochemistry of core samples from an ultra-deep hot well (300 Deg. C., 7 km). *Chem. Geol.,* 37: 215 – 228.

Price, P.E., Kyle, J.R. and Wessel, G.R., 1983. Salt dome related lead – zinc deposits. In: G. Kisvarsanyi, S.K. Grant, W.P. Pratt and J.W. Koenig (Editors), *Proc. Int. Conf. Mississippi Valley Type Lead-Zinc Deposits.* Rolla, Mo., pp. 558 – 577.

Prikryl, J.D., Posey, H.H. and Kyle, J.R., 1988. A petrographic and geochemical model for the origin of calcite cap rock at Damon Mound salt dome, Texas, U.S.A. In: H.H. Posey and J.R. Kyle (Editors), *Fluid – Rock Interactions in the Salt Dome Environment. Chem. Geol.,* 74: 67 – 97.

Rezak, R., 1985. Local carbonate production on a terrigenous shelf. *Gulf Coast Assoc. Geol. Soc., Trans.,* 35: 477 – 484.

Ruckmick, J.C., Wimberly, B.H. and Edwards, A.F., 1979. Classification and genesis of biogenic sulfur deposits. *Econ. Geol.,* 74: 469 – 474.

Russel, M.J., 1976. Incipient plate separation and possible related mineralization in lands bordering the North Atlantic. In: D.F. Strong (Editor), *Metallogeny and Plate Tectonics. Geol. Assoc. Can., Spec. Pap.,* 14: 339 – 349.

Salvador, A. and Buffler, R.T., 1982. The Gulf of Mexico Basin. In: A.R. Palmer (Editor), *Perspectives in Regional Geological Synthesis. Geol. Soc. Am. Decade of N. Am. Geol., Spec. Publ.,* pp. 157 – 162.

Sassen, R., 1987. Organic geochemistry of salt dome cap rocks, Gulf Coast Basin. In: I. Lerche and J.J. O'Brien (Editors), *Dynamical Geology of Salt and Related Structures.* Academic Press, Orlando, Fla., pp. 631 – 649.

Schlumberger, 1976. *Natural Gamma Ray Spectrometry Interpretation.* 112 pp.

Schlumberger, 1981. *Litho Density Tool Interpretation.* 62 pp.

Schmidt, V. and McDonald, D.A., 1979. Texture and recognition of secondary porosity in sandstones. In: P.A. Scholle and P. Schluger (Editors), *Aspects of Diagenesis. Soc. Econ. Paleontol. Mineral. Spec. Publ.,* 26: 175 – 207.

Schmidt, V.D., McDonald, D.A. and Platt, R.L., 1977. Pore geochemistry and reservoir aspects of secondary porosity in sandstones. *Bull. Can. Pet. Geol.,* 25: 271 – 290.

Selley, R.C., 1979. Concepts and methods of subsurface facies analysis. *Am. Assoc. Pet. Geol.,* Continuing Education Course, Note Series, 9: 82 pp.

Seni, S.J., 1987. Evolution of Boling dome cap rock with emphasis on included terrigenous clastics, Fort Bend and Wharton Counties, Texas. In: I. Lerche and J.J. O'Brien (Editors), *Dynamical Geology and Salt Related Structures.* Academic Press, Orlando, Fla., pp. 543 – 591.

Seni, S.J. and Jackson, M.P.A., 1983. Evolution of salt, East Texas diapir province, part 1: Sedimentary record of halokinesis. *Bull. Am. Assoc. Pet. Geol.,* 67: 1219 – 1244.

Seni, S.J., Mullican, W.F., III and Hamlin, H.S., 1984. *Texas Salt Domes – Aspects Affecting Disposal of Toxic-Chemical Waste in Solution-Mined Caverns. The University of Texas at Austin, Bureau of Economic Geology,* Rep. prepared for Texas Dep. Water Resour., inter-agency contract IAC (84-85)-1019, 94 pp.

Seni, S.J., Kreitler, C.W., Mullican, W.F., III and Hamlin, H.S., 1985. Utilization of salt domes for chemical waste disposal. In: E.T. Smerdon and W.R. Jordan (Editors), *Issues in Groundwater Management. Water Resources Symp.,* 12. Center for Research in Water Resources, The University of Texas at Austin, pp. 419 – 460.

Sommer, F., 1975. Histoire diagénétique d'une série gréseuse de Merdu Nord – datation de l'introduction des hydrocarbures. *Rev. Inst. Fr. Pét.,* 30: 729 – 741.

Sommer, F., 1978. Diagenesis of Jurassic sandstones in Viking graben. *Geol. Soc. London J.,* 135: 63 – 67.

Stesky, R.M. and Brace, W.F., 1973. Electrical conductivity of serpentinised rocks to 6 kilobars. *J. Geophys. Res.,* 78: 7614 – 7621.

Surdam, R.C., Crossey, L.J., Hagen, E.S. and Heasler, H.P., 1989. Organic – Inorganic interactions and sandstone diagenesis. *Bull. Am. Assoc. Pet. Geol.,* 73: 1 – 23.

Tade, M.D., 1967. Helium storage in Cliffside Field. *J. Pet. Technol.,* 19(7): 885 – 888.

Talbot, C.J. and Jackson, M.P.A., 1987. Salt tectonics. *Sci. Am.,* 255(8): 70 – 79.

Taliyev, S.D., Gribkov, V.V., Kozlov, V.F., Astafyev, V.P., Norenkova, I.K., Svechina, P.M. and Yaroshenko, V.N., 1978. Combined biogeochemical study of bottom sediments in hydrocarbon prospecting on shelf areas. *8th Int. Congr. Organic Geochemistry,* 1: 93 – 94.

Taylor, R.E., 1938. Origin of the cap rock of Louisiana salt domes. *La. Geol. Surv. Bull.,* 11: 191 pp.

Taylor, J.C.M., 1981. Zechstein facies and petroleum prospects in Central and Northern North Sea. In: L.V. Illing and G.D. Hobson (Editors), *Petroleum Geology of the Continental Shelf of North-West Europe.* Inst. Petroleum, Heyden and Son, London, pp. 176 – 185.

Tissot, B.P. and Welte, D.H., 1984. *Petroleum Formation and Occurrence.* Springer, Berlin, 538 pp.

Tolstikin, I.N., 1978. A review of some recent advances in isotope geochemistry of light rare gases. In: E.C. Alexander, Jr. and M. Ozima (Editors), *Terrestrial Rare Gases.* Japan Scientific Societies Press, Tokyo, pp. 33 – 62.

Touret, J., 1986. Fluid inclusions in rocks from the lower crust. In: J.B. Dawson, D.A. Carswell, J. Hall and K.H. Wedepohl (Editors), *The Nature of the Lower Continental Crust. Geol. Soc., Spec. Publ.,* 24: 161 – 172.

Trommsdorff, V. and Skippen, G., 1986. Vapour loss ("boiling") as a mechanism for fluid evolution in metamorphic rocks. *Contrib. Mineral. Petrol.,* 94: 317 – 322.

Trusheim, F., 1960. Mechanism of salt migration in northern Germany. *Bull. Am. Assoc. Pet. Geol.,* 44: 1519 – 1540.

Ulrich, M.R., Kyle, J.R. and Price, P.E., 1984. Metallic sulfide deposits in the Winnfield salt dome, Louisiana – evidence for episodic introduction of metalliferous brines during cap rock formation. *Gulf Coast Assoc. Geol. Soc., Trans.,* 34: 435 – 442.

Van den Bark, E. and Thomas, O.D., 1980. Ekofisk; first of the Giant oil fields in Western Europe. In: M.T. Halbouty (Editor), *Giant Oil and Gas Fields of the Decade,* 1968 – 1978. *Am. Assoc. Pet., Geol. Mem.,* 30: 195 – 224.

Van der Vlis, A.C., Duns, H. and Fernandez Luque, R., 1979. Increasing well productivity in tight chalk reservoirs. *10th World Petroleum Congr.,* Bucharest, Pap. P.D. 7: 3 pp.

Van Wijhe, D.H., 1987. Structural evolution of inverted basins in the Dutch offshore. *Tectonophysics,* 137: 171 – 219.

Van Wijhe, D.E., Lutz, M. and Kaasschieter, J.P.H., 1980. The Rotliegendes in the Netherlands and its gas accumulation. *Geologie Mijnb.,* 59: 3 – 24.

Wakita, H., Nakamura, Y., Kita, I., Fujii, N. and Notsu, K., 1980. Hydrogen release, new indicator of fault activity. *Science,* 210: 188 – 190.

Warner, M. and McGeary, S., 1987. Seismic reflection coefficients from mantle fault zones. In: Deep seismic reflection profiling of the continental lithosphere. *Geophys. J. R. Astron. Soc.,* 89: 223 – 230.

Watts, N.L., 1983. Microfractures in chalks of the Albuskjell Field, Norwegian Sector, North Sea: Possible origin and distribution. *Bull. Am. Assoc. Pet. Geol.,* 67(2): 201 – 234.

Wickham, S.M. and Taylor, H.P., 1987. Stable isotope constraints on the origin and depth of penetration of hydrothermal fluids associated with Hercynian regional metamorphism and crustal anatexis in the Pyrenees. *Contrib. Mineral. Petrol.,* 95: 255 – 268.

Windley, B.F., 1986. *The Evolving Continents.* Wiley, Chichester, 399 pp.

Winkler, H.G.F., 1976. *Petrogenesis of Metamorphic Rocks.* Springer, New York, N.Y., 4, 320 pp.

Woidt, W.D., 1978. Finite element calculations applied to salt dome analysis. In: M.N. Toksoz (Editor), *Numerical Modelling in Geodynamics. Tectonophysics,* 50 (2/3): 369 – 386.

Workman, A.L. and Hanor, J.S., 1985. Evidence for large-scale vertical migration of dissolved fatty acid in Louisiana oil field brines. *Gulf Coast Assoc. Geol. Soc., Trans.,* 35: 293 – 300.

Reprinted from: *Diagenesis, III. Developments in Sedimentology, 47.*
Edited by K.H. Wolf and G.V. Chilingarian.

Chapter 11

PRECAMBRIAN IRON-FORMATIONS: NATURE, ORIGIN, AND MINERALOGIC EVOLUTION FROM SEDIMENTATION TO METAMORPHISM

HAROLD L. JAMES

INTRODUCTION

Because of the economic significance of Precambrian iron-formations as the world's major source of iron, and the unique physical and chemical attributes that make them prime targets for scientific research, investigations into these unusual and distinctive rocks have generated a vast literature over the past 100 years. Several major compendia of recent or relatively recent vintage are available: "Genesis of Precambrian Iron and Manganese Deposits" (UNESCO, 1973); "Precambrian Iron-Formations of the World" (James and Sims, 1973); "Iron-Formations: Facts and Problems" (Trendall and Morris, 1983); and "Precambrian Iron-Formations" (Appel and LaBerge, 1987). These volumes are indispensable sources of data and theory for the researcher, but they lack authoritative syntheses that would meet the needs of the nonspecialist reader. It is the aim of the present chapter to fill that gap – to provide a brief but comprehensive summary of the occurrence, geologic associations, age distributions, physical and chemical properties, probable origin, and mineralogic evolution of Precambrian iron-formations.

The summary review presented in this chapter draws heavily from the voluminous published record, but it is grounded in field studies by the writer of a number of iron-formation districts, carried out intermittently over the past 45 years. Predictably, therefore, the approach taken to general problems of classification, genesis, and significance of iron-formation relies more on field evidence and field associations than on laboratory research and theoretical analysis. Nevertheless, it must be granted that the results of experimental and theoretical research have had, and will continue to have, an important role in the development of concepts of iron-formation genesis. Of particular importance are the relationships between precipitation of iron and certain physicochemical attributes of the hydrous medium, notably oxidation potential (Eh) and acidity – alkalinity (pH), which were outlined in a series of papers published in the 1950's by geochemist R.M. Garrels and his associates at Northwestern University (Castaño and Garrels, 1950; Krumbein and Garrels, 1952; Huber and Garrels, 1953; and Huber, 1958). Subsequent laboratory and theoretical research (for example, Garrels and Christ, 1965; Eugster and Chou, 1973; Holland, 1984; LaBerge et al., 1987) has continued to expand the range of geochemical knowledge that can be invoked to help interpret the complex relationships observed in the rocks themselves.

The virtual limitation of iron-formation to the Precambrian – and of the bulk of it to strata older than 1900 m.y. – is a powerful indication in itself that the deposits originated under conditions appreciably different from those of later periods in the earth's history. Inasmuch as no recent or modern-day counterparts can be cited (iron-rich sediments are being formed at present in some rift environments in response to exhalative processes but the deposits bear little similarity otherwise to iron-formation), almost any theory of origin will involve untested (and perhaps untestable) postulates. Presently accepted genetic models doubtless will be modified, or even rejected, as further research provides new clues and constraints. But is is not likely that the elements of conjecture and speculation can ever be eliminated in considering the origin of these ancient and unusual deposits.

BACKGROUND REVIEW

Definitions and nomenclature

The term iron-formation* evolved, through common usage, from the designation "iron-bearing formation" used in early reports on the Lake Superior iron ranges, and it has come to replace or be accepted as equivalent to other regional designations, such as "itabirite" (Brazil), "banded ironstone" (South Africa), "quartz-banded ores" (Sweden), "banded hematite quartzite" (India), and "jasper" or "jaspilite" (Lake Superior and Australia). It has been defined formally (if incompletely) as "a chemical sediment, typically thin-bedded or laminated, containing 15% or more iron of sedimentary origin, commonly but not necessarily containing layers of chert" (James, 1954).

This definition has been widely quoted and used, but not without criticism. Some have considered the 15% cutoff value to be too limiting – Trendall (1983a, p. 8), for example, would substitute "an anomalously high content of iron" – but, in fact, the value is not arbitrarily chosen. As Lepp (1987, fig. 3) has shown diagrammatically, the iron content of Precambrian sedimentary rocks has a bimodal distribution pattern, and a 15% value approximates the low-frequency trough that separates the common rock average (about 4%) from the iron-formation average (about 29%). Others have considered the definition too broad in scope and would apply the term only to chert-banded rocks, or, even further, to those chert-banded rocks in which the iron-rich layers consist mainly of iron oxides, magnetite or hematite (Trendall, 1983a, p. 2). Though it is recognized that it is indeed the chert-banded rocks that are the most distinctive and the most readily identified of the strata assigned the term "iron-formation", it must also be noted that virtually every stratigraphic unit so designated (e.g., Brockman Iron Formation) will contain interlayered non-cherty iron-rich strata, on a scale ranging from centimeters to meters, that are equally iron-rich and share the depositional history. The broader

* In this chapter the convention of hyphenation is retained, in order to emphasize the fact that iron-formation is a rock name, not a stratigraphic designation. It can, of course, be used as part of a formal stratigraphic term, such as Riverton Iron-formation, in which case it should be capitalized. It is to be noted, however, that many authors dispense with the hyphen in this usage.

definition of the term is utilized in this review, both for the sake of consistency and for usefulness in discussion of genesis and regional variations. In some descriptive sections, the modified term *banded iron-formation* will be applied, where appropriate, to iron-formation with rhythmically interlayered chert.

Taconite* is a term used in the Lake Superior region and elsewhere to iron-formation suitable for processing as a low-grade iron ore. Though at times used loosely ("slaty taconite", for example) the typical material so designated is metamorphosed, relatively coarse-grained, chert-banded iron-formation, in which most of the iron is in the form of crystalline hematite and/or magnetite that can be recovered magnetically or by flotation.

By intent, if not by strict reading, the definition of iron-formation here adopted excludes the (generally) oolitic iron-rich rocks of (mostly) Phanerozoic age, typified by the Jurassic-age "minette" ores of Europe. These rocks are classed as *ironstone* (James, 1966; Gross, 1973). Though, generally containing considerably more than 15% iron of sedimentary origin, these non-cherty beds lack the laminated or thin-bedded structure typical of iron-formation, and chemically they are marked by distinctly lower contents of silica, as shown in Fig. 11-1.

The proposal by Kimberly (1978; see also 1989a,b) that *all* iron-rich sedimentary rocks be called ironstone and that the stratigraphic units marked by such rock be called iron formations, though by no means illogical, is not likely to be adopted, and for good reasons. Considering past practice, and the voluminous literature affected, use of such nomenclature would cause widespread confusion; further, it would result in the grouping together two quite dissimilar classes of rock (each of which consists of a variety of rock types) that have little in common in terms of physical structure, age, occurrence, and genesis. The iron-formation/ironstone terminology employed here, though far from ideal, does assist in communicating these differences, thus meeting one important criterion of appropriate classification and nomenclature.

Classification

Inasmuch as iron-formation is a term for an entire class of rocks rather than an individual rock type, further subdivision is needed both for orderly description and for consideration of regional variations and genesis. Within individual districts, subdivision schemes commonly are based on gross structural and textural differences that can be used for local correlation. In the Mesabi district, for example, the Biwabik Iron Formation was early subdivided into four units: Lower Cherty, Lower Slaty, Upper Cherty, and Upper Slaty (Wolff, 1917). In the Hamersley region of Western Australia, the Dales Gorge member of the Brockman Iron Formation has been subdivided into BIF (banded iron-formation) and S (shale) units, the latter

* The term has a curious etymology. It is the last remnant of a grossly erroneous stratigraphic correlation made late in the late 1800's, when Proterozoic strata of the Gunflint and Mesabi districts were designated "Taconic Series" on the basis of supposed equivalency with strata of Paleozoic age in the Taconic Mountains of northeastern United States.

referring to dark-colored strata that are stilpnomelane- and siderite-rich (Trendall, 1983b, p. 85).

Three general schemes of classification now have current status: (1) a *facies* classification that is based on the dominant iron mineral present (oxide, silicate, carbonate, sulfide); (2) a *type* classification based on geologic association and analogy (Superior, Algoma, Rapitan); and, finally (3) a *textural* classification patterned

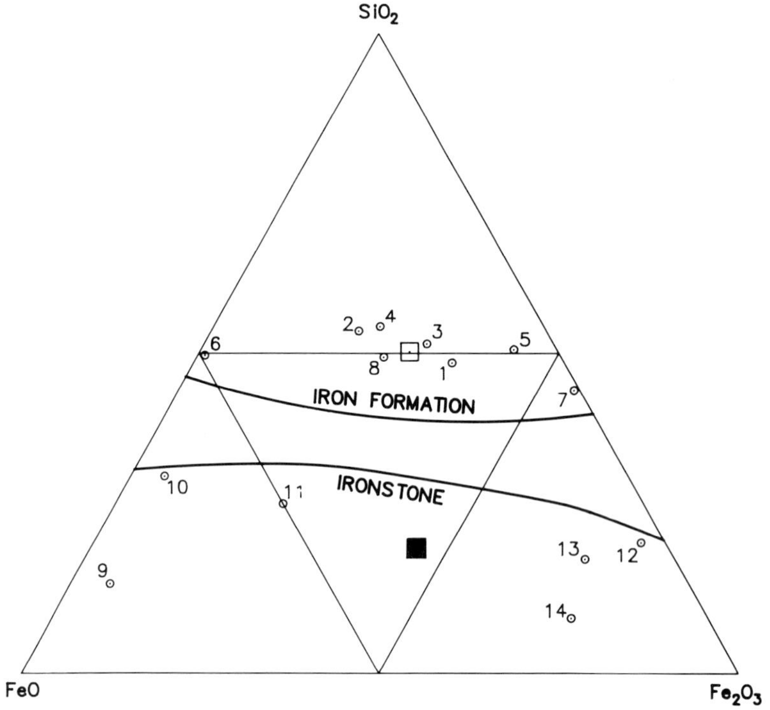

Fig. 11-1. Compositions of iron-formation and ironstone, expressed as $SiO_2 + Fe_2O_3 + FeO = 100\%$.
(1 – 8) Iron-formation, district composites:
(1) Atlantic City, Wyoming (Bayley and James, 1973);
(2) Yilgarn Block, Australia (Gole, 1981);
(3) Hamersley, Australia (Gole, 1981);
(4) Mesabi, U.S.A. (Lepp, 1972);
(5) Labrador, Canada (Gross, 1968);
(6) Iron River, U.S.A. (James, 1954);
(7) Minas Gerais, Brazil (Dorr, 1973); and
(8) Transvaal, South Africa (Beukes, 1973).
□ Average of district iron-formation composites.
(9 – 14) Ironstone, district composites. (Data from James, 1966, as indicated.)
(9) Northampton sideritic ironstone; average of analyses B – G, table 16;
(10) Cleveland siderite – chamosite ironstone; average of analyses H – K, table 16;
(11) Chamosite oolite, Chamosentze, Switzerland; average of analyses F – H, table 13;
(12) Luxembourg oolitic limonite; average of analyses I – M, table 8;
(13) Clinton oolitic hematite, New York; analysis C, table 9; and
(14) Wabana oolitic hematite and chamosite, Newfoundland; average of analyses G – H, table 9.
■ Average of district ironstone composites.

after limestone descriptive schemes (orthochemical, or lutitic; and allochemical, or arenaceous). The bases for these classifications (which are not mutually exclusive), and the limits to their usefulness, are reviewed in the paragraphs that follow. Of these three, the facies classification is the most basic and most useful and will be discussed in greatest detail.

Classification according to mineralogy (facies)

Unlike common elements of invariant valency, such as calcium and magnesium, iron (like manganese) is multivalent, capable of forming a variety of minerals in response to particular environmental conditions. This attribute provides an immediate and fundamental basis for objective classification of iron-rich sediments and important clues as to environment of origin.

The nature of iron precipitates (or reaction products) formed at low temperature in aqueous media is governed in large part by the pH and Eh (oxidation potential) of the depositional or diagenetic environment. Given appropriate thermodynamic data, stability fields of relevant iron minerals can be defined in terms of these two variables for particular compositions of the participating fluid. An early but still instructive attempt at such analysis is given in Fig. 11-2, one of many such diagrams that have been prepared (see Garrels and Christ, 1965, for diagrams assuming a

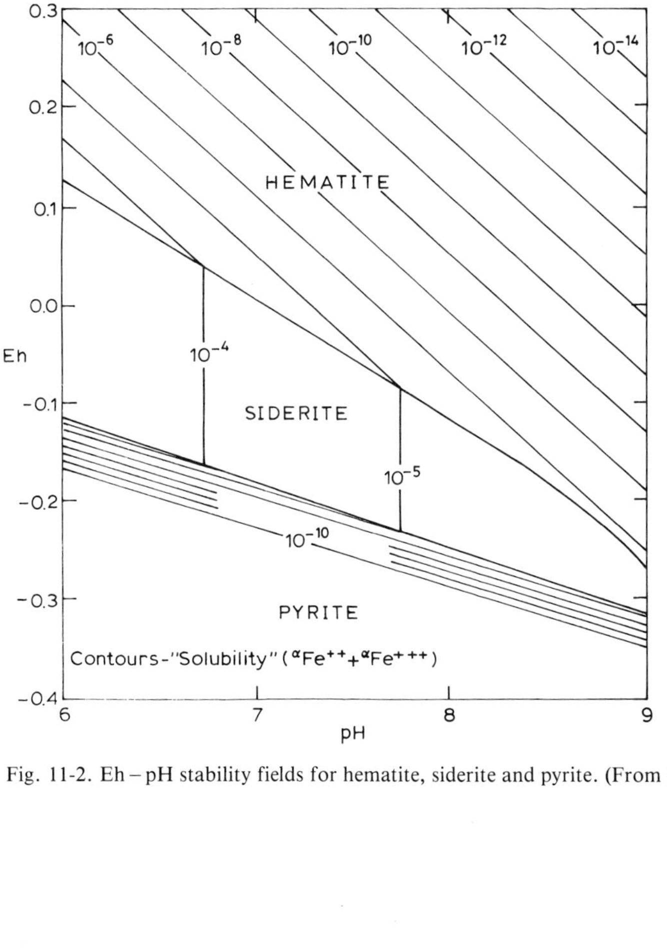

Fig. 11-2. Eh – pH stability fields for hematite, siderite and pyrite. (From Krumbein and Garrels, 1952.)

variety of compositional situations). These diagrams are helpful in appraisal of natural assemblages that in turn provide some limits on depositional and diagenetic conditions, but they are to be used as general guidelines only. Aside from the all-important necessity to specify other properties of the system, the actual first-formed materials are not likely to be stable compounds, such as hematite and magnetite but, as stated by Ewers (1983, p. 494), rather " . . . ill-defined metastable compounds for which the thermodynamic data are either unknown or only very approximately known".

TABLE 11-1

Principal features of iron-formation facies

Facies	Iron minerals*	Fe contents (wt%)	Lithology	Distinctive features
Oxide				
Hematitic	Specular hematite (Magnetite)	30–40	Thin-bedded to wavy bedded alternation of bluish black hematite and gray or reddish chert	Specularite content; oolitic in many districts
Magnetitic	Magnetite (Minnesotaite) (Stilpnomelane) (Greenalite) (Siderite)	25–35	Heavy dark rock, evenly bedded to irregularly bedded; layers of magnetite alternate with dark chert and mixtures of silicates and siderite	Strongly magnetic
Silicate	Minnesotaite Stilpnomelane (Magnetite) (Siderite) (Greenalite) (Chamosite) (Chlorite)	20–30	Light to dark green rock, generally laminated or even-bedded, but in some districts wavy to irregularly bedded; commonly interlayered with magnetitic oxide facies or carbonate facies	Commonly magnetic; may be granule-bearing
Carbonate	Siderite (Stilpnomelane) (Minnesotaite) (Magnetite) (Pyrite)	20–30	Light to dark gray, evenly bedded or laminated alternation of siderite and chert	Generally non-magnetic; stylolites common
Sulfide	Pyrite (Siderite) (Greenalite)	15–25	Laminated to thinly layered black carbonaceous argillite; chert rare	"Graphitic"

* In rocks of low metamorphic grade; common but non-essential minerals in parentheses.

That mineralogy of iron provides a ready, objective basis for classification of iron-rich sedimentary rocks, was recognized early (e.g., Harder, 1919; and Teodorovich, 1947). For iron-formation, it leads to recognition of four major facies: *oxide, silicate, carbonate,* and *sulfide* (James, 1954). The principal properties of these facies are summarized in Table 11-1, in which the oxide facies is further subdivided into a hematitic and magnetitic subfacies.

Estimates for average chemical compositions of end-member oxide, silicate, and carbonate facies are given in Table 11-2. Data are too fragmentary to provide a valid basis for a compositional estimate for the sulfide facies, but two representative analyses of carbonaceous pyritic iron-formation are given in Table 11-3. Estimates of average compositions of the various facies have also been made by Eichler (1976, table VI).

Oxide facies. Although complete gradation between hematitic and magnetitic iron-formation is commonplace in some districts, in others two clearly defined subfacies are recognizable, and in fact may be interbedded on a scale of meters.

The *hematitic subfacies,* in its most distinctive (though not necessarily the quantitatively most abundant) habit, is a wavy bedded to irregularly bedded rock, commonly oolitic in some layers with transitions into contemporaneous breccia or even iron-rich conglomerate – all features indicative of accumulation in a shallow-water, high-energy environment. Algal structures (stromatolites), illustrated in Fig. 11-3, are not uncommon in such rocks.

TABLE 11-2

Average chemical compositions of oxide, silicate and carbonate facies (adapted, with slight modifications, from Gross, 1965)

	Oxide			
	Hematitic	Magnetitic	Silicate	Carbonate
SiO_2	40.10	48.51	51.30	34.51
Al_2O_3	0.80	0.96	2.83	1.09
Fe_2O_3	50.10	26.39	7.14	1.84
FeO	1.60	15.70	24.80	31.72
MgO	2.00	2.21	3.76	3.15
CaO	1.40	1.17	0.53	1.16
Na_2O	n.d.	n.d.	1.18	0.04
K_2O	n.d.	n.d.	2.09	0.20
H_2O^-	–	–	0.44	–
H_2O^+	–	–	3.96	–
H_2O	–	1.45	–	0.88
CO_2	2.6	3.91	4.91	22.96
P_2O_5	0.07	0.07	0.49	0.51
S	–	–	–	0.05
MnO	0.20	–	2.08	1.37
TiO_2	–	–	0.52	0.07
C	–	0.06	0.21	1.88
Total Fe	36.28	30.66	24.27	25.92

TABLE 11-3

Chemical analyses of two samples of sulfide facies

	A	B
SiO_2	36.67	18.6
Al_2O_3	6.90	13.9
Fe_2O_3	–	–
FeO	2.35	0.44
FeS_2	38.70	36.1
MnO	0.002	0.0
CaO	0.13	0.8
MgO	0.65	0.0
Na_2O	0.26	0.16
K_2O	1.81	2.9
TiO_2	0.39	1.1
V_2O_5	0.15	–
P_2O_5	0.20	0.047
H_2O^-	0.55	1.5
H_2O^+	1.25	2.9
SO_3	2.60	1.77
S	–	0.11
C + organic matter	7.60	20.2
Total	100.21	99.81

A: Wauseca Pyritic Member of Dunn Creek slate, Iron River, Michigan. Early Proterozoic. (James et al., 1968, p. 42.)
B: Carbonaceous pyritic rock associated with Soudan Iron-formation of northern Minnesota. Archean. (Cloud et al., 1965.)

Not all hematitic iron-formations display evidence of turbulent bottom conditions during sedimentation. The uppermost part of the Negaunee iron-formation of the Marquette district, Michigan, though wavy bedded and oolitic in the western part of the present structural trough, is nonoolitic and even-bedded in the eastern part, as shown in Fig. 11-4. The most spectacular example of nonoolitic even-bedded hematitic iron-formation, however, is the great Cauê Itabirite of Brazil, with an average thickness of 250 m (Dorr, 1973; Eichler, 1976), formation of which would appear to demand the improbable combination of long-lived highly-oxidizing conditions in a shallow, but slowly subsiding, basin of utmost quiescence.

Most hematitic iron-formations have appreciable amounts of magnetite, often demonstrably of diagenetic origin, and most contain minor quantities of carbonate minerals – ankerite, dolomite, and calcite; rarely siderite. The Cauê Itabirite is reported to be strongly dolomitic in some places (Dorr, 1973, p. 1012).

The *magnetitic subfacies* is probably the most common type of iron-formation and to many workers is the type example of banded iron-formation. The rock, as seen in outcrop, is thinly layered and evenly bedded (Fig. 11-5) and, in detail, commonly is laminated on a centimeter to millimeter scale (Fig. 11-6). Rocks of similar character form the BIF (banded iron-formation) units of the Brockman Iron-formation of Western Australia, for which Trendall (1973a) has demonstrated

A

B

Fig. 11-3. Algal structures in hematitic chert. Biwabik Iron-formation, Mesabi district. Polished surfaces. (A) parallel to bedding; (B) approximately normal to bedding.

Fig. 11-4. Even-bedded hematitic iron-formation, Marquette district, Michigan. Consists of interlayered reddish jasper (dark in photograph) and silvery specularite (white). (From James and Sims, 1973.)

Fig. 11-5. Magnetitic iron-formation (Archean), Wind River Range, Wyoming. (From Bayley and James, 1973.)

Fig. 11-6. Layering in Negaunee Iron-formation, Michigan. (Left) Thin-bedded chert (white), siderite (gray), and minnesotaite – stilpnomelane (dark gray). (Right) Interbedded chert (white), magnetite (black), and silicate – carbonate (gray).

astonishing lateral continuity on a millimeter scale over distances of many kilometers.

Iron silicates and siderite are common constituents of magnetitic iron-formation, either interstitial or as discrete layers, and in some districts ankerite is present as relatively coarse-grained patches of late diagenetic origin. Riebeckite, the soda-iron amphibole, is present in minor quantity in many magnetitic iron-formations of both Proterozoic and Archean age, but it is known in moderate abundance only in the

Proterozoic iron-formations of the Transvaal Supergroup of South Africa and the Hamersley Group of Western Australia; its origin and significance will be discussed later in this chapter.

Silicate facies. Much of the rock classed here as silicate facies of iron-formation is described in district reports as "ferruginous slate", "slaty iron-formation", or simply "slate" or "shale". Lithologic types range from flinty rocks to fissile shale, finely laminated to irregularly bedded. Mineral contents and paragenetic relations are as diverse as structures and textures. Minnesotaite and stilpnomelane, in some areas demonstrably replacements of earlier greenalite, are the typical iron silicates, but chlorite is common and chamosite is not rare. Much of the chert is interstitial rather than in discrete layers. Magnetite is ubiquitous, in places in major proportions so that the rock becomes a mixed oxide – silicate facies. Similarly, siderite commonly is abundant and may form separate layers alternating with silicate and chert to yield a mixed silicate – carbonate facies, illustrated in Fig. 11-6.

Irregular or wavy bedding is characteristic of some silicate-facies iron-formation in several Lake Superior districts and in the Labrador belt. Internally these rocks commonly are granular or oolitic in texture, reflecting wave and current action that disrupted the initial bottom sediments. As shown in Fig. 11-7, the granules (allochems) range in form from spherical to angular, and most have been modified by some degree by diagenetic replacements.

Clastic material, in the form of silt-size grains of quartz and feldspar, is recognizable in some varieties of silicate iron-formation, and angular shards, purportedly of volcanic origin, are reported from major iron-formations of South Africa and Australia (LaBerge, 1966a,b). It is entirely likely that most silicate-rich

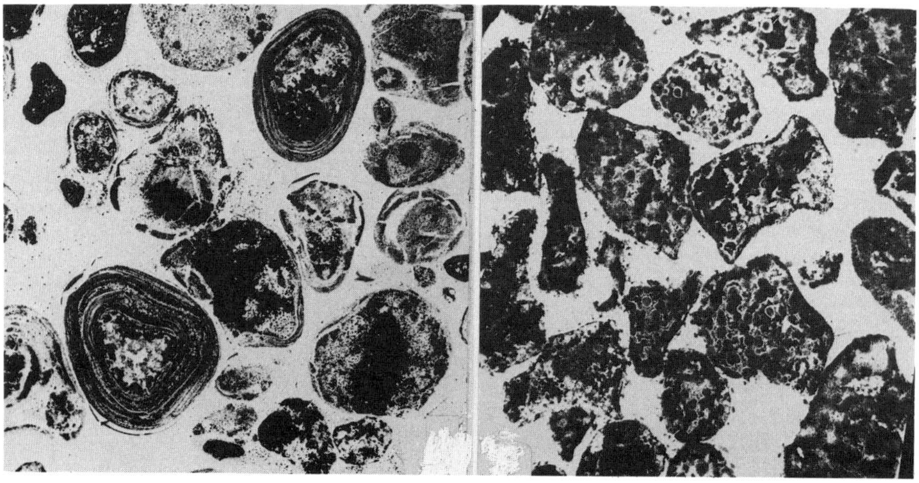

Fig. 11-7. Oolitic and granular textures in Gunflint Iron-formation. (Left) Ooliths of hematite and granules of greenalite – chert, in chert matrix. (Right) Greenalite granules in chert matrix. (Field diameter = 3 mm.)

horizons in iron-formation do, in fact, reflect introduction of fine volcanic ash (or simply terrestrial dust) from distant sources, but shard structure in itself is not necessarily indicative of volcanic origin. As noted by Trendall and Blockley (1970, p. 290), also by Dimroth and Chauvel (1973, p. 113), such forms could also have been produced by breakup of iron-rich bottom sediments.

Carbonate facies. Ideally, the carbonate facies of iron-formation consists almost entirely of interlayered siderite and chert, and in fact some thick stratigraphic units do conform to this simple make-up – for example, the Riverton Iron-formation of the Iron River – Crystal Falls district of Michigan (James et al.,

Fig. 11-8. Carbonate iron-formation, Iron River, Michigan. Polished surface, etched with HCl. Siderite – gray; chert – dark. Note contemporaneous deformation structures. (Width of specimen = 8 inches.)

1968), illustrated in Fig. 11-8. The principal carbonate of this district, like the bedded carbonate of other districts, is a single phase, the composition of which can be expressed almost entirely in terms of four components: $FeCO_3$, $MnCO_3$, $MgCO_3$, $CaCO_3$. Variations within Lake Superior districts are illustrated in Table 11-4. Manganese is the most variable constituent, ranging from low values (or even zero) in bedded carbonate associated with oxide facies to as much as 15% where associated with sulfide facies. Miyano (1978, p. 204) presents somewhat similar data for siderite in the Dales Gorge Member of the Brockman Iron-formation of Western Australia; here, in association with magnetite-rich rocks, the manganese content is very low. Manganese-rich carbonate is an important constituent in several iron-formation units of the Transvaal Supergroup of South Africa (Beukes, 1983, pp. 175, 182, 191 – 193), and in places the carbonate horizons are overlain by oxidic manganese beds.

Interbedded transitions from carbonate iron-formation into silicate or silicate – magnetite iron-formation (illustrated in Fig. 11-6), on the one hand, and into sulfide facies on the other are recognized in many districts. Transitions of the latter type, are described for the basal part in the Kuruman Iron-formation of South Africa (Beukes, 1973, p. 983)) and in the Riverton Formation of Michigan (James et al., 1968). The author knows of no example of sideritic facies of iron-formation being transitional into, or interbedded with, hematitic facies; in all examples observed of carbonate interbedded with hematitic rock, the carbonate has proved to be calcite, ankerite, or dolomite.

In a number of districts, bedded siderite has been shown to be composed of tiny spherules, generally 10 – 40 microns in diameter, that evidently are primary sedimentary features. Their origin and significance, still uncertain, are discussed at some length by LaBerge (1973, pp. 1105 – 1108) and LaBerge et al. (1987, p. 75).

Sulfide facies. Most strata assigned to the sulfide facies are black carbonaceous shales or slates that occur either alone or as minor components of iron-formation suites. This is not only the least abundant (and least typical) of the various facies of iron-formation, but it also is the most likely to be overlooked. Even when present, the rock rarely is found in well-preserved outcrop and where intersected by mine workings or drill core may go unrecognized because of the typically fine grain size of the pyrite. In the type example of carbonaceous sulfide facies, illustrated in Fig. 11-9, the pyrite has an average grain size of 0.005 millimeters and, despite ap-

TABLE 11-4

Compositions of bedded siderite from Lake Superior districts (in wt%; from James, 1966)

	Iron River – Crystal Falls	Gogebic	Gunflint	Mesabi
$FeCO_3$	81.0	80.9	80.2	89.0
$MnCO_3$	10.2	4.1	0.5	–
$MgCO_3$	7.9	12.9	12.6	9.4
$CaCO_3$	0.9	2.1	6.7	1.6

pearance in photographed polished surface, is entirely invisible in ordinary hand specimen or drill core (the high pyrite content, nearly 40%, of this particular stratigraphic unit went unrecognized for more than 60 years in an active mining district). Chemical analyses of two samples of pyritic rock have been given in Table 11-3; noteworthy are the high contents of carbon – 20% in the Archean-age sample from the Vermilion district of Minnesota.

An entirely different type of sulfide iron-formation is present in the Michipicoten district of Ontario (Goodwin, 1962, and later papers). In the central and eastern parts of the district, the 2700-m.y. Helen Iron-formation, overlain and underlain by volcanic strata, consists of a basal siderite member, an intermediate thin pyritic

Fig. 11-9. Sulfide iron-formation, Iron River – Crystal Falls District, Michigan. Underlies and is interbedded with sideritic iron-formation. Polished surface; width of specimen = 7 inches. (From James et al., 1968.)

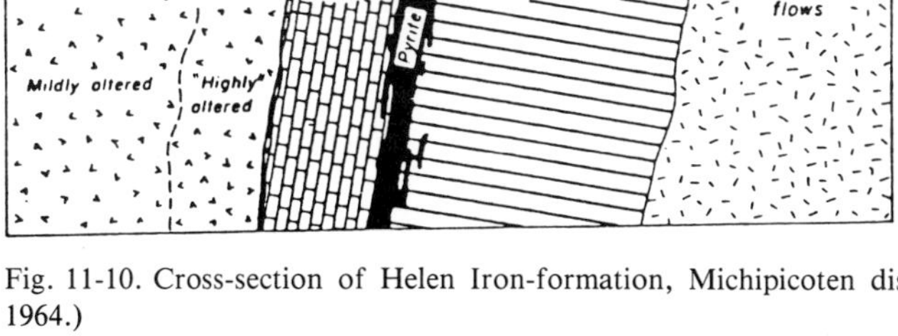

Fig. 11-10. Cross-section of Helen Iron-formation, Michipicoten district, Ontario. (From Goodwin, 1964.)

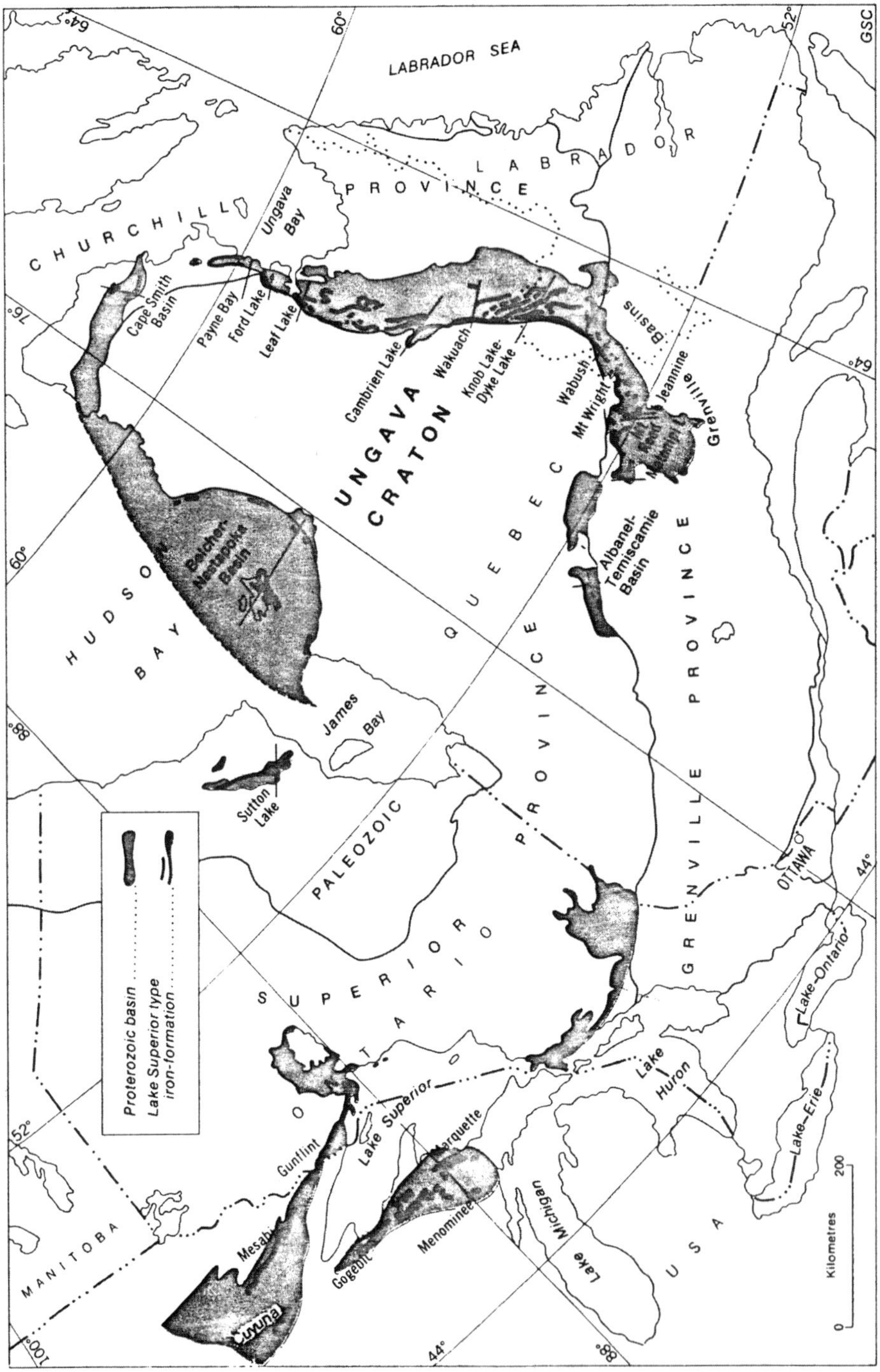

Fig. 11-11. Distribution of early Proterozoic basins and iron-formations marginal to Ungava craton, eastern Canada and north-central United States. (From Gross and Zajac, 1983.)

zone, and a thick upper banded chert (Fig. 11-10). The lower sideritic and pyritic units are lensy, with considerable variation from place to place. Detailed geochemical data, obtained from study of drill core from the western part of the district, are reported by Goodwin et al. (1985). The iron-formation at this locality has a stratigraphic thickness of 342 m, divided into a lower siderite – pyrite unit (50 m) and an upper banded chert (292 m). The siderite – pyrite unit is further subdivided into four zones, data for which are summarized in Table 11-5. The iron-formation is attributed to deposition from seawater charged with volcanic exhalations in a cauldron subsidence basin. The pyrrhotite content of the pyritic unit is considered to be of primary origin: " . . sulfur deposition was rhythmical, with three main concentrations, each comprising an earlier pyrite peak and a later pyrrhotite peak" (Goodwin et al., 1985, p. 79). The Helen Iron-formation is in detail an unusual, possibly unique, deposit, but volcanogenic iron-formation of various facies is abundantly represented in the Archean greenstone belts of the Canadian Shield (Goodwin, 1973) and elsewhere in the world.

TABLE 11-5

Chemical and mineralogic data for the siderite – pyrite member of the Helen Iron-formation, Michipicoten district (summarized from tables in Goodwin et al., 1985); zones listed in stratigraphic order, upper to lower

Zone	Stratigraphic thickness (m)	SiO_2 (wt%)	Fe (wt%)	S (wt%)	Mineralogy
"Pyrite"	19	4.50	41.5	21.67	**Pyrite**, siderite, pyrrhotite **Chlorite**, chamosite, sauconite
"Siderite"	12	7.1	37.8	6.70	**Siderite**, pyrite, quartz **Chlorite**, chamosite, sauconite
"Central silica"	5	53.4	18.8	4.38	**Quartz**, siderite, arsenopyrite
"Siderite"	14	11.1	35.4	4.55	**Siderite**, quartz, chlorite

Classification according to geologic setting (type)

In contrast to facies and textural classifications, which, in a given stratigraphic unit of iron-formation are applicable at all scales down to that of individual layers, the type classification applies to the stratigraphic unit as a whole. The concept was introduced by Gross (1965), who classed the iron-formations of the Canadian Shield as *Superior-type,* exemplified by the remarkable chain of early Proterozoic deposits that are preserved around the 3000-km margin of the Ungava craton of eastern Canada and north-central United States (Fig. 11-11), and *Algoma-type,* typified by the many occurrences of iron-formation in late Archean greenstone belts that transect the craton. A third category, *Rapitan-type,* was later added to distinguish those enigmatic late Proterozoic iron-formations associated with coarse clastic deposits, some of glaciogenic origin (Button, 1982, p. 260). The comparative aspects

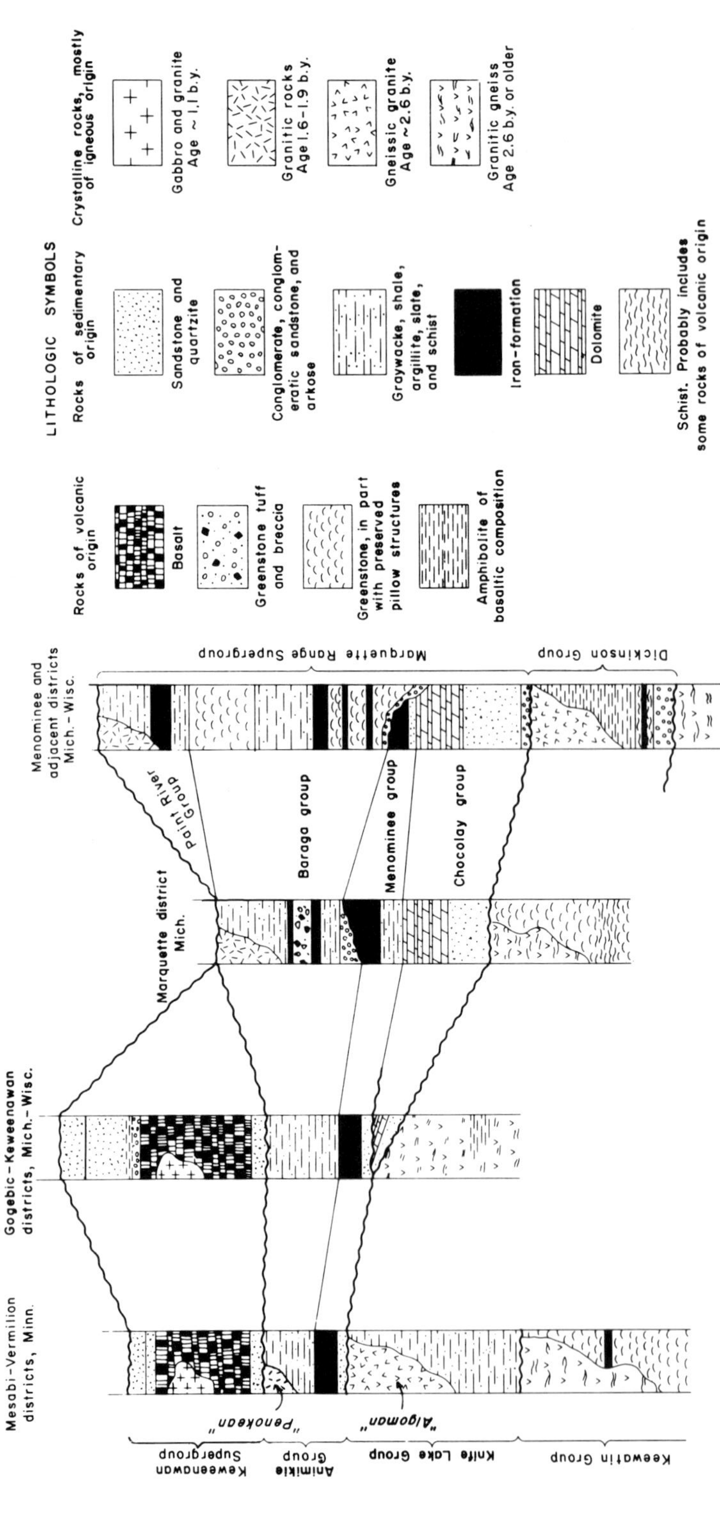

Fig. 11-12. Stratigraphic positions of iron-formations in Lake Superior districts. Thicknesses not to scale. (From Bayley and James, 1973.)

of these classes of deposits are summarized in Table 11-6. The type classification is a valuable device, but not all iron-formations can be fitted into this scheme. Many deposits of major dimensions, notably those of Archean age, e.g., the Anshan belt of northern China and the Cerro Bolivar deposits of Venezuela, do not lend themselves readily to categorization, and some assignments, including that of the great Hamersley deposits of Western Australia to the Superior-type (Gross, 1980, p. 219), have been vigorously rejected (Trendall, 1983a, p. 8 – 9). Nevertheless, within cratonic areas at least, the twofold Superior – Algoma split does provide a ready, if incomplete, means of separating deposits of distinctly different associations (and perhaps of genesis). The Lake Superior area itself, for example, contains identifiable deposits of both types, as well as some that do not fit comfortably in either category. As illustrated in Fig. 11-12, the Archean-age iron-formation of the Vermilion district (lowest iron-formation unit in far left column) is volcanic-hosted

TABLE 11-6

Comparative features of Superior-, Algoma- and Rapitan-types of iron formations (adapted in part from Eichler, 1976)

	Superior	Algoma	Rapitan
Depositional environment	Anorogenic "miogeosynclinal"-stable shelf, marginal basins, generally nonvolcanic	Orogenic, "eugeosynclinal"-submarine volcanic greenstone belts	Non-volcanic rift zones at continental margins
Associated strata	Quartzite, dolomite, black shale; iron-formation generally low in sequence	Volcanogenic rocks, graywacke; iron-formation irregularly distributed	Mudstone, shale, conglomerate, diamictite; position of iron-formation uncertain
Dimensions	Commonly 100 m or more thick; laterally continuous for tens to hundreds of kilometers	Lenticular, meters to tens of meters thick; rarely continuous for more than a few tens of kilometers	Tens to hundreds of meters thick; laterally continuous for as much as hundreds of kilometers
Age	Mostly early Proterozoic (1900 m.y. – 2500 m.y.)	Majority are Archean, particularly late Archean, but some are as young as Paleozoic	Known deposits are late Proterozoic
Facies	All facies represented, order of abundance: oxide – carbonate – silicate – sulfide	All facies represented, but hematitic oxide facies rare	Hematitic oxide facies only
Examples	Sokomon Iron-formation, Labrador (Gross and Zajac, 1983); Kuruman and Penge Iron-formations, Transvaal, Supergroup, South Africa (Beukes, 1983)	Helen Iron-formation, Michipicoten district, Ontario (Goodwin, 1973); Late Proterozoic iron-formations of Eastern Desert, Egypt (Sims and James, 1984)	Iron-formation of the Rapitan Group, McKenzie Mountains, Canada (Young, 1976); iron-formation of Jacadigo Group, Urucum, Brazil (Hoppe et al., 1987)

and clearly of Algoma-type, as are the several iron-formation units of the Proterozoic Baraga Group (two right-hand columns), whereas most of the major iron-formations of the region, in the Menominee Group and its equivalents, are of the classic Superior-type. But the Riverton Iron-formation of the Iron River – Crystal Falls district (uppermost iron-formation in the far right column), entirely enclosed in nonvolcanic sedimentary strata, but broadly within a "eugeosynclinal" sequence, differs significantly in association from both groups and thus is not readily classifiable.

Deposits of the Rapitan-type are the least common of the several varieties of iron-formation and the least studied. All are of late Proterozoic age, and all are marked by association with coarse clastic rocks, among which are some of probable or certain glaciogenic origin. The type area is in the Mackenzie Mountains of western Canada, where iron-formation is present as a distinctive member of the Rapitan Group, with an estimated age of about 700 my (Yeo, 1986). The stratigraphic succession is illustrated in Fig. 11-13; elsewhere in the district, the iron-formation is considerably thicker, as much as 150 m (Yeo, 1986, p. 144).

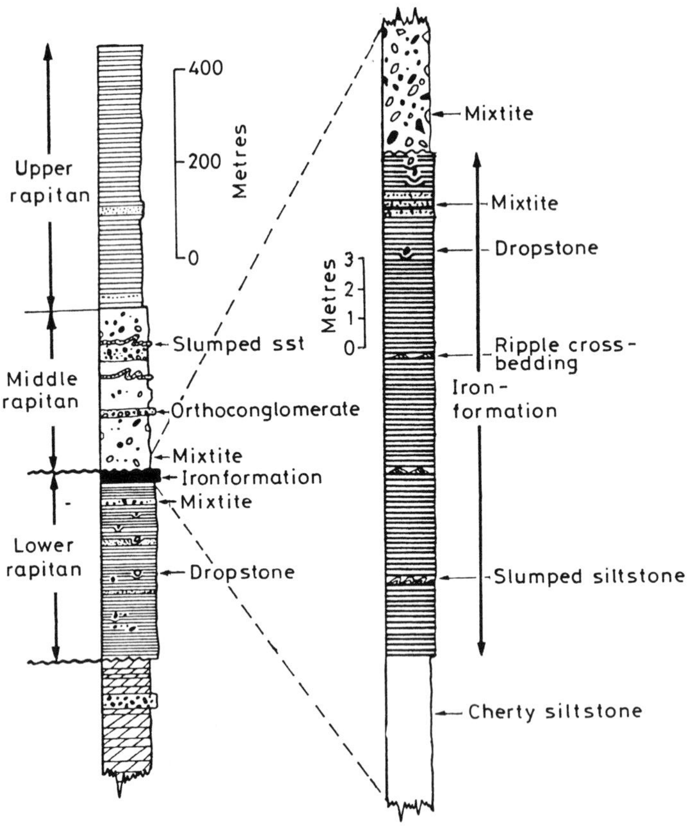

Fig. 11-13. Stratigraphic position of iron-formation in the late Proterozoic Rapitan Group, Mackenzie Mountains, Canada. (From Young, 1976.)

Iron-formations of similar age, character, and association are known in several other parts of the world; notable examples are those of the Jacadigo Group of Urucum, Brazil, as much as 300 m thick (Dorr, 1973, p. 1016; Hoppe et al., 1987), and of the Damara Supergroup of southern Africa (Beukes, 1973; Breitkopf, 1988).

Early reports on the iron-formations of the Mackenzie Mountains and of the Urucum district indicated an unusually high content of iron, as much as 50% or more (Gross, 1973, p. 17; Dorr, 1973, p. 1015). A more recent comparative study by Yeo (Yeo, 1986) fails to bear this out; in ten districts, selected worldwide, average values for iron range between 25 and 42%. The calculated overall average, as shown in Table 11-7, is about 31%, entirely comparable to that of older iron-formations.

Classification according to texture

Many textural features of iron-formation owe their origin to the interplay between chemical and physical processes in the depositional environment, and to diagenetic reactions. Dimroth (1968) and, later, Dimroth and Chauvel (1973) called attention to similarities between these textures and those of limestone, and introduced terms – femicrite, for example – corresponding to those used in a widely accepted textural classification of limestone. Beukes later expanded these suggestions

TABLE 11-7

Average chemical composition of Rapitan-type iron-formation, compared with typical iron-formation of Superior and Algoma types

	1	2	3
SiO_2	44.30	40.71	44.13
Al_2O_3	3.18	2.32	3.85
Fe_2O_3	44.30*	21.04	30.75
FeO		20.10	15.49
MgO	1.24	3.15	2.80
CaO	1.79	2.63	1.31
Na_2O	0.28	1.05	1.22
K_2O	0.45	0.57	0.36
TiO_2	0.27	0.12	n.r.
P_2O_5	0.35	0.25	n.r.
MnO	0.23	0.29	0.09
CO_2	n.r.	6.56	n.r.
H_2O, LOI	3.36	0.98	n.r.
Total	99.80	99.77	100.00
Fe	30.98	30.34	33.45

* Total iron as Fe_2O_3.
n.r. - not reported.
1: Rapitan-type iron-formation, late Proterozoic, average of analyzed samples from 11 districts (calculated from data in Yeo, 1986, p. 151).
2: Superior-type iron-formation, early Proterozoic. Kuruman Iron-formation, South Africa (Beukes, 1973, p. 959).
3: Algoma-type iron-formation, late Archean. Wind River Range, Wyoming (Bayley and James, 1973, p. 952).

into a rather elaborate classification scheme for iron-formation (Beukes, 1983, p. 141). Simonson (1985) also has provided further discussion (and nomenclature) on textural classification.

The descriptive terminology for textural analysis, perhaps already overly complex for general usage, is summed up as follows: A major division is made between *orthochemical* (or *lutitic*) iron-formation, which represents essentially undisturbed chemically precipitated muds, and *allochemical* (or *arenaceous*) iron-formation, which consists mainly of reworked materials that reflect more energetic bottom environments. Terms such as *femicrite, felutite,* and *ferhythmite* are applied to the fine-grained muds. Reworked constituents (allochems) are labeled according to standard limestone terminology – pellets, pelloids, ooliths (or ooids), intraclasts, and shards, depending on size, shape and structure.

Though not given formal definition until recently, the basic division between orthochemical and allochemical (or lutitic and arenaceous) varieties of iron-formation is one that has long been recognized by field workers, at least implicitly. It was, for example, essentially the basis for the stratigraphic subdivision of the Ironwood Iron-formation of the Gogebic range in the Lake Superior district, introduced more than 60 years ago (Aldrich, 1929): the five members of the 125-m thick unit were defined principally by relative dominance of thin-bedded or slaty structure (characteristic of orthochemical iron-formation) and coarse, wavy bedding (typical of allochemical iron-formation).

Textural analysis is a valuable tool for precise description and interpretation of certain mineral relationships in iron-formation, and in reconstructing sedimentary environment of deposition in individual basins, but it does not appear to bear heavily on the more general problems of iron-formation genesis. It does permit conclusions to be drawn concerning factors such as depth of water and degree of turbulence in a particular basin of deposition, but these aspects are of secondary importance to those that control or permit the initial accumulation of chemically precipitated iron- and silica-rich sediments.

Abundance and distribution in space and time

Iron-formation, though a highly distinctive component of Precambrian metasedimentary sequences in all continents, is quantitatively subordinate to other sedimentary rock types. Only in a few major districts does it account for as much as 10% of the sedimentary succession. Nevertheless, the Precambrian iron-formations do constitute a vast storehouse for iron and silica. Estimates of initial tonnage for a number of districts, including most of those of major importance are summarized in Table 11-8 and Fig. 11-14. The aggregate amount of iron-formation in deposits listed here is in excess of 10^{15} m.t.; with a reasonably generous allowance for the many deposits not appraised, the amount of iron-formation contained initially in deposits now in part preserved in the earth's crust is $n \times 10^{16}$ m.t. of rock containing about 30% iron and 35 – 50% silica.

The oldest iron-formation known is that at Isua, Greenland, with an age of about 3800 m.y. (Baadsgaard et al., 1984; Dymek and Klein, 1988), and the youngest of significant size probably are the Algoma-type deposits of the Maly Khinghan and Uda areas of far-eastern U.S.S.R., with an apparent Early Cambrian age (Nalivkin,

TABLE 11-8

Estimated initial size and age of selected deposits of cherty banded iron formation (from James and Trendall, 1982)

Ref No*	Area (may include more than one stratigraphic unit)	Class	Estimated age, in m.y.**
Africa			
AF1	Damara Belt, Namibia	Moderate	650 (590 – 720)
AF2	Shushong Group, Botswana	Small	1875 (1750 – 2000)
AF3	Ijil Group, Mauritania	Moderate	2100 (1700 – 2500)
AF4	Transvaal-Griquatown, S. Africa	Very large	2263 (2095 – 2643)
AF5	Witwatersrand, S. Africa	Small	2720 (2643 – 2800)
AF6	Liberian Shield, Liberia – Sierra Leone	Large	3050 (2750 – 3350)
AF7	Pongola beds, Swaziland – S. Africa	Moderate	3100 (2850 – 3350)
AF8	Swaziland Supergroup, Swaziland – S. Africa	Small	3200 (3000 – 3400)
Australia			
AU1	Nabberu Basin	Large	2150 (1700 – 2600)
AU2	Middleback Range	Moderate	2200 (1780 – 2600)
AU3	Hamersley Range	Very large	2500 (2350 – 2650)
Eurasia			
EU1	Altai region, Kazakhstan – W. Siberia	Moderate	375 (350 – 400)
EU2	Maly Khinghan – Uda, Far East USSR	Large	550 (500 – 800(?)
EU3	Central Finland	Moderate	2085 ± 45
EU4	Krivoy Rog – KMA, USSR	Very large	2250 (1900 – 2600)
EU5	Bihar – Orissa, India	Large	3025 (2900 – 3150)
EU6	Belozyorsky – Konski, Ukraine USSR	Moderate	3250 (3100 – 3400)
North America			
NA1	Rapitan Group, NWT Canada	Moderate	700 (550 – 850)
NA2	Yavapai Series, Southwest USA	Small	1795 (1775 – 1820)
NA3	Lake Superior, USA	Large	1975 (1850 – 2100)
NA4	Labrador Trough and extensions, Canada	Very large	2175 (1850 – 2500)
NA5	Michipicoten, Canada – Vermilion, USA	Moderate	2725 (2700 – 2750)
NA6	Beartooth Mountains, Montana, USA	Small	2920 (2700 – 3140)
NA7	Isua, Greenland	Small	3760 ± 70
South America			
SA1	Morro du Urucum – Mutun, Brazil – Bolivia	Moderate	600(?) (450 – 900)
SA2	Minas Gerais, Brazil	Very large	2350 (2000 – 2700)
SA3	Imataca Complex, Venezuela	Large	3400 (3100 – 3700)

* See Fig. 11-14.

** In absence of other data, the assigned age is arithmetic mean of age limits given in brackets.

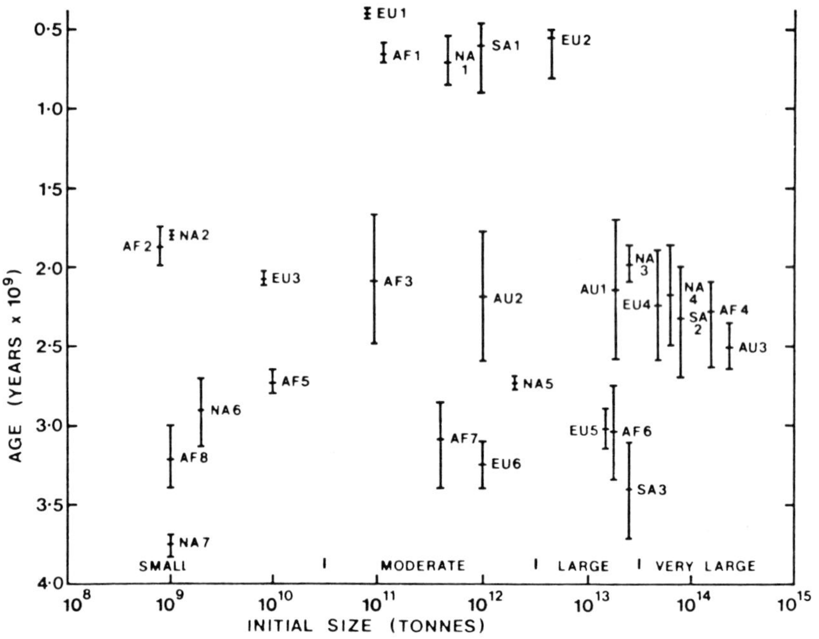

Fig. 11-14. Estimated age and initial tonnage of selected deposits of Precambrian iron-formation. Designations refer to Table 8. (From James and Trendall, 1982.) (Additional note: estimate for deposit AF4 should be increased to 10^{15} tonnes on basis of later data from Beukes, 1983.)

1960, p. 135). But the distribution over the 3300 m.y. time spread represented by these two extremes is far from being uniform; indeed there may be gaps as long or longer than all of Phanerozoic time during which no iron-formation was deposited. A previous survey of time distribution (James, 1983) indicated several peak periods of iron-formation deposition: mid-Archean, late Archean, early Proterozoic, and late Proterozoic – early Phanerozoic. The significance – perhaps even the reality – of some of these peaks is somewhat uncertain, but a few positive statements can be made:

(1) The early Proterozoic, 2500 to 1900 m.y., doubtless was the time of greatest accumulation of sedimentary iron in earth's history. Fully 75% of all known iron-formation preserved in the earth's crust is contained in eight districts of this age bracket: Lake Superior and Labrador, North America; Quadrilatero Ferrifero, Brazil; Transvaal, South Africa; Krivoy Rog and Kursk Magnetic Anomaly (KMA), Ukraine; and Nabberu Basin and Hamersley, Western Australia.

Few of these deposits are dated precisely, but the age data now available suggest a considerable spread in time of deposition. The oldest fairly well-dated iron-formations of the group is that of the Hamersley district, with an apparent age bracketed between 2650 m.y. (underlying basalt) and 2350 m.y. (intrusive sills) and an internal zircon age of 2490 m.y. from a contemporaneous tuff (Trendall, 1983b). At the younger end of the early Proterozoic spectrum are the iron-formations of the Lake Superior district, bracketed by an apparent 2100 m.y. age for basement rocks and a metamorphic event at about 1850 m.y., with an internal zircon age of 1920 m.y. from a rhyolite in the sequence (Morey, 1983).

This great range in ages – perhaps 600 m.y. – suggests that the early Proterozoic peak in iron-formation deposition reflects, not some general worldwide atmospheric, biospheric, or hydrospheric event, but rather the regionally controlled development of continental shelves that followed the strongly diachronous Archean cratonization of late Archean time.

(2) Although Algoma-type deposits are found in greenstone belts of many ages, ranging from mid-Archean to Paleozoic, the vast majority are in volcanogenic sequences of late Archean age. These late Archean deposits, most of which are of relatively small dimensions, constitute a distinct numerical peak in occurrences of iron-formation in the geologic record.

(3) Iron-formations of Rapitan-type lack precise age data, but it is evident that most, if not all, are of late Proterozoic age. Their appearance at this time does not coincide with any particular known stage in evolution of the earth's hydrosphere or atmosphere, but it does coincide with a widespread structural event, namely the rifting and breakup of Proterozoic supercontinents in late Precambrian time (Yeo, 1986, p. 151).

Many large Archean-age deposits are difficult both to classify and to date with any reasonable assurance, and it is by no means certain that any peak Archean period of iron-formation deposition exists, other than that related to formation of late Archean greenstone belts. Indeed, the record can be interpreted to mean that the major period of iron-formation deposition extended without break from mid-Archean time into the early Proterozoic (Gole and Klein, 1981; Walker et al., 1983, p. 279).

Included among the Archean age deposits that are uncertainly classified and dated are a number of major dimensions. Perhaps the most extensively known are those of the Imataca Complex of the Guyana Shield of South America, with an age in the 3000 – 3400 m.y. range (Hurley et al., 1972), and the apparently once-linked counterparts in the Liberian Shield of western Africa (Fig. 11-15). Also of major size are those of Bihar and Orissa, India, with an apparent age in excess of 3100 m.y. (Chakraborty and Majumder, 1986). Others include the economically important deposits of the Anshan belt of northeastern China (Qiusheng, 1987) and of the Serro dos Carajos area of Brazil (Hoppe et al., 1987), both classed as Algoma-type by the cited researchers but lacking many of the features characteristic of volcanic-related deposits.

ORIGIN OF IRON-FORMATION

Considering the range of physical character, dimensions, ages, and geological associations represented by the various types of iron-formation, it is entirely unlikely that a single genetic model will hold for all. Before addressing the various possibilities, it is worthwhile to outline some general propositions that must be considered in evaluating any proposed model.

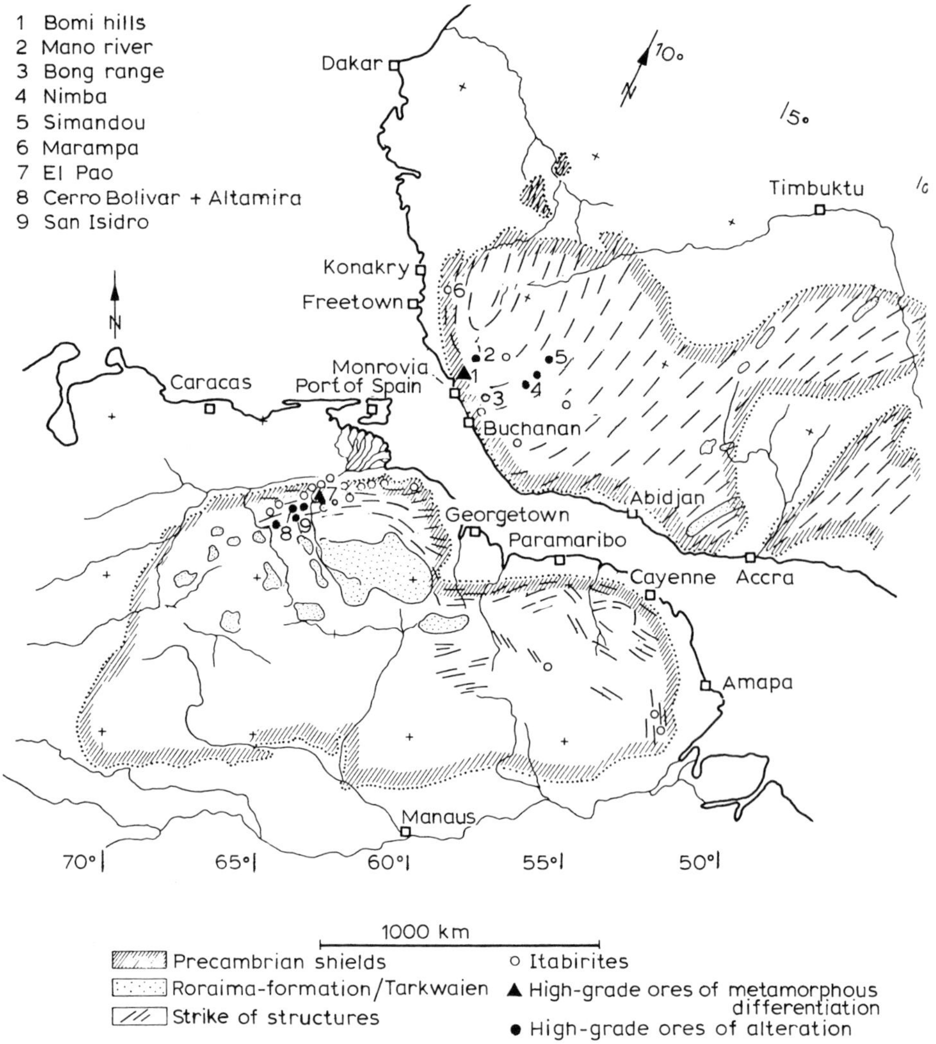

Fig. 11-15. Reconstruction of the relative positions of iron deposits in the Guyana Shield of South America and the Liberian Shield of western Africa, prior to continental separation. (From Gross, 1973.)

(1) Iron-formations were deposited on continental shelves or marginal marine basins of relatively limited size

The assumption of a marine environment cannot be proved and in fact has been challenged (Hough, 1958; Eugster, 1969), but is supported by the nature of associated deposits, such as extensive thick beds of dolomite in Superior-type sequences and oceanic-type basalts in Algoma associations. As to dimensions, even the most extensive of the known iron-formations, that of the Hamersley district with an estimated initial distribution of about 150,000 km^2, is not truly extensive on a continental or oceanic scale – the area of Hamersley deposition, for example, is equal to about one-third that of the present Black Sea.

(2) Chert-banded iron-formation of the oxide and carbonate facies consisted initially of sea-bottom accumulates of essentially the same composition as the preserved material

These initial deposits doubtless underwent considerable later reorganization, notably dehydration, as stable assemblages evolved. But the bulk composition, specifically in terms of iron and silica, remained without significant change.

The hypothesis advanced by Dimroth (1977, fig. 1), following a suggestion by Kimberley (1974), that the initial precipitate was aragonite, which then was replaced diagenetically by silica and iron compounds, is refuted by the very scale of the deposits and the completeness of the supposed transformation. The same holds true for the thesis that the chert layers represent replaced initial non-iron carbonate, as suggested by Lepp and Goldich (1964) and again by Lepp (1987).

(3) Iron-formation of the silicate facies generally reflects introduction of particulate material from external sources

The actual bottom accumulates are largely products of reaction between an externally derived fraction and iron-rich seawater at the depositional site. The scarcity (but not complete absence) of recognizable clastic or volcanic material suggest that the introduced fraction consisted typically of fine wind-blown dust of either volcanic or terrestrial origin. Chemically, this contribution to the system is evident in the relatively high contents of Al_2O_3 and TiO_2 in silicate-facies rocks (see Table 11-2).

(4) Iron-formation of the black-shale sulfide facies is the product of complex sea bottom and diagenetic reactions between precipitated iron compounds, locally produced organic material, and pore waters rich in iron and sulfur

The ultimate source of the sulfur is assumed to be dissolved sulfate in the marine reservoir, from which reactive H_2S is produced by bacterial sulfate reduction in the bottom or burial environment. The existence of oxidized sulfur in the marine reservoir may be considered questionable for Archean-age systems, but the inference is supported by the discovery of bedded gypsum in strata of 3500 m.y. age in Australia (Groves et al., 1981).

Conclusions

These propositions (and it is to be remembered that they are propositions, not necessarily verities) apply to iron-formations of all types and ages, and although they are not without at least modest exceptions, they do set some limits on genetic concepts. We turn now to a key question that has been attached to the iron-formation issue ever since the rock was first described over a hundred years ago – namely, the source of the iron and silica.

SOURCES OF IRON AND SILICA

This question was greatly broadened by the recognition of three specific types of iron-formation (Algoma, Superior, and Rapitan) that originated in distinctly different geologic environments. It is a reasonable assumption that the sources of element supply also may have been different, so that the question may have three separate answers, not just one.

Much of the earlier debate on the source issue concerned iron-formation of what is now referred to as Superior-type; that is, deposits of major dimension lacking in immediate volcanic association. It focussed essentially on two polarized alternatives: the iron and silica were derived either from weathering processes, or from volcanic contributions. The quantitative aspects of these alternatives finally were examined critically by Holland (1973), specifically with respect to the great early Proterozoic iron-formation of the Hamersley district; each was found to have serious, possibly fatal shortcomings. The major problem with the weathering hypothesis, aside from the enormous volume of crust that would have to be eroded (about 10^6 km^3, according to earlier calculations; Trendall and Blockley, 1970), is the absence of any credible mechanism that would so efficiently separate the chemical contributions from the clastic sediment load. The volcanic-source hypothesis encounters difficulties of equal magnitude: using a rate of iron discharge equal to that of the most productive volcanic source now known (Ebeko volcano in the Kurile Islands). Holland (1973, p. 1170) estimated that, to supply the necessary quantities of iron in the specified time, the Hamersley basin would have had to have been rimmed with active, iron-productive volcanoes at one-mile intervals. Such a setting clearly would be grossly at odds with the structural quiescence indicated by uniformity of layering within the chemical sediments in a basin and by the general absence of recognizable volcanic debris. Faced with these inadequacies in both the weathering and volcanic sources of supply, Holland (1973, p. 1171) introduced formally the proposition that the source of iron and silica for the great iron-formations was the deep ocean itself and that deposition was the result of upwelling of these charged oceanic waters into the moderately oxygenated shallow waters of continental shelves and marginal basins. In the same volume, Cloud (1973) approaching the problem from a paleoecological point of view concerned particularly with biologic generation of oxygen, indirectly reached a similar conclusion. Although the idea of an oceanic source was not entirely new – Strakhov (1959) and Borchert (1960) had suggested that iron might be mobilized locally from sediment in anoxic bottoms and transferred to shallow depositional sites, and James (1966, p. 49) noted the possibility on a larger scale – the analyses convincingly presented by Holland and by Cloud added great strength to what had been an undocumented speculation.

In the decade following, the concept of a deep ocean source was developed further by Drever (1974) and by a 1980 Dahlem workshop (Button, 1982), so that by 1983 it could be written: "Though not yet at the level of complete consensus, there is now widespread acceptance, either implicit or explicit, of the assumption that a controlling factor was the particular composition of ocean waters during Archean and early Proterozoic time – a composition that differed significantly from that of later eras with respect to pH and oxidation potential. In a word, many now believe that the early oceans were major reservoirs for dissolved iron and silica, the ultimate source of which was diverse – volcanic, terrestrial, and even cosmic'' (Foreword by H.L. James, in: Trendall and Morris, 1983). This conclusion, it is to be noted, applies not only to iron-formation of recognized Superior-type but also to other less readily classified iron-formations of both Archean and early Proterozoic age. It carries the implication that the composition of these deep ocean waters was established during early Precambrian time in equilibrium with a dominantly reducing atmosphere and that during later Archean and early Proterozoic time the near-surface layers became increasingly oxygenated in equilibrium with an evolving oxidic atmosphere. Holland (1984, p. 408) estimates that during the time of major iron-formation deposition, essentially the interval between 3 and 2 billion years ago, the oxygen content of the atmosphere attained a level of about 1/50 that of the present day, or about 4×10^{-3} atm.

The material sources of supply for precipitation of iron-formations of the Algoma-type may have been complex, at least for those with ages of 2 billion years or more. But for those of later Precambrian and Phanerozoic age, for which the oceanic source would not be available (oceanic waters by this time can be expected to have come into equilibrium with a more completely evolved oxygenic atmosphere and thus be rather comparable to that of the present-day oceans, in which the content of iron in solution is virtually zero), the source doubtless consisted of emissions from the spatially and temporally associated volcanic sources. For those of older age the volcanic contributions of iron and silica may simply have augmented that already at a high level in the oceanic basin, providing enough excess to trigger precipitation. The relative balance between the two sources would depend on local factors such as depth of water, shape of the depositional basin, and intensity of hydrothermal activity.

The poorly understood Rapitan-type deposits pose special supply problems. The deep ocean source postulated for Superior-type deposits of Archean and early Proterozoic age was not available in the late Proterozoic (about 700 m.y.), the time of deposition of Rapitan-type deposits. Further, the deposits show no particular relation to volcanism, so that the concept of an immediate exhalative source is not readily applicable. The glaciogenic association adds further to the puzzle; the attempt (Young, 1976) to link iron-formation deposition directly to glaciation by invoking build-up of iron and silica by freezing of marine water on an ice shelf (essentially a fractional crystallization process) has not proved quantitatively adequate. Attention, therefore, has returned to inferences that can be drawn from the structural setting of the deposits on rifted continental margins, and there now is tacit acceptance of the thesis advanced earlier by Gross (1973) that iron-formation deposition was related to hydrothermal fluids that rose along rift breaks to mix with normal seawater of the rift trough (Yeo, 1986; Young, 1988). The association with glaciation may be simply fortuitous (James, 1983), or it may be related to a glacially induced dearth of clastic contributions, which permitted accumulation of undiluted chemical sediments (Young, 1988).

Thus, the following material sources of supply for the iron and silica of iron-formations have been established, at least conceptually: an anoxic pregnant deep ocean during Archean and early Proterozoic time, to furnish materials for extensive shelf and marginal basin deposits lacking a significant volcanic association; exhalative emissions within areas of active submarine volcanism for deposits of Algoma-type; and rift-related hydrothermal fluids, unaccompanied by volcanism, for deposits of Rapitan-type. We turn now to an issue of broad significance, namely, the role of organisms in the depositional process.

Role of organisms in iron-formation genesis

This is an issue on which views vary greatly. LaBerge and his co-workers regard biota as essential participants: "In our model iron-formations originated almost entirely as a biological precipitate . . ." (LaBerge et al., 1987, p. 69); whereas others take a more conservative view: according to Walter and Hofmann (1983, p. 394), " . . . there is no convincing evidence that microorganisms had a direct, major role in precipitating iron minerals (except perhaps through the photosynthetic production of oxygen). In addition, there is no convincing evidence that silica-secreting organisms contributed to the deposition of chert of iron-formations . . . ". The latter, more negative, view is supported by the fact that although some major Superior-type iron-formations – those classified texturally as allochemical – contain abundant evidence of microbiota, others of comparable age and even greater dimension (notably those of the Hamersley district) appear to be barren of fossils, which implies that no necessary relation exists between biologic activity and iron-formation deposition.

As in previous review of the biologic role (James, 1966, p. 16–17), three general situations can be considered:

(1) The organism acts as a catalyst to cause or speed up precipitation of a thermodynamically appropriate compound

The abundance of bacterial populations in iron oxide deposits formed where emerging ground waters carrying ferrous iron in solution react with oxygen-rich surface waters (or air) is an everyday example. Such reactions possibly may have taken place on a larger scale in iron-formation deposition, but, even so, the biologic role would be incidental and of secondary importance: the amount and nature of the end product is determined primarily by the chemistry of the system, not its biology.

(2) The organism selectively secretes a compound as part of its cell wall or skeletal structure

This is a process of major importance in the silica budget of present-day oceans, and in the origin of chert deposits of Phanerozoic age, some of which consist largely of tests of radiolaria or diatoms. It is of particular interest in that silica extraction may occur at concentration levels well below that of saturation in the physicochemical sense, and in that it may flourish periodically or seasonally in response to certain environmental conditions.

The chert in many Precambrian iron-formations contains abundant spherules averaging about 0.03 mm diameter that LaBerge (1973) suggests are of biologic origin. It is, however, known that such forms can be produced abiotically (Oehler,

1976; Schopf and Walter, 1983). Cloud (1973) and Walter and Hofmann (1983) also have discounted biologic genesis on the grounds that no silica-secreting organisms have been identified in Precambrian-age rocks. This interesting possibility, with its potential for explaining at least some of the rhythmic layering in banded iron-formation, must, therefore, remain largely speculative.

(3) The organism significantly modifies the physicochemical nature of the depositional environment through its life processes

It now appears certain that biologically induced environmental changes were of major significance in the deposition of at least some types of iron-formation. Almost surely the creation of near-surface strongly oxygenic conditions in marine waters during late Archean and early Proterozoic time is to be attributed mainly to photosynthetic production of oxygen. Biologic activity, therefore, was the critical, if indirect, factor in the precipitation of oxide-facies iron-formation, as outlined by Cloud (1973). In a more complex manner, precipitation of pyrite in sulfide-facies iron-formation involved interactions between accumulated organic material and H_2S produced by sulfate-reducing bacteria (Berner, 1989, p. 105).

Conclusion

On the basis of these considerations a general conclusion can be drawn, namely: that biologic activity was of first-order importance to the formation of iron-formation of certain types and ages, but the role was largely that of environmental modification, not direct precipitation. This conclusion, however, would have to be modified significantly if it could be shown that silica-secreting organisms did in fact exist in late Archean and early Proterozoic time.

Depositional models

As noted at the outset of this discussion, it is unlikely that a single depositional model could meet the requirements of the several iron-formation types, with their diverse associations and geologic settings. Moreover, after review of the probable sources of supply of the major elements, it is evident that a fundamental distinction exists between the conditions that yielded deposits of Algoma- and Rapitan-types, on the one hand, and the quantitatively far more extensive deposits of Superior- and related types, on the other. For the volcano- and rift-related systems the key aspect is supply; concentrations of iron and silica can (conceptually) be increased to whatever levels are necessary to induce precipitation. For the much larger shelf and marginal basin deposits, the critical aspect of the depositional model must be those processes and conditions under which upwelling charged deep ocean waters release their load of iron and silica from solution.

Algoma- and Rapitan-types

Once granting the source and nature of the supply systems for these deposits, the further requirements for iron-formation accumulation may be relatively few, consisting mainly of a suitable marine basin into which charged solutions are fed and retained until precipitation occurs, plus a timing that would allow for lulls in volcanic or erosional activity so that chemical precipitates would not be diluted by introduced fragmental material. The nature of the actual accumulate would depend upon specific bottom conditions. For the late Archean Algoma-type deposits of the Canadian Shield, Goodwin (1973) has outlined ten generalized basins of deposition within broad areas of submarine volcanism, and in the most intensely studied area – the Michipicoten district – he has shown a systematic distribution of facies, ranging from sulfide to oxide. Rapitan-type deposits appear to be uniformly of oxide facies, consistent with deposition in relatively shallow, rift-bounded basins at continental margins under a later Precambrian oxygenic atmosphere.

There are of course problems with these simple and generalized models, but they concern chiefly secondary issues, such as suppression of the precipitation of other metals that might be expected to be present in exhalative or hydrothermal fluids. Resolution of such problems, however, is not likely to call for change in the broad outlines of the model.

Superior- and related types

The key element in concepts of origin of the extensive shelf and marginal basin deposits that constitute the bulk of known iron-formation is the assumption of an anoxic deep ocean during Archean and early Proterozoic time. It is further assumed that, even at early stages in the earth's history, surface waters were at least mildly oxidizing (indicated by the 3800 m.y. iron-formation at Isua and 3500 m.y. evaporitic sulfate deposits in Western Australia), due to photolytic dissociation of water by ultraviolet radiation in the absence of a protective ozone layer in the atmosphere (Towe, 1983). During later Archean and early Proterozoic time – essentially, between about 3000 m.y. and 2000 m.y. ago – the atmosphere and surface waters became increasingly oxidic as a result of photosynthetic production of oxygen by prokaryotic organisms (Holland, 1984, p. 420).

This progressive buildup in oxygen content of the atmosphere and hydrosphere coincided with – perhaps was symbiotically related to – the late Archean stabilization of cratonic areas and the ensuing development of widespread stable continental shelves, a lengthy, diachronous, but worldwide structural event of the first magnitude. It was the combination, in time and space, of these great events, one biochemical, the other structural, that set the stage for deposition of the world's great iron-formations.

The range of physical configurations for the postulated upwelling system is considerable, ranging from simple broad shelves, open to the sea, to discrete marginal basins of varying size, depth, and degree of restriction. Two general situations are portrayed diagrammatically in Fig. 11-16.

The composition of the upwelling deep ocean water can only be inferred. Drever (1974) suggests the not unreasonable values of about 10 ppm for ferrous iron and 120 ppm for silica; other nutrients, notably phosphorus and nitrate, can be assumed to have been present in significant amounts. The effects consequent on the entry of this charged solution into the warmer shallows of a shelf or marginal basin environment would be profound: immediate drop in pressure and increase in temperature, which would lead to CO_2 outgassing and decreased carbonate solubility; rapid increase in biologic activity in response to greater availability of nutrients, possibly

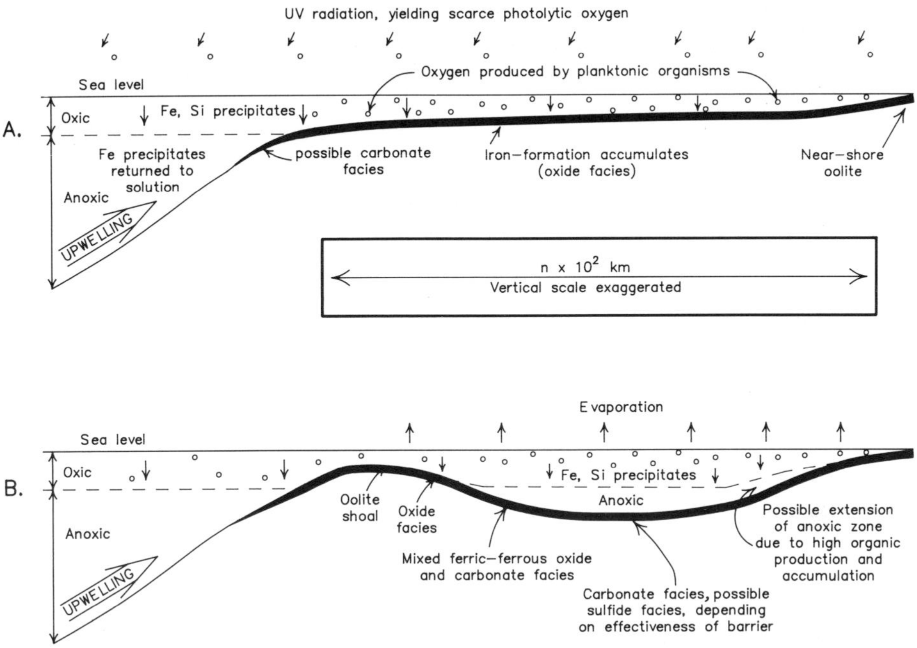

Fig. 11-16. Examples (conceptual) of possible configurations for upwelling-related, shelf and marginal marine basin deposition of iron-formation. (A) Simple broad shelf, no restriction; (B) marginal basin; partly to wholly restricted, depending upon height of barrier at any particular time. (Adapted from Button, 1982.)

with some direct organic oxidation of ferrous iron; and a sharp increase in the level of ambient oxygen as a result of activity by photosynthetic organisms, with consequent inorganic oxidation of ferrous iron.

Ferric hydrate would constitute the bulk of initial iron precipitates in the postulated system. The nature of the *accumulates,* however, would be dependent upon the particular bottom conditions prevailing. Referring to Fig. 11-16, the accumulates would remain wholly ferric on the open shelf of configuration A and on the shoal and nearshore parts of configuration B, some to be reconstituted as oolite in areas of vigorous wave and current action. In the restricted marginal basin, however, other alternatives are not only possible but probable. If, in the water column, the oxic zone is underlain by a thick anoxic layer, ferric hydrate showered down from the near-surface environment may return entirely to solution, increasing the ferrous iron content of bottom waters to a level that would permit direct precipitation of siderite. If the anoxic layer is thin, the result may be partial reduction of the ferric hydrate to yield a mixture of ferrous and ferric material, a possible precursor of magnetite. And in areas of extremely high biologic productivity, the accumulated organic material may be preserved, ultimately to be involved in the complex process by which iron precipitates, oxide or carbonate, are converted to sulfide.

The mechanism for extraction of silica from the upwelling waters is not known. In the absence of a verifiable biologic agent, it is assumed that the process was inorganic – that factors such as temperature rise and evaporation were adequate to cause precipitation. Another possibility mentioned by Drever (1974, p. 1102) is that because the kinetics of silica gel precipitation are pH-sensitive, precipitation might be an indirect effect of hydrogen ion increase due to oxidation of ferrous iron.

Layering in the accumulates can be attributed to variations in precipitation rates related to changes in the level of biologic activity and the amount and timing of upwelling. A not unreasonable model would involve a fairly steady rain of dominantly silica precipitates interrupted at intervals by heavy iron precipitation during cyclic, perhaps seasonally induced, explosive growth of photosynthetic organisms. A further consideration in the origin of layering would be the intermittent introduction of externally derived dust or ash, now represented by silicate-rich layers.

The precipitation of iron and silica from waters trapped in a restricted basin doubtless would be accelerated by evaporation, and indeed some workers, notably Eugster and Chou (1973) and Garrels (1987), have invoked this as a major process. Eugster and Chou postulate an "evaporitic playa lake complex", in which silica is precipitated as magadiite ($NaSi_7O_{13}(OH)_3 \cdot 3H_2O$), which then converts to chert, as at Lake Magadi, Kenya (Eugster, 1969). The presence of abundant riebeckite, the soda amphibole, in major iron-formations of the Transvaal and Hamersley areas is cited as support for such a concept. Garrels (1987) draws attention to the similarity of layering in Hamersley iron-formation to that of evaporites of the Permian-age Castile Formation of southwestern United States, and postulates seasonal precipitation in a closed continental basin. Garrels' basic assumptions and conclusions have been vigorously challenged by Morris and Trendall (1988), who take care to note, however, that their objections do not extend to the general concept of an evaporative factor in iron-formation deposition.

No final conclusions can be drawn as to the importance of evaporation in the genesis of major iron-formations, but a purely continental setting is entirely unlikely, considering the nature of associated strata. Eugster and Chou (1983, p. 1166) do suggest the possibility of a semi-marine model, in which the depositional basin was a playa-like marginal lagoon into which oceanic waters periodically had access. It would appear, therefore, that the Eugster – Chou model is not fundamentally in conflict with that portrayed in Fig. 11-16. The difference is perhaps mainly one of emphasis on degree of evaporation and volume of oceanic contribution. Riebeckite-rich zones in iron-formation may in fact reflect intervals of maximum evaporation, in which some of the silica was precipitated as magadiite.

MINERALOGIC EVOLUTION OF IRON-FORMATION

Iron-formation exposed in the earth's crust today is the product of a sequence of processes that involved initial sedimentation, diagenesis, and one or more epochs of metamorphism. Each of these stages left its imprint on the rock, texturally or mineralogically.

Complete treatment of this topic, in all its complexity, would require a work of monographic proportions (e.g., Melnik, 1973). Various segments of the evolutionary process have been analyzed with some success, notably by Dimroth (1968, 1977a,b) on the physical aspects of sedimentation and early diagenesis, by Berner (1984, 1989) on formation of pyrite, by Ewers (1983) on initial precipitation, by Klein (1983) and Miyano (1987) on diagenesis and metamorphism, and of course by many authors in reports dealing with specific deposits (e.g., Trendall, 1983b, p. 120; Belevstev et al., 1983, p. 241 – 245). Here an attempt will be made to summarize and synthesize concepts and conclusions that have been developed. These, it is to be noted, relate specifically to extensive deposits of the (general) Superior-type (see Fig. 11-16), which in fact constitute the great bulk of known iron-formation.

For convenience, the review of mineralogic evolution will be under the headings of sedimentation, diagenesis, and metamorphism. Conceptually these are distinct processes, but in fact no clearcut dividing lines separate either the processes or the products. Certainly in at least some respects the processes of sedimentation and diagenesis represent a continuum, and the products of diagenesis are not easily distinguished, compositionally or texturally, from those of low-grade metamorphism.

Sedimentation

As noted in a previous section dealing with depositional models, the nature of iron-rich *accumulates* will be governed to a large degree by the bottom environment prevailing. Some accumulates, ferric hydrate in oxidic environments and siderite in certain relatively deep anoxic sites, will represent unchanged initial precipitates. But, as noted in a previous section, in stratified basins, in which the oxic zone is underlain by an anoxic bottom layer of modest thickness, it is likely that the accumulate will be a mixture of ferric and ferrous materials: that is, some initially precipitated ferric hydrate would survive, but some would be reduced to the ferrous state, possibly to combine as "hydromagnetite". Introduced particulate material – volcanic or terrestrial dust – would react with iron-rich bottom waters to produce, particularly in chemically active anoxic sites, an iron-rich silicate mud or gel (in

TABLE 11-9

Principal minerals and compounds of iron and silica involved in sedimentation and diagenesis

Ferric hydrate	$Fe(OH)_3$
Ferrous hydrate	$Fe(OH)_2$
Hematite	Fe_2O_3
"Hydromagnetite"	$Fe_3O_4.H_2O$
Magnetite	Fe_3O_4
Siderite	(Fe, Mg, Ca, Mn) CO_3
Greenalite	(Fe, Mg)$_3Si_2O_5$ $(OH)_4$ (approx.)
Chamosite	Fe-rich septachlorite
Silica gel	$Si(OH)_4$
Chert	SiO_2
Magadiite	$NaSi_7O_{13}(OH)_3.3H_2O$
Riebeckite	$Na_2Fe_5Si_8O_{22}(OH)_2$
Pyrite	FeS_2

shallower oxic environments the tendency would be for this introduced material to be winnowed out and transferred to deeper sites). Organic mud, mixed with initial iron precipitates (carbonate or oxide) would accumulate on highly anoxic bottoms overlain by organically productive surface waters. All of these iron-rich deposits would be mixed with, or interlayered with precipitated silica gel. The compositions of the inferred initial accumulates and the diagenetic products are listed in Table 11-9. The paragenetic relationships of these minerals and compounds are illustrated diagrammatically in Fig. 11-17.

Iron-formations of classic Superior-type preserve abundant evidence of the in-

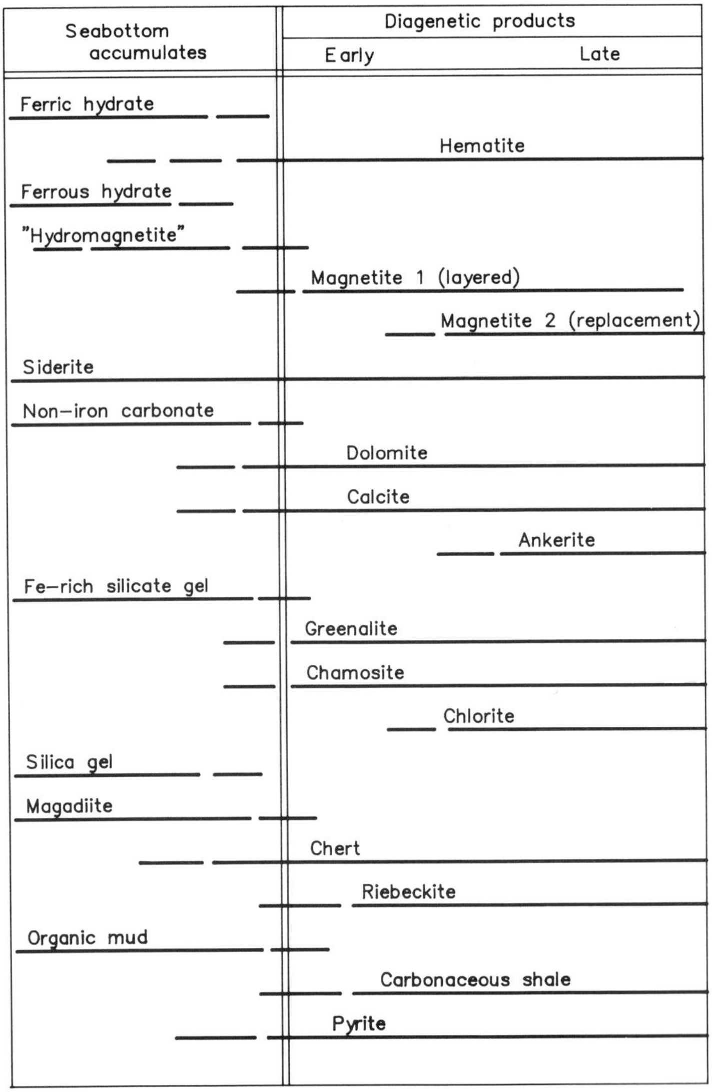

Fig. 11-17. Paragenesis of initial accumulates and diagenetic products.

terplay between chemical and physical processes in sedimentation. On shoals and near-shore positions, ferric hydrate was agglomerated by wave and current action and accumulated as oolite, ultimately to be converted to hematite. In places, deeper-water sediments, including some formed in anoxic bottom environments, were also subject to disruption by wave action, producing peloids, shards, and granules that were transferred as clastic particles to sites of accumulation; they may have been significantly different chemically from those of the initial deposition. Many of the resultant clastic accumulates were chemically heterogeneous (see Fig. 11-7, which shows mixed assemblages of ferric ooliths and ferrous granules), subject to a profound change during diagenesis and metamorphism as mineralogic equilibrium was established.

Somewhat surprisingly, the evidence is clear that the first **layered** material to lithify was chert. Breccias consisting of sharp-edged angular fragments of chert in fine-grained ferruginous matrices are common in many districts; examples from the Kuruman Iron-formation of Transvaal are described by Beukes (1973, p. 984), from the carbonate-facies (Riverton Iron-formation of Michigan, by James et al., 1968, fig. 34), and from the Sokoman Iron-formation of Labrador by Dimroth and Chauvel (1973, fig. 1; see also later papers by Dimroth).

Diagenesis

The processes involved in diagenesis include dewatering and compaction, dehydration and crystallization of gels, biochemical reduction of dissolved sulfate to yield reactive H_2S, and mineralogic reactions in response to the physicochemistry of the burial environment. In general, the tendency in formation of new minerals and replacements will be toward products lower in the Eh scale, because of entrapped organic material. This reducing tendency is readily recognized texturally in iron-formation of various types: hematite is commonly replaced by magnetite or even siderite; siderite is replaced by pyrite. The extent of such replacements is highly variable, because it depends quantitatively on the capacity of the system – that is, the availability of reactants – not just by the chemical tendency set by intensive variables such as Eh.

The principal minerals produced by, or involved in, diagenesis are shown paragenetically in Fig. 11-17 and are discussed individually in the following paragraphs.

Hematite is generally assumed to have formed early from ferric hydrate, perhaps in part during the sedimentation process. Though not explicitly recognized, it is entirely likely that the transition involved an intermediate oxide hydrate phase.

Magnetite has several different modes of formation. Some unquestionably is a diagenetic replacement product, most obviously where magnetite crystals transect primary sedimentary structures such as oolite layering. But magnetite also forms dense fine-grained layers interbedded with unaltered layers of possible precursor minerals – hematite, siderite, silicates – and the conclusion is inescapable that this magnetite (Magnetite 1 in Fig. 11-17) formed very early in the sedimentation – diagenesis sequence. The actual mode of formation of layered magnetite is not known, but it is assumed to have developed from a precursor ferric–ferrous hydrate ("hydromagnetite"). The distribution and amount of replacement magnetite (Magnetite 2 of Fig. 11-17) varies widely from district to district and even within a formation, but it may be extensive. In parts of the Ironwood Iron-formation of the Gogebic Range, for example, in strata originally hematitic (as shown by the preservation of fine-grained hematite trapped in interlayered jasper and by relict oolitic structures), the initial hematite is almost completely replaced by magnetite (Huber, 1959). To further complicate the magnetite paragenesis record, the tendency for hematite to be reduced to magnetite persists during metamorphism, leading some workers to conclude that this is a major mode of magnetite formation (Floran and Papike, 1978).

Siderite is probably the only layered material in iron-formation assemblages that represents an initial precipitate, essentially unchanged except for lithification. A clear demonstration of early origin is provided by the Riverton Iron-formation of Michigan. As shown in Fig. 11-8, layered and laminated siderite, together with interbedded chert, participated in soft-rock deformation. In the same formation, a primary sedimentary control is also indicated by significant layer-to-layer variations in siderite composition (James et al., 1968, p. 56). Additionally, though not shown separately on the chart (Fig. 11-17), some siderite is of diagenetic replacement origin, participating with magnetite in the reduction of hematite.

Non-iron carbonates, i.e., dolomite, calcite, and ankerite, are common minor constituents in many iron-formations. Dolomite and calcite may represent original accumulates, but ankerite, where present, is clearly of late diagenetic origin, typically appearing as clots or clusters of relatively coarse grains.

The iron-rich silicates (greenalite, chamosite, and chlorite) formed by reconstitution of amorphous iron-rich silicate gel of variable composition. Iron-formations that are least metamorphosed, as in the Gunflint Range in its eastern extension, contain greenalite as a major constituent. Greenalite, however, is readily altered to minnesotaite and stilpnomelane, even within zones of very low-grade metamorphism, and preserved only locally. Chamosite and chlorite, scarce but recognizable constituents in some iron-formations, presumably also formed diagenetically from silicate residues.

The origin of chert is a widely researched problem to which little of consequence can be added here. As previously stated, the concept of biogenic precipitation, attractive though it is, does not appear to be applicable, because no silica-secreting organisms are known to have existed at the time of deposition of Precambrian iron-formations (Walter and Hofmann, 1983). Concepts of hydrothermal or replacement origin, the one advanced by Simonson (1985) and the other by Lepp and Goldich (1964) and Lougheed (1983), are rejected as earlier stated, on the basis of scale and uniformity of final product. It is concluded, therefore, that most iron-formation chert evolved from precipitated silica gel, and, on the evidence from contemporaneous breccias, much of it formed early in the depositional phase. The evolution from gel to the crystalline state doubtless was complex, as described by Dimroth and Chauvel (1973, p. 121) for chert of the Sokoman Iron-formation of Labrador. A lesser amount of chert may have formed directly from magadiite under strongly evaporative conditions, as noted in the next section.

Riebeckite, an observed but generally scarce mineral in most iron-formations of oxide facies, is relatively abundant in parts of the early Proterozoic iron-formations of the Hamersley district, Australia, and of the Transvaal, South Africa, where it occurs as separate discontinuous layers, made up of felted masses of interlocking fibers, of almost incredible toughness. In detail, as described by Trendall and Blockley (1970, p. 300), riebeckite transects earlier sedimentary structures, so it evidently is of diagenetic origin. Eugster (1969, p. 26) illustrates lateral transitions, with uninterrupted lamination, from chert to riebeckite, and links the formation of riebeckite to the behavior of sodium in the magadiite – chert transition; " . . . massive riebeckite has formed by reaction of magadiite with the iron minerals in those portions of a mesoband, in which the magadiite – chert conversion had not yet begun or had not been completed". In both the South African and Australian occurrences, riebeckite locally has been transformed in later diagenetic stages to cross-fiber *crocidolite*, the blue asbestos.

The origin of *pyrite* is intimately tied up with the biochemical cycles of carbon and sulfur. The key factor is production of reactive H_2S by reduction of dissolved sulfate, which " . . . can occur only in the presence of organic compounds that can be metabolized by the appropriate set of bacteria" (Berner, 1989, p. 105). Berner (p. 109) further notes that the limiting factor of pyrite formation in most euxinic sediments is availability of iron, not organic matter or dissolved sulfate. But in an iron-formation depositional environment, ample iron (in the form of either oxide or carbonate) is present in the bottom accumulates, so that pyrite forms to levels of abundance (see Table 11-3) rarely if ever observed in non-iron-formation strata. The layered structure typical of sulfide iron-formation (see Fig. 11-9) is inherited from initial iron-rich precipitates. Probably most layered pyrite formed early in the depositional – diagenetic sequence, but – like magnetite and siderite – pyrite also occurs as a later diagenetic mineral replacing siderite or other earlier formed iron minerals.

Isotopic data on carbon and sulfur are not entirely unambiguous but they tend to support the model of biochemical sulfate reduction, even for deposits of Archean age. Goodwin et al. (1976) provide data on the iron-formations of the Michipicoten district and conclude (p. 87): "The sulfur and carbon isotope data provide strong evidence for the existence of autotrophic organisms and reducing bacteria in Archean times . . . " Analyses of carbon from Precambrian iron-formations of Brazil (Schidlowski et al., 1976) indicate an average composition of $-24.7‰$ for $^{13}C_{org}$, compared with $-0.9‰$ for $^{13}C_{carb}$, a spread consistent with biologic origin. Cameron (1983), however, interpreted data on sulfur isotope distribution of various iron-formations to indicate a magmatic rather than a biologic derivation.

Metamorphism

All iron-formation at the earth's surface today has been modified to some degree by metamorphic processes: there is no example known of wholly unmetamorphosed iron-formation. As has been mentioned earlier in this review, the nearest approach to the unmetamorphosed state is represented by the fossiliferous Gunflint Iron-formation in its eastern extension (Barghoorn and Tyler, 1965), but even here early formed greenalite is partly replaced by metamorphic stilpnomelane and minnesotaite (Floran and Papike, 1978). At the other end of the metamorphic spectrum are the coarsely crystalline iron-formations in contact with large mafic intrusions such as the Duluth Gabbro and the Stillwater Complex (Vaniman et al., 1980), and the products of high-grade regional metamorphism, as in the Archean-age iron-formation of the Yilgarn Block of Western Australia (Gole and Klein, 1981).

The mineralogic evolution of iron-formation, from primary sedimentary – diagenetic deposits to high-grade metamorphic assemblages, comprehensively reviewed by Klein (1983), is summarized in simplified form in Fig. 11-18, in which the degree of metamorphism is divided, somewhat arbitrarily, into low, medium, and high categories. Low grade corresponds approximately to chlorite – biotite zones of regional metamorphism (greenschist facies), medium grade to garnet – staurolite zones (epidote amphibolite and low amphibolite facies), and high grade to the sillimanite zone (upper amphibolite and granulite facies) and to the contact-

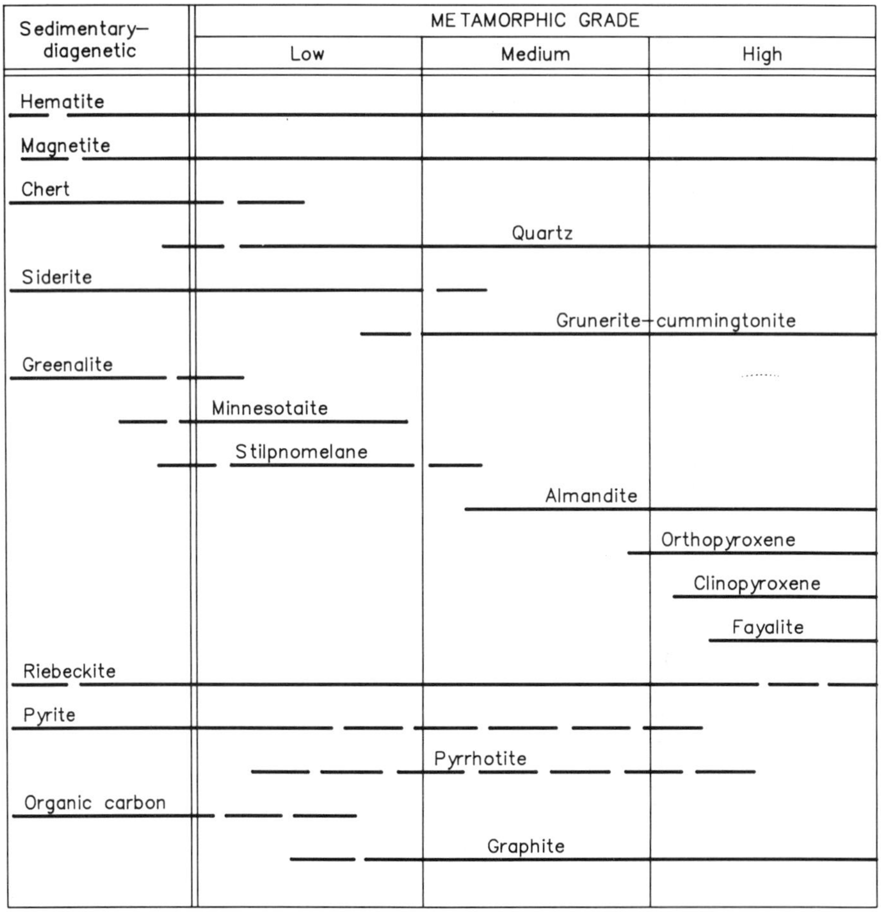

Fig. 11-18. Stability ranges of principal constituents of iron-formation, from sedimentary – diagenetic stage to high-grade metamorphic stage. (Adapted from James, 1954; and Klein, 1983.)

metamorphosed hornfels facies. The constituents listed are only those of common occurrence or significance. Actual iron-formation assemblages contain many other minerals, including various chlorites (such as chamosite and ripidolite), additional amphiboles (hornblende, anthophyllite, and ferro-actinolite), the carbonates (calcite, dolomite, and ankerite), and, in some iron-formations, manganese-rich carbonate, manganese silicates, and manganese oxides. Apatite is a minor, but ubiquitous, constituent in iron-formations of all metamorphic grades.

The temperature (and pressure) ranges represented by low-, medium-, and high-grade metamorphic categories cannot be given with any degree of precision, but information is available for particular minerals and assemblages. Miyano (1987, p. 157) estimates the temperature of the greenalite to minnesotaite transition to be 130°C at 1 kbar pressure, which probably is an approximate minimum for the onset of significant metamorphism. Oxygen isotope ratios in coexisting quartz, dolomite, and calcite in the nearly flat-lying Brockman Iron-formation of the Hamersley Range, Western Australia, indicate equilibration in a temperature range of 270° – 310°C (Becker and Clayton, 1976) – suprisingly high for an iron-formation that contains minnesotaite and stilpnomelane and which often has been considered to be essentially unmetamorphosed.

The regionally metamorphosed iron-formation of the Ruby Range, Montana (James, 1990) ranges areally from amphibolite facies to granulite facies. On the basis of Fe/Mg distributions in co-existing garnet and pyroxene, Dahl (1979) concludes that the iron-formation was metamorphosed under conditions ranging from about 675°C, 6 kbar (amphibolite facies) to about 745°C, 7 kbar (granulite facies). A comparable estimate (700° – 750°C; 10 – 11 kbar) for granulite-facies metamorphism has been made by Klein (1978) for the highly metamorphosed iron-formation of the southwestern Labrador Trough.

Fayalite-bearing assemblages occur mainly in iron-formation of hornfels facies within contact-metamorphic zones of major mafic intrusions. On the basis of inferred and observed pyroxene relations (prograde pigeonite inverted to orthopyroxene with two sets of Ca-rich pyroxene exsolution lamellae), Vaniman et al. (1980) estimate that the fayalite-bearing iron-formation in the contact zone of the Stillwater Complex was metamorphosed at a temperature of at least 800°C and 2kbar minimum pressure.

The mineralogic changes that result from metamorphism are many and varied. Siderite, a major iron-bearing mineral of many iron-formations of low metamorphic grade, is unstable above approximately the garnet zone of regional metamorphism; at higher levels it is consumed entirely, mostly by reaction with quartz to produce gruneritic amphibole. Primary silicates undergo a series of metamorphic transformations with sequential appearance of minnesotaite and stilpnomelane, amphiboles of the grunerite – cummingtonite series, orthopyroxene, clinopyroxene, and, at highest levels, fayalite. Organic carbon initially present is either lost by reaction or converted (probably with some intermediate amorphous phase) to graphite. Pyrite persists as a stable phase in some rocks of relatively high metamorphic grade, but the paragenesis of pyrrhotite is uncertain. Some – as in the sulfidic iron-formation of the Michipicoten district – occurs as early-formed mineral in certain environments, and pyrrhotite reportedly coexists with pyrite iron-formation of low metamorphic grade in the Hamersley district of Western Australia (Klein, 1983, p. 436). But pyrrhotite may also form by breakdown of pyrite; French (1968) suggests a possible pyrrhotite isograd in the contact-metamorphosed iron-formation of the Mesabi Range of Minnesota.

Iron-formation of primary oxide facies retains much of its initial character even under intense metamorphic conditions, and indeed is the only variety of iron-formation that remains readily recognizable even in metamorphic assemblages of the highest grade. The principal minerals (hematite, magnetite, and quartz) persist as stable phases, although there doubtless exists a constant tendency toward reduction of hematite to magnetite. The original layered structure of the rock may be modified somewhat by the progressive increase in grain size and mineralogic reactions involving initial silicate or carbonate content, but in general much of the characteristic thin bedding of iron-formation is retained, and the bulk composition (in terms of major elements) remains essentially unchanged. This remarkable stability of mineral content, bulk composition, and structure of oxide iron-formation accounts for the fact that iron-formation is a recognized metasedimentary constituent even in the most highly metamorphosed sequences and in the most ancient of terranes, as at Isua, Greenland.

EPILOGUE

The Precambrian iron-formations are quantitatively relatively minor components in the Precambrian sedimentary record, but – in addition to their great economic value – they are significant indicators of certain large-scale aspects of the earth's evolution.

But the relationships are not simple. As with many economic mineral deposits, iron-formation genesis involved coincidence in time and space of different processes and events, none of which individually would have resulted in a deposit of such particular character. For deposits of the Algoma- and Rapitan-types, the simplified models that have been presented here have three principal requirements: appropriate depositional sites, absence (at least periodic) of clastic or fragmentary volcanic contributions to the basins of deposition, and deep-seated or volcanic sources of iron- and silica-rich fluids. In terms of the earth's history, Algoma-type iron-formations simply reflect that particular type of marine volcanism that produced greenstone belts (of all ages), in which there had been a coincidence of the needed depositional factors. The Rapitan-type deposits, though still not well understood, do provide substantial support for the concept of a (possibly) worldwide epoch of continental rifting in late Proterozoic time.

The great iron-formations of Superior-type were the ultimate consequence of a series of conditions and events, some of which extended far back in the earth's history. In the immediate sense, the requirements for deposition are similar to those for iron-formation of other types: suitable depositional site, little or no clastic input, and an abundant supply of iron and silica in solution. But the supply system envisioned in the model that has been presented involves a necessary assumption, indeed a conclusion, of truly major importance concerning prior history of the earth's oceans. The iron-formations constitute positive evidence that, despite indications that some oxygen was present in earth's early atmosphere (Towe, 1983), the oceanic deeps were anoxic throughout Archean time, reservoirs in which iron and silica attained levels far beyond those possible in later geologic eras. The structural evolution of continental margins that followed the worldwide epoch of mid- to early-Archean cratonization provided the necessary structural setting for upwelling of these charged solutions into basins where they could react with near-surface marine waters and an evolving oxygenic atmosphere. The existence of these depositional sites was an essential factor in the process whereby large-scale equilibrium ultimately was established between the anaerobic deep ocean waters and an increasingly oxygenated atmosphere; this process culminated in early Proterozoic time and was irreversible. It was halted, in a world-wide sense, only when the oceans were essentially depleted of their content of iron in solution, about 2 billion years ago.*

The reactive nature of iron-formation assemblages and the tantalizing preservation of biologic activity in some assure that research on the petrochemistry and biochemistry of these rocks will continue long in the future. It is to be hoped that such future research will provide clues to questions that now remain unanswered as well as to those that are yet to be asked. One of the most puzzling questions is why the bulk compositions and iron–silica ratios of iron-formation vary within such small limits for rocks of widely different mineralogic facies and of widely different ages. Such relative constancy of composition is not a predictable consequence of the depositional models that have been here described. The fact itself suggests mutual interaction in iron and silica precipitation, but if so it must be an interaction that prevailed over a wide range of environmental conditions, seemingly far greater than those known to induce coprecipitation under controlled laboratory conditions (see Ewers, 1983, p. 505, for discussion of possible reactions).

Other issues concerning iron-formation genesis remain to be fully explored, notably the role of biologic activity and the importance of the evaporative factor, issues on which there is little agreement at the present time. The genetic concepts that have been presented in this chapter incorporate judgments on these and other debatable points, and though they appear reasonable on the basis of present knowledge of the earth's history and geochemical processes, it would be foolhardy to claim more than interim status for them: Final answers are yet to come.

REFERENCES

Aldrich, H.R., 1929. The geology of the Gogebic Iron range of Wisconsin. *Geol. Nat. Hist. Surv., Bull.*, 71, 279 pp.

Appel, P.W.U. and LaBerge, G.L. (Editors), 1987. *Precambrian Iron-Formations*. Theophrastus Publ., Athens, 674 pp.

Baadsgaard, H., Nutman, A.P., Bridgwater, D., Rosing, M., McGregor, V.R. and Allaart, J.H., 1984. The zircon geochronology of the Akilia association and Isua supracrystal belt, West Greenland. *Earth*

* It is recognized, of course, that oceanic stagnation has occurred periodically during the earth's late history; for example, in the Pacific realm during Aptian time (Sliter, 1989). The difference between such anoxic events and that invoked here is one of scale, particularly that of time. Oceanic anoxia was probably the norm for 1000–2000 my of Archean time, whereas time spans for later events appear to have been on the order of 1 my or less.

Planet. Sci. Lett., 68: 221–228.
Barghoorn, E.S. and Tyler, S.A., 1965. Microorganisms from the Gunflint chert. *Science*, 147: 563–577.
Bayley, R.W. and James, H.L., 1973. Precambrian iron-formations of the United States. *Econ. Geol.*, 68: 934–959.
Becker, R.H and Clayton, R.N., 1976. Oxygen isotope study of a Precambrian banded iron-formation in Western Australia. *Geochim. Cosmochim. Acta*, 40: 1153–1165.
Belevtsev, Ya.N., Belevtsev, R.Ya. and Siroshtan, R.I., 1983. The Krivoy Rog Basin. In: A.F. Trendall and R.C. Morris (Editors), *Iron-Formation: Facts and Problems*. Elsevier, Amsterdam, pp. 211–252.
Berner, R.A., 1984. Sedimentary pyrite formation: An update. *Geochim. Cosmochim. Acta*, 48: 605–615.
Berner, R.A., 1989. Biogeochemical cycles of carbon and sulfur and their effect on atmospheric oxygen over Phanerozoic time. *Palaeogeogr. Palaeoclimatol. Palaeoecol.*, 75: 97–122.
Beukes, N.J., 1973. Precambrian iron-formations of southern Africa. *Econ. Geol.*, 68: 960–1004.
Beukes, N.J., 1983. Palaeoenvironmental setting of iron-formations in the depositional basin of the Transvaal Supergroup, South Africa. In: A.F. Trendall and R.C. Morris (Editors). *Iron-Formation: Facts and Problems*. Elsevier, Amsterdam, pp. 131–210.
Borchert, H., 1960. Genesis of marine sedimentary iron ore. *Inst. Min. Metall. Trans.*, 69: 261–279.
Breitkopf, J.H., 1988. Iron-formations related to mafic volcanism and ensialic rifting in the southern margin zone of the Damara Orogen, Namibia. *Precambrian Res.*, 38: 111–130.
Button, A., 1982. Sedimentary iron deposits, evaporites, and phosphorites. State of the art report. In: H.D. Holland and M. Schidlowski (Editors), *Mineral Deposits and the Evolution of the Biosphere*. Springer, Berlin, pp. 259–273.
Cameron, E.M., 1983. Genesis of Proterozoic iron-formation: sulphur isotope evidence. *Geochim. Cosmochim. Acta,* 47: 1069–1074.
Castaño, J.R. and Garrels, R.M., 1950. Experiments on the deposition of iron with special reference to the Clinton iron deposits. *Econ. Geol.*, 45: 755–770.
Chakraborty, K.L. and Majumder, T., 1986. Geological aspects of the banded iron-formation of Bihar and Orissa. *J. Geol. Soc. India*, 28: 109–133.
Cloud, P., 1973. Paleo ecological significance of Precambrian banded iron-formations. *Econ. Geol.*, 68: 1135–1143.
Cloud, P.E., Gruner, V.W. and Hagen, H., 1965. Carbonaceous rocks of the Soudan Iron-formation (Early Precambrian). *Science*, 148: 178–1716.
Dahl, P.S., 1979. Comparative geothermometry based on major-element and oxygen isotope distributions in Precambrian metamorphic rocks from southwestern Montana. *Am. Mineral.*, 64: 1280–1293.
Dimroth, E., 1968. Sedimentary textures, diagenesis, and sedimentary environment of certain Precambrian ironstones. *Neues Jahrb. Geol. Paleontol.*, 130: 247–274.
Dimroth, E., 1977a. Facies models. 5. Models of physical sedimentation of iron-formations. *Geosci. Can.*, 4: 23–30.
Dimroth, E., 1977b. Facies models. 6. Diagenetic facies of iron-formation. *Geosci. Can.*, 4: 83–88.
Dimroth, E. and Chauvel, J.J., 1973. Petrography of the Sokoman Iron Formation in part of the Central Labrador Trough, Quebec, Canada. *Geol. Soc. Am., Bull.*, 84: 111–134.
Dorr, J.V.N. II, 1973. Iron-formation in South America. *Econ. Geol.*, 68: 1005–1022.
Drever, J.L., 1974. Geochemical model for the origin of Precambrian banded iron-formations. *Geol. Soc. Am., Bull.*, 85: 1099–1106.
Dymek, R.F. and Klein, C., 1988. Chemistry, petrology and origin of banded iron-formation lithologies from the 3800 Ma Isua Supracrustal Belt, West Greenland. *Precambrian Res.*, 39: 247–302.
Eichler, J., 1976. Origin of Precambrian banded iron-formations. In: K.H. Wolf (Editor), *Handbook of Strata-Bound and Stratiform Ore Deposits*, 7. Elsevier, Amsterdam, pp. 157–201.
Eugster, H., 1969. Inorganic bedded cherts from the Magadi area, Kenya. *Contrib. Mineral. Petrol*, 22: 1–31.
Eugster, H. and Chou, I-Ming, 1973. The depositional environments of Precambrian banded iron-formations. *Econ. Geol.*, 68: 1144–1168.
Ewers, W.E., 1983. Chemical factors in the deposition and diagenesis of banded iron-formations. In: A.F. Trendall and R.C. Morris (Editors), *Iron-Formation: Facts and Problems*. Elsevier, Amsterdam, pp. 491–512.

Floran, R.J. and Papike, J.J., 1978. Mineralogy and petrology of the Gunflint Iron-formation, Minnesota – Ontario: Correlation of compositional and assemblage variations at low to moderate grade. *J. Petrol.*, 19: 215 – 288.

French, B.M., 1968. Progressive contact metamorphism of the Biwabik Iron-formation, Mesabi Range, Minnesota. *Minn. Geol. Surv., Bull.*, 45: 103 pp.

Garrels, R.M., 1987. A model for the deposition of microbanded Precambrian iron-formations. *Am. J. Sci.*, 287: 81 – 106.

Garrels, R.M. and Christ, C., 1965. *Solutions, Minerals and Equilibria*. Harper and Row, New York, N.Y., 450 pp.

Gole, M.J., 1981. Archean banded iron-formations, Yilgarn Block, western Australia. *Econ. Geol.*, 76: 1954 – 1975.

Gole, M.J. and Klein, C., 1981. Banded iron-formations through much of Precambrian time. *J. Geol.*, 89: 169 – 183.

Goodwin, A.M., 1962. Structure, stratigraphy, and origin of iron-formations, Michipicoten area, Algoma district, Ontario, Canada. *Geol. Soc. Am., Bull.*, 73: 581 – 586.

Goodwin, A.M., 1964. Geochemical studies at the Helen iron range. *Econ. Geol.*, 59: 684 – 718.

Goodwin, A.M., 1973. Archean iron-formations and tectonic basins of the Canadian Shield. *Econ. Geol.*, 68: 915 – 933.

Goodwin, A.M., Monster, V. and Thode, H.G., 1976. Carbon and sulfur isotope abundances in Archean iron-formations and early Precambrian life. *Econ. Geol.*, 71: 870 – 891.

Goodwin, A.M., Thode, H.G., Chou, G.L. and Karkhansis, S.H., 1985. Chemostratigraphy and origin of the late Archean siderite – pyrite-rich Helen Iron Formation, Michipicoten belt, Canada. *Can. J. Earth Sci.*, 22: 72 – 84.

Gross, G.A., 1965. Geology of iron deposits in Canada. General geology and evaluation of iron deposits. *Geol. Surv. Can., Econ. Geol.*, 22, 1: 181 pp.

Gross, G.A., 1968. Geology of iron deposits in Canada. Iron ranges of the Labrador geosyncline. *Geol. Surv. Can., Econ. Geol.*, 22, 3: 179 pp.

Gross, G.A., 1973. The depositional environments of principal types of Precambrian iron-formations. In: *Genesis of Precambrian Iron and Manganese Deposits*. Proc. Kiev Symp., 1970, UNESCO Earth Sci. Rep., 9: 15 – 21.

Gross, G.A., 1980. A classification of iron formations based on depositional environments. *Can. Mineral.*, 18: 215 – 222.

Gross, G.A., 1983. Tectonic systems and deposition of iron formation. *Precambrian Res.*, 20: 171 – 187.

Gross, G.A. and Zajac, I.S., 1983. Iron-formation in fold belts marginal to the Ungava Craton. In: A.F. Trendall and R.C. Morris (Editors), *Iron-Formation: Facts and Problems*. Elsevier, Amsterdam, pp. 253 – 294.

Groves, D.I., Dunlop, V.S.R. and Buick, R., 1981. An early habitat of life. *Sci. Am.*, 245: 64 – 73.

Gruss, H., 1973. Itabirite iron ores of the Limberia and Guyanan Shields. In: *Genesis of Precambrian Iron and Manganese Deposits*. Proc. Kiev Symp. 1970, UNESCO Earth Sci. Rep., 9: 335 – 359.

Harder, E.C., 1919. Iron depositing bacteria and their geologic relations. *U.S. Geol. Surv., Prof. Pap.*, 113: 89 pp.

Holland, H.D., 1973. The Oceans: A possible source of iron in iron-formations. *Econ. Geol.*, 68: 1169 – 1172.

Holland, H.D., 1984. *The Chemical Evolution of the Atmosphere and the Oceans*. Princeton Univ. Press, Princeton, N.J., 582 pp.

Hoppe, A., Schobbenhaus, C. and Walde, D.H.G., 1987. Precambrian iron-formation in Brazil. In: P.W.U. Appel and G.L. LaBerge (Editors), *Precambrian Iron-Formations*. Theophrastus Publ., Athens, pp. 347 – 392.

Hough, J.L., 1958. Fresh-water environment of deposition of Precambrian banded iron-formations. *J. Sediment. Petrol.*, 28: 414 – 430.

Huber, N.K., 1958. The environmental controls of sedimentary iron minerals. *Econ. Geol.*, 53: 123 – 140.

Huber, N.K., 1959. Some aspects of the origin of the Ironwood Iron-Formation of Michigan and Wisconsin. *Econ. Geol.*, 54: 82 – 118.

Huber, N.K. and Garrels, R.M., 1953. Relation of pH and oxidation potential to sedimentary iron mineral formation. *Econ. Geol.*, 48: 337 – 357.

Hurley, R.M., Kalliokoski, J., Fairbairn, H.W. and Pinson, W.H., 1972. Progress report on the age of granulite facies rocks in the Imataca Complex, Venezuela. *Proc. IX Interguyanas Geol. Conf.*, Ciudad Bolivar, p. 10.
James, H.L., 1946. Chromite deposits near Red Lodge, Carbon County, Montana. *U.S. Geol. Surv., Bull.*, 945-F: 151 – 189.
James, H.L., 1954. Sedimentary facies of iron-formation. *Econ. Geol.*, 49: 235 – 293.
James, H.L., 1966. Chemistry of the iron-rich sedimentary rocks. *U.S. Geol. Surv., Prof. Pap.*, 440W: 60 pp.
James, H.L., 1981. Bedded Precambrian iron deposits of the Tobacco Root Mountains, southwestern Montana. *U.S. Geol. Surv., Prof. Pap.*, 1187: 16 pp.
James, H.L., 1983. Distribution of banded iron-formation in space and time. In: A.F. Trendall and R.C. Morris (Editors), *Iron-Formation: Facts and Problems.* Elsevier, Amsterdam, pp. 471 – 490.
James, H.L., 1990. Precambrian geology and bedded iron deposits of the southwestern Ruby Range, Montana. *U.S. Geol. Surv., Prof. Pap.*, 1495: 39 pp.
James, H.L. and Sims, P.K. (Editors), 1973. Precambrian iron-formations of the world. *Econ. Geol.*, 68 (7): 913 – 1179.
James, H.L. and Trendall, A.F., 1982. Banded iron-formation: Distribution in time and paleoenvironmental significance. In: H.D. Holland and M. Schidlowski (Editors), *Mineral Deposits and the Evolution of the Biosphere.* Springer, Berlin, pp. 199 – 218.
James, H.L., Dutton, C.E., Pettijohn, F.J. and Wier, K.L., 1968. Geology and ore deposits of the Iron River – Crystal Falls district, Iron County, Michigan. *U.S. Geol. Surv., Prof. Pap.*, 570: 134 pp.
Kimberley, M.M., 1974. Origin of iron ore by replacement of calcareous oolite. Ph. D. thesis, Princeton Univ., Princeton, N.J., 386 pp.
Kimberly, M.M., 1978. Paleoenvironmental classification of iron-formations. *Econ. Geol.*, 73: 215 – 229.
Kimberley, M.M., 1989a. Nomenclature for iron ore formations. *Ore Geol. Rev.,* 5 (1/2): 1 – 12.
Kimberley M.M., 1989b. Exhalative origin of iron formations. *Ore Geol. Rev.,* 5 (1/2): 13 – 146.
Klein, C., 1978. Regional metamorphism of Proterozoic iron-formation, Labrador Trough, Canada. *Am. Mineral.*, 63: 898 – 912.
Klein, C., 1983. Diagenesis and metamorphism of Precambrian banded iron-formations. In: A.F. Trendall and R.C. Morris (Editors), *Iron-Formation: Facts and Problems.* Elsevier, Amsterdam, pp. 417 – 470.
Krumbein, W.C. and Garrels, R.M., 1952. Origin and classification of chemical sediments in terms of pH and oxidation – reduction potentials. *J. Geol.*, 60: 1 – 33.
LaBerge, G.L., 1966a. Altered pyroclastic rocks in the Hamersley Range, Western Australia. *Econ. Geol.*, 61: 147 – 161.
LaBerge, G.L., 1966b. Pyroclastic rocks in South African iron-formations. *Econ. Geol.*, 61: 572 – 581.
LaBerge, G.L., 1973. Possible biological origin of Precambrian iron-formations. *Econ. Geol.*, 68: 1098 – 1109.
LaBerge, G.L., Robbins, E.I. and Han, T.M., 1987. A model for the biological precipitation of Precambrian iron-formations. In: P.W.U. Appel and G.L. LaBerge (Editors), *Precambrian Iron-Formations.* Theophrastus Publ., Athens, pp. 69 – 96.
Lepp, H., 1972. Normative mineral composition of the Biwabik Formation: A first approach. In: B.R. Doe and D.K. Smith (Editors), *Studies in Mineralogy and Precambrian Geology.* Geol. Soc. Am., Mem., 135: 265 – 280.
Lepp, H., 1987. Chemistry and origin of Precambrian iron-formations. In: P.W.U. Appel and G.L. LaBerge (Editors), *Precambrian Iron-Formations.* Theophrastus Publ., Athens, pp. 3 – 30.
Lepp, H. and Goldich, S.S., 1964. Origin of Precambrian iron-formations. *Econ. Geol.*, 59: 1025 – 1060.
Lougheed, M.S., 1983. Origin of Precambrian iron-formations in the Lake Superior region. *Geol. Soc. Am., Bull.*, 94: 325 – 340.
Melnik, Yu. P., 1982. *Precambrian Banded Iron-Formations: Physicochemical Conditions of Formation.* Elsevier, Amsterdam, 310 pp.
Miyano, T., 1978. Effects of CO_2 on mineralogic differences in some low-grade metamorphic iron-formations. *Geochem. J.*, 12: 201 – 211.
Miyano, T., 1987. Diagenetic to low-grade metamorphic conditions of Precambrian iron-formations. In: P.W.U. Appel and G.L. LaBerge (Editors), *Precambrian Iron-Formations.* Theophrastus Publ.,

Athens, pp. 155 – 186.

Morey, G.B., 1983. Animikie Basin, Lake Superior Region, U.S.A. In: A.F. Trendall and R.C. Morris (Editors), *Iron-Formation: Facts and Problems*. Elsevier, Amsterdam, pp. 13 – 68.

Morris, R.C. and Trendall, A.F., 1988. A model for the deposition of the microbanded Precambrian iron-formations (Discussion of paper by R.M. Garrels). *Am. J. Sci.*, 288: 664 – 673.

Nalivkin, D.V., 1960. *The Geology of the U.S.S.R.* Pergamon Press Int. Ser., Monogr. Earth Sci., vol. 8. Pergamon, New York, N.Y., 170 pp.

Oehler, J.H., 1976. Hydrothermal crystallization of silica gel. *Gel. Soc. Am., Bull.*, 87: 1143 – 1152.

Qiusheng, Z., 1987. Banded iron-formations in China. In: P.W.U. Appel and G.L. LaBerge (Editors), *Precambrian Iron-Formations*. Theophrastus Publ., Athens, pp. 423 – 448.

Schidlowski, M., Eichmann, R. and Fiebiger, W., 1976. Isotopic fractionation between organic carbon and carbonate carbon in Precambrian banded ironstone series from Brazil. *Neues Jahrb. Mineral. Monatsh.*, 8: 344 – 353.

Schopf, J.W. and Walter, M.R., 1983. Archean microfossils: New evidence of ancient microbes. In: J.W. Schopf (Editor), *Earth's Earliest Biosphere*. Princeton Univ. Press, Princeton, N.J., pp. 214 – 239.

Simonson, B.M., 1985. Sedimentological constraints on the origins of Precambrian iron-formations. *Geol. Soc. Am., Bull.*, 96: 244 – 252.

Sliter, W.V., 1989. Aptian anoxia in the Pacific Basin. *Geology,* 17: 909 – 912.

Strakhov, N.M., 1959. Schema de la diagenèse des depots marins. *Eclogae Geol. Helvetiae,* 51: 761 – 767.

Teodorovich, G.I., 1947. Sedimentary geochemical cycles: *Soc. Naturalistes Moscou Bull.*, 52: *Sec. Geol.*, 22 (1): 3 – 24 (in Russian with English summary).

Towe, K.M., 1983. Precambrian atmospheric oxygen and banded iron-formations: A delayed ocean model. *Precambrian Res.*, 20: 161 – 170.

Trendall, A.F., 1983a. Introduction. In: A.F. Trendall and R.C. Morris (Editors), *Iron-Formation: Facts and Problems.* Elsevier, Amsterdam, pp. 1 – 12.

Trendall, A.F., 1983b. The Hamersley Basin. In: A.F. Trendall and R.C. Morris (Editors), *Iron-Formation: Facts and Problems.* Elsevier, Amsterdam, pp. 69 – 130.

Trendall, A.F. and Blockley, J.G., 1970. The iron-formations of the Precambrian Hamersley Group, Western Australia. *Geol. Surv. W. Austr., Bull.* 119: 366 pp.

Trendall, A.F. and Morris, R.G. (Editors), 1983. *Iron-Formation: Facts and Problems.* Elsevier, Amsterdam, 558 pp.

UNESCO, 1973. *Genesis of Precambrian Iron and Manganese Deposits.* Proc. Kiev Symp., 1970, UNESCO Earth Sci. Rep., 9: 382 pp.

Vaniman, D.T., Papike, J.J. and Labotka, T., 1980. Contact-metamorphic effects of the Stillwater Complex, Montana: The concordant iron formation. *Am. Mineral.*, 65: 1087 – 1102.

Walker, J.C.G., Klein, C., Schidlowski, M., Schopf, J.W., Stevenson, D.J. and Walter, M.R., 1983. Environmental evolution of the Archean – Early Proterozoic Earth. In: J.W. Schopf (Editor), *Earth's Earliest Biosphere.* Princeton University Press, Princeton, N.J., pp. 260 – 290.

Walter, M.R. and Hofmann, H.J., 1983. The palaeontology and palaeoecology of Precambrian iron-formations. In: A.F. Trendall and R.C. Morris (Editors), *Iron-Formation: Facts and Problems.* Elsevier, Amsterdam, pp. 373 – 400.

Wolff, J.F., 1917. Recent geologic developments on the Mesabi range, Minnesota. *Am. Inst. Min. Metall. Eng.*, 56: 142 – 169.

Yeo, G.M., 1986. Iron-formation in the late Proterozoic Rapitan Group, Yukon and Northwest Territories. In: J.A. Morin (Editor), *Mineral Deposits of the Northern Cordillera. Can. Inst. Min. Metall., Spec. Vol.*, 37: 142 – 153.

Young, G.M., 1976. Iron-formation and glaciogenic rocks of the Rapitan Group, Northwest Territories, Canada. *Precambrian Res.,* 3: 137 – 158.

Young, G.M., 1988. Proterozoic plate tectonics, glaciation, and iron-formations. *Sed. Geol.*, 58: 127 – 144.

Young, T.P. and Taylor, W.E.G. (Editors), 1989. Phanerozoic Ironstones. *Geol. Soc., Spec. Publ.,* 46: 251 pp.

Chapter 12

PALEOSOL RECOGNITION: A GUIDE TO EARLY DIAGENESIS IN TERRESTRIAL SETTINGS

V. P. WRIGHT

INTRODUCTION

A wide spectrum of early diagenetic processes take place in terrestrial settings. Weathering is the most obvious of these and leads to the breakdown of material (see reviews in Martini and Chesworth, 1991). Much of the reorganization of surface materials, however, takes place within soil profiles. Extensive alteration can also take place in shallow groundwater zones (Nesbitt and Young, 1989). Whereas paleo-weathering horizons have been recognized, even widely in the Precambrian (e.g., Holland et al., 1989; Palmer et al., 1989), the abundance of paleosols in the non-marine stratigraphic record is only now becoming clear. There is a very good reason for the abundance of such fossil soils. The average sedimentation rates in terrestrial settings are low, typically in the order of a few millimeters or less per year. As a consequence, sediments will have a "residence time" of tens of years to hundreds of thousands of years within the upper part of the weathering profile. Within this zone, effectively the zone of soil formation, the sediments can be radically modified by a veriety of biological, chemical and physical processes associated with pedogenesis. These constitute a distinctive type of early diagenesis and, as such, should be regarded as an integral part of the diagenetic spectrum. It is essential that such early diagenetic effects be recognized and differentiated from other types of diagenesis. Recently Nesbitt and Young (1989) have reviewed the chemistry of weathering profiles and geochemical differences between groundwater and weathering processes.

Appreciating the role of pedogenically-induced early diagenesis will depend on the recognition of paleosols in the geological record. The aim of this chapter is to review the criteria for their recognition, and also to discuss some of the early, post-pedogenic but shallow-burial alteration processes to which paleosols are susceptible. The chapter will stress those aspects relevant to studies of older, generally lithified paleosols and takes a sedimentological rather than a pedological approach.

The problems associated with using various criteria will be discussed and aspects which are in urgent need of a more thorough evaluation will be stressed. The basis of this chapter owes much to the review of Quaternary paleosol recognition by Fenwick (1985).

PALEOSOLS

Paleosols, that is, soils that formed on a landscape of the past (Valentine and

Dalrymple, 1976), are now being widely recognized in the geological record and are potentially very powerful tools for a variety of paleoenvironmental purposes. They are especially common in ancient alluvial sequences but are also widespread in lacustrine, deltaic and marginal marine deposits, including shallow-water carbonates. Besides their intrinsic importance of indicating phases of subaerial exposure and early modification, they are critical for deciphering paleoclimates and paleogeomorphology, and have also been put to other uses such as time resolution (Kraus and Bown, 1986) and basin analysis (Allen, 1974). They have also been used as indicators of porosity evolution in ancient limestones (Wright, 1988).

The bulk of the geological studies of paleosols have concentrated on calcareous forms (caliche or calcretes), which are especially common in red-bed alluvial deposits. Other types of paleosols have been recognized for some time but it is only relatively recently that they have been studied in detail. Criteria for recognizing paleosols in pre-Quaternary deposits are not widely appreciated but have been reviewed briefly by Retallack (1981, 1983a, 1988, 1990), who has done much to stimulate interest in the study of such paleosols.

Three main types of paleosols can be recognized: relict, buried, and exhumed. The former type represents a soil formed under one set of environmental conditions and now found in a landscape influenced by different conditions. Buried paleosols are the type encountered by geologists working in the rock record. These are soils buried by younger deposits. If such soils become re-exposed and incorporated into younger soils they are referred to as exhumed paleosols.

It is relevant here to consider what is meant by a soil in the definition of a paleosol. A useful definition has been provided by Birkeland (1984): "A soil is a natural body consisting of layers of horizons of mineral and/or organic constituents of variable thicknesses, which differ from the parent material in their morphological, physical, chemical and mineralogical properties and their biological characteristics; at least some of these properties are pedogenic". It is a matter of emphasis when deciding at what point to distinguish weathering profiles from soil profiles. Some workers may feel that evidence of biological activity would be essential in order to recognize a soil profile, but the fossilization potential of many organic structures (e.g., roots) is low in many soil types so this criterion is not the best.

The subdiscipline of paleopedology has received much attention in recent years and a series of reviews have appeared (Yaalon, 1971; Boardman, 1984; Wright, 1986a; Reinhardt and Sigleo, 1988) and specific papers on terminology, concepts and paleosol stratigraphy have been provided by Finkl (1980), Morrison (1978) and Pawluk (1978).

PALEOPEDOGENESIS – THE NORM OR THE EXCEPTION

The degree of pedogenic modification that a terrestrial sediment will undergo depends on the residence time of the material within the zone of active soil formation, and the rate of pedogenesis. The residence time depends on the sedimentation rate and the thickness of that zone.

If we take a floodplain which has an "average" sedimentation rate of 2 mm y^{-1}

(Bridge, 1984), and assume the zone of significant pedogenesis is 2 m thick (Fig. 12-1), then the average floodplain sediment will have a residence time of 1000 years in the soil before it is progressively buried and isolated. This is enough time to radically alter some sediment properties, under certain soil conditions. By this argument most floodplain sediments should have received some pedogenic modification resulting in rudimentary soil profiles. This time period, however, is not enough for more mature profiles to form and the abundance of mature paleosols in alluvial sequences is evidence of the punctuated nature of sedimentation (Kraus and Bown, 1986).

It is usual for sedimentologists to be able to recognize the more strongly developed profiles, but most floodplain deposits should be regarded as having been pedogenically modified to lesser or greater degrees. The absence of pedogenic overprinting in such deposits may reflect high rates of sedimentation, low rates of pedogenesis, the loss of evidence by diagenetic overprinting, or simply that the evidence has been missed by the observer or misinterpreted as a burial diagenetic effect. The opposite case is where floodplain deposits contain evidence of pedogenic processes but which actually represent reworked soil material. The widespread parna and related deposits of Australia are a striking example of this (Nanson et al., 1986; Nanson and Rust, 1988; Rust and Nanson, 1989).

Figure 12-1 is a residence graph for terrestrial deposits and shows the time periods required for the formation of some of the commonest features found in paleosols.

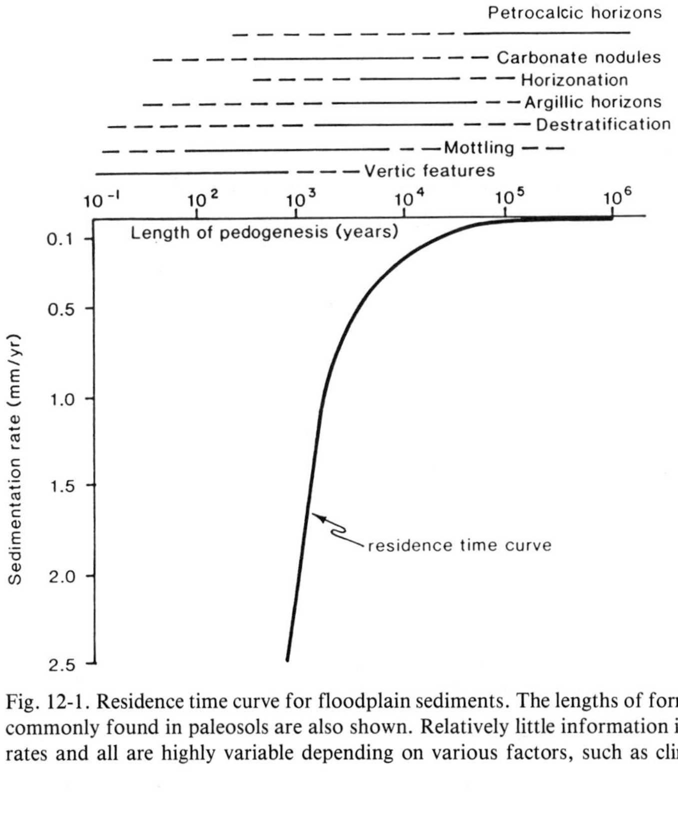

Fig. 12-1. Residence time curve for floodplain sediments. The lengths of formation of pedogenic features commonly found in paleosols are also shown. Relatively little information is available on many of these rates and all are highly variable depending on various factors, such as climate, parent material, etc.

These are further discussed in the text. It must be stressed that all the rates are highly variable in most cases with at least one order of magnitude variation. Birkeland (1984) has reviewed many of the rates of these processes and the reader is referred to that text for further details.

CRITERIA FOR RECOGNIZING PALEOSOLS

A variety of criteria is used for recognizing paleosols. Some are not replicated by sedimentological or other diagenetic processes and so provide unique and diagnostic evidence. These are: biological features, color, destratification, horizonation and boundaries, granulometrics, mineral assemblages, macrostructures, and micromorphology.

Biogenic features

A diverse suite of biogenic features have been used to identify and interpret paleosols. Retallack (1984a) provided a review of some of the features and discussed the taphonomic aspects of both soils and paleosols. The most commonly seen features are rootlets, root moulds or casts, or rhizocretions (permineralized or mineral-encrusted roots), and such occurrences provide clear evidence that rooted vegetation once grew in that soil. Root structures, however, have a low preservation potential in many soil types, and Lower Palaeozoic and older paleosols, formed before vasular plants evolved, will lack such features. The absence of roots in post-Lower Palaeozoic paleosols is principally the result of their decomposition by microbial activity. It is usually the case that well-preserved roots with organic remains occur only in paleosols formed where hydromorphism led to a decrease in the rate of decomposition, allowing the organic matter to be buried before complete decay. As a generality, it is reasonable to state that most paleosols do not exhibit obvious fossil root structures. Types of evidence for an original vegetation cover which can be found include phytoliths, which are opaline silica structures particularly associated with grasses (Wilding and Drees, 1972). Their study is still at an early stage but they may prove useful as general indicators of vegetation types in late Tertiary paleosols, when grasses evolved. Other evidence of vegetation in paleosols includes features such as the endocarps of the hackberry (*Celtis occidentalis*) which are relatively abundant in Tertiary paleosols in North America (Thomassen, 1979; Retallack, 1986).

Evidence for faunal activity has also been widely reported and includes remains of terrestrial molluscs (gastropods), vertebrates and arthropods. Trace fossil evidence is also useful and again reported examples include calcified pupal cases of insects, burrows, and even whole termite nests (e.g., Bown and Kraus, 1983; Wright, 1983; Retallack, 1984b). Coprolites occur in many paleosols and even calcified humus layers have been recorded from early Carboniferous paleosols allowing speculations on the different soil communities based on the different types of paleohumus found (Wright, 1987). Traces of paleosol microbes have been described from material as old as the early Carboniferous (Wright, 1986c) and even the Cambrian (Southgate, 1986).

The surface layers of a biologically active soil are enriched with organic matter. Such material has a low preservation potential in many soil environments but the presence of organically-enriched horizons, showing darkening by organic staining (melanization), especially when associated with other pedogenic criteria, is a useful guide to identifying paleosols.

A detailed review of these biological criteria lies outside the framework of this study, and such features are usually relatively obvious.

Color of paleosols

Many paleosols are identified initially at outcrop because of their distinctive coloring (e.g., red coloration), or because of localized color variations such as mottling. Reddening is commonly associated with soil processes and results from the oxidation of iron to form hematite. This process, known as rubifaction, is widespread in many soil types (Duchaufour, 1982; Fitzpatrick, 1987; Schwertmann, 1987), but in older deposits it can also result from early diagenetic processes (Walker et al., 1978). The degree of reddening caused by rubifaction will be dependent on the length of exposure and on the climate during formation, especially temperature (Birkeland, 1984). Many ancient red paleosols have had their red pigmentation formed by the diagenetic alteration of iron hydroxides (such as goethite) to hematite. One characteristic of pedogenic reddening is that it typically increases upwards within a profile and has diffuse lower boundaries (Fenwick, 1985). This also applies to organic staining (melanization). In temperate-zone soils, however, hematite is replaced by goethite in the upper horizons so that the "reddening" decreases upward (Schwertmann, 1987). In general, the interpretation of Fe-coloration in paleosols must be made with extreme care.

Other colors occur in paleosols. Grey to white coloration is typically associated with eluvial (leached) horizons such as albic horizons (Percival, 1986). In these cases, the removal of clays and free iron oxides results in the horizon's color being determined by the color of the coarser quartzose sediment component. In highly reducing settings, such as waterlogged (hydromorphic) or gley soils, reduced iron will impart a darker coloration (greens, blues and blue-greys) (e.g., Lehman, 1989), as will organic matter which can accumulate due to anoxia.

One of the most distinctive aspects of some paleosols is color mottling reflecting localized changes in oxidation and reduction (Duchaufour, 1982). Periodic waterlogging can create reduction within the soil profile and, in contrast, a lowering of the local water table can result in oxidation. Localized zones of oxidation can occur in zones of increased porosity in the soil, such as natural fracture surfaces between soil aggregates (peds), resulting in "gley" mottling with local zones of oxidation within a groundmass of "reduced" soil. The opposite situation also exists, where local zones of waterlogging occur where waters penetrate into cracks. In this case colors indicative of iron reduction occur around the permeable areas, in a groundmass of oxidized soil (Fig. 12-2). Such horizons are referred to as pseudogleys. Many burrows and root moulds in paleosols exhibit such "drab haloes" (Retallack, 1988). These zones result from enhanced microbial reduction around the

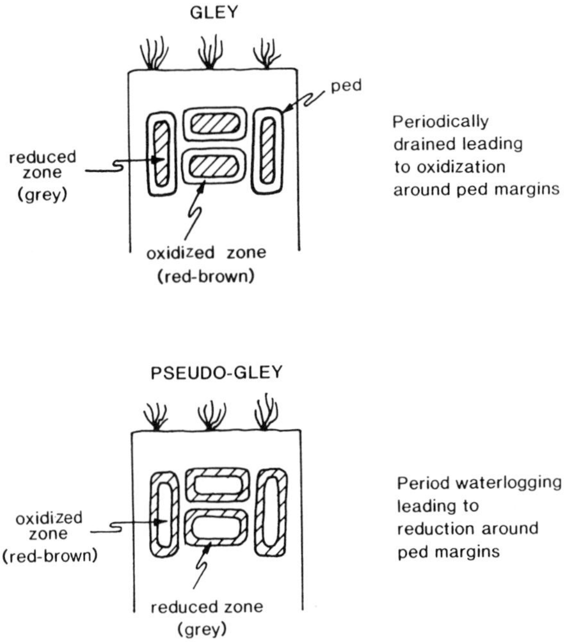

Fig. 12-2. Mottling styles in gleys and pseudo-gleys.

organic-rich tubule zone during hydromorphy, probably caused by rising water tables during floodplain aggradation (accumulative hydromorphy, see below).

It must be stressed that mottling is a very common feature of many paleosols; yet, as pointed out by Yaalon (1971), such features form by reversible processes in soils and are readily susceptible to alteration. It appears that mottling is so commonly seen in paleosols because it develops relatively rapidly and so even short periods of pedogenesis will create mottles (Fig. 12-1). The remarkable persistence of such features, even in Precambrian paleosols, therefore, requires an explanation. Dr. Bernard Besly (pers. commun., 1989) has offered the suggestion that mottling is well preserved in paleosols in mud-rocks because burial and compaction isolate the pigments from reaction with groundwaters, and also because most paleosols rapidly lose their organic matter on burial, and so subsequent organic-matter-driven reduction does not take place.

Destratification

Several processes control the reorganization of sedimentary materials in soils, but from a sedimentological point of view, destratification is one of the most easily recognized. It can, of course, occur in a variety of settings and is most usually due to bioturbation. In pedology, this physical destratification is called pedoturbation. A number of processes are responsible for this mixing (or haploidizing) of the soil. Bioturbation results from the direct action of infaunal or semi-faunal animals or to root activity. Tree uprooting (heave) by tree decay and wind activity, also causes mixing of the soil (Seminiuk, 1986). The shrinking and swelling of expandable clays (argilliturbation) can create rapid mixing in some soils and results in a variety of macro-structures commonly seen in paleosols (see below). Salt crystal growth (crystalturbation) is especially important in soils where calcium carbonate or gypsum forms. Freezing and thawing (cryoturbation) is significant in cold regions and disturbance caused by soil gas movement (aeroturbation) can also modify surface layers (Buol et al., 1980).

The net result of these processes will be the destruction of the original sedimentary anisotropy in the upper part of the profile (the upper meter or so). Thus the sedimentary layering, structures and grain-size trends will be destroyed.

An understanding of the rates of destratification in soils would be of considerable importance in interpreting ancient profiles, but relatively little information is available. Studies of subaqueous to intertidal marine and lacustrine sediments, show that bioturbation can be a rapid process, as is the rate of argilliturbation in some soils (Yaalon and Kalmar, 1978). Harden (1982), in a study of a sequence of alluvial terrace soils in central California, found that some soils required 10,000 years to lose primary sedimentary layering. This is a surprisingly long period of time when compared to seemingly equally-biologically active marine sediments. For example Gagan et al. (1988) recorded the complete obliteration of offshore storm deposits, centimeters thick, on the central Great Barrier Reef shelf within only one year of deposition. Further work on the rates of biological destratification are urgently needed, although applying these results to, for example, pre-Carboniferous paleosols will be difficult. Considering the average residence times of floodplain deposits within the soil zone (1000 years or so), some signs of bioturbation should be present.

Horizonation and boundaries

Paleosols are commonly recognized in the field because of the development of prominent horizons with different properties (typically color or degree of cementation). Horizonation (Hole, 1961) refers to processes of soil formation which lead to the differentiation of the soil profile, with various distinctive horizons definable because of different biological, physical and chemical properties. This differentiation is controlled by mainly vertically-directed reorganization processes.

Horizonation is the opposite to haploidization (mixing) but both sets of processes take place at the same time (Johnson and Watson-Stegner, 1987). In younger soils the latter dominates and destratification occurs but, generally speaking, with time, the processes of differentiation become dominant and horizonation takes place. The time scale for horizon differentiation varies with soil factors (Birkeland, 1984). Harden (1982), studying alluvial terrace soils of California, found that mixing processes dominated in soils younger than 40,000 years old, after which time horizonation became dominant (Fig. 12-3). In some soils mixing processes dominate over the long term, as in Vertisols (Soil Survey Staff, 1975), which are clay-rich soils with smectitic layer silicates that undergo seasonal movements caused by wetting and drying cycles.

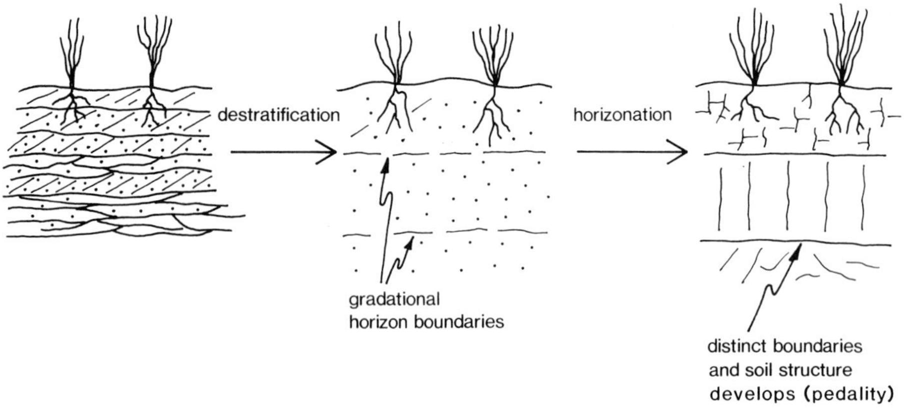

Fig. 12-3. Structural changes during pedogenesis. Initially the sediment is destratified (haploidized) but anisotropy later develops as horizonation occurs. With time, horizon distinctness typically increases. Distinctive soil structure (pedality) also develops. In some soils, such as Vertisols, mixing processes are dominant. (Modified from Harden, 1982.)

The degree of horizon development can be used to estimate the period of pedogenesis. Birkeland (1984) has recognized three main categories of soil development, depending on the degrees and types of horizons, whereas Retallack (1988) has defined five categories.

Pedogenic indices have been developed by soil scientists (Bilzi and Ciolkosz, 1977; Harden, 1982) which stress the differences between the parent material or C horizon (base of soil profile still possessing some inherited features of the weathered parent material) and the overlying horizons (Relative Profile Development or RPD indices).

This type of approach has been widely used in ancient "red-bed" alluvial sequences possessing paleocalcrete horizons, of which the studies by Allen (1986) on the Old Red Sandstone of southern Britain, and Steel (1974) on the New Red Sandstone of Scotland represent classic studies. Calcrete soils develop in a series of stages (Gile et al., 1966; Machette, 1985) from immature weakly calcareous forms to ones with prominent, indurated and impermeable carbonate horizons (Fig. 12-4). Such horizons are widely recognized in ancient alluvial sequences (Fig. 12-5).

A similar approach, that of comparing the degree of C horizon (parent material) modification, has been used by Wright (1987) to recognize time-related sequences in early Carboniferous xero-rendzina calcretes.

Another aspect of horizonation useful for sedimentological analysis is that of relative horizon distinctness (RHD; Bilzi and Ciolkosz, 1977; Meixner and Singer, 1981). In this case, two adjacent horizons are compared using quantitative indices. Distinctness reflects the length of pedogenesis but may also indicate discontinuities in the profile with different horizons separated by periods of deposition or erosion. In the author's experience such sedimentological hiatuses are very common in paleosol sequences and should be differentiated from purely pedogenic horizon distinctness. For example, in a study of apparently thick paleosol profiles in early

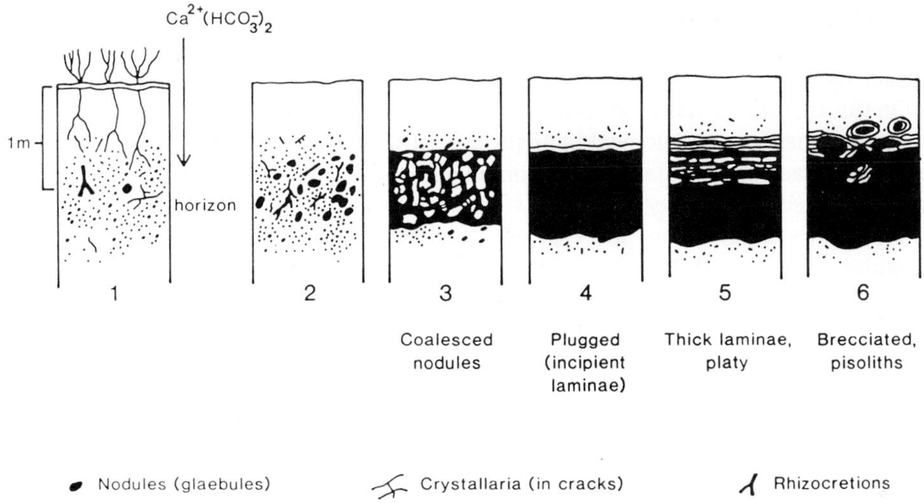

Fig. 12-4. Stages of calcrete development. (Based on Machette, 1985.)

Fig. 12-5. Paleo-calcrete. This is a stage 2 – 3 calcrete profile developed in alluvial fan deposits, Upper Jurassic, Porto Novo, Santa Cruz, Portugal. The overlying, erosive-based, channel sandstone shows soft-sediment deformation features.

Carboniferous paleosols in South Wales, Wright and Robinson (1988), using clay mineral analyses, were able to show that two sharply-defined paleosol horizons were in fact parts of two distinctly different profiles, superimposed by erosion.

On this theme, the nature of boundary types in paleosols is a critical factor to consider. Sedimentary lithological changes are either sharply defined, especially with erosive contacts, or regularly gradational (such as in graded bedding). Bioturbation can, of course, create irregular and diffuse boundaries. Whereas horizon boundaries can be sharp in soils, they are typically gradational or diffuse, and can be highly irregular in form. Mottling or vertical interdigitations are commonly seen. Soil boundaries will cross-cut primary lithological boundaries indicating their secondary origin, and such non-depositional contacts are a particularly useful criterion for identifying alteration zones in lower Palaeozoic or older paleosols. The paleosol, if developed on already-indurated material, may pass into the bedrock via a distinctive regolith or saprolite zone with evidence of progressive loss of the original lithological structure up through the transition horizon. As a result of subsequent erosion, only the regolith or saprolite horizon may be left as evidence of subaerial exposure. In many cratonic regions these weathering zones may be tens of meters thick or more and are probably very common, though rarely recognized, at major unconformity surfaces in the geological record.

Granulometrics

Pedogenic processes (transformations, pedoturbations and translocations) create physical (particle size), mineralogical and chemical gradients in soil profiles. The removal of materials from one horizon (eluviation) and their subsequent concentration at a lower horizon (illuviation) is an important process in many soils. Several processes operate to create particle-size gradients including the preferential breakdown of grains at certain levels in the soil profiles and also by the formation of authigenic clays (Duchaufour, 1982). The movement (translocation) of the finer particles in suspension (a process termed lessivage) creates horizons in the profile of greater concentration of clay-grade materials to form illuvial argillic "B" horizons. This illuviation is most usefully detected using the ratio of fine ($< 1\ \mu$m) to coarse ($> 1\ \mu$m) clay fraction (Khalifa and Buol, 1968), although coarser particles, up to silt-size, can also be translocated in suspension. Argillic horizons form slowly, 10,000 years or more typically being required (Birkeland, 1984), and their presence in paleosols indicates prolonged pauses in sedimentation (Fig. 12-1).

In soils where mixing (haploidizing) processes dominate, no grain-size trends will develop. Vertisols, which occur relatively commonly as paleosols, are soils developed on clay-rich parent materials with an original high shrink – swell capacity, or on materials with the capacity to produce, by weathering, clays with such a capacity, and, therefore, undergo often intense mixing by argilliturbation. Surprisingly, such soils may apparently show an upward movement of the coarser fraction (fine sand – silt) because of the upwardly-oriented stress caused by differential wetting (Yaalon and Kalmar, 1978).

Grain-size profiles might prove useful in recognizing illuvial clay horizons in paleosols, especially in finer grained materials, but there are many problems in looking for such gradients. There are two basic methods which have been used: disaggregation and point counting. Totally disaggregating lithified paleosols may be difficult and it is rarely possible to separate diagenetic phases. Point counting has been used in lithified materials (Retallack, 1983a,b; Percival, 1986), but problems arise in estimating the finer fractions (Murphy and Kemp, 1984) and it may be impossible to differentiate soil materials from later diagenetic phases, such as clays formed during burial diagenesis.

There are additional problems associated with the recognition of particle-size gradients in paleosols (Fenwick, 1985). Firstly, additions of sediment, by wind action or flooding, can mask the pedogenic trends. Secondly, not all deposits are initially homogeneous in particle-size distribution; for example, marked grain-size variations occur across floodplains and as a result of environmental changes the floodplain sediments may exhibit a variety of purely depositional vertical trends (Ruhe, 1975). Similar trends, attributable to sedimentary and not pedogenic processes, have been recorded from early Carboniferous floodplain paleosols by Wright and Robinson (1988).

Other grain properties can be used, although not for field recognition. Grain morphology (Jenkins, 1985) is a powerful tool for identifying weathering processes in Quaternary paleosols, but does not appear to have been used in older paleosols.

Mineralogy and geochemistry

Mineralogy and geochemistry are particularly useful criteria, especially for detecting "weathering" gradients, and Jenkins (1985) has reviewed the use of these techniques in Quaternary paleopedology.

Both the mineral content and chemistry can be used to detect the previous activity of pedogenic processes. The basic factor controlling these gradients is the rate of mineral decomposition, which is usually greatest in the upper soil horizons and decreases with depth. During weathering, various cations are released. Their distribution within a profile can be used to assess both the nature and degree of weathering: Fe, Al, P, Mn, Na, K, Ca and Si are measured, commonly as the oxides and oxyhydroxides. Either the actual amounts of these cations or oxides within the profile can be plotted against depth, or the ratios of mobile to immobile elements can be used. In the case of the latter (Fig. 12-6A), this ratio is lower in the upper part of the soil where leaching is greatest (Smith and Buol, 1968). Immobile elements include Ti and Zr, although in the case where the Ti is mostly derived from the weathering of biotite, some may be lost by leaching (Chartres et al., 1988). Some workers, however, have questioned the validity of using these "immobile" elements (Milnes and Fitzpatrick, 1989).

The ratio of mobile to immobile elements will also be affected by changes in the parent material with depth. In an aggrading soil, if material is being added from more than one source, marked variations in chemical gradient can occur due to changes in parent mineralogy (Fig. 12-6B).

Immobile elements can also be used to test the uniformity of parent materials. Zirconium and titanium are relatively immobile, and their ratio in the soil should provide an indication of their ratio in the parent sediment. Any variations in their

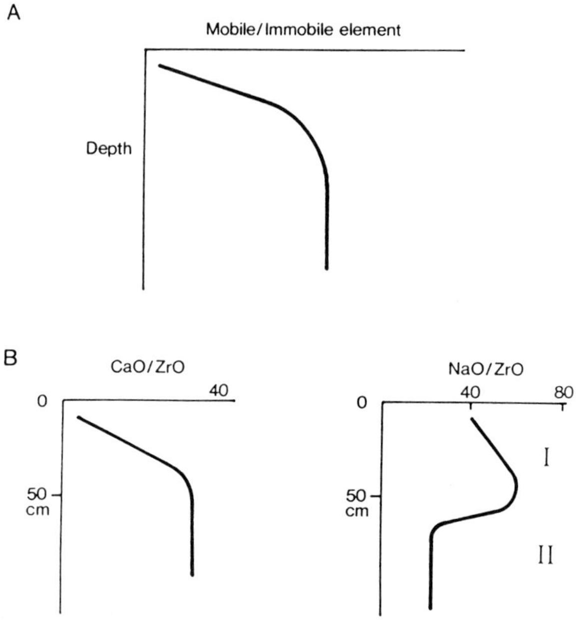

Fig. 12-6. Chemical profiles in soils. (A) The typical relationship for a mobile/immobile element ratio in a soil. The loss of the mobile element in the upper part reflects leaching. (B) Ratios from complex soils from Australia where the CaO/ZrO ratio shows a typical trend. Whereas the NaO/ZrO shows a similar trend in the upper part of the profile (*I*), it has a higher NaO content than the less leached lower part of the solum (*II*). This is believed to reflect the addition of higher amounts of sodic feldspar in the upper part of the profile from aeolian input. All graphs are for the 50 – 60 μm component. (Based on data in Chartres and Walker, 1988.)

ratios may reflect changes in the source sediment (Birkeland, 1984; Chartres et al., 1988).

If weathering conditions and the parent material composition remain constant, the loss of mobile species will be related to the length of pedogenesis. This is used in studies of "chronosequences" (sequences of soils whose differences relate to the length of pedogenesis: Vreeken, 1975), to gauge the relative ages of the component soils (Birkeland, 1984; Harden, 1987).

Applications of the general use of chemical gradients in studies of pre-Quaternary paleosols have been provided by Retallack (1983b, 1986) and Bown and Kraus (1987). In the case of older paleosols, lacking diagnostic biogenic features, such chemical "weathering" gradients are a powerful tool for recognition. They are typically sought in soils with silicate-rich parent materials, but they can also be recognized on carbonate parent materials. In this case, a substantial literature exists devoted to the recognition of subaerial exposure surfaces in limestone sequences, using trace element (Mg, Sr, Na) and stable isotope ($\delta^{18}O$, $\delta^{13}C$) gradients. James and Choquette (1984) have reviewed the uses. In these cases, Mg, Sr and Na are progressively lost from the parent material as the unstable aragonite (Sr-rich) and high magnesian calcite are leached out or replaced by low magnesian calcite. During these changes the marine $\delta^{18}O$ signature in the parent marine sediments is replaced by a meteoric signature causing enrichment in the lighter ^{16}O isotope. Light organic carbon is also introduced into the replacement calcites, derived from soil CO_2 and soil acids, as the meteoric water percolated through the overlying soils. As a result the new carbonate has a lighter $\delta^{13}C$ value, although this trend is generally limited to the uppermost part of the profile.

During weathering silicates are transformed to a variety of secondary products, but especially to clay minerals (Nesbitt and Young, 1989). These processes and products have been intensively studied by soil scientists and an enormous literature exists (see volume edited by Dixon and Weed, 1989). Clay mineralogy has been widely used to detect paleosols, in particular by recognizing highly-leached clays such as kaolinite (e.g., Goldbery, 1982a; Meyers, 1981). Palygorskite has also been found in paleosols and is a useful palaeoenvironmental indicator (e.g., Singer 1985; Thiry, 1989). Smectite can be a useful mineral in paleosol interpretation, but diagenetic overprinting is a particular problem, whereby progressive illitization is caused by increased depth and heating. Two recent studies (Robinson and Wright, 1987; Deconinck and Strasser, 1987), however, have documented a potentially useful ordered illite/smectite interlayered clay, from Carboniferous and Jurassic paleosols from South Wales and mainland Europe, respectively. These pedogenically-illitized smectites formed by K-fixation caused by wetting and drying cycles, and not by burial illitization. Robinson and Wright (1987) found this clay-type in well-developed paleo-Vertisols, confirming the interpretation of formation under a soil moisture regime with wetting and drying cycles. Recently, Singer (1988) has also stressed the importance of pedogenic illitization related to such cycles in Quaternary soils.

Iron and manganese compounds can also be used to indicate specific soil processes (Fitzpatrick, 1987). Mineral concentrations through pedogenic processes are most strikingly developed in duricrusts. These are prominent accumulations of iron and aluminium oxides and oxyhydroxides (laterites and bauxites), silica (silcretes), calcium carbonate (calcrete) or gypsum (gypcrete). These various types have been reviewed in chapters in Martin and Chesworth (1991), and Goudie and Pye (1983). Fossil duricrusts have been widely documented in paleosols and examples are to be found in several volumes of case studies (Wilson, 1983; Wright, 1986a; Reinhardt and Sigleo, 1988). Although similar concentrations can occur in other geological situations, there are usually sufficient indications for pedogenic activity to confirm their origins as duricrusts.

Many near-surface mineral concentrations, however, are not, strictly speaking, pedogenic but are formed in the shallow phreatic zone or within the lower part of the vadose zone, within the capillary fringe. These groundwater-related concentrations are discussed below. This setting represents the transition realm between pedogenic and shallow burial diagenesis and is a much neglected environment.

Macro-structural features

There are a variety of distinctive structures which form in soils and have been recognized in paleosols. These can be reliable indicators of pedogenesis because

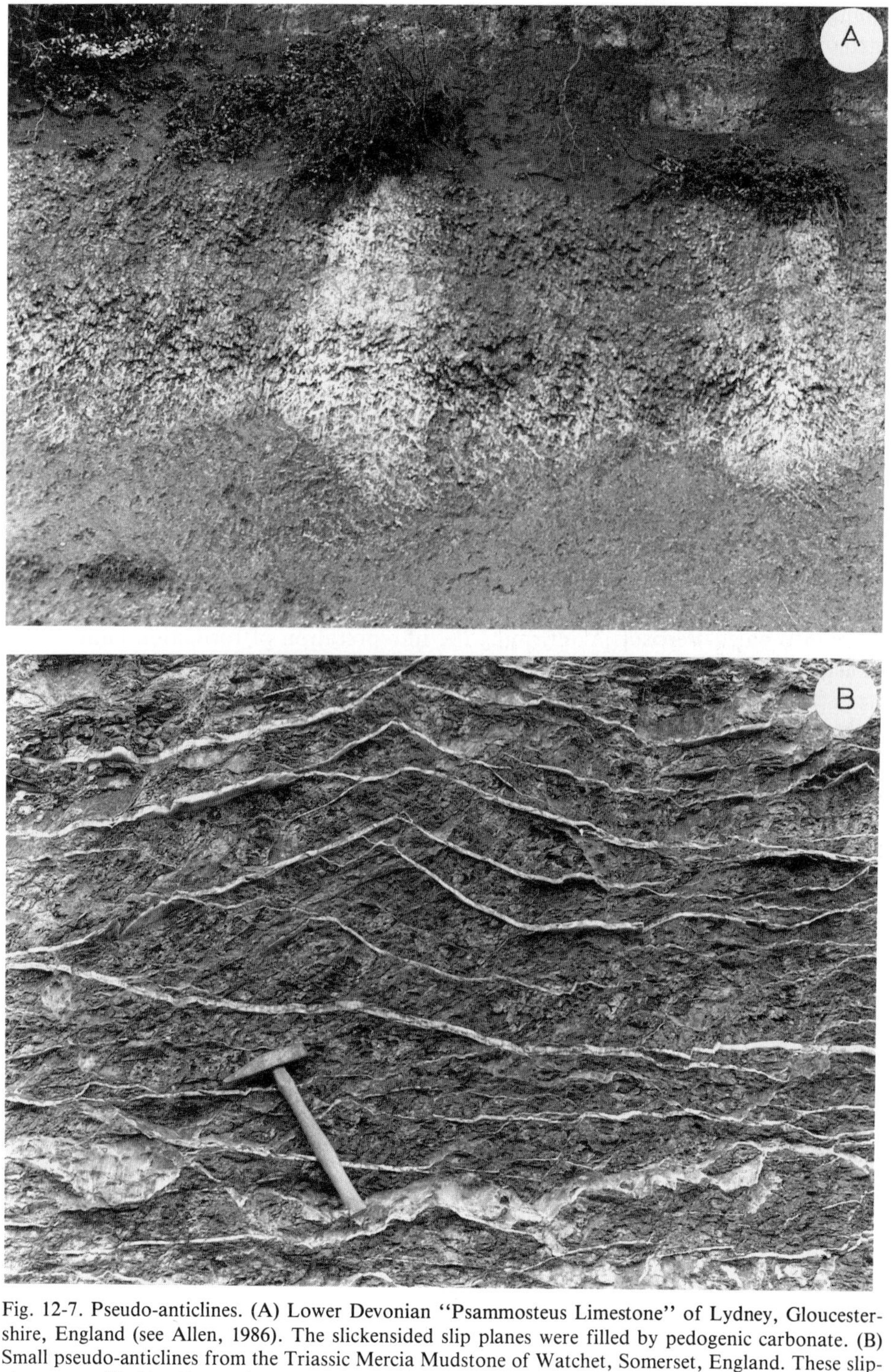

Fig. 12-7. Pseudo-anticlines. (A) Lower Devonian "Psammosteus Limestone" of Lydney, Gloucestershire, England (see Allen, 1986). The slickensided slip planes were filled by pedogenic carbonate. (B) Small pseudo-anticlines from the Triassic Mercia Mudstone of Watchet, Somerset, England. These slip-planes were filled by diagenetic gypsum which became mobilized during Tertiary uplift. In this case, these paleo-Vertisols developed on playa flats and were only wetted during rare floods. See also Fig. 12-9B.

similar features do not seem to form by sedimentological or diagenetic processes. In addition, many of these features are readily identifiable at outcrop. The typical ped structures seen in present day soils, however, are not preserved in most paleosols as a consequence of compaction. In soils, as a general rule, ped size increases with depth, e.g., from finely granular to blocky to prismatic. Remnants of such vertical differentiation might be found in some paleosols, especially where lithification occurred before burial. Calcrete horizons from the early Carboniferous of South Wales show a vertical change in ped shape and size (Wright, 1982).

Pseudo-anticlinal structures have been described from a variety of paleosols (Goldbery, 1982b; Wright, 1982; Blodgett, 1984; Allen, 1986; Retallack, 1986; Wright and Robinson, 1988). These structures should not be confused with carbonate tepees in peritidal or lake margin deposits with artesian flows (Kendall and

Fig. 12-8. Pedogenic slickensides developed within early Carboniferous paleo-Vertisols, Llanelly Quarry, South Wales. (See Wright and Robinson, 1988.)

Warren, 1987), or with calcrete (or caliche) tepees which mainly result from the displacive deformation of indurated soil carbonate layers (Esteban and Klappa, 1983). These pseudo-anticlinal features are defined by sets of parallel, often slickensided, fracture surfaces (commonly later infilled by calcite) arranged in broad, gently sloping synclines and steep, cuspate anticlines (Figs. 12-7 and 12-8). The wavelengths range from 2 to 10 meters, with amplitudes of indivivudal sets being over a meter. Such structures have been extensively documented from present-day soils which have a high content of swelling clays (smectite) and classified as Vertisols (Yaalon and Kalmar, 1978; Wilding and Tessier, 1988, and references therein). The occurrence of such features in soil paleosols indicates a high swelling-clay content in the original soil and formation under a strongly soil moisture regime, with locally impeded drainage (Ahmad, 1983).

The reports of a surprisingly large number of paleosols with pseudo-anticlinal structures requires comment. Yaalon (1971) and Wilding and Tessier (1988) amongst others, have stressed that these vertic structural features can form very rapidly (5 – 200 years). In rapidly aggrading situations (Fig. 12-1), these features will form and be buried rapidly, and thus have a high preservation potential.

Other distinctive features include columnar and prismatic structures (Fig. 12-9). These are vertically elongate features, typically a few centimeters in diameter and up to a meter high. They occur in a wide variety of soil types and paleosols. Both varieties mainly result from expansion and contraction associated with wetting and drying (Fitzpatrick, 1983).

Calcretes commonly exhibit columnar and prismatic structures as well as a host of other features (Netterberg, 1980; Goudie, 1983), such as crystallaria (early, crystal-filled fractures), honey-comb structure, laminar cap layers, and pisoids. All of these have been widely recognized in paleo-calcrete horizons (Hubert, 1977; McPherson, 1979; Freytet and Plaziat, 1982; Wright, 1982; Allen, 1986).

Some of the most striking features described from paleosols are the crescentic clay structures in Upper Carboniferous paleosols of northern England (Percival, 1986). These consist of low-amplitude (usually several millimeters) but long wavelength (up to tens of centimeters) structures with a convex-up top surface and convex-up to flat lower surface. These structures are up to a few centimeters thick, are composed of clays and are typically parallel to the bedding. The laminae occur in present-day soils which have had clays removed (translocated) to lower horizons.

Soil microstructures (micromorphology)

Micromorphology (soil petrography) has proved to be a very powerful technique for recognizing paleosols and is one readily grasped by geologists used to examining thin-sections of rocks. The technique has been very successful in identifying a variety of paleosols in both calcareous (for example, Wright, 1983, 1987; Wright and Wilson, 1987) and non-calcareous settings (for example, Fastovsky and McSweeney, 1987; McSweeney and Fastovsky, 1987).

A well-illustrated review of calcrete-soil micromorphology is provided by Esteban and Klappa (1983), and texts on the micromorphology of soils include Brewer (1976), Bullock et al. (1985), and Fitzpatrick (1984).

Fig. 12-9. Columnar structures in paleosols. Columnar structures are a striking feature of many paleosols, especially paleo-calcretes. (A) Miocene calcretes showing columnar structure, Benues Formation, Benues, Spanish Pyrenees. (B) Columnar structure developed in saline paleo-Vertisols, from playa lake deposits of Triassic Mercia Mudstone, Watchet, Somerset (see also Fig. 12-7B).

Many microfabrics in soils have no known sedimentological or diagenetic analogues and appear, at least at present, to be criteria of true "diagnostic" reliability. Sepic fabrics are commonly reported from paleosols (e.g., Goldbery, 1982b; McSweeney and Fastovksy, 1987). They consist of patches or zones of soil material (typically clays) with various aligned and oriented extinction patterns, typically caused by stresses within the soil. Vertisols exhibit prominent stress fabrics of this type (Nettleton and Sleeman, 1985). It is possible, however, to imagine burial and structural modifications of clay-rich sediments which might mimic sepic stress fabrics, and there is a need to check for such features in non-pedified ancient mudstones.

Micromorphology can be compared directly to the use of petrographic analysis in sedimentary petrology. The aim of petrography in diagenetic studies is to formulate paragenetic sequences, where all the diagenetic features are related to one another in a "stratigraphic" context. Once such a sequence of events has been established, it can then be related to the burial history of the sediment and, if necessary, porosity-forming or occluding events can be timed in relation to phases of hydrocarbon generation and migration. The development of geothermometric techniques, such as the use of oxygen isotopes and fluid inclusions, has enabled actual depth values to be placed on stages in paragenetic sequences. In exactly the same way, studies of Quaternary soils have revealed similar pedo-"paragenetic" sequences reflecting environmental changes during the soil's history (Chartres, 1980; Rose et al., 1985). Micromorphology has been used to detect climatic changes during the Pleistocene of Britain, with alternating phases of clay illuviation (interglacials) and clay disruption (glacial phases with cryoturbation; Chartres, 1980). One of the most remarkable examples of the use of the technique is quoted by Macphail (1986) with reference to some French soils. These exhibit grain coatings (or cutans) and pore fills with three growth zones. The first consists of clear coats of tangentially oriented fine clay (limpid clay), followed by "dusty", silty clay coatings, and finally by poorly-sorted impure clays with charcoal and organic matter. These three stages are believed to represent clay illuviation under undisturbed woodland (limpid clay), followed by woodland clearance and a greater flow-through of water ("dusty" layer), followed by tillage and soil slaking (poorly-sorted pore fills).

Clearly, detecting anthropogenic effects like these is not the concern of the geologist, but it serves to illustrate the use of the technique to detect environmental change.

As soils evolve and as environmental changes occur, some soil properties, such as chemical gradients and mineral accumulations, may be remobilized and earlier patterns overprinted. Many micromorphological features, however, appear to have high preservation potentials and survive sufficiently to allow earlier soil phases to be recognized. It is the author's experience that most paleosols reveal a polygenetic history using careful petrographic techniques. For example, a common feature of early Carboniferous paleosols in South Wales are duplex soils, in which two distinctly different soil horizons are superimposed indicating at least two phases of soil development (Wright, 1987). Micromorphology confirms these relationships where calcified mineral humus (A) horizons of the early surface soil are incorporated as pedorelicts in the later petrocalcic (C) horizon (Fig. 12-10).

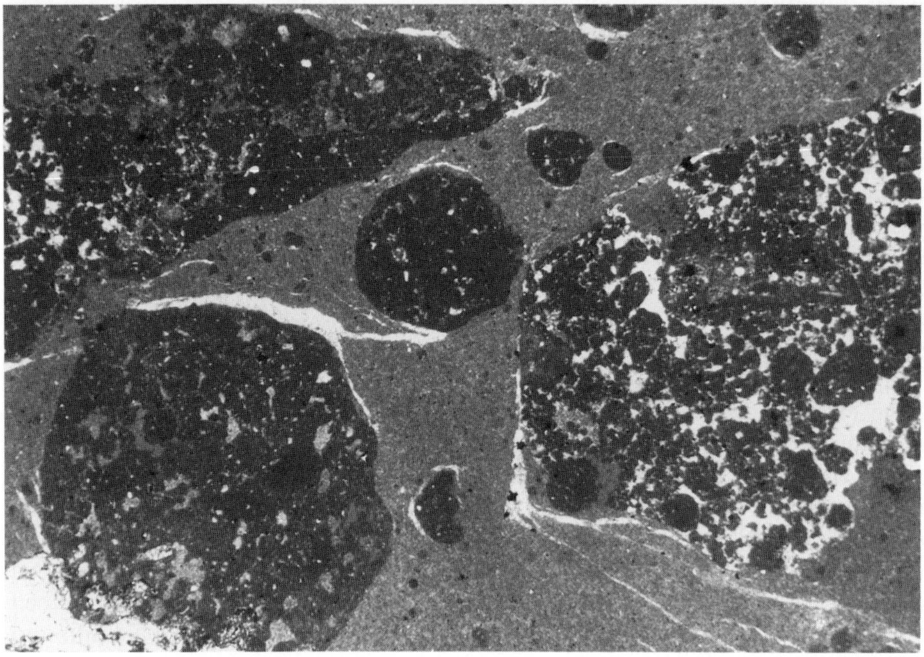

Fig. 12-10. Pedorelicts. These peloidal clasts represent calcified surface humus material from a xero rendzina-type paleosol. This soil was buried and overprinted by a thick petrocalcic horizon calcrete represented by the dense micritic matrix. The calcified humus pedorelicts have been partially impregnated by the calcrete matrix. There are circum-granular cracks around the peloid clasts; this is a common feature of present-day and ancient calcretes (Wright, 1982). The paleosol is the early Carboniferous Heatherslade Geosol from Miskin, South Wales (see Wright, 1987). Field of view is 12 mm wide.

PALEOSOL DIAGENESIS

Many of the features used to recognize paleosols are subtle and prone to alteration during diagenesis, but surprisingly little work has been carried out on these problems. The physical effects of burial diagenesis, such as compaction and pressure solution, are relatively easily recognized as overprints. As stated above, however, the possible effects of burial and structural overprinting in forming sepic-like fabrics in non-pedified materials require study. The effects of diagenesis on paleosol mineralogy is an even less well-understood aspect.

Many major changes can occur very early in the burial history of the paleosol, and two common situations require consideration. The first is that of an aggrading floodplain and second, a paralic setting during a transgression.

During floodplain aggradation any sediment deposited on the surface will pass through several phases (Fig. 12-11). Firstly, the sediment is incorporated into the "top soil". It will be destratified, particularly by bioturbation. It will become progressively buried and, under suitable conditions, may pass through an eluvial horizon into an illuvial one. Each phase will alter the sediment and overprint earlier phases. Eventually the level will become buried beneath the zone of active soil processes and will become a "buried" soil. Hayward (1985) refers to such a sequence as the "sediment to topsoil to subsoil to deposit" succession.

As a consequence of aggradation, the floodplain surface and the water table will rise. Initially, in areas near the alluvial ridge, the addition of sediment will cause the surface of the floodplain to rise more quickly than the mean water table. Soil drainage may actually improve with water-logging being replaced by better drained, possibly leached soils (Hayward, 1985; Farrell, 1987). With continued aggradation and compaction, however, the water table will rise and the buried soils will become water saturated (Fig. 12-11). This hydromorphic phase has been shown to be important in modifying the Eocene alluvial paleosols of Wyoming (Bown and Kraus, 1987), and the term "accumulative hydromorphy" was introduced by Bown and Kraus to describe such a situation.

The effects of such hydromorphism includes the loss of primary coloration, especially by Fe reduction, and the formation of drab haloes (see above). A fluctuating water table could result in pseudo-gley or gley features developing in the buried soil, later overprinted by gley features with continued burial. In the upper part of the phreatic zone, where lateral flow may occur, groundwater "weathering" can also take place. The processes and products associated with these phenomena are discussed in the next section, but the effects on paleosols have been described by Goldbery (1982a) from the Jurassic of Israel. In this case, both the paleosols and associated fluvial sediments were modified to bauxite at depth by early interstratal, groundwater-related weathering. Furthermore, groundwater ferricrete formation (not pedogenic ferricrete formation) has been noted in the Triassic Budra Formation of Sinai by Goldbery and Beyth (1984).

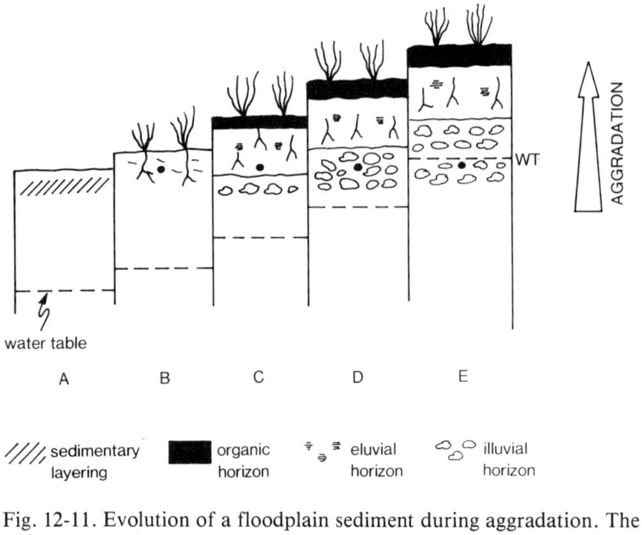

Fig. 12-11. Evolution of a floodplain sediment during aggradation. The initial sediment (*A*) becomes incorporated into the top soil (*B*) and is destratified. It may then pass through an eluvial layer (*C*) into an illuvial horizon (*D*). With continued aggradation, the water table will rise leading to accumulative hydromorphy (*E*).

In marginal marine sequences, such as deltaic or peritidal sequences paleosols are also relatively common. In such settings, rises in sea level will lead to a variety of hydrological changes. One initial effect is for the local water table to rise creating hydromorphic conditions. Further rises will result in the development of a brackish water phase, followed by a marine swamp or marsh, ahead of the transgressing sea. Such changes have been documented in Carboniferous deltaic and peritidal sequences by Love et al. (1983) and Wright (1986b), respectively. The complex geochemical changes which occur during these events have been reviewed by Curtis and Coleman (1986) and Buurman (1975) has described the general pedological changes in soils during drowning. Figure 12-12 shows the effects of a slow transgression on early Carboniferous floodplain paleosols from South Wales, described by Wright and Robinson (1988). The paleosols initially developed on the floodplains were calcrete-bearing Vertisols. These were replaced, at least locally, by leached kaolinite – smectite – chlorite-like clays also with vertic characteristics, such as deep desiccation cracks and pseudo-anticlinal slip planes. These clays represent a phase of ferrolytic weathering (in sense of Brinkman, 1977) caused by prolonged waterlogging and reducing conditions (Robinson and Wright, 1988). This may have been due to a change in climate or to a raised water table. Above this ferrolytic weathering unit there is a ferroan dolomite, whose isotopic composition indicates precipitation under brackish, reducing conditions (Wright et al., 1988). Pyrite is also abundant indicating sulphate-rich waters. The paleosol unit is overlain by low-energy marine limestones. Similar features occur in other marine-drowned paleosols, and

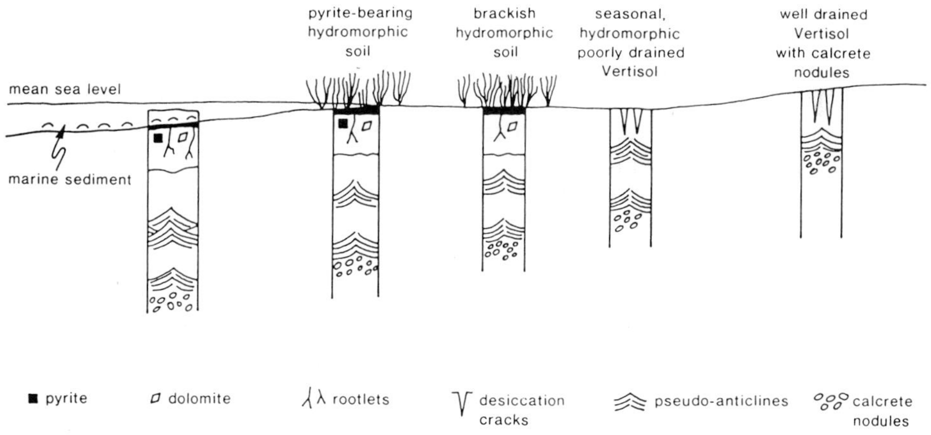

Fig. 12-12. Progressive marine hydromorphy related to a marine transgression. Such sequences are common in paralic deposits (see text). This sequence is based on the early Carboniferous Gilwern Clay paleosols of South Wales (Wright and Robinson, 1988). The initial floodplain soils were calcrete-bearing Vertisols. These are overlain by paleo-Vertisols showing evidence of ferrolytic weathering, which was caused by prolonged seasonal hydromorphism. This might reflect either a change to a more humid climate, a rise in the local water table, or to increased flooding. The latter two effects may have been caused by a marine transgression. These ferrolytic paleo-Vertisols are overlain by horizons with ferroan dolomite formed in brackish waters (Wright et al., 1988). Subsequent drowning led to the formation of pyrite in more marine waters, and to the covering of the paleosol complex by marine limestones. The geochemistry of such changes has been discussed in detail by Curtis and Coleman (1986).

this "progressive marine hydromorphy" sequence should be a common feature of paleosols in paralic sequences.

SHALLOW PHREATIC DIAGENESIS

Terrestrial sediments can be modified by active, shallow phreatic groundwaters (see previous section). These processes are sufficiently widespread in some settings to warrant further consideration, especially because some result in features very similar to those of soils. Duricrust phenomena, in particular, can be mimicked by shallow groundwater cementation (Arakel and McConchie, 1982; Goldbery 1982a).

In the case of groundwater (syn. phreatic, channel or valley) calcretes, which are strikingly like pedogenic forms (Arakel and McConchie, 1982), very large volumes of sediment can be cemented. Some of these deposits are enormous; for example, there are cemented Plio–Pleistocene alluvial fan gravels in Oman (the Wahiba Sands) which have been diagenetically altered (replaced) to dolomitic clays to depths of 200 m (Maizels, 1987). In Australia (Mann and Deutscher, 1978; Mann and Horwitz, 1979; Carlisle, 1983; Arakel, 1986) cemented alluvial deposits are commonly kilometers wide and tens of kilometers long. The maximum recorded size is 10 km wide by 100 km long, and their average thickness in Australia is 10 m. On a local scale they may be lensoid, and locally thickened zones occur as mounds or domes. They form from carbonate-rich, mobile groundwaters which become progressively concentrated during down-dip flow. The carbonate is precipitated mainly in the capillary fringe directly above moving subsurface water, but can also be precipitated below the water table. The precipitation of carbonate is due to several factors. The losses of CO_2 by degassing and evaporation are the major causes. As a result, cementation preferentially occurs at "highs" in the phreatic zone, where basement irregularities bring groundwaters near to the surface where degassing and evaporation occur. Where Ca and Mg carbonate-bearing waters mix with Ca or Mg sulphate- or chloride-rich playa groundwaters, precipitation occurs because of the common ion effect. They also preferentially form where drainages converge, flow gradients decrease, saline waters mix, or where permeabilities are low. The major features of the carbonate profile are shown in Fig. 12-13. Pedogenic calcretes may cap the sequence.

The resulting carbonate (calcrete or dolocrete) is typically micritic and densely crystalline, and authigenic silica, clays (sepiolite, palygorskite) and gypsum are commonly present. The growth of carbonate is both displacive and replacive so that nodular and massive forms occur. In some cases, shrinkage cracks and dissolution features are abundant. As cementation progresses, the zones become plugged and flow shifts laterally, thus creating wide, ribbon-like bodies. These cementation zones are responsible for relief inversion in many regions; the drainage systems become cemented, but later erosion (deflation) removes the uncemented interfluve sediments and the groundwater calcretes are left as areas of positive relief.

The formation of groundwater calcrete is only one part of the diagenetic spectrum documented from arid alluvial basins. As groundwaters move down regional flow lines, their ionic concentrations increase due to evaporation, and various mineral

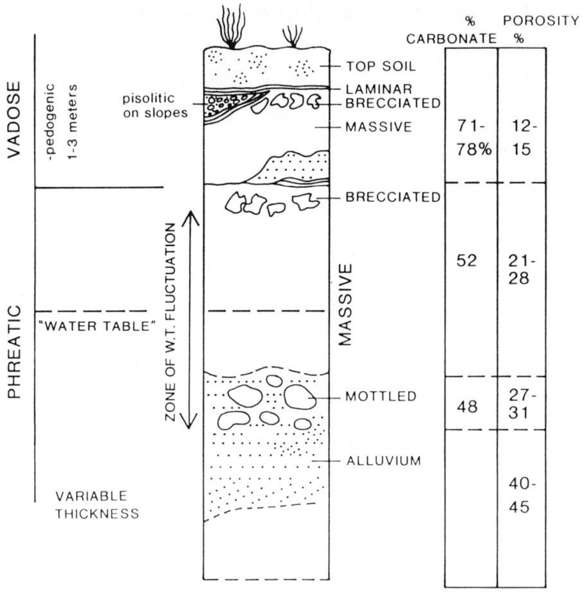

Fig. 12-13. Groundwater calcrete profile, based on Arakel (1986) and other sources (see text). Laminar calcretes also develop in the capillary fringe, associated with phreatophytic vegetation (Seminiuk and Meagher, 1981). As a consequence of water table fluctuations and phases of lowering by erosion, complex overprinting of the various zones can occur. Some workers (e.g., Carlisle, 1980) recognized two zones in the massive phreatic unit: an upper earthy zone, with remnant soil and alluvium, and a lower "porcellaneous" dense zone with abundant cracks and cavities.

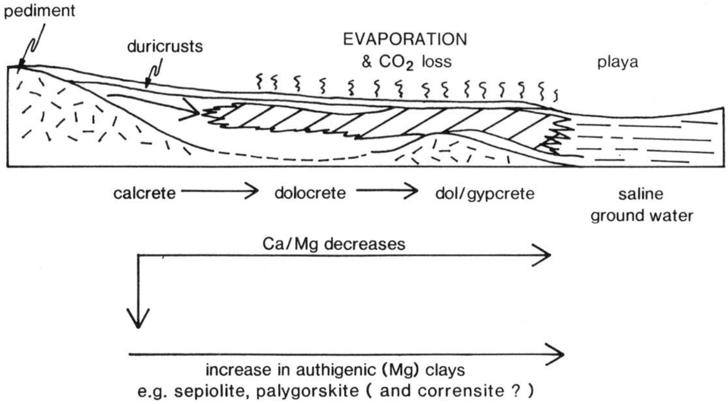

Fig. 12-14. Evolution of groundwaters and their precipitates across an arid alluvial basin (see text).

phases precipitate out. Total salinities, pH, and the concentrations of Ca, Mg, Na, K, U, V, SO_4 and Cl all increase down the drainage system (Carlisle, 1983; Arakel, 1986). In particular these groundwaters are able to both dolomitize and silicify the host sediment as well as precipitate Mg-rich clays such as sepiolite and palygorskite (Arakel, 1986; Fig. 12-14). These diagenetic zones are also commonly associated with economic mineral concentrations (Carlisle et al., 1978; Mann and Deutsher, 1978; Carlisle, 1980; 1983).

From such late Cenozoic deposits, it appears that very shallow groundwater diagenesis is a major factor affecting continental sediments in semi-arid to arid regions. Similar diagenetic processes must have operated in the past and represent part of the pedogenetic – eogenetic spectrum.

CONCLUSIONS

There are many paleosols which exhibit complex, polyphase histories, and as such provide unique store-houses of information if their histories can be deciphered. Arguably, no other sedimentological phenomena can provide so much, and such diverse, data as paleosols.

Paleosols can also be radically altered during shallow burial in zones of active shallow phreatic diagenesis. This setting has been neglected by sedimentologists studying the geological record and forms a transition zone between pedogenesis and diagenesis. The similarity of some of the processes in this zone and the soil results in similar products, especially in the case of duricrusts.

Pedogenesis should be seen as an integral aspect of a terrestrial sediment's post-depositional history, and not as some quirk of nature. To ignore pedogenesis, and likewise shallow phreatic alteration, when reconstructing diagenetic histories, is to omit the phase of most rapid change that a sediment undergoes during its post-depositional history.

ACKNOWLEDGEMENTS

I especially wish to thank the reviewers of the earlier version of this paper for their incisive and constructive comments: Colin Chartres (CSIRO Canberra), R. W. Fitzpatrick (CSIRO, Glen Osmond), Jane Francis (University of Adelaide) and Tony Milnes (CSIRO, Glen Osmond). Janet Binns and Elizabeth Wyeth skilfully typed the manuscript; Alan Cross and James Watkins prepared the illustrations. This chapter (Reading University, PRIS Contribution 024) is partly based on a chapter in PRIS Short Course Series no. 1, prepared for a course held at PRIS, University of Reading, May 1989.

REFERENCES

Ahmad, N., 1983. Vertisols. In: L.P. Wilding, N.E. Smeck and G.F. Hall (Editors), *Pedogenesis and Soil Taxonomy II. The Soil Orders.* Elsevier, Amsterdam, pp. 91 – 123.

Allen, J.R.L., 1974. Studies in fluviatile sedimentation: implications of pedogenic carbonate units, Lower Old Red Sandstone, Anglo-Welsh outcrop. *Geol. J.,* 9: 181 – 208.

Allen, J.R.L., 1986. Pedogenic calcretes in the Old Red Sandstone facies (Late Silurian – Early Carboniferous) of the Anglo-Welsh area, southern Britain. In: V.P. Wright (Editor), *Paleosols: Their Recognition and Interpretation.* Blackwell Scientific, Oxford, pp. 58 – 86.

Arakel, A.V., 1986. Evolution of calcrete in palaeodrainages of the Lake Napperby area, central Australia. *Palaeogeogr. Palaeoclimatol. Palaeoecol.,* 54: 283 – 303.

Arakel, A.V. and McConchie, D., 1982. Classification and genesis of calcrete and gypsite lithofacies in paleodrainage systems of inland Australia and their relationship to carnotite mineralization. *J. Sediment. Petrol.,* 52: 1149 – 1170.

Bilzi, A.F. and Ciolkosz, E.J., 1977. A field morphology rating scale for evaluating pedologic development. *Soil Sci.,* 124: 45 – 48.

Birkeland, P.W., 1984. *Soils and Geomorphology.* Oxford Univ. Press, New York, N.Y., 372 pp.

Blodgett, R.H., 1984. Non-marine depositional environments and paleosol development in the Upper Triassic Dolores Formation, southwestern Colorado. In: D.C. Brew (Editor), *Field Trip Guidebook.* Geol. Soc. Am. Rocky Mountain Section, 37th Annu. Meet., Durango, Colo., Fort Lewis Coll., pp. 46 – 92.

Boardman, J. (Editor), 1985. *Soils and Quaternary Landscape Evolution.* John Wiley, Chichester, 391 pp.

Bown, T.M. and Kraus, M.J., 1983. Ichnofossils of the alluvial Willwood Formation (lower Eocene), Bighorn Basin, northwest Wyoming, USA. *Palaeogeogr. Palaeoclimatol. Palaeoecol.,* 43: 95 – 128.

Bown, T.M. and Kraus, M.J., 1987. Integration of channel and floodplain scutes in aggrading alluvial systems. I. Developmental sequence and lateral relations of Lower Eocene alluvial paleosols, Willwood Formation, Bighorn Basin, Wyoming. *J. Sediment. Petrol.,* 57: 587 – 601.

Brewer, R., 1976. *The Fabric and Mineral Analysis of Soils.* Krieger Publ., New York, N.Y., 482 pp.

Bridge, J.S., 1984. Large scale facies sequences in alluvial overbank environments. *J. Sediment. Petrol.,* 54: 583 – 588.

Brinkman, R., 1977. Surface-water gley soils in Bangladesh: genesis. *Geoderma,* 17: 111 – 144.

Bullock, P., Fedoroff, N., Jongerius, A., Stoops, G. and Tursina, T., 1985. *Handbook for Soil Thin Section Description.* Waine Research Publ., U.K., 152 pp.

Buol, S.W., Hole, F.D. and McCracken, R.J., 1980. *Soil Genesis and Classification.* 2nd ed. Iowa State Univ. Press, Ames, Iowa, 404 pp.

Buurman, P. 1975. Submarine soil formation changing fossil terrestric soils. *Soil Sci.,* 119: 24 – 27.

Carlisle, D., 1980. Possible variations on the calcrete-gypcrete uranium model. *Open-file Rep. U.S. Dep. Energy,* GJBX-53(80): 38 pp.

Carlisle, D., 1983. Concentration of uranium and vanadium in calcretes and gypcretes. In: R.C.L. Wilson (Editor), *Residual Deposits. Geol. Soc. London, Spec. Publ.,* 11: 185 – 195.

Carlisle, D., Merifield, P.M., Orme, A.R., Kohl, M.S. and Kolker, O., 1978. The distribution of calcretes and gypcretes in southwestern United States and their uranium favorability based on a study of deposits in Western Australia and South West Africa (Namibia). *Open-file Rep., U.S. Dep. Energy,* GJBX-29(78): 274 pp.

Chartres, C.J., 1980. A Quaternary soil sequence in the Kennet Valley, central southern England. *Geoderma,* 23: 125 – 146.

Chartres, C.J. and Walker, P.H., 1988. The effect of aeolian accessions on soil development on granitic rocks in south-eastern Australia. III. Micromorphological and geochemical evidence of weathering and soil development. *Aust. J. Soil. Res.,* 26: 33 – 53.

Chartres, C.J., Chivas, A.R. and Walker, P.H., 1988. The effect of aeolian accessions on soil development on granitic rocks in south-eastern Australia. II. Oxygen-isotope, mineralogical and geochemical evidence of aeolian deposition. *Aust. J. Soil. Res.,* 26: 17 – 31.

Curtis, C.D. and Coleman, M.L., 1986. Controls on the precipitation of early diagenetic calcite, dolomite and siderite concentrations in complex depositional sequences. In: D.L. Gautier (Editor), *Roles of Organic Matter in Sediment Diagenesis. Soc. Econ. Paleontol. Mineral., Spec. Publ.,* 38: 23 – 33.

Deconinck, J.-F. and Strasser, A., 1987. Sedimentology, clay mineralogy and depositional environment of Purbeckian green marls (Swiss and French Jura). *Eclogae. Geol. Helv.,* 80: 753 – 772.

Dixon, J.B. and Weed, S.B. (Editors), 1989. *Minerals in Soil Environments,* 2nd ed. Soil Sci. Soc. Am., Madison, Wisc.

Duchaufour, P., 1982. *Pedology.* Allen and Unwin, London, 448 pp.

Esteban, M. and Klappa, C.F., 1983. Subaerial exposure environment. In: P.A. Scholle, D.G. Bebout and C.H. Moore (Editors), *Carbonate Depositional Environments.* Am. Assoc. Pet. Geol., Mem., 33: 1 – 54.

Farrell, K.M., 1987. Sedimentology and facies architecture of the overbank deposits of the Mississippi River, False River Region, Louisiana. In: F.G. Etheridge, R.M. Flores and M.D. Harvey (Editors), *Recent Developments in Fluvial Sedimentology. Soc. Econ. Paleontol. Mineral., Spec. Publ.,* 39: 112 – 120.
Fastovky, D.E. and McSweeney, K., 1987. Paleosols spanning the Cretaceous-Paleogene transition, eastern Montana and western North Dakota. *Geol. Soc. Am. Bull.,* 99: 66 – 77.
Fenwick, I., 1985. Paleosols: problems of recognition and interpretation. In: J. Boardman (Editor), *Soils and Quaternary Landscape Evolution.* Wiley, Chichester, pp. 3 – 21.
Finkl, F.W., 1980. Stratigraphic principles and practices as related to soil mantles. *Catena,* 7: 169 – 194.
Fitzpatrick, E.A., 1983. *Soils: their formation, classification and distribution.* Longmans, Harlow, 353 pp.
Fitzpatrick, E.A., 1984. *Micromorphology of Soils.* Chapman and Hall, London, 433 pp.
Fitzpatrick, R.W., 1987. Iron compounds as indicators of pedogenic processes. Examples from the Southern Hemisphere. In: J.W. Stucki, A. Goodman and U. Schwertmann (Editors), *Iron in Soils and Clay Minerals.* Reidel, Dordrecht, pp. 351 – 389.
Freytet, P. and Plaziat, J.C., 1982. Continental carbonate sedimentation and pedogenesis – Late Cretaceous and early Tertiary of southern France. *Contrib. Sedimentol.,* 12: 213 pp.
Gagan, M.K., Johnson, D.P. and Carter, R.M., 1988. The cyclone Winifred storm bed, central Great Barrier Reef shelf, Australia. *J. Sediment. Petrol.,* 58: 845 – 856.
Gile, L.H., Peterson, F.F. and Grossman, R.B., 1966. Morphological and genetic sequences of carbonate accumulation in desert soils. *Soil Sci.,* 100: 347 – 360.
Goldbery, R., 1982a. Paleosols in the Lower Jurassic Mishor and Ardon Formations ("laterite-derivative facies"), Makhtesh Ramon, Israel. *Sedimentology,* 29: 669 – 690.
Goldbery, R., 1982b. Structural analysis of soil microrelief in paleosols of the Lower Jurassic "laterite derivative facies" (Mishor and Ardon Formations), Makhtesh Ramon, Israel. *Sediment. Geol.,* 31: 119 – 140.
Goldbery, R. and Beyth, M., 1984. Laterization and groundwater alteration phenomena in the Triassic Budra Formation, southwestern Sinai. *Sedimentology,* 31: 575 – 594.
Goudie, A.S., 1983. Calcrete. In: A.S. Goudie and K. Pye (Editors), *Chemical Sediments and Geomorphology: Precipitates and Residua in the Near Surface Environment.* Academic Press, London, pp. 93 – 131.
Goudie, A.S. and Pye, K., 1983. *Chemical Sediments and Geomorphology: Precipitates and Residua in the Near Surface Environment.* Academic Press, London, 439 pp.
Harden, J.W., 1982. A quantitative index of soil development from field descriptions: examples from a chronosequence in central California. *Geoderma,* 28: 1 – 28.
Harden, J.W., 1987. Soils developed in granitic alluvium near Merced, California. *U. S. Geol. Surv. Bull.,* 1590-A: 26 pp.
Hayward, M., 1985. Soil development in Flandrian floodplains: River Severn case-study. In: J. Boardman (Editor), *Soils and Quaternary Landscape Evolution.* Wiley, Chichester, pp. 281 – 299.
Hole, F.D., 1961. A classification of pedoturbations and some other processes and factors of soil formation in relation to isotropism and anisotropism. *Soil Sci.,* 91: 375 – 377.
Holland, H.D., Feakes, C.R. and Zbinden, E.A., 1989. The Flin Flon paleosol and the composition of the atmosphere 1.8 BYBP. *Am. J. Sci.,* 97: 362 – 389.
Hubert, J.F., 1977. Paleosol caliche in the New Haven Arkose, Newark Group, Connecticut. *Palaeogeogr. Palaeoclimatol. Palaeoecol.,* 24: 151 – 168.
James, N.P. and Choquette, P.W., 1984. Diagenesis of limestones – the meteoric diagenetic environment. *Geoscience Canada,* 11: 161 – 194.
Jenkins, D.A., 1985. Chemical and mineralogical composition in the identification of palaeosols. In: J. Boardman (Editor), *Soils and Quaternary Landscape Evolution.* Wiley, Chichester, pp. 23 – 43.
Johnson, D.L. and Watson-Stegner, D., 1987. Evolution model of pedogenesis. *Soil Sci.,* 143: 349 – 366.
Kendall, C.G.St.C. and Warren, J., 1987. A review of the origin and setting of tepees and their associated fabrics. *Sedimentology,* 34: 1007 – 1027.
Khalifa, E.M. and Buol, S.W., 1968. Studies of clay skins in a Cecil (Typic Hapludult) soil: I Composition and genesis. *Soil Sci. Soc. Am. Proc.,* 32: 857 – 861.
Kraus, M.J. and Bown, T.M., 1986. Paleosols and time resolution in alluvial stratigraphy. In: V.P.

Wright (Editor), *Paleosols: Their Recognition and Interpretation.* Blackwell Scientific, Oxford, pp. 180–207.

Lehman, T.M., 1989. Upper Cretaceous (Maastrichtian) paleosols in Trans-Pecos Texas. *Bull. Geol. Soc. Am.,* 101: 188–203.

Love, L.G., Coleman, M.L. and Curtis, C.D., 1983. Diagenetic pyrite formation and sulphur isotope fractionation associated with a Westphalian marine incursion, northern England. *Trans. R. Soc. Edinburgh, Earth Sci.,* 74: 165–182.

Machette, M.N., 1985. Calcic soils of the southwestern United States. In: D.L. Wiede (Editor), *Soils and Quaternary Geology of the Southwestern United States. Geol. Soc. Am., Spec. Pap.,* 203: 1–21.

Macphail, R.I., 1986. Paleosols in archaeology: their role in understanding Flandrian pedogenesis. In: V.P. Wright (Editor), *Paleosols: Their Recognition and Interpretation.* Blackwell Scientific, Oxford, pp. 263–290.

Maizels, J.K., 1987. Plio–Pleistocene raised channel systems of the western Sharqiya (Wahiba), Oman. In: L. Frostick and I. Reid (Editors), *Desert Sediments: Ancient and Modern. Geol. Soc. London, Spec. Publ.,* 35: 31–50.

Mann, A.W. and Deutscher, R.L., 1978. Genesis principles for the precipitation of carnotite in calcrete drainages in Western Australia. *Econ. Geol.,* 73: 1724–1737.

Mann, A.W. and Horwitz, R.C., 1979. Groundwater calcrete deposits in Australia. Some observations from Western Australia. *J. Geol. Soc. Aust.,* 26: 293–303.

Martini, I.P. and Chessworth, W., 1991. Weathering, Soil and Paleosols. Elsevier, Amsterdam.

McPherson, J.G., 1979. Calcrete (caliche) paleosols in fluvial redbeds of the Aztec Siltstone (Upper Devonian), southern Victoria Land, Antarctica. *Sediment. Geol.,* 22: 267–285.

McSweeney, K. and Fastovsky, D.E., 1987. Micromorphological and SEM analysis of Cretaceous-Paleogene petrosols from eastern Montana and western North Dakota. *Geoderma,* 40 (1,2): 49–63.

Meixner, R.E. and Singer, M.J., 1981. Use of a field morphology rating system to evaluate soil formation and discontinuities. *Soil Sci.,* 131: 114–123.

Meyers, R., 1981. Rôle de paléoaltération, de la paléopédogenèse et de la diagenèse précoce au cours de l'élaboration des sérces continentales. *Thèse Univ. de Nancy.*

Milnes, A.R. and Fitzpatrick, R.W., 1989. Titanium and zirconium minerals. In: J.B. Dixon and S.B. Weed (Editors), *Minerals in Soil Environments,* 2nd ed. Soil Sci Soc. Am., Madison, Wisc.

Morrison, R.B., 1978. Quaternary soil stratigraphy – concepts, methods, and problems. In: W.C. Mahaney (Editor), *Quaternary Soils.* Geoabstracts, Norwich, pp. 77–108.

Murphy, C.P. and Kemp, R.A., 1984. The over-estimation of clay and the underestimation of pores in soil thin sections. *J. Soil Sci.,* 35: 481–495.

Nanson, G.C. and Rust, B.R., 1988. Stratigraphy, sedimentology and late Quaternary chronology of the Channel Country of Western Queensland. In: R.F. Warner (Editor), *Fluvial Geomorphology of Australia.* Academic Press, Australia, pp. 151–175.

Nanson, G.C., Rust, B.R. and Taylor, G., 1986. Coexistent mud braids and anastomosing channels in an arid-zone river: Cooper Creek, central Australia. *Geology,* 14: 175–178.

Nesbitt, H.W. and Young, G.M., 1989. Formation and diagenesis of weathering profiles. *Am. J. Sci.,* 97: 129–147.

Netterberg, F., 1980. Geology of southern African calcretes: 1. Terminology, description, macrofeatures, and classification. *Trans. Geol. Soc. S. Afr.,* 83: 255–283.

Nettleton, W.D. and Sleeman, J.R., 1985. Micromorphology of Vertisols. In: L.A. Douglas and M.L. Thompson (Editors), *Soil Micromorphology and Soil Classification.* SSSA Spec. Publ., 15: 165–196.

Palmer, J.A., Phillips, G.N. and McCarthy, T.S., 1989. Paleosols and their relevance to Precambrian atmospheric compositions. *Am. J. Sci.,* 97: 77–92.

Pawluk, S., 1978. The pedogenic profile in the stratigraphic section. In: W.C. Mahaney (Editor), *Quaternary Soils.* Geoabstracts, Norwich, pp. 61–75.

Percival, C.J., 1986. Paleosols containing an albic horizon: examples from the Upper Carboniferous of northern England. In: V.P. Wright (Editor), *Paleosols: Their Recognition and Interpretation.* Blackwell Scientific, Oxford, pp. 87–111.

Reinhardt, J. and Sigleo, W.R., 1988. Paleosols and weathering through geologic time: principles and applications. *Geol. Soc. Am., Spec. Pap.,* 216: 181 pp.

Retallack, G.J., 1981. Fossil soils: indicators of ancient terrestrial environments. In: K.J. Niklas

(Editor), *Paleobotany, Paleoecology and Evolution,* Vol. 1. Praeger Publ., New York, N.Y., pp. 52 – 102.

Retallack, G.J., 1983a. A paleopedological approach to the interpretation of terrestrial sedimentary rocks: the mid-Tertiary paleosols of Badlands National Park, South Dakota. *Bull. Geol. Soc. Am.,* 94: 823 – 840.

Retallack, G.J., 1983b. Late Eocene and Oligocene paleosols from Badlands National Park, South Dakota. *Geol. Soc. Am., Spec. Pap.,* 193: 82 pp.

Retallack, G.J., 1984a. Completeness of the rock and fossil record: some estimates using fossil soils. *Paleobiology,* 10: 59 – 78.

Retallack, G.J., 1984b. Trace fossils of burrowing beetles and bees in an Oligocene paleosol, Badlands National Park, South Dakota. *J. Paleontol.,* 58: 571 – 592.

Retallack, G.J., 1986. The fossil record of soils. In: V.P. Wright (Editor), *Paleosols: Their Recognition and Interpretation.* Blackwell Scientific, Oxford, pp. 1 – 57.

Retallack, G.J., 1988. Field recognition of paleosols. In: J. Reinhardt and W.R. Sigles (Editors), *Paleosols and Weathering through Geologic Time: Principles and Applications. Geol. Soc. Am., Spec. Pap.,* 216: 1 – 20.

Retallack, G.J., 1990. *Soils of the Past: An Introduction to Palaepedology.* Unwin Hyman, London, 520 pp.

Robinson, D. and Wright, V.P., 1987. Ordered illite/smectite and kaolinite/smectite as possible primary minerals in a Lower Carboniferous paleosol sequence, South Wales? *Clay Miner.,* 22: 109 – 118.

Robinson, D. and Wright, V.P., 1988. Ferrolytic weathering in Lower Carboniferous paleosols. *Abstract British Sedimentary Research Group Meeting, Cambridge; Br. Antarct. Surv., Publ.,* Cambridge.

Rose, J., Allen, P., Kemp, R.A., Whiteman, C.A. and Owen, N., 1985. The early Anglian Barham Soil of eastern England. In: J. Boardman (Editor), *Soils and Quaternary Landscape Evolution.* Wiley, Chichester, pp. 197 – 229.

Ruhe, R.V., 1975. *Geomorphology: Geomorphic Processes and Surficial Geology.* Houghton Mifflin, Boston, Mass., 246 pp.

Rust, B.R. and Nanson, G.C., 1989. Bedload transport of mud as pedogenic aggregates in modern and ancient streams. *Sedimentology,* 36: 291 – 306.

Schwertmann, U., 1987. Occurrence and formation of oxides in various pedoenvironments. In: J.W. Stucki, A. Goodman and U. Schwertmann (Editors), *Iron in Soils and Clay Minerals.* Reidel, Dordrecht, pp. 267 – 308.

Seminiuk, V., 1986. Calcrete breccia floatstone in Holocene sand developed by storm-uprooted trees. *Sediment. Geol.,* 48: 183 – 192.

Seminiuk, V. and Meagher, T.D., 1981. Calcrete in Quaternary coastal dunes in southwestern Australia: a capillary rise phenomenon associated with plants. *J. Sediment. Petrol.,* 51: 47 – 68.

Singer, A., 1985. Pedogenic palygorskite in the arid environment. In: A. Singer and E. Galan (Editors), *Palygorskite – Sepiolite: Occurrences, Genesis and Uses.* Developments in Sedimentology, 37. Elsevier, Amsterdam, pp. 169 – 176.

Singer, A., 1988. Illite in arid soils, desert dusts and desert loess. *Sediment. Geol.,* 59: 251 – 259.

Smith, B.R. and Buol, S.W., 1968. Genesis and relative weathering studies in three semi-arid soils. *Soil Sci. Soc. Am., Proc.,* 32: 261 – 265.

Soil Survey Staff, 1975. *Soil Taxonomy Handbook.* U.S. Dep. Agriculture, 436: 754 pp.

Southgate, P.N., 1986. Cambrian phoscrete profiles, coated grains and microbial processes in phosphogenesis: Georgina Basin, Australia. *J. Sediment. Petrol.,* 56: 429 – 441.

Steel, R.J., 1974. Cornstone (fossil caliche): its origin, stratigraphic and sedimentological importance in the New Red Sandstone, W. Scotland. *J. Geol.,* 82: 351 – 369.

Thiry, M., 1989. Geochemical evolution and paleoenvironments of the Eocene continental deposits in the Paris Basin. *Palaeogeogr. Palaeoclimatol. Palaeoecol.,* 70: 153 – 163.

Thomassen, J.R., 1979. Late Cenozoic grasses and other angiosperms from Kansas, Nebraska and Colorado: biostratigraphy and relationship to living taxa. *Bull. Geol. Surv. Kansas,* 218: 68 pp.

Valentine, K.W.G. and Dalrymple, J.B., 1976. Quaternary buried palaeosols: a critical review. *Quat. Res.,* 6: 209 – 222.

Vreeken, W.J., 1975. Principal kinds of chronosequences and their significance in soil history. *J. Soil Sci.,* 26: 378 – 394.

Walker, T.R., Waugh, B. and Crone, A.J., 1978. Diagenesis in first cycle desert alluvium of Cenozoic age, southwestern United States and northwestern Mexico. *Bull. Geol. Soc. Am.*, 89: 19 – 32.

Wilding, L.P. and Drees, L.R., 1972. Biogenic opal in Ohio soils. *Soil Sci. Soc. Am., Proc.*, 35: 1004 – 1010.

Wilding, L.P. and Tessier, D., 1988. Genesis of Vertisols: shrink-swell phenomena. In: L.P. Wilding and R. Puentes (Editors), *Vertisols: Their Distribution, Properties, Classification and Management.* Tech. Monogr. 18 SMSS, 55 – 81, Texas A & M Univ., College Station, Tex.

Wilson, R.C.L. (Editor), 1983. *Residual Deposits: Surface Related Weathering Processes and Materials. Geol. Soc. Lond., Spec., Publ.*, 11. Blackwell Scientific, Oxford, 258 pp.

Wright, V.P., 1982. Calcrete paleosols from the Lower Carboniferous. Llanelly Formation, South Wales. *Sediment. Geol.*, 33: 1 – 33.

Wright, V.P., 1983. A rendzina from the Lower Carboniferous of South Wales. *Sedimentology*, 30: 159 – 179.

Wright, V.P. (Editor), 1986a. *Paleosols: Their Recognition and Interpretation.* Blackwell Scientific, Oxford, 315 pp.

Wright, V.P., 1986b. Pyrite formation and the drowning of a palaeosol. *Geol. J.*, 21: 139 – 149.

Wright, V.P., 1986c. The role of fungal biomineralization in the formation of Early Carboniferous soil fabrics. *Sedimentology*, 33: 831 – 838.

Wright, V.P., 1987. The ecology of two early Carboniferous paleosols. In: J. Miller, A.E. Adams and V.P. Wright (Editors), *European Dinantian Environments.* Geol. J., Spec. Publ. 12, Wiley, Chichester, pp. 345 – 358.

Wright, V.P., 1988. Paleokarsts and paleosols as indicators of paleoclimate and porosity evolution: a case-study from the Carboniferous of South Wales. In: N.P. James and P.W. Choquette (Editors), *Paleokarst.* Springer, New York, N.Y., pp. 329 – 341.

Wright, V.P. and Robinson, D., 1988. Early Carboniferous floodplain deposits from South Wales: a case study of the controls on palaeosol development. *J. Geol. Soc. London*, 145: 847 – 857.

Wright, V.P. and Wilson, R.C.L., 1987. A terra rossa-like paleosol complex from the Upper Jurassic of Portugal. *Sedimentology*, 34: 259 – 273.

Wright, V.P., Marshall, J.D., Robinson, D. and Kemp, A., 1988. Stable isotope analyses of pedogenic carbonates from the Lower Carboniferous of South Wales. *Abstract British Sedimentary Research Group, Annual Meeting. Brit. Antarct. Survey., Publ.* Cambridge.

Yaalon, D.H., 1971. Soil-forming processes in time and space. In: D.H. Yaalon (Editor), *Paleopedology – Origin, Nature and Dating in Paleosols.* Israel Univ. Press, Jerusalem, pp. 29 – 39.

Yaalon, D.H. and Kalmar, D., 1978. Dynamics of cracking and swelling of clay soils: displacement of skeletal grains, optimum depth of slickensides, and rate of intra-pedonic turbation. *Earth Surf. Prof. Pap.*, 3: 31 – 42.

Reprinted from: *Diagenesis, III. Developments in Sedimentology, 47.*
Edited by K.H. Wolf and G.V. Chilingarian.

Chapter 13

SILICA-CEMENTED PALEOSOLS (GANISTERS) IN THE PENNSYLVANIAN WADDENS COVE FORMATION, NOVA SCOTIA, CANADA

MARTIN R. GIBLING and BRIAN R. RUST[1]

INTRODUCTION

Ganisters are tough, silica-cemented sandstones which are found in alluvial and deltaic, commonly coal-bearing strata, where they form a local source of refractory silica bricks. They are economically significant in underground coal mining, because sparks generated when machinery strikes tough, siliceous rocks adjacent to the seams can ignite methane (Houseknecht and Iannacchione, 1982). Forgeron et al. (1986) recognized siliceous seatearths as incendiary mining hazards in the Sydney Basin of Nova Scotia, the site of the present study. Some highly siliceous units probably reflect erosion of quartz-rich source areas and/or selective accumulation of quartz during weathering and sediment transport (Houseknecht, 1980; Donaldson and Shumaker, 1981). Many ganisters, however, have been interpreted as paleosols, the predominance of silica reflecting surficial modification of an originally more argillaceous sandstone (Williamson, 1967, pp. 60 – 61). Ganisters that originated as paleosols have been documented from the Australian Trias (McDonnell, 1974; Retallack, 1976, 1977) and the British Carboniferous (Percival, 1983a,b, 1986; Fielding et al., 1988; Besly and Fielding, 1989), but relatively little information is available concerning their petrography and geochemistry.

Silica-cemented paleosols, similar in many respects to those of the WCF, have been described from Albian strata of British Columbia, Canada, by Leckie et al. (1989).

The present study concerns the origin and diagenesis of a suite of Pennsylvanian ganister-bearing paleosols in the Sydney Basin of Nova Scotia, Canada. The influence of the ganisters on the development of associated channel-sandstone bodies (Gibling and Rust, 1990) provides strong confirmatory evidence for a near-surface, pedogenetic origin for the ganisters. The writers document the megascopic and microscopic nature and geochemistry of the ganisters (a) to determine the diagenetic (pedogenetic) conditions involved in their formation, and (b) to compare them with siliceous soils and silcretes in modern landscapes and siliceous paleosols in the geological record.

---

[1] Brian died in the summer of 1990 in Ottawa of malaria, contracted during a trip to Africa.

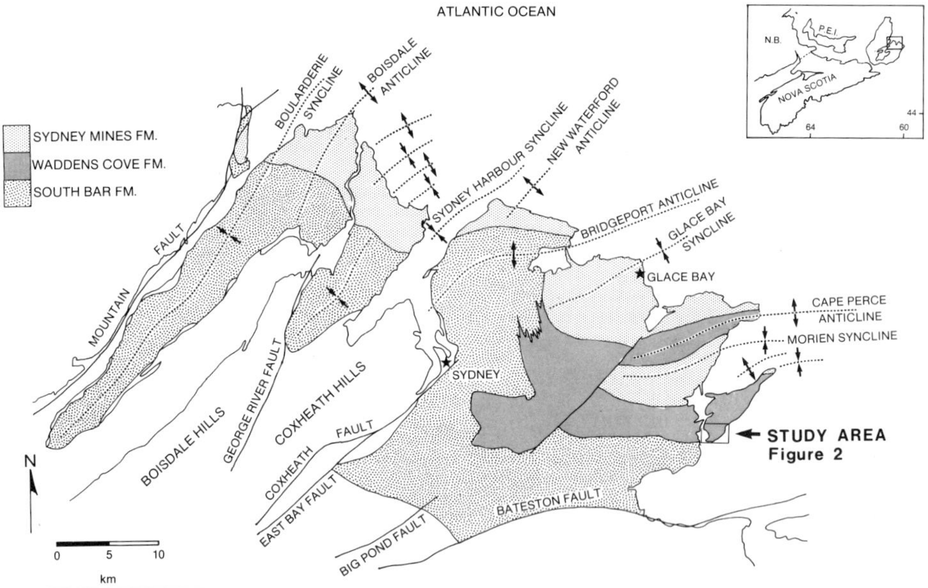

Fig. 13-1. Outcrop area of the Morien Group in the Sydney Basin, Nova Scotia. The areal extent of the Waddens Cove Formation in the southeastern part of the basin is approximate due to complex facies changes and poor inland exposure. (Modified from Boehner and Giles, 1986.)

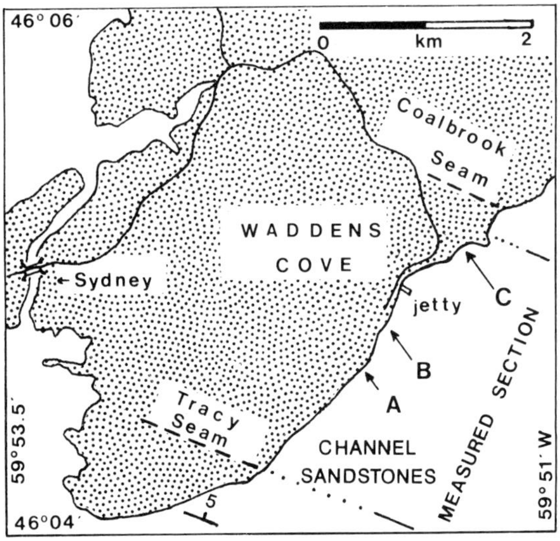

Fig. 13-2. Location of stratigraphic section measured in the Waddens Cove Formation (Fig. 13-3). The formation underlies all the onshore area (stippled). *A*, *B* and *C* indicate the location of channel sandstones designated in Fig. 13-3. See Fig. 13-1 for location of the area within the basin.

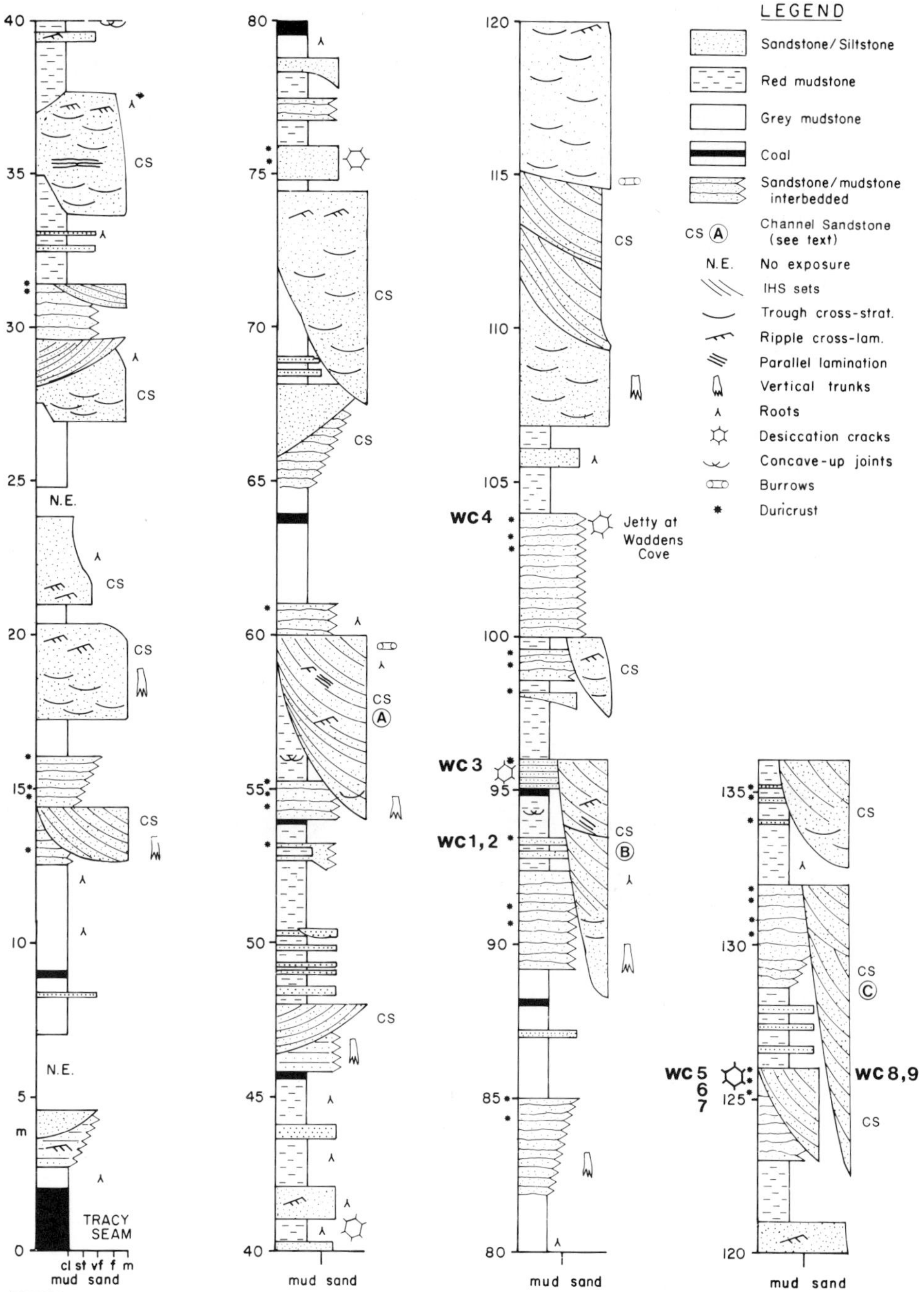

Fig. 13-3. Stratigraphic column for the Waddens Cove Formation from the Tracy Seam to the limit of accessibility northeast of Waddens Cove (Fig. 13-2). The Coalbrook Seam lies shortly above the top of the section. Where sediment bodies change in thickness and character laterally in the cliffs, their relationships are indicated. The duricrusts indicated are ganisters. The stratigraphic position of ganister samples used in this study is shown (WC 1 – 9); samples WC 8 and 9 were taken from slumpblocks derived from the same ganister beds as those represented by samples WC 5 – 7. Channel sandstones *A, B* and *C* were described in detail by Gibling and Rust (1990).

GEOLOGICAL SETTING

The Sydney Basin forms the northeast corner of Nova Scotia (Fig. 13-1) and extends under the sea as far as Newfoundland (King and MacLean, 1976). Paleomagnetic evidence (Scotese et al., 1979) indicates that the basin lay a few degrees south of the equator during the Pennsylvanian. The coal-bearing Morien Group, of Westphalian B to Stephanian age, is the uppermost bedrock unit in the Sydney Basin. It includes in the southeastern onshore part of the basin a redbed unit about 840 m thick, mapped as the Waddens Cove Formation (Boehner and Giles, 1986; Rust et al., 1987) (Fig. 13-1). The formation comprises channel sandstones and floodplain deposits of sandstones and mudstones with thin coals and ganisters. A section 136 m thick was measured in the vicinity of Waddens Cove (Figs. 13-2 and 13-3), and contains several distinctive facies.

Channel-sandstone bodies are up to 14 m thick and are single- or multistoreyed. Where both sides are exposed, the bodies show width : thickness ratios < 15:1 (ribbon sandstones in the terminology of Friend et al., 1979). They show inclined heterolithic stratification (IHS: Thomas et al., 1987) sets of medium- to very fine-grained sandstone interbedded with mudstone (Fig. 13-4), with numerous high-angle erosional surfaces. The sandstones show large-scale trough cross-stratification, with ripple cross-lamination and planar stratification. The margins of the bodies dip at up to 34° and show steps over ganister-bearing sheet (floodplain) sandstones, as well as slumpblocks of ganisters that can be matched with those in the adjacent

Fig. 13-4. Channel sandstone A (Fig. 13-3). The sandstone body shows a width : thickness ratio of 13:1 and contains IHS sets separated by erosional surfaces. The basal erosional surface transects red mudstones and a composite sheet sandstone bearing ganisters. A composite sheet sandstone capped by a ganister overlies the channel sandstone. Ganisters are arrowed.

Fig. 13-5. Pale colored ganisters (arrowed) in the Waddens Cove Formation. The lowermost three layers are developed in the top of a channel sandstone (*CS 1*). Higher layers are developed in sheet sandstones intercalated with mudstones. A channel sandstone (*CS 2*: channel sandstone C in Fig. 13-3)) shows a stepped profile over the ganisters and contains ganister slumpblocks (*GSB*) (see Figs. 13-6 and 13-11 for close-up photos). Note the locally convex-upward surfaces of the lowermost arrowed ganisters, especially evident just to the left of the lowermost arrow (see also Fig. 13-6). Strata illustrated form the 124 – 136 m interval in Fig. 13-3.

Fig. 13-6. Close-up of margin of channel sandstone (*CS-2* in Fig. 13-5). Margin is indicated by arrows. Note the stepped profile of the margin over the three superimposed ganisters (*g*) from which a slump-block (*GSB* in Fig. 13-5) was derived. Note also the convexities of the ganister layers to the right. About 8 m of strata are shown.

floodplain strata (Figs. 13-5 and 13-6). *Calamites* in growth position, roots and rhizoliths were observed in the channel sandstones. Gibling and Rust (1990) interpreted the bodies as the fills of incised paleovalleys within which channels followed a more sinuous course between bank-attached bars. The low width : thickness ratios and tendency for storeys to be superimposed were attributed, at least in part, to confinement of the channels by the resistant ganisters. The abundant vegetation at low levels in the channel deposits, coupled with their heterolithic nature and contained erosional surfaces, indicates episodic, probably seasonal, flow.

Sheet sandstone bodies have planar bounding surfaces and width : thickness ratios in excess of 100:1 (Fig. 13-5). A few thicken and pass laterally into channel bodies. The sandstone bodies range from single beds 20 cm thick to composite sheets several meters thick with mudstone interbeds. Grain size of the sheet sandstones varies from fine to silty, very fine sand. Many composite sheets show a tendency for the sandstone beds to thicken and coarsen upward, but some sheets show more complex trends, such as thickening up followed by thinning up. Erosional hollows, which cut through the sheets to a depth of 2 m, contain inclined layers of sandstone and mudstone. The sheets show sand- and mud-filled desiccation cracks up to 1 m deep, *Stigmaria* and small carbonized roots, sideritic rhizoliths (Klappa, 1980), and upright *Sigillaria* trunks. The composite sheets commonly show lensoid and contorted beds, probably due to disruption by and compaction around vegetation. Individual sandstone beds are interpreted as crevasse splays, composite sheets as stacked splay deposits or levees, and the erosional hollows as crevasse channels. The sheets were vegetated during and/or after deposition. Modern splay and levee deposits with similar features have been described by Coleman and Gagliano (1965), Smith (1983), and Farrell (1987).

Mudstones form sheet-like units up to 4 m thick (Fig. 13-5), weakly stratified and with roots and rhizoliths. Red mudstone predominates over grey, especially in the upper part of the measured section (Fig. 13-3). Several red mudstone units show intersecting, concave-upward joint sets, which resemble pseudo-anticlines described from the Old Red Sandstone (Allen, 1973, 1986); the latter contain calcareous glaebules and veins, unlike examples from the Lower Carboniferous (Wright, 1982) and the Waddens Cove Formation. Allen compared the pseudo-anticlines to gilgai formed in some clay-rich, alluvial soils, and attributed to pedogenesis under conditions of seasonal, deep-soil wetting (Hallsworth and Beckman, 1969; Knight, 1980). Smectite is usually a prominent component of modern soils with gilgai, but is absent or present in trace amounts only in mudstones of the Waddens Cove Formation and other formations of the Morien Group (Dilles, 1983; Gibling et al., 1985 and unpublished data). This scarcity may reflect alteration of smectite to illite during burial metamorphism (Hower, 1981) of sufficient intensity to have converted associated peats to bituminous coals.

Coals, apart from the 2-m-thick Tracy Seam, are less than 30 cm thick. They are underlain by rooted seatearths, but not by ganisters, and vertical trunks are rooted in some seams.

GANISTERS

Thirty-six ganister beds from 0.1 to 1 m in thickness were observed in the measured section. They are present within sheet sandstone bodies composed of single beds, at numerous levels within (but especially at the tops of) composite sheet sandstones, and at the tops of a few channel-sandstone bodies (Fig. 13-3).

The upper surfaces of the ganisters form wave-cut platforms where, because of their resistant nature and the low dips ($<10°$), they are exposed over areas in excess of 2000 m^2 (Figs. 13-7 and 13-8). Many ganisters may be traced laterally for at least 150 m, the limit of exposure of individual beds. Upper surfaces are uneven, with up to 30 cm of relief, and show nodular ridges and raised rings 10 – 50 cm in diameter, which probably developed around tree trunks and roots. Similar features were observed on ganister surfaces in New South Wales by Retallack (1976) and attributed by him to "cradle knolls" formed in modern soils around trees that were blown down by the wind (Buol et al., 1973, p. 253). *Stigmaria* roots are common on some surfaces. The surfaces are broadly planar, but in cross-section some show gentle, convex-upward segments with about 5 m spacing, separated by narrow concavities (Fig. 13-6). This topography may reflect displacive cementation (N.L. Watts, 1977), but more probably reflects disruption of splays and levees by vegetation. Thin (a few millimeters thick) carbonaceous layers overlie some ganisters, but in no case does a coal rest upon or closely overlie a ganister (Fig. 13-3).

Fig. 13-7. Ganister developed at the top of a sheet sandstone and exposed on wave-cut platform. Note the nodular appearance and ring-like forms (arrowed). 104-m position in Fig. 13-3.

Fig. 13-8. Ganister exposed on wave-cut platform. Relief of the nodular surface is up to 30 cm, with a ring-like form 50 cm in diameter (behind hammer 30 cm long). 16-m position in Fig. 13-3.

The ganisters are white, light red or mottled in color and exhibit weak or no stratification (Fig. 13-9). Vertically oriented rhizoliths, 1 – 10 cm in diameter and up to several decimeters long, show sharply defined margins and a tuberose, downward-branching form. The rhizoliths are composed of dark siderite and some show pale, siliceous centers (Fig. 13-10). A few ganisters show vertically oriented, white siliceous pipes 20 cm in diameter and up to 1 m long, in some cases penetrating the entire unit (Figs. 13-11 and 13-12); such pipes correspond to the hollows within raised rings on the ganister surface, and probably indicate the position of former root channels. Diffuse, vertical streaks in some mottled ganisters (Fig. 13-13A), correspond on the upper surfaces to polygonal desiccation cracks with white, siliceous fills (Fig. 13-13B).

The lower surfaces of the ganisters are generally planar to undulating (Figs. 13-9, 13-11, 13-13A), but some show lobate projections up to 20 cm long into the underlying mudstones (Figs. 13-14A,B). Percival (1986) noted similar lobes beneath some British ganisters. Similar features in modern soils may have originated by flow of soil materials and water down tree-root channels (Buol et al., 1973, p. 257).

Retallack (1983, 1988) and Fenwick (1985) discussed criteria for recognizing paleosols. The existence of a paleocatena of soil types within the ancient landscape, a soil characteristic not replicable by processes of sedimentation and diagenesis (Valentine and Dalrymple, 1975), cannot be demonstrated for the Waddens Cove Formation because of scarcity of inland outcrop and subsurface data. Within more localized paleosol exposures, Retallack identified root traces, soil horizons and soil structures as important indicators of pedogenesis. The Waddens Cove ganisters show abundant roots and rhizoliths. Soil horizons are poorly developed, but numerous structures present in modern soils (Brewer, 1976) were observed in thin-section (see below). Pedogenesis in associated mudstones is indicated by pseudoanticlines, resembling the gilgai of modern soils, and by rooted seatearths beneath coals.

GANISTER PETROGRAPHY AND GEOCHEMISTRY

Analytical methods

Six ganisters were sampled (WC 1, 2, 4 – 7, Fig. 13-3), and represent four composite sheet sandstones from the uppermost 43 m of the section. The samples were analyzed for major elements and selected trace elements using X-ray fluorescence (Table 13-1). Duplicate analyses of a laboratory standard differed by 0.13% for silica, 2.0% average for major elements, and 4.6% average for trace elements, apart from Th which showed a much greater divergence. The ganisters' mineral constituents were determined quantitatively from 300 point counts each in thin-section (Table 13-2), and petrographic features are illustrated in Fig. 13-15. Two samples were examined under the scanning electron microscope (SEM) equipped with an EDX system for phase identification (Welton, 1984) (Fig. 13-16). One sample was examined by X-ray diffraction (XRD), using the semi-quantitative method of Cook et al. (1975).

Fig. 13-9. Ganister developed at the top of a sheet sandstone. Note vertically oriented sideritic rhizoliths, which branch downwards, and weak stratification. Underlying sandstone and red mudstone beds are intensely rooted, but lack ganisters. Scale is 1 m long. 85-m position in Fig. 13-3.

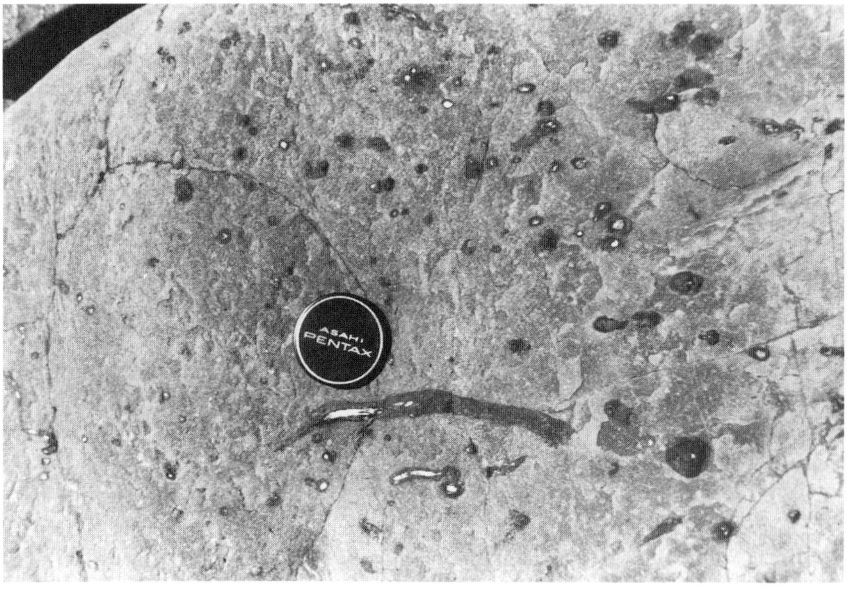

Fig. 13-10. Rhizoliths in ganister (bedding-plane view). The bulk of the rhizoliths is dark siderite with a central zone of pale microquartz. Lens cap is 5 cm in diameter. 104-m position in Fig. 13-3.

Fig. 13-11. Three superimposed ganisters within a sheet sandstone. Note the vertical, pale pipes and general mottling (red/green). The base of Channel Sandstone C (Fig. 13-3) is located at the top of the uppermost ganister, but steps down to the left of the scale (1 m). Photo is a close-up of the three lowermost arrowed ganisters at 10-m position in Fig. 13-5 (see also Fig. 13-6).

Fig. 13-12. Ganister with white, vertically oriented siliceous pipes. Graduations on scale at 10-cm intervals. 32-m position in Fig. 13-3.

Fig. 13-13. Vertical section (A) and upper surface (B) of ganister. Note in (A) the siliceous vertical streaks which extend for about 1 m, and correspond to polygonal desiccation cracks on the upper surface (B). Lens cap 5 cm in diameter in (B). Ganister is at 126-m position in Fig. 13-3, and at 130-m position (just above sea level) in Fig. 13-5.

Fig. 13-14. (A) Ganister 20 cm thick with nodular form, exhumed on wave-cut platform. Scale is 1 m long. 104-m position, Fig. 13-3. (B) Red mudstone underlying the ganister in (A), with siliceous lobate extensions from the base of the ganister. Hammer is 30 cm long.

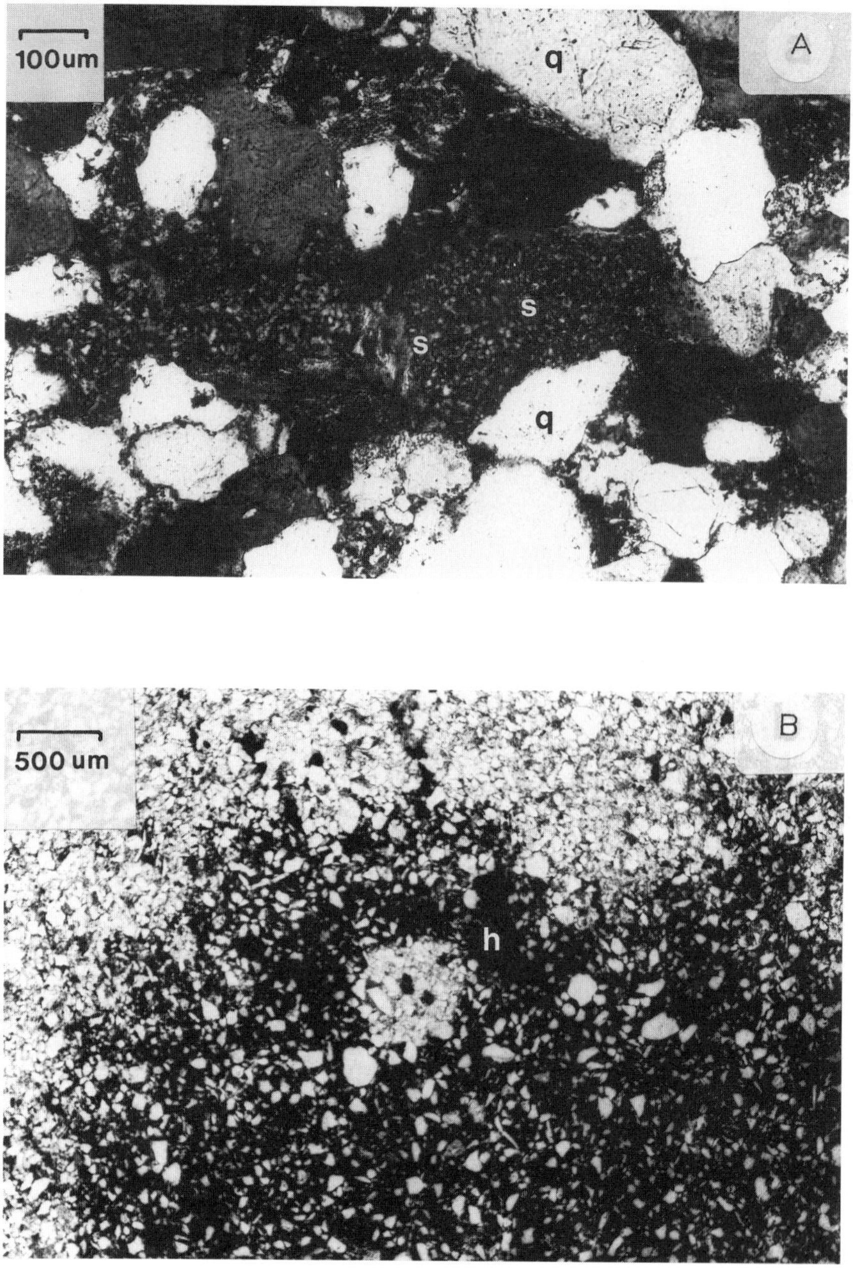

Fig. 13-15. Photomicrographs of ganisters.
(A) Microquartz (*s*) filling interstitial area (orthovugh?) between quartz grains (*q*). Crossed polarizers.
(B) Hematite glaebule (*h*) enclosing quartz grains and showing sharp contact with hematite-free sandstone. Ordinary light.

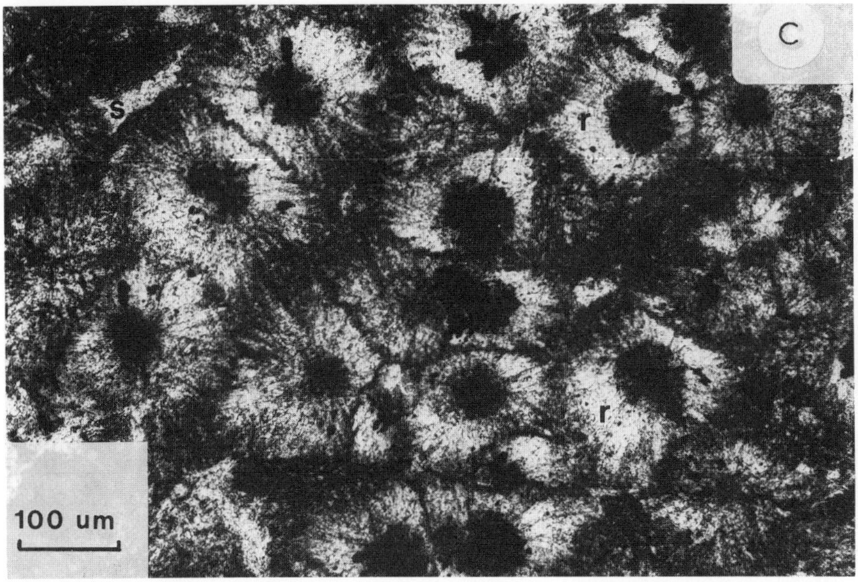

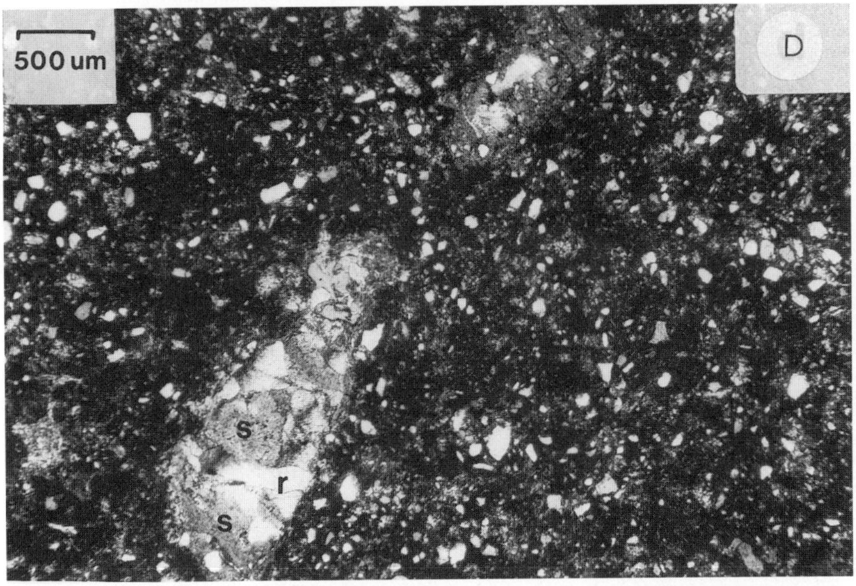

Fig. 13-15 (continued).
(C) Pale siderite rosettes (*r*) within a large rhizolith. The dark material at the centers of the rosettes and between them is also sideritic. Microquartz (*s*) fills voids. Ordinary light.
(D) Crystallaria within a rhizolith, filled with cryptocrystalline silica (*s*) and pale siderite (*r*). Note the floating fabric of quartz grains cemented by darker siderite. Ordinary light.

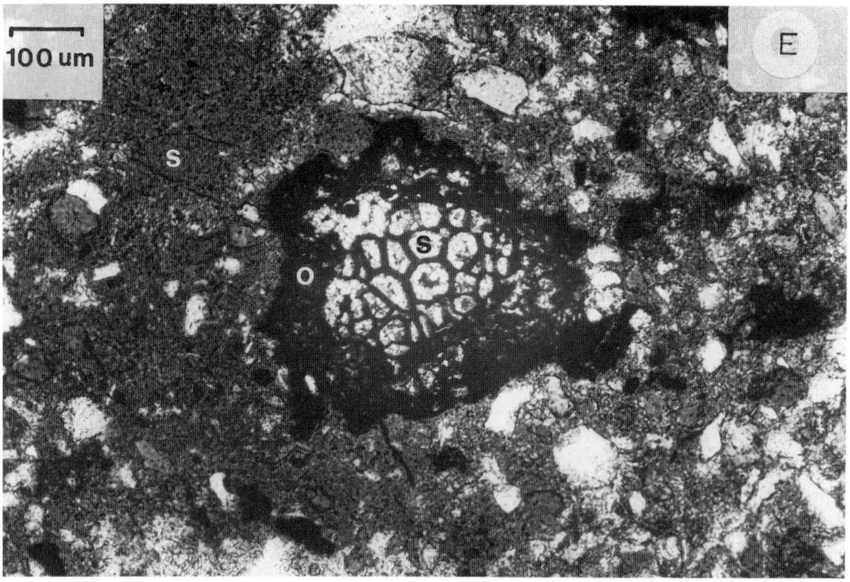

Fig. 13-15 (continued).
(E) Silicified root. Dark organic material (*o*) shows cellular forms that contain microquartz (*s*), which also encases the root. Ordinary light.

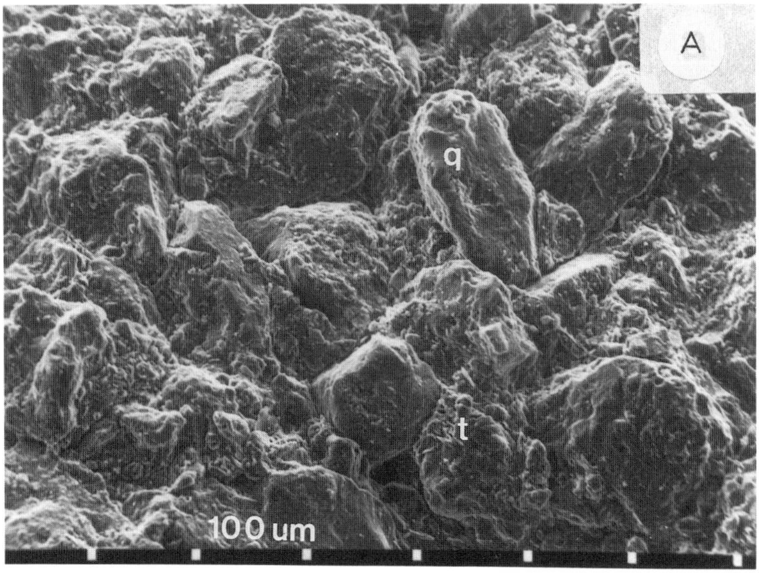

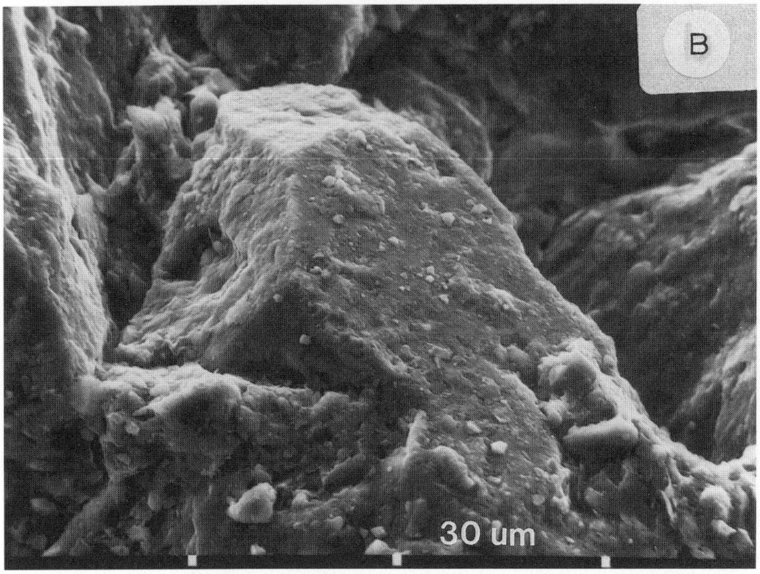

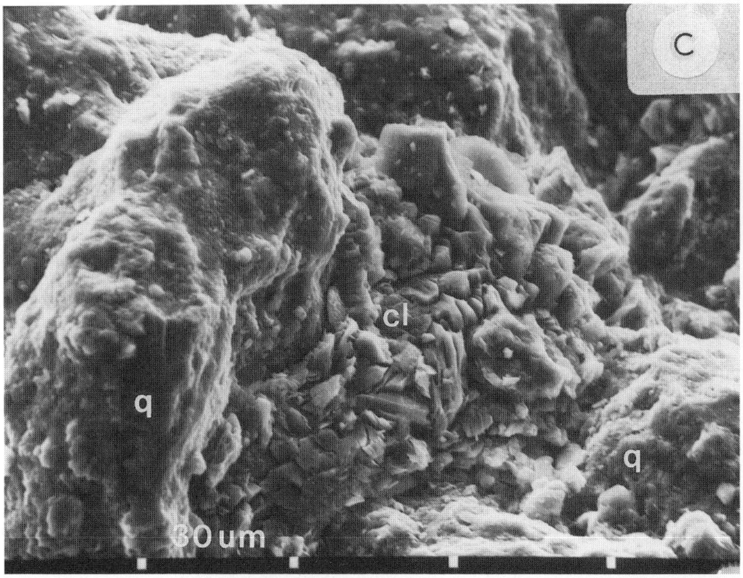

Fig. 13-16. Ganisters under the scanning electron microscope. Scale is indicated between adjacent white marks.

(A) Grain-supported fabric of subrounded quartz grains (*q*) with clay matrix between and wrapped around the grains. A rare overgrown quartz grain (embedded grain quartzan) is present to left of center. Titaniferous grains are present (*t*).

(B) Close-up of overgrown quartz grain in (A) to show euhedral form.

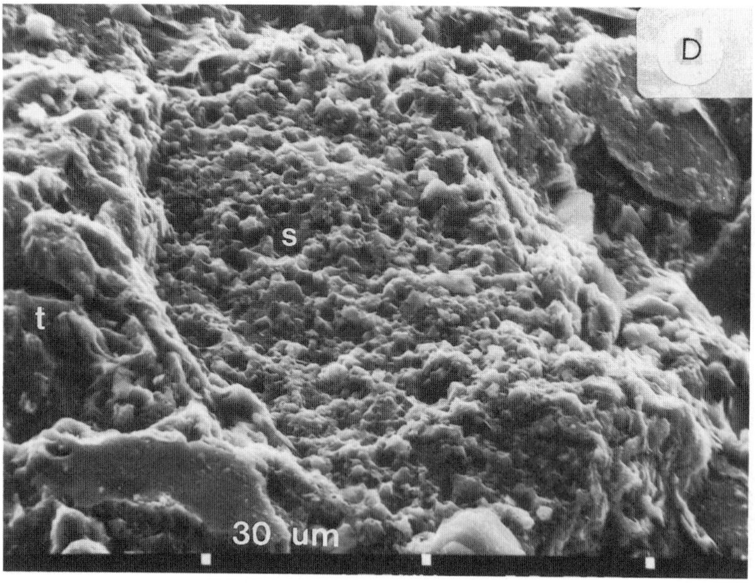

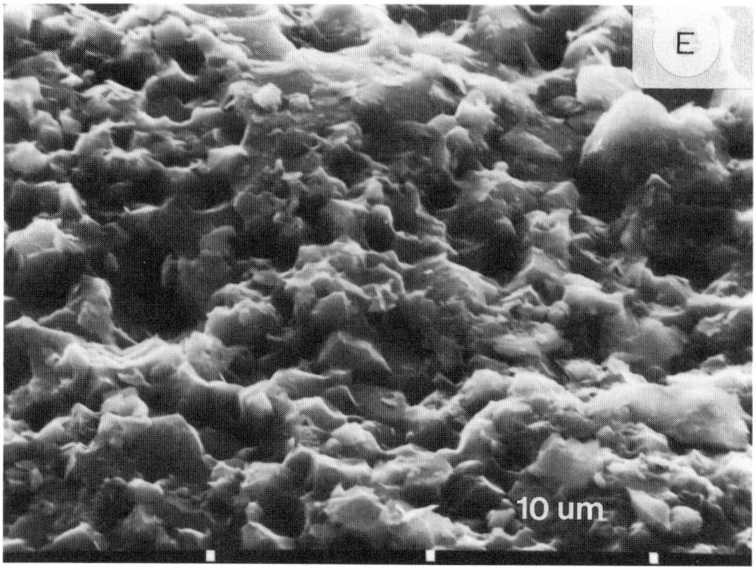

Fig. 13-16 (continued).
(C) Close-up of illitic clay (*cl*) in lower left of (A). The clay forms moderately well-developed blocky crystals up to 20 μm in diameter in the interstices between quartz grains (*q*).
(D) Microquartz (*s*) in interstices 1 mm long between framework grains. One titaniferous grain is present (*t*).

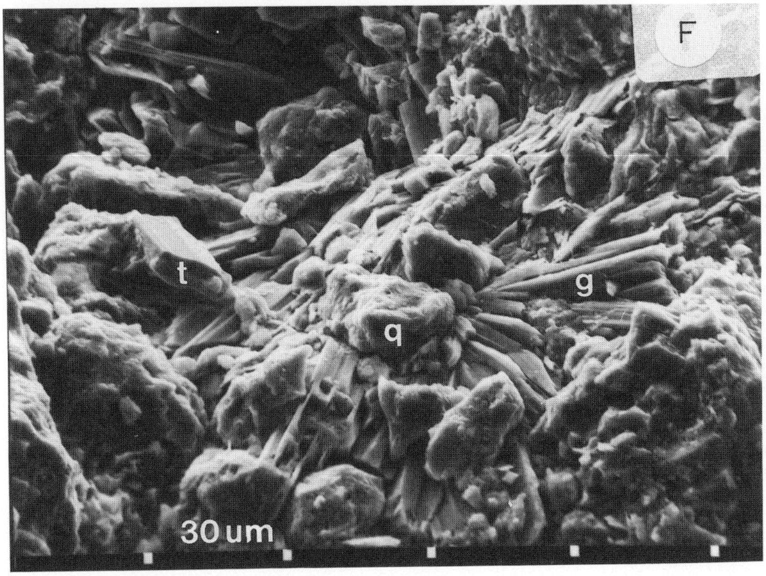

Fig. 13-16 (continued).
(E) Close-up of microquartz in (D). Individual crystals are less than 5 μm in diameter, with subhedral form.
(F) Radiate gypsum crystals (*g*), apparently surrounding a quartz grain (*q*) with an illitic coating. A well-formed titaniferous grain (*t*) shows striated faces.

TABLE 13-1
Contents of major elements (%;top) and selected trace elements (ppm; bottom) in Waddens Cove ganisters, determined from X-ray fluorescence analysis

Sample	SiO_2	Al_2O_3	Fe_2O_3	MgO	CaO	Na_2O	K_2O	TiO_2	MnO	P_2O_5	L.O.I.
WC 1	83.91	8.47	2.70	0.94	0.06	0.50	1.09	1.02	0.02	0.04	2.3
WC 2	84.39	7.54	3.72	1.04	0.28	0.59	0.91	0.75	0.04	0.03	1.9
WC 4	81.66	9.95	2.59	1.16	0.13	0.59	1.56	0.88	0.04	0.04	2.5
WC 5	85.03	8.16	2.11	0.96	0.18	0.63	1.19	0.79	0.03	0.04	2.0
WC 6	82.74	8.07	4.22	1.20	0.11	0.38	1.14	0.80	0.04	0.06	2.2
WC 7	86.51	5.88	3.47	1.04	0.42	0.29	0.75	0.60	0.06	0.03	1.9

Sample	Ba	Rb	Sr	Y	Zr	Nb	Th	Pb	Ga	Zn	Cu	Ni	V	Cr
WC 1	214	60	32	34	457	17	7	14	12	50	13	28	73	67
WC 2	136	38	29	21	238	12	n.d.	7	14	47	2	14	62	54
WC 4	192	66	55	23	348	14	4	13	18	46	n.d.	20	81	81
WC 5	139	50	50	24	384	13	3	8	15	38	n.d.	12	71	80
WC 6	136	46	44	23	402	13	3	9	17	51	n.d.	12	66	64
WC 7	96	29	33	20	343	11	8	11	13	47	n.d.	10	46	49

Total iron was determined as Fe_2O_3.
Loss on ignition (L.O.I.) was determined after heating sample to 1050°C.
n.d. indicates not detected.

TABLE 13-2

Mineral composition of Waddens Cove ganisters. Data in % from point counts of 300 grains per thin-section

	WC 1	*WC 2*	*WC 4*	*WC 5*	*WC 6*	*WC 7*
Framework grains						
Quartz	75.3	70.6	60.3	81.3	60.3	74.0
Lithic grains				0.3	3.0	0.3
Muscovite						0.3
Zircon				0.6		0.3
Tourmaline	0.3					
Ferruginized grains	0.3		0.3		2.6	2.3
Calcitized grains	1.6	2.3	1.6	1.3	0.6	1.3
Matrix/cement						
Clay minerals	9.6	17.0	21.0	7.0	21.3	13.6
Microquartz	10.0	8.0	16.0	7.0	9.0	6.3
Opaque minerals	2.6	2.0	0.6	2.3	3.0	1.3

Four ganister samples were analyzed using an electron microprobe (EMP), a JEOL 733 Superprobe, in the WDS mode, with operating conditions of 15 kV at 5 nA and a 2 – 3 micron spot size. Counting time was 40 seconds, or one standard deviation, and mineral standards of calcite, hornblende, and Mn_2O_3 were employed. Background-corrected peak intensities were analyzed using a Tracor Northern ZAF matrix correction programme. The accuracy of the reported analyses is estimated to be about 5%, and is better for major elements.

Additional samples from a sideritic rhizolith (WC 3) (Fig. 13-15C – E) and from ganister slumpblocks in channel deposits (WC 8, 9) were examined in thin-section. Some descriptive petrographic terms used below follow soil-fabric terminology (Brewer, 1976) as the ganisters are considered to form part of paleosols.

Constituents

The ganisters are siliceous rocks with 81 – 86% silica, 6 – 10% alumina, 2 – 4% iron oxides, and less than 1.5% each of other major elements (Table 13-1). Titanium oxides form 0.6 – 1.0% of the rocks. Zr and Ba are the major trace elements at 100 – 450 ppm, with less than 100 ppm of the remaining elements analyzed. Contents of all trace elements analyzed lie within the ranges recorded for modern soils by Levinson (1974, table 2-1) and Aubert and Pinta (1977), but in most cases below quoted average values. (These authors reported no values for Y and Nb.) Exceptions are Ti (discussed by Aubert and Pinta, 1977, as a trace element) and Zr, which are present at above-average levels. Titanium is concentrated in many silcretes, and Hutton et al. (1978) noted that Zr and Nb were concentrated in some Australian silcretes. Levels of Y and Nb in the Waddens Cove ganisters are similar to those for most silcretes documented by Hutton et al. (1978, table 1).

The ganisters show a grain-supported fabric, being composed of a framework of coarse silt to fine sand-sized grains (Figs. 13-15A and 13-16A). Quartz forms from 60 to 81% of grains counted in thin-section (Table 13-2), the remainder comprising phyllitic and schistose lithic fragments, muscovite, zircon, and tourmaline. Quartz grains are commonly elongate, subrounded, and exhibit tangential and point contacts with a few sutured contacts (Sippel, 1968). Small numbers of embayed quartz grains are present in most samples.

Subhedral grains composed principally of calcite with small, angular quartz inclusions form 0.6 – 2.3% of the ganisters in thin-section. The EMP analysis indicates the presence in some calcitic grains of Si, Al, K and Fe. The grains are probably pseudomorphs after detrital grains, probably alkali feldspar and quartz. Unaltered feldspar grains are very rare. Ferruginized (hematitic) grains, of unknown original composition, form up to 2.6% of the rocks.

Clay minerals form 7 – 21% of the ganisters in thin-section. The clay is associated with fine quartz silt and muscovite in interstices between framework grains and is inferred to be a predominantly detrital matrix (Fig. 13-16A), with grain size less than 30 μm. The XRD analysis and the examination of many flakes using the EDX system show that illite predominates over kaolinite, with minor chlorite identified in one thin-section. The clays exhibit a blocky form, moderately to poorly crystallized (Fig. 13-16C).

Based on the proportion of grain types and the abundance of detrital matrix, the ganisters are classified as quartz arenites and quartz wackes according to the system of Pettijohn et al. (1987, fig. 5-1).

The ganisters show relatively little fabric development, with a flecked extinction pattern and very few plasmic separations (asepic or silasepic soil fabrics: Brewer, 1976). One sample shows two sets of short, discontinuous plasma separations, composed predominantly of clay minerals, approximately at right angles to each other (a lattice-like pattern termed lattisepic fabric). Grain coatings (embedded grain cutans) are prominent, discontinuously developed around framework grains and moderately to strongly separated from the adjoining plasma. They are composed of clay minerals (argillans) and include chlorite in one sample, with some iron oxide coatings (sesquans or hematans). Quartz overgrowths (embedded grain quartzans) are present on a few framework grains only (Fig. 13-16B).

Channel cutans composed of well-oriented clay minerals associated with iron oxides (argillans and ferri-argillans) were observed in several samples. The cutans are up to 30 μm in cross-sectional thickness, are strongly separated from the adjoining rock, and are simple, lacking compositional layering. Framework grains project into the cutans, indicating that the cutans formed before the sediment was lithified. The strong degree of separation and the good orientation of the clays indicates an illuvial origin for the cutans (Brewer, 1976).

Microcrystalline quartz (or microquartz: crystals less than 20 μm in diameter; Summerfield, 1983a) forms 6 – 16% of the ganisters in thin-section. The microquartz forms equidimensional, subhedral crystals about 3 μm in diameter (Figs. 13-16D,E). Microquartz and clay minerals form interstitial patches up to 0.25 mm in apparent size between framework grains. The two constituents are intimately mixed and commonly difficult to distinguish, and the microquartz is inferred to have formed by silicification of the detrital clay matrix. In a minority of cases, the microquartz appears relatively clay-free (Fig. 13-15A), is associated with minor calcite, and the adjoining framework grains and plasma are coated with clay or iron oxide. These cases are probably microquartz fills of orthovugs lined with vug cutans (argillans or sesquans), of meso- to macrovoid dimensions (>30 μm in size: Brewer, 1976). Where mixed with clays, the microquartz may have filled smaller voids.

Opaque grains form 0.6 to 3% of the ganisters in thin-section, as equidimensional or acicular, subhedral grains 10 – 25 μm in diameter. Both EMP and EDX data confirm the predominance of a relatively pure, crystalline phase of TiO_2, crystals of which were observed with the SEM (Figs. 13-16A, D, F). Five EMP point counts in three samples indicate that the titania grains contain 0.22 – 1.71% of iron and up to 0.16% of manganese (as oxides). The presence of these impurities may account for the opacity of the anisotropic titanium oxide grains. Finely divided titanium oxide is also associated with illitic clay and/or microquartz in interstitial patches and vug fillings, as indicated by both thin-section and EMP analysis. A similar occurrence of titanium oxide in some silcretes was observed by Hutton et al. (1978, figs. 6 and 7). The grains are too small to be analyzed individually with the EMP, but titanium enrichment was noted in most clay-rich areas analyzed. The occurrence of finely disseminated titanium oxide and of well-formed titaniferous crystals (Fig. 13-16F) strongly suggests that much of the titanium oxide is authigenic. Rutile, anatase, and brookite can all form authigenically (Morad, 1986).

Hematite glaebules several centimeters in diameter are present in a few samples (Fig. 13-15B). They consist of quartz and other framework grains in a hematitic matrix, and show an undifferentiated fabric. Their external boundaries are rather sharp (Brewer, 1976), with glaebular halos narrow or absent (Fig. 13-15B). Minor pyrite is associated with plant material.

A rhizolith (sample WC 3) from a ganister consists predominantly of coalescing sheaves of siderite with "rosettes" up to 0.2 mm in cross-sectional thickness (sphaerosiderite) (Fig. 13-15C). The siderite has up to 4.2% MnO and less than 0.06% TiO_2. Silt-sized quartz grains form 22% of the rhizolith in thin-section (300 point counts), appear to be floating in siderite (Fig. 13-15D), and are highly fractured and embayed. Microquartz and cryptocrystalline silica form 6% of the rhizolith, filling voids up to 2 mm long (Fig. 13-15D). The void fills (compound crystallaria) commonly show a rim of microquartz with a mammillated inner surface with coarser crystals, and a central fill of coarsely crystalline siderite. Siderite veins also cut the microquartz. Void- and vein-fill siderite resembles the rosette siderite in MnO content (3.3%). A few void fills show partial linings of oriented clay, suggesting that the voids originated in association with cutans. Microquartz also fills the interstices between the sheaves and rosettes (Fig. 13-15C) and preserves the cellular structure of roots within the rhizolith (Fig. 13-15E). A single rosette of gypsum was noted with the SEM in one sample (Fig. 13-16F).

The channel sandstones are coarse grained and consist predominantly of quartz. Microcline and perthitic feldspar are present in small amounts (<1%), some partially altered to calcite. Grains of microquartz-cemented sandstone closely resemble the ganisters, and are interpreted as intraclasts derived from them. Clay matrix forms about 10% of thin-sections and a few cutans are present. Calcite and microquartz cements are present in patches up to 1 mm in diameter, and hematitic and titaniferous material forms up to 3.3% of the samples. Sideritic rhizoliths are common in the channel sandstones and show well-developed banding. The channel sandstones thus show some features in common with the ganisters, including the presence of microquartz, titaniferous material and rhizoliths. They differ from them mineralogically principally in the greater abundance of feldspar and the presence of ganister intraclasts.

ORIGIN OF GANISTER-BEARING PALEOSOLS

Paleosol profiles

The Waddens Cove ganisters form components of poorly-developed soil profiles, and their composition and fabric reflect early diagenesis associated with shallow, terrestrial groundwaters. Thin, organic-rich layers are present above a few ganisters, but are probably organic coatings of diagenetic (subaqueous) origin. The ganisters probably correspond to eluvial (E or A2) horizons, as suggested by Retallack (1976, 1977) and Percival (1986). They have undergone pervasive silica cementation, associated with some dissolution of other constituents, but petrographic evidence suggests a limited degree of pedogenetic alteration. Fabrics are predominantly asepic, the ganisters contain considerable amounts of alumina and iron, associated with up to 21% of clay (Tables 13-1, 13-2), and evidence of illuviation is restricted to local channel argillans and ferri-argillans, and to embedded grain argillans and (rarely) sesquans. Alteration of some detrital grains and clay minerals, coupled with authigenic crystallization of silica, iron and titanium oxides, and siderite, appears to have been the major pedogenic process.

Because ganister development has eliminated the original texture of their sheet-sandstone hosts, the nature of the original parent material in any profile is difficult to ascertain. Percival (1986) noted illuvial clay accumulation below ganisters, but no enrichment of clays, sesquioxides, or organic matter was observed below the Waddens Cove ganisters, possibly in part because such accumulations are difficult to distinguish from the primary mudstone beds with which the ganister-bearing sandstones are intercalated. In some cases (e.g., Fig. 13-11), ganisters directly overlie well-stratified beds with primary sedimentary structures and little indication of pedogenetic modification. Where several ganisters are superimposed in composite sheet-sandstones, little vertical differentiation is evident. Samples WC 5 – 7, which represent three superimposed ganisters (Fig. 13-11), show little consistent vertical trend in geochemical and petrographic features (Tables 13-1 and 13-2). Embedded grain sesquans are restricted to the lowermost two layers and embedded grain argillans rich in chlorite to the lowermost layer, suggesting some downward translocation of iron oxides into, and possible neoformation of clay within, the lower ganisters.

Profile immaturity is probably a function of the aggradational, alluvial setting, especially in proximal, near-channel regions of the floodplain (Bown and Kraus, 1987). Such a setting is represented by much of the Waddens Cove section. The tendency for developing alluvial soils to receive new influxes of parent material results in cumulative soil profiles which are difficult to characterize (Birkeland, 1984). The fine intercalation of sandstone and mudstone in the floodplain strata could also have prevented good profile development.

Ganister-bearing paleosols in the British Carboniferous were ascribed to soil types akin to modern podzoluvisols, albic luvisols, and planosols (Percival, 1986), and those in the Australian Triassic to modern gleyed podzolic and alluvial types, principally immature aquods and aquents (Retallack, 1977). All these modern soil types show albic, eluviated horizons. In view of the importance of soil chemistry in classifying modern soils (see discussion in Besly and Fielding, 1989) and of the burial which the Waddens Cove Formation has undergone during basinal subsidence, the correlation of Waddens Cove paleosols with modern soil types is considered uncertain. There is, however, a general similarity of Waddens Cove paleosols to some tropical soils described by Klinge (1965) and Andriesse (1969, 1970) (a humid tropical setting is inferred for the Waddens Cove Formation, see below). Spodosols (aquods) in Sarawak (Andriesse, 1969) develop commonly from alluvial parent materials with strong textural contrast between adjacent strata, rich in quartz but poor in clays and weatherable minerals. Organic-rich surficial horizons are locally thick. The soils show a well-developed eluvial (A2) horizon up to about 3 m thick, the base of which is commonly silica-rich and indurated when dry but capable of flowage when wet and thus not fully cemented. Clay minerals, humus, and sesquioxides accumulate in the underlying B horizons as a result of illuviation from overlying horizons and neoformation of clays, but such accumulations are commonly poorly developed on account of the low clay and labile mineral content of the parent materials. In addition, much illuvial material, including humus colloids, is removed through lateral percolation (see also Klinge, 1965). Fitzpatrick (1980) noted that many soils with albic horizons are acidic, with pH as low as 3.5 at some levels, and that silica cementation in the lower parts of eluvial horizons is common. Some Recent soils with albic horizons formed rapidly, within a few hundred years (Buol et al., 1973, p. 254).

Much of the silica cement (microquartz) in the Waddens Cove ganisters is intimately mixed with clay minerals, and the writers suggest that the silica was derived predominantly from clay-mineral alteration, as suggested by Summerfield (1983a,b) for some silcretes. Summerfield (1983a) noted that microquartz occurs almost to the exclusion of other silica forms in weathering-profile silcretes developed by replacement of clay. The presence of embayed quartz grains in many ganister samples, especially within rhizoliths, suggests that dissolution of quartz grains was an additional source of silica locally, although the floating fabric of the rhizoliths suggests displacive carbonate crystallization (N.L. Watts, 1978) rather than dissolution of framework quartz grains. Dissolution of feldspars, which are less abundant in the ganisters than in coeval channel sandstones and in the Waddens Cove Formation as a whole (Dilles, 1983), may also have contributed silica to groundwaters. This process was suggested by Percival (1986) for some British ganisters in which the proportion of feldspar is lower in the ganisters than in relatively unaltered parent material below. Altered grains presently of ferruginous composition may also have yielded silica. The formation of kaolinite from illite (both present in the Waddens

Cove ganisters) also could have resulted in silica release. Some silica could have been derived from plant litter, phytoliths and siliceous microorganisms. The above discussion suggests that most of the silica cement could have been derived from sources within and closely adjacent to the paleosol profile.

Topographic and climatic conditions

The predominance of ganisters in the upper parts of channel, levee, and splay deposits suggests that the ganisters formed after the temporary or permanent abandonment of these landforms, under conditions of minimal net sedimentation. Such sites of pedogenesis were among the highest and best-drained parts of the floodplain. Direct evidence that the ganisters were exposed on the floodplain surface is lacking. Thin carbonaceous layers overlying some ganisters could represent surface organic accumulations but could also be of subsurface origin. The ganisters were exposed, however, by channel incision into the floodplain strata, as indicated by channel margins stepped over the ganisters and by ganister slumpblocks within the channel fills (Figs. 13-5 and 13-6). A stepped profile over ganisters 3 m below the top of one channel fill (Fig. 13-5) indicates that the ganisters were substantially lithified at depths of no more than a few meters below the floodplain surface. The restriction of ganisters to sandstone beds suggests dependence upon the availability of sand-rich parent material.

Valentine and Dalrymple (1975) noted that buried paleosols are difficult to identify on account of diagenesis during burial. In the case of the Waddens Cove ganisters, however, early lithification implies that much of the observed fabric is original, leaving relatively little scope for subsequent diagenetic fabric alteration, although some mineralogical change may have taken place during burial.

Paleomagnetic evidence (Scotese et al., 1979) and the local abundance of vegetation, including coal, in the Waddens Cove Formation indicate a tropical setting close to the equator. A strongly seasonal climate is suggested by pseudo-anticlines in the associated mudstones and by evidence in channel deposits for episodic flow (see above), as well as by the nature of the paleosols themselves, as discussed below. The abundant crevasse-splay deposits indicate periodic overbank flooding, but the presence of vertical root systems (Mount and Cohen, 1984) and desiccation cracks up to 1 m deep suggest that a substantial vadose groundwater zone existed during drier seasons.

Geochemical conditions during paleosol formation

Geochemical conditions for silica solution and precipitation have been discussed by Krauskopf (1956) and Summerfield (1983c). Although many complex variables play a part, pH and temperature are believed to be major factors. The solubility of amorphous silica is proportional to pH, with a heightened rate of increase above pH 9 under laboratory conditions. Solubility is also proportional to temperature, and rates of geochemical reactions should increase markedly as temperature increases. Solubility, however, probably changes little over the pH (about 2 – 8) and temperature range of most surficial settings (Summerfield, 1983c; Morris and Fletcher, 1987), although pH may be a significant factor where soil waters are dilute (Duchaufour, 1982, pp. 75 – 76).

Seasonal variation in groundwater level may have been an important factor in silica diagenesis on the vegetated Waddens Cove floodplain. Dissolution of silica-bearing minerals would have been enhanced as water tables dropped during the dry season and the vadose zone was affected by organic acids derived from decaying vegetation (Fitzpatrick, 1980, p. 244). In this connection, Bennett and Siegel (1987) found that quartz solubility increased markedly in natural waters contaminated by biodegraded petroleum, and proposed that silica was complexed and mobilized by organic acids in waters with close to neutral pH. Morris and Fletcher (1987) showed that redox reactions involving iron result in rapid dissolution of quartz in laboratory experiments, probably because oxidation causes breakdown of a ferrous iron/silica complex with accompanying liberation of silica to solution. They suggested that seasonal wetting and drying of soils could provide fluctuating redox conditions suitable for such dissolution. The presence of siderite rhizoliths and hematite glaebules in the Waddens Cove ganisters testifies to the importance of reactions involving iron during pedogenesis.

Rising water tables would have bathed the sediment in waters of higher pH, possibly allowing the silica content of these groundwaters to increase. Groundwaters saturated with respect to amorphous silica, even in relatively dilute solutions, were documented in Western Australian calcretes by Mann and Deutscher (1978) and Mann and Horwitz (1979). Amorphous silica was inferred to be precipitating below the water table under neutral to slightly alkaline conditions. The silica probably originated from rapid weathering of granitic rocks at depth. The setting of these deposits is different from that inferred for the Waddens Cove paleosols: the calcretes are forming in an arid region where the water table lies 2 – 5 m below ground surface and is subject to only minor seasonal fluctuation. It is likely, however, that silica dissolution and cementation was especially active between the levels of seasonal fluctuation of the water table beneath the Waddens Cove floodplain surface.

Evaporative concentration of soil waters could have aided silica precipitation within the soil (Duchaufour, 1982, p. 75 – 76), especially in the vadose zone. Silica cementation, however, need not imply saline porewaters: microquartz was inferred to form at low concentrations of pore-water silica and low to moderate concentrations of foreign ions (Summerfield, 1983a). Cryptocrystalline silica is believed to form with higher concentrations of silica and foreign ions (Summerfield, 1983a), and its presence in Waddens Cove rhizoliths may be attributable to the high activity of carbonate in the groundwater during rhizolith formation. The scarcity of overgrowths on detrital quartz grains in the Waddens Cove ganisters may be explained by the presence of clay coatings on the grains (Fig. 13-16A). Heald and Larese (1974) showed that clay coatings effectively inhibit the development of quartz overgrowths.

Titanium is mobile at pH of less than about 4 (S.H. Watts, 1977), and the presence of authigenic titanium oxide in the Waddens Cove ganisters suggests highly acidic conditions, at least periodically. Ferruginous and manganiferous accumulations are common, especially in association with roots, in the Waddens Cove ganisters. Alhonen et al. (1975) invoked alternating oxidizing and reducing (wet and dry) conditions as an important factor in the formation of such accumulations. Gypsum (present in one sample) may have formed from local oxidation of pyrite (Bullock, 1985), rather than from evaporative concentration of sulphate within the groundwaters.

Siderite is abundant in deposits associated with poorly drained Louisiana swamps under low Eh and slightly alkaline conditions (pH about 7 – 8) (Coleman, 1966; Ho and Coleman, 1969). Geochemical conditions for siderite formation include low Eh (reducing conditions), commonly associated with the decay of organic material, and limited availability of sulphur (Berner, 1971; Curtis and Spears, 1968). Such conditions probably prevailed at least seasonally in the vicinity of roots on the Waddens Cove floodplains, both before and after death of the plants. The relatively low Mn content of carbonate phases is consistent with good surface drainage (Mount and Cohen, 1984).

In summary, acidic conditions (pH < 4) probably prevailed periodically during ganister formation and more alkaline conditions (pH 7 – 8) affected the sediments, at least in the vicinity of roots. A fluctuating water table in association with a strongly seasonal climate could explain this inferred geochemical variation. There is little geochemical or petrologic evidence for saline groundwaters during pedogenesis.

The occurrence in channel-sandstone bodies of microquartz and titanium oxide precipitates suggests that the early diagenetic geochemical environment of the channel sandstones was similar in many respects to that of the floodplain sheet sandstones. This may in part reflect virtual drying up of the channels during dry seasons, accompanied by vegetative growth, weathering, and incipient pedogenesis within the channel sediments themselves. It is probable, however, that microquartz cementation in the channel deposits was associated with phreatic, silica-rich groundwaters at shallow depth rather than with pedogenesis. These points of similarity between the early diagenesis of channel sandstones and pedogenetic ganisters suggests that lithification of the channel deposits took place at shallow depth.

Relation to other siliceous deposits

The Waddens Cove ganisters differ both in setting and petrography from three other groups of ganisters described in the literature (Table 13-3). The "type" Sheffield Blue and two other British ganisters of Late Carboniferous age, described by Percival (1983a,b, 1986), lie at the top of progradational deltaic sequences and are overlain by coals, which in turn are overlain by marine strata. Percival (1986) suggested that the ganisters formed prior to a relative base-level rise that led to the subsequent accumulation of thick peat (see also Gardner et al., 1988). Five ganisters for which analyses were reported by Williamson (1967, table 15) show 93 to $>$ 99% silica. The Sheffield Blue ganister shows quartz overgrowths on framework grains (Percival, 1983a). The present authors suggest that the British ganisters form part of relatively well-developed paleosols, probably formed under prolonged conditions of minimal net sedimentation during which virtually all clay and labile minerals in the eluvial horizons were dissolved or removed by illuviation (the paleosols show

TABLE 13-3
Composition and geological setting of some ganisters reported in the literature

Age and location	Nova Scotia	U.K.		Australia
	(Lower Carboniferous)	Early Carboniferous	Lower Carboniferous	(Triassic)
Form of silica	Microquartz replaced clay matrix; overgrowths and cryptocrystalline silica rare	n.d.	Quartz overgrowths	Cryptocrystalline silica between framework grains
Fabric	GS	GS	GS	GS
SiO_2 (%)	81 – 86	77, 88 – 98	93 – 99	n.d.
TiO_2 (%)	0.6 – 1.0	0.1 – 0.5	0.1 – 0.5	n.d.
Paleosol profile	Poorly developed, minor illuviation	Well-developed		Well-developed, pronounced illuvation
Association with coal	Coal associated but rarely caps ganisters	Mainly overlain by coal		Associated with thin, organic layers
Geological setting	Alluvial; cap sandstone splay, levee and channel deposits	In sandstoned at top of progradational deltaic sequences		Alluvial; in sandstone splay deposits
Climate	Humid tropical, seasonal	Humid tropical		Cool temperate
References	This paper; Gibling and Rust (1990)	Fielding et al. (1988)	Percival (1983a,b, 1986) Besly and Fielding (1989)	McDonell (1974) Retallack (1976, 1977)

GS = grain-supported textures.
n.d. = no data reported.

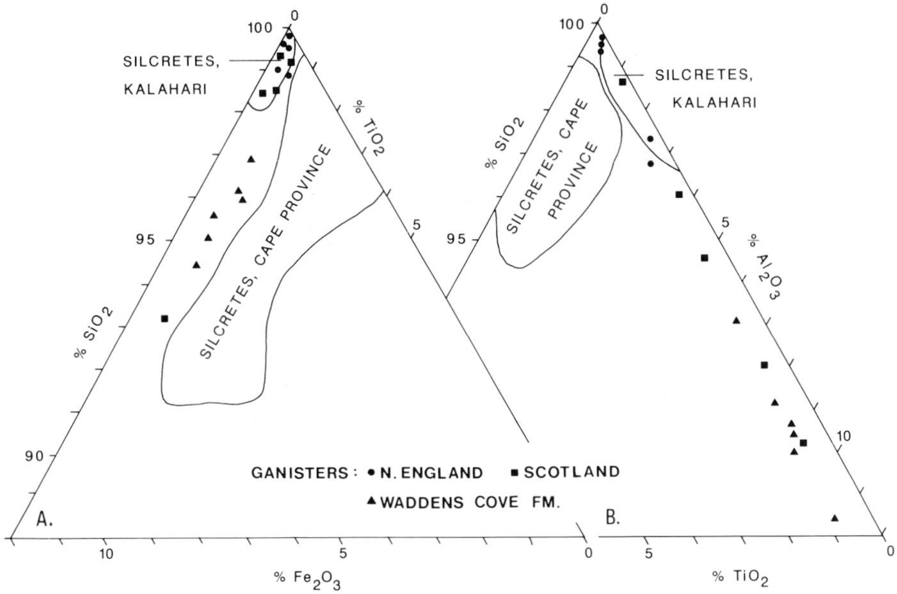

Fig. 13-17. Geochemical variation diagrams for some silcretes and ganisters. The two diagrams show the samples, with the three oxide contents recalculated to 100% in each case. Data for "weathering profile" silcretes from the Cape Province and for "non-weathering profile" silcretes of the Kalahari Basin, southern Africa, from Summerfield (1983c,d). Data for some English Carboniferous ganisters from Williamson (1967, table 15) and for some Scottish Carboniferous ganisters from C. Fielding (pers. commun., 1989; see also Fielding et al., 1988). Some Australian silcretes (Young, 1985, fig. 3) show a similar range to southern African silcretes in Fig. 13-17B (see text).

prominent cutanic features in lower horizons). Additionally, acid leaching associated with the overlying peat may have aided soil development. The removal of clay may have contributed to the development of overgrowths (cf., Heald and Larese, 1974). The region lay at near-equatorial latitudes at this time, and probably experienced a humid, tropical climate (Fielding, 1984; Besly and Fielding, 1989).

Some Early Carboniferous paleosols of southeast Scotland show similarity to the ganisters described by Percival (Fielding et al., 1988). The paleosols, designated as Type III, contain tough, white siliceous layers with vertical rootlet casts, and some exceed 2 m in thickness. They form part of progradational deltaic sequences and most are overlain by coals. Geochemical and XRD data (C. Fielding, pers. commun., 1989) show that the silica content of four samples ranges from 88 to > 98%, with one sample at about 77%. Quartz is the predominant mineral, with detectable quantities of orthoclase, plagioclase, clay minerals (illite, kaolinite and chlorite) and siderite in most samples. The paleosols were attributed to freely drained soils developed in sandy substrates. An additional type of paleosol (Type IV) shows a bleached, siliceous upper unit accompanied by clay, carbonate and iron oxide-enriched horizons in the lower part of the profile. These paleosols were likewise attributed to free drainage, but developed in more lithologically mixed substrates where minerals accumulated at permeability barriers during downward translocation.

In contrast, the Waddens Cove ganisters are not overlain by coal and formed in aggradational, proximal floodplain settings where paleosol development was interrupted by renewed alluvial sedimentation. Gibling and Rust (1990) noted evidence for progressive lowering of base-level (allocyclic events) upward in the measured section, based on changes in the proportion of red strata and coals, and channel-deposit characteristics. Ganisters at higher levels in the measured section tend to be thicker, show more pronounced desiccation cracks, and occur stacked in composite sheets (Fig. 13-3). The authors believe, however, that ganister formation in the Waddens Cove Formation primarily reflects autocyclic events related to channel avulsion and splay abandonment. The ganisters are relatively silica-poor (81 – 86% silica) in comparison with most of the British ganisters, and contain relatively large amounts of other elements. The Waddens Cove paleosols probably reflect an early stage of pedogenesis at which considerable amounts of clay were still retained in eluvial horizons. The microquartz cementation probably reflects in part the presence of the clays.

The Triassic ganisters of New South Wales formed in alluvial, crevasse-splay deposits under a cool, temperate climate (McDonnell, 1974; Retallack, 1976, 1977). The paleosols with which they are associated show well-developed horizons, underlie thin organic-rich layers and are framework-supported with a cryptocrystalline quartz cement. In certain respects, they resemble the Waddens Cove ganisters, but no quantitative petrographic data were presented.

Silcretes are widespread in modern terrains, especially in Australia and southern Africa (Langford-Smith, 1978; Summerfield, 1983a). Most silcretes are 1 – 3 m thick, locally in excess of 5 m. Although many are associated with present-day landforms, most appear to be relict deposits from earlier geological periods, so that their conditions of formation must be inferred. Summerfield (1983d) divided South African silcretes into two groups (Fig. 13-17). Weathering profile (WP) silcretes are found in coastal areas of the Cape Province, South Africa, in association with deep weathering of landscape surfaces under humid conditions, commonly with abundant vegetation and low pH. They show Ti enrichment, with titanium oxides generally > 1% and up to 3.4% of the rock (Summerfield, 1983b). In contrast, non-weathering profile (NWP) silcretes are present inland in the Kalahari Desert, lack weathering profiles, and are commonly associated with calcretes. They originated under relatively alkaline conditions, in arid to semi-arid settings (N.L. Watts, 1980), and lack Ti enrichment (< 0.5% of titanium oxides). The use of titanium content as a tool in silcrete classification has recently been questioned by Young (1985). Although ganisters have been considered analogous to silcretes (McDonnell, 1974; Wopfner, 1978), detailed comparisons have yet to be drawn (Summerfield, 1983c). The following section provides a brief petrographic and geochemical comparison of the Waddens Cove ganisters with some silcretes from modern settings.

The Waddens Cove ganisters are thinner than most silcretes. The ganisters show grain-supported (GS) fabrics similar to those of some silcretes, but lack the F (floating), M (silicified fine-grained matrix) and C (conglomeratic) fabrics recognized in silcretes by Summerfield (1983a,c). Replacement of a fine-grained matrix by microquartz was noted in both the ganisters and many silcretes (Summerfield,

1983a). The GS fabrics are apparently rare in South African and Australian silcretes, but are known from silicified horizons in North Africa, France and southern England, where a microquartz cement is very common (Summerfield, 1983c). The ganisters lack colloform and other Ti-rich authigenic structures (Williamson, 1957, Summerfield, 1983a), but show good evidence for authigenic precipitation of titanium oxides and are relatively rich in Ti compared with many modern soils.

The Waddens Cove ganisters lie geochemically outside the range of both WP and NWP silcretes (Fig. 13-17). They are much more aluminous than most of the plotted silcretes, but show equivalent iron oxide and silica contents to many WP silcretes. Their titanium oxide contents are equivalent to titanium-poor WP silcretes of southern Africa and to many silcretes of eastern, coastal Australia (Fig. 13-17). Most British ganisters plot within the field for NWP silcretes, although some are considerably more alumina- and iron-rich (Fig. 13-17). Despite their geochemical similarity to NWP silcretes, they evidently formed in humid climates as indicated by their association with coal. Relatively humid climatic conditions are inferred both for the formation of the Waddens Cove and other ganisters (Table 13-3) and for some silcretes (Young, 1978, 1985; Wopfner, 1978; Summerfield, 1983d). Further work is clearly required to document the relationship between ganisters and silcretes.

CONCLUSIONS

(1) Ganisters in the Waddens Cove Formation of the Sydney Basin represent pedogenetic (early diagenetic) accumulation of silica associated with continental groundwaters. They form components of immature paleosols within crevasse-splay, levee, and channel deposits. Ganister development followed channel avulsion and splay abandonment, when the elevated surfaces of sandy landforms underwent pedogenesis under a seasonal, tropical climate. Poor paleosol horizonation probably reflects the aggradational floodplain setting and cumulative profile development.

(2) Channel margins stepped over ganisters, and ganister slumpblocks and fragments in channel deposits indicate that the ganisters were substantially lithified at or just below the floodplain surface. Consequently, the authors believe that the ganister fabrics originated at subsurface depths of less than a few meters and probably within a few hundred years after deposition, with little modification during later burial.

(3) The ganisters are moderately siliceous, with 81 – 86% silica. Fabric is grain-supported, and microquartz replaced detrital clay and filled orthovugs. The presence of embedded-grain cutans may explain the rarity of overgrowths on quartz grains. Illuviation cutans testify to clay and iron oxide eluviation, and authigenic titanium oxides are disseminated within clay and microquartz patches.

(4) Geochemical considerations for the ganisters, as well as sedimentary features of the channel and floodplain deposits, indicate a seasonally fluctuating groundwater level. Periodic deep desiccation and abundant vegetation are also indicated.

These factors would have influenced soil water pH, Eh, and temperature during pedogenesis, and contributed to the soil-profile development.

(5) Probable sources of silica for ganister cementation include clay minerals, quartz grains (commonly embayed, especially within rhizoliths) and feldspars.

(6) Siliceous seatearths underlie coals in the Sydney Mines Formation of the Sydney Basin and in the British Carboniferous, where they form a potential fire hazard due to spark generation during underground mining. Ganister-bearing paleosols in the studied section, however, are not overlain by coals. A subsequent rise of relative base-level may have been required for peat accumulation to follow ganister formation.

(7) The ganisters show petrographic features analogous to those of some silcretes.

ACKNOWLEDGEMENTS

The authors thank Chris Fielding and Bob Young for discussion and for their thoughtful comments on an earlier version of this paper. K.H. Wolf and G.V. Chilingar gave helpful assistance with editing the manuscript. Chris Fielding kindly provided unpublished data for some British ganisters. Gordon Brown, Kevin Cameron, Bob Mackay and Patricia Stoffyn assisted with technical analysis, and Ferenc Stefani and Max Perkins with photography. The authors gratefully acknowledge financial support provided by the Natural Science and Engineering Research Council of Canada, Grants A8437 and A2672. M.R.G. expresses his debt of gratitude to his co-author and thesis supervisor, Brian Rust, who died on June 22, 1990.

REFERENCES

Alhonen, P., Koljonen, T., Lahermo, P. and Uusinoka, R., 1975. Ferruginous concretions around root channels in clay and fine sand deposits. *Bull. Geol. Soc. Finl.*, 47: 175 – 181.

Allen, J.R.L., 1973. Compressional structures (patterned ground) in Devonian pedogenic limestones. *Nature,* 243: 84 – 86.

Allen, J.R.L., 1986. Pedogenic calcretes in the Old Red Sandstone facies (Late Silurian – Early Carboniferous) of the Anglo – Welsh area, southern Britain. In: V.P. Wright (Editor), *Paleosols, Their Recognition and Interpretation.* Blackwell, Oxford, pp. 58 – 86.

Andriesse, J.P., 1969. A study of the environment and characteristics of tropical podzols in Sarawak (East-Malaysia). *Geoderma,* 2: 201 – 227.

Andriesse, J.P., 1970. The development of the podzol morphology in the tropical lowlands of Sarawak (Malaysia). *Geoderma,* 3: 261 – 279.

Aubert, H. and Pinta, M., 1977. *Trace Elements in Soils. Developments in Soil Science, 7.* Elsevier, Amsterdam, 395 pp.

Bennett, P. and Siegel, D.I., 1987. Increased solubility of quartz in water due to complexing by organic compounds. *Nature,* 326: 684 – 686.

Berner, R.A., 1971. *Principles of Chemical Sedimentology.* McGraw-Hill, New York, N.Y., 240 pp.

Besly, B.M. and Fielding, C.R., 1989. Palaeosols in Westphalian coal-bearing and red-bed sequences, Central and Northern England. *Palaeogeogr. Palaeoclimatol. Palaeoecol.,* 70: 303 – 330.

Birkeland, P.W., 1984. *Soils and Geomorphology.* Oxford Univ. Press, Oxford, 372 pp.

Boehner, R.C. and Giles, P.S., 1986. *Geological map of the Sydney Basin.* Nova Scotia Dep. Mines and Energy, Map 86-1.

Bown, T.M. and Kraus, M.J., 1987. Integration of channel and floodplain suites, I. Developmental se-

quence and lateral relations of alluvial paleosols. *J. Sediment. Petrol.,* 57: 587 – 601.
Brewer, R., 1976. *Fabric and Mineral Analysis of Soils.* Krieger, Huntington, New York, N.Y., 482 pp.
Bullock, P., 1985. The role of micromorphology in the study of Quaternary soil processes. In: J. Boardman (Editor), *Soils and Quaternary Landscape Evolution.* Wiley, New York, N.Y., pp. 45 – 68.
Buol, S.W., McCracken, R.J. and Hole, F.D., 1973. *Soil Genesis and Classification,* 1nd ed. Iowa State Univ. Press, Ames, Iowa, 360 pp.
Coleman, J.M., 1966. Ecological changes in a massive fresh-water clay sequence. *Trans. Gulf Coast Assoc., Geol. Soc.,* 15: 159 – 174.
Coleman, J.M. and Gagliano, S.M., 1965. Sedimentary structures: Mississippi River deltaic plain. In: G.V. Middleton (Editor), *Primary Sedimentary Structures and Their Hydrodynamic Interpretation. Soc. Econ. Paleontol. Mineral., Spec. Publ.,* 12: 133 – 148.
Cook, H.E., Johnson, P.D., Matti, J.C. and Zemmels, I., 1975. Methods of sample preparation and X-ray diffraction data analysis. X-Ray Mineralogy Lab., Deep Sea Drilling Project, Univ. California, Riverside, Calif. In: D.E. Hayes et al. (Editors), *Initial Reports of Deep Sea Drilling Project,* 28: 999 – 1007.
Curtis, C.D. and Spears, D.A., 1968. The formation of sedimentary iron minerals. *Econ. Geol.,* 63: 257 – 270.
Dilles, S.J., 1983. *Sedimentology of the lower part of the Morien Group in the southeast portion of the Sydney Coal Basin, Nova Scotia.* M.Sc. thesis, Univ. Ottawa, Ottawa, Ont., 196 pp.
Donaldson, A.C. and Shumaker, R.C., 1981. Late Paleozoic molasse of Central Appalachians. In: A.D. Miall (Editor), *Sedimentation and Tectonics in Alluvial Basins. Geol. Assoc. Can., Spec. Pap.,* 23: 99 – 124.
Duchaufour, P., 1982. *Pedology.* Geo. Allen and Unwin, London, 448 pp.
Farrell, K.M., 1987. Sedimentology and facies architecture of overbank deposits of the Mississippi River, False River region, Louisiana. In: F.G. Ethridge, R.M. Flores and M.D. Harvey (Editors), *Recent Developments in Fluvial Sedimentology. Soc. Econ. Paleontol. Mineral., Spec. Publ.,* 39: 111 – 120.
Fenwick, I., 1985. Paleosols: problems of recognition and interpretation. In: J. Boardman (Editor), *Soils and Quaternary Landscape Evolution.* Wiley, New York, N.Y., pp. 3 – 21.
Fielding, C.R., 1984. Upper delta plain lacustrine and fluviolacustrine facies from the Westphalian of the Durham coalfield, N.E. England. *Sedimentology,* 31: 547 – 567.
Fielding, C.R., Al-Rubaii, M. and Walton, E.K., 1988. Deltaic sedimentation in an unstable tectonic environment – the Lower Limestone Group (Lower Carboniferous) of East Fife, Scotland. *Geol. Mag.,* 125: 241 – 255.
Fitzpatrick, E.A., 1980. *Soils.* Longman, London, 353 pp.
Forgeron, S., Mackenzie, B. and Macpherson, K., 1986. The effects of geological features on coal mining, Sydney Coalfield, Nova Scotia. *Bull. Can. Inst. Min.,* 79: 79 – 87.
Friend, P.F., Slater, M.J. and Williams, R.C., 1979. Vertical and lateral building of river sandstone bodies, Ebro Basin, Spain. *J. Geol. Soc. London,* 136: 39 – 46.
Gardner, T.W., Williams, E.G. and Holbrook, P.W., 1988. Pedogenesis of some Pennsylvanian underclays: ground-water, topographic, and tectonic controls. *Geol. Soc. Am., Spec. Pap.,* 216: 81 – 101.
Gibling, M.R. and Rust, B.R., 1990. Ribbon sandstones in the Pennsylvanian Waddens Cove Formation, Sydney Basin, Atlantic Canada: the influence of siliceous duricrusts on channel-body geometry. *Sedimentology,* 37: 45 – 65.
Gibling, M.R., Rust, B.R. and Aksu, A.E., 1985. Mudrocks of the Pennsylvanian Morien Group, Sydney Coalfield: mud deposition and early diagenesis on an alluvial plain. *Geol. Assoc. Can., Annu. Meet.,* Fredericton, N.B., Abstr., p. 21.
Hallsworth, E.G. and Beckman, G.G., 1969. Gilgai in the Quaternary. *Soil Sci.,* 107: 409 – 420.
Heald, M.T. and Larese, R.E., 1974. Influence of coatings on quartz cementation. *J. Sediment. Petrol.,* 44: 1269 – 1274.
Ho, C. and Coleman, J.M., 1969. Consolidation and cementation of Recent sediments in the Atchafalaya Basin. *Bull. Geol. Soc. Am.,* 80: 183 – 192.
Houseknecht, D.W., 1980. Comparative anatomy of a Pottsville lithic arenite and quartz arenite of the Pocahontas Basin, southern West Virginia: petrogenetic, depositional, and stratigraphic implications. *J. Sediment. Petrol.,* 50: 3 – 20.
Houseknecht, D.W. and Iannacchione, A.T., 1982. Anticipating facies-related coal mining problems in

Hartshorne Formation, Arkoma Basin. *Bull. Am. Assoc. Pet. Geol.*, 66: 923 – 930.
Hower, J., 1981. Shale diagenesis. In: F.J. Longstaffe (Editor), *Clays and the Resource Geologist.* Min. Assoc. Can., Short Course Handbook, Vol. 7: 60 – 80.
Hutton, J.T., Twidale, C.R. and Milnes, A.R., 1978. Characteristics and origin of some Australian silcretes. In: T. Langford-Smith (Editor), *Silcretes in Australia.* Dep. Geography, Univ. New England, Armidale, N.S.W., pp. 19 – 39.
King, L.H. and MacLean, B., 1976. Geology of the Scotian Shelf and adjacent areas. *Geol. Surv. Can.,* Pap. 74 – 31: 31 pp.
Klappa, C.F., 1980. Rhizoliths in terrestrial carbonates: classification, recognition, genesis and significance. *Sedimentology,* 27: 613 – 629.
Klinge, H., 1965. Podzol soils in the Amazon Basin. *J. Soil Sci.,* 16: 95 – 103.
Knight, M.J., 1980. Structural analysis and mechanical origins of gilgai at Boorook, Victoria, Australia. *Geoderma,* 23: 245 – 283.
Krauskopf, K.B., 1956. Dissolution and precipitation of silica at low temperatures. *Geochim. Cosmochim. Acta,* 10: 1 – 26.
Langford-Smith, T. (Editor), 1978. *Silcrete in Australia.* Dep. Geography, Univ. New England, Armidale, N.S.W., 304 pp.
Leckie, D., Fox, C. and Tarnocai, C., 1989. *Sedimentology,* 36: 307 – 323.
Levinson, A.A., 1974. *Introduction to Exploration Geochemistry.* Applied Publ., Maywood, Ill., 614 pp.
Mann, A.W. and Deutscher, R.L., 1978. Hydrogeochemistry of a calcrete-containing aquifer near Lake Way, Western Australia. *J. Hydrol.,* 38: 357 – 377.
Mann, A.W. and Horwitz, R.C., 1979. Groundwater calcrete deposits in Australia: some observations from Western Australia. *J. Geol. Soc. Aust.,* 26: 293 – 303.
McDonnell, K.L., 1974. Depositional environments of the Triassic Gosford Formation, Sydney Basin. *J. Geol. Soc. Aust.,* 21: 107 – 132.
Morad, S., 1986. SEM study of authigenic rutile, anatase and brookite in Proterozoic sandstones from Sweden. *Sediment. Geol.,* 46: 77 – 89.
Morris, R.C. and Fletcher, A.B., 1987. Increased solubility of quartz following ferrous – ferric iron reactions. *Nature,* 330: 558 – 561.
Mount, J.F. and Cohen, A.S., 1984. Petrology and geochemistry of rhizoliths from Plio – Pleistocene fluvial and marginal lacustrine deposits, East Lake Turkana, Kenya. *J. Sediment. Petrol.,* 54: 263 – 275.
Percival, C.J., 1983a. A definition of the term ganister. *Geol. Mag.,* 120: 187 – 190.
Percival, C.J., 1983b. The Firestone Sill ganister, Namurian, northern England – the A2 horizon of a podzol or podzolic palaeosol. *Sediment. Geol.,* 36: 41 – 49.
Percival, C.J., 1986. Paleosols containing an albic horizon: examples from the Upper Carboniferous of northern England. In: V.P. Wright (Editor), *Paleosols, Their Recognition and Interpretation.* Blackwell, Oxford, pp. 87 – 111.
Pettijohn, F.J., Potter, P.E. and Siever, R., 1987. *Sand and Sandstone,* 2nd ed. Springer, New York, N.Y., 553 pp.
Retallack, G.J., 1976. Triassic palaeosols in the Upper Narrabeen Group of New South Wales. Part I. Features of the palaeosols. *J. Geol. Soc. Aust.,* 23: 383 – 399.
Retallack, G.J., 1977. Triassic palaeosols in the Upper Narrabeen Group of New South Wales. Part II. Classification and reconstruction. *J. Geol. Soc. Aust.,* 24: 19 – 36.
Retallack, G.J., 1983. Late Eocene and Oligocene paleosols from Badlands National Park, South Dakota. *Geol. Soc. Am., Spec. Pap.,* 193: 82 pp.
Retallack, G.J., 1988. Field recognition of paleosols. *Geol. Soc. Am., Spec. Pap.,* 216: 1 – 20.
Rust, B.R., Gibling, M.R., Best, M.A., Dilles, S.J. and Masson, A.G., 1987. A sedimentological overview of the coal-bearing Morien Group (Pennsylvanian), Sydney Basin, Nova Scotia, Canada. *Can. J. Earth Sci.,* 24: 1869 – 1885.
Scotese, C.R., Bambach, R.K., Barton, C., van der Voo, R. and Ziegler, A.M., 1979. Paleozoic base maps. *J. Geol.,* 87: 217 – 277.
Sippel, R.F., 1968. Sandstone petrology, evidence from luminescence petrography. *J. Sediment. Petrol.,* 38: 530 – 554.
Smith, D.G., 1983. Anastomosed fluvial deposits: modern examples from western Canada. In: J.D. Col-

linson and J. Lewin (Editors), *Modern and Ancient Fluvial Systems. Int. Assoc. Sediment., Spec. Publ.,* 6: 155 – 168.
Summerfield, M.A., 1983a. Petrography and diagenesis of silcrete from the Kalahari Basin and Cape Coastal Zone, southern Africa. *J. Sediment. Petrol.,* 53: 895 – 909.
Summerfield, M.A., 1983b. Geochemistry of weathering profile silcretes, southern Cape Province, South Africa. In: R.C.L. Wilson (Editor), *Residual Deposits: Surface Related Weathering Processes and Material.* Blackwell, Oxford, pp. 167 – 178.
Summerfield, M.A., 1983c. Silcrete. In: A.S. Goudie and K. Pye (Editors), *Chemical Sediments and Geomorphology.* Academic Press, London, pp. 59 – 91.
Summerfield, M.A., 1983d. Silcrete as a palaeoclimatic indicator: evidence from southern Africa. *Palaeogeogr., Palaeoclimatol., Palaeoecol.,* 41: 65 – 79.
Thomas, R.G., Smith, D.G., Wood, J.M., Visser, J., Calverley-Range, E.A. and Koster, E.H., 1987. Inclined heterolithic stratification – terminology, description, interpretation and significance. *Sediment. Geol.,* 53: 123 – 179.
Valentine, K.W.G. and Dalrymple, J.B., 1975. The identification, lateral variation, and chronology of two buried paleocatenas at Woodhall Spa and West Runton, England. *Quat. Res.,* 5: 551 – 590.
Watts, N.L., 1977. Pseudo-anticlines and other structures in some calcretes of Botswana and South Africa. *Earth Surf. Processes,* 2: 63 – 74.
Watts, N.L., 1978. Displacive calcite: evidence from recent and ancient calcretes. *Geology,* 6: 699 – 703.
Watts, N.L., 1980. Quaternary pedogenic calcretes from the Kalahari (southern Africa): mineralogy, genesis and diagenesis. *Sedimentology,* 27: 661 – 686.
Watts, S.H., 1977. Major element geochemistry of silcrete from a portion of inland Australia. *Geochim. Cosmochim. Acta,* 41: 1164 – 1167.
Welton, J.E., 1984. *SEM Petrology Atlas. Am. Assoc. Pet. Geol.,* Tulsa, Okla., 237 pp.
Williamson, I.A., 1967. *Coal Mining Geology.* Oxford Univ. Press, Oxford, 266 pp.
Williamson, W.O., 1957. Silicified sedimentary rocks in Australia. *Am. J. Sci.,* 255: 23 – 42.
Wopfner, H., 1978. Silcretes of northern South Australia and adjacent regions. In: T. Langford-Smith (Editor), *Silcretes in Australia.* Dep. Geography, Univ. New England, Armidale, N.S.W., pp. 93 – 141.
Wright, V.P., 1982. Calcrete palaeosols from the Lower Carboniferous Llanelly Formation, South Wales. *Sediment. Geol.,* 33: 1 – 33.
Young, R.W., 1978. Silcrete in a humid landscape: the Shoalhaven Valley and adjacent coastal plains of southern New South Wales. In: T. Langford-Smith (Editor), *Silcretes in Australia.* Dep. Geography, Univ. New England, Armidale, N.S.W., pp. 195 – 207.
Young, R.W., 1985. Silcrete distribution in eastern Australia. *Z. Geomorphol.,* 29: 21 – 36.

SUBJECT INDEX

ERRATUM

Iijima, A., 1988. Diagenetic transformations of minerals as exemplified by zeolites and silica minerals – A Japanese view. In: G.V. Chilingarian and K.H. Wolf (Editors), Diagenesis, II. Developments in Sedimentology, 43. Elsevier, pp. 147–211.

Please note that Fig. 3-38 on p. 202 should be replaced by the figure below.

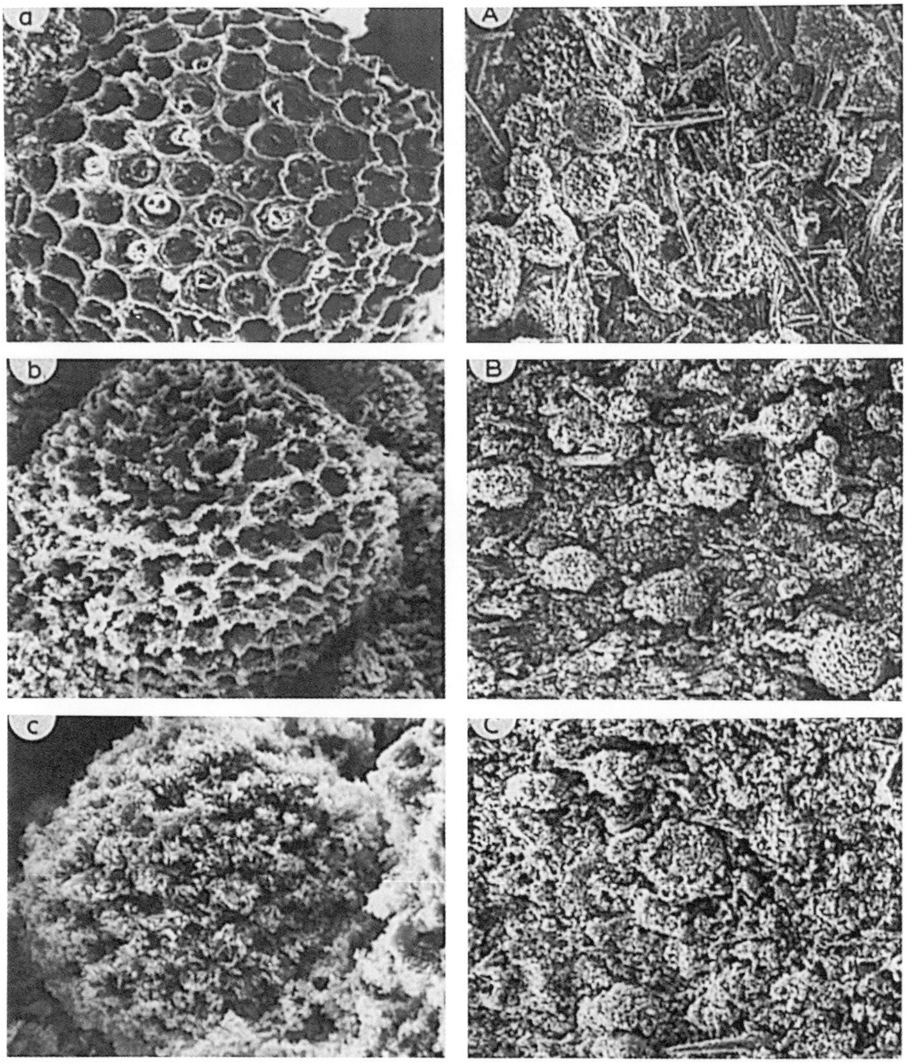

Fig. 3-38. For caption see next page.

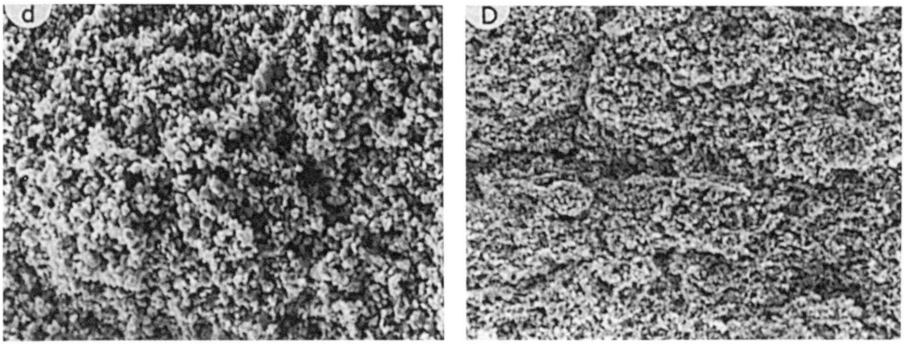

Fig. 3-38. Scanning electron micrographs showing the degree of preservation of radiolarian skeletons in quartzose bedded cherts. Degree of preservation of individual skeletons is on the left side. Degree of preservation of collective skeletons in chert is on the right side. (After Kakuwa, 1984, plates 2 – 7, pp. 56 – 61.)